Infosys Science Foundation Series

Infosys Science Foundation Series in Mathematical Sciences

The **Infosys Science Foundation Series in Mathematical Sciences,** a Scopus-indexed book series, focuses on high-quality content in the domain of pure and applied mathematics, biomathematics, financial mathematics, operations research and theoretical computer science. With this series, Springer hopes to provide readers with monographs, textbooks, edited volume, handbooks and professional books of the highest academic quality on current topics in relevant disciplines. Literature in this series will appeal to a wide audience of researchers, students, educators and professionals across the world.

Gopal Prasad · Andrei Rapinchuk ·
Balasubramanian Sury · Aleksy Tralle
Editors

Zariski Dense Subgroups, Number Theory and Geometric Applications

Mini Courses and Expository Articles

Springer

Editors
Gopal Prasad
Department of Mathematics
University of Michigan
Ann Arbor, MI, USA

Andrei Rapinchuk
Department of Mathematics
University of Virginia
Charlottesville, VA, USA

Balasubramanian Sury
Statistics and Mathematics Unit
Indian Statistical Institute
Bengaluru, Karnataka, India

Aleksy Tralle
Faculty of Mathematics and Computer Science
University of Warmia and Mazury in Olsztyn
Olsztyn, Poland

ISSN 2363-6149 ISSN 2363-6157 (electronic)
Infosys Science Foundation Series
ISSN 2364-4036 ISSN 2364-4044 (electronic)
Infosys Science Foundation Series in Mathematical Sciences
ISBN 978-981-95-3436-4 ISBN 978-981-95-3437-1 (eBook)
https://doi.org/10.1007/978-981-95-3437-1

Mathematics Subject Classification: 14G05, 20G15, 20G30, 50C35, 20H05, 20H25, 53C30, 11E73, 11F06, 14G25, 20G10, 20G35

Responsible Editor: Shamim Ahmad

This Springer imprint is published by the registered company Springer Nature Singapore Pte Ltd.
The registered company address is: 152 Beach Road, #21-01/04 Gateway East, Singapore 189721, Singapore

Preface

We are very pleased to present to the mathematical community the two volumes of the proceedings of the program on *Zariski-Dense Subgroups, Number Theory and Geometric Applications* that was held at the International Center for Theoretical Sciences (ICTS) in Bangalore/Bengaluru (India) during the 2-week period January 1–12, 2024.

We would like to begin this preface with a brief historical account of the investigation of Zariski-dense subgroups and of the previous meetings on this subject. While the Zariskidensity of certain subgroups of algebraic/Lie groups (such as arithmetic subgroups or lattices) has always been considered as one of their important properties, about 15–20 years ago Zariski-dense subgroups themselves, whether arithmetic or not (in other words, so-called "thin" subgroups), became the focus of intense investigation. Interest in Zariski-dense subgroups was fueled by their spectacular applications in various areas, one of the first of these being the construction of expander graphs. The resulting new research direction that focused on the analysis of the spectral gap for Zariski-dense subgroups became known as the theory of superstrong approximation. This was the subject of the workshop on *Thin Groups and Superstrong Approximation* held at MSRI in February 2012[1], which histrorically was one of the first meetings on the subject of Zariski-dense subgroups. In parallel, results on Zariski-dense subgroups were used to resolve some longstanding problems in differential and spectral geometry concerning isospectral and lengthcommensurable locally symmetric spaces. The classical problem in algebraic geometry, going back to D. Mumford, to determine fake projective planes was solved using Prasad's volume formula explained in Thilmany's article in Volume I. These results initiated a new trend of using techniques from number theory, the arithmetic theory of algebraic groups, and arithmetic geometry in the theory of Lie groups and differential and algebraic geometry. They, along with many other related results, were featured in the program of the workshop on Zariski-dense Subgroups hosted by IPAM (UCLA) in February 2015. Among other things, this workshop affirmed the

[1] The proceedings of this workshop were published in Superstrong Approximation and Thin Groups, edited by E. Breuillard and H. Oh, MSRI Publications No. 61, Cambridge Univ. Press, 2014.

ambition of extending some of the geometric results to nonarithmetically defined spaces by developing the arithmetic theory of algebraic groups over fields more general than global fields.

In the years that followed, many new results in this direction have been obtained, so the necessity of a new meeting in the broad and rapidly developing area involving Zariski-dense subgroups and related topics became evident. The ICTS kindly agreed to host a 2-week long program on Zariski-dense subgroups in March 2020. Unfortunately, due to COVID, the full-scale program was cancelled, and in its place, a one-day event was held online (icts.res.in/program/ZDS2020). When normal research activities resumed, the ICTS rescheduled the program for January 2024. It should be noted, however, that in the four years following the cancellation of the original program, many new results have been obtained, so the content of the program that was held in January 2024 was rather different from what we had planned for March 2020. Thus, it would be fair to say that the new program was not just a re-run of the old one, but rather the second edition of it.

The January 2024 program brought together over 70 mathematicians, from graduate students and postdocs, to research leaders in the respective fields, who, while investigating a variety of subjects, were led to the analysis of Zariski-dense subgroups of algebraic/Lie groups, their applications, and related issues. In fact, the diversity of the mathematical areas dealing in some way with Zariski-dense subgroups, which by itself is quite amazing, was well reflected in the scientific program of the meeting (included, for the reader's convenience in the Appendices to Volume I). The variety of topics presented in the mini-courses and research talks given at the program resulted in a very stimulating environment for active discussions, exchanges, and interactions. As the program progressed, it became obvious that the materials that it generated would be beneficial to a broad group of research mathematicians working in different areas. This sentiment was the starting point for the preparation of these proceedings. While not all speakers/participants were able to contribute to this project, many of them (including all of those who delivered minicourses) endorsed the idea with enthusiasm and prepared their manuscripts in a timely manner, in some cases significantly expanding the material from their presentations. As a result, we have decided to divide all the submissions into two volumes, with Volume I comprising the materials from all four mini-courses and two more expository articles and Volume II containing predominantly research articles. The reader can also find the titles and abstracts of all presentations (mini-courses, talks and short communications) delivered during the program as well as its schedule at the end of this volume.

For the reader's convenience, we will now give an overview of the articles included in the current Volume I, starting with those based on mini-courses. In his article, Brian Conrad provides an extensive introduction to the new approach to Bruhat-Tits theory of algebraic groups over local fields developed by T. Kaletha and G. Prasad, and includes some arithmetic applications of the theory. The article of Pietro Corvaja focuses on the use of techniques from Diophantine approximation in the investigation of linear groups, which recently led to a resolution of old problems concerning groups with bounded generation. The article by Kazaki Kannaka and Toshiyuki Kobayashi, which is an expanded version of the mini-course delivered by Kobayashi at the ICTS program, addresses many aspects of the classical problem of studying actions of discrete subgroups on homogeneous spaces of Lie groups, with special attention to the deformation of such actions and the associated notion of (local) rigidity. The article by Bharatram Rangarajan is based on the mini-course he gave jointly with Alex Lubotzky. It presents recent results on the new form of rigidity of higher-rank arithmetic groups—uniform stability—asserting that in an appropriate set-up every "almost representation" of such a group is close to a genuine representation.

Volume I also contains two expository articles based on the talks given at the program. In his article, Dave Witte Morris focuses on a recent proof (by B. Deroin and S. Hurtado) of the conjecture that higher-rank lattices in semisimple Lie groups are not left-orderable (equivalently, do not admit nontrivial, orientation preserving actions on a real line). He explains the main ideas of the proof and then uses them to establish the analogous results for semisimple *p*-adic groups. The article by François Thilmany gives a detailed account of Prasad's formula for the covolume of an arithmetic subgroup of a simple algebraic group, starting with historical remarks and the necessary background material. The author also presents many important applications of the volume formula including the classification of fake projective planes, counting lattices in terms of their covolume, determination of lattices of minimal covolume, and others.

We conclude this preface with a few acknowledgements. We would like to thank the ICTS Director Professor Rajesh Gopakumar and the ICTS staff for funding and hosting the program and for providing excellent working conditions to its participants. We are also grateful to Springer Executive Editor Shamim Ahmad for all his help associated with the publication of these proceedings. But first and foremost we would like to thank the participants and especially the speakers of the program for making it a success, with special thanks to the contributors of these proceedings.

Gopal Prasad (Ann Arbor, USA)
Andrei Rapinchuk (Charlottesville, USA)
Balasubramanian Sury (Bengaluru, India)
Aleksy Tralle (Olsztyn, Poland)
June 2025

As these proceedings are being sent into production, our colleague and friend Gopal Prasad turns 80. We all wish him a Happy Birthday and many happy returns of the day!

Contents

Bruhat-Tits Theory: A Structural Introduction

Brian Conrad

Abstract This article expands on a 3-lecture overview of Bruhat-Tits theory given in Bangalore. It conveys the new approach by Kaletha and Prasad giving the first axiomatization of the theory, thereby organizing its many features (affine root systems, buildings, integral models, etc.) in a coherent manner.

Keywords Reductive algebraic group · Local fields · Bruhat-Tits buildings

Bruhat-Tits theory illuminates connected reductive groups G over non-Archimedean local fields k and both the open subgroup structure and representation theory of $G(k)$. The original papers [BT1, BT2, BT3] by Bruhat and Tits are a formidable mixture of intricate group-theoretic calculations and combinatorial-geometric structures that are challenging to visualize beyond a handful of cases.

An initial aim of the theory is to come to grips with the multiplicity of conjugacy classes of maximal compact groups of $G(k)$ (in contrast with connected Lie groups, whose maximal compact subgroups are conjugate). We also want to understand the open subgroup structure of the maximal compact subgroups. These aims are achieved by analyzing point-stabilizers $G(k)_x$ for an isometric $G(k)$-action on a specific contractible and complete non-positively curved metric space called the *Bruhat-Tits building* $\mathcal{B}(G)$. The building is covered by Euclidean spaces that overlap in an extremely complicated manner. For many classical G there is a concrete interpretation of $\mathcal{B}(G)$ in terms of chains of integral lattices. (For split G a uniform construction of $\mathcal{B}(G)$ was known prior to Bruhat-Tits.) Bruhat-Tits theory allows a

Acknowledgements I am grateful to Quanlin Chen for live-TEXing the Stanford lectures, the students in the course for lively questions, and Lernik Asserian for a masterful job with the pictures and efficiently diagnosing TEXnical issues. I am also grateful to Kaletha and Prasad for many illuminating discussions during the preparation of the lectures, and to the organizers of the Bangalore conference for the opportunity to promote a conceptual approach to Bruhat-Tits theory.

B. Conrad (✉)
Stanford University, Stanford, USA
e-mail: bdconrad@stanford.edu

G. Prasad et al. (eds.), *Zariski Dense Subgroups, Number Theory and Geometric Applications*, Infosys Science Foundation Series,
https://doi.org/10.1007/978-981-95-3437-1_1

uniform approach with general G, for which a concrete interpretation of $\mathcal{B}(G)$ is lacking.

For arithmetic applications, a crucial feature of Bruhat-Tits theory is the relation to algebraic geometry over the valuation ring $\mathfrak{o}$ of k: it provides an array of smooth affine $\mathfrak{o}$-group schemes $\mathscr{G}$ with generic fiber $\mathscr{G}_k = G$ for which the compact open subgroup $\mathscr{G}(\mathfrak{o}) \subset G(k)$ has properties that can be understood via affine root systems and the geometry of the building. These *Bruhat-Tits $\mathfrak{o}$-group schemes* are central to the deeper uses of Bruhat-Tits theory, such as by relating the open subgroup structure of $G(k)$ to properties of special fibers $\mathscr{G}_{\mathfrak{f}}$ over the residue field $\mathfrak{f}$. For example, the calculation of $\mathrm{H}^1(k, G)$ and the construction of k-anisotropic maximal k-tori in G for non-Archimedean local k and semisimple G are results over k with (uniform) proofs that use Bruhat-Tits $\mathfrak{o}$-group schemes. Both proofs are carried out here based on the Kaletha-Prasad axioms for Bruhat-Tits theory as in [KP], not requiring knowledge of how the theory is constructed.

Substantial use of smooth affine $\mathfrak{o}$-group schemes far beyond how Bruhat and Tits used $\mathfrak{o}$-groups enables the axiomatic approach in [KP] to significantly streamline both the construction of the theory over the maximal unramified extension k^{un} (harnessing that $G_{k^{\mathrm{un}}}$ is quasi-split and k^{un} has algebraically closed residue field) and the "unramified descent" that constructs Bruhat-Tits theory over k from its availability over k^{un}. (Another aspect of the theory, valuations on a root datum, was intertwined with unramified descent in the original approach of Bruhat and Tits. The use of group schemes over the valuation ring in [KP] bypasses needing valuations on a root datum beyond the quasi-split case; that structure is recovered in general in [KP, Sect. 9.6] after unramified descent is done).

This article is a polished set of notes from a 10-week course at Stanford University that had four aims: (i) convey with proofs and examples a sense of tools in the theory (Henselian fields, Tits systems, affine root systems, buildings in combinatorial and metric guises, etc.), (ii) formulate the axiomatic approach and provide arithmetic applications, (iii) carry out unramified descent, (iv) ease the path into reading the monumental [KP]. Knowledge of reductive groups over fields [C3] and of schemes (e.g., smooth and étale maps) is assumed. Time ran out for (iii), but the course seemed to succeed in the other aspects. Anyone who reads these notes is well-prepared to enjoy the elegant treatment of unramified descent in [KP, Sects. 9.1–9.5].

There are 11 core axioms for Bruhat-Tits theory, labeled in [KP, Sect. 4.1] as 4.1.x for 11 integers $1 \le x \le 27$. We label the axioms by the same set of non-consecutive integers x. It is convenient to rearrange the order of a few axioms relative to the order in [KP] (without introducing any circularity), so the labels sometimes occur in here in non-increasing order (see Sects. 27 and 30); hopefully this is not disorienting.

In the final lecture Sect. 30, the axioms of Bruhat-Tits theory yield a striking arithmetic application of Bruhat-Tits $\mathfrak{o}$-group schemes: the k-anisotropic connected semisimple G that are absolutely simple and simply connected are inner type A (i.e., $\mathrm{SL}_1(D)$ for a central division algebra D over k), in contrast with $\mathbf{R}$ over which there is a compact form for every Dynkin type. Appendices B–D provide an array of additional interesting applications, and the later chapters in [KP] supply very many more.

1 Compact Open Subgroups

Let k be a non-Archimedean (complete) local field, $\mathfrak{o}$ the valuation ring, $\mathfrak{m}$ the maximal ideal of $\mathfrak{o}$, and $\mathfrak{f}$ the (finite) residue field. Let G be a connected reductive group over k (e.g, $\mathrm{SL}_n(D)$ for a finite-dimensional central division algebra D over k, a symplectic group Sp_{2g}, $\mathrm{SO}(q)$ for a non-degenerate quadratic space (V, q) over k, $\mathrm{U}(h)$ for a non-degenerate Hermitian space (W, h) relative to a quadratic Galois extension k'/k, the automorphism scheme G_2 of the split octonion algebra over k, ...). Let's start with a basic lemma.

Lemma 1.1 *Let H be an affine k-group scheme of finite type. Then $H(k)$ has a neighborhood basis of* 1 *consisting of compact open subgroups.*

We only need this in the smooth case, and the general case immediately reduces to the smooth case anyway: H can be replaced with the Zariski closure of $H(k)$ in H, the latter being a geometrically reduced closed k-subgroup, hence k-smooth (since geometrically reduced k-schemes of finite type have a dense open k-smooth locus).

Proof We may choose a closed k-subgroup inclusion $H \hookrightarrow \mathrm{GL}_{n,k}$, so $H(k) \subset \mathrm{GL}_n(k)$ is a closed subgroup (with the subspace topology). Thus it's sufficient to treat the case $H = \mathrm{GL}_n$.

We have $\mathrm{GL}_n(k) \supset \mathrm{GL}_n(\mathfrak{o}) \supset 1 + \mathfrak{m}^i \mathrm{Mat}_n(\mathfrak{o})$ for $i \geq 1$. For any affine scheme X of finite type over $\mathfrak{o}$, the subset $X(\mathfrak{o}) \subset X(k)$ is compact open because we can use a closed immersion of X into an affine space $\mathbf{A}^m_{\mathfrak{o}}$ over $\mathfrak{o}$ to reduce to the case $X = \mathbf{A}^m_{\mathfrak{o}}$ that is clear (as $\mathfrak{o}$ is compact open in k). Taking X to be $\mathrm{GL}_{n,\mathfrak{o}}$ then yields that $\mathrm{GL}_n(\mathfrak{o})$ is compact open in $\mathrm{GL}_n(k)$ (as we could have seen directly too), and $1 + \mathfrak{m}^i \mathrm{Mat}_n(\mathfrak{o})$ is a neighborhood basis of 1 even in $\mathrm{Mat}_n(k)$ (in which $\mathrm{GL}_n(k)$ is an open subset with the subspace topology). □

We will often write $\mathcal{K}$ to denote a compact open subgroup of $G(k)$.

Remark 1.2 There is much interest in (generally infinite-dimensional) $\mathbf{C}$-linear representations V of $G(k)$ that satisfy both of the following two conditions:

(1) Smoothness: This means that for each $v \in V$, the stabilizer $\mathrm{Stab}_{G(k)}(v)$ is an open subgroup. It then follows via Lemma 1.1 that this stabilizer contains some compact open subgroup $\mathcal{K}$, so smoothness is equivalent to the condition

$$V = \bigcup_{\mathcal{K}} V^{\mathcal{K}}.$$

This is a directed union since $V^{\mathcal{K}}, V^{\mathcal{K}'} \subset V^{\mathcal{K} \cap \mathcal{K}'}$ with $\mathcal{K} \cap \mathcal{K}'$ also compact open in $G(k)$.

(2) Admissibility: this means $\dim_{\mathbf{C}} V^{\mathcal{K}} < \infty$ for all $\mathcal{K}$.

When classifying such irreducible V, one often focuses on the "biggest" $\mathcal{K}$ such that $V^{\mathcal{K}} \neq 0$. This leads us to want to analyze possibilities for "large" compact open $\mathcal{K}$ in $G(k)$.

Proposition 1.3 (Langlands] *Each compact subgroup of $G(k)$ lies in a maximal one and all maximal compact subgroups are open.*

The proof will be given later in wider generality (Proposition 4.4), with k assumed to be *Henselian*. This is a technically useful weakening of completeness of the valuation and finiteness of the residue field that will be defined in Sect. 3.

It is a classical and important fact that for a smooth affine $\mathbf{R}$-group H, the Lie group $H(\mathbf{R})$ (which may be disconnected for the analytic topology but always has finite component group), every compact subgroup lies in a maximal one. This is not quite obvious because after the dimension of a rising chain of such subgroups stops growing, perhaps the finite component groups keep growing.

Equally important, all such maximal compact subgroups meet every connected component of $H(\mathbf{R})$, and are all $H(\mathbf{R})$-conjugate to each other. These stated properties hold for any Lie group with finite component group; one reference is [Hoch, Chap. 15] (which directly tackles disconnected cases), and another is [Bor, Chap. 7] (which bootstraps from the connected case).

For non-Archimedean local k, in contrast, we'll see that often there are *multiple $G(k)$-conjugacy classes* of maximal compact open subgroups, though always only finitely many. Bruhat-Tits theory gives an elegant way to understand this (see Corollary 29.9, and especially its proof).

Let's now turn to some basic examples.

Example 1.4 Consider $G = \mathrm{GL}_n$. All maximal compact subgroups of $\mathrm{GL}_n(k)$ are conjugate to $\mathrm{GL}_n(\mathfrak{o})$. There is a classical proof using that the subgroup $\mathcal{K} \cap \mathrm{GL}_n(\mathfrak{o}) \subset \mathcal{K}$ has finite index (as it is open in a compact group, and its cosets constitute a pairwise-disjoint open cover). We will give a proof in Sect. 5 more generally for Henselian k (replacing "compactness" with a property that makes sense for Henselian k and is equivalent to compactness when k is complete with finite residue field).

By an $\mathfrak{o}$-*structure* on a smooth affine k-group H we mean a smooth affine $\mathfrak{o}$-group $\mathscr{H}$ *equipped with* an isomorphism $\alpha : \mathscr{H}_k \xrightarrow{\sim} H$. In other words, an $\mathfrak{o}$-structure is a smooth Hopf $\mathfrak{o}$-subalgebra $\mathcal{A} \subset k[H]$ such that $\mathcal{A}_k = k[H]$. Given an $\mathfrak{o}$-structure $(\mathscr{H}, \alpha)$, we get a (compact open!) subgroup

$$\mathscr{H}(\mathfrak{o}) \subset \mathscr{H}(k) = \mathscr{H}_k(k) \simeq H(k).$$

For instance, since the generic point $\mathrm{Spec}(k)$ is an open subscheme of $\mathrm{Spec}(\mathfrak{o})$, for $g \in G(k)$ the composite open immersion

$$G = \mathrm{GL}_{n,k} \xrightarrow[\mathrm{Conj}(g^{-1})]{\sim} \mathrm{GL}_{n,k} \hookrightarrow \mathrm{GL}_{n,\mathfrak{o}} \tag{1.1}$$

is an $\mathfrak{o}$-structure $\mathscr{G}$ for G with $\mathscr{G} \simeq \mathrm{GL}_{n,\mathfrak{o}}$ as $\mathfrak{o}$-groups but the induced isomorphism $\alpha : \mathscr{G}_k \simeq G$ is not the identity morphism on $\mathrm{GL}_{n,k}$. Unraveling this, the subgroup $\mathscr{G}(\mathfrak{o}) \subset G(k) = \mathrm{GL}_n(k)$ is exactly $g\mathrm{GL}_n(\mathfrak{o})g^{-1}$ (which is usually not $\mathrm{GL}_n(\mathfrak{o})$).

Remark 1.5 If $(\mathscr{G}, \alpha)$, $(\mathscr{G}', \alpha')$ are two $\mathfrak{o}$-structures on a smooth affine k-group G, there is *at most one* isomorphism $\theta : \mathscr{G} \simeq \mathscr{G}'$ of $\mathfrak{o}$-groups such that over k it respects α and α': the diagram

$$\begin{array}{ccc} \mathscr{G}_k & \xrightarrow{\theta_k} & \mathscr{G}'_k \\ & {\scriptstyle\alpha}\searrow \quad \swarrow{\scriptstyle\alpha'} & \\ & G & \end{array}$$

commutes. Due to such uniqueness (when θ exists), it is reasonable to speak of two $\mathfrak{o}$-structures being "the same" when such a θ exists. This is exactly the condition of *equality* of $\mathfrak{o}$-subalgebras $\mathfrak{o}[\mathscr{G}] = \mathfrak{o}[\mathscr{G}']$ inside $k[G]$.

If $\mathscr{G}(\mathfrak{o}) \neq \mathscr{G}'(\mathfrak{o})$ as subgroups of $G(k)$ then the $\mathfrak{o}$-structures are certainly not the "same." In Proposition 6.8 we will see that when the *residue field* $\mathfrak{f}$ *is algebraically closed* a converse also holds: if $\mathfrak{f} = \overline{\mathfrak{f}}$ and $\mathscr{G}(\mathfrak{o}) = \mathscr{G}'(\mathfrak{o})$ inside $G(k)$ then $\mathscr{G}' \simeq \mathscr{G}$ as $\mathfrak{o}$-structures (smoothness over the valuation ring is crucial here). Bruhat-Tits theory is not just a theory of certain topological groups, but rather has crucial algebro-geometric aspects involving $\mathfrak{o}$-structures.

Let's look at some special cases of these conjugation calculations in $\mathrm{GL}_n(k)$. In these, we write π to denote a uniformizer of $\mathfrak{o}$.

(1) For $r \in \mathbf{Z}$ we have

$$\begin{pmatrix} 1 & 0 \\ 0 & \pi^r \end{pmatrix} \mathrm{GL}_2(\mathfrak{o}) \begin{pmatrix} 1 & 0 \\ 0 & \pi^r \end{pmatrix}^{-1} = \left\{ \begin{pmatrix} \mathfrak{o} & \mathfrak{m}^{-r} \\ \mathfrak{m}^r & \mathfrak{o} \end{pmatrix} \right\} = \left\{ \begin{pmatrix} x & \pi^{-r} y \\ \pi^r z & w \end{pmatrix} : x, y, z, w \in \mathfrak{o} \right\} \subset \mathrm{GL}_2(k).$$

By inspection, the associated $\mathfrak{o}$-structures $\mathscr{G}_r$ via our preceding construction correspond to pairwise distinct $\mathfrak{o}$-subalgebras of $k[\mathrm{GL}_2]$ (so in particular different from $\mathfrak{o}[\mathrm{GL}_{2,\mathfrak{o}}]$ when $r \neq 0$). The $\mathfrak{o}$-subalgebras $\mathfrak{o}[\mathscr{G}_r]$ are each isomorphic to the Hopf algebra for $\mathrm{GL}_{2,\mathfrak{o}}$ (using x, y, z, w and $1/(xw - yz)$ as algebra generators giving the usual Hopf algebra structure, so to speak) but the way $k[\mathscr{G}_r]$ is identified with $k[\mathrm{GL}_{2,k}]$ involves the intervention of $\pi^{\pm r}$.

(2) Consider $0 \leq m \leq n - 1$, and define the diagonal matrix

$$\gamma_m = \begin{pmatrix} \mathrm{Id}_{n-m} & 0 \\ 0 & \pi \mathrm{Id}_m \end{pmatrix}.$$

One checks that

$$\gamma_m \mathrm{GL}_n(\mathfrak{o}) \gamma_m^{-1} = \left\{ \begin{pmatrix} \mathfrak{o} & \mathfrak{m}^{-1} \\ \mathfrak{m} & \mathfrak{o} \end{pmatrix} \right\} \subset \mathrm{GL}_n(k)$$

where the top-left block is $(n - m) \times (n - m)$ and the bottom-right block is $m \times m$. As we vary such m, we get pairwise distinct (maximal) compact open subgroups of $\mathrm{GL}_n(k)$ and thus pairwise different $\mathfrak{o}$-structures.

When $\mathcal{K} = \mathscr{G}(\mathfrak{o})$ for an $\mathfrak{o}$-structure $\mathscr{G}$ of G, we get some interesting compact open subgroups $\mathcal{H}$ of $\mathcal{K}$ via the subgroup structure of the special fiber $\mathscr{G}_{\mathfrak{f}}$ over the residue field (so this is a mechanism by which algebraic geometry illuminates the subgroup structure):

$$\begin{array}{ccccc} \mathrm{GL}_n(k) & \hookleftarrow & \mathcal{K} = \mathrm{GL}_n(\mathfrak{o}) & \twoheadrightarrow & \mathrm{GL}_n(\mathfrak{f}) \\ & & \| & & \| \\ & & \mathrm{GL}_{n,\mathfrak{o}}(\mathfrak{o}) & \longrightarrow & \mathrm{GL}_{n,\mathfrak{f}}(\mathfrak{f}) \\ & & \uparrow & & \uparrow \\ 1 + \mathfrak{m}\mathrm{Mat}_n(\mathfrak{o}) & \hookrightarrow & \mathcal{H} & \longrightarrow & P(\mathfrak{f}) \end{array}$$

where P is a parabolic $\mathfrak{f}$-subgroup of $\mathrm{GL}_{n,\mathfrak{f}}$ and the bottom square is cartesian (i.e., $\mathcal{H}$ is defined to be the full preimage of $P(\mathfrak{f})$). The surjectivity of $\mathcal{K} \twoheadrightarrow \mathrm{GL}_n(\mathfrak{f})$ is a manifestation of $\mathfrak{o}$-smoothness of $\mathrm{GL}_{n,\mathfrak{o}}$ (this role for smoothness will be more visible when we move beyond GL_n), and it ensures that different P's give rise to different $\mathcal{H}$'s since $P(\mathfrak{f}) \neq Q(\mathfrak{f})$ for distinct parabolic $\mathfrak{f}$-subgroups $P, Q \subset \mathrm{GL}_{n,\mathfrak{f}}$ (as for parabolic subgroups of a connected reductive group over any field, allowing finite fields).

Remark 1.6 For probing the structure of $\mathscr{G}_{\mathfrak{f}}$, we often need to work with an algebraic closure $\overline{\mathfrak{f}}$ when carrying out group-theoretic calculations in proofs. This is one reason not to limit our focus to finite residue fields.

Example 1.7 Consider $G = \mathrm{SL}_n$. If $\mathcal{K} \subset G(k)$ is a compact subgroup then by viewing it as a compact subgroup of $\mathrm{GL}_n(k)$ we know $\mathcal{K}$ is contained in a maximal compact open subgroup of $\mathrm{GL}_n(k)$. The latter has the form $\gamma_0 \mathrm{GL}_n(\mathfrak{o})\gamma_0^{-1}$ for some $\gamma_0 \in \mathrm{GL}_n(k)$.

By normality of SL_n in GL_n (even as group schemes over $\mathbf{Z}$), we conclude that

$$\mathcal{K} \subset \mathrm{SL}_n(k) \cap (\gamma_0 \mathrm{GL}_n(\mathfrak{o})\gamma_0^{-1}) = \gamma_0(\mathrm{SL}_n(k) \cap \mathrm{GL}_n(\mathfrak{o}))\gamma_0^{-1} = \gamma_0 \mathrm{SL}_n(\mathfrak{o})\gamma_0^{-1}.$$

This ambient subgroup is certainly compact open in $\mathrm{SL}_n(k)$, and so if $\mathcal{K}$ is *maximal* as a compact subgroup of $\mathrm{SL}_n(k)$ then this inclusion has to be an equality. But any k-group automorphism of $\mathrm{SL}_{n,k}$ (such as conjugation by $\gamma_0^{-1} \in \mathrm{GL}_n(k)$) induces a topological automorphism of $\mathrm{SL}_n(k)$ and thus preserves maximality of compact open subgroups. Thus, by choosing $\mathcal{K}$ to be maximal (as we may) we see $\mathrm{SL}_n(\mathfrak{o}) = \gamma_0^{-1}\mathcal{K}\gamma_0$ *must* be maximal as a compact open subgroup of $\mathrm{SL}_n(k)$.

To summarize: (i) $\mathrm{SL}_n(\mathfrak{o})$, and hence *every* $\mathrm{GL}_n(k)$-conjugate of it, is a maximal compact open subgroup of $\mathrm{SL}_n(k)$, and (ii) these conjugates *exhaust* the set of maximal compact open subgroups of $\mathrm{GL}_n(k)$. Hence, to determine the maximal compact open subgroups of $\mathrm{SL}_n(k)$ up to $\mathrm{SL}_n(k)$-conjugacy (rather than just $\mathrm{GL}_n(k)$-conjugacy) amounts to determining $\mathrm{SL}_n(k)$-conjugacy among the subgroups

$$\gamma \mathrm{SL}_n(\mathfrak{o})\gamma^{-1} \subset \mathrm{SL}_n(k)$$

for $\gamma \in \mathrm{GL}_n(k)$.

Clearly such γ only matters through its image in the quotient $\mathrm{GL}_n(k)/\mathrm{SL}_n(k) \simeq k^\times$ (identification via the determinant). In this quotient the center $k^\times \subset \mathrm{GL}_n(k)$ (which has *no effect* under conjugation) has image $(k^\times)^n$. Since

$$k^\times = \mathfrak{o}^\times \times \pi^{\mathbf{Z}},$$

we have $\det(\gamma) = u\pi^m$ for some $u \in \mathfrak{o}^\times$ and $m \in \mathbf{Z}$, so by writing $m = nq + r$ with $0 \le r \le n-1$ we can replace γ with its multiple by the central $\pi^{-q} \in \mathrm{GL}_n(k)$ (leaving γ-conjugation unaffected) to arrange that $0 \le m \le n-1$.

But the matrix γ_m as in our $\mathrm{GL}_{n,k}$-discussion (Example 1.4) satisfies $\det(\gamma_m) = \pi^m$, so $\gamma_m^{-1}\gamma$ has determinant in $\mathfrak{o}^\times = \det(\mathrm{GL}_n(\mathfrak{o}))$. In other words,

$$\gamma \in \mathrm{SL}_n(k)\gamma_m g$$

for some $g \in \mathrm{GL}_n(\mathfrak{o})$ and $0 \le m \le n-1$. Writing $\gamma = h\gamma_m g$ with $h \in \mathrm{SL}_n(k)$, we have

$$\gamma \mathrm{SL}_n(\mathfrak{o})\gamma^{-1} = h(\gamma_m(g\mathrm{SL}_n(\mathfrak{o})g^{-1})\gamma_m^{-1})h^{-1}.$$

We are only interested in working up to $\mathrm{SL}_n(k)$-conjugacy, so the outer h-conjugation can be dropped. In other words, the $\mathrm{SL}_n(k)$-conjugacy classes among $\mathrm{GL}_n(k)$-conjugates of $\mathrm{SL}_n(\mathfrak{o})$ are represented by

$$\gamma_m(g\mathrm{SL}_n(\mathfrak{o})g^{-1})\gamma_m^{-1}$$

with $g \in \mathrm{GL}_n(\mathfrak{o})$ and $0 \le m \le n-1$. Next we show the conjugation by $g \in \mathrm{GL}_n(\mathfrak{o})$ can be dropped.

For any ring R (such as $\mathfrak{o}$), $\mathrm{GL}_n(R)$-conjugation on the R-group $\mathrm{GL}_{n,R}$ preserves the R-subgroup $\mathrm{SL}_{n,R}$ (as we see via functor-of-points on R-algebras), so

$$g\mathrm{SL}_n(\mathfrak{o})g^{-1} = \mathrm{SL}_n(\mathfrak{o}).$$

Thus, the $\mathrm{SL}_n(k)$-conjugacy classes of maximal compact open subgroups of $\mathrm{SL}_n(k)$ are represented by

$$\mathcal{K}_m = \gamma_m \mathrm{SL}_n(\mathfrak{o})\gamma_m^{-1} \tag{1.2}$$

with $0 \le m \le n-1$.

Might there be some repetition in this final list; i.e., a way we could whittle it down even further by discovering $\mathrm{SL}_n(k)$-conjugacy among some of these n subgroups?

No: general features of Bruhat-Tits theory for simply connected semisimple groups that we will discuss in Sect. 2 ensure that $\mathrm{SL}_n(k)$ has exactly n conjugacy classes of such subgroups, so the final list (1.2) has no repetitions among such conjugacy classes. (The general theory also provides a more direct way to see that the n subgroups in (1.2) exhaust the conjugacy classes without repetition).

2 Actions on Buildings

Bruhat-Tits theory provides a metric space $\mathcal{B}(G)$ (called the *building* of G) with an isometric action of $G(k)$ having open stabilizers $G(k)_x$ (for all $x \in \mathcal{B}(G)$) and satisfying a "non-positive curvature" property that ensures a fixed point theorem for the action of compact subgroups. This will all be discussed in Sects. 13, 14 and 28, and implies that each compact subgroup $\mathcal{K}$ of $G(k)$ lies in $G(k)_x$ for some $x \in \mathcal{B}(G)$. Beware $G(k)_x$ is *not* generally compact when G is not semisimple.

If $S \subset G$ is a maximal split k-torus and $S' := (S \cap \mathscr{D}G)^0_{\mathrm{red}}$ is the corresponding maximal split k-torus in $G' = \mathscr{D}G$ then the building $\mathcal{B}(G)$ contains a subset $\mathcal{A}(S)$—called the *apartment* attached to S—that is an affine space for the finite-dimensional $\mathbf{R}$-vector space $\mathrm{X}_*(S')_{\mathbf{R}}$ (i.e., it is non-empty and equipped with a simply transitive $\mathrm{X}_*(S')_{\mathbf{R}}$-action). The labeling by such S's is $G(k)$-equivariant in the sense that for $g \in G(k)$ we have $g.\mathcal{A}(S) = \mathcal{A}(gSg^{-1})$ inside $\mathcal{B}(G)$. In particular, $N_G(S)(k)$ preserves $\mathcal{A}(S)$. When the metric on $\mathcal{B}(G)$ is restricted to the apartment, it is the translation-invariant metric on that affine space induced by a choice of inner product on $\mathrm{X}_*(S')_{\mathbf{R}}$ invariant under the finite relative Weyl group $W(G, S) = N_G(S)(k)/Z_G(S)(k)$. These matters are discussed in Sect. 24ff.

For G semisimple (e.g., SL_n, not GL_n), the stabilizers $G(k)_x$ *are* compact and so since every compact subgroup lives in one of those we have that the maximal compact subgroups are among these $G(k)_x$'s when G is semisimple.

Example 2.1 When G is also simply connected (e.g., SL_n, Sp_{2g}, *not* $\mathrm{SO}(q)$, *not* PGL_n), then maximal compact subgroups turn out to be *exactly* the $G(k)_x$'s for x that are *vertices* in the apartment (a notion to be defined in Sect. 24, and explored more fully in Sects. 26 and 27).

From the structure of reductive groups, when G is simply connected we have $G = \prod_i G_i$ for G_i that are simply connected and k-simple (in fact, $G_i = \mathrm{R}_{k'_i/k} G'_i$ where k'_i/k is finite separable and G'_i is absolutely simple over k'_i and simply connected). We have $S = \prod_i S_i$ where S_i is a maximal split k-torus in G_i, and the root system $\Phi(G_i, S_i)$ is irreducible or empty (corresponding to $S_i = 1$). Let $r_i = \dim(S_i) = \dim(\Phi(G_i, S_i))$ be the k-rank of G_i. The maximal compact subgroups $\mathcal{K}$ of $G(k)$ are clearly exactly $\mathcal{K} = \prod_i \mathcal{K}_i$ where $\mathcal{K}_i$ is maximal compact in $G_i(k) = G'_i(k'_i)$.

Bruhat-Tits theory will yield (see Corollary 29.10) that the number of conjugacy classes of maximal compact subgroups in $G(k)$ is $\prod_i (r_i + 1)$ when G is simply connected (it is often false otherwise).

Example 2.2 For $G = \mathrm{SL}_n$ we have $r = n - 1$, so $r + 1 = n$. In Example 1.7 the n subgroups of $\mathrm{SL}_n(k)$ in (1.2) were seen to account for all conjugacy classes of maximal compact subgroups of $\mathrm{SL}_n(k)$, so they must be pairwise non-conjugate.

Now let's discuss the visual meaning of the formula $\prod(r_i + 1)$ for the number of conjugacy classes via buildings, with an eye toward the special case of SL_n. Consider G that is k-split, so S is also a maximal k-torus on G (e.g., $G = \mathrm{SL}_n$, PGL_n, $\mathrm{Sp}_{2g}, \dots$). For $G' = \mathscr{D}G$ we have $\Phi(G, S) = \Phi(G', S') \subset \mathrm{X}^*(S')_{\mathbf{R}}$ and $\mathcal{A}(S)$ is identified with $\mathrm{X}_*(S')_{\mathbf{R}}$ upon choosing a Chevalley $\mathfrak{o}$-model $(\mathscr{G}, \mathscr{S})$ for (G, S). By definition, the *affine Weyl group* W_{aff} is generated by reflections in the hyperplanes $\{x \in \mathcal{A}(S) : a(x) = k\}$ for $k \in \mathbf{Z}$ and $a \in \Phi(G, S)$. In Sect. 19 we'll consider fundamental domains for the action of W_{aff} on $\mathcal{A}(S)$, giving a kind of "triangulation" whose maximal open (poly)simplicies $\mathcal{C} \subset \mathcal{A}(S)$ are called *chambers*. This is shown in Fig. 1 for SL_2 (a line with uniformly-spaced vertices) and SL_3 (a plane tiled by equilateral triangles).

The "corners" of chamber closures in the apartments are called vertices. For G *simply connected*, maximal compact subgroups turn out to be exactly $G(k)_x$ for $x \in \mathcal{B}(G)$ a vertex. For such G, if $\mathcal{C} \subset \mathcal{A}(S)$ is a choice of chamber then the stabilizers $\{G(k)_x\}_{x \in \mathrm{vert}(\mathcal{C})}$ are representatives of $G(k)$-conjugacy classes of maximal compact subgroups without repetition. In the special case $G = \mathrm{SL}_n$, the groups $\mathcal{K}_m = \gamma_m \mathrm{SL}_n(\mathfrak{o}) \gamma_m^{-1}$ we have discussed are exactly the stabilizers of the vertices of a specific chamber, and each chamber in $\mathcal{B}(\mathrm{SL}_n)$ is an open simplex in $\mathbf{R}^{n-1}$, so it has n vertices.

Remark 2.3 By definition, we say the derived group $G' = \mathscr{D}G$ is *k-anisotropic* when its maximal split k-tori are trivial, which is to say $\mathrm{X}_*(S')_{\mathbf{R}} = 0$ or equivalently $\mathcal{A}(S)$ is a point. That is equivalent to $\mathcal{B}(G)$ being a point (see Remark 13.5). It was proved by Bruhat-Tits and Rousseau, with another proof later by G. Prasad, that the algebro-geometric condition of k-anisotropicity for a connected reductive k-group H is equivalent to compactness of the group $H(k)$ (we'll discuss this in a broader context in Theorem 4.3). Hence, $\mathcal{B}(G)$ is a point precisely when $G'(k)$ is compact.

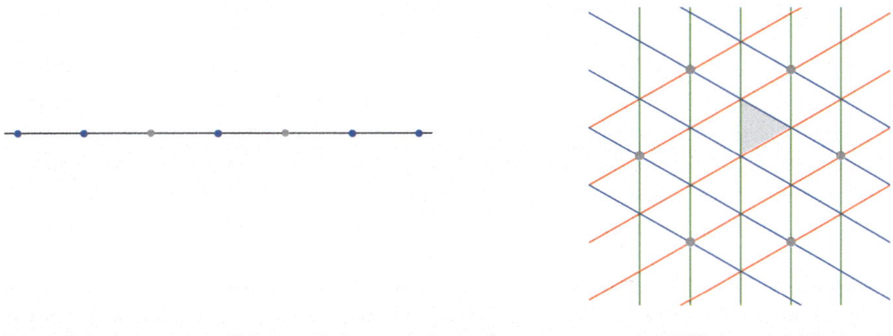

Fig. 1 The "triangulated" line and plane are respective apartments in $\mathcal{B}(\mathrm{SL}_2)$ and $\mathcal{B}(\mathrm{SL}_3)$, with roots darkened (using a choice of Weyl-invariant inner product)

Bruhat-Tits theory will also yield that the building is unaffected by passage to a central quotient, and more specifically that there is a compatible identification

$$\begin{array}{ccc} \mathcal{B}(G) & = & \mathcal{B}(G^{\mathrm{ad}} = G/Z_G) \\ \circlearrowleft & & \circlearrowleft \\ G(k) & \longrightarrow & G^{\mathrm{ad}}(k) \end{array}$$

where $G(k) \to G^{\mathrm{ad}}(k)$ is generally not surjective (think about $G = \mathrm{SL}_n$, so $G^{\mathrm{ad}} = \mathrm{PGL}_n$). Hence, the action of $G^{\mathrm{ad}}(k)$ can be regarded as an enrichment of the $G(k)$-action on the building for G.

When $G = \mathrm{SL}_n$ with $n \geq 2$, we have compatible actions

$$\begin{array}{ccc} \mathcal{B}(\mathrm{SL}_n) & = & \mathcal{B}(\mathrm{PGL}_n) \\ \circlearrowleft & & \circlearrowleft \\ \mathrm{SL}_n(k) & \longrightarrow & \mathrm{PGL}_n(k), \end{array}$$

where the cokernel of $q : \mathrm{SL}_n(k) \to \mathrm{PGL}_n(k)$ (which has normal image) is the group $k^\times/(k^\times)^n$ (via determinants); this cokernel is non-trivial, an extension of $\mathbf{Z}/n\mathbf{Z}$ by $\mathfrak{o}^\times/(\mathfrak{o}^\times)^n$.

The natural map $\mathrm{SL}_n(k) \to \mathrm{PGL}_n(k)$ carries each of the maximal compact subgroups $\mathcal{K}_m \subset \mathrm{SL}_n(k)$ (see (1.2)) into a conjugate of $\mathrm{PGL}_n(\mathfrak{o})$, and $\mathrm{SL}_n(\mathfrak{o}) \to \mathrm{PGL}_n(\mathfrak{o})$ is usually not surjective (its image is normal and the cokernel is $\mathfrak{o}^\times/(\mathfrak{o}^\times)^n$). We'll see in Proposition 4.1(4) that the map

$$q : \mathrm{SL}_n \to \mathrm{PGL}_n$$

on k-points has compact preimage for any compact subset of $\mathrm{PGL}_n(k)$, but the *failure* of surjectivity of q on k-points prevents us from using knowledge of maximal compact subgroups of $\mathrm{SL}_n(k)$ (up to conjugacy) to deduce the possibilities for maximal compact subgroups of $\mathrm{PGL}_n(k)$ up to conjugacy.

The problem is that if $\mathcal{K} \subset \mathrm{PGL}_n(k)$ is a maximal compact subgroup then $q^{-1}(\mathcal{K})$ is compact open in $\mathrm{SL}_n(k)$ but might not map *onto* $\mathcal{K}$. Hence, if $q^{-1}(\mathcal{K})$ is perhaps not maximal (so we don't have a sense of what it is), by picking a maximal compact subgroup $\widetilde{\mathcal{K}} \supset q^{-1}(\mathcal{K})$ maybe $q(\widetilde{\mathcal{K}})$ leaks outside $\mathcal{K}$ (this does not cause an inconsistency when $q(q^{-1}(\mathcal{K})) \subsetneq \mathcal{K}$)! Thus, knowledge of the maximal compact subgroups of $\mathrm{SL}_n(k)$ may fail to control the maximal compact subgroups of $\mathrm{PGL}_n(k)$. This failure is not just a hypothetical possibility, but really does happen, as we now sketch.

Example 2.4 Consider the case $n = 2$. The left picture in Fig. 1 is the apartment for the diagonal maximal torus of SL_2, with its vertices as shown, and chambers C as the open intervals between successive vertices. For a suitable chamber $\widetilde{C}$, with endpoints x_0 and x_1, we have

$$\mathrm{PGL}_2(k)_{x_0} = \mathrm{PGL}_2(\mathfrak{o}), \quad \mathrm{PGL}_2(k)_{x_1} = \begin{pmatrix} 1 & 0 \\ 0 & \pi \end{pmatrix} \mathrm{PGL}_2(\mathfrak{o}) \begin{pmatrix} 1 & 0 \\ 0 & \pi \end{pmatrix}^{-1},$$

so in contrast with SL_2 the $\mathrm{PGL}_2(k)$-stabilizers of the vertices of a single chamber wind up being *conjugate* subgroups. These do turn out to be maximal.

Now comes a new phenomenon for PGL_2 not seen with SL_2 (related to the failure of surjectivity of $\mathrm{SL}_2(k) \to \mathrm{PGL}_2(k)$): the element $\begin{pmatrix} 0 & 1/\pi \\ -1 & 0 \end{pmatrix}$ (whose determinant has odd valuation) flips the segment $\widetilde{\mathcal{C}}$, swapping x_0 and x_1 but fixing its midpoint ξ. The stabilizer $\widetilde{\mathcal{K}} := \mathrm{PGL}_2(k)_\xi$ (same as the stabilizer of $\widetilde{\mathcal{C}}$, not to be confused with the pointwise-stabilizer of $\widetilde{\mathcal{C}}$) turns out to be

$$\left\langle \begin{pmatrix} \mathfrak{o} & \mathfrak{o} \\ \mathfrak{m} & \mathfrak{o} \end{pmatrix}, \begin{pmatrix} 0 & 1/\pi \\ -1 & 0 \end{pmatrix} \right\rangle$$

that is a "new" maximal compact subgroup (up to conjugacy).

Bruhat-Tits theory provides an $\mathfrak{o}$-structure $\widetilde{\mathscr{G}}$ for PGL_2 such that $\widetilde{\mathscr{G}}(\mathfrak{o}) = \widetilde{\mathcal{K}}$ and $\widetilde{\mathscr{G}}_{\mathfrak{f}}$ is disconnected (its component group has order 2) with identity component $\mathbf{G}_m \ltimes (\mathbf{G}_a \times \mathbf{G}_a)$, where the semi-direct product is given by the action $t.(x, y) = (tx, t^{-1}y)$. Note via the special fiber that $\widetilde{\mathcal{K}} = \widetilde{\mathscr{G}}(\mathfrak{o})$ is prosolvable, whereas the maximal compact subgroups of $\mathrm{SL}_2(k)$ are all non-solvable when $\mathfrak{f} \neq \mathbf{F}_2, \mathbf{F}_3$ (since $\mathrm{SL}_2(\mathfrak{f})$ is non-solvable when $\#\mathfrak{f} > 3$). Bruhat-Tits theory also shows that the only conjugacy classes of maximal compact subgroups in $\mathrm{PGL}_2(k)$ are those of $\mathrm{PGL}_2(\mathfrak{o})$ and $\widetilde{\mathcal{K}}$.

Although there are 2 maximal compact subgroups in each of $\mathrm{SL}_2(k)$ and $\mathrm{PGL}_2(k)$, there is no natural bijection between the sets of conjugacy classes of the two groups; getting the same number 2 in both cases is a coincidence. For all *prime* n there are exactly two conjugacy classes of maximal compact subgroups of $\mathrm{PGL}_n(k)$: the class of $\mathrm{PGL}_n(\mathfrak{o})$ and of the stabilizer of the "metric center" of a chamber. (The maximality of $\mathrm{PGL}_n(\mathfrak{o})$ for all n is not obvious; it is a special case of Theorem B.3.) For composite n the number of such conjugacy classes is influenced by the number of divisors of n.

The main takeaway is that the elegant formula $\prod(r_i + 1)$ for the number of $G(k)$-conjugacy classes of maximal compact subgroups of $G(k)$ when G is simply connected does *not* generally hold for G merely connected semisimple. Hence, the group-theoretic information that we learn about $G(k)$ from the building $\mathcal{B}(G)$ can be extremely sensitive to passing to a central quotient of G even though the building (equipped with its structure of apartments and so on) is insensitive to such change in G. The failure of $G(k) \to (G/Z)(k)$ to be surjective for central $Z \subset G$ underlies rather different properties for the subgroup structure of these two groups of k-points.

3 Henselian Fields and Boundedness

We need to go beyond local fields for a variety of reasons. One reason is that having an *algebraically closed* residue field is sometimes convenient for analyzing special fibers of $\mathfrak{o}$-structures (much as passing to an algebraically closed base field is often convenient for some calculations in proofs with algebraic groups).

Let's review in a concrete setting how to arrive at a Galois extension of a local field k with the same uniformizer but algebraically closed residue field $\bar{\mathfrak{f}}$ at the cost of losing completeness but retaining a "shadow" of completeness.

Example 3.1 For $k = \mathbf{Q}_p$ (so $\mathfrak{o} = \mathbf{Z}_p$), we have the unramified extensions $\mathbf{Q}_p(\zeta_n)$ for $p \nmid n$, with valuation rings $\mathbf{Z}_p[\zeta_n]$. A cofinal system of these is given by $\mathbf{Q}_p(\zeta_{p^r-1})$ (with valuation ring $\mathbf{Z}_p[\zeta_{p^r-1}]$) for $r \geq 0$. The directed union $\mathbf{Z}_p^{\text{un}} := \bigcup_{p\nmid n} \mathbf{Z}_p[\zeta_n] = \bigcup_{r>0} \mathbf{Z}_p[\zeta_{p^r-1}]$ is a discrete valuation ring with uniformizer p and residue field $\overline{\mathbf{F}}_p$, and its fraction field $\mathbf{Q}_p^{\text{un}} := \bigcup_{p\nmid n} \mathbf{Q}_p(\zeta_n) = \bigcup_{r>0} \mathbf{Q}_p(\zeta_{p^r-1})$ is *not* complete. For example, the infinite series

$$\sum_{r\geq 0} \zeta_{p^r-1}\, p^r$$

has partial sums that form a Cauchy sequence but (by looking at the partial sums mod p^N for sufficiently large N) as an element of the completion equipped with the isometric action of $\mathrm{Gal}(\mathbf{Q}_p^{\text{un}}/\mathbf{Q}_p)$ we see it is not fixed by any open subgroup of the Galois group. Hence, as a point in the completion it cannot lie in $\mathbf{Q}_p(\zeta_{p^r-1})$ for any $r \geq 0$, so the infinite series really does not converge in $\mathbf{Q}_p^{\text{un}}$.

Although the discrete valuation ring $R = \mathbf{Z}_p^{\text{un}}$ is not complete, let's now see that it satisfies Hensel's lemma in a strong form (as for *any* complete discrete valuation ring): for any monic $f \in R[x]$ with reduction $f_0 = g_0 h_0$ for monic *coprime* g_0, h_0 over the residue field, there exist unique monic lifts $g, h \in R[x]$ satisfying $gh = f$. Note that the more classical instance of unique lifting of simple roots over the residue field is the special case $\deg(g_0) = 1$.

As a first step, we recall why all complete discrete valuation rings (or even complete local Noetherian rings) $(R, \mathfrak{m})$ with residue field κ satisfy the strong form of Hensel's Lemma. Write $A = R[x]/(f)$, a finite free R-module, so the map of κ-algebras

$$A/\mathfrak{m}A \to \kappa[x]/(f_0) \simeq \kappa[x]/(g_0) \times \kappa[x]/(h_0)$$

is an isomorphism. By successive approximation using that A is $\mathfrak{m}$-adically separated and complete, we may uniquely lift the non-trivial idempotent $e_0 = (1, 0) \in \kappa[x]/(g_0) \times \kappa[x]/(h_0)$ (as a solution to $Y^2 - Y = 0$) to an idempotent $e \in A$ and this yields a unique R-algebra decomposition $A = A_1 \times A_2$ lifting the κ-algebra decomposition of $A/\mathfrak{m}A$. Some careful thought yields that as $R[x]$-algebras, $A_1 = R[x]/(g)$ and $A_2 = R[x]/(h)$ for unique monic $g, h \in R[x]$ lifting g_0 and h_0 respectively and satisfying $gh = f$ (see [EGA, IV$_4$, Lemme 18.5.13.1] for the general equivalence between monic factorization of a monic $f \in C[x]$ for local C and ring-theoretic decompositions of $C[x]/(f)$).

To bootstrap from the complete case to infer that the strong form of Hensel's Lemma holds for the discrete valuation ring $\mathbf{Z}_p^{\mathrm{un}}$ despite its non-completeness, note that any monic $f \in \mathbf{Z}_p^{\mathrm{un}}$ arises "at finite level" over $\mathbf{Z}_p$, (i.e., $f \in \mathbf{Z}_p[\zeta_n][x]$ for some n not divisible by p), and we can run the complete case *there* to get the existence aspect of the strong form of Hensel's Lemma. The uniqueness at all higher levels ensures the uniqueness aspect over $\mathbf{Z}_p^{\mathrm{un}}$.

All the above works similarly for any complete discrete valuation ring $\mathfrak{o}$ with fraction field k, uniformizer π, and arbitrary residue field $\mathfrak{f}$ in place of $\mathbf{Z}_p$ and $\mathbf{Q}_p$, setting aside that $\overline{\mathbf{F}}_p$ has an explicit description in terms of roots of unity over $\mathbf{F}_p$. This is due to the categorical equivalence between finite unramified extensions k'/k and finite separable extensions $\mathfrak{f}'/\mathfrak{f}$ of the residue field.

The key point is that if $\mathfrak{f}'/\mathfrak{f}$ is finite separable and we then write $\mathfrak{f}' = \mathfrak{f}[x]/(g_0)$ for monic *separable* g_0 then we may pick a monic lift $g \in \mathfrak{o}[x]$ of g_0, so the module-finite flat $\mathfrak{o}$-algebra $\mathfrak{o}' := \mathfrak{o}[x]/(g)$ is a discrete valuation ring with π as a uniformizer (since $\mathfrak{o}'/\pi\mathfrak{o}' = \mathfrak{f}[x]/(g_0) = \mathfrak{f}'$ is a field, so [Ser1, Chap. 1, Sect. 2, Proposition 2] applies). For any finite unramified extension $\mathfrak{o}''$ of $\mathfrak{o}$, Hensel's Lemma in $\mathfrak{o}''$ for roots of g yields bijectivity of the natural map

$$\mathrm{Hom}_{\mathfrak{o}}(\mathfrak{o}', \mathfrak{o}'') \to \mathrm{Hom}_{\mathfrak{f}}(\mathfrak{f}', \mathfrak{f}'').$$

Such bijectivity gives a characterization of $\mathfrak{o}'$ intrinsic to $\mathfrak{f}'$ over $\mathfrak{f}$ without reference to the non-canonical choice of primitive element of $\mathfrak{f}'$ over $\mathfrak{f}$ or monic lift g of g_0.

Now if we *fix* a separable closure $\mathfrak{f}_s/\mathfrak{f}$ and for each $\mathfrak{f}' \subset \mathfrak{f}_s$ of finite $\mathfrak{f}$-degree build a corresponding $\mathfrak{o}'/\mathfrak{o}$ (unique up to $\mathfrak{o}$-algebra isomorphism) then we get a well-posed direct limit $\mathfrak{o}^{\mathrm{un}} := \varinjlim \mathfrak{o}'$ that is readily seen (by working at "finite level" in the limit process) to be a discrete valuation ring with the same uniformizer as for $\mathfrak{o}$ and residue field $\mathfrak{f}_s$ over $\mathfrak{f}$.

The $\mathfrak{o}$-algebra $\mathfrak{o}^{\mathrm{un}}$ is called the *maximal unramified extension* of $\mathfrak{o}$, and is generally *not* complete (much as for $\mathbf{Z}_p^{\mathrm{un}}$) but satisfies the strong form of Hensel's lemma. One checks that for the inertia subgroup $I_k \subset \mathrm{Gal}(k_s/k)$, we have $(k_s)^{I_k} = \mathrm{Frac}(\mathfrak{o}^{\mathrm{un}}) =: k^{\mathrm{un}}$.

A deep theorem by Steinberg on the non-abelian Galois cohomology of such extensions k^{un} ensures that connected reductive groups over such fields are always *quasi-split* (i.e., have a Borel subgroup over the base field) when $\mathfrak{f}$ is perfect; see Theorem 10.1 and Example 10.2. Constructing Bruhat-Tits theory in the quasi-split case is far simpler than in the general case (though still highly non-trivial). This is a fundamental reason that we allow for passage to k^{un} when setting up Bruhat-Tits theory.

A crucial merit of unramifiedness is that "étale descent over $\mathfrak{o}$" can be expressed in essentially Galois-theoretic terms, whereas finite flat descent through ramified extensions $\mathfrak{o}'$ over $\mathfrak{o}$ cannot be expressed in such terms. Such étale descent is used to construct Bruhat-Tits theory over the original base field k when it is in hand over k^{un}. If we were only to work with the (typically non-algebraic) completion $\widehat{k^{\mathrm{un}}}$, it would be very hard to descend constructions down to k; this is why we shall avoid limiting ourselves to complete base fields.

Having reviewed where k^{un} comes from, why it is usually not complete, and why it satisfies the strong form of Hensel's Lemma, we now put this into a broader setting:

Proposition 3.2 *Let R be a discrete valuation ring with fraction field F. The following are equivalent:*

(1) the valuation on F extends uniquely to all algebraic (equivalently, finite separable) extensions F'/F;
(2) R satisfies the strong form of Hensel's lemma (existence and uniqueness for monic factorization over R lifting a coprime one over the residue field);
(3) the local ring R is Henselian in sense of [EGA, IV_4, Theorem 18.5.11].

Proof The equivalence of (1) and (2) is [Ber, Proposition 2.4.3]. The equivalence of (2) and (3) is [EGA, IV_4, Remark 18.5.13]. □

We emphasize that in Proposition 3.2, property (1) is purely field-theoretic in flavor, whereas (2) and (3) focus on the integral structure R. Moreover, if we weaken the requirement on the valuation to be merely rank-1 rather than a discrete valuation then the Proposition holds with the same proof (i.e., the cited references work in that extra generality). Finally, with more work it can be shown that the strong form of Hensel's Lemma as discussed above is unaffected if we remove the uniqueness requirement on the lifted factorization and if we limit the hypothesis only to the existing of lifting of simple roots over the residue field: see [Stacks, Tag 04GG].

Definition 3.3 A discretely-valued field F is called *Henselian* when the equivalent conditions in Proposition 3.2 hold (using the given valuation ring R).

Via Proposition 3.2(2), for any Henselian discretely-valued F we can construct F^{un} as in our earlier discussion for complete discretely-valued fields. This is easily seen to be a Galois extension of F, with associated closed normal subgroup $I_F \subset G_F$ consisting of those $\gamma \in G_F$ inducing the identity on the residue field of F_s. The valuation ring R^{un} of F^{un} is a Henselian discrete valuation ring with residue field $\mathfrak{f}_s$ and any uniformizer of R is a uniformizer of R^{un} exactly as for complete F. The R-algebra R^{un} coincides with the *strict henselization* of R in the sense of [EGA, IV_4, Definition 18.8.7] due to the universal property [EGA, IV_4, Proposition 18.8.8(ii)] characterizing strict henselizations.

Remark 3.4 The discreteness condition on F in the definition of "Henselian" can be removed since Proposition 3.2 is valid for rank-1 valuations in general. In that generality we can say that every algebraic extension of a Henselian field is Henselian, though the construction of F^{un} requires more care without discreteness (since the notion of "ramification index" is more subtle to define).

We will have no use for such generality, but you can find a discussion of it in [Ber, Sect. 2.4] if you are interested.

Sometimes when working with Henselian fields it is useful to reduce problems to the complete setting. This is often achieved via the following handy fact:

Proposition 3.5 *For F Henselian, $L \mapsto L \otimes_F \widehat{F}$ is an equivalence of categories from "finite separable fields over F" to "finite separable fields over $\widehat{F}$." In particular,* $\mathrm{Gal}(F_s/F) \simeq \mathrm{Gal}((\widehat{F})_s/\widehat{F})$

Proof This is [Ber, Proposition 2.4.1]. □

In the Henselian setting, the base field k is generally not locally compact and so we need to find an appropriate concept to replace compactness for subgroups of $G(k)$ (with finite type affine k-groups G). The replacement arises from the following fact:

Lemma 3.6 *Let F be a discretely-valued field. Let X be an affine F-scheme of finite type and $j : X \hookrightarrow \mathbf{A}^n_F$ a closed immersion over F. For any subset $B \subset X(F)$, the following are equivalent*

(1) the subset $j(B) \subset F^n$ is bounded (for the valuation topology),
(2) for any h in the coordinate ring $F[X]$ of X over F, the image $h(B) \subset F$ is bounded.

Proof The coordinate ring $F[X]$ is generated as an F-algebra by the component functions of j. □

Definition 3.7 A subset $B \subset X(F)$ satisfying the equivalent conditions in Lemma 3.6 is called *bounded* (in X).

We will often speak of boundedness without saying "in X", but the ambient affine scheme is always implicit in such considerations. For instance, as we'll see in Example 3.9 (with $n = 1$), a subset B of $F^\times$ is bounded in GL_1 if and only if B and $1/B$ are each bounded in $\mathbf{G}_a$ (e.g., $R - \{0\}$ is bounded in $\mathbf{G}_a$ but not in GL_1).

The concept of boundedness can be defined beyond the affine case, for any F-scheme of finite type (e.g., if X is F-proper then $X(F)$ is always bounded in X), but we will not have any need for such generality. See [BLR, Sect. 1.1, Def. 2(b)] and the discussion following it for a development of boundedness beyond the affine case.

Example 3.8 If k is a complete discretely-valued field with finite residue field then a subset $B \subset X(k)$ is bounded if and only if its topological closure $\overline{B}$ for the natural topology on $X(k)$ is compact.

This holds because a closed subset of k^n is compact if and only if it is bounded for the sup-norm topology relative to the standard basis.

Since open subgroups of a topological group are always closed (as cosets for an open subgroup are open too), it follows that for the study of open subgroups of $G(k)$ for Henselian k and affine k-groups G of finite type, the boundedness property coincides with compactness when k is complete with finite residue field. This is the reason that our focus in Bruhat-Tits theory will be on bounded subgroups of $G(k)$ for Henselian k.

Example 3.9 An instructive special case for making boundedness concrete is $X = \mathrm{GL}_n$ over a discretely-valued field F. We claim that a subset $B \subset \mathrm{GL}_n(F)$ is bounded in GL_n if and only if both subsets $B, B^{-1} \subset \mathrm{Mat}_n(F)$ are bounded for the usual topology of the matrix algebra as a finite-dimensional F-vector space.

The implication "$\Rightarrow$" is readily verified via Cramer's formula, since $F[\mathrm{GL}_n]$ is generated over F by the "matrix entry" functions x_{ij} and $1/\det$. For the converse, we have to show that if the subsets $B, B^{-1} \subset \mathrm{Mat}_n(F)$ are bounded in the usual sense then $1/\det : B \to F$ is bounded. For this it is enough to show that $F[\mathrm{GL}_n]$ is generated as a F-algebra by the matrix entry functions x_{ij} along with the matrix entries of the inverse of (x_{ij}). In more concrete terms, this is saying that the natural map of F-schemes $\varphi : \mathrm{GL}_n \to \mathrm{Mat}_{2n}$ defined by

$$x \mapsto \begin{pmatrix} x & 0 \\ 0 & x^{-1} \end{pmatrix}$$

(with $n \times n$ blocks on the right side) is a closed immersion. That closed immersion property holds because φ is an isomorphism onto the closed subscheme

$$\left\{ \begin{pmatrix} x & y \\ z & w \end{pmatrix} \in \mathrm{Mat}_{2n} : xw = 1_n, y = 0, z = 0 \right\}$$

of Mat_{2n} (with x, y, z, w denoting $n \times n$ blocks).

When $n = 1$, this characterization of boundedness in GL_n says that a subset $B \subset \mathrm{GL}_1(F) = F^\times$ is bounded in GL_1 if and only if both B and B^{-1} are bounded in F, which is to say that B is "bounded away from 0 and ∞".

For a *subgroup* $B \subset \mathrm{GL}_n(k)$ we have $B^{-1} = B$, so B is bounded in GL_n if and only if B is bounded as a subset of $\mathrm{Mat}_n(F)$ in the usual sense (i.e., each "matrix entry" function on B is bounded). This will be used all the time without comment.

4 More on Bounded Subgroups

Now k is assumed to be a Henselian (discretely-valued) field, with a general residue field $\mathfrak{f}$. From now on, by *Henselian* we always mean Henselian discretely-valued (with non-trivial valuation, so the valuation ring is a discrete valuation ring rather than k itself). Here are some basic properties of boundedness for subsets of the rational points of affine k-schemes of finite type:

Proposition 4.1 *The following properties of boundedness hold:*

(1) If $f : X \to Y$ is a k-map between affine k-schemes of finite type and $B \subset X(k)$ is bounded (in X) then the image $f(B) \subset Y(k)$ is bounded (in Y).

(2) *For G an affine k-group of finite type and B, $B' \subset G(k)$ bounded, the subset of products*

$$BB' = \{bb' : b \in B, b' \in B'\} \subset G(k)$$

is bounded.

(3) *For the Weil restriction $X = \mathrm{R}_{k'/k}(X')$ with finite separable k'/k (so k' is also Henselian) and X' affine of finite type over k', the bijection $X(k) = X'(k')$ preserves boundedness both ways.*

(4) *If $f : X \to Y$ is a finite k-map between affine k-schemes of finite type and $B \subset Y(k)$ is bounded (in Y) then the preimage $f^{-1}(B) \subset X(k)$ is bounded (in X).*

Remark 4.2 The importance of Proposition 4.1(3) is that for connected semisimple G that is either simply connected or adjoint type we have $G = \prod_i \mathrm{R}_{k'_i/k}(G'_i)$ for G'_i that are absolutely simple over k'_i (and simply connected, resp. adjoint type).

Proof of Proposition 4.1 (1) For $h \in k[Y]$, we want $h(f(B)) \subset k$ to be bounded. But $h(f(B)) = (f^*(h))(B)$ for $f^*(h) \in k[X]$, and B is bounded in X.

(2) Apply (1) to $B \times B' \subset (G \times G)(k)$ and $f : G \times G \to G$ the multiplication map.

(3) Pick a closed immersion $\gamma : X' \hookrightarrow \mathbf{A}^n_{k'}$ over k' to get $X = \mathrm{R}_{k'/k}(X') \to \mathrm{R}_{k'/k}(\mathbf{A}^n_{k'}) \simeq \mathbf{A}^{nd}_k$ that is a closed immersion over k. Thus we reduce to the case $X' = \mathbf{A}^n_{k'}$.

A subset $B' \subset X'(k') = (k')^n$ is bounded if and only if its image in each factor is bounded. More generally, a subset of $(Y_1 \times Y_2)(k')$ is bounded (in $Y_1 \times Y_2$) if and only if its images in $Y_1(k')$ and $Y_2(k')$ are bounded (since $k'[Y_1 \times Y_2] = k'[Y_1] \otimes_{k'} k'[Y_2]$), and similarly over k. Hence, without loss of generality we may and do assume $X' = \mathbf{A}^1_{k'}$.

Pick a k-basis for k' to define a k-isomorphism $X \simeq \mathbf{A}^d_k$, so our task reduces to showing a subset of k' is bounded for that valuation topology if and only if it is bounded for a k-vector space topology defined using a choice of k-basis (and the valuation topology of k).

We will show that the valuation norm on k' is "equivalent" to the sup-norm with respect to a k-basis, in the sense that each is bounded by a positive constant multiple of the other (viewing the valuation as an absolute value). For complete k this is true because there is only *one* complete norm up to equivalence on a finite-dimensional k-vector space (for k complete); that is seen via the Banach Open Mapping Theorem or other means.

To reduce to the complete case, consider the diagram

$$\begin{array}{ccc} \widehat{k} \otimes_k k' & \xrightarrow{\sim} & \widehat{k'} \\ \uparrow & & \uparrow \\ k' & = & k' \end{array}$$

where the isomorphism along the top is as $\widehat{k}$-algebras (here we are using Proposition 3.5 and separability of k'/k). For the inclusion along the left, the sup-norms defined using a k-basis of k' (which is also a $\widehat{k}$-basis of $\widehat{k} \otimes_k k'$) are compatible. For the inclusion along the right, the field valuations are compatible. Hence, the settled "uniqueness up to equivalence" along the top from the complete base field case gives the desired result in the Henselian case.

(4) The map $f^* : k[Y] \to k[X]$ is module-finite, so each of a finite set of k-algebra generators of $k[X]$ satisfies a *monic* polynomial relation over $k[Y]$. The usual bound on the absolute values of roots of a monic polynomial of specified degree d with a given bound on the absolute value of its coefficients then does the job. □

Let's now turn to bounded subgroups of $G(k)$ for affine k-groups G of finite type. We begin with a fundamental result that explains why in the study of bounded subgroups the presence of non-trivial (and especially non-central) split tori corresponds to interesting settings. This was mentioned in Remark 2.3 for complete k (with finite residue field) in the language of compactness, but here it is stated in a more natural form with Henselian k:

Theorem 4.3 (Bruhat-Tits, Rousseau) *If G is connected reductive over a Henselian k, $G(k)$ is bounded in G if and only if G is k-anisotropic (i.e., there are no non-trivial k-split tori in G).*

This result relates a topological property (boundedness) to an algebraic property (having no non-trivial split k-tori). Recall also that having non-central split k-tori is equivalent to having proper parabolic k-subgroups.

Proof There is an elegant proof due to G. Prasad given in [KP, Theorem 2.2.9]. The key case is when G is semisimple (which Prasad treats via the Jacobson Density Theorem), since:

(1) Boundedness is preserved under preimage for finite maps (Proposition 4.1(4)).
(2) Isogenies are finite, and there is a canonical (central) isogeny $G \to G^{\mathrm{ad}} \times (G/\mathscr{D}G)$ (where $G^{\mathrm{ad}} := G/Z_G$ is semisimple and $G/\mathscr{D}G$ is a torus) and a canonical isogeny $\mathscr{D}(G) \times (Z_G^0)_{\mathrm{red}} \to G$ (where $\mathscr{D}(G)$ is semisimple and $(Z_G^0)_{\mathrm{red}}$ is a torus).
(3) Tori can be treated directly because every k-torus T is an "isogeny factor" of a power of $\mathrm{R}_{k'/k}(\mathbf{G}_m)$ for some finite separable k'/k (one writes the geometric character lattice of T as a quotient of a direct sum of copies of $\mathbf{Z}[\mathrm{Gal}(k'/k)]$ for a splitting field k'/k of T).

□

Proposition 4.4 (Langlands) *Let G be a connected reductive k-group. Then every bounded subgroup $\mathcal{K} \subset G(k)$ lies in a maximal one, and maximal ones are open.*

Proof We'll first show that $\mathcal{K}$ lies in a bounded *open* subgroup, so then we can focus on *open* $\mathcal{K}$. Indeed, once this general statement is proved then any maximal bounded subgroup also lies in a bounded open subgroup which must (by maximality) coincide with itself, and so such maximal bounded subgroups would be automatically open.

Pick a closed immersion of k-groups $j : G \hookrightarrow \mathrm{GL}_n$, so $\mathcal{K}$ is bounded in GL_n (via $\mathcal{K} \subset G(k) \subset \mathrm{GL}_n(k)$). If $\mathcal{K} \subset \mathrm{GL}_n(k)$ lies in an *open* bounded subgroup $\mathcal{U} \subset \mathrm{GL}_n(k)$ then $G(k) \cap \mathcal{U} = j^{-1}(\mathcal{U})$ is *bounded* open in $G(k)$ containing $\mathcal{K}$, where boundedness for this j-preimage rests on finiteness of j. Hence, for finding a bounded open subgroup containing $\mathcal{K}$ we may suppose $G = \mathrm{GL}_n$.

We will show any bounded subgroup $\mathcal{K} \subset \mathrm{GL}_n(k)$ in GL_n preserves some $\mathfrak{o}$-lattice Λ, yielding the containment $\mathcal{K} \subset \mathrm{GL}(\Lambda) \subset \mathrm{GL}_n(k)$ where $\mathrm{GL}(\Lambda)$ is open and bounded in GL_n (since after a change of k-basis of k^n, $\mathrm{GL}(\Lambda)$ becomes $\mathrm{GL}_n(\mathfrak{o}) \subset \mathrm{GL}_n(k)$). Such a Λ will be found at the start of Sect. 5, and then we will show that every *open* bounded subgroup lies in a maximal bounded subgroup. □

5 Maximal Bounded Subgroups

We continue the proof of Proposition 4.4. Remember that without further notice, the field k is always assumed to be Henselian (with a non-trivial discrete valuation).

Proof of Proposition 4.4 (***continued***) Let $\mathcal{K} \subset \mathrm{GL}_n(k)$ be a bounded subgroup. We seek an $\mathfrak{o}$-lattice $\Lambda \subset k^n$ stable under $\mathcal{K}$, since then $\mathcal{K} \subset \mathrm{GL}(\Lambda) \subset \mathrm{GL}_n(k)$, where $\mathrm{GL}(\Lambda) \subset \mathrm{GL}_n(k)$ is bounded open because it is conjugate to $\mathrm{GL}_n(\mathfrak{o})$ (since $\Lambda = \gamma(\mathfrak{o}^n)$ for some $\gamma \in \mathrm{GL}_n(k)$). Once this is done, it was seen near the end of Sect. 4 that for any connected reductive k-group G and bounded subgroup $\mathcal{K} \subset G(k)$ there exists a *bounded open* subgroup $\mathcal{U} \subset G(k)$ such that $\mathcal{K} \subset \mathcal{U}$. Thus, it would then suffice to show that any bounded *open* subgroup of such $G(k)$ is contained in a maximal bounded subgroup.

Let $\Lambda_0 = \mathfrak{o}^n \subset k^n$. Consider the usual action map

$$\mathrm{GL}_n \times \mathbf{A}_k^n \xrightarrow{\alpha} \mathbf{A}_k^n$$

over k. Since the subset $\mathcal{K} \times \Lambda_0 \subset \mathrm{GL}_n(k) \times k^n$ is bounded in $\mathrm{GL}_n \times \mathbf{A}_k^n$, the image $\mathcal{K}\Lambda_0 = \alpha(\mathcal{K} \times \Lambda_0) \subset k^n$ is *bounded* and so is contained in $(1/\pi^N)\Lambda_0$ for some $N \geq 0$. Thus, the $\mathfrak{o}$-linear span

$$\Lambda := \mathfrak{o}[\mathcal{K}\Lambda_0] \subset \frac{1}{\pi^N}\Lambda_0$$

is also bounded, yet it contains Λ_0 too (since $1 \in \mathcal{K}$). This implies Λ is an $\mathfrak{o}$-lattice in k^n which is $\mathcal{K}$-stable by design. That completes the proof that any bounded subgroup is contained in an open bounded subgroup.

Returning to connected reductive G over k and a bounded *open* subgroup $\mathcal{K} \subset G(k)$, we seek a maximal bounded subgroup $\widetilde{\mathcal{K}} \subset G(k)$ containing $\mathcal{K}$ (note that any such $\widetilde{\mathcal{K}}$ inherits openness from $\mathcal{K}$ due to being a subgroup containing an open neighborhood of 1).

By Zorn's lemma, it suffices to show for a *chain* $\{\mathcal{K}_i\}$ of bounded subgroups of $G(k)$ containing $\mathcal{K}$, the union $\bigcup_i \mathcal{K}_i =: \mathcal{H}$ (which is a subgroup!) is bounded. Note that $\mathcal{H}$ is open since $\mathcal{K} \subset \mathcal{H}$. This ensures that $\mathcal{H}$ is Zariski-dense in G by the following lemma.

Lemma 5.1 *For smooth connected affine X over k, any non-empty open $\mathcal{U} \subset X(k)$ is Zariski-dense in X.*

This lemma holds with "affine" removed, by reduction to the affine case, but we stick to the affine case since it is all we shall need.

Proof We first prove the lemma for k complete, and then explain how to reduce the problem for general k to the problem over $\widehat{k}$. Since X is smooth and connected, so it is irreducible and reduced, the coordinate ring $k[X]$ is a domain. For Zariski-density, it suffices to show that if $f \in k[X]$ satisfies $f(\mathcal{U}) = \{0\}$ then $f = 0$. Pick $u_0 \in \mathcal{U}$, so the natural map $k[X] \to k[X]_{\mathfrak{m}_{u_0}} = \mathscr{O}_{X,u_0}$ is injective.

By the Zariski-local structure of smooth maps applied to $X \to \mathrm{Spec}(k)$ (i.e., Zariski-locally factors as an étale map to an affine space) and the Zariski-local structure of étale maps, for complete k (so there is a good theory of k-analytic manifolds: recall that the k-analytic Inverse and Implicit Function Theorems rest on the contraction mapping theorem for *complete* metric spaces) the topological space $X(k)$ has the structure of a k-analytic manifold X^{an} (not to be confused with the associated rigid-analytic space X^{rig} that usually has many non-rational points).

For any $x \in X(k)$, the natural local map $\mathscr{O}_{X,x} \to \mathscr{O}_{X^{\mathrm{an}},x}$ of local k-algebras induces an isomorphism of max-adic completions

$$\mathscr{O}^{\wedge}_{X,x} = \mathscr{O}^{\wedge}_{X^{\mathrm{an}},x}$$

(this is seen by inspection for X an affine space, and in general it is deduced from affine spaces via an étale k-map $V \to \mathbf{A}^n_k$ for a Zariski-open $V \subset X$ around x).

The hypothesis $f|_{\mathcal{U}} = 0$ implies that f has vanishing image in $\mathscr{O}_{X^{\mathrm{an}},u_0} \subset \mathscr{O}^{\wedge}_{X^{\mathrm{an}},u_0} = \mathscr{O}^{\wedge}_{X,u_0}$, but this is the same as the image of the algebraic u_0-stalk of f under $\mathscr{O}_{X,u_0} \to \mathscr{O}^{\wedge}_{X,u_0}$. This latter map is injective, so f has vanishing algebraic stalk at u_0 and hence $f = 0$.

Having treated the case of complete k, for general Henselian k we argue as follows. Consider $X(k)$ as a subset of $X(\widehat{k}) = X_{\widehat{k}}(\widehat{k})$. It has the subspace topology (and this can all be viewed inside the metric space $\widehat{k}^N$ via a closed immersion $X \hookrightarrow \mathbf{A}^N_k$ over k), so there is an open subset $\mathcal{U}' \subset X(\widehat{k})$ such that $\mathcal{U}' \cap X(k) = \mathcal{U}$. We now apply the following fact that is useful in a variety of contexts for reducing problems over a Henselian field to the case of a complete base field.

Proposition 5.2 *For any smooth scheme Y over a Henselian discretely-valued field k, the subset $Y(k) \subset Y(\widehat{k})$ is dense.*

Proof This is [KP, Proposition 2.3.4(1)]. The idea of its proof is to use the Zariski-local description of smooth schemes as étale over an affine space to reduce to étale Y, which is settled using Proposition 3.5. □

By Proposition 5.2, every point in the open subset $\mathcal{U}' \subset X(\widehat{k})$ is a limit of points in $X(k) \cap \mathcal{U}' = \mathcal{U}$, so any $f \in k[X] \subset \widehat{k}[X_{\widehat{k}}]$ vanishing on $\mathcal{U}$ also vanishes on $\mathcal{U}'$. Thus, the settled Zariski-density over $\widehat{k}$ would do the job *provided* $X_{\widehat{k}}$ is also connected. But X is connected with $X(k)$ non-empty, so X is *geometrically* connected over k [EGA, IV$_2$, 4.5.13] and hence $X_{\widehat{k}}$ is indeed connected. □

Now we assume $\mathcal{H}$ is unbounded and seek a contradiction. In G. Prasad's proof [KP, Lemma 2.2.11] of the Bruhat-Tits-Rousseau Theorem (Proposition 4.3), reductivity is used to deduce via the Jacobson Density Theorem that any unbounded *Zariski-dense* subgroup of $G(k)$ contains an element γ having a non-integral eigenvalue $\lambda \in \overline{k}$ with respect to a fixed choice of closed immersion of k-groups $G \hookrightarrow \mathrm{GL}_{N,k}$.

Conversely, any subgroup $\mathcal{H}' \subset G(k)$ containing such a γ *must* be unbounded. Indeed, if such an $\mathcal{H}'$ were bounded in G, hence in GL_N, we would get uniform bounds on the absolute values of the coefficients of the degree-N characteristic polynomial at all points of $\mathcal{H}'$ and hence uniform bounds on the absolute values of all eigenvalues in $\overline{k}$ of all elements of $\mathcal{H}'$. But $\gamma^m \in \mathcal{H}'$ has λ^m as an eigenvalue and $|\lambda^m| = |\lambda|^m$ approaches to ∞ as $m \to \infty$ since $|\lambda| > 1$, violating such uniform bounds. Thus, any such $\mathcal{H}'$ must indeed be unbounded.

By design $\mathcal{H} = \bigcup_i \mathcal{K}_i$, so the "bad" $\gamma \in \mathcal{H}$ arising from unboundedness (and Zariski-density) lies in some $\mathcal{K}_{i_0}$. Thus, the preceding converse implies $\mathcal{K}_{i_0}$ is unbounded, a contradiction! □

Corollary 5.3 *Maximal bounded subgroups of* $\mathrm{GL}_n(k)$ *are exactly the conjugates of* $\mathrm{GL}_n(\mathfrak{o})$.

Proof We know that a maximal bounded subgroup $\mathcal{K} \subset \mathrm{GL}_n(k)$ exists. But $\mathcal{K} \subset \mathrm{GL}(\Lambda)$ for some $\mathfrak{o}$-lattice $\Lambda \subset k^n$, so $\mathcal{K} = \mathrm{GL}(\Lambda)$ by the maximality of $\mathcal{K}$. Thus $\mathrm{GL}(\Lambda)$ is *maximal* for that Λ, and maximality then follows for all of its conjugates, which are precisely the conjugates of $\mathrm{GL}_n(\mathfrak{o})$. This argument shows that all maximal bounded subgroups arise in this way too. □

Observe that for $\mathscr{X}$ an affine (flat) $\mathfrak{o}$-scheme of finite type with generic fiber X over k, the subset $\mathscr{X}(\mathfrak{o}) \subset X(k)$ is bounded and open. Indeed, make a choice of closed immersion $\mathscr{X} \hookrightarrow \mathbf{A}^N_{\mathfrak{o}}$ over $\mathfrak{o}$ to define an associated choice of closed immersion $X \hookrightarrow \mathbf{A}^N_k$, and then note that $\mathscr{X}(\mathfrak{o}) = X(k) \cap \mathbf{A}^N_{\mathfrak{o}}(\mathfrak{o}) = X(k) \cap \mathfrak{o}^k \subset k^N$ with $\mathfrak{o}^N$ bounded open in k^N and $X(k)$ having the subspace topology from k^N.

In particular, if $\mathscr{G}$ is a flat affine $\mathfrak{o}$-group of finite type with k-fiber denoted G then $\mathscr{G}(\mathfrak{o})$ is a bounded open subgroup of $G(k)$. But which bounded open subgroups arise in this way, particularly with *smooth* $\mathscr{G}$? As an application of Corollary 5.3, we now record that this construction at least accounts for all *maximal* bounded subgroups (when G is connected reductive). This will be the basis for the role of $\mathfrak{o}$-group schemes in the development of Bruhat-Tits theory.

Proposition 5.4 *If G is connected reductive over k and $\mathcal{K} \subset G(k)$ is a maximal bounded subgroup then $\mathcal{K} = \mathscr{G}(\mathfrak{o})$ for a (smooth!) $\mathfrak{o}$-structure $\mathscr{G}$ of G.*

The only role of reductivity in the proof, which is taken up at the start of Sect. 6, is that Proposition 4.4 provides a plentiful supply of maximal bounded subgroups in $G(k)$ for connected reductive G.

Remark 5.5 Although for any $\mathfrak{o}$-structure $\mathscr{G}$ of G the subgroup $\mathscr{G}(\mathfrak{o}) \subset G(k)$ is bounded and open, such subgroups are often *not* maximal (and this will be rather important for Bruhat-Tits theory). Indeed, given such a $\mathscr{G}$, consider $\mathcal{K}_1 = \ker(\mathscr{G}(\mathfrak{o}) \to \mathscr{G}(\mathfrak{f}))$ (usually a proper subgroup of $\mathscr{G}(\mathfrak{o})$). This is readily seen to coincide with $\mathscr{G}^+(\mathfrak{o})$ for

$$\mathscr{G}^+ := \mathrm{Spec}\,\left(\mathfrak{o}[\mathscr{G}][(1/\pi)\mathscr{I}_{e_0}]\right)$$

the (smooth!) "dilatation" of $\mathscr{G}$ along the trivial subgroup $e_0 \in \mathscr{G}_\mathfrak{f}$ (this dilatation is an affine open in the blow-up $\mathrm{Bl}_{e_0}(\mathscr{G})$ along the identity point e_0 with associated ideal $\mathscr{I}_{e_0} \subset \mathfrak{o}[\mathscr{G}]$).

It is clear that the affine flat finite type $\mathfrak{o}$-scheme $\mathscr{G}^+$ has k-fiber $\mathscr{G}_k = G$, and it is easily verified that $\mathscr{G}^+$ has an $\mathfrak{o}$-group structure compatible with the k-group structure on its generic fiber G. It is less evident that $\mathscr{G}^+$ inherits $\mathfrak{o}$-smoothness from $\mathscr{G}$; we will return to dilatations in Example 6.7, including how they interact with smoothness.

Remark 5.6 An important class of cases when $\mathscr{G}(\mathfrak{o})$ *is* guaranteed to be maximal as a bounded subgroup of $G(k)$ is $\mathscr{G}$ a *reductive* $\mathfrak{o}$-group, which means (by definition, given that $\mathscr{G}_k = G$ is connected reductive) the special fiber $\mathscr{G}_\mathfrak{f}$ is connected reductive. Maximality in such cases is not obvious (not even for Sp_{2g}, for example), and is established as part of Bruhat-Tits theory (see Theorem B.3). Such maximal bounded subgroups are called *hyperspecial* (for reasons explained in Appendix B). Even for a *split* connected reductive G, there can be maximal bounded subgroups that are *not* hyperspecial (e.g., this happens for Sp_4 due to Example 20.11 for $\mathrm{B}_2 = \mathrm{C}_2$, Corollary 29.10, and Corollary B.5).

In the definition of reductivity for $\mathfrak{o}$-structures, it suffices to require $\mathscr{G}_\mathfrak{f}^0$ is reductive (i.e., this forces $\mathscr{G}_\mathfrak{f}$ to be connected). Indeed, more generally for any scheme S and smooth affine S-group $G \to S$ with all fibral identity components G_s^0 reductive, the locus of $s \in S$ for which G_s is connected is always *closed* [C2, Proposition 3.1.12]. Thus, when S is irreducible and the generic fiber G_η is connected then that closed set must be the entirety of S and so all fibers are connected.

6 Boundedness and Integral Models

Let's recall the statement of Proposition 5.4, which we shall now prove:

Proposition 6.1 *Let G be a connected reductive group over k and $\mathcal{K} \subset G(k)$ a maximal bounded subgroup. Then $\mathcal{K} = \mathscr{G}(\mathfrak{o})$ for some (smooth!) $\mathfrak{o}$-structure $\mathscr{G}$ of G.*

Remark 6.2 It is non-obvious but true that when $\mathfrak{f}$ is *finite*, the action of $G^{\mathrm{ad}}(k)$ on G is transitive on the set of all reductive $\mathfrak{o}$-structures on G (if any exist; often none exist!) by [C2, Theorem 7.2.16] (this reference is written with the hypothesis there that the ground field is also complete, but only its Henselian property is used in the proof, along with Lang's theorem that *connected* smooth groups over finite fields have no non-trivial torsors).

Beware however that $G(k) \to G^{\mathrm{ad}}(k)$ is usually *not* surjective (e.g., $\mathrm{SL}_n(k) \to \mathrm{PGL}_n(k)$), so this $G^{\mathrm{ad}}(k)$-action on G tends to go *beyond* $G(k)$-conjugation. We have seen this phenomenon before in our discussion of $G = \mathrm{SL}_n$ in Example 1.7, where (for instance) with $n = 2$ the action by $\begin{pmatrix} \pi & 0 \\ 0 & 1 \end{pmatrix} \in \mathrm{PGL}_2(k) = \mathrm{GL}_2(k)/k^\times$ (which can be interpreted as actual conjugation inside $\mathrm{GL}_2(k)$) gives rise to the two conjugacy classes of maximal bounded subgroups of $\mathrm{SL}_2(k)$.

(The discussion of maximal bounded subgroups of $\mathrm{SL}_n(k)$ in Example 1.7 was written for k locally compact because at that early stage we had not developed notions for Henselian fields. Now you can look back to see that it applies verbatim with Henselian k and "bounded" in place of "compact", thanks to Corollary 5.3 over Henselian fields).

Proof of Proposition 6.1 Pick a closed immersion of k-groups $\rho : G \hookrightarrow \mathrm{GL}_{n,k}$, so $\mathcal{K} \subset G(k) \hookrightarrow \mathrm{GL}_n(k)$ is a bounded subgroup. Thanks to Corollary 5.3, by replacing ρ with some $\mathrm{GL}_n(k)$-conjugate we may and do arrange $\mathcal{K} \subset \mathrm{GL}_n(\mathfrak{o})$.

Consider the schematic closure $\mathcal{G}$ of $G \hookrightarrow \mathrm{GL}_{n,k} \hookrightarrow \mathrm{GL}_{n,\mathfrak{o}}$ (a composition of a closed immersion with an open immersion). Concretely, $\mathfrak{o}[\mathcal{G}]$ is the image of $\mathfrak{o}[\mathrm{GL}_n] \to k[G]$; since $\mathfrak{o}$ is Dedekind, $\mathcal{G}$ is $\mathfrak{o}$-flat and thus an $\mathfrak{o}$-subgroup of $\mathrm{GL}_{n,\mathfrak{o}}$ (in general for any Dedekind domain R with fraction field F and any flat R-scheme $\mathscr{X}$, the set of closed subschemes Z of $X = \mathscr{X}_F$ is in natural bijective correspondence with the set of R flat closed subschemes $\mathscr{Z}$ of $\mathscr{X}$, via the operations of schematic closure and generic fiber; to prove this we may work Zariski-locally on $\mathscr{X}$ to reduce to affine $\mathscr{X}$).

By design, $\mathcal{G}_k = G$ inside $\mathrm{GL}_{n,k}$. We have $\mathcal{G}(\mathfrak{o}) = \mathrm{GL}_n(\mathfrak{o}) \cap G(k)$ inside $\mathrm{GL}_n(k)$ yet $\mathrm{GL}_n(\mathfrak{o}) \cap G(k) \supset \mathcal{K}$ inside $G(k)$ where $\mathcal{K}$ is *maximal* bounded, so the bounded subgroup $\mathcal{G}(\mathfrak{o})$ in $G(k)$ cannot be any larger than $\mathcal{K}$: in other words, $\mathcal{K} = \mathcal{G}(\mathfrak{o})$ inside $G(k)$. This doesn't complete the proof, since the affine $\mathfrak{o}$-scheme $\mathcal{G}$ of finite type is merely $\mathfrak{o}$-flat rather than $\mathfrak{o}$-smooth (so it isn't an $\mathfrak{o}$-structure in our sense, as we require $\mathfrak{o}$-smoothness for these).

By "group smoothening" over discrete valuation rings in the theory of Néron models [BLR, Sect. 7.1, Lemma 4–Theorem 5] (this doesn't use reductivity or connectedness of G), there exists an affine finite type $\mathfrak{o}$-group map $q : \mathscr{G} \to \mathcal{G}$ such that $\mathscr{G}$ is $\mathfrak{o}$-smooth and $q_k : \mathscr{G}_k \to \mathcal{G}_k = G$ is an isomorphism (so $\mathscr{G}$ is an $\mathfrak{o}$-structure on G) with $\mathscr{G}(\mathfrak{o}) = \mathcal{G}(\mathfrak{o})$ inside $G(k)$ (even equality for sets of $\mathfrak{o}^{\mathrm{un}}$-points inside $G(k^{\mathrm{un}})$). This is obtained by finitely many iterations of a construction called "dilatation" (a specific affine open subscheme in the blow-up along a closed subscheme Z of the special fiber; see (6.1)). □

Remark 6.3 The $\mathfrak{o}$-structure $\mathscr{G}$ in Proposition 6.1 may have disconnected special fiber $\mathscr{G}_{\mathfrak{f}}$. In Bruhat-Tits theory such disconnectedness over $\mathfrak{f}$ is a ubiquitous phenomenon when studying maximal bounded subgroups for G that is semisimple but not simply connected (e.g., PGL_2). However, "parahoric" $\mathfrak{o}$-structures (see Axiom 20ff. in Sect. 30) have connected special fiber by definition.

Remark 6.4 If $\mathscr{G} \to \mathrm{Spec}\ \mathfrak{o}$ is a flat group scheme of finite type and the generic fiber $\mathscr{G}_k$ is connected then $\mathscr{G}$ is connected as a topological space even if $\mathscr{G}_{\mathfrak{f}}$ is *disconnected*. Indeed, by $\mathfrak{o}$-flatness all generic points of $\mathscr{G}$ must lie in $\mathscr{G}_k$, so no connected component of $\mathscr{G}$ can lie entirely within $\mathscr{G}_{\mathfrak{f}}$ (otherwise it would have a generic point in the special fiber, violating $\mathfrak{o}$-flatness). Hence, disconnectedness of the total space implies disconnectedness of the generic fiber.

Let's now grapple more directly with the possibility of a disconnected special fiber, focusing on what will matter for us: a *smooth* (not merely flat of finite type) affine group $\mathscr{G} \to \mathrm{Spec}\ \mathfrak{o}$ for which we assume the generic fiber $G = \mathscr{G}_k$ is connected (reductivity will not be relevant for the moment).

Consider the union of fibral identity components: this amounts to removing from the closed fiber $\mathscr{G}_{\mathfrak{f}} \subset \mathscr{G}$ the *closed* locus of non-identity connected components of the special fiber. Having removed such a closed set from the total space, this union is *open* in $\mathscr{G}$, denoted $\mathscr{G}^0 \subset \mathscr{G}$ and equipped with its open subscheme structure that is clearly (by topological considerations) an $\mathfrak{o}$-*subgroup*. Somewhat less evident is that this is *affine*. To explain that, we record the ultimate result in this direction and then will see how to understand it in concrete terms for what we need:

Proposition 6.5 (Raynaud) *If R is a discrete valuation ring, a* separated *flat group scheme $\mathscr{H} \to$* Spec *R of finite type is affine if and only if its generic fiber is affine.*

Proof This is [PY, Proposition 3.1], but we only need the special case in which R-flatness is strengthened to R-smoothness. In the R-smooth case, the hard input in the proof from [PY, Appendix A] is not needed, so the argument becomes much simpler.

But let's simplify matters even further by giving a direct proof of affineness for $\mathscr{G}^0$ when $\mathscr{G}$ is a smooth affine R-group with connected generic fiber. We give two proofs.

Let $\pi \in R$ be a uniformizer and $\kappa = R/\pi R$ the residue field. Observe that for any R-algebra A we have

$$\mathscr{G}^0(A) = \left\{ g \in \mathscr{G}(A) : g \bmod \pi \in \mathscr{G}^0_\kappa(A/\pi) \text{ inside } \mathscr{G}_\kappa(A/\pi A) \right\},$$

and for R-flat A this description of the functor of points of $\mathscr{G}^0$ can be expressed (check!) as $\mathscr{Y}(A)$ for the visibly affine R-scheme $\mathscr{Y} := \mathrm{Spec}\ (R[\mathscr{G}][(1/\pi)\mathscr{I}_{\mathscr{G}^0_\kappa}])$ (where $\mathscr{I}_{\mathscr{G}^0_\kappa}$ is the ideal in $R[\mathscr{G}]$ for the closed subscheme $\mathscr{G}^0_\kappa \subset \mathscr{G}$). But $\mathscr{Y}$ has R-flat coordinate ring (as it lies inside the coordinate ring of the generic fiber of $\mathscr{G}_F$ and torsion-free R-modules are flat), so both the R-smooth $\mathscr{G}^0$ and R-flat $\mathscr{Y}$ have the same functor of points on flat R-algebras. Hence, as R-schemes they are isomorphic, yielding the affineness of $\mathscr{G}^0$.

Alternatively, it suffices to show the open immersion $j : \mathscr{G}^0 \to \mathscr{G}$ is an affine morphism. We check Zariski-locally on the target: over the open $\mathscr{G}^0 \subset \mathscr{G}$ the map j restricts to an isomorphism, and over the open $\mathscr{G} - \mathscr{G}^0_\kappa$ it holds by separatedness of $\mathscr{G}$ since the j-preimage is the (connected) affine generic fiber of $\mathscr{G}$. □

It is natural to wonder if the $\mathfrak{o}$-structure we have built for a connected reductive k-group G realizing a given maximal bounded subgroup $\mathcal{K}$ as its group of $\mathfrak{o}$-points is unique (given $\mathcal{K}$). Let's pose this question more broadly:

Question 6.6 Given a smooth affine k-scheme X and $\mathfrak{o}$-structures $\mathscr{X}$, $\mathscr{X}'$ for X (so these are $\mathfrak{o}$-smooth and affine with $\mathscr{X}_k = X = \mathscr{X}'_k$), if $\mathscr{X}(\mathfrak{o}) = \mathscr{X}'(\mathfrak{o})$ inside $X(k)$ then is it necessarily true that $\mathscr{X} = \mathscr{X}'$ (i.e., $\mathfrak{o}[\mathscr{X}] = \mathfrak{o}[\mathscr{X}']$ inside $k[X]$)?

The answer is negative, even for X connected (and $\mathscr{X}(\mathfrak{o})$ non-empty, to keep matters interesting).

Example 6.7 Consider $\mathfrak{o}$ with *finite* residue field and $\mathscr{X}$ a smooth affine $\mathfrak{o}$-curve whose special fiber has some $\mathfrak{f}$-points. To make these in abundance, pick a smooth projective plane curve over $\mathfrak{f}$ with some rational points (so it is geometrically connected), and lift a defining homogeneous polynomial over $\mathfrak{o}$ to get an $\mathfrak{o}$-flat projective plane curve $\mathscr{C}$. This lift must be smooth (due to $\mathfrak{o}$-properness, $\mathfrak{o}$-flatness, and smoothness of the special fiber), and its generic fiber is geometrically connected (hence irreducible) due to $\mathscr{C} \to \operatorname{Spec} \mathfrak{o}$ being proper smooth with geometrically connected special fiber (use Stein factorization or [EGA, IV_3, Theorem 12.2.4(vi)]). Then take $\mathscr{X}$ to be an affine open in $\mathscr{C}$ around some rational point in the special fiber.

Having built such an $\mathscr{X}$, let $Z \subset \mathscr{X}_{\mathfrak{f}}$ be the reduced non-empty closed set of its finitely many rational points, and consider the "dilatation" along Z:

$$\mathscr{X}' = \operatorname{Spec}\,(\mathfrak{o}[\mathscr{X}]\left[\frac{\mathscr{I}_Z}{\pi}\right]) \tag{6.1}$$

representing the functor of points of $\mathscr{X}$ whose reduction lies in Z. The $\mathfrak{o}$-smoothness of $\mathscr{X}'$ is due to $\mathfrak{o}$-smoothness of $\mathscr{X}$ and $\mathfrak{f}$-smoothness of Z by [KP, Lemma A.5.10]. By inspection, the $\mathfrak{o}$-structures $\mathscr{X}$ and $\mathscr{X}'$ on X are distinct ($\mathscr{I}_Z \not\subset \pi\mathfrak{o}[\mathscr{X}]$ since $Z \neq \emptyset$), yet by design $\mathscr{X}'(\mathfrak{o}) = \mathscr{X}(\mathfrak{o})$ inside $X(k)$.

The preceding class of many examples in which Question 6.6 has a negative answer (with connected X and finite $\mathfrak{f}$) relies crucially on the set of rational points in the special fiber of $\mathscr{X}$ not being Zariski-dense. That motivates the following criterion:

Proposition 6.8 *Let $\mathscr{X}$, $\mathscr{Y}$ be smooth affine $\mathfrak{o}$-schemes with k-fibers X, Y respectively. Assume $\mathscr{X}(\mathfrak{f}) \subset \mathscr{X}_{\mathfrak{f}}$ is Zariski-dense (such as when $\mathfrak{f}$ is separably closed). Then a k-map $\varphi : X \to Y$ extends to an $\mathfrak{o}$-map $\varphi : \mathscr{X} \to \mathscr{Y}$ if and only if $\varphi(\mathscr{X}(\mathfrak{o})) \subset \mathscr{Y}(\mathfrak{o})$ inside $Y(k)$.*

In particular, for $Y = X$, any two $\mathfrak{o}$-structures on X with a dense set of rational points in their special fibers are the same when their sets of $\mathfrak{o}$-points coincide in $X(k)$.

Before proving the result, we make some remarks to give context.

Remark 6.9 For $\mathfrak{f} = \mathfrak{f}_s$, the Zariski-density of $\mathscr{X}(\mathfrak{f})$ in $\mathscr{X}_{\mathfrak{f}}$ is *automatic* by smoothness, so then the result takes on a more useful form: k-maps extend to $\mathfrak{o}$-maps if and only if they carry all $\mathfrak{o}$-points to $\mathfrak{o}$-points. In particular, when $\mathfrak{f}$ is separably closed it is natural to say (for a smooth affine k-scheme X) that a bounded subset $B \subset X(k)$ is *schematic* if $B = \mathscr{X}(\mathfrak{o})$ for an $\mathfrak{o}$-structure $\mathscr{X}$ of X since such an $\mathscr{X}$ is *unique* and even *functorial* in the pair (X, B).

Remark 6.10 Another important case when the Zariski-density hypothesis holds is if $\mathscr{X}$ is unirational over $\mathfrak{f}$ and the residue field is *infinite*. This applies in particular to reductive $\mathfrak{o}$-groups, since connected reductive groups over a field are always unirational over the field.

Remark 6.11 By working Zariski-locally on $\mathscr{X}$ we can remove the affineness hypothesis on $\mathscr{X}$ in Proposition 6.8 (though we have no need for such generality). The affineness hypothesis on $\mathscr{Y}$ *cannot* be removed: a counterexample over any $\mathfrak{o}$ (with affine $\mathscr{X}$) is given in [KP, Ex. 2.10.14]

Proof of Proposition 6.8 The implication "$\Rightarrow$" is obvious. For the converse, given that $\varphi(\mathscr{X}(\mathfrak{o})) \subset \mathscr{Y}(\mathfrak{o})$ inside $Y(k)$ we want to show the map $\varphi^* : k[Y] \to k[X]$ carries $\mathfrak{o}[\mathscr{Y}]$ into $\mathfrak{o}[\mathscr{X}]$.

For $h \in \mathfrak{o}[\mathscr{Y}]$, consider $\varphi^*(h) \in k[X]$. For $\widetilde{x} \in \mathscr{X}(\mathfrak{o}) \subset X(k)$, we see that the value $(\varphi^* h)(\widetilde{x}) = h_k(\varphi(\widetilde{x})) \in k$ is inside $\mathfrak{o}$ because $\varphi(\widetilde{x}) \in \mathscr{Y}(\mathfrak{o})$ by our hypothesis on φ. This reduces us to verifying a property solely about $\mathscr{X}$: if $\alpha \in k[X]$ satisfies $\alpha(\mathscr{X}(\mathfrak{o})) \subset \mathfrak{o}$ then $\alpha \in \mathfrak{o}[\mathscr{X}]$.

We can write

$$\alpha = \frac{A}{\pi^n}$$

for some $A \in \mathfrak{o}[\mathscr{X}]$ and $n \geq 0$ since $k[X] = \mathfrak{o}[\mathscr{X}][1/\pi]$. If $n = 0$, we are done. Suppose $n \geq 1$. Since $\mathscr{X}$ smooth and $\mathfrak{o}$ is Henselian, the reduction map $\mathscr{X}(\mathfrak{o}) \to \mathscr{X}(\mathfrak{f})$ is surjective [EGA, IV$_4$, Theorem 18.5.17]. Thus, for the reduction $A_0 \in \mathfrak{f}[\mathscr{X}_{\mathfrak{f}}]$ we see that the values of A_0 on $\mathscr{X}(\mathfrak{f})$ are *exactly* the reductions of the values of A on $\mathscr{X}(\mathfrak{o})$. But $A = \pi^n \alpha$ is $\pi^n \mathfrak{o}$-valued since $\alpha(\mathscr{X}(\mathfrak{o})) \subset \mathfrak{o}$, so (since $n \geq 1$) we see that A_0 *vanishes* on $\mathscr{X}(\mathfrak{f})$. However, this set of rational points is *Zariski-dense* in the smooth (hence reduced) $\mathscr{X}_{\mathfrak{f}}$, so $A_0 = 0$ in $\mathfrak{f}[\mathscr{X}_{\mathfrak{f}}] = \mathfrak{o}[\mathscr{X}]/(\pi)$. We conclude that $A = \pi A'$ for some $A' \in \mathfrak{o}[\mathscr{X}]$, so

$$\alpha = \frac{A}{\pi^n} = \frac{A'}{\pi^{n-1}}.$$

Now we are done if $n = 1$, and otherwise we keep repeating the argument (if $n - 1 \geq 1$). □

In the special case $\mathscr{Y} = \mathbf{A}^1_{\mathfrak{o}}$, this result includes the key property verified in the proof:

$$\mathfrak{o}[\mathscr{X}] = \{\alpha \in k[X] : \alpha(\mathscr{X}(\mathfrak{o})) \subset \mathfrak{o}\}.$$

This latter property of $\mathscr{X}$, which rests crucially on the Zariski-density of $\mathscr{X}(\mathfrak{f})$ in $\mathscr{X}_{\mathfrak{f}}$, is called *étoffé* in the work of Bruhat and Tits (it means "substantial").

This characterization of $\mathfrak{o}[\mathscr{X}]$ inside $k[X]$ breaks down whenever the set of rational points in the special fiber is *not* Zariski-dense (e.g., $\mathfrak{f}$ finite and $\dim \mathscr{X}_{\mathfrak{f}} > 0$). Indeed, when Zariski-density fails there is some nonzero $\varphi_0 \in \mathfrak{f}[\mathscr{X}_{\mathfrak{f}}]$ that vanishes at the rational points, so any lift $\varphi \in \mathfrak{o}[\mathscr{X}]$ of φ_0 satisfies $\varphi(\mathscr{X}(\mathfrak{o})) \subset \pi\mathfrak{o}$. The fraction $(1/\pi)\varphi \in k[X]$ is then $\mathfrak{o}$-valued on $\mathscr{X}(\mathfrak{o})$, but it is *not* contained in $\mathfrak{o}[\mathscr{X}]$. Indeed if $(1/\pi)\varphi \in \mathfrak{o}[\mathscr{X}]$ then $\varphi \in \pi \cdot \mathfrak{o}[\mathscr{X}]$ and so its reduction $\varphi_0 \in \mathfrak{f}[\mathscr{X}_{\mathfrak{f}}]$ would *vanish*, contrary to how φ_0 was chosen.

Remark 6.12 In view of the preceding failures with finite residue fields, it is a remarkable fact that for *reductive groups* over $\mathfrak{o}$ with a given generic fiber G, equality of their sets of $\mathfrak{o}$-points inside $G(k)$ forces equality of those $\mathfrak{o}$-structures (in the language of Appendix B, hyperspecial bounded subgroups determine the corresponding $\mathfrak{o}$-structure) without restriction on the residue field, particularly allowing a finite residue field. For an infinite residue field this is not remarkable, since it is a consequence of Proposition 6.8 (unrelated to groups) and Remark 6.10. But for a finite residue field, the only proof known to me (and to Kaletha and Prasad) is via the full force of Bruhat-Tits theory [KP, Proposition 9.3.25, Theorem 9.9.3] (which applies to any $\mathfrak{o}$ whose residue field is perfect; such perfectness is a requirement for the development of Bruhat-Tits theory in [KP]).

7 The Subgroup $G(k)^1$

Let G be a connected reductive group over k. The *valuation homomorphism*

$$\omega_G : G(k) \to \mathrm{Hom}_{\mathbf{Z}}(\mathrm{X}^*_k(G), \mathbf{Z}) \subset \mathrm{Hom}_{\mathbf{Q}}(\mathrm{X}^*_k(G^{\mathrm{ab}})_{\mathbf{Q}}, \mathbf{Q}) \tag{7.1}$$

of G is defined by

$$\omega_G(g) = (\chi \mapsto \mathrm{ord}(\chi(g)))$$

where $G^{\mathrm{ab}} := G/\mathscr{D}G$ is the maximal torus quotient of G and $\mathrm{X}^*_k(G) := \mathrm{Hom}_k(G, \mathbf{G}_m) = \mathrm{X}^*_k(G^{\mathrm{ab}}) = \mathrm{X}^*(G^{\mathrm{ab}}_{k_s})^{\Gamma_k}$ is the fixed-point locus under the action of $\Gamma_k = \mathrm{Gal}(k_s/k)$ on the geometric character lattice of the k-torus G^{ab}. The kernel denoted as

$$G(k)^1 := \ker \omega_G = \{g \in G(k) : |\chi(g)| = 1 \text{ for all } \chi \in \mathrm{X}^*_k(G)\}$$

is certainly normal in $G(k)$, and it is open in $G(k)$ due to openness of $\mathfrak{o}^\times$ in $k^\times$ ($\mathrm{X}_k^*(G)$ is finitely generated as a $\mathbf{Z}$-module, with each of finitely many generators $\chi_i : G \to \mathbf{G}_m$ inducing a continuous homomorphism $G(k) \to k^\times$, so if some g_0 is carried into $\mathfrak{o}^\times$ by all χ_i then so are all nearby $g \in G(k)$).

The subgroup $G(k)^1$ of $G(k)$ is also functorial in G, since ω_G is functorial in G (an easy verification). Note that $G(k)^1 = G(k)$ when $\mathrm{X}_k^*(G) = 0$, which is to say G has no $\mathbf{G}_m$ quotient, or equivalently no central $\mathbf{G}_m$ as a k-subgroup (equivalence since $(Z_G)^0_{\mathrm{red}} \to G^{\mathrm{ab}}$ is an isogeny; e.g., for $G = \mathrm{GL}_n$ this is $\mathbf{G}_m \xrightarrow{z^n} \mathbf{G}_m$). All bounded subgroups $\mathcal{K} \subset G(k)$ lie in $G(k)^1$ since for any homomorphism $\chi : G \to \mathbf{G}_m$ the image $\chi(\mathcal{K}) \subset k^\times$ is a bounded subgroup of GL_1 and hence $\chi(\mathcal{K}) \subset \mathfrak{o}^\times$. However, $G(k)^1$ is usually *not* bounded, since $G(k)^1 = G(k)$ for semisimple G (so boundedness fails for non-trivial split connected semisimple k-groups). The following resulting question seems hopeless in general beyond the quasi-split case, though for k complete with finite residue field an affirmative answer holds due to the validity of the Kneser-Tits conjecture over such fields:

Question 7.1 Is $G(k)^1$ generated by the bounded subgroups of $G(k)$?

Since the definition of $G(k)^1$ is very much tied to the use of characters defined over k, it is not immediately apparent how the formation of $G(k)^1$ interacts with separable algebraic extension on k (say preserving discreteness of the valuation, such as k^{un}/k). To understand this, we'll first focus on the case of tori since from the definition of $G(k)^1$ we see that for the quotient map $q : G \to G^{\mathrm{ab}}$ the subgroup $G(k)^1 \subset G(k)$ is the preimage of $(G^{\mathrm{ab}})(k)^1$ under q on k-points. Loosely speaking, $G(k)^1$ removes "unbounded parts" from $(Z_G)^0_{\mathrm{red}}(k)$. From now on, we'll write $Z(G) := (Z_G)^0_{\mathrm{red}}$ to streamline notations.

Suppose $G = T$ is a k-torus, and consider

$$\omega_T : T(k) \to \mathrm{Hom}(\mathrm{X}_k^*(T), \mathbf{Z}) \subset \mathrm{Hom}_{\mathbf{Q}}(\mathrm{X}_k^*(T)_{\mathbf{Q}}, \mathbf{Q}) = \mathrm{Hom}_{\mathbf{Q}}(\mathrm{X}^*(T_{k_s})_{\mathbf{Q}}^{\Gamma_k}, \mathbf{Q}) = \mathrm{X}_{*,k}(T)_{\mathbf{Q}},$$

where the final equality uses that formation of "invariants" under the action of a finite group on a $\mathbf{Q}$-vector space commutes with duality thanks to averaging in $\mathbf{Q}$-vector spaces. Beware that such averaging does *not* generally work as the level of lattices, and more specifically the natural composition pairing

$$\langle \cdot, \cdot \rangle : \mathrm{X}_k^*(T) \times \mathrm{X}_{*,k}(T) \to \mathbf{Z} \tag{7.2}$$

is usually *not* a perfect pairing (even though the analogous pairing at the level of geometric character and co-character lattices is perfect over $\mathbf{Z}$).

Example 7.2 Consider $T = \mathrm{R}_{k'/k}(\mathbf{G}_m)$ for k'/k finite separable of degree d. There is an evident natural cocharacter

$$j : \mathbf{G}_m \hookrightarrow T, \qquad A^\times \to (k' \otimes_k A)^\times$$

and a natural character

$$\mathrm{N}_{k'/k} : T = \mathrm{R}_{k'/k}(\mathbf{G}_m) \to \mathbf{G}_m, \qquad (k' \otimes_k A)^\times \xrightarrow{\text{norm}} A^\times.$$

It's not hard to check that these span the respective rank-1 lattices of cocharacters and characters defined over k: $\mathrm{X}_k^*(T) = \mathrm{N}_{k'/k}^{\mathbf{Z}}$, $\mathrm{X}_{*,k}(T) = j^{\mathbf{Z}}$. However, $\langle \mathrm{N}_{k'/k}, j\rangle = d$ (this says $\mathrm{N}_{k'/k} \circ j : z \mapsto z^d$), so for $d > 1$ perfectness fails.

By using the canonical isogeny between a general k-torus T and a product of split and k-anisotropic tori, one can show (see [KP, Lemma 2.5.7]) that the image of ω_T *contains* $\mathrm{X}_{*,k}(T)$. A key special case in this verification for the image of ω_T is the case of a split torus $T = \mathbf{G}_m^N$, for which ω_T is identified with the map $(k^\times)^N \to \mathbf{Z}^N$ given by $\mathrm{ord}^{\oplus N}$ that is surjective with kernel $(\mathbf{G}_m^N)(k)^1 = (\mathfrak{o}^\times)^N \subset (k^\times)^N = T(k)$. The containment of $\mathrm{X}_{*,k}(T)$ in the image of ω_T has finite index for general T since by design the image is contained in the $\mathbf{Z}$-dual of $\mathrm{X}_k^*(T)$ and (7.2) is rationally perfect.

Proposition 7.3 *Let T be a k-torus.*

(1) The subgroup $T(k)^1 \subset T(k)$ is bounded, hence is the unique maximal bounded subgroup;
(2) $T(k)^1 = \{t \in T(k) : |\chi(t)| = 1$ for all $\chi \in \mathrm{X}^*(T_{k_s})\}$, where $\mathrm{X}^*(T_{k_s}) = \mathrm{X}^*(T_{k'})$ for any finite separable k'/k splitting T.

Note that for part (2) to make sense we need that k_s has exactly one absolute value extending the one given on k, and that holds because k is Henselian.

Proof Let's first see that (1) implies (2). The natural k-map $j : T \to \mathrm{R}_{k'/k}(T_{k'})$ induces on k-points a map $T(k) \to T(k')$ that carries bounded subgroups to bounded subgroups. Inside $T(k')$ the notion of boundedness using the k-basis topology on k' is the same as using the valuation topology on k' (see Proposition 4.1(3)). Hence, if (1) holds in general (such as over k and k') then $T(k)^1 \subset T(k) \cap T(k')^1$, and the reverse equality holds since finiteness of the closed immersion j forces $T(k) \cap T(k')^1 = j^{-1}(T(k')^1)$ to be a bounded subgroup of $T(k)$ (see Proposition 4.1(4)) and thus it is contained in $T(k)^1$. Hence, the description of $T(k')^1$ in terms of characters of $T_{k'}$ yields the asserted description of $T(k)^1$.

To prove (1), consider the short exact sequence

$$1 \to T_0 \to T \to S \to 1$$

with T_0 the maximal k-anisotropic subtorus and $S = T/T_0$ the maximal split quotient. We then have $\mathrm{X}_k^*(T) = \mathrm{X}_k^*(S)$, so $T(k)^1$ is the preimage of $S(k)^1 = (\mathfrak{o}^\times)^d$ via a choice of k-group isomorphism $S \simeq \mathbf{G}_m^d$. Now consider the *isogeny* (hence a finite map!)

$$T \to S \times (T/T^{\mathrm{split}})$$

for the maximal split k-subtorus $T^{\text{split}} \subset T$. Since T/T^{split} is k-anisotropic, so its group of k-points is bounded, the preimage of

$$S(k)^1 \times (T/T^{\text{split}})(k)^1 = S(k)^1 \times (T/T^{\text{split}})(k)$$

in $T(k)$ is bounded. But this preimage is $T(k)^1$, so (1) is proved. □

By applying Proposition 7.3(2) to $T = G^{\text{ab}}$ and using the relationship between $G(k)^1$ and $G^{\text{ab}}(k)^1$ we get

Corollary 7.4 *For connected reductive G over k, $G(k)^1 = \{g \in G(k) : |\chi(g)| = 1$ for all $\chi \in X^*(G_{k_s})\}$. In particular, for separable algebraic k'/k with $|k'^{\times}| \subset \mathbf{R}_{>0}$ discrete (e.g., $k' = k^{\text{un}}$) we have*

$$G(k')^1 \cap G(k) = G(k)^1.$$

Remark 7.5 For $\mathfrak{f}$ perfect, $|k'^{\times}|$ is discrete if and only if $[(k')^{\text{un}} : k^{\text{un}}] < \infty$ since $\mathfrak{f}_s = \overline{\mathfrak{f}}$.

Let's briefly discuss why the subgroup $G(k)^1$ of $G(k)$ is of interest, beyond that it contains all bounded subgroups. We'll later see for perfect $\mathfrak{f}$ (such perfectness is needed for the development of Bruhat-Tits theory) that all maximal bounded subgroups of $G(k)$ are given by point-stabilizers $G(k)^1_x$ of $G(k)^1$ for certain points $x \in \mathcal{B}(G)$ (see Corollary 29.7), and that all point-stabilizers $G(k)^1_x$ are bounded (Theorem 12.1 and Axiom 2 in Sect. 25).

To get deeper insight into the structure of such stabilizers, for all $x \in \mathcal{B}(G)$ we will build a *canonical* $\mathfrak{o}$-structure $\mathscr{G}^1_x$ for G such that $\mathscr{G}^1_x(\mathfrak{o}) = G(k)^1_x$ and likewise over k^{un} (so the point-stabilizers of $G(k)^1$ are always "schematic"); $\mathcal{B}(G_{k^{\text{un}}})$ is rather huge and contains $\mathcal{B}(G)$ as a subset compatibly with actions by $G(k)$ on each, but often $(\mathscr{G}^1_x)_{\mathfrak{f}}$ is *disconnected*. We want to get a handle on $(\mathscr{G}^1_x)^0$. Using Néron models of k-tori, in Definition 10.7 we will define a finite index subgroup $G(k)^0 \subset G(k)^1$ that is open normal in $G(k)$ and will turn out to satisfy $G(k)^0_x = (\mathscr{G}^1_x)^0(\mathfrak{o})$ for all $x \in \mathcal{B}(G)$ (see Theorem 30.7 and Axiom 20 in Sect. 30); note that $G(k)^0$ has nothing to do with x! For simply connected G we will have $G(k)^0 = G(k)^1$, but often not otherwise (this underlies in why $\mathrm{SL}_n(k)$ has a simpler description for its maximal bounded subgroups than does $\mathrm{PGL}_n(k)$).

8 The Subgroup $T(k)^0$

Let A_G be the maximal split central k-torus in a connected reductive k-group G; this is the maximal split torus in $Z(G) := (Z_G)^0_{\text{red}}$. Recall the valuation homomorphism

$$\omega_G : G(k) \to \mathrm{Hom}_{\mathbf{Z}}(X^*_k(G), \mathbf{Z}) \subset \mathrm{Hom}_G(X^*_k(G^{\text{ab}})_{\mathbf{Q}}, \mathbf{Q}) = X_{*,k}(G^{\text{ab}})_{\mathbf{Q}}.$$

Since $Z(G) \to G^{\mathrm{ab}} = G/\mathscr{D}G$ is an *isogeny*, we have a natural isomorphism

$$X_{*,k}(G^{\mathrm{ab}})_{\mathbf{Q}} \xleftarrow{\sim} X_{*,k}(Z(G))_{\mathbf{Q}} = X_{*,k}(A_G)_{\mathbf{Q}}$$

and $G(k)^1 := \ker(\omega_G)$. For a parabolic k-subgroup $P \subset G$ we can write $P = M \ltimes U$ for the split k-unipotent radical $U = \mathscr{R}_{u,k}(P)$ and a Levi k-subgroup M (i.e., $M \to P/U$ is an isomorphism). By the structure theory of connected reductive groups over fields, there exists a k-homomorphism $\lambda : \mathbf{G}_m \to G$ such that $M = Z_G(\lambda)$ and $P = P_G(\lambda)$.

Proofs with reductive groups often proceed via induction with such Levi k-subgroups, so let's now relate $M(k)^1$ to $G(k)^1$. Since $A_G \subset Z_G(\lambda) = M$, we have $A_G \subset A_M$. For any k-split torus $S \subset G$ with $A_G \subset S$ (e.g., $S = A_M$) we claim $A_G \times S' \to S$ is an isogeny for $S' := (S \cap \mathscr{D}G)^0_{\mathrm{red}}$ (the maximal k-torus of S that is inside $\mathscr{D}G$). To prove this, consider the isogeny $Z(G) \times \mathscr{D}G \to G$ via multiplication, so likewise we have an isogeny

$$Z(G) \times (Z(G) \cdot S)' \to Z(G) \cdot S.$$

The maximal split torus in $Z(G) \cdot S$ is S since $A_G \subset S$, so by passing to maximal split k-subtori we get that

$$A_G \times S' \to S$$

is an isogeny. Using $S = A_M$ we thus have

$$A_M = A_G \cdot A'_M$$

as an almost direct product and so

$$X_*(A_M)_{\mathbf{Q}} = X_*(A_G)_{\mathbf{Q}} \oplus X_*(A'_M)_{\mathbf{Q}},$$

which yields a projection map

$$\mathrm{pr} : X_*(A_M)_{\mathbf{Q}} \to X_*(A_G)_{\mathbf{Q}}.$$

Lemma 8.1 *The diagram*

$$\begin{array}{ccc} M(k) & \xrightarrow{\omega_M} & X_*(A_M)_{\mathbf{Q}} \\ \downarrow & & \downarrow \mathrm{pr} \\ G(k) & \xrightarrow[\omega_G]{} & X_*(A_G)_{\mathbf{Q}} \end{array}$$

commutes and we have a natural injection

$$(M(k) \cap G(k)^1)/M(k)^1 \hookrightarrow \ker(\mathrm{pr}) = X_*(A'_M)_{\mathbf{Q}}.$$

Proof The commutativity is definition-chasing, and that implies the final assertion. For more details, see [KP, Lemma 2.6.19]. □

Example 8.2 Consider a minimal parabolic k-subgroup $P_0 \subset G$, so the Levi k-subgroups of P_0 have the form $M_0 = Z_G(S) =: Z$ for maximal split k-tori $S \subset P_0$ (in particular, $A_{M_0} = S$). Thus, we have an injection

$$(Z(k) \cap G(k)^1)/Z(k)^1 \hookrightarrow \mathrm{X}_*(S')_{\mathbf{Q}}$$

where $S' = (S \cap \mathscr{D}G)^0_{\mathrm{red}}$. This underlies why $G(k)^0 \subset G(k)^1$ with finite index (see Definition 10.7, Lemma 10.14, Theorem 12.1).

We define $G(k)^0$ for general G in Sect. 10, building on the case of k-tori that occupies the rest of this section and part of Sect. 9. The key tool we use is Néron models of tori [BLR, Sect. 10.1, Propositions 4–6]: for any k-torus T, there exists a smooth *separated* locally finite type (commutative) $\mathfrak{o}$-group $\mathscr{T}$ with $\mathscr{T}_k = T$ (we call $\mathscr{T}$ the *Néron model* of T) such that for all smooth $\mathfrak{o}$-schemes $\mathscr{X}$, the natural map

$$\mathrm{Hom}_{\mathfrak{o}}(\mathscr{X}, \mathscr{T}) \to \mathrm{Hom}_k(\mathscr{X}_k, T)$$

defined by $\varphi \mapsto \varphi_k$ is bijective. The formation of Néron models clearly commutes with direct products, and for $\mathscr{X} = \mathrm{Spec}\,\mathfrak{o}'$ with finite *unramified* k'/k (equivalently, $\mathfrak{o} \to \mathfrak{o}'$ is finite étale) we have

$$\mathscr{T}(\mathfrak{o}') = T(k').$$

Proposition 8.3 *The following hold:*

(1) The open $\mathfrak{o}$-subgroup $\mathscr{T}^0 \subset \mathscr{T}$ is affine.
(2) The base change $\mathscr{T} \otimes_{\mathfrak{o}} \mathfrak{o}^{\mathrm{un}}$ is *the Néron model of $T_{k^{\mathrm{un}}}$.*
(3) The group $\pi_0(\mathscr{T}_{\mathfrak{f}_s}) = \pi_0(\mathscr{T}_{\overline{\mathfrak{f}}}) = \mathscr{T}_{\overline{\mathfrak{f}}}/\mathscr{T}^0_{\overline{\mathfrak{f}}}$ is finitely generated, hence has finite *torsion subgroup.*

In particular, there is a unique open subgroup $\mathscr{T}^{\mathrm{ft}} \subset \mathscr{T}$ for which $\pi_0(\mathscr{T}^{\mathrm{ft}}_{\overline{\mathfrak{f}}}) = \pi_0(\mathscr{T}_{\overline{\mathfrak{f}}})_{\mathrm{tor}}$, and it is finite type.

Proposition 8.3(1) follows from Proposition 6.5. For Proposition 8.3(2), we seek to show that the separated smooth $\mathfrak{o}^{\mathrm{un}}$-group $\mathscr{T} \otimes_{\mathfrak{o}} \mathfrak{o}^{\mathrm{un}}$ with generic fiber $T_{k^{\mathrm{un}}}$ satisfies the Néron mapping property:

$$\mathrm{Hom}_{\mathfrak{o}^{\mathrm{un}}}(X, \mathscr{T} \otimes_{\mathfrak{o}} \mathfrak{o}^{\mathrm{un}}) \to \mathrm{Hom}_{k^{\mathrm{un}}}(X_{k^{\mathrm{un}}}, T_{k^{\mathrm{un}}})$$

is bijective for smooth $\mathfrak{o}^{\mathrm{un}}$-schemes X. This is the same as bijectivity of the natural map

$$\mathrm{Hom}_{\mathfrak{o}}(X, \mathscr{T}) \to \mathrm{Hom}_k(X_{k^{\mathrm{un}}}, T) \tag{8.1}$$

(for $\mathfrak{o}^{\mathrm{un}}$-smooth X). Injectivity of (8.1) is immediate from $\mathfrak{o}$-flatness of X and separatedness of $\mathscr{T}$ (a similar argument will be given in Example 9.1). To prove surjectivity of (8.1), we can use an affine open cover of X to reduce to the case $X = \mathrm{Spec}(A)$ for $\mathfrak{o}^{\mathrm{un}}$-smooth A.

By design $\mathfrak{o}^{\mathrm{un}}$ is a direct limit of local finite étale $\mathfrak{o}$-algebras $\mathfrak{o}'$, so for sufficiently big $\mathfrak{o}'$ (with finite unramified fraction field k'/k) we can write $A = \mathfrak{o}^{\mathrm{un}} \otimes_{\mathfrak{o}'} A'$ for a smooth $\mathfrak{o}'$-algebra A'. By increasing $\mathfrak{o}'$ further if necessary, a chosen k-map $f : X \to T$ descends to a k-map $f' : \mathrm{Spec}(A'_{k'}) \to T$. But $A'_{k'} = A' \otimes_{\mathfrak{o}} k$ and A' is $\mathfrak{o}$-smooth (since $\mathfrak{o}'$ is $\mathfrak{o}$-étale), so f' extends to an $\mathfrak{o}$-map $\mathrm{Spec}(A') \to \mathscr{T}$. Composing with $\mathrm{Spec}(A) \to \mathrm{Spec}(A')$ then does the job (recovering f over k).

The proof of Proposition 8.3(3), which occupies most of Sect. 9, uses the construction of Néron models, starting from $\mathbf{G}_m$ whose Néron model has special fiber with component group $\mathbf{Z}$ (so the Néron model $\mathscr{T}$ of $T = \mathbf{G}_{m,k}$ satisfies $\mathscr{T}^{\mathrm{ft}} = \mathscr{T}^0 = \mathbf{G}_{m,\mathfrak{o}}$). The non-quasi-compactness of $\mathscr{T}$ (in contrast with Néron models of abelian varieties) is related to the fact that $\mathbf{G}_m(k) = k^\times$ is unbounded; explicitly, $k^\times = \bigcup_{n \in \mathbf{Z}} \mathfrak{o}^\times \pi^n$ for a uniformizer π of $\mathfrak{o}$.

Definition 8.4 We define $T(k)^0 := \mathscr{T}^0(\mathfrak{o})$, an open bounded subgroup of $(\mathscr{T}^0)_k(k) = T(k)$.

Note that $T(k)^0 = T(k) \cap T(k^{\mathrm{un}})^0$ since $k \cap \mathfrak{o}^{\mathrm{un}} = \mathfrak{o}$ and $T(k^{\mathrm{un}})^0 = \mathscr{T}^0(\mathfrak{o}^{\mathrm{un}})$ (use (1) and (2) in Proposition 8.3).

Proposition 8.5 *For T and $\mathscr{T}^{\mathrm{ft}}$ as above, the following hold:*

(i) The containment $\mathscr{T}^{\mathrm{ft}}(\mathfrak{o}) \subset T(k)^1$ is an equality, so $T(k)^1/T(k)^0 \subset \pi_0(\mathscr{T}^{\mathrm{ft}}_{\mathfrak{f}_s})$ and hence $T(k)^1/T(k)^0$ is finite.

(ii) If $\mathfrak{f}$ is algebraically closed and k'/k is a finite separable extension splitting T then the norm map $\mathrm{N}_{k'/k} : T(k') \to T(k)$ arising from the commutative k-group norm $N : \mathrm{R}_{k'/k}(T_{k'}) \to T$ on k-points carries $T(k')^1 = T(k')^0$ onto $T(k)^0$. More broadly, still with $\mathfrak{f}$ algebraically closed, if $1 \to C' \to C \to C'' \to 1$ is a short exact sequence of k-tori then $C(k)^0 \to C''(k)^0$ is surjective.

(iii) If $\mathfrak{f}$ is perfect and $T_{k^{\mathrm{un}}}$ is an induced torus (such a T is said to have "induced ramification" in [KP] because the condition takes place over k^{un}) then $T(k)^0 = T(k)^1$.

Before we discuss the proof, we make some observations. Since the formation of $\mathscr{T}$ commutes with scalar extension to $\mathfrak{o}^{\mathrm{un}}$, when $\mathfrak{f}$ is perfect (so $\mathfrak{f}_s$ is algebraically closed) we see from (ii) applied over $\mathfrak{o}^{\mathrm{un}}$ that if L/k^{un} is any finite separable extension splitting T then $T(k^{\mathrm{un}})^0$ is the image of $T(L)^1$ under the norm map $T(L) \to T(k^{\mathrm{un}})$. This gives a description of $T(k^{\mathrm{un}})^0$ in purely "Galois-theoretic" terms for perfect $\mathfrak{f}$ without reference to Néron models, and so likewise for $T(k)^0 = T(k) \cap T(k^{\mathrm{un}})^0$.

The Galois-theoretic formula for $T(k) \cap T(k^{\mathrm{un}})^0$ is the *definition* of $T(k)^0$ for perfect $\mathfrak{f}$ in [KP, Definition 2.5.13], where its independence from L/k^{un} and openness of the image of the norm map $T(L) \to T(k^{\mathrm{un}})$ have to be directly proved. That is done in [KP, Lemma 2.5.12], resting on the surjectivity of the norm map on integral units $\mathcal{O}'^{\times} \to \mathcal{O}^{\times}$ for any finite separable extension K'/K with Henselian K having algebraically closed residue field. This surjectivity is [Ser1, Chap. 5, Sect. 6, Cor. 4] when K is *complete*. What if we don't assume completeness? Since $K' \otimes_K \widehat{K} = \widehat{K}'$ (Proposition 3.5), the *smooth* surjective norm map $N : \mathrm{R}_{K'/K}(\mathbf{G}_m) \to \mathbf{G}_m$ viewed on $\widehat{K}$-points has non-empty fiber over any $u \in \mathcal{O}^{\times} \subset \widehat{\mathcal{O}}^{\times}$. But the *smooth* fiber scheme $N^{-1}(u)$ then has a $\widehat{K}$-point, so by density of the set of K-points in the set of $\widehat{K}$-points for smooth K-schemes (Proposition 5.2) we get a K-point $u' \in N^{-1}(u)$. In other words, $u' \in K'^{\times}$ has norm $u \in \mathcal{O}^{\times}$, so $u' \in \mathcal{O}'^{\times}$ (with norm u) as desired.

Now let's prove Proposition 8.5:

Proof To prove (i), the desired equality is clear when $T \simeq \mathbf{G}_m^N$ is split since then $\mathscr{T}^{\mathrm{ft}}(\mathfrak{o}) = (\mathfrak{o}^{\times})^N$ inside $T(k) = (k^{\times})^N$. In general, let k'/k be a finite separable extension splitting T, and $\mathscr{T}'$ the Néron model of $T_{k'} = \mathbf{G}_{m,k'}^N$, so $\mathscr{T}'^{\mathrm{ft}} = \mathscr{T}'^0$ and the closed immersion $j : T \hookrightarrow \mathrm{R}_{k'/k}(T_{k'})$ (which induces the identification $T(k)^1 = T(k) \cap T(k')^1$ on k-points) gives rise to a map of Néron models

$$\varphi : \mathscr{T} \to \mathrm{R}_{\mathfrak{o}'/\mathfrak{o}}(\mathscr{T}').$$

(In Sect. 9 we'll discuss the existence of $\mathrm{R}_{\mathfrak{o}'/\mathfrak{o}}(\mathscr{T}')$ for the locally finite type Néron model $\mathscr{T}'$ of a power $T_{k'}$ of $\mathbf{G}_{m,\mathfrak{o}'}$. This Weil restriction is $\mathfrak{o}$-smooth by the infinitesimal criterion and hence is the Néron model of $\mathrm{R}_{k'/k}(T_{k'})$ by the universal property of Weil restriction.)

The map φ carries $\mathscr{T}^{\mathrm{ft}}$ (with special fiber that is torsion in the geometric component group) into the open subgroup $\mathrm{R}_{\mathfrak{o}'/\mathfrak{o}}(\mathscr{T}'^0)$. Indeed, the universal property of Weil restriction reduces this to showing any $\mathfrak{o}'$-homomorphism $(\mathscr{T}^{\mathrm{ft}})_{\mathfrak{o}'} \to \mathscr{T}'$ lands in the open subgroup $\mathscr{T}'^0$, and that in turn holds since the special fiber of the Néron model $\mathscr{T}'$ of some $\mathbf{G}_{m,\mathfrak{o}'}^N$ has *torsion-free* geometric component group.

Any point $t \in T(k)^1 = T(k) \cap T(k')^1$ can be viewed in $T(k) = \mathscr{T}(\mathfrak{o})$ and is carried by φ into $T(k')^1 = \mathscr{T}'^0(\mathfrak{o}')$, so it suffices to show that the containment $\mathscr{T}^{\mathrm{ft}} \subset \varphi^{-1}(\mathrm{R}_{\mathfrak{o}'/\mathfrak{o}}(\mathscr{T}'^0))$ is an equality (as then on $\mathfrak{o}$-points this forces $t \in \mathscr{T}^{\mathrm{ft}}(\mathfrak{o})$ as desired). This comparison of open $\mathfrak{o}$-subgroups of $\mathscr{T}$ is a topological question in the special fiber of $\mathscr{T}$. The inclusion j of k-tori admits an left isogeny-inverse: a k-homomorphism $q : \mathrm{R}_{k'/k}(T_{k'}) \to T$ such that $q \circ j : T \to T$ is multiplication by some $d \in \mathbf{Z} - \{0\}$. The map q induces $\mathrm{R}_{\mathfrak{o}'/\mathfrak{o}}(\mathscr{T}') \to \mathscr{T}$ between Néron models, so it carries the open $\mathrm{R}_{\mathfrak{o}'/\mathfrak{o}}(\mathscr{T}'^0)$ with connected special fiber into $\mathscr{T}^0$. Hence, $\varphi^{-1}(\mathrm{R}_{\mathfrak{o}'/\mathfrak{o}}(\mathscr{T}'^0))$ is carried by d-multiplication into $\mathscr{T}^0$, so we get the reverse containment $\varphi^{-1}(\mathrm{R}_{\mathfrak{o}'/\mathfrak{o}}(\mathscr{T}'^0)) \subset \mathscr{T}^{\mathrm{ft}}$. This settles (i).

For (ii), the norm map $N : \mathrm{R}_{k'/k}(T_{k'}) \to T$ has kernel that is also a torus (check!), so it suffices to show that for *any* short exact sequence of k-tori

$$1 \to C' \to C \xrightarrow{q} C'' \to 1$$

the induced map of Néron models $\mathcal{C} \to \mathcal{C}''$ induces a surjection between $\mathfrak{o}$-points of relative identity components when $\mathfrak{f}$ is algebraically closed. Beware that the diagram of Néron models

$$\mathcal{C}' \to \mathcal{C} \to \mathcal{C}''$$

generally has poor exactness properties; e.g., $\mathcal{C}' \to \mathcal{C}$ might not be a closed immersion and $\mathcal{C} \to \mathcal{C}''$ might not be surjective on special fibers.

The key point is that $C(k) \to C''(k)$ is surjective because $\mathrm{H}^1(k, C') = 1$ (this torus case of Steinberg's theorem has a direct proof [KP, Lemma 2.5.4], and uses that $\mathfrak{f}$ is algebraically closed). Hence, for any $x'' \in \mathcal{C}''^0(\mathfrak{o}) \subset C''(k)$ the fiber $q^{-1}(x'')$ is non-empty and k-smooth (as q is smooth since $\ker(q) = C'$ is smooth), so its set of k-points is dense in its set of $\widehat{k}$-points (Proposition 5.2). But $\mathcal{C}(\mathfrak{o}) = \mathcal{C}(\widehat{\mathfrak{o}}) \cap C(k)$ inside $C(\widehat{k})$ since $\widehat{\mathfrak{o}} \cap k = \mathfrak{o}$ inside $\widehat{k}$, so it suffices to show $q^{-1}(x'')$ has an $\widehat{\mathfrak{o}}$-point in $\mathcal{C}^0$ (as such a point viewed in $C(\widehat{k})$ can be arbitrarily approximated by $C(k)$, yet any k-point near a point in the *open* subset $\mathcal{C}^0(\widehat{\mathfrak{o}}) \subset C(\widehat{k})$ must itself lie in $\mathcal{C}^0(\widehat{\mathfrak{o}})$ and so comes from $\mathcal{C}^0(\widehat{\mathfrak{o}}) \cap C(k) = \mathcal{C}^0(\mathfrak{o})$). This reduces our task to the *complete* case.

Now with k complete, to prove (ii) we can argue via successive approximation as follows. The known surjectivity of $C(k) \to C''(k)$ can be reformulated as surjectivity of $\mathcal{C}(\mathfrak{o}) \to \mathcal{C}''(\mathfrak{o})$ by the Néron mapping property. Because $\mathcal{C} \to \mathcal{C}''$ is a map between smooth separated $\mathfrak{o}$-groups for which $\pi_0(\mathcal{C}_{\mathfrak{f}})$ is finitely generated, the surjectivity of $\mathcal{C}(\mathfrak{o}) \to \mathcal{C}''(\mathfrak{o})$ implies the desired surjectivity of $\mathcal{C}^0(\mathfrak{o}) \to \mathcal{C}''^0(\mathfrak{o})$ by [BLR, Sect. 9.6, Lemma 2], also presented as [KP, Lemma A.4.25] (whose proof uses successive approximation arguments via completeness of $\mathfrak{o}$, along with a mixed-characteristic analogue of artinian Weil restriction called the Greenberg functor when $\mathfrak{o}$ has mixed characteristic). This settles (ii).

Finally, to prove (iii), it suffices to show $T(k^{\mathrm{un}})^0 = T(k^{\mathrm{un}})^1$ since then intersecting with $T(k)$ gives the desired equality $T(k)^0 = T(k)^1$. Hence, we may rename k^{un} as k so that now the perfect residue field $\mathfrak{f}$ is algebraically closed. But then T is induced, hence a product of k-tori of the form $T_i = \mathrm{R}_{k_i'/k}(\mathbf{G}_m)$ for finite separable k_i'/k. Our task is compatible with direct products, so we may assume $T = \mathrm{R}_{k'/k}(\mathbf{G}_m)$ for a finite separable extension k'/k. Hence $\mathscr{T} = \mathrm{R}_{\mathfrak{o}'/\mathfrak{o}}(\mathcal{N}')$ for the Néron model $\mathcal{N}'$ of $\mathbf{G}_m$ over $\mathfrak{o}'$, so $\mathscr{T}^0 = \mathrm{R}_{\mathfrak{o}'/\mathfrak{o}}(\mathcal{N}'^0) = \mathrm{R}_{\mathfrak{o}'/\mathfrak{o}}(\mathbf{G}_m)$ (the first of these two equalities will be justified in Sect. 9) and thus $T(k)^0 := \mathscr{T}^0(\mathfrak{o}) = \mathrm{R}_{\mathfrak{o}'/\mathfrak{o}}(\mathbf{G}_m)(\mathfrak{o}) = \mathbf{G}_m(\mathfrak{o}') = \mathfrak{o}'^{\times}$ inside $T(k) = k'^{\times}$. But $\mathrm{X}_k^*(T)$ is generated by the norm $N : T = \mathrm{R}_{k'/k}(\mathbf{G}_m) \to \mathbf{G}_m$ that induces the usual norm $k'^{\times} \to k^{\times}$ on k-points, so $T(k)^1$ is the group of points $t' \in k'^{\times}$ such that $\mathrm{N}_{k'/k}(t') \in \mathfrak{o}^{\times}$. These are exactly the points in $\mathfrak{o}'^{\times} = T(k)^0$, so we are done. □

9 Construction of Néron Models of Tori

Now we discuss how the Néron model $\mathscr{T}$ of a general k-torus T is built, and use this to verify Proposition 8.3, particularly that the geometric component group of its special fiber is finitely generated. As was noted in Sect. 8 via the Néron mapping property, if T_1 and T_2 have Néron models $\mathscr{T}_1$ and $\mathscr{T}_2$ then $\mathscr{T}_1 \times \mathscr{T}_2$ is a Néron model of $T_1 \times T_2$. We will first focus on the construction of Néron models for split T, so this amounts to treating the case $T = \mathbf{G}_{m,k}$.

Example 9.1 When $T = \mathbf{G}_{m,k}$, we build $\mathscr{T}$ as the gluing of $\mathfrak{o}$-schemes $\{U_n\}_{n \in \mathbf{Z}}$ for $U_n = \mathbf{G}_{m,\mathfrak{o}}$ as follows. For all $n, m \in \mathbf{Z}$, we glue U_n and U_m along their open generic fibers via the k-scheme isomorphism $(U_n)_k \simeq (U_m)_k$ defined to be multiplication by $\pi^{n-m} \in k^\times$ on $\mathbf{G}_{m,k}$. The cocycle condition for gluing is readily verified since $\pi^{n-n} = 1$ and $\pi^{m-r}\pi^{n-m} = \pi^{n-r}$ for all $n, m, r \in \mathbf{Z}$. Informally, we imagine U_n as an $\mathfrak{o}$-scheme version of the subset $\pi^n \mathfrak{o}^\times$ of the partition

$$k^\times = \coprod_{m \in \mathbf{Z}} \pi^m \mathfrak{o}^\times,$$

where $U_n(\mathfrak{o}) = \mathfrak{o}^\times$ is identified with the subset of k-points of $(U_n)_k \simeq (U_0)_k = \mathbf{G}_{m,k}$ given by $\pi^n \mathfrak{o}^\times$.

Now let's check that $\mathscr{T}$ is separated, which is to say that the relative diagonal $\Delta_{\mathscr{T}/\mathfrak{o}} : \mathscr{T} \to \mathscr{T} \times \mathscr{T}$ is a closed immersion. An open affine covering of the target is given by the open subschemes $U_n \times U_m$ for $n, m \in \mathbf{Z}$, so we have to show that

$$\Delta^{-1}(U_n \times U_m) \to U_n \times U_m$$

is a closed immersion.

When $n = m$, this map is $\Delta_{U_n/\mathfrak{o}}$ that is a closed immersion (as U_n is affine). When $n \neq m$, we have

$$\Delta^{-1}(U_n \times U_m) = \left\{(u, u') \in \mathbf{G}_{m,k} \times \mathbf{G}_{m,k} : u' = \pi^{n-m} u\right\}$$

over k and we want to show that this closed subscheme of the open generic fiber of $\mathbf{G}_{m,\mathfrak{o}} \times \mathbf{G}_{m,\mathfrak{o}}$ is a closed subscheme of the product $\mathbf{G}_{m,\mathfrak{o}} \times \mathbf{G}_{m,\mathfrak{o}}$ over $\mathfrak{o}$. Over k, $u' = \pi^{n-m}u$ if and only if $\pi^{m-n}u' = u$. Using the formulation with a positive exponent for π (so $u' = \pi^{n-m}u$ when $n > m$ and $u = \pi^{m-n}u'$ when $n < m$), as a condition on units u, u' in an $\mathfrak{o}$-algebra A this forces A to be a k-algebra! Hence, the same defining equation over $\mathfrak{o}$ ($u' = \pi^{n-m}u$ when $n > m$, and $u = \pi^{m-n}u'$ when $n < m$) cuts out the *same* locally closed subscheme of $\mathbf{G}_{m,\mathfrak{o}} \times \mathbf{G}_{m,\mathfrak{o}}$, hence a closed subscheme by inspection. This completes the verification of separatedness of the gluing.

We define an $\mathfrak{o}$-group structure on $\mathscr{T}$ recovering the one on its generic fiber $\mathscr{T}_k = (U_0)_k = \mathbf{G}_{m,k}$ as follows. The identity section is $1 \in U_0(\mathfrak{o}) = \mathbf{G}_{m,\mathfrak{o}}(\mathfrak{o}) = \mathfrak{o}^\times$, the multiplication law $\mathscr{T} \times \mathscr{T} \to \mathscr{T}$ is defined by $U_n \times U_m \to U_{n+m}$ that is usual multiplication on $\mathbf{G}_{m,\mathfrak{o}}$ (since $\pi^n u \cdot \pi^m u' = \pi^{n+m}(uu')$ for units $u, u' \in \mathfrak{o}^\times$), and inversion $i : \mathscr{T} \to \mathscr{T}$ is defined by $U_n \simeq U_{-n}$ that is usual inversion on $\mathbf{G}_{m,\mathfrak{o}}$ (since $(\pi^n u)^{-1} = \pi^{-n}u^{-1}$ for units $u \in \mathfrak{o}^\times$).

To check this data constitutes an $\mathfrak{o}$-group law, first note that the diagrams we need to commute over $\mathfrak{o}$ at least commute over k. Thus, they commute over $\mathfrak{o}$ because if R is a discrete valuation ring with fraction field F and $f, f' : X \rightrightarrows Y$ are two R-maps from a flat R-scheme to a separated R-scheme such that $f_F = f'_F$ then $f = f'$. Indeed, for $(f, f') : X \to Y \times Y$ we want the *closed* subscheme $Z = (f, f')^{-1}(\Delta_{Y/R}) \subset X$ to be X, which is to say the corresponding quasi-coherent ideal sheaf $\mathscr{I}_Z \subset \mathcal{O}_X$ vanishes. But $Z_F = X_F$ since $f_F = f'_F$, so $\mathscr{I}_Z|_{X_F} = 0$, forcing $\mathscr{I}_Z = 0$ since $\mathcal{O}_X$ is R-flat.

For any finite unramified k'/k (so finite étale $\mathfrak{o} \to \mathfrak{o}'$), π is also a uniformizer of $\mathfrak{o}'$ and so the gluing recipe yields

$$\mathscr{T}(\mathfrak{o}') = \bigcup_{n \in \mathbf{Z}} (\mathfrak{o}')^{\times} \pi^n = k'^{\times}$$

as groups, where

$$(\mathfrak{o}')^{\times} = U_n(\mathfrak{o}') \subset (U_n)_k(k') \xrightarrow[\cdot \pi^n]{\sim} (U_0)_k(k') = \mathbf{G}_{m,k}(k') = k'^{\times}$$

is multiplication by π^n. Also, $\mathscr{T}_{\mathfrak{f}} = \coprod_{n \in \mathbf{Z}} U_{n,\mathfrak{f}} = \mathbf{G}_{m,\mathfrak{f}} \times \mathbf{Z}$ as $\mathfrak{f}$-groups.

The $\mathfrak{o}$-group $\mathscr{T}$ is covered by $\pi^{\mathbf{Z}}$-translates of the open $\mathfrak{o}$-subgroup $\mathscr{T}^0 = \mathbf{G}_{m,\mathfrak{o}}$ with a uniformizer $\pi \in k^{\times} = T(k) = \mathscr{T}(\mathfrak{o})$. Here $\pi_0(\mathscr{T}_{\mathfrak{f}}) = \mathbf{Z}$ as groups. The verification of the Néron mapping property is given in [BLR, Sect. 10.1, Ex. 5]. This mapping property makes no mention of the choice of π, so in that way $\mathscr{T}$ is independent of the choice of π (though one can also make the $\mathfrak{o}$-group identification $\mathscr{T}_{\pi} \simeq \mathscr{T}_{\pi'}$ for uniformizer choices $\pi, \pi' \in \mathfrak{o}$ explicit: it is multiplication by $(\pi'/\pi)^n \in \mathfrak{o}^{\times}$ on each open subscheme $U_n = \mathbf{G}_{m,\mathfrak{o}}$).

To build the smooth and *separated* Néron model $\mathscr{T}$ for a general k-torus T (and to verify Proposition 8.3), pick a finite separable extension k'/k such that $T_{k'}$ is split: $T_{k'} \simeq \mathbf{G}_{m,k'}^d$ as k'-groups. We have a Néron model $\mathscr{T}'$ over $\mathfrak{o}'$ for $T_{k'}$ as the dth power of the Néron model of $\mathbf{G}_{m,k'}$. It satisfies $(\mathscr{T}')^0 = \mathbf{G}_{m,\mathfrak{o}'}^d$, $\pi_0(\mathscr{T}_{\mathfrak{f}'}) = \mathbf{Z}^d$, and $\mathscr{T}'_{k'} = T_{k'}$.

We have a natural closed immersion of k-groups $T \hookrightarrow \mathrm{R}_{k'/k}(T_{k'})$, with $T_{k'}$ the open generic fiber of $\mathscr{T}'$. As functors on $\mathfrak{o}$-algebras (or $\mathfrak{o}$-schemes), we see that $\mathrm{R}_{k'/k}(T_{k'})$ is the k-fiber of $\mathrm{R}_{\mathfrak{o}'/\mathfrak{o}}(\mathscr{T}')$. Thus, $\mathrm{R}_{k'/k}(T_{k'})$ is an open subscheme of $\mathrm{R}_{\mathfrak{o}'/\mathfrak{o}}(\mathscr{T}')$ provided this Weil restriction along the finite flat Spec $(\mathfrak{o}') \to$ Spec $(\mathfrak{o})$ is representable.

See [BLR, Sect. 7.6] for a general discussion of Weil restriction $\mathrm{R}_{S'/S}(X')$ relative to a finite locally free map $S' \to S$ of base schemes, applied to arbitrary S'-schemes X'. This includes a representability criterion and the behavior of properties of morphisms of schemes under the functor $\mathrm{R}_{S'/S}$. We now use that to show the existence of $\mathrm{R}_{\mathfrak{o}'/\mathfrak{o}}(\mathscr{T}')$ as a *separated* smooth $\mathfrak{o}$-scheme (generally not quasi-compact, merely locally of finite type over $\mathfrak{o}$).

The Néron model $\mathscr{T}'$ of $\mathbf{G}_{m,k'}^d$ is a product of d copies of the Néron model of $\mathbf{G}_{m,k'}$, and Weil restriction (as a functor) commutes with direct products. Since $\mathscr{T}'$ is smooth and separated over $\mathfrak{o}'$, for the existence of $\mathrm{R}_{\mathfrak{o}'/\mathfrak{o}}(\mathscr{T}')$ as a smooth separated $\mathfrak{o}$-scheme it suffices by [BLR, 7.6/4, 7.6/5(b,h)] to show:

Proposition 9.2 *For the Néron model $\mathscr{T}'$ of $\mathbf{G}_{m,k'}$, each finite subset of a fiber of* $\mathscr{T}' \to \mathrm{Spec}\,(\mathfrak{o}') \to \mathrm{Spec}\,(\mathfrak{o})$ *is contained in an affine open subscheme.*

The argument for this on [BLR, pp. 290–1] refers to [BLR, Sect. 6.4, Theorem 1] that is for group objects of finite type rather than locally of finite type as we need (to have a group structure over $\mathfrak{o}'$).

Proof The fibral topological spaces are the same as the fibers of $\mathscr{T}' \to \mathrm{Spec}\,(\mathfrak{o}')$, and the *open* generic fiber $\mathscr{T}'_{k'} = T_{k'}$ is affine. Hence, we just need to verify any finite subset of $\mathscr{T}'_{\mathfrak{f}'}$ is contained in an affine open subscheme of $\mathscr{T}'$.

For a uniformizer π' of $\mathfrak{o}'$, using the $\mathfrak{o}'$-scheme automorphism of translation by $\pi'^N \in k'^\times = T_{k'}(k') = \mathscr{T}'(\mathfrak{o}')$ for sufficiently large N reduces our task to considering finite subsets of $\mathscr{T}'$ contained in the union V_N of the open subschemes $\{U'_n\}_{0\le n\le N}$ for some $N \ge 1$, where U'_n is as in Example 9.1 (now over k' rather than k). Thus, it suffices to show that each such V_N is quasi-projective.

Let $X_0 = \mathbf{A}^1_{\mathfrak{o}} = \mathrm{Spec}\,(\mathfrak{o}[t_0])$, so $(X_0)_k = \mathbf{A}^1_k$, and define $X_1 = \mathrm{Bl}_{\overline{x}_0}(X_0)$ to be the blow-up of X_0 at the point $\overline{x}_0$ in the special fiber that is the reduction of $x_0 := 0 \in \mathfrak{o} = X_0(\mathfrak{o})$. This blow-up is covered by two affine open charts $D_+(\pi)$ and $D_+(t_0)$. The special fiber of X_1 has an exceptional divisor $\mathbf{P}^1_{\mathfrak{f}}$ over $\overline{x}_0$ in which the special fiber of $D_+(t_0)$ is an affine line whose complementary point at infinity is the reduction $\overline{x}_1$ of the point x_1 corresponding to $x_0 \in X_0(\mathfrak{o}) = X_1(\mathfrak{o})$ (equality by the valuative criterion of properness for $X_1 \to X_0$).

The affine open $D_+(\pi)$ is an affine line $\mathrm{Spec}\,(\mathfrak{o}[t_1])$ (for $t_1 = t_0/\pi$) in which the origin of the special fiber is $\overline{x}_1$. The maximal open smooth locus $(X_1 - \{x_1\})^{\mathrm{sm}}$ is exactly V_1.

Now iterate by defining $X_2 = \mathrm{Bl}_{\overline{x}_1}(X_1)$ and using the equality $X_0(\mathfrak{o}) = X_1(\mathfrak{o}) = X_2(\mathfrak{o})$ to define $x_2 \in X_2(\mathfrak{o})$ over x_1. Since $D_+(\pi) = \mathrm{Spec}(\mathfrak{o}[t_1])$ (for $t_1 = t_0/\pi$) is an open affine line $\mathbf{A}^1_{\mathfrak{o}}$ containing x_1, the restriction of $X_2 \to X_1$ over that open locus around x_1 is identical to $X_1 \to X_0$ equipped with $x_0 \in X_0(\mathfrak{o})$. Thus, $(X_2 - \{x_2\})^{\mathrm{sm}} = V_2$, and we can keep repeating this process to eventually arrive at X_N that is projective over X_0 (hence quasi-projective over $\mathfrak{o}$) with $x_N \in X_N(\mathfrak{o}) = X_0(\mathfrak{o})$ corresponding to x_0 such that the quasi-projective $(X_N - \{x_N\})^{\mathrm{sm}}$ is identified with V_N. □

Since open immersions are the same as étale monomorphisms [EGA, IV_4, Theorem 17.9.1], by using the infinitesimal criterion for étaleness we see that the locally finitely presented $\mathfrak{o}$-group map

$$\mathrm{R}_{\mathfrak{o}'/\mathfrak{o}}(\mathbf{G}^d_{m,\mathfrak{o}'}) = \mathrm{R}_{\mathfrak{o}'/\mathfrak{o}}(\mathscr{T}'^0) \to \mathrm{R}_{\mathfrak{o}'/\mathfrak{o}}(\mathscr{T}')$$

is an open immersion. Passing to the special fiber, we claim the open $\mathfrak{f}$-subgroup $\mathrm{R}_{\mathfrak{o}'_{\mathfrak{f}}/\mathfrak{f}}(\mathscr{T}'^0_{\mathfrak{o}'_{\mathfrak{f}}})$ (with artin local $\mathfrak{o}'_{\mathfrak{f}} := \mathfrak{o}' \otimes_{\mathfrak{o}} \mathfrak{f}$) is *connected*, so it is the identity component. In other words, we could then conclude that

$$\mathrm{R}_{\mathfrak{o}'/\mathfrak{o}}(\mathscr{T}')^0 = \mathrm{R}_{\mathfrak{o}'/\mathfrak{o}}(\mathscr{T}'^0) \tag{9.1}$$

(so in particular this is *affine* since $\mathscr{T}'^0$ is affine). More broadly:

Lemma 9.3 *For any field κ and finite local κ-algebra A, $\mathrm{R}_{A/\kappa}(\mathbf{G}_{m,A}^d)$ is geometrically connected.*

Proof Since Weil restriction commutes with direct products, we can focus on the case $d = 1$. Letting $N : \mathrm{R}_{A/\kappa}(\mathbf{A}_A^1) \to \mathbf{A}_\kappa^1$ be the norm map (i.e., on the functor of points, for any κ-algebra R, this is the ring-theoretic norm map $R \otimes_\kappa A \to R$), we have $\mathrm{R}_{A/\kappa}(\mathbf{G}_m) = N^{-1}(\mathbf{G}_m)$. Thus, if $r = \dim_\kappa A$ and we use a κ-basis of A to identify $\mathrm{R}_{A/\kappa}(\mathbf{A}_A^1)$ with $\mathbf{A}_\kappa^r$ then N is identified with a nonzero polynomial function on $\mathbf{A}_\kappa^r$ and so $\mathrm{R}_{A/\kappa}(\mathbf{G}_{m,A})$ is its non-empty open non-vanishing locus. As a non-empty open subset of an affine space over κ, it is geometrically connected.

[A more conceptual but heavier alternative argument is to show that for any *smooth* affine A-scheme V with geometrically connected special fiber, the Weil restriction $\mathrm{R}_{A/\kappa}(V)$ is geometrically connected. This is [CGP, Proposition A.5.9]. Beware that such preservation of geometric connectedness in the smooth setting is *false* without smoothness when A has residue field not separable over κ: the discussion immediately after the statement of [CGP, Proposition A.5.9] gives a geometrically integral regular plane curve smooth away from one point whose Weil restriction through a purely inseparable field extension is *disconnected*, with an entire connected component that is non-reduced.] □

Lemma 9.4 *The geometric component group of the special fiber of $\mathrm{R}_{\mathfrak{o}'/\mathfrak{o}}(\mathscr{T}')$ is a free $\mathbf{Z}$-module of finite rank. More specifically, if $\delta = [\mathfrak{f}' : \mathfrak{f}]_s$ is the separability degree for the extension $\mathfrak{f}'/\mathfrak{f}$ of residue fields for $\mathfrak{o} \to \mathfrak{o}'$ then this geometric component group is $\mathbf{Z}^{d\delta}$, where d is the dimension of the split k'-torus $T_{k'}$.*

Proof It suffices to treat the case $d = 1$ (we can then pass to powers). Apply the left-exact functor $\mathrm{R}_{\mathfrak{o}'_\mathfrak{f}/\mathfrak{f}}$ to the short exact sequence

$$0 \to \mathscr{T}'^0_{\mathfrak{o}'_\mathfrak{f}} \to \mathscr{T}'_{\mathfrak{o}'_\mathfrak{f}} \to \mathbf{Z} \to 0$$

of $\mathfrak{o}'_\mathfrak{f}$-groups to get that the smooth affine identity component $\mathfrak{f}$-subgroup

$$\mathrm{R}_{\mathfrak{o}'_\mathfrak{f}/\mathfrak{f}}(\mathscr{T}'^0_{\mathfrak{o}'_\mathfrak{f}}) \subset \mathrm{R}_{\mathfrak{o}'_\mathfrak{f}/\mathfrak{f}}(\mathscr{T}'_{\mathfrak{o}'_\mathfrak{f}})$$

has corresponding étale quotient

$$\pi_0(\mathrm{R}_{\mathfrak{o}'_\mathfrak{f}/\mathfrak{f}}(\mathscr{T}'_{\mathfrak{o}'_\mathfrak{f}})) \subset \mathrm{R}_{\mathfrak{o}'_\mathfrak{f}/\mathfrak{f}}(\mathbf{Z}).$$

We will show that this is an equality and that over $\mathfrak{f}_s$ it is $\mathbf{Z}^\delta$.

Applying scalar extension to $\mathfrak{f}_s$ amounts to replacing $\mathfrak{o}'$ with $\mathfrak{o}' \otimes_\mathfrak{o} \mathfrak{o}^{\mathrm{un}}$, a normal finite $\mathfrak{o}'$-algebra which must be the product of the valuation rings of the factor fields of the finite étale k^{un}-algebra $k' \otimes_k k^{\mathrm{un}}$. The latter is a product of δ finite separable extension fields of k^{un}, as we may check after scalar extension to $\widehat{k^{\mathrm{un}}}$ and using the local factor ring decomposition of $\mathfrak{o}'_\mathfrak{f} \otimes_\mathfrak{f} \mathfrak{f}_s$ that is detected modulo nilpotents (i.e., $\mathfrak{f}' \otimes_\mathfrak{f} \mathfrak{f}_s$ is a product of δ field extensions of $\mathfrak{f}_s$).

Since $\mathrm{R}_{(\coprod S_i')/S}(\coprod Y_i') = \prod \mathrm{R}_{S_i'/S}(Y_i')$ as functors (for S_i'-schemes Y_i'), by passing to a product of δ Weil restrictions we may replace $\mathfrak{o}$ with $\mathfrak{o}^{\mathrm{un}}$ and separately replace $\mathfrak{o}'$ with the valuation ring of a factor field of $k' \otimes_k k^{\mathrm{un}}$ to reduce to the case where $\mathfrak{f}$ is *separably closed*. Now $\mathfrak{o}'_{\mathfrak{f}}$ is an artin local $\mathfrak{f}$-algebra whose residue field $\mathfrak{f}'$ is *purely inseparable* over $\mathfrak{f}$. Thus, if Y is an $\mathfrak{f}$-scheme then $Y \otimes_{\mathfrak{f}} \mathfrak{o}'_{\mathfrak{f}}$ has the **same** underlying topological space as Y, so the sets of locally constant $\mathbf{Z}$-valued functions on each are naturally identified. This shows $\mathrm{R}_{\mathfrak{o}'_{\mathfrak{f}}/\mathfrak{f}}(\mathbf{Z}) = \mathbf{Z}$. Moreover, the containment

$$\pi_0(\mathrm{R}_{\mathfrak{o}'_{\mathfrak{f}}/\mathfrak{f}}(\mathscr{T}'_{\mathfrak{o}'_{\mathfrak{f}}})) \subset \mathrm{R}_{\mathfrak{o}'_{\mathfrak{f}}/\mathfrak{f}}(\mathbf{Z}) = \mathbf{Z}$$

of (constant) $\mathfrak{f}$-groups is an equality because for a uniformizer π' of $\mathfrak{o}'$ the point

$$\pi' \in k'^{\times} = \mathscr{T}'(\mathfrak{o}')$$

has reduction in the geometric component group equal to $1 \in \mathbf{Z}$. □

We can now form a commutative diagram

$$\begin{array}{ccc} T & \xrightarrow{\text{closed}} & \mathrm{R}_{k'/k}(T_{k'}) \\ {\scriptstyle\text{open}}\downarrow & & \downarrow{\scriptstyle\text{open}} \\ \mathscr{T}^{\mathrm{fl}} & \longrightarrow & \mathrm{R}_{\mathfrak{o}'/\mathfrak{o}}(\mathscr{T}') \end{array}$$

where $\mathscr{T}^{\mathrm{fl}}$ is defined to be the $\mathfrak{o}$-flat closed subscheme of $\mathrm{R}_{\mathfrak{o}'/\mathfrak{o}}(\mathscr{T}')$ given by the schematic closure of the locally closed subscheme T in $\mathrm{R}_{\mathfrak{o}'/\mathfrak{o}}(\mathscr{T}')$. The generic fiber of $\mathscr{T}^{\mathrm{fl}}$ is T as a closed subscheme of the open generic fiber $\mathrm{R}_{k'/k}(T_{k'})$ of $\mathrm{R}_{\mathfrak{o}'/\mathfrak{o}}(\mathscr{T}')$. The natural inverse bijections (via "generic fiber" and "schematic closure") between the set of $\mathfrak{o}$-flat closed subschemes of an $\mathfrak{o}$-flat $\mathfrak{o}$-scheme $\mathscr{X}$ and the set of closed subschemes of $\mathscr{X}_k$ is functorial in $\mathscr{X}$ (in an evident manner) and is compatible with direct products, so it preserves the subgroup scheme property over the base when $\mathscr{X}$ is an $\mathfrak{o}$-group. Thus, $\mathscr{T}^{\mathrm{fl}}$ is a closed $\mathfrak{o}$-subgroup of $\mathrm{R}_{\mathfrak{o}'/\mathfrak{o}}(\mathscr{T}')$ with generic fiber T.

The $\mathfrak{o}$-flat $\mathscr{T}^{\mathrm{fl}}$ generally won't be the Néron model of its generic fiber T because it is merely $\mathfrak{o}$-flat (and locally of finite type) rather than $\mathfrak{o}$-smooth. The group smoothening process applied to $\mathscr{T}^{\mathrm{fl}}$ will wind up being the sought-after Néron model, but we need to keep track of component groups of geometric special fibers to be sure that this smoothening has finitely generated component group for its geometric special fiber.

Since we have shown that $\mathrm{R}_{\mathfrak{o}'/\mathfrak{o}}(\mathscr{T}')$ has special fiber with geometric component group that is a finitely generated $\mathbf{Z}$-module (it is even free), the same goes for its closed $\mathfrak{o}$-subgroup $\mathscr{T}^{\mathrm{fl}}$ by passing to special fibers and applying the following general lemma:

Lemma 9.5 *For a field κ, if G is a κ-group locally of finite type and H is a closed κ-subgroup scheme then the map of étale component groups $\pi_0(H) := H/H^0 \to G/G^0 =: \pi_0(G)$ has finite kernel.*

Proof We may assume κ is algebraically closed and replace G and H with their underlying reduced schemes so each is smooth. The map of component groups $H/H^0 \to G/G^0$ has kernel $(H \cap G^0)/H^0$. This is the component group of the closed subgroup $H \cap G^0$ of G^0 (note that H^0 is a clopen subscheme of H and $H \cap G^0$ is a clopen subscheme of H, as G^0 is clopen in G, so H^0 is clopen in $H \cap G^0$), so we want to show that $H \cap G^0$ has finite component group.

Since G^0 is a connected κ-group locally of finite type, it is of finite type [SGA3, Exp. $\mathrm{VI_A}$, Proposition 2.4] (for our case of interest G^0 is affine, so this general citation is then not needed). Thus, its closed subscheme $H \cap G^0$ is finite type and hence has finite component group. □

Since $\pi_0(\mathscr{T}^{\mathrm{fl}}_{\overline{\mathfrak{f}}})$ is a finitely generated $\mathbf{Z}$-module, the quotient $\mathscr{T}^{\mathrm{fl}}(\mathfrak{o}^{\mathrm{un}})/(\mathscr{T}^{\mathrm{fl}})^0(\mathfrak{o}^{\mathrm{un}})$ is too since this quotient is naturally a subgroup of $\pi_0(\mathscr{T}^{\mathrm{fl}}_{\mathfrak{f}_s}) = \pi_0(\mathscr{T}^{\mathrm{fl}}_{\overline{\mathfrak{f}}})$. By [BLR, Sect. 10.1, Proposition 4], the group smoothening $q : \mathscr{T} \to \mathscr{T}^{\mathrm{fl}}$ (an affine map of finite type) *is* a Néron model of T. The special fiber of this smoothening inherits finite generatedness for its geometric component group from that of $\mathscr{T}^{\mathrm{fl}}_{\overline{\mathfrak{f}}}$ (in particular, the torsion subgroup $\pi_0(\mathscr{T}_{\overline{\mathfrak{f}}})_{\mathrm{tor}}$ is *finite*) by the following lemma, settling Proposition 8.3(3).

Lemma 9.6 *If $q : G \to G'$ is a homomorphism between locally finite type group schemes over a field κ such that q is of finite type (e.g., affine) then $\pi_0(G) \to \pi_0(G')$ has finite kernel.*

Proof We may pass to the case that κ is algebraically closed and then replace G and G' with their underlying reduced schemes so that each is smooth. We aim to show the kernel of $G/G^0 \to G'/G'^0$ is finite. This kernel is $q^{-1}(G'^0)/G^0$, with G^0 clopen in G and $q^{-1}(G'^0)$ also clopen in G since G'^0 is clopen in G'. Hence, G^0 is the identity component of $q^{-1}(G'^0)$, so it suffices to show $q^{-1}(G'^0)$ is of finite type. Since q is of finite type, it suffices to show G'^0 is of finite type. But G'^0 is a connected κ-group locally of finite type, so it is of finite type by [SGA3, Exp. $\mathrm{VI_A}$, Proposition 2.4] (for our case of interest G'^0 is affine, so this general citation is then not needed). □

10 The Subgroup $G(k)^0$

The following result, on which we'll give some commentary in Sect. 12, is one of the key reasons for requiring a perfect residue field in the development of Bruhat-Tits theory.

Theorem 10.1 (Steinberg) *If $\mathfrak{f}$ is algebraically closed then* $\mathrm{H}^1(k, G) = 1$ *for any connected reductive k-group G. In particular, when $\mathfrak{f}$ is perfect we have* $\mathrm{H}^1(k^{\mathrm{un}}, G) = 1$ *for any connected reductive k-group G.*

Example 10.2 The structure theory of connected reductive groups over fields gives that G over k has a (unique up to isomorphism) quasi-split inner form. The set of inner forms is the image of $\mathrm{H}^1(k, G^{\mathrm{ad}})$ in $\mathrm{H}^1(k, \mathrm{Aut}_{G/k})$, where $\mathrm{Aut}_{G/k}$ is a smooth locally finite type k-group representing the automorphism functor of G (on the category of k-schemes). This Aut-scheme fits into a short exact sequence

$$1 \to G^{\mathrm{ad}} \to \mathrm{Aut}_{G/k} \to \Gamma \to 1$$

with étale Γ (i.e., the adjoint central quotient $G^{\mathrm{ad}} = G/Z_G$ with its natural action on G is the identity component of $\mathrm{Aut}_{G/k}$).

Now assume $\mathfrak{f}$ is perfect, so Theorem 10.1 applied to G^{ad} shows $G_{k^{\mathrm{un}}}$ is its own quasi-split inner form; i.e., *G becomes quasi-split over k^{un}*. For non-Archimedean local fields k, there are *many* connected semisimple k-groups that are not quasi-split (equivalently, maximal split k-tori S have centralizer $Z_G(S)$ not a torus). The most basic examples are the anisotropic k-groups of units of reduced norm 1 in central division algebras, but there are many isotropic examples too, such as: special orthogonal groups $\mathrm{SO}(q)$ for non-degenerate quadratic spaces (V, q) whose anisotropic part (V_0, q_0) has dimension at least 3 (see [C3, Sect. T.3], especially near the end) and special unitary groups $\mathrm{SU}(h)$ in [C3, Sect. Q.4] for $h = h' \perp h_0$ with "split" Hermitian h' and anisotropic Hermitian h_0 with dimension at least 2.

Remark 10.3 Another reason for the utility of requiring $\mathfrak{f}$ to be perfect in the development of Bruhat-Tits theory is that the equivalence of the Néron model definition of $T(k)^0$ and Galois-theoretic definition in [KP, Definition 2.5.13] (see Proposition 8.5(ii) applied over k^{un}) assumes the perfectness of $\mathfrak{f}$.

For a short while longer, we continue to allow general $\mathfrak{f}$. Afterwards, we will assume that $\mathfrak{f}$ is perfect.

Definition 10.4 Let $G(k)^\natural := \mathrm{im}(G^{\mathrm{sc}}(k) \to G(k))$ for the simply connected central cover $G^{\mathrm{sc}} \to \mathscr{D}G$.

We have $(G^{\mathrm{sc}})^{\mathrm{ad}} = G^{\mathrm{ad}}$, so G^{ad} acts compatibly on the source and target of the map $G^{\mathrm{sc}} \to G$ and hence G does too via $G \twoheadrightarrow G^{\mathrm{ad}}$. In particular, the $G(k)$-action on itself via conjugation naturally lifts to a $G(k)$-action on $G^{\mathrm{sc}}(k)$, so $G(k)^\natural \lhd G(k)$ and $G(k)^\natural$ is *functorial* in G (since the central cover $G^{\mathrm{sc}} \to \mathscr{D}(G)$ is functorial in $\mathscr{D}G$ and hence in G). A further very useful property is

Lemma 10.5 *The quotient group $G(k)/G(k)^\natural$ is commutative, and the $G^{\mathrm{ad}}(k)$-action on $G(k)$ becomes trivial on this quotient.*

Proof The commutator map $G \times G \to G$ factors uniquely through $G \times G \to G^{\mathrm{ad}} \times G^{\mathrm{ad}}$ via a map $\rho : G^{\mathrm{ad}} \times G^{\mathrm{ad}} \to G$. But applying this with G^{sc} in place of G and using the equality $(G^{\mathrm{sc}})^{\mathrm{ad}} = G^{\mathrm{ad}}$ thereby provides $\widetilde{\rho} : G^{\mathrm{ad}} \times G^{\mathrm{ad}} \to G^{\mathrm{sc}}$ lifting ρ, visibly uniquely (since $G^{\mathrm{ad}} \times G^{\mathrm{ad}}$ is smooth and connected whereas $G^{\mathrm{sc}} \to G$ is an isogeny). Passing to k-points, it follows that the image of the "commutator" map $G^{\mathrm{ad}}(k) \times G^{\mathrm{ad}}(k) \to G(k)$ lands inside $G(k)^\natural$. This makes the $G^{\mathrm{ad}}(k)$-action on $G(k)/G(k)^\natural$ trivial (check!), and in particular makes all commutators in this quotient group trivial. □

Proposition 10.6 *We have $G(k) = G(k)^{\natural} M_0(k)$ for any Levi k-subgroup $M_0 = Z_G(S)$ of a minimal parabolic k-subgroup of G (here $S \subset G$ is a maximal k-split torus). In particular, this equality holds with M_0 a Levi k-subgroup of any parabolic k-subgroup of G.*

Proof Let $M_0 = Z_G(S)$ be a Levi k-subgroup of a minimal parabolic k-subgroup $P_0 = M_0 \ltimes U_0$ (we'll often say "minimal Levi"), where S is a maximal split k-torus in G and U_0 is the split k-unipotent radical of P_0. We have the Bruhat decomposition

$$G(k) = \bigcup_{w \in W(G,S)} P_0(k)\dot{w}P_0(k)$$

where $W(G, S) := N_G(S)(k)/Z_G(S)(k)$ and $\dot{w} \in N_G(S)(k)$ denotes any choice of representative of w.

Let $S' = (S \cap \mathscr{D}G)^0_{\mathrm{red}}$ be the maximal k-subtorus of S contained in $\mathscr{D}G$ (a maximal k-split torus in $\mathscr{D}G$), so S is the almost direct product of S' and the maximal split central torus A_G in G. Let $S^{\mathrm{sc}} \subset G^{\mathrm{sc}}$ the maximal k-split torus in G^{sc} for which the central isogeny $G^{\mathrm{sc}} \to \mathscr{D}G$ carries S^{sc} onto S'. The structure theory for connected reductive groups over fields tells us that the natural map $W(G^{\mathrm{sc}}, S^{\mathrm{sc}}) \to W(G, S)$ is an isomorphism, and likewise that the minimal parabolic k-subgroup $P_0^{\mathrm{sc}} \subset G^{\mathrm{sc}}$ given by the preimage of P_0 under $G^{\mathrm{sc}} \to G$ has k-unipotent radical mapping *isomorphically* onto U_0. Hence, $U_0(k) \subset G(k)^{\natural}$ and the representatives $\dot{w}$ can be chosen from $G(k)^{\natural}$ too. Thus, inside the *commutative* quotient $G(k)/G(k)^{\natural}$, the image of $M_0(k)$ is full. □

From now on $\mathfrak{f}$ is perfect unless specified otherwise. We discussed at the end of Sect. 7 the aim of defining a finite-index subgroup $G(k)^0 \subset G(k)^1$ so that for all $x \in \mathcal{B}(G)$ the stabilizer $G(k)^0_x$ coincides with $(\mathscr{G}^1_x)^0(\mathfrak{o})$ for a specific smooth affine $\mathfrak{o}$-group model $\mathscr{G}^1_x$ for G. Let's motivate how we shall define it.

In accordance with the philosophy (as in [Tits, Sect. 1.13]) that G being simply connected corresponds to $G(k)$ behaving like a "connected" object for Bruhat-Tits theory, we want $G(k)^0 = G(k)^1$ for simply connected G. Thus, the desire for $G(k)^0$ to be functorial in G leads us to want $G(k)^{\natural} \subset G(k)^0$ in general. Proposition 10.6 thereby motivates wanting $G(k)^0 = G(k)^{\natural} M_0(k)^0$ for minimal Levi k-subgroups $M_0 \subset G$, but then $M_0(k)^0$ needs to be defined.

Since M_0^{ad} is k-anisotropic, let's focus on how to define $G(k)^0$ when G^{ad} is k-anisotropic. The principle that Bruhat-Tits theory over k should be a synthesis of the theory over k^{un} along with unramified descent leads us to want $G(k)^0 = G(k) \cap G(k^{\mathrm{un}})^0$ in general (see Corollary 7.4 with $k' = k^{\mathrm{un}}$ for a $G(k)^1$-analogue). To use this recipe as a *definition* when G^{ad} is k-anisotropic, we need a definition over k^{un}. By Example 10.2 (applicable since $\mathfrak{f}$ is perfect!), over k^{un} all connected reductive groups are quasi-split. Thus, let's focus on defining $G(k)^0$ whenever G is *quasi-split*. Proposition 10.6 in the quasi-split case has M_0 a *torus*, and for k-tori T the work in Sects. 8 and 9 provides a good definition of a finite-index subgroup $T(k)^0 \subset T(k)^1$ via the smooth affine relative identity component $\mathscr{T}^0$ of the (locally finite type) smooth separated Néron model $\mathscr{T}$ of T! Hence, for quasi-split G a natural candidate for $G(k)^0$ is $G(k)^\natural T(k)^0$ for a maximal k-torus T in a Borel k-subgroup of G.

Definition 10.7 Define $G(k)^0 \subset G(k)$ as follows:

(1) If G is quasi-split (as happens over k^{un}, by Theorem 10.1 and Example 10.2), define $G(k)^0 := G(k)^\natural T(k)^0$ for a maximal k-torus T in a Borel k-subgroup (T is a minimal Levi in such G, hence all choices of this T are $G(k)$-conjugate).
(2) If G^{ad} is anisotropic (e.g., $G = M_0$ is a minimal Levi $Z_H(S)$ for a connected reductive k-group H and maximal k-split torus $S \subset H$), define $G(k)^0 := G(k) \cap G(k^{\text{un}})^0$ (where $G(k^{\text{un}})^0$ makes sense by applying (1) over k^{un}).
(3) In the *general case*, define $G(k)^0 = G(k)^\natural M_0(k)^0$ for a minimal Levi M_0 (where $M_0(k)^0$ makes sense by applying (2) to M_0).

Remark 10.8 Let's discuss why (1) is independent of the choice of minimal Levi T and (3) is independent of the choice of minimal Levi M_0. The reason is that all such choices in each case are in a single orbit under the action by $G^{\text{ad}}(k)$ on $G(k)$ (by the structure theory of reductive groups over fields) yet this action induces the identity on $G(k)/G(k)^\natural$ by Lemma 10.5.

Let's also check that $G(k)^0 \supset G(k)^\natural$ in all cases. This is ensured by definition for (1) and (3). For (2) we see that $G(k)^\natural \subset G(k) \cap G(k^{\text{un}})^\natural \subset G(k) \cap G(k^{\text{un}})^0 =: G(k)^0$. By the commutativity of $G(k)/G(k)^\natural$, it follows that $G(k)^0 \lhd G(k)$.

Remark 10.9 See [KP, Remark 2.6.25] for a straightforward verification of consistency among overlaps of cases (1)–(3) in Definition 10.7 (e.g., cases (1) and (2) overlap exactly when G is a k-torus, in which case the consistency amounts to the equality $T(k)^0 = T(k) \cap T(k^{\text{un}})^0$ that follows from Proposition 8.3(2), as this turns the claimed equality into the assertion $\mathscr{T}^0(\mathfrak{o}) = \mathscr{T}(\mathfrak{o}) \cap \mathscr{T}^0(\mathfrak{o}^{\text{un}})$ for the Néron model $\mathscr{T}$ of T, which is easily verified).

Remark 10.10 The multi-case aspect of Definition 10.7 can be expressed by a single recipe (that is however not well-suited to be used directly): $G(k)^0 = G(k)^\natural(M_0(k) \cap M_0(k^{\text{un}})^\natural T(k^{\text{un}})^0)$ where $T \subset (M_0)_{k^{\text{un}}}$ is a maximal k^{un}-torus in a Borel k^{un}-subgroup of $(M_0)_{k^{\text{un}}}$.

Lemma 10.11 *The normal subgroup $G(k)^0 \subset G(k)$ is open.*

Proof We treat each of the cases in Definition 10.7 in turn. For case (1), consider the open cell

$$\Omega = U^- \times T \times U^+ \subset G$$

via multiplication, where $U^\pm$ are the split k-unipotent radicals of opposite minimal parabolic k-subgroups of G containing T. Both U^- and U^+ lift isomorphically into G^{sc}, and the resulting open subspace of k-points (via multiplication)

$$U^-(k) \times T(k) \times U^+(k) \subset G(k)$$

implies openness of $U^-(k)T(k)^0U^+(k) \subset G(k)$ (as a subset of triple products) since $T(k)^0$ is open in $T(k)$. In the quotient group $G(k)/G(k)^\natural$ the images of $U^\pm(k)$ are trivial, so $T(k)^0$ has open image in that quotient. Hence, $G(k)^\natural T(k)^0$ is open in $G(k)$ as desired for (1).

For (2), by (1) applied over k^{un} we see that $G(k^{\mathrm{un}})^0$ is open in $G(k^{\mathrm{un}})$, so $G(k) \cap G(k^{\mathrm{un}})^0$ is open in the subspace $G(k) \subset G(k^{\mathrm{un}})$ (which has the subspace topology, just as k does in k^{un}).

Finally, (3) is settled by the exact same argument as (1) with the minimal Levi M_0 in the role of T there (a minimal Levi of the quasi-split case) provided that $M_0(k)^0$ is open in $M_0(k)$ (an openness property that for tori was known via Néron models). But the settled openness for case (2) applies to the k-group M_0. □

Example 10.12 If G is k-split then by definition $G(k)^0 = G(k)^\natural S(k)^0$ for $G(k)^\natural$ the image of $G^{\mathrm{sc}}(k)$ and S a maximal split k-torus in G, where $S(k)^0 = \mathscr{S}(\mathfrak{o})$ for the unique $\mathfrak{o}$-torus $\mathscr{S}$ with generic fiber S (relative identity component of the Néron model of S).

For $G = \mathrm{PGL}_2$, $G(k)^0 = (\text{image of } \mathrm{SL}_2(k))\mathscr{S}(\mathfrak{o})$ for $\mathscr{S} = \left\{\left(\begin{smallmatrix} * & 0 \\ 0 & 1 \end{smallmatrix}\right)\right\} \subset \mathrm{PGL}_{2,\mathfrak{o}}$. But $\mathrm{PGL}_2(k) \overset{\det}{\twoheadrightarrow} k^\times/(k^\times)^2$ has kernel equal to the image of $\mathrm{SL}_2(k)$ (check!), and the image of $\mathscr{S}(\mathfrak{o})$ in this quotient is the subgroup $\mathfrak{o}^\times/(\mathfrak{o}^\times)^2 \subset k^\times/(k^\times)^2$ that has index 2 because of the commutative diagram

$$\begin{array}{ccc} k^\times/\mathfrak{o}^\times & \xrightarrow{\sim} & \mathbf{Z} \\ \uparrow & & \uparrow \\ (k^\times)^2/(\mathfrak{o}^\times)^2 & \xrightarrow{\sim} & 2\mathbf{Z} \end{array}$$

This shows the open normal subgroup $\mathrm{PGL}_2(k)^0 \subset \mathrm{PGL}_2(k)$ has index 2.

Beware that in contrast with $G(k)^1$, the functoriality of $G(k)^0$ in G is *not* obvious (beyond the case when G is a torus, for which it follows from functoriality of Néron models of tori). This property only comes much later, using Bruhat-Tits theory in a serious way; see Theorem 30.7. (An alternative approach is to *define* $G(k)^0$ in a more abstract way via the so-called Kottwitz homomorphism [KP, Sects. 11.1–11.5], but this is harder to access for concrete calculations early on in the theory.)

There is a special case in which the functoriality of $G(k)^0$ can be verified directly now:

Proposition 10.13 *If $\mathfrak{f}$ is algebraically closed then $G(k)^0$ is functorial in G; i.e., if $f : G' \to G$ is a homomorphism between connected reductive k-groups then $G'(k) \to G(k)$ carries $G'(k)^0$ into $G(k)^0$.*

In particular, for general perfect $\mathfrak{f}$ the group $G(k^{\mathrm{un}})^0$ is functorial in G.

Proof Since $\mathfrak{f}$ is algebraically closed, so G' and G are quasi-split by Example 10.2, we have $G'(k)^0 = G'(k)^\natural T'(k)^0$ and $G(k)^0 = G(k)^\natural T(k)^0$ for maximal k-tori $T' \subset G'$ and $T \subset G$ in Borel k-subgroups. We know that $G'(k) \to G(k)$ carries $G'(k)^\natural$

into $G(k)^\natural$, so the task is to show that $T'(k)^0$ is carried into $G(k)^\natural T(k)^0$. Since $f(T') \subset G$ is a k-torus, so it is contained in a maximal k-torus $\mathbb{T} \subset G$, we have $f(T'(k)^0) \subset \mathbb{T}(k)^0$ by functoriality with respect to the map of k-tori $T' \to \mathbb{T}$.

Now we can forget about the general G' and are reduced to a problem entirely about G: for any maximal k-torus $\mathbb{T} \subset G$, show $\mathbb{T}(k)^0 \subset G(k)^\natural T(k)^0$ for a maximal k-torus T in a Borel k-subgroup $B \subset G$. For this purpose we now build a central extension

$$1 \to Z \to \widetilde{G} \xrightarrow{q} G \to 1 \tag{10.1}$$

in which Z is a torus and $\widetilde{G}$ is a connected reductive k-group whose derived group $\mathscr{D}\widetilde{G}$ is simply connected. (A more refined and very useful construction called a *z-extension* can be made in which Z is even an induced torus, as explained in [KP, Sect. 11.4], but we don't need this here.)

Let μ be the kernel of the central isogeny $H := Z(G) \times G^{\mathrm{sc}} \to G$ (where $Z(G) := (Z_G)^0_{\mathrm{red}}$ is the maximal central k-torus) and pick a maximal k-torus Z of H, so $\mu \subset Z$ (since Z is its own centralizer in H). Thus, we can form the "central pushout" of

$$1 \to \mu \to H \to G \to 1$$

along the inclusion $\mu \hookrightarrow Z$ to get (10.1) by using $\widetilde{G} := (Z \times H)/\mu$ in which the quotient rests on μ being anti-diagonally embedded: precompose with inversion for one of the two inclusions $\mu \rightrightarrows Z, H$.

The preimage $\widetilde{B} := q^{-1}(B)$ is a Borel k-subgroup of $\widetilde{G}$, and likewise $\widetilde{\mathbb{T}} := q^{-1}(\mathbb{T})$ is a maximal k-torus of $\widetilde{G}$ and $\widetilde{T} := q^{-1}(T)$ is a maximal k-torus in $\widetilde{B}$. To reduce our problem for G to the analogue for $\widetilde{G}$, we now assume $\widetilde{\mathbb{T}}(k)^0 \subset \widetilde{G}(k)^\natural \widetilde{T}(k)^0$ and deduce that $\mathbb{T}(k)^0 \subset G(k)^\natural T(k)^0$. The crucial point is that we have a short exact sequence of k-tori

$$1 \to Z \to \widetilde{\mathbb{T}} \xrightarrow{q} \mathbb{T} \to 1,$$

so $q : \widetilde{\mathbb{T}}(k)^0 \to \mathbb{T}(k)^0$ is surjective by Proposition 8.5(ii) (this uses that $\mathfrak{f}$ is algebraically closed). Hence, applying q to both sides of the *assumed* containment $\widetilde{\mathbb{T}}(k)^0 \subset \widetilde{G}(k)^\natural \widetilde{T}(k)^0$ yields

$$\mathbb{T}(k)^0 \subset q(\widetilde{G}(k)^\natural) q(\widetilde{T}(k)^0) \subset G(k)^\natural T(k)^0$$

as desired.

Now we may and do replace G with $\widetilde{G}$, so $\mathscr{D}G$ is simply connected and hence $G(k)^\natural = (\mathscr{D}G)(k)$. Our aim is to show

$$\mathbb{T}(k)^0 \subset (\mathscr{D}G)(k) T(k)^0$$

for any maximal k-torus $\mathbb{T} \subset G$ and any maximal k-torus T in a Borel k-subgroup B of G. The intersection $T' = T \cap (\mathscr{D}G)$ is a maximal k-torus in $\mathscr{D}G$ that is contained in a Borel k-subgroup of $\mathscr{D}G$. The quotient $\overline{T} := G/(\mathscr{D}G) = G^{\text{ab}}$ is a k-torus fitting into a short exact sequence of k-tori

$$1 \to T' \to T \to \overline{T} \to 1.$$

Once again appealing to Proposition 8.5(ii), the natural map $T(k)^0 \to \overline{T}(k)^0$ is surjective, so via the inclusion $G(k)/(\mathscr{D}G)(k) \hookrightarrow \overline{T}(k)$ the subgroup $G(k)^0/(\mathscr{D}G)(k)$ coincides with $\overline{T}(k)^0$.

In other words, $G(k)^0$ is the *preimage* of $\overline{T}(k)^0$ under the map $G(k) \to \overline{T}(k)$. Hence, our task reduces to verifying that the composite map of k-groups $\mathbb{T} \to G \to \overline{T}$ carries $\mathbb{T}(k)^0$ into $\overline{T}(k)^0$. But the construction $T(k)^0$ for general k-tori T is functorial in T, so we are done. □

Next we want to relate $G(k)^0$ to $G(k)^1$, generalizing what we have seen in the case of tori.

Lemma 10.14 *As subgroups of $G(k)$, we have $G(k)^0 \subset G(k)^1$.*

Proof We check cases (1)–(3) in Definition 10.7 using that $H(k)^1$ is functorial in H for connected reductive k-groups H and that $G(k)^\natural \subset G(k)^1$ (since $G^{\text{sc}}(k) = G^{\text{sc}}(k)^1$):

(1) $G(k)^0 = G(k)^\natural T(k)^0 \subset G(k)^1 T(k)^1 = G(k)^1$;
(2) $G(k)^0 = G(k) \cap G(k^{\text{un}})^0 \subset G(k) \cap G(k^{\text{un}})^1 = G(k)^1$ (see Corollary 7.4);
(3) $G(k)^0 = G(k)^\natural M_0(k)^0 \subset G(k)^1 M_0(k)^1 = G(k)^1$. □

The quotient group $G(k)^1/G(k)^0$ is commutative since it is a subquotient of the commutative $G(k)/G(k)^\natural$. In Sect. 11 we'll show the index of $G(k)^0$ in $G(k)^1$ is always finite, generalizing the established finiteness in the case of tori that is part of Proposition 8.5(i).

11 Refined Aspects of $G(k)^0$

We have defined $G(k)^0 \subset G(k)$ that is an open normal subgroup contained in $G(k)^1$ and containing the image $G(k)^\natural$ of $G^{\text{sc}}(k)$, so $G(k)^1/G(k)^0$ is commutative since $G(k)/G(k)^\natural$ is commutative. From the definition it is *not* evident that $G(k)^0$ is functorial in G. However, over k^{un} we *do* have functoriality by a *direct* (non-obvious) argument (see Proposition 10.13). The main difficulty is to show if $\mathbb{T} \subset G$ is a maximal k-torus then $\mathbb{T}(k)^0 \subset G(k)^0$. We encourage the reader to try proving this directly (to be more appreciative of Proposition 10.13). As a special case, $G(k^{\text{un}})^0$ is functorial in G.

Proposition 11.1 *We have $G(k)^0 \subset G(k) \cap G(k^{\text{un}})^0$.*

The reverse containment lies much deeper; see Theorem 30.7. From now on, we write $K := k^{\text{un}}$.

Proof We check $G(k)^0 \subset G(K)^0$ casewise via the definition of $G(k)^0$:

(1) if G is quasi-split then $G(k)^0 := G(k)^\natural \cdot T(k)^0 \subset G(K)^\natural \cdot T(K)^0 = G(K)^0$ for T a maximal torus in a Borel k-subgroup $B \subset G$, where the inclusion follows from the fact that for the Néron model $\mathscr{T}$ of T, $\mathscr{T}_{\mathfrak{o}^{\mathrm{un}}}$ *is* the Néron model of T_K (hence $T(k)^0 \subset T(K)^0$). The last equality follows from that T_K is maximal in the Borel K-subgroup $B_K \subset G_K$.
(2) if G^{ad} is k-anisotropic then $G(k)^0 = G(k) \cap G(K)^0$, by *definition*.
(3) the general case: for a minimal Levi M_0 in G, $G(k)^0 = G(k)^\natural M_0(k)^0 \subset G(K)^\natural M_0(K)^0 \subset G(K)^0$ where the first inclusion follows from (2) for M_0 and the last inclusion holds since $G(K)^\natural \subset G(K)^0$ by definition and $M_0(K)^0 \subset G(K)^0$ by functoriality over K (Proposition 10.13) applied to $M_0 \hookrightarrow G$.

□

Now we discuss two results concerning $M_0(k)^0$ for a minimal Levi k-subgroup $M_0 \subset G$. The first result feeds into the proof of the second, and the second is used in Appendix D to prove the Cartan decomposition for $G(k)$ (an important decomposition for representation-theoretic purposes) via Bruhat-Tits theory. The reader may skip them on a first reading.

Proposition 11.2 *If G is simply connected then $M_0(K)^0 = M_0(K)^1$. Upon intersecting with $M_0(k)$, we have $M_0(k)^0 = M_0(k)^1$.*

Example 11.3 If G is *quasi-split* and simply connected then $M_0 = T$ is an *induced* torus (this uses that G is simply connected), so T_K is also induced and hence $T(K)^0 = T(K)^1$ by Proposition 8.5(iii).

Proof We want the inclusion $M_0(K)^0 \subset M_0(K)^1$ to be an equality. For a maximal K-torus $T \subset (M_0)_K$ in a Borel K-subgroup of $(M_0)_K$, by definition $M_0(K)^0 = M_0(K)^\natural T(K)^0$ where $M_0(K)^\natural$ is the image on K-points of $M_0^{\mathrm{sc}} \to \mathscr{D}M_0$. But derived groups of torus centralizers (such as Levi subgroups) of simply connected groups are simply connected (see [C3, Cor. 9.5.11]), so $M_0(K)^0 = (\mathscr{D}M_0)(K)T(K)^0$.

Since M_0 is a Levi subgroup of G, likewise $(M_0)_K \subset G_K$ is a Levi subgroup (perhaps not minimal), so maximal tori of Borel K-subgroups of $(M_0)_K$ are *also* maximal tori of Borel K-subgroups of G_K: a maximal torus in a Borel subgroup is the same as a maximal torus that is maximally split, and $(M_0)_K$ and G_K have maximal *split* tori of *same* dimension (as for Levi subgroups of connected reductive groups over any field). The dynamic description of Borel subgroups and Levi subgroups over a field then yields that if $B \subset G_K$ is a Borel K-subgroup containing T then $B \cap (M_0)_K$ is a Borel K-subgroup of $(M_0)_K$ containing T and hence $B' := B \cap \mathscr{D}(M_0)_K$ is a Borel K-subgroup of $\mathscr{D}(M_0)_K$ containing the maximal K-torus $T' = T \cap \mathscr{D}(M_0)_K$ of $\mathscr{D}(M_0)_K$.

Since T is a maximal torus in a Borel subgroup of the K-group $G_K = G_K^{\mathrm{sc}}$ that is simply connected, it is *induced*. Likewise T' is a maximal torus of a Borel K-subgroup of the K-group $\mathscr{D}(M_0)_K$ that is simply connected (since G is), so T' is also induced. Since $(M_0)_K = T\,\mathscr{D}(M_0)_K$, we also have $T/T' = (M_0/\mathscr{D}M_0)_K$.

We next show $\overline{T} := T/T'$ is induced. For this we use how *quasi-split* connected reductive groups (such as G_K equipped with a Borel K-subgroup containing T, and $\mathscr{D}(M_0)_K$ equipped with a Borel K-subgroup containing T') are built via Galois-twisting of a split form via "diagram automorphisms". The Galois action on $\Phi(G_{K_s}, T_{K_s})$ is a permutation action on the set Δ simple positive coroots for B, and this preserves the subset Δ' of simple positive coroots in $\Phi((\mathscr{D}M_0)_{K_s}, T'_{K_s})$ for B', so the Galois action on the complementary set $\Delta - \Delta'$ of simple positive coroots over K_s is also a permutation. The latter set is a basis for the geometric cocharacter lattice of $\overline{T} := T/T'$, so $\overline{T}$ is induced.

Since $\overline{T}$ is induced, we have $\overline{T}(K)^1 = \overline{T}(K)^0$ (Proposition 8.5(iii)). By definition of $M_0(K)^0$ for the quasi-split $(M_0)_K$, we have $T(K)^0 \subset M_0(K)^0$ and thus obtain the commutative diagram

$$\begin{array}{ccccc} & & M_0(K)^1 & & \\ & \nearrow & & \searrow & \\ M_0(K)^0 & & & & \overline{T}(K)^1 = \overline{T}(K)^0 \\ & \nwarrow & & \nearrow & \\ & & T(K)^0 & & \end{array}$$

where the bottom-right morphism is surjective due to the short exact sequence of K-tori $1 \to T' \to T \to \overline{T} \to 1$ (i.e., $T \to \overline{T}$ is surjective with a torus kernel) and vanishing of $\mathrm{H}^1(K, T')$ (as T' is induced). Hence $M_0(K)^1/M_0(K)^0$ comes from $\ker(M_0(K)^1 \to \overline{T}(K)^1) \subset \mathscr{D}(M_0)(K) = M_0(K)^{\natural} \subset M_0(K)^0$. □

Proposition 11.4 *For a connected reductive k-group G and minimal Levi k-subgroup $M_0 \subset G$,*

$$G(K)^0 \cap M_0(k)^1 = M_0(k)^0$$

Proof We have $M_0(k)^0 := M_0(k) \cap M_0(K)^0$ by definition (as M_0^{ad} is k-anisotropic), and $M_0(k)^1 = M_0(k) \cap M_0(K)^1$ by Corollary 7.4. Thus, it is enough to show that $G(K)^0 \cap M_0(K)^1 = M_0(K)^0$, since intersecting with $M_0(k)$ then yields the desired result.

By functoriality of $H(K)^0$ in H over K from Proposition 10.13 (applied to $M_0 \hookrightarrow G$) we have

$$M_0(K)^0 \subset G(K)^0 \cap M_0(K)^1,$$

where $G(K)^0 = G(K)^{\natural} T(K)^0$ for a maximal K-torus T in a Borel K-subgroup of G_K. We want to show this containment is an equality.

Such a T can be *chosen* to be contained in a Borel K-subgroup of $(M_0)_K$, so $T(K)^0 \subset M_0(K)^0$ by definition of $M_0(K)^0$ for the quasi-split $(M_0)_K$. Thus, we have

$$G(K)^0 \cap M_0(K)^1 = (G(K)^{\natural} T(K)^0) \cap M_0(K)^1 = (G(K)^{\natural} \cap M_0(K)^1) T(K)^0,$$

so to show this is contained in $M_0(K)^0$ it is the same as showing

$$G(K)^\natural \cap M_0(K)^1 \subset M_0(K)^0.$$

We will verify this using Proposition 11.2.

The natural map $q : G^{\mathrm{sc}} \to G$ is the composition of the central isogeny $G^{\mathrm{sc}} \to \mathscr{D}G$ and the inclusion $\mathscr{D}G \hookrightarrow G$, so the formation of q-preimages sets up a bijection between sets of parabolic k-subgroups, and then likewise for Levi k-subgroups. Thus, the q-preimage M_0' of M_0 is a minimal Levi k-subgroup of G^{sc}, so $M_0'(K)^0 = M_0'(K)^1$ by Proposition 11.2 applied to G^{sc}.

Points in $G(K)^\natural \cap M_0(K)$ come from $M_0'(K)$. Since $M_0' \to M_0$ carries M_0' onto a smooth connected *normal* k-subgroup of the connected reductive M_0 (namely, onto $M_0 \cap \mathscr{D}G$) via a central isogeny, the maximal central torus of M_0' maps isogenously onto a subtorus of the maximal central torus of M_0. Hence, $M_0'(K)^1$ is the preimage of $M_0(K)^1$, so points in $G(K)^\natural \cap M_0(K)^1$ come from $M_0'(K)^1 = M_0'(K)^0$. The functoriality of $H(K)^0$ in H (Proposition 10.13) implies $M_0' \to M_0$ carries $M_0'(K)^0$ into $M_0(K)^0$. This yields $G(K)^\natural \cap M_0(K)^1 \subset M_0(K)^0$ as desired. □

12 Finiteness Result and Steinberg's Theorem

Here is an important fact:

Theorem 12.1 *The inclusion $G(k)^0 \subset G(k)^1$ has finite index.*

For $G = T$ a torus with Néron model $\mathscr{T}$, this finiteness follows from Proposition 8.5(i): since $T(k)^1 = \mathscr{T}^{\mathrm{ft}}(\mathfrak{o})$ and $T(k)^0 = \mathscr{T}^0(\mathfrak{o})$, we have $T(k)^1/T(k)^0 \hookrightarrow \pi_0(\mathscr{T}_{\mathfrak{f}})_{\mathrm{tor}}$ which is finite.

Proof The plan is

(1) reduce the task for G to the same for a minimal Levi $Z = Z_G(S)$;
(2) reduce the task for Z to the case of Z_K. That in turn is quasi-split, so by (1) over K this reduces to tori over K, which is known (since $T(k)^0$ has finite index in $T(k)^1$ for all k-tori T).

Carrying out step (1), we temporarily grant the finiteness for H for which H^{ad} anisotropic over k (e.g., $H = Z = Z_G(S)$). By definition $G(k)^0 = G(k)^\natural Z(k)^0$, and $Z(k)^0$ has finite index in $Z(k)^1$ by our temporary hypothesis. We have

$$G(k)/G(k)^\natural \twoheadrightarrow G(k)/G(k)^0 \supset G(k)^1/G(k)^0,$$

where $G(k)/G(k)^\natural$ is *commutative*. Thus, when representatives of elements of $G(k)/G(k)^0$ are described via the Bruhat decomposition

$$G(k) = \coprod_{w \in W(G,S)} P_0(k)\dot{w}P_0(k)$$

(for a minimal parabolic k-subgroup $P_0 = Z \ltimes U_0$ having Levi Z, and $\dot{w} \in N_G(S)(k)$ a representative of w), the order of multiplication in the "decomposition" for the representative doesn't matter.

Since $W(G, S) = W(G^{\mathrm{sc}}, S^{\mathrm{sc}})$, we can *choose* $\dot{w}$ to come from $N_{G^{\mathrm{sc}}(k)}(S^{\mathrm{sc}}) \subset G^{\mathrm{sc}}(k)$. Likewise the minimal parabolic k-subgroup $P_0^{\mathrm{sc}} \subset G^{\mathrm{sc}}$ corresponding to P_0 (this is the preimage of P_0 under $G^{\mathrm{sc}} \to G$) has $\mathscr{R}_u(P_0^{\mathrm{sc}}) \simeq U_0$ via $G^{\mathrm{sc}} \to G$, so $U_0(k) \subset G(k)^{\natural}$. Hence $G(k)/G(k)^0$ is represented by $Z(k)$, so

$$Z(k) \cap G(k)^1 \to G(k)^1/G(k)^0$$

is surjective. Thus,

$$(Z(k) \cap G(k)^1)/(Z(k) \cap G(k)^0) \to G(k)^1/G(k)^0$$

is an isomorphism.

Clearly $Z(k) \cap G(k)^0 = (G(k)^{\natural} \cap Z(k))Z(k)^0$. Since we are assuming temporarily that the inclusion $Z(k)^0 \subset Z(k)^1$ has finite index, and $Z(k)^1 \subset Z(k) \cap G(k)^1$, it suffices to show

$$(Z(k) \cap G(k)^1)/((G(k)^{\natural} \cap Z(k))Z(k)^1) \tag{12.1}$$

is finite.

Let's first consider the quotient

$$(Z(k) \cap G(k)^1)/Z(k)^1.$$

In our earlier study of the relationship between ω_G and ω_M for Levi subgroups M (such as $M = Z = Z_G(S)$), we saw in Example 8.2 that there is a natural inclusion

$$\frac{Z(k) \cap G(k)^1}{Z(k)^1} \hookrightarrow \mathrm{X}_*(S')_{\mathbf{Q}}$$

(for $S' = (S \cap \mathscr{D}G)^0_{\mathrm{red}}$) whose source is finitely generated (as $Z(k)/Z(k)^1$ is even a finite free $\mathbf{Z}$-module, since $H(k)/H(k)^1 \subset \mathrm{Hom}_{\mathbf{Z}}(\mathrm{X}_k^*(H), \mathbf{Z})$ for any connected reductive k-group H). This subgroup of $\mathrm{X}_*(S')_{\mathbf{Q}}$ has full rank since it contains $\mathrm{X}_*(S')$: $S'(k) \subset Z(k) \cap G(k)^1$ because $S'(k) \subset (\mathscr{D}G)(k) = (\mathscr{D}G)(k)^1$, and this is carried onto $\mathrm{X}_*(S')$ because (i) S' is a product of copies of $\mathbf{G}_m$ and (ii) the normalized valuation $\mathrm{ord} : k^\times \to \mathbf{Q}$ has image $\mathbf{Z}$.

Having shown that $(Z(k) \cap G(k)^1)/Z(k)^1$ is finite free of rank $\dim S'$, we will show the same holds for the image of $G(k)^{\natural} \cap Z(k)$ in this quotient, so the cokernel (12.1) by that image is *finite*, as desired.

The point is that if we repeat our preceding considerations with G, S, and $Z = Z_G(S)$ for G^{sc}, S^{sc}, and $Z' := Z_{G^{\mathrm{sc}}}(S^{\mathrm{sc}})$ then since $G^{\mathrm{sc}}(k) = G^{\mathrm{sc}}(k)^1$ we conclude that the finite free $\mathbf{Z}$-module

$$(Z'(k) \cap G^{\mathrm{sc}}(k))/Z'(k)^1$$

has rank $\dim S^{\mathrm{sc}} = \dim S'$ and is naturally contained in $\mathrm{X}_*(S^{\mathrm{sc}})_{\mathbf{Q}}$. Moreover, the preimage of Z under $G^{\mathrm{sc}} \to G$ is Z' (since $S^{\mathrm{sc}} \to S'$ is an isogeny and $A_G \times S' \to S$ is an isogeny with A_G central in G), so $Z'(k) \cap G^{\mathrm{sc}}(k) \to G(k)^\natural \cap Z(k)$ is surjective.

Since the isogeny $S^{\mathrm{sc}} \to S'$ of split k-tori induces an isomorphism between rationalized cocharacter spaces, it follows that

$$(Z'(k) \cap G^{\mathrm{sc}}(k))/Z'(k)^1 \to (Z(k) \cap G(k)^1)/Z(k)^1$$

is an inclusion between finite free $\mathbf{Z}$-modules of the same rank and its image is the same as the image of $G(k)^\natural \cap Z(k)$ in the target quotient. This completes the reduction of our finiteness problem for G to the analogue for Z.

Turning attention to Z for which Z^{ad} is k-anisotropic, by *definition* $Z(k)^0 = Z(k) \cap Z(K)^0$. (The analogous equality for general G only comes much later in the development of Bruhat-Tits theory; see Theorem 30.7.) We have $Z(k)^1 = Z(k) \cap Z(K)^1$ by Corollary 7.4, so $Z(k)^1/Z(k)^0 \hookrightarrow Z(K)^1/Z(K)^0$. Hence, the finiteness problem for Z reduces to the same for Z_K (a reduction we can't carry out directly for G because we don't yet know the containment $G(k)^0 \subset G(k) \cap G(K)^0$ is an equality).

But Z_K is quasi-split (as for any connected reductive K-group), so its minimal Levi K-subgroups are tori! Now rerun the reduction step "$G \rightsquigarrow Z$ over k" for "$Z_K \rightsquigarrow$ torus over K" to reduce our finiteness problem to the settled case of tori over K. □

As a final topic before we introduce affine root systems and non-positively curved metric spaces, we discuss Steinberg's Theorem 10.1 that has been mentioned and used earlier: $\mathrm{H}^1(K, G) = 1$ for connected reductive G. This is a special case of a broader vanishing result for $\mathrm{H}^1(F, G)$ with connected reductive groups G over a certain class of fields F (to be defined in Sect. 13) that includes those K for which $\widehat{K}/K$ is separable (equivalently, $\mathcal{O}_K$ is excellent).

We don't want such a separability hypothesis on K to be imposed on Bruhat-Tits theory (there are non-excellent discrete valuation rings R [BLR, Sect. 3.6, Ex. 11], and the henselization R^{h} of any such R is also not excellent since $\widehat{R^{\mathrm{h}}} = \widehat{R}$ and $\mathrm{Frac}(R^{\mathrm{h}})/\mathrm{Frac}(R)$ is separable). Thus, first we want to reduce the triviality of $\mathrm{H}^1(K, G)$ to that of $\mathrm{H}^1(\widehat{K}, G)$. This follows from a more general fact:

Lemma 12.2 *The natural map of sets* $\mathrm{H}^1(k, G) \to \mathrm{H}^1(\widehat{k}, G)$ *is injective for any Henselian field* k *and smooth affine* k*-group* G.

We are not only saying that the kernel is trivial as a map of pointed sets, but that all non-empty fibers have only one element.

Proof The task is to show that if E and E' are G-torsors over k that become isomorphic over $\widehat{k}$ then $E \simeq E'$ as G-torsors over k. Using Galois descent (smoothness of such torsors ensures each splits over a finite Galois extension of k), the Isom-functor $\mathrm{Isom}_G(E, E')$ of G-equivariant isomorphisms is represented by a smooth affine k-scheme $\mathscr{I}$ (if L/k is a finite Galois extension splitting both torsors, this

Isom-functor over L becomes $\mathrm{Isom}_G(G, G)$ that is identified with G via right multiplication through inversion). We want to show that $\mathscr{I}(k)$ is non-empty, and we're given that $\mathscr{I}(\widehat{k})$ is non-empty. By smoothness of $\mathscr{I}$, we have that $\mathscr{I}(k)$ is even dense in $\mathscr{I}(\widehat{k})$ (Proposition 5.2). □

13 Fields of Dimension ≤ 1 and Non-positively Curved Metric Spaces

By [Ser2, Chap. 2, Sect. 3.1, Proposition 5], for a general field F the following are equivalent:

(1) $\mathrm{cd}(\mathrm{Gal}(F_s/F)) \leq 1$ (where $\mathrm{cd}(\Gamma)$ for a profinite group Γ is the least integer $n \geq 0$ such that $\mathrm{H}^i(\Gamma, \cdot)$ vanishes on torsion discrete Γ-modules for all $i > n$), and when $\mathrm{char}(F) = p > 0$ then also $\mathrm{Br}(E)[p] = 0$ for all finite separable E/F;
(2) $\mathrm{Br}(E) = 0$ for all finite separable E/F;
(3) for all finite separable E/F and finite Galois L/E, the norm map $L^\times \to E^\times$ is surjective.

Such fields F are said to be *of dimension at most* 1, denoted "$\dim(F) \leq 1$".

Example 13.1 A key result in Lang's thesis is that if K is a Henselian discretely-valued field with algebraically closed residue field and $\widehat{K}/K$ is separable then $\dim(K) \leq 1$. A reference for a proof is given in [Ser2, Chap. 2, Sect. 3.3]; we only need it for $K = \widehat{K}$, which is proved as [Ser1, Chap. 5, Sect. 4, Proposition 7].

Our reason for interest in these matters is

Theorem 13.2 (Steinberg) *If F is a field with* $\dim(F) \leq 1$ *and G is a connected reductive F-group then* $\mathrm{H}^1(F, G) = 1$. *In particular G is quasi-split.*

The quasi-split conclusion comes from Example 10.2. The proof of Theorem 13.2 as written in [Ste] assumes F is perfect, which rules out discretely-valued fields of positive characteristic. The method works without perfectness, as indicated in [BS, Sect. 8.6].

Remark 13.3 In the special case of F-tori T, the vanishing of $\mathrm{H}^i(F, T)$ for all $i \geq 1$ when $\dim(F) \leq 1$ is directly proved in [KP, Lemma 2.5.4].

Let's see why Theorem 13.2 implies $\mathrm{H}^1(K, G) = 1$ for all connected reductive K-groups G. Since $\widehat{K}$ is a *complete* discretely-valued field with algebraically closed residue field, Example 13.1 implies $\dim(\widehat{K}) \leq 1$ and so $\mathrm{H}^1(\widehat{K}, G) = 1$ by Theorem 13.2. By Lemma 12.2 we then have $\mathrm{H}^1(K, G) = 1$.

Next we develop some aspects of metric spaces. When $\mathcal{B}(G)$ is constructed, it will be a metric space with isometric $G(k)$-action and covered by subsets $\mathcal{A}(S)$ (called *apartments*) indexed by the maximal split k-tori S in G, with $g.\mathcal{A}(S) = \mathcal{A}(gSg^{-1})$ for

$g \in G(k)$. Each $\mathcal{A}(S)$ is an affine space for $\mathrm{X}_*(S')_{\mathbf{R}}$, with $S' := (S \cap \mathscr{D}G)^0_{\mathrm{red}}$, where an *affine space* for a finite-dimensional vector space V over a field L is a non-empty set equipped with a simply transitive V-action. For example if $V := \{\sum a_i x_i = 0\} \subset L^n$ with a_i's not all 0 then $\{\sum a_i x_i = c\}$ is an affine space for V via translation for any $c \in L$.

Under the $G(k)$-action on $\mathcal{B}(G)$, the effect of $N_G(S)(k)$ on $\mathcal{A}(S)$ is compatible with the affine space structure for $\mathrm{X}_*(S')_{\mathbf{R}}$ via the quotient map $q : N_G(S)(k) \twoheadrightarrow W(G, S)$:

$$\begin{array}{ccc} \mathcal{A}(S) & & \mathrm{X}_*(S')_{\mathbf{R}} \\ \circlearrowright & & \circlearrowright \\ N_G(S)(k) & \twoheadrightarrow & W(G, S). \end{array}$$

(i.e., $n.(v + a) = q(n).v + n.a$). We will discuss these actions further in Sects. 26 and 27.

Remark 13.4 The metric on $\mathcal{B}(G)$ restricted to each $\mathcal{A}(S)$ comes from the metric arising from a choice of $W(G, S)$-invariant inner product on $\mathrm{X}_*(S')_{\mathbf{R}}$ (this makes sense since the identification of $\mathcal{A}(S)$ with $\mathrm{X}_*(S')_{\mathbf{R}}$ via a choice of base point is well-defined up to a translation, and translations on $\mathrm{X}_*(S')_{\mathbf{R}}$ leave invariant the metric arising from any inner product). Moreover, any two $x, y \in \mathcal{B}(G)$ will turn out to lie in a common $\mathcal{A}(S)$, so this makes the metric on $\mathcal{B}(G)$ tangible.

Remark 13.5 When $\mathscr{D}G$ is k-anisotropic, so there is only one S and $\mathrm{X}_*(S')_{\mathbf{R}} = 0$, $\mathcal{B}(G)$ is a point.

What is particularly important is that certain group actions on the metric space $\mathcal{B}(G)$ have a fixed point. More specifically, we will show that for any bounded subgroup $\mathcal{K} \subset G(k)$ (so $\mathcal{K} \subset G(k)^1$), its action on $\mathcal{B}(G)$ has a fixed point x and so $\mathcal{K}$ is contained in the x-stabilizer $G(k)^1_x$ of $G(k)^1$. Such stabilizers are bounded for all x (by Theorem 12.1 and Axiom 2 in Sect. 25), so $G(k)^1_x$ accounts for the maximal bounded subgroups of $G(k)$ as we vary x, though for many x it is generally not maximal.

Likewise, a natural isometric action by $\Gamma = \mathrm{Gal}(K/k)$ on $\mathcal{B}(G_K)$ will have fixed points, and one *constructs* $\mathcal{B}(G)$ as $\mathcal{B}(G_K)^\Gamma$ (which is non-empty!) and deduces this locus has the desired structures. For instance, to make the apartments $\mathcal{A}(S) \subset \mathcal{B}(G)$ we note that S_K is generally *not* maximal split in G_K, so one has to do some work to (i) find a k-torus $T \subset Z_G(S)$ such that T_K is a maximal K-split torus in G_K, (ii) show the subset $\mathcal{A}(T_K)^\Gamma \subset \mathcal{B}(G)$ is independent of T and does the job to be $\mathcal{A}(S)$.

Now we turn to a general discussion with metric spaces.

Definition 13.6 A *curve* in a metric space (X, ρ) is a continuous map $c : [a, b] \to X$, and we say c is *rectifiable* if $\ell(c) := \sup_{a=t_0 \le \cdots \le t_n \le b} \sum_{i=0}^{n-1} \rho(c(t_i), c(t_{i+1})) < \infty$ (which we call the *length* of c).

Remark 13.7 It is easy to see that if c is rectifiable then so is $c|_{[a',b']}$ for $a \le a' \le b' \le b$.

Definition 13.8 Say $c : [a, b] \to X$ is a *geodesic* if it is rectifiable and satisfies

$$\ell(c|_{[t_0,t_1]}) = \rho(c(t_0), c(t_1)) = \rho(c(a), c(b))|t_0 - t_1|$$

for all $a \le t_0 \le t_1 \le b$ (the final equality is a strong form of "parameterized by arc length").

We often write $[x, y]$ to denote a geodesic from x to y (if one exists). Call (X, ρ) a *geodesic space* if all $x, y \in X$ are joined by a geodesic, and say (X, ρ) is *uniquely geodesic* if for all $x, y \in X$ there exists a unique such $[x, y]$. We emphasize again that a geodesic is a "parameterized path".

Example 13.9 Uniquely geodesic spaces include $\mathbf{R}^n$, affine spaces over $\mathbf{R}^n$, and the building $\mathcal{B}(G)$. Spheres are geodesic but not uniquely so, due to antipodal points.

Remark 13.10 If (X, ρ) is geodesic then $\rho(x, y)$ is the infiumum of $\ell(c)$ over all $c : [a, b] \to X$ joining x to y.

Definition 13.11 Say (X, ρ) is *non-positively curved* if for all $x, y \in X$, there exists $m \in X$ such that

$$\rho(x, z)^2 + \rho(y, z)^2 \ge 2\rho(m, z)^2 + \frac{1}{2}\rho(x, y)^2 \tag{13.1}$$

for all $z \in X$ (this is an equality for $X = \mathbf{R}^n$ by taking m to be the midpoint of $[x, y]$, due to the parallelogram law).

Lemma 13.12 *A non-positively curved geodesic space is uniquely geodesic and m in* (13.1) *is unique: it is the midpoint $m = c(1/2)$ for $c = [x, y]$.*

Proof See [KP, Lemma 1.1.11]. □

Remark 13.13 By [BH, Part II, Exercise 1.9], for a geodesic space X the condition (13.1) is equivalent to being a CAT(0)-space, and also equivalent to the property that for $\overline{p}, \overline{q}, \overline{r} \in \mathbf{R}^2$ whose respective distances between points are the same as the distances between p, q, r (forming a "reference triangle" in $\mathbf{R}^2$) the respective midpoints m and $\overline{m}$ of $[q, r]$ and $[\overline{q}, \overline{r}]$ satisfy $\rho(p, m) \le d_{\mathbf{R}^2}(\overline{p}, \overline{m})$.

Lemma 13.14 *For a non-positively curved geodesic space X, all $\overline{B}(z, r)$ are convex: if $x, y \in \overline{B}(z, r)$ then $[x, y] \subset \overline{B}(z, r)$.*

Proof From (13.1) with m the midpoint of $[x, y]$,

$$2r^2 \ge \rho(x, z)^2 + \rho(y, z)^2 \ge 2\rho(m, z)^2 + \frac{1}{2}\rho(x, y)^2 \ge 2\rho(m, z)^2,$$

so $\rho(m, z) \le r$. This shows the midpoint of $[x, y]$ is also in the same closed ball. Now iterate, using $[x, m]$ and $[m, y]$! The resulting "iterated bisection points" in $[x, y]$ are then all contained in $\overline{B}(z, r)$ but the set of these is dense in $[x, y]$, so $[x, y] \subset \overline{B}(z, r)$. □

Any non-empty bounded subset M in a non-positively curved geodesic space X is contained in some closed ball $\overline{B}(z, r)$, and the latter is convex, closed, and bounded. Hence

$$M \subset \bigcap_{C \supset M \text{convex,closed,bounded}} C =: \mathrm{conv}(M),$$

where the right side is called the *convex hull* of M; it is convex, closed, and bounded. The next order of business with non-positively curved geodesic spaces is to use the preceding lemma to prove the Bruhat-Tits fixed point lemma (which will ultimately be applied to bounded sets of the form $M = \mathcal{K}.x_0$ for certain subgroups $\mathcal{K} \subset \mathrm{Isom}(X)$ in the isometry group of X; note that $\mathcal{K}$ preserves any such M in the sense that $k.M = M$ for any $k \in \mathcal{K}$).

14 Fixed-Point Theorems

Recall that a metric space (X, ρ) is *non-positively-curved* if for all $x, y \in X$, there exists $m \in X$ such that for all $z \in X$ we have

$$\rho(x, z)^2 + \rho(y, z)^2 \geq 2\rho(m, z)^2 + \frac{1}{2}\rho(x, y)^2.$$

We've seen that non-positively curved geodesic spaces are uniquely geodesic and m is unique: it is the midpoint of the (parametrized!) geodesic curve $[x, y]$.

Theorem 14.1 *Let X be a* **complete** *non-positively curved geodesic space.*

(1) Let $M \subset X$ be a non-empty *closed* bounded *convex subset and define $f : M \to \mathbf{R}$ by*

$$y \mapsto \mathrm{diam}(y, M) = \sup_{y' \in M} \rho(y, y')$$

and the "radius" $r(M) = \inf_{y \in M} f(y) \geq 0$. There exists a unique metric center *$c \in M$: this means $\mathrm{diam}(c, M) = r(M)$ (i.e., c minimizes the diameter function f).*
In particular, c is a fixed point of $\mathrm{Stab}_{\mathrm{Isom}(X)}(M) := \{\varphi \in \mathrm{Isom}(X) : \varphi(M) = M\}$, where $\mathrm{Isom}(X)$ is the isometry group of X.

(2) (Bruhat-Tits fixed point lemma) For any non-empty bounded subset $M \subset X$, $\mathrm{Stab}_{\mathrm{Isom}(X)}(M)$ has a fixed point $x \in X$.

Remark 14.2 For $M \subset \mathbf{R}^n$ a closed n-simplex (a case of interest for Bruhat-Tits theory), its metric center is usually *not* the barycenter (i.e., average of the vertices). For example, if M is a right triangle T in $\mathbf{R}^2$ then the midpoint m of the hypotenuse has a common distance r from all three vertices of T since a triangle inscribed along the diameter of a circle is a right triangle, and m is the metric center because the three disks of radius r centered at the vertices of T have only the point m in common.

Example 14.3 The building $X = \mathcal{B}(G)$ (equipped with a $G(k)$-action via isometries) will be seen to be a complete non-positively curved geodesic space. A useful class of M's in such X will be "facets" in each apartment $\mathcal{A}(S)$ relative to a specific hyperplane arrangement (when $\mathcal{B}(G)$ is not a point).

Proof of Theorem 14.1 The implication (1) $\Rightarrow$ (2) is easy: $\mathrm{Stab}_{\mathrm{Isom}(X)}(M) \subset \mathrm{Stab}_{\mathrm{Isom}(X)}(\mathrm{conv}(M))$ for the closed bounded convex non-empty convex hull $\mathrm{conv}(M)$ of M (since any isometry of X carrying M onto itself bijectively permutes the set of all closed bounded convex subsets of X containing M), so we can apply (1) to $\mathrm{conv}(M)$ to obtain (2).

Now we prove (1). The informal idea is to consider points $y \in M$ for which $f(y) = \mathrm{diam}(y, M) \approx r(M)$ and to show such y's "are collectively close". Then c will be built as the Cauchy limit of a sequence of such y's.

To carry out a proper argument, first we dispose of the special case when $r(M) = 0$. This forces the non-empty M to be a point because for $x, x', y \in M$ the upper bound

$$\rho(x, x') \leq \rho(y, x) + \rho(y, x') \leq \mathrm{diam}(y, M) + \mathrm{diam}(y, M) = 2f(y)$$

forces

$$\frac{1}{2}\rho(x, x') \leq f(y)$$

for all $y \in M$ and hence

$$\frac{1}{2}\rho(x, x') \leq r(M) = 0$$

(so $x = x'$).

Now suppose $r := r(M) > 0$. Pick $0 < \epsilon < r$. Since r is defined as an infimum, we are led to consider $y, y' \in M$ such that $f(y), f(y') < r + \epsilon$ (there are such points for every $\epsilon > 0$). We claim that $\rho(y, y')^2 < 16r\epsilon$. To prove this claim, note that M is convex, so $[y, y'] \subset M$ and hence the midpoint m of $[y, y']$ is in M too. But $\mathrm{diam}(m, M) = f(m) \geq r$, so from the definition of the diameter as a supremum there exists $z \in M$ such that $\rho(m, z) \geq f(m) - \epsilon \geq r - \epsilon$. We thereby get

$$2(r+\epsilon)^2 > f(y)^2 + f(y')^2 \geq \rho(y, z)^2 + \rho(y', z)^2 \geq 2\rho(m, z)^2 + \frac{1}{2}\rho(y, y')^2 \geq 2(r-\epsilon)^2 + \frac{1}{2}\rho(y, y')^2.$$

Therefore $\rho(y, y')^2 < 2(2(r+\epsilon)^2 - 2(r-\epsilon)^2) = 16r\epsilon$. This proves our intermediate claim.

Pick a strictly deceasing sequence $\epsilon_n \to 0^+$ inside $(0, r)$, and (as we may do) pick $y_n \in M$ with $f(y_n) < r + \epsilon_n$. For $n' \geq n$, $\rho(y_{n'}, y_n)^2 < 16r\epsilon_n$. Thus $\{y_n\}$ is Cauchy, so there exists $c := \lim_{n\to\infty} y_n \in M$ (by completeness of X and closedness of M in X). The function $f : y \mapsto \mathrm{diam}(y, M)$ is continuous (since $|f(y) - f(y')| \leq \rho(y, y')$), so $f(c) = r$. This constructs the desired metric center.

Now we show the uniqueness of the metric center. If $c' \in M$ also satisfies $f(c') = r$, then $f(c), f(c') < r + \epsilon$ for all $0 < \epsilon < r$ and hence $\rho(c, c')^2 < 16r\epsilon \to 0$. This forces $c' = c$. □

Example 14.4 For $X = \mathcal{B}(G)$ and $\mathcal{K} \subset G(k)$ a *bounded* subgroup, the orbit $M = \mathcal{K}.x_0 \subset \mathcal{B}(G)$ will be seen to be *bounded* (Theorem 29.5), and $\mathcal{K} \to \mathrm{Isom}(X)$ lands inside $\mathrm{Stab}_{\mathrm{Isom}(X)}(M)$ since $\mathcal{K}$ is a group, so $\mathcal{K}$ has a fixed point in $\mathcal{B}(G)$! As we noted in the discussion preceding Remark 13.5, this explains why the point-stabilizers $G(k)^1_x$ as x varies in $\mathcal{B}(G)$ account for *all* maximal bounded subgroups of $G(k)$ (though the points x for which $G(k)^1_x$ is maximal bounded turn out to be rather specific).

Likewise, $\Gamma := \mathrm{Gal}(K/k)$ acting on $\mathcal{B}(G_K)$ via isometries has orbits that are compact (even finite: Proposition 30.5), and hence bounded. The fixed-point lemma then ensures that $\mathcal{B}(G_K)^\Gamma \neq \emptyset$ (and this will be the *construction* of $\mathcal{B}(G)$, with apartments $\mathcal{A}(S) \subset \mathcal{B}(G_K)^\Gamma$ likewise built as Γ-fixed point sets in carefully-chosen apartments in $\mathcal{B}(G_K)$).

Proposition 14.5 *Let (X, ρ) be a non-positively curved complete geodesic space, and $Y \subset X$ a closed convex non-empty subset.*

(1) For all $x \in X$, the function $Y \to \mathbf{R}$ defined by $y \mapsto \rho(x, y)$ is minimized at a unique point $\pi(x) \in Y$.
(2) The retraction function $\pi : X \to Y$ is continuous.
(3) The function π is equivariant for the actions on X and on Y by $\mathrm{Stab}_{\mathrm{Isom}(X)}(Y)$.

Remark 14.6 The retraction $\pi : \mathcal{B}(G) \to A$ onto an apartment in (2) is *not* the retraction $\rho_{A,C} : \mathcal{B}(G) \to A$ "centered" at a chamber $C \subset A$ (see Lemma A.1) used to study the metric on $\mathcal{B}(G)$ in Appendix A.

Proof It is clear that (1) and (2) together imply (3). The proof of (1) is similar to the Cauchy argument in the proof of Theorem 14.1(1) by replacing the role of $r(M)$ with

$$r(x) := \inf_{y \in Y} \rho(x, y) \geq 0,$$

so we just sketch the idea. If $r(x) = 0$ then $x \in Y$ by closedness of Y, so $\pi(x) := x$ works and is unique. If $r(x) > 0$ then for $0 < \epsilon < r(x)$ one shows that if $\rho(x, y), \rho(x, y') < r(x) + \epsilon$ then $\rho(y, y') < 16r(x)\epsilon$. In this way, we build $\pi(x)$ as a limit of a Cauchy sequence in Y and show its uniqueness.

The proof of (2) is more involved than one would expect, as we now explain. We use the sequential continuity criterion: if $x_n \to x$ in X we want to show $\pi(x_n) \to \pi(x)$. Letting $r = \rho(x, \pi(x))$ (this is $r(x)$), we separately treat the cases $r = 0$ (this is exactly the case $x \in Y$) and $r > 0$. By the defining minimizing properties of $\pi(x)$ and $\pi(x_n)$ we have

$$\begin{aligned} r = \rho(x, \pi(x)) \le \rho(x, \pi(x_n)) \le \rho(x, x_n) + \rho(x_n, \pi(x_n)) &\le \rho(x, x_n) + \rho(x_n, \pi(x)) \\ &\le 2\rho(x, x_n) + \rho(x, \pi(x)) \\ &= 2\rho(x, x_n) + r, \end{aligned}$$

so as $n \to \infty$ we have $\rho(x, \pi(x_n)) \to r$. In particular, if $r = 0$ then $\pi(x_n) \to x = \pi(x)$. If $r > 0$ then the convergence $\rho(x, \pi(x_n)) \to r$ forces the sequence $\{\pi(x_n)\}$ in the closed subset Y to be Cauchy by the method of proof of (1), so $\pi(x_n) \to y$ for some $y \in Y$ and hence $\rho(x, y) = r$. The uniqueness in (1) then forces $y = \pi(x)$. □

15 Affine Spaces and Affine Root Systems

For a connected reductive group G over a field F with a maximal split F-torus $S \subset G$, the S-representation on $\mathrm{Lie}(G)$ is completely reducible and the set $\Phi(G, S) \subset \mathrm{X}^*(S) - \{0\}$ of non-trivial S-weights that arise on $\mathrm{Lie}(G)$ constitutes a root system in the vector space $\mathrm{X}^*(S')_{\mathbf{R}}$ for $S' = (S \cap \mathscr{D}G)^0_{\mathrm{red}}$. Let's recall what this means:

Definition 15.1 A *root system* consists of a pair (V, Φ) where V is a finite-dimensional $\mathbf{R}$-vector space and $\Phi \subset V - \{0\}$ is a finite subset satisfying

(1) $\mathbf{R}\Phi = V$;
(2) for all $a \in \Phi$, there exists $a^\vee \in V^*$ such that

- $\langle b, a^\vee \rangle \in \mathbf{Z}$ for all $b \in \Phi$ and $\langle a, a^\vee \rangle = 2$;
- the reflection $r_a : V \to V$ defined by $x \mapsto x - \langle x, a^\vee \rangle a$ carries Φ into itself.

Remark 15.2 The coroot $a^\vee$ is uniquely determined in these axioms, and $b^\vee = a^\vee$ if and only if $b = a$. For any $a \in \Phi$, if $\lambda a \in \Phi$ for some $\lambda \in \mathbf{R}$ then $\lambda \in \{\pm 1/2, \pm 1, \pm 2\}$.

Remark 15.3 The pair $(V^*, \Phi^\vee = \{a^\vee\}_{a \in \Phi})$ is a root system, called the *dual* of (V, Φ).

The *Weyl group* $W(\Phi) \subset \mathrm{GL}(V)$ is the finite subgroup generated by the reflections $\{r_a\}_{a \in \Phi}$. The corresponding subgroup generated by the dual reflections in $\mathrm{GL}(V^*)$ is $W(\Phi^\vee)$.

There exists a subset $\Delta \subset \Phi$ such that every $a \in \Phi$ can be uniquely written as

$$a = \sum_{a \in \Delta} \lambda_i a_i$$

with either all $\lambda_i \in \mathbf{Z}_{\geq 0}$ or all $\lambda_i \in \mathbf{Z}_{\leq 0}$. Such a set Δ is called a *set of simple roots*.

For any two root systems (V_1, Φ_1) and (V_2, Φ_2), the pair

$$(V_1 \oplus V_2, (\Phi_1 \times \{0\}) \cup (\{0\} \times \Phi_2) =: \Phi)$$

is also a root system, called the *product* of (V_1, Φ_1) and (V_2, Φ_2).

Definition 15.4 A root system is called *irreducible* if it is not the *trivial* root system $(0, \emptyset)$ and is *not* the product of two non-trivial root systems.

Definition 15.5 A root system is called *reduced* if $\mathbf{R}a \cap \Phi = \{\pm a\}$ for all $a \in \Phi$.

There is a complete classification of irreducible root systems: types A, B, C, D, E, F, G in the reduced case, and one non-reduced irreducible root system called BC_n for each rank $n \geq 1$.

Returning now to a connected reductive group G over a discretely-valued Henselian field k (with a perfect residue field) and S a maximal split k-torus in G, each root group $U_a(k)$ for $a \in \Phi(G, S)$ will admit a filtration by bounded open subgroups satisfying certain properties. The members of the filtration are parametrized by *affine roots*, so we need the notion of *affine root system*. We now define and explore this latter notion, and will have much more to say about it (including examples) in subsequent sections.

Definition 15.6 An *affine space* A for a finite-dimensional vector space V over a field L is a V-torsor; i.e., a non-empty set equipped with a simply transitive action of V. We call $\dim V$ the *dimension* of A.

An *affine subspace* B of A is a W-torsor inside A for a linear subspace $W \subset V$ (uniquely determined: W is the set of elements of V translating some element of B to another element of B).

Definition 15.7 If A is an affine space for V then A^* denotes the L-vector space of affine-linear functions on A; i.e., A^* consists of functions $f : A \to L$ such that

$$f(x + v) = f(x) + \ell(v)$$

for all $x \in A$ and $v \in V$ with $\ell : V \to L$ a linear functional (uniquely determined: $\ell(v) = f(x_0 + v) - f(x_0)$ for all $v \in V$ and a fixed choice of $x_0 \in A$). The functional ℓ is denoted $\dot{f}$ or ∇f, and is called the *derivative* of f.

We have a short exact sequence

$$0 \to L \to A^* \stackrel{\nabla}{\to} V^* \to 0$$

where the first inclusion identifies L with the set of constant L-valued functions on A.

Definition 15.8 Let (V_1, A_1) and (V_2, A_2) be two affine spaces. An *affine transformation* $F : A_1 \to A_2$ is a map of sets such that for some linear $\dot{F} : V_1 \to V_2$ we have $F(x + v) = F(x) + \dot{F}(v)$ for all $x \in A_1$ and $v \in V_1$. We call the uniquely-determined $\dot{F}$ the *derivative* of F; it is also denoted ∇F.

If we choose $(V_1, A_1) = (V_2, A_2)$, we then have an exact sequence of groups

$$1 \to V \to \mathrm{Aff}(A) \to \mathrm{GL}(V) \to 1,$$

where the middle group consists of affine automorphisms of A (in which V is included via its translation action on A). When $\dim A > 0$, for non-constant $\psi \in A^*$ (so $\dot{\psi} \neq 0$) the subset

$$H_\psi = \{x \in A : \psi(x) = 0\}$$

is an affine space under the action of the linear hyperplane $\ker(\dot{\psi}) \subset V$. Upon choosing a base point $x_0 \in A$ to identify A with V, we can write $\psi = \dot{\psi} + c$ for some scalar $c \in L$.

Definition 15.9 An *affine root system* is (A, Ψ) where A is an affine space over a finite-dimensional $\mathbf{R}$-vector space V and $\Psi \subset A^* - \mathbf{R}$ is a collection of *non-constant* affine functions such that

(AR1) Ψ spans A^* (so $\Psi \neq \emptyset$ since $A^* \neq 0$, as $\mathbf{R} \subset A^*$, so $\dim A > 0$),
(AR2) $\Phi := \nabla(\Psi) \subset V^* - \{0\}$ is finite (necessarily a spanning set, by (AR1)),
(AR3) for all $\psi \in \Psi$, there exists a $\dot{\psi}^\vee \in V = (V^*)^*$ such that $\langle \dot{\psi}, \dot{\psi}^\vee \rangle = 2$ and Ψ is preserved by the reflection $r_{\psi,\dot{\psi}^\vee} : A^* \simeq A^*$ defined by $x \mapsto x - \dot{x}(\dot{\psi}^\vee)\psi$ (this negates ψ and is the identity on the linear hyperplane of points $x \in A^*$ such that $\dot{x} \in V^*$ vanishes on the nonzero $\dot{\psi}^\vee \in V$, a hyperplane containing the constant functions; in particular $-\Psi = \Psi$),
(AR4) $\langle \dot{\eta}, \dot{\psi}^\vee \rangle \in \mathbf{Z}$ for all $\psi, \eta \in \Psi$,
(AR5) for all $a \in \Phi$, the subset $\{\psi \in \Psi \mid \dot{\psi} = a\} \subset A^*$ has at least two members and no accumulation point (i.e., the parallel hyperplanes H_ψ for $\psi \in \Psi$ with $\dot{\psi} = a$ don't get arbitrarily close).

Since $r_{\psi,\dot{\psi}^\vee} : A^* \simeq A^*$ is the identity on the line $\mathbf{R}$ of constant functions, it induces a reflection on $A^*/\mathbf{R} = V^*$. Under the preceding axioms, (V^*, Φ) is a root system for which the associated reflection $r_{\dot{\psi}} \in \mathrm{GL}(V^*)$ is induced by $r_{\psi,\dot{\psi}^\vee}$, so (V^*, Φ) is a root system with corresponding coroots $\dot{\psi}^\vee \in V$. In particular, the vector $\dot{\psi}^\vee \in V$ is uniquely determined due to the theory of usual root systems.

Remark 15.10 If we strengthen axiom (AR2) to require Φ be non-empty and span V^* then (AR1) can be dropped. Indeed, there are nonzero constants in $\mathrm{span}(\Psi)$ by the first part of (AR5) since $\Phi \neq \emptyset$, and $\mathrm{span}(\Psi)$ maps onto V^* by (AR2), so (AR1) follows.

Since $\nabla\Psi \subset V^*$ is a root system and $\nabla : A^* \to V^*$ is linear, the theory of root systems ensures that $\Psi \cap \mathbf{R}\psi \subset \{\pm\psi, \pm 2\psi, \pm(1/2)\psi\}$ for $\psi \in \Psi$.

Definition 15.11 An affine root system (A, Ψ) is *reduced* if $\Psi \cap \mathbf{R}\psi = \{\pm\psi\}$ for each $\psi \in \Psi$; otherwise it is called *non-reduced*.

The *affine Weyl group* $W(\Psi) \subset \mathrm{GL}(A^*)$ is the subgroup generated by the reflections $r_{\psi,\dot{\psi}^\vee}$ for $\psi \in \Psi$. By design, everything in the affine Weyl group induces the identity on the line $\mathbf{R}$ of constant functions and thereby induces a linear automorphism on $A^*/\mathbf{R} \overset{\nabla}{\simeq} V^*$. The resulting map $W(\Psi) \to \mathrm{GL}(V^*)$ has image $W(\Phi)$ and kernel consisting of automorphisms of A^* having the form $\ell \mapsto \ell + \dot{\ell}(v)$ for $v \in \mathrm{Hom}(V^*, \mathbf{R}) = V$.

Here is a useful way to interpret $W(\Psi)$ via affine automorphisms of A (rather than linear automorphisms of A^*). For each $\psi \in \Psi$, the affine transformation $r_\psi : A \simeq A$ defined by $x \mapsto x - \psi(x)\dot{\psi}^\vee$ is an *affine reflection*: it is the identity on the affine hyperplane H_ψ but not the identity on A (since $\dot{\psi}^\vee \neq 0$) and has order 2 (since $\dot{\psi}(\dot{\psi}^\vee) = 2$). It is a pleasant exercise to check that the affine reflection r_ψ on A induces on A^* exactly $r_{\psi,\dot{\psi}^\vee}$, so the injective (!) homomorphism of groups $\mathrm{Aff}(A) \to \mathrm{GL}(A^*)$ defined by $w \mapsto (w^{-1})^*$ carries $\langle r_\psi \rangle_{\psi \in \Psi} \subset \mathrm{Aff}(A)$ isomorphically onto $W(\Psi)$. Under this interpretation, the quotient map $W(\Psi) \twoheadrightarrow W(\Phi) \subset \mathrm{GL}(V^*)$ whose kernel we identified above with a subgroup of V is identified with the quotient map

$$\nabla : W(\Psi) \twoheadrightarrow W(\Phi^\vee) \subset \mathrm{GL}(V)$$

via differentiation, so its kernel consists of *translations* on A from V (this is the same embedding of $\ker(W(\Psi) \twoheadrightarrow W(\Phi))$ into V).

Remark 15.12 If $\Phi := \nabla\Psi$ is *irreducible* then the first part of (AR5) (i.e., each affine hyperplane H_ψ admits another H_η parallel to it) is a consequence of the rest of the axioms for an affine root system, as we now explain. Since Φ is irreducible, the $W(\Phi)$-action is transitive on the set of roots in Φ with a common length. By surjectivity of $\nabla : W(\Psi) \to W(\Phi)$, it follows that for all roots $a \in \Phi$ with some common length, (AR5) holds for all of them if it holds for one of them.

The set Ψ of affine roots must be infinite. Indeed, otherwise (A^*, Ψ) would be a usual root system with $W(\Psi) \subset \mathrm{GL}(A^*)$ as its Weyl group, but then by the theory of root systems we'd have $(A^*)^{W(\Psi)} = 0$, contradicting that the line $\mathbf{R} \subset A^*$ of constant functions on A has trivial $W(\Psi)$-action. The surjection $\nabla : \Psi \twoheadrightarrow \Phi$ must then have some infinite fiber, so there is a pair of affine roots differing by a nonzero constant. The corresponding root hyperplanes in A are parallel and distinct, so the first part of (AR5) holds for *some* $a \in \Phi$ and hence for all roots in the Weyl-orbit of that a. In particular if Φ is reduced and simply laced then we are done.

Now assume the irreducible Φ is not simply laced, so we can pick non-orthogonal linearly independent roots $a, b \in \Phi$ with distinct root lengths such that the first part of (AR5) holds for all roots in the Weyl-orbit of a. We'll show it holds for b too, hence for all roots in Φ if Φ is reduced (as then there are only two root lengths). Pick

$\psi \in \Psi$ with derivative a and a *nonzero* constant k such that $\psi + k \in \Psi$. For $x \in A$ we have $r_\psi(x) = x - \psi(x)\dot{\psi}^\vee = x - \psi(x)a^\vee$, so

$$(r_{\psi+k} \circ r_\psi)(x) = r_\psi(x) - (\psi + k)(r_\psi(x))a^\vee = x - \psi(x)a^\vee - (-\psi(x) + k)a^\vee$$

(the final equality uses that $\psi \circ r_\psi = -\psi$). Thus, $(r_{\psi+k} \circ r_\psi)(x) = x - ka^\vee$ for all $x \in A$.

Let $\eta \in \Psi$ be an affine root with derivative b, so $\eta' := \eta \circ (r_{\psi+k} \circ r_\psi)$ is an affine root with derivatve $r_a^2(b) = b$ too. The constant difference $\ell = \eta' - \eta$ is the value of η' on H_η. For $x_0 \in H_\eta$ this value is $\eta(x_0 - ka^\vee) = \eta(x_0) - k\langle\dot{\eta}, a^\vee\rangle = -k\langle b, a^\vee\rangle$, which is *nonzero*. Thus, the first part of (AR5) holds for b.

We are done if Φ is reduced, so now suppose Φ is non-reduced and hence is BC_n for some n. In particular $\Phi = \Phi^{\mathrm{nm}} \cup \Phi^{\mathrm{nd}}$ is a union of two root systems that are irreducible and reduced: the set Φ^{nm} of non-multipliable roots (this is type C_n), and the set Φ^{nd} of non-divisible roots (this is type B_n). If a is both non-divisible and non-multipliable (i.e., intermediate root length) then we can pick b above to be either divisible (longest length) or multipliable (shortest root length) as we please, so again we are done. If a is either divisible or multipliable then we can pick b to be of intermediate length to see the first part of (AR5) holds for the Weyl-orbit of b. Then we can run the preceding argument with the non-divisible non-multipliable b in the role of a to conclude.

Remark 15.13 Apart from [KP, Sect. 1.3], it seems all other literature on affine root systems (including [Mac, Sect. 2] and Wikipedia) omits the first part of (AR5). This is irrelevant when $\Phi := \nabla\Psi$ is irreducible, as we saw in Remark 15.12, so for the purpose of classifying possibilities with irreducible Φ it doesn't matter. It is an error in general because the first part of (AR5) can fail when the rest of the axioms hold due to the fact (which we'll discuss in Sect. 16) that the formation of the affine-dual A^* fails to commute with direct sums in A. We now show that the existence of "irreducible decomposition" as stated in [Mac, Proposition 3.3] without the first part of (AR5) is false (its existence proof preceding the statement overlooks that an "irreducible factor" S_i there can fail to contain a nonzero constant). The definition of "affine root system" in [Mac, Sect. 2] should be changed to require the full strength of (AR5).

Let (A, Ψ) be an affine root system (in Sect. 17 we'll make many of these) with A an affine space for V, and let (V'^*, Φ') be a non-trivial usual root system. Consider $A \times V'$ as an affine space for $V \oplus V'$. The affine dual $(A \times V')^*$ is naturally identified with $A^* \oplus V'^*$, where A^* is an affine dual and V'^* is a *linear* dual (there is only one line of constant functions!). The disjoint union $\Psi \coprod \Phi' \subset A^*$, together with the associated coroots in $(V^* \oplus V'^*)^* = V \oplus V'$ and resulting reflections then satisfies the axioms for an affine root system *except* the first part of (AR5).

The preceding construction satisfies the definition of "affine root system" in [Mac, Sect. 2] but contradicts the statement of irreducible decomposition in [Mac, Proposition 3.3]. This is a possibility we will rule out from arising when we build "irreducible components" of general affine root systems and verify that such "factors" are indeed affine root systems (rather than some such "factors" being usual root systems; i.e., finite). The end of the proof of Lemma 16.3 takes up this matter.

16 Decomposition for Affine Root Systems

Let's start by tying up loose ends on affine spaces. Let A be an affine space for a finite-dimensional vector space V over a field L. The L-vector space $A^* = \{A \xrightarrow{\text{affine}} L\}$ of affine L-valued functions on A has dimension $1 + \dim V = 1 + \dim A$ due to the canonical line of constant L-valued functions on A, and we have a short exact sequence of L-vector spaces

$$0 \to L \to A^* \overset{\nabla}{\to} V^* \to 0$$

where $\nabla : \psi \mapsto \dot{\psi}$.

Here are some important aspects of affine spaces, all of which can be checked as exercises:

(1) For any $\psi \in A^*$ that is non-constant (equivalently, $\dot{\psi} \in V^*$ is nonzero), $H_{\psi+s} \cap H_\psi = \emptyset$ for $s \in L^\times$ ("parallel"), where H_ψ is the affine hyperplane in A of points x satisfying $\psi(x) = 0$.

(2) For $\psi, \eta \in A^* - L$, the equality of hyperplanes $H_\psi = H_\eta$ is equivalent to $\eta = s\psi$ for some $s \in L^\times$. Also, $\nabla(s\psi) = s\nabla\psi$, so $s\psi$ is characterized by the pair of properties:

   (a) $H_{s\psi} = H_\psi$;
   (b) $\nabla(s\psi) = s\nabla\psi$ (which determines $s\psi$ up to *adding* an element of L, and (a) removes that ambiguity) thanks to (1).

(3) The linear dual of the affine dual $(A^*)^*$ is *not* A'^* for any naturally associated affine space A' for $V^{**} = V$. Thus to define a notion of dual affine root system $\Psi^\vee$ for an affine root system (A, Ψ) we will have to make an auxiliary choice (of an inner product on V) and wind up defining $\Psi^\vee \subset A^*$ as a subset (this is discussed with proper definitions in Sect. 18).

(4) Consider an affine space (V, A) and a linear decomposition $V = V_1 \oplus V_2$ (e.g., for an affine root system (A, Ψ) we decompose the root system $(V^*, \nabla\Psi = \Phi)$ into irreducible constituents, and then decompose V accordingly). Then $A_1 := A/V_2$ equipped with the natural action of $V/V_2 = V_1$ and $A_2 := A/V_1$ equipped with the natural action of $V/V_1 = V_2$ are affine spaces and we have an evident isomorphism

$$\begin{array}{ccc} A & \xrightarrow{\sim} & A_1 \times A_2 \\ \circlearrowleft & & \circlearrowleft \\ V & \xrightarrow{\sim} & V_1 \oplus V_2. \end{array}$$

The same holds for more general linear decompositions $V = \bigoplus V_i$ yielding a canonically associated decomposition $A = \prod A_i$ as affine spaces.

To define the notion of *direct sum* for affine root systems, we need to be attentive to how the formation of the "affine dual" A^* (a vector space of dimension $1 + \dim A$) interacts with direct products in A. The canonical line of constant functions requires care. Using composition with the projections $A = A_1 \times A_2 \to A_j$ over $V \twoheadrightarrow V_j$ for $j = 1, 2$ we get linear maps $A_j^* \hookrightarrow A^*$ and then a short exact sequence

$$\begin{array}{ccccccccc} 0 & \longrightarrow & L & \longrightarrow & A_1^* \oplus A_2^* & \stackrel{q}{\twoheadrightarrow} & A^* & \longrightarrow & 0 \\ & & \| & & \uparrow & & \uparrow & & \\ & & \{(c, -c) \in L \times L : c \in L\} & & L \oplus L & \stackrel{+}{\longrightarrow} & L & & \end{array}$$

In particular, $A_1^* \cap A_2^*$ inside A^* is the line L of constant functions, so $\Psi_1 \cap \Psi_2 = \emptyset$ inside A^* since each $\psi \in \Psi_j$ is non-constant. The surjectivity of q ensures the spanning axiom for $(A_1 \times A_2, \Psi_1 \coprod \Psi_2)$, and the other axioms are readily verified (e.g., if $\psi_1 \in \Psi_1$ then its hyperplane of zeros in $A_1 \times A_2$ is $H_{\psi_1} \times A_2$), so such a pair is indeed an affine root system; we call it the *direct sum* and denote it $\Psi_1 \oplus \Psi_2$.

For a direct sum of n affine root systems (as we'll need for decomposing a general affine root systems into irreducible components), we likewise have a natural surjection

$$\bigoplus_{j=1}^{n} A_j^* \twoheadrightarrow \left(\prod_{j=1}^{n} A_j\right)^*$$

whose kernel is the hyperplane $\{\sum x_j = 0\} \subset L^n$.

Now we let $L = \mathbf{R}$.

Definition 16.1 An affine root system (A, Ψ) is *reducible* if $\Psi = \Psi_1 \coprod \Psi_2$ with $\Psi_j \neq \emptyset$ such that for any $\psi \in \Psi_1$ and $\eta \in \Psi_2$ we have the "orthogonality" properties $\langle \dot{\psi}, \dot{\eta}^\vee \rangle = 0$ and $\langle \dot{\eta}, \dot{\psi}^\vee \rangle = 0$ (evaluation pairings between V^* and $V^{**} = V$). Equivalently, for all $\psi \in \Psi_1$ the reflection $r_{\psi, \dot{\psi}^\vee} : A \simeq A$ is the identity on Ψ_2 and likewise for all $\eta \in \Psi_2$ the reflection $r_{\eta, \dot{\eta}^\vee}$ is the identity on Ψ_1.

We say (A, Ψ) is *irreducible* if it is not reducible.

Remark 16.2 Whereas for root systems there is the "trivial" (or "zero") object $(0, \emptyset)$, there is no such thing for affine root systems since the affine dual A^* always contains the line of constant functions. More specifically, a one-point affine space has affine dual $\mathbf{R}$, and that is not spanned by $\emptyset$.

For the notion of reducibility of affine root systems, we now verify that the orthogonal decomposition $\Psi = \Psi_1 \coprod \Psi_2$ arises from a product decomposition $A = A_1 \times A_2$ (with regard to a direct sum decomposition $V = V_1 \oplus V_2$) such that the resulting subspaces $A_1^*, A_2^* \subset A^*$ satisfy $\Psi_j \subset A_j^*$ inside A^* with each pair (A_j, Ψ_j) an affine root system.

This requires more work than for usual root systems. To see the issue, note that with usual root systems we can decompose the ambient vector space into a direct sum by forming the subspaces *spanned* by the subsets in the partition of the root system. There is no concept of "span" for affine spaces! Fortunately we can build the desired A_j's as follows.

Look at the root system (V^*, Φ) for

$$\Phi := \nabla\Psi = \Phi_1 \cup \Phi_2 \subset V^* - \{0\}$$

where $\Phi_j := \nabla(\Psi_j)$. The two sets Φ_1 and Φ_2 of nonzero points in V^* are mutually orthogonal to each other (in the sense of the associated reflections) and hence are *disjoint* sets, so (V, Φ) is *reducible* in the sense of root systems. By the theory of root systems, the spans V_j^* of the Φ_j's yield $V^* = V_1^* \oplus V_2^*$, making $(V^*, \Phi) = (V_1^*, \Phi_1) \times (V_2^*, \Phi_2)$. This gives that $V = V_1 \oplus V_2$, so we get a corresponding decomposition of affine spaces $A = A_1 \times A_2$ and $\Psi_j \subset A_j^* \subset A^*$ because Ψ_j is the preimage of $\Phi_j \subset \Phi$ under $\nabla : \Psi \twoheadrightarrow \Phi$.

Lemma 16.3 *Each (A_j, Ψ_j) is an affine root system, so $\Psi = \Psi_1 \oplus \Psi_2$ via the identification $A = A_1 \times A_2$.*

Proof Axiom (AR2) on $\nabla(\Psi_j)$ spanning V_j^* holds since this set of derivatives is Φ_j. For axiom (AR3), if $\psi \in \Psi_1$ then the reflection $r_{\psi,\dot{\psi}^\vee}$ acting on A respects $A_1 \times A_2$ via a reflection on $A_1 = A/V_2$ and the identity on $A_2 = A/V_1$ since (i) $\dot{\psi}^\vee \in V_1$ and (ii) the functional $\dot{\psi} \in V^*$ vanishes on V_2. The same holds with the roles of A_1 and A_2 swapped, so (AR3) holds. The integrality axiom (AR4) clearly holds. Axiom (AR5) concerning parallel hyperplanes also holds since for each $\psi \in \Psi_j$ its hyperplane of zeros in $A = A_1 \times A_2$ is the direct product of A_{3-j} and its hyperplane of zeros in A_j.

The spanning axiom (AR1) holds for (A_j, Ψ_j) thanks to (AR5) being expanded beyond what is in the literature, as explained in Remark 15.13. □

The same reasoning yields the *canonical* decomposition of affine root systems as a direct sum of irreducible ones, as stated in the next result. This decomposition is unique in the same strong sense as for usual root systems, as a consequence of the uniqueness for usual root systems.

Proposition 16.4 *Let (A, Ψ) be an affine root system, where A is an affine space for V, and let (V_j^*, Φ_j) be the irreducible components of $\Phi = \nabla(\Psi)$.*

Letting $\prod A_j$ be the corresponding decomposition of A arising from the decomposition $\prod V_j$ of V, and $\Psi_j \subset \Psi$ be the preimage of Φ_j under $\nabla : \Psi \twoheadrightarrow \Phi$, each pair (A_j, Ψ_j) is an irreducible affine root system and $\oplus\Psi_j = \Psi$ (this identification is called the **irreducible decomposition** *of (A, Ψ)).*

The notion of *isomorphism* among affine root systems has a subtlety, as we'll soon see. To start off, it is natural to make the following definition (implicit at the end of Proposition 16.4).

Definition 16.5 A *strong isomorphism* $(A, \Psi) \simeq (A', \Psi')$ is an isomorphism of affine spaces $f : A \simeq A'$ such that the backwards linear map $A^* \xleftarrow{\sim} A'^* : f^*$ (this is $\ell' \mapsto \ell' \circ f$) carries Ψ' onto Ψ.

There is no need to demand compatibility with the reflections in the axioms for affine root systems in this definition since the uniqueness of such reflections makes this compatibility automatic.

The notion of strong isomorphism turns out to be too restrictive, ultimately because the effect of f^* between the canonical lines $\mathbf{R} \subset A^*$ and $\mathbf{R} \subset A'^*$ of constant functions is the identity map and hence f^* *cannot* be scaling by any $s \in \mathbf{R}^\times - \{1\}$.

To see why the property $f^*|_{\mathbf{R}} = \mathrm{id}_{\mathbf{R}}$ makes the notion of "strong isomorphism" too restrictive, consider the pair $(A, s\Psi)$ for $s \in \mathbf{R}^\times - \{\pm 1\}$. For an irreducible (A, Ψ), we claim the affine root system $(A, s\Psi)$ (with the same reflections on A and the same collection of affine hyperplanes in A as zero loci of the affine roots that are obtained from (A, Ψ)) satisfies

$$(A, s\Psi) \not\simeq (A, \Psi) \tag{16.1}$$

(i.e., not strongly isomorphic) when $s \neq \pm 1$; we avoid $s = -1$ since $-\Psi = \Psi$ because $r_{\psi,\psi^\vee}(\psi) = \psi - 2\psi = -\psi$. (A broader notion of isomorphism to bypass this issue will be defined in Sect. 18.) Observe that in contrast, for root systems (V, Φ) we have $(V, s\Phi) \simeq (V, \Phi)$ using s-multiplication on V; this has no analogue for affine root systems since s-multiplication makes no sense on affine spaces!

To verify (16.1), it seems necessary to invoke deeper structural features of irreducible affine root systems. More specifically, using the notation of "basis" to be discussed in Sect. 18, in the irreducible case there is a specific $\mathbf{Z}^+$-linear combination of the affine roots in a choice of "basis" $\Delta \subset \Psi$ that is a nonzero *constant* function independent of the choice of Δ (see [KP, Proposition 1.3.47]); this constant is denoted $c(\Psi)$. By [KP, Proposition 1.3.54(2)], the condition $f^*|_{\mathbf{R}} = \mathrm{id}_{\mathbf{R}}$ for strong isomorphisms f forces $c(s\Psi) = c(\Psi)$ if $(A, s\Psi)$ is strongly isomorphic to (A, Ψ). But it can be directly computed that $c(s\Psi) = |s|c(\Psi)$ for all $s \in \mathbf{R}^\times$ [KP, Ex. 1.3.55], so we obtain $|s| = 1$ yet we assumed $s \neq \pm 1$.

In Sect. 17 we'll build *many* reduced affine root systems: given a non-trivial root system (V, Φ) (possibly non-reduced!) we'll build a *reduced* affine root system (A, Ψ_Φ) satisfying $\nabla(\Psi_\Phi) = \Phi$. In Sect. 18 we'll see (for an appropriate notion of dual affine root system) that $(\Psi_{\Phi^\vee})^\vee$ has derivative root system $\Phi^{\vee\vee} \simeq \Phi$. For the right broader notion of isomorphism of affine root systems, we'll have $(\Psi_{\Phi^\vee})^\vee \not\simeq \Psi_\Phi$ when Φ is reduced but not simply laced (i.e., Φ of type B_n $(n \geq 2)$, C_n $(n \geq 2)$, G_2, or F_4). In particular, the classification of irreducible reduced affine root systems (using a broader notion of isomorphism) contains multiple members with the same irreducible reduced derivative root system.

17 Construction of Affine Root Systems

In Sect. 16 we defined the notion of *reducibility* for an affine root system (A, Ψ), and directly from that definition we see Ψ is reducible precisely when its derivative root system $\Phi := \nabla\Psi$ is reducible. The same also holds for irreducibility, since the trivial root system $(0, \emptyset)$ never arises as the derivative of an affine root system.

The irreducible root systems Φ are classified by connected Dynkin diagrams in the reduced case, and by their dimension (for type BC_n) in the non-reduced case. By Proposition 16.4, to find the possibilities for Ψ it is natural to fix an irreducible Φ and seek to classify all Ψ with $\nabla\Psi \simeq \Phi$.

In this section we'll construct a *reduced* affine root system Ψ_Φ with any desired (possibly non-reduced!) non-trivial root system Φ as its derivative root system. This is inspired by [Bour, Chap. 6, Sect. 2.1]. Let (V^*, Φ) be a root system that is not $(0, \emptyset)$ (it is possibly non-reduced). Let $A = V$, define the subset $\Psi \subseteq A^* = V^* \oplus \mathbf{R}$ as

$$\Psi = \{a + n : a \in \Phi, n \in I_a\}, \quad \text{where } I_a = \begin{cases} \mathbf{Z} & a \notin 2\Phi, \\ 2\mathbf{Z}+1 & a \in 2\Phi, \end{cases}$$

and define the associated coroot $(\nabla(a+n))^\vee := a^\vee \in \Phi^\vee \subset V = (V^*)^*$ for all $n \in I_a$. (Note that Ψ is non-empty since Φ is non-empty, and its elements are non-constant since $V^* \neq 0$.) The corresponding affine reflection $r_{a+n}\colon A \to A$ is $x \mapsto (x - a(x)a^\vee) - na^\vee$, where the affine reflection $r_a : x \mapsto x - a(x)a^\vee$ (for $n = 0$) is just the linear reflection $r_{a^\vee}$ on V attached to the dual root system $(V, \Phi^\vee)$. The reader can check that (A, Ψ) is an affine root system and $\nabla\Psi = \Phi \subset V^*$.

Remark 17.1 The choice $I_a = 2\mathbf{Z}+1 = \mathbf{Z} - 2\mathbf{Z}$ for $a \in 2\Phi$ makes the affine root system Ψ be reduced. For $a = 2b \in \Phi$ with $b \in \Phi$ (so $b \in \Phi^{\rm nd}$), we have $a + 2k = 2(b+k)$ for any $k \in \mathbf{Z} = I_b$. If we had used $I_a = \mathbf{Z}$ for all $a \in \Phi$ then we would still get an affine root system but it would be non-reduced when Φ is non-reduced.

The construction using $I_a = \mathbf{Z}$ for all $a \in \Phi$ is given in [Bour, Chap. 6, Sect. 2.1] where Φ is assumed to be reduced. In the preceding construction Ψ is always reduced (check!) and its derivative root system is Φ (which we allow to be non-reduced).

Remark 17.2 Let's discuss why replacing the role of $\mathbf{Z}$ (and $\mathbf{Z} - 2\mathbf{Z} = 2\mathbf{Z}+1$ for non-reduced Φ) with a general infinite cyclic discrete subgroup of $\mathbf{R}$ leaves the isomorphism class of Ψ_Φ unchanged upon broadening the notion of "isomorphism" beyond strong isomorphisms.

Replace I_a with sI_a for a choice of $s \in \mathbf{R}^\times$. The construction then yields $s\Psi_{(1/s)\Phi}$ since $a + ns = s((1/s)a + n)$. Here, for $s \in \mathbf{R}^\times$ and an affine root system Ψ, we define the *rescaling* $s\Psi$ to consist of the affine roots $s\psi$ for $\psi \in \Psi$ with associated coroots $\nabla(s\psi)^\vee = (1/s)\dot{\psi}^\vee$ (to ensure the pairing $\langle\nabla(s\psi), \nabla(s\psi)^\vee\rangle$ remains equal to 2).

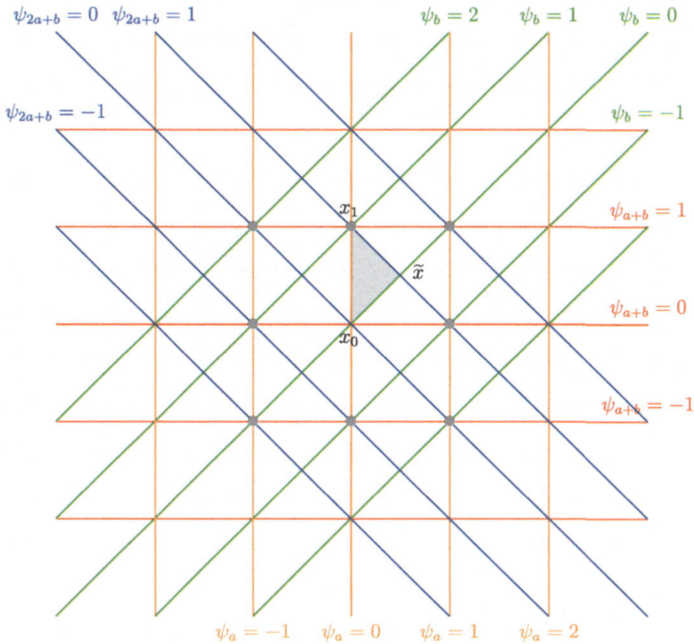

Fig. 2 The affine hyperplanes (lines) for Ψ_{B_2}, with roots darkened

When $s \neq \pm 1$ and Φ is irreducible (so Ψ_Φ is also irreducible), $s\Psi_{(1/s)\Phi}$ is not strongly isomorphic to $\Psi_{(1/s)\Phi}$. In Sect. 18 we'll introduce a broader notion of *isomorphism* for affine root systems so that rescaling preserves the isomorphism class and hence $s\Psi_{(1/s)\Phi} \simeq \Psi_{(1/s)\Phi} \simeq \Psi_\Phi$.

Example 17.3 Let's draw the collection of affine hyperplanes H_ψ where the affine roots ψ each vanish for the case $\Psi = \Psi_\Phi$ with $\Phi = \mathrm{B}_2 = \mathrm{C}_2$. Pick a basis $\Delta = \{a, b\}$ where a is shorter (equivalently, we pick a positive system of roots Φ^+). We use the standard dot product isomorphism $V = \mathbf{R}^2 \simeq (\mathbf{R}^2)^* = V^*$ and rescale the roots so $\|a\| = 1$, and hence $\|b\| = \sqrt{2}$. The affine roots are of the form $\psi_{c+n} = \langle c, \cdot \rangle + n$ for $n \in \mathbf{Z}$ (since Φ is reduced).

The affine hyperplanes are given by the equations $\psi_c = -n$ for $c \in \Phi$ and $n \in \mathbf{Z}$. For example, $\psi_c = 0$ is the line orthogonal to c through the origin (and $\psi_c = -n$ is a parallel line).

The connected components of the open subset $A - \bigcup_{\psi \in \Psi} H_\psi \subset A$ are called *chambers* (much like for usual root systems). This is where it is crucial to work over $\mathbf{R}$ (rather than $\mathbf{Q}$, for instance), so we can make definitions and arguments via connectedness considerations (rather than having to do purely algebraic work with collections of inequalities on affine roots).

The shaded chamber C as shown in Fig. 2 is defined by the three simultaneous inequalities

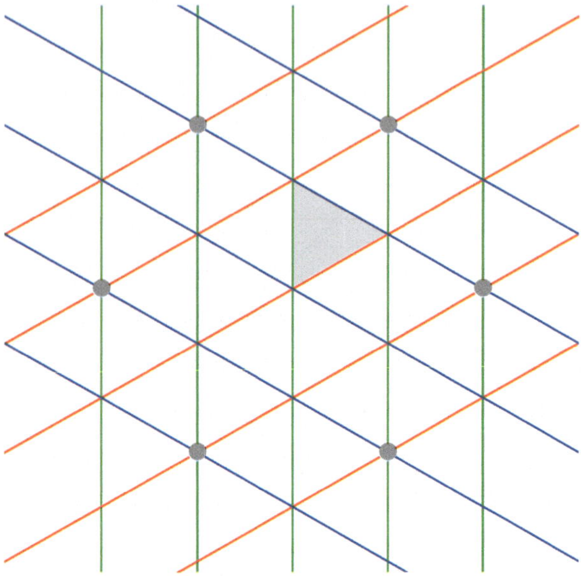

Fig. 3 The affine hyperplanes (lines) for Ψ_{A_2}, with roots darkened

$$\psi_a > 0, \quad \psi_b > 0, \quad 1 - \psi_{2a+b} > 0,$$

where $2a + b$ is the highest root. This is the general picture: a chamber is cut out by the usual positive strict inequalities for a Weyl chamber along with one further strict inequality involving the highest positive root (this is formulated in general in Sect. 18). We will have a notion of *basis* for affine root systems, and $\{\psi_a, \psi_b, 1 - \psi_{2a+b}\}$ is a basis of Ψ in that sense.

Remark 17.4 In Fig. 2 the closure triangle $\overline{C}$ is not equilateral, unlike for Ψ_{A_2} as in Fig. 3.

For a connected reductive group G over a Henselian discretely-valued field k (with perfect residue field), when $\mathscr{D}(G)$ is k-isotropic the apartment $\mathcal{A}(S) \subset \mathcal{B}(G)$ as an affine space for the nonzero $X_*(S')_{\mathbf{R}}$ will support an affine root system $\Psi(G, S)$ whose derivative root system is the non-empty $\Phi = \Phi(G, S)$. For split G that is semisimple and simply connected, this will turn out to be exactly Ψ_Φ, so in the case $G = \mathrm{Sp}_4$ it is $\Psi_{C_2} = \Psi_{B_2}$ for which an apartment is shown in Fig. 2.

The two vertices x_0, x_1 for the shaded chamber in Fig. 2 will turn out to be "hyperspecial", due to which each corresponding bounded open stabilizer $\mathcal{K}_j := G(k)^1_{x_j}$ (for $G = \mathrm{Sp}_4$) is the group of $\mathfrak{o}$-points of a reductive $\mathfrak{o}$-model $\mathscr{G}_{x_j}$, with $\mathcal{K}_1, \mathcal{K}_2$ not $G(k)$-conjugate. On the other hand, $\widetilde{x}$ will turn out to not be hyperspecial, and $\widetilde{\mathcal{K}} := G(k)^1_{\widetilde{x}}$ is the group of $\mathfrak{o}$-points of an $\mathfrak{o}$-model $\mathscr{G}_{\widetilde{x}}$ of G with connected but *non-reductive* special fiber satisfying $(\mathscr{G}_{\widetilde{x}})_{\mathfrak{f}}/\mathscr{R}_u((\mathscr{G}_{\widetilde{x}})_{\mathfrak{f}}) \simeq \mathrm{SL}_2 \times \mathrm{SL}_2$.

18 Duality, Isomorphisms, and Classification

Let (A, Ψ) be an affine root system, where A is an affine space for a vector space V. To define the dual $\Psi^\vee$, we will find it inside A^* because the linear dual $(A^*)^*$ of the affine dual A^* does not seem to be the affine dual $(A')^*$ for a "natural" affine space A' for V^*. The derivative root system $\Phi = \nabla(\Psi)$ lies in V^*.

An inner product on V (equivalently, a positive-definite quadratic form) yields a translation-invariant metric on V and hence a well-defined metric on A (namely $d(a, a') = \|a - a'\|$, where $a - a' \in V$ is the unique $v \in V$ satisfying $v + a' = a$). Pick an inner product $\langle\cdot,\cdot\rangle$ on V that is *invariant under* $W(\Phi^\vee) = W(\Phi)^*$ (linear duals of operators from $W(\Phi)$), and use it to identify V^* with V. This will be used to define $\Psi^\vee$, so let's first look into how the choice of inner product may vary.

If Φ is irreducible then $W(\Phi)$ acting on V^* is *absolutely irreducible*, so by Schur's Lemma the positive-definite $\langle\cdot,\cdot\rangle$ is unique up to $\mathbf{R}_{>0}$-scaling. In general, if $\Phi = \coprod \Phi_i (= \oplus\Phi_i)$ is the irreducible decomposition then correspondingly $V^* = \oplus V_i^*$ and so we get a decomposition $V = \oplus V_i$. Then $W(\Phi) = \prod W(\Phi_i)$ acts on $\prod V_i^*$ absolutely irreducibly on each factor, so $\langle\cdot,\cdot\rangle$ *must* be given by $\sum\langle\cdot,\cdot\rangle_i$, unique up to separate $\mathbf{R}_{>0}$-scalings on each $\langle\cdot,\cdot\rangle_i$.

By Proposition 16.4 we have naturally $(A, \Psi) = \prod(A_i, \Psi_i)$ where each factor is an irreducible affine root system having derivative (V_i^*, Φ_i). Our definition below of the dual $\Psi^\vee$ in general will (from the construction) be the direct sum of the duals $\Psi_i^\vee$, and the latter will depend on the choice of $\langle\cdot,\cdot\rangle_i$ up to an overall scaling factor. In other words, the dependence of $\Psi^\vee$ on our choice of $\langle\cdot,\cdot\rangle$ will amount to $\mathbf{R}_{>0}$-scaling applied to each $\Psi_i^\vee$.

Let's finally define $\Psi^\vee$ using our choice of $\langle\cdot,\cdot\rangle$. As motivation, recall that for a usual root system (V', Φ'), upon choosing a Weyl-invariant $\langle\cdot,\cdot\rangle$ to define an isomorphism $V'^* \simeq V'$ we get

$$a'^\vee = \frac{2a'}{\langle a', a'\rangle}$$

for all $a' \in \Phi'$. We now adapt that recipe to define an *affine* coroot in A^* attached to each $\psi \in \Psi$.

Define

$$\psi^\vee = \frac{2\psi}{\langle\dot{\psi}, \dot{\psi}\rangle} \in A^*$$

for each $\psi \in \Psi$; this makes sense since $\dot{\psi} \in V^* \simeq V$ is nonzero. Note that this has derivative

$$\nabla(\psi^\vee) = \frac{2}{\langle\dot{\psi}, \dot{\psi}\rangle}\dot{\psi} = (\dot{\psi})^\vee;$$

that is, this recovers the usual coroot for Φ (from the axioms for Ψ!). Clearly $\langle\dot{\psi}, \nabla(\psi^\vee)\rangle = 2$, so we can use $\dot{\psi} \in V^*$ as the "coroot" attached to $\psi^\vee$ in axioms for a dual affine root system: we have the linear reflection

$$r_{\psi^\vee,\dot{\psi}} : A^* \to A^*, \qquad y \mapsto y - \dot{y}(\dot{\psi})\psi^\vee$$

where $\dot{y}, \dot{\psi} \in V^*$ are paired using the inner product (or identifying V^* with V to view $\dot{\psi}$ in V). Hence, plugging in the definition of $\psi^\vee$ yields

$$r_{\psi^\vee,\dot{\psi}}(y) = y - \langle \dot{y}, \dot{\psi}\rangle\psi^\vee = y - \langle \dot{y}, \frac{2\dot{\psi}}{\langle\dot{\psi}, \dot{\psi}\rangle}\rangle\psi = r_{\psi,\dot{\psi}^\vee}(y).$$

In other words, the reflection $r_{\psi^\vee} := r_{\psi^\vee,\dot{\psi}}$ coincides with the reflection $r_\psi := r_{\psi,\dot{\psi}^\vee}$ of A^* attached to $\psi \in \Psi$. Define

$$\Psi^\vee = \{\psi^\vee : \psi \in \Psi\}.$$

To verify this is an affine root system using $\dot{\psi}$ as the "coroot" attached to $\psi^\vee$, we check:

Lemma 18.1 $r_{\psi^\vee,\dot{\psi}} : A^* \simeq A^*$ *carries* $\Psi^\vee$ *onto itself.*

Proof For $\psi, \eta \in \Psi$ we have

$$r_{\eta^\vee}(\psi^\vee) = \frac{2}{\langle\dot{\psi}, \dot{\psi}\rangle} r_{\eta^\vee}(\psi) = \frac{2}{\langle\dot{\psi}, \dot{\psi}\rangle} r_\eta(\psi)$$

so it suffices to show

$$\frac{2}{\langle\dot{\psi}, \dot{\psi}\rangle} r_\eta(\psi) = r_\eta(\psi)^\vee.$$

This is the same as verifying

$$\langle\dot{\psi}, \dot{\psi}\rangle = \langle\nabla(r_\eta(\psi)), \nabla(r_\eta(\psi))\rangle.$$

But $\nabla(r_\eta(\psi)) = r_{\dot{\eta}}(\dot{\psi})$ with $r_{\dot{\eta}} \in W(\Phi)$, and by design $W(\Phi)$ leaves $\langle\cdot, \cdot\rangle$ unchanged. □

Since $H_{\psi^\vee} = H_\psi$ for all ψ, it is now clear that $(A, \Psi^\vee)$ satisfies the axioms for an affine root system.

Definition 18.2 The pair $(A, \Psi^\vee)$ is called the *dual* affine root system.

By construction, $\nabla(\Psi^\vee) = \Phi^\vee \subset V^* \simeq V$. We have noted before that if we change $\langle\cdot, \cdot\rangle$ then the effect is to rescale each $\Psi_i^\vee$ for each irreducible component Ψ_i of Ψ.

Proposition 18.3 *Using the same* $\langle\cdot, \cdot\rangle$, *we have* $(\Psi^\vee)^\vee = \Psi$ *inside* A^*.

Proof We have

$$\nabla(\psi^\vee) = \frac{2\dot{\psi}}{\langle\dot{\psi}, \dot{\psi}\rangle}$$

so

$$(\psi^\vee)^\vee = \frac{2\psi^\vee}{(4\langle\dot{\psi},\dot{\psi}\rangle)/\langle\dot{\psi},\dot{\psi}\rangle^2} = \frac{\langle\dot{\psi},\dot{\psi}\rangle}{2}\psi^\vee = \psi.$$

□

Example 18.4 Consider $\Psi = \Psi_\Phi$ for an irreducible *reduced* (V^*, Φ), so there are at most two root lengths. Choose the unique Weyl-invariant $\langle\cdot,\cdot\rangle$ scaled so the longest root length is $\sqrt{2}$. The reader can check that inside A^* we have

$$(\Psi_{\Phi^\vee})^\vee = \{a + n : a \in \Phi, n \in J_a\}, \qquad \text{with} \quad J_a := \begin{cases} \mathbf{Z}, & a \text{ of longest length} \\ (1/\ell)\mathbf{Z}, & a \text{ of shortest length,} \end{cases}$$

where $\ell \in \{1, 2, 3\}$ is the square of the quotient of the longest root length by the shortest root length. Note also that the dual $(\Psi_{\Phi^\vee})^\vee$ has derivative root system $(\Phi^\vee)^\vee = \Phi$, so $(\Psi_{\Phi^\vee})^\vee$ is reduced and irreducible.

For $\ell = 1$ ("simply laced") this shows $(\Psi_{\Phi^\vee})^\vee = \Psi_\Phi$ inside A^*, or equivalently (by Proposition 18.3) $\Psi_{\Phi^\vee} = (\Psi_\Phi)^\vee$. But for $\ell = 2, 3$ (i.e., B_n $(n \geq 2)$, C_n $(n \geq 2)$, F_4, G_2) that equality fails and even better we get *new* reduced irreducible affine root systems. The reason we know it is "new" is that the set of *constant differences* among affine roots with a given derivative a isn't the same discrete subset of $\mathbf{R}$ independently of $a \in \Phi$, in contrast with construction Ψ_Φ for reduced Φ: for some roots $a \in \Phi$ we get $\mathbf{Z} \subset \mathbf{R}$ and for others $a \in \Phi$ we get $(1/\ell)\mathbf{Z} \subset \mathbf{R}$.

To make precise that the preceding construction is genuinely new when Φ is not simply laced, we have to define "isomorphism" for affine root systems in a manner that unifies the notion of strong isomorphism from Definition 16.5 and the scaling operation on irreducible components. First we give the general definition and then we discuss how it achieves the desired unification.

Definition 18.5 Let (A, Ψ) and (A', Ψ') be two affine root systems. An *isomorphism* $(A, \Psi) \simeq (A', \Psi')$ is a bijection $\varphi : \Psi \simeq \Psi'$ such that for any $\psi, \eta \in \Psi$ and $r, s \in \mathbf{Z}$ for which $r\psi + s\eta \in \Psi$ we have $\varphi(r\psi + s\eta) = r\varphi(\eta) + s\varphi(\eta)$, and likewise for φ^{-1}.

Note that for every isomorphism there is an inverse isomorphism, and a composition of isomorphisms is an isomorphism. Likewise, direct sum carries isomorphisms to isomorphisms.

Example 18.6 Let's explain how strong isomorphisms can be viewed as special types of isomorphisms. Any affine map $f : A \to A'$ is determined by the associated map of affine duals $f^* : A'^* \to A^*$ (easily checked), and if f is a strong isomorphism then f^* is determined by its restriction φ to a bijection $\Psi' \simeq \Psi$ (since Ψ' spans A'^*) that visibly satisfies the limited $\mathbf{Z}$-linearity condition in the definition of "isomorphism" in Definition 18.5.

Example 18.7 For any affine root system (A, Ψ) and $s \in \mathbf{R}^\times$, the bijection $\varphi : \Psi \to s\Psi$ via s-multiplication is an isomorphism.

Fortunately an isomorphism of affine root systems always arises from a combination of strong isomorphisms and Example 18.7 applied to irreducible components:

Proposition 18.8 *If* $\varphi : (A, \Psi) \simeq (A', \Psi')$ *is an isomorphism of affine root systems and* $\bigoplus_{i=1}^n \Psi_i$ *is the irreducible decomposition of* Ψ *as in Proposition 16.4 then there is a strong isomorphism* $f : (A, \oplus(s_i\Psi_i)) \simeq (A', \Psi')$ *with* $s_i \in \mathbf{R}^\times$ *such that* $f^{-1} \circ \varphi : \Psi \simeq \oplus(s_i\Psi_i)$ *is* s_i*-multiplication on* Ψ_i *for each* i*.*

Proof This is [KP, Proposition 1.3.56(2)]. □

Having set the stage, we can finally deliver the punchline from [KP, Theorem 1.3.63] whose proof (which we omit) requires a *lot* of work with the structure of affine root systems (particularly with special features of the irreducible case):

Theorem 18.9 *Up to isomorphism, the irreducible reduced affine root systems are* exactly *the* Ψ_Φ*'s for irreducible* Φ*'s and the duals* $\Psi_\Phi^\vee$ *for irreducible* Φ *that are* reduced *and not simply laced.*

In particular, Ψ_{BC_n} *is self-dual for all* $n \geq 1$.

The self-duality of Ψ_Φ for non-reduced irreducible Φ (i.e., Φ of type BC) is established in the course of the proof of [KP, Theorem 1.3.63].

In Sect. 19 we'll record the possibilities up to isomorphism for *non-reduced* irreducible affine root systems Ψ (there are four possibilities in each dimension $n \geq 2$, and three possibilities in dimension 1), and introduce the notions of *basis* and *affine Dynkin diagram* that help us to organize information in the structure of apartments in buildings.

19 Bases of Affine Root Systems

Now we classify *non-reduced* irreducible affine root systems Ψ; these arise in Bruhat-Tits theory for $\mathrm{SU}(h)$ with even-dimensional quasi-split Hermitian spaces (W, h) [KP, Theorem 6.4.8, Table 6.4.1]. Necessarily $\Phi = \nabla(\Psi)$ is irreducible and non-reduced (if $\psi = 2\eta$ then $\dot{\psi} = 2\dot{\eta}$), so $\Phi = \mathrm{BC}_n$ where $n = \dim A = \dim V^*$. Since multiplication by 4 never carries affine roots to affine roots, we have

$$\Psi = \Psi^{\mathrm{nd}} \cup \Psi^{\mathrm{nm}}$$

where $\Psi^{\mathrm{nd}} = \{\psi \in \Psi : \psi \notin 2\Psi\}$ is the set of non-divisible affine roots and $\Psi^{\mathrm{nm}} = \{\psi \in \Psi : 2\psi \notin \Psi\}$ is the set of non-multipliable affine roots (indeed, if an affine root

ψ is divisible, so $\psi = 2\eta$ for some $\eta \in \Psi$ then ψ cannot be multipliable, as otherwise $4\eta \in \Psi$). Both of these subsets are readily seen to be affine root systems with the *same* hyperplanes and reflections as for Ψ (hence the same Weyl group) and thus are irreducible and clearly *reduced*.

The possibilities for "irreducible and reduced" affine root systems are known up to isomorphism, organized by derivative root systems, so we can apply that to constrain the possibilities for Ψ^{nd} and Ψ^{nm} by first controlling the possibilities for their derivative root systems. We have

$$\Phi^{\mathrm{nd}} \subset \nabla(\Psi^{\mathrm{nd}}) \subset \nabla(\Psi) = \Phi = \mathrm{BC}_n \tag{19.1}$$

where the first inclusion is because $a = \dot{\psi} \in \Phi^{\mathrm{nd}}$ implies that $\psi \notin 2\Psi$. We similarly have

$$\Phi^{\mathrm{nm}} \subset \nabla(\Psi^{\mathrm{nm}}) \subset \nabla(\Psi) = \Phi = \mathrm{BC}_n \tag{19.2}$$

of rank n. By inspection of BC_n as a union of B_n and C_n, $\Phi^{\mathrm{nd}} = \mathrm{B}_n$ and $\Phi^{\mathrm{nm}} = \mathrm{C}_n$.

The containments of root systems in (19.1) and (19.2) each have all terms with the *same* Weyl group $W(\Phi)$ (by consideration of reflections), so by transitivity of $W(\Phi)$ on roots of Φ with a common length (such as the divisible roots in BC_n, and the multipliable roots in BC_n) we see the two containments in each of (19.1) and (19.2) cannot both be strict; i.e., $\nabla(\Psi^{\mathrm{nd}})$ is equal to either Φ (type BC_n) or Φ^{nd} (type B_n) and $\nabla(\Psi^{\mathrm{nm}})$ is equal to Φ (type BC_n) or Φ^{nm} (type C_n). Hence, by looking at the derivative root system in the reduced irreducible classification we deduce up to isomorphism

$$\Psi^{\mathrm{nd}} \in \{\Psi_{\mathrm{BC}_n}, \Psi_{\mathrm{B}_n}, \Psi_{\mathrm{C}_n}^{\vee}\}, \quad \Psi^{\mathrm{nm}} \in \{\Psi_{\mathrm{BC}_n}, \Psi_{\mathrm{C}_n}, \Psi_{\mathrm{B}_n}^{\vee}\}.$$

Note that for $n > 1$ the three possibilities for each are pairwise non-isomorphic, whereas for $n = 1$ we have a repetition since $\mathrm{B}_1 = \mathrm{C}_1 = \mathrm{A}_1$ and Ψ_{A_1} is self-dual.

The preceding is a first step toward controlling the possibilities for Ψ. The full result is as follows (stated in [KP, Theorem 1.3.69], and resting on many details given in the errata [KPerr] for [KP]).

Theorem 19.1 *Any non-reduced irreducible Ψ is determined up to isomorphism by the pair of isomorphism classes of Ψ^{nd} and Ψ^{nm}. The possibilities for $(\Psi^{\mathrm{nd}}, \Psi^{\mathrm{nm}})$ are $(\Psi_{\mathrm{B}_n}, \Psi_{\mathrm{B}_n}^{\vee})$, $(\Psi_{\mathrm{C}_n}^{\vee}, \Psi_{\mathrm{C}_n})$, $(\Psi_{\mathrm{BC}_n}, \Psi_{\mathrm{C}_n})$, and $(\Psi_{\mathrm{C}_n}^{\vee}, \Psi_{\mathrm{BC}_n})$.*

In particular, the latter two are dual to each other and the only repetition up to isomorphism among these 4 cases are the first two when $n = 1$.

Proof One first gives an explicit construction of all 4 pairs arising, and then one proves exhaustiveness (via a more refined version of the type of argument given above). □

Now we introduce the notion of "basis" for affine root systems. Let (A, Ψ) be an affine root system, and let $C \subset A$ be a *chamber*: a connected component of the open subset $A - \bigcup_{\psi \in \Psi} H_\psi$ of A. For $\psi \in \Psi$ we have $H_\psi \cap C = \emptyset$, so $C \subset A^{\psi > 0}$

or $A^{\psi<0}$. We say ψ is *positive* (resp. *negative*) with respect to $\mathcal{C}$ if $\psi(\mathcal{C}) \subset \mathbf{R}_{>0}$ (resp. $\psi(\mathcal{C}) \subset \mathbf{R}_{<0}$). This gives rise to a partition $\Psi = \Psi(\mathcal{C})^+ \coprod \Psi(\mathcal{C})^-$. Associated to $\mathcal{C}$, we define the subset $\Delta_{\mathcal{C}}$ of Ψ called a *basis* to be

$$\{\psi \in \Psi^{\mathrm{nd}} : \psi \in \Psi(\mathcal{C})^+, H_\psi \text{ is a wall of } \mathcal{C}\},$$

where we note that $\mathcal{C}$ is on the "postive side" of such a wall. We call the elements of $\Delta_{\mathcal{C}}$ the *simple affine roots* with respect to $\mathcal{C}$.

Remark 19.2 We aren't giving a combinatorial characterization of which subsets of Ψ arise as such bases $\Delta_{\mathcal{C}}$, but such a subset does determine the chamber from which it is made because we can reconstruct $\mathcal{C}$ from $\Delta_{\mathcal{C}}$ (so it is reasonable to call a subset $\Delta \subset \Psi$ a *basis* if it has the form $\Delta_{\mathcal{C}}$ for some chamber $\mathcal{C}$):

$$\mathcal{C} = \bigcap_{\psi \in \Delta_{\mathcal{C}}} A^{\psi>0}.$$

This equality is a general feature of connected components of *any* locally finite collection of hyperplanes in an affine space for a finite-dimensional $\mathbf{R}$-vector space, not specific to affine root systems, as the reader may check.

We will soon make $\Delta_{\mathcal{C}}$ rather explicit for a chamber $\mathcal{C}$ in Ψ_Φ for reduced irreducible Φ, but first let's discuss how the preceding notions behave with respect to the irreducible decomposition $\Psi = \bigoplus \Psi_i$: we have correspondingly $V = \prod V_i$, $\Phi = \bigoplus \Phi_i$, $A = \prod A_i$, and $\mathcal{C} = \prod \mathcal{C}_i$, and for $\psi \in \Psi_{i_0}$ the hyperplane $H_\psi \subset A$ is $(\prod_{i \neq i_0} A_i) \times \{\psi = 0 \text{ in } A_{i_0}\}$. Moreover, $W(\Phi_i)$ acts on V_i absolutely irreducibly and $W(\Psi_i)$ is infinite (since for $v \perp H$, $r_{H+v} \circ r_H$ is translation by $2v$), so by [Bour, Chap. 4, Sect. 3.9, Proposition 8] each $\mathcal{C}_i$ is an *open simplex* in A_i and hence $\mathcal{C} = \prod \mathcal{C}_i$ is an open "polysimplex" in $A = \prod A_i$. Moreover, $\overline{\mathcal{C}}$ is a fundamental domain for the $W(\Psi)$-action on A: the natural map $\overline{\mathcal{C}} \to A/W(\Psi)$ is bijective [Bour, Chap. 5, Sect. 3.3, Theorem 9].

Since any open simplex in $\mathbf{R}^N$ can be carried via an affine automorphism onto the standard simplex in the positive orthant (with closure having vertices at the origin and the standard basis), we deduce:

Proposition 19.3 *If* Ψ *is* irreducible *then* $\Delta_{\mathcal{C}} = \{\psi_0, \ldots, \psi_n\}$ *is a basis of* A^* *and the set of vertices of* $\overline{\mathcal{C}}$ *can be uniquely labeled as* $\{x_0, \ldots, x_n\}$ *where* $\psi_i(x_j) \in \delta_{ij}\mathbf{R}_{>0}$.

Example 19.4 Let $\Psi = \Psi_\Phi$ for (V^*, Φ) reduced and irreducible. Pick a basis $\Delta_0 = \{a_1, \ldots, a_n\}$ of Φ, thereby defining a positive system of roots and associated Weyl chamber open cone $C \subset V$ (defined by positivity of all a_i's). By [Bour, Chap. 6, Sect. 2.2, Proposition 10], there is a unique chamber $\mathcal{C}$ of Ψ with $\mathcal{C} \subset C$ and $0 \in \overline{\mathcal{C}}$, and explicitly $\mathcal{C}$ has corresponding basis $\Delta_{\mathcal{C}} = \{a_1, \ldots, a_n, 1 - a_0\}$ where a_0 is the highest root in Φ^+.

Remark 19.5 For root systems Φ, negation carries a Weyl chamber cone to another one (likewise the negative of a positive system of roots is another such) and so the negative of a basis of Φ is a basis of Φ. But beware that for affine root systems (A, Ψ) *the negative* $-\Delta$ *of a basis* Δ *of* Ψ *is* **never** *a basis of* Ψ.

This fact is illustrated by Example 19.4, for which negation of $a_1, \ldots, a_n$ brings one to the negated Weyl chamber cone that is disjoint from the negative side $A^{1-a_0<0} = A^{a_0>1}$ of the hyperplane H_{1-a_0}. To prove with general Ψ that $-\Delta$ is not a basis it suffices (via irreducible decomposition) to treat irreducible Ψ, for which it follows from [KP, Proposition 1.3.47(1,2,5)].

Theorem 19.6 *Let* (A, Ψ) *be an affine root system, and* $W(\Psi) := \langle r_{H_\psi} : \psi \in \Psi \rangle$ *the affine Weyl group.*

(1) The group $W(\Psi)$ *acts simply transitively on the set of chambers (equivalently, on the set of bases);*
(2) $W(\Psi) = \langle r_H \rangle_{H \in \mathrm{wall}(\mathcal{C})}$ *for any chamber* $\mathcal{C}$.

Proof This is part of [KP, Props. 1.3.20, 1.3.22] (amounting to references to [Bour]). □

Corollary 19.7 *We have* $\Psi^{\mathrm{nd}} = W(\Psi).\Delta$ *where* $\Delta = \Delta_{\mathcal{C}}$ *for a choice of chamber* $\mathcal{C}$.

Proof Pick $\psi \in \Psi^{\mathrm{nd}}$. The hyperplane H_ψ has lots of chambers on either side with H_ψ as a wall (as for any locally finite collection of hyperplanes in a Euclidean space). Pick a chamber $\mathcal{C}'$ in the positive half-space $A^{\psi>0}$ of $A - H_\psi$ with $H_\psi \in \mathrm{wall}(\mathcal{C}')$, so $\psi \in \Delta_{\mathcal{C}'}$. The Weyl group $W(\Psi)$ moves $\mathcal{C}'$ to $\mathcal{C}$ by transitivity, so it moves $\Delta_{\mathcal{C}'}$ onto $\Delta_{\mathcal{C}} = \Delta$ and hence carries ψ into Δ. □

Now we can finally establish an affine analogue of a key feature of bases of affine root systems, justifying the use of the word "basis":

Corollary 19.8 *Let* Δ *be a basis of* Ψ. *Each* $\psi \in \Psi(\mathcal{C})^+$ *is in* $\mathbf{Z}_{\geq 0} \cdot \Delta$ *(with a unique such expression if* Ψ *is* irreducible).

Proof The uniqueness in the irreducible case is immediate from Δ being a basis of A^* in such cases (Proposition 19.3). To prove such an expression exists for each ψ positive with respect to $\mathcal{C}$, we can certainly assume ψ is non-divisible (if we have the result for some η then we get it for 2η).

Let $\mathcal{C}$ be the chamber giving rise to Δ. If $\bigoplus \Psi_i$ is the irreducible decomposition of Ψ then we get a corresponding decomposition $\mathcal{C} = \prod \mathcal{C}_i$ and hence bases $\Delta_i = \Delta_{\mathcal{C}_i}$

of the Ψ_i's with $\Delta = \coprod \Delta_i$ (we do *not* claim linear independence among this disjoint union when Ψ is reducible, as each A_i^* contains the line of constant functions in A^*). We have $\psi \in \Psi_{i_0}$ for some i_0, so it is harmless to pass to Ψ_{i_0} and Δ_{i_0} so that we may assume Ψ is irreducible. In particular, $\Delta \subset A^*$ is a basis.

Let $L = \mathbf{Z} \cdot \Delta \subset A^*$ (a lattice). We have $\psi \in \Psi^{\mathrm{nd}} = W(\Psi).\Delta \subset W(\Psi).L = \langle r_H \rangle_{H \in \mathrm{wall}(\mathcal{C})} \cdot (\mathbf{Z} \cdot \Delta)$. For $\psi_i \in \Delta$ and $\ell \in L$ we have

$$r_{\psi_i}(\ell) = \ell - \langle \dot{\ell}, \dot{\psi}_i^\vee \rangle \psi_i$$

with $\langle \dot{\ell}, \dot{\psi}_i^\vee \rangle \in \mathbf{Z}$, so $W(\Psi)$ preserves L. This shows $\psi \in L = \mathbf{Z} \cdot \Delta$; in other words,

$$\psi = \sum n_i \psi_i$$

with $n_i \in \mathbf{Z}$. Evaluating this on the vertex x_j of the closed simplex $\overline{\mathcal{C}}$ then yields

$$0 \leq \psi(x_j) = \sum n_i \psi_i(x_j) = n_j \psi_j(x_j)$$

(where $\psi(x_j) \geq 0$ since $\psi(\mathcal{C}) > 0$). But $\psi_j(x_j) > 0$, so $n_j \geq 0$ for all j as desired. □

20 Special Points and Extra Special Points

Let (A, Ψ) be an affine root system, with A an affine space for a vector space V.

Definition 20.1 For $x \in A$, let $\Psi_x = \{\psi \in \Psi : \psi(x) = 0\}$ (this is empty precisely when x is not in any of the affine hyperplanes H_ψ; i.e., when x belongs to some (open) chamber $\mathcal{C}$).

Let $A_x^* \subset A^*$ be the subspace of affine functionals vanishing at x; this is a linear complement to the line $\mathbf{R}$ of constant functionals, so $\nabla : A_x^* \to V^*$ is an isomorphism. Any two distinct $\psi, \eta \in \Psi_x$ cannot have nonzero constant difference (as both affine functionals vanish at x), so $\nabla : \Psi_x \to \Phi$ induced by $\nabla : A_x^* \simeq V^*$ is injective.

The injection of $\Psi_x \hookrightarrow \Phi$ intertwines the action on $\Psi_x \subset A_x^*$ by the reflection r_ψ^* for $\psi \in \Psi_x$ (if $\eta \in A_x^*$ then $r_\psi^*(\eta) = \eta - \langle \dot{\eta}, \dot{\psi}^\vee \rangle \psi \in A_x^*$, so this belongs to $\Psi \cap A_x^* = \Psi_x$ when $\eta \in \Psi_x$ too) and the action on Φ by the derivative reflection $\nabla(r_\psi^*) = r_{\dot{\psi}}$. Thus, Ψ_x is a root system in its span $V_x' \subset A_x^* = V^*$ by using the restriction to V_x' of reflections on A^* through affine roots in Ψ_x.

The subgroup $W(\Psi_x) \subset W(\Psi) \subset \mathrm{GL}(A)$ generated by reflections r_ψ for $\psi \in \Psi_x$ consists of affine transformations that fix x, so it contains no non-trivial translations and hence via ∇ injects into the finite group $W(\Phi)^* = W(\Phi^\vee) \subset \mathrm{GL}(V)$. Thus, if we equip V^* with a $W(\Phi)$-invariant inner product then the resulting inner product on $A^* = A_x^* \oplus \mathbf{R}$ (upon identifying A_x^* with V^*) makes $W(\Psi_x)$ act through orthogonal transformations. Hence, the decomposition $A^* = V'_x \oplus (V'_x)^\perp$ and the root system property of (V'_x, Ψ_x) imply $V'_x \oplus (A^*)^{W(\Psi_x)} = A^*$. The subgroup $W(\Psi_x) \subset W(\Psi)$ is thereby identified via restriction to V'_x with the Weyl group of (V'_x, Ψ_x). By using a $W(\Phi)$-invariant inner product on V^*, it is checked that the subset $\{\psi^\vee : \psi \in \Psi_x\} \subset A^*_x = V^*$ is $(\Psi_x)^\vee$.

The locally finite collection of hyperplanes $\mathfrak{h} = \{H_\psi\}_{\psi \in \Psi}$ decomposes A into *facets*: equivalence classes for the equivalence relation on A defined by declaring that $x, y \in A$ are equivalent when for every $H \in \mathfrak{h}$ either $x, y \in H$ or x and y are in the same open half-space of $A - H$ (equivalently: for all $\psi \in \Psi$, $\psi(x)$ and $\psi(y)$ are either both 0, both positive, or both negative). By [Bour, Chap. 5, Sect. 1.3], the facets not contained in any $H \in \mathfrak{h}$ are exactly the connected components of $A - (\cup_{H \in \mathfrak{h}} H)$, called *chambers* (so these are the open facets), and every facet is in the closure of a chamber. An explicit description of the facets in the closure of a chosen chamber is given in [Bour, Chap. 5, Sect. 1.6].

If $\mathcal{F} \subset A$ is the facet containing x then the subset $\Psi_x \subset \Psi$ depends only on $\mathcal{F}$ because for any $\psi \in \Psi_x$ we have H_ψ meets $\mathcal{F}$ (namely both contain x) and thus $\mathcal{F} \subset H_\psi$ (as for facets of any locally finite hyperplane arrangement in an affine space). For this reason, we'll often write $\Psi_{\mathcal{F}}$ to denote Ψ_x.

The identification of $W(\Psi_{\mathcal{F}})$ as subgroup of $W(\Psi)$ can be refined as follows:

Lemma 20.2 *For $x \in \mathcal{F} \subset A$,*

$$W(\Psi)_x = \{w \in W(\Psi) : w|_{\mathcal{F}} = \mathrm{id}_{\mathcal{F}}\} = \langle r_H \rangle_{H = H_\psi \supset \mathcal{F}}.$$

In particular, $W(\Psi)_x = W(\Psi_x)$ as subgroups of $W(\Psi)$.

Proof The pair of displayed equalities of subgroups of $W(\Psi)$ is part of [KP, Lemma 1.3.17]. The subgroup $W(\Psi_x) \subset W(\Psi)$ is by definition generated by the reflections r_H for hyperplanes $H = H_\psi$ containing x. Such H are exactly those which contain $\mathcal{F}$, so we get the final assertion too. □

Remark 20.3 The axioms in Bruhat-Tits theory include Axiom 6 in Sect. 27 that if $\mathscr{D}G$ is isotropic (i.e., $\mathcal{B}(G)$ is not a point) then each apartment $A = \mathcal{A}(S) \subset \mathcal{B}(G)$ is equipped with a natural affine root system Ψ with derivative $\Phi = \Phi(G, S)$, and Axiom 20 in Sect. 30 that for each facet $\mathcal{F} \subset A$ the subset $\Psi_{\mathcal{F}} \hookrightarrow \Phi$ is the root system of the maximal reductive quotient $\mathsf{G}_{\mathcal{F}}$ of the connected special fiber of a Bruhat-Tits $\mathfrak{o}$-group $\mathscr{G}^0_{\mathcal{F}}$ that satisfies $\mathscr{G}^0_{\mathcal{F}}(\mathfrak{o}) = G(k)^0_{\mathcal{F}}$ (the pointwise $\mathcal{F}$-stablizer in $G(k)^0$, which coincides with the $\mathcal{F}$-stabilizer in $G(k)^0$ by Theorem 24.11(2) and Axiom 9 in Sect. 28). Further combinatorics with $\Psi_{\mathcal{F}}$ keeps track of proper parabolic subgroups of $\mathsf{G}_{\mathcal{F}}$ (Axiom 22 in Sect. 30).

Let's illustrate the Ψ_x's for various vertices x in $\Psi = \Psi_\Phi$ for some rank-2 irreducible root systems Φ. (In each of the following examples, for any x in a hyperplane that isn't a vertex we'll have $\Psi_x = \mathrm{A}_1$.) (Fig. 4)

Example 20.4 For $\Psi = \Psi_\Phi$ with Φ of type A_2, every vertex x has $\Psi_x \overset{\nabla}{\simeq} \Phi = \mathrm{A}_2$.

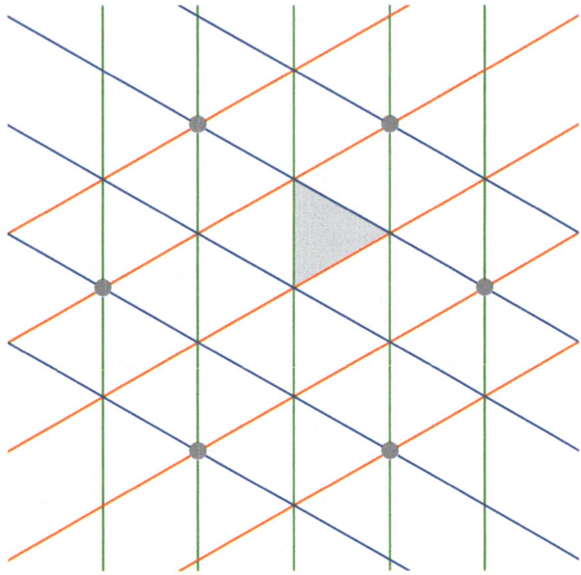

Fig. 4 The affine hyperplanes (lines) for Ψ_{A_2}, with roots darkened

Example 20.5 For $\Psi = \Psi_\Phi$ with Φ of type B_2, the vertices y in the Weyl-orbits of x_0 and x_1 (i.e., triangle vertices at a 45-degree angle) have $\Psi_y \overset{\nabla}{\simeq} \Phi = B_2$ whereas vertices z in the Weyl-orbit of $\widetilde{x}$ (i.e., triangle vertices at a right angle) have $\Psi_z \overset{\nabla}{\simeq} A_1 \times A_1 \subset \Phi$ (Fig. 5)

Example 20.6 For $\Psi = \Psi_\Phi$ with $\Phi = \{\pm e_i^*, \pm e_1^* \pm e_2^*, \pm 2e_i^*\}$ of type BC_2, we have the explicit construction $\Psi = \{\pm e_i^* + \mathbf{Z}, \pm e_1^* \pm e_2^* + \mathbf{Z}, \pm 2e_i^* + (2\mathbf{Z}+1)\}$ where the vanishing hyperplanes $H_\psi \subset \mathbf{R}^2$ are shown in Fig. 6: the black hyperplanes where some $\pm e_i^* + n$ vanishes ($n \in \mathbf{Z}$), the blue hyperplanes where some $\pm e_1^* \pm e_2^* + m$ vanishes ($m \in \mathbf{Z}$), and the red dashed hyperplanes where some $\pm 2e_i^* + k$ vanishes ($k \in 2\mathbf{Z}+1$); the latter are H_ψ's for which $\dot{\psi} \in 2\Phi$ (i.e., $\dot{\psi}$ is divisible).

For $\Psi = \Psi_{BC_2}$ there are three types of vertices up to the action of $W(\Phi)$, and we can describe them in terms of chamber-closure triangles: vertices x at right angles of the triangles, vertices y at 45-degree angles through which there is no dashed hyperplane (i.e., $\Psi_y \simeq \nabla(\Psi_y) \subset \Phi^{\mathrm{nd}}$), and vertices z at 45-degree angles through which there is a dashed hyperplane (i.e., $\Psi_z \simeq \nabla(\Psi_z)$ meets 2Φ).

We have $\Psi_x \overset{\nabla}{\simeq} A_1 \times A_1 \subset \Phi$, $\Psi_y \overset{\nabla}{\simeq} \Phi^{\mathrm{nd}} = B_2 \subset BC_2 = \Phi$, and $\Psi_z \overset{\nabla}{\simeq} \Phi^{\mathrm{nm}} = C_2 \subset BC_2 = \Phi$. To make the latter of these explicit we take z to be the point $(1/2, 1/2)$, so then

$$\Psi_z = \{\pm(2e_1^* - 1), \pm(2e_2^* - 1), \pm(e_1^* - e_2^*), \pm(e_1^* + e_2^* - 1)\} = \{\pm\varepsilon_1, \pm\varepsilon_2, \pm(\varepsilon_1 + \varepsilon_2), \pm(\varepsilon_1 - \varepsilon_2)\}$$

where $\varepsilon_1 = e_1^* + e_2^* - 1$ and $\varepsilon_2 = e_1^* - e_2^*$.

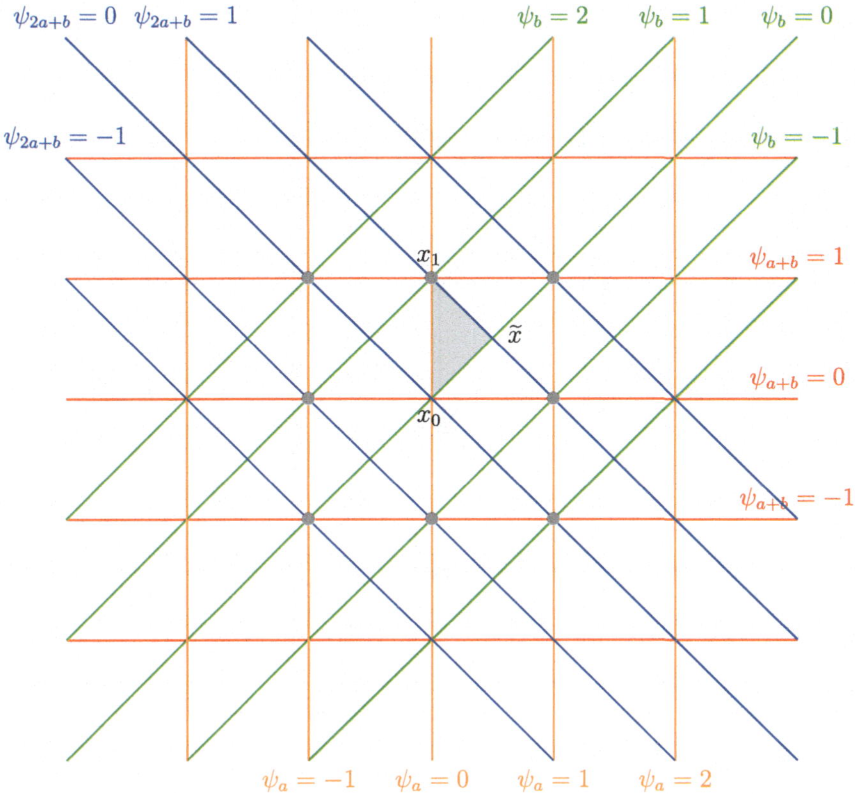

Fig. 5 The affine hyperplanes (lines) for Ψ_{B_2}, with roots darkened

Remark 20.7 When $\mathcal{F} \subset \overline{\mathcal{C}}$, the subset $\Delta_{\mathcal{C}} \cap \Psi_{\mathcal{F}} \subset \Psi$ is a basis of $\Psi_{\mathcal{F}}$: see [KP, Proposition 1.3.35(6)] for a proof.

Definition 20.8 Say $x \in A$ is *special* if each H_ψ (for $\psi \in \Psi$) is parallel to some $H_{\psi'} \ni x$ (i.e., $\psi' \in \Psi_x$).

For special x we have $W(\Psi_x) = W(\Phi)$ because reflections generating Φ arise from derivatives of reflections through the hyperplanes $H_{\psi'}$ for $\psi' \in \Psi_x$ (due to the parallelism classes of the $H_{\psi'}$'s exhausting all the hyperplanes H_ψ by the meaning of "special"). The property that the subgroup $W(\Psi_x) \subset W(\Phi)$ is full even *characterizes* when x is special: see Proposition 20.12.

Definition 20.9 Say x is *extra special* if Φ has a *basis* of the form $\{\dot{\psi}_1, \dots, \dot{\psi}_\ell\}$ for $\psi_i \in \Psi_x$ (equivalently $\Psi_x \overset{\nabla}{\hookrightarrow} \Phi$ contributes a basis, forcing $\Psi_x = \Phi$ when Φ is reduced due to the equality $W(\Phi).\Delta = \Phi$ in the reduced case and $W(\Phi)$ being generated by reflections in the roots from Δ).

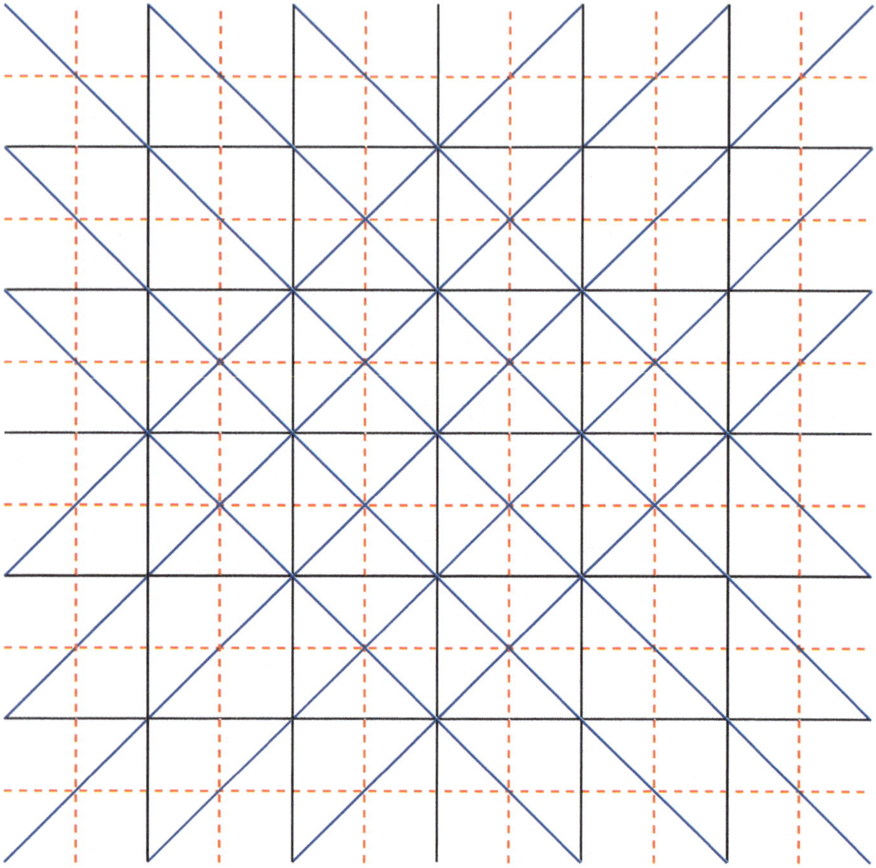

Fig. 6 The affine hyperplanes (lines) for Ψ_{BC_2}, with the hyperplanes H_ψ for $\dot{\psi} \in 2\Phi$ as dashed (in red)

Example 20.10 Consider $\Psi = \Psi_\Phi$ for (V^*, Φ) irreducible and reduced. Pick a basis $\Delta_0 = \{a_1, \ldots, a_n\}$ of Φ; this specifies a positive system of roots and an associated open cone Weyl chamber C.

Take $\mathcal{C}$ to be the chamber of Ψ contained in C with $0 \in \overline{\mathcal{C}}$. This $\mathcal{C}$ is explicitly described in Example 19.4, where we saw that

$$\Delta_{\mathcal{C}} = \{a_1, \ldots, a_n, 1 - a_0\}$$

for the highest positive root a_0. The hyperplanes H_{a_i} for $1 \leq i \leq n$ pass through 0 and are walls of $\mathcal{C}$ with $a_i(\mathcal{C}) > 0$, so 0 is thereby seen to be extra special. We view "extra special" in general as a generalization of this example.

Example 20.11 Here is a summary of some observations we made about Ψ_Φ for the rank-2 irreducible Φ in Examples 20.4–20.6. For A_2, all vertices are special and extra special. For B_2, both x_0 and x_1 are special and extra special while $\widetilde{x}$ is neither. For BC_2, x is neither special nor extra special, y is *both*, and z is special but *not* extra special.

The preceding example illustrates the next result:

Proposition 20.12 *The following hold for affine root systems:*

(1) Extra special points always exist, and are vertices (i.e., 1-point facets).
(2) A point $x \in A$ is special if and only if the inclusion $W(\Psi_x) \hookrightarrow W(\Phi)$ is an equality.
(3) If x is extra special then it is special, and the converse holds if Φ is reduced.

Proof To prove the first part of (1), pick a basis of Φ, so it can be written as $\{\dot{\psi}_1, \ldots, \dot{\psi}_\ell\}$ (where $\ell = \dim A = \dim V^*$) with some $\psi_i \in \Psi$. The affine hyperplanes $H_{\psi_i} \subset A$ are defined by affine linear equations with *linearly independent* derivatives. Thus, as we intersect successively more H_{ψ_i}'s the dimension drops by 1 each time, so $\bigcap H_{\psi_i}$ is a single point. From the definitions, such a point is extra special.

Assertion (2) is [KP, Lemma 1.3.42].

To prove the first part of (3), by (2) it suffices to show $W(\Psi_x) \hookrightarrow W(\Phi)$ is an equality when x is extra special. But $W(\Psi_x)$ contains reflections through the hyperplanes defined by $\dot{\psi}$ for each $\psi \in \Psi_x$, and by the meaning of "extra special" there are such ψ's for which the derivatives $\dot{\psi}$ sweep out a basis of Φ. But the reflections in the hyperplanes defined by a basis of Φ generate $W(\Phi)$, so the desired equality holds.

Now we assume Φ is reduced and aim to show that if x is special then it is extra special. Pick a basis of Φ, and denote its elements as $\{\dot{\psi}_1, \ldots, \dot{\psi}_\ell\}$ for $\psi_i \in \Psi$. Since x is special, for each i there is a $\psi_i' \in \Psi_x$ such that $H_{\psi_i'} \parallel H_{\psi_i}$, so the derivatives $\dot{\psi}_i'$ and $\dot{\psi}_i$ are $\mathbf{R}^\times$ multiples of each other. But Φ is reduced, so $\dot{\psi}_i' = \pm\dot{\psi}_i$. By replacing ψ_i' with $-\psi_i' \in \Psi_x$ if necessary, we thereby arrange that $\dot{\psi}_i' = \dot{\psi}_i$. Hence, Φ has a basis of the form $\{\dot{\psi}_1', \ldots, \dot{\psi}_\ell'\}$ for affine roots $\psi_i' \in \Psi_x$. This says x is extra special. □

The proof of Proposition 20.12(3) can be used to show a special point $x \in A$ is *always a vertex* (i.e., a 1-point facet), as follows. For $\ell = \dim A = \dim \Phi$, pick a basis of Φ; its members can be written as $\dot{\psi}_1, \ldots \dot{\psi}_\ell$ for some $\psi_1, \ldots, \psi_\ell \in \Psi$. The speciality of x provides $\psi_1', \ldots \psi_\ell' \in \Psi_x$ such that $H_{\psi_i'} \parallel H_{\psi_i}$ for all i. This implies $\dot{\psi}_i' \in \mathbf{R}^\times \dot{\psi}_i$, so the functionals $\dot{\psi}_i' \in V^*$ are linearly independent (as the $\dot{\psi}_i$'s are). Hence, $H_{\psi_1'}, \ldots, H_{\psi_\ell'}$ are affine hyperplanes in A through x whose defining affine-linear equations have linearly independent derivatives. As in the proof of Proposition 20.12, the intersection $\bigcap H_{\psi_i'}$ must then be a single point (necessarily then a vertex!) yet x is in that intersection, so the point is x.

21 Affine Diagrams

In Sect. 20 we introduced the notions of special and extra special points for an affine root system (A, Ψ), and saw these are always vertices. We now use that to prove:

Proposition 21.1 *Let (A, Ψ) be an irreducible affine root system, $\Phi = \nabla(\Psi)$, and $\Delta \subset \Psi$ a basis. At most one $\psi \in \Delta$ can have derivative $\dot{\psi} \in \Phi$ that is divisible.*

As an example, $\Psi_{\mathrm{BC}_1} = \{\pm a + \mathbf{Z}, \pm 2a + (2\mathbf{Z} + 1)\}$ has $\Delta = \{a, 1 - 2a\}$ as a basis, where a has derivative that is a basis of $\Phi = \mathrm{BC}_1$ and $1 - 2a$ has derivative $-2a$ that is divisible in Φ.

Proof In principle we could prove this by checking all irreducible Ψ in the classification for which Φ is non-reduced (namely $\Psi = \Psi_{\mathrm{BC}_n}$ and the 4 families of non-reduced Ψ), looking at one basis in each such possibility. Instead we shall give a uniform argument not requiring the classification.

Let $\mathcal{C} \subset A$ be the chamber such that $\Delta = \Delta_{\mathcal{C}}$. We know there is an extra special vertex $x_0 \in \overline{\mathcal{C}}$, so Ψ_{x_0} contains affine roots $\psi_1, \ldots, \psi_\ell$ such that $\{\dot{\psi}_1, \ldots, \dot{\psi}_\ell\}$ is a basis of Φ. This latter basis defines positive systems of roots for Φ and its dual, so in particular a positive Weyl chamber cone $C \subset V$.

The canonical isomorphism of pointed spaces $(A, x_0) \simeq (V, 0)$ carries each H_{ψ_i} over to the vanishing linear hyperplane $H_{\dot{\psi}_i}$ that varies through the walls of C. The action of $W(\Psi)_{x_0} = W(\Psi_{x_0}) = W(\Phi) = W(\Phi^\vee)^*$ is transitive on the set of Weyl chamber cones of Φ, and so can be used to bring us to the case that the affine chamber $\mathcal{C}$ is contained in C.

But $x_0 \mapsto 0$, so the open $\mathcal{C}$ inside C has the origin in its closure and hence (check by reasoning with locally finite collections of hyperplanes) the walls H_{ψ_i} of C are among the walls of $\mathcal{C}$. Since $\psi_i(\mathcal{C}) \subset \dot{\psi}_i(C) \subset \mathbf{R}_{>0}$ and $\psi_i \in \Psi^{\mathrm{nd}}$ (as its derivative $\dot{\psi}_i$ is in Φ^{nd} due to being part of a basis of Φ), we conclude that each ψ_i belongs to

$$\{\psi \in \Psi(\mathcal{C})^+ \cap \Psi^{\mathrm{nd}} : H_\psi \in \mathrm{wall}(\mathcal{C})\} = \Delta_{\mathcal{C}} = \Delta.$$

We have thereby built affine roots $\psi_1, \ldots, \psi_\ell \in \Delta$ for which $\dot{\psi}_i \in \Phi^{\mathrm{nd}}$ for all i. Here $\ell = \dim \Phi = \dim A = \#\Delta - 1$, so there is exactly *one* affine root $\psi \in \Delta$ not among the ψ_i's. This is the only one which could possibly have derivative divisible in Φ, so we are done. □

The preceding result will be used in one aspect of the definition of Dynkin diagrams for affine root systems, as we now discuss. The Dynkin diagram $\mathrm{Dyn}(\Psi)$ will be a weighted ("multiplicity on edges") graph whose edges with multiplicity > 1 are assigned a direction (except for some edges with multiplicity 4, which never arise for Dynkin diagrams of usual root systems). The *set of vertices* in $\mathrm{Dyn}(\Psi)$ is a choice of basis Δ. For *distinct* $\psi, \eta \in \Delta$, no edge joins them if $\dot{\psi}$ and $\dot{\eta}$ are orthogonal in Φ (equivalently $\langle \dot{\psi}, \dot{\eta}^\vee \rangle = 0$, and equivalently $\langle \dot{\eta}, \dot{\psi}^\vee \rangle = 0$).

For the irreducible components Ψ_i of Ψ, we have $\Delta = \coprod \Delta_i$ for bases Δ_i of Ψ_i (since $\mathcal{C} = \prod \mathcal{C}_i$ for chambers $\mathcal{C}_i$ of Ψ_i, and $\mathrm{wall}(\mathcal{C}) = \coprod \mathrm{wall}(\mathcal{C}_i)$). The irreducible decomposition $\Psi = \bigoplus \Psi_i$ has the irreducible Ψ_i's as the preimages of the irreducible components Φ_i of Φ, with roots in Φ_i orthogonal to roots in Φ_j when $i \neq j$. Thus, there is no edge in $\mathrm{Dyn}(\Psi)$ joining a vertex in $\mathrm{Dyn}(\Psi_i)$ to a vertex in $\mathrm{Dyn}(\Psi_j)$ when $i \neq j$, so we may and do focus the rest of the definition of affine Dynkin diagrams (the edges, their multiplicities and directions, etc.) on the case when Ψ is *irreducible*, so $\#\Delta = 1 + \dim A = 1 + \dim \Phi$.

By irreducibility, a $W(\Phi)$-invariant inner product on V (or V^*) is unique up to scaling, so the ratio $\|\dot{\psi}\|/\|\dot{\eta}\|$ for *distinct* $\psi, \eta \in \Delta$ is intrinsic (as is the angle $\measuredangle(\dot{\psi}, \dot{\eta})$). Furthermore, by irreducibility and Proposition 21.1 *at most one* of ψ, η has derivative divisible in Φ, so we can arrange $\dot{\psi} \in \Phi^{\mathrm{nd}}$.

Suppose $\dot{\psi}$ and $\dot{\eta}$ are *not* orthogonal, so

$$f(\psi, \eta) := \langle \dot{\psi}, \dot{\eta}^{\vee} \rangle \langle \dot{\eta}, \dot{\psi}^{\vee} \rangle = 4\cos^2(\measuredangle(\dot{\psi}, \dot{\eta})) \in \{1, 2, 3, 4\},$$

where the last equality comes from the formula (using a $W(\Phi)$-invariant inner product)

$$\langle a, b^{\vee} \rangle = \frac{2(a \cdot b)}{b \cdot b} = 2\frac{\|a\|}{\|b\|}\cos(\measuredangle(a, b))$$

for $a, b \in \Phi$. Note that the value $f(\psi, \eta)$ is equal to 4 if and only if the angle between $\dot{\psi}$ and $\dot{\eta}$ is 0 or π, which is to say these derivatives are linearly dependent. We have arranged that $\dot{\psi} \in \Phi^{\mathrm{nd}}$, so the possibilities for linear dependence are $\dot{\eta} = \pm\dot{\psi}$ or $\dot{\eta} = \pm 2\dot{\psi}$; the case $\dot{\eta} = \pm(1/2)\dot{\psi}$ is ruled out since $\dot{\psi} \in \Phi^{\mathrm{nd}}$. (The phenomenon of edges with multiplicity 4 never arises for Dynkin diagrams of usual root systems, since for those there is never linear dependence among distinct roots in a basis.)

By staring at possibilities for pairs of non-orthogonal roots in a usual root system (as summarized early in [Bour, Chap. 6, Sect. 1.3]), when $\|\dot{\eta}\| \neq \|\dot{\psi}\|$ we have

$$f(\psi, \eta) = \frac{\|\mathrm{long}\|^2}{\|\mathrm{short}\|^2}$$

(the ratio of squared lengths of the longer among $\dot{\psi}$ and $\dot{\eta}$ to the shorter among the two). We assign an edge between ψ and η the multiplicity $f(\psi, \eta)$, and when $f(\psi, \eta) > 1$ with $\|\dot{\eta}\| \neq \|\dot{\psi}\|$ we put an arrow from long to short (the arrow points to whichever of ψ or η has shorter derivative).

Remark 21.2 From the list of possibilities early in [Bour, Chap. 6, Sect. 1.3], we have $\|\dot{\eta}\| = \|\dot{\psi}\|$ with $f(\psi, \eta) > 1$ exactly when $\dot{\eta} = \pm\dot{\psi}$ (in which case $f(\psi, \eta) = 4$). But this says H_{η} and H_{ψ} are parallel, and a simplex (such as $\overline{C}$, by irreducibility of Ψ) has parallel distinct walls exactly in the 1-dimensional case. Hence, this can only possibly occur for Ψ_{A_1}, Ψ_{BC_1}, and the three non-reduced irreducible Ψ's with dimension 1. We'll see those for which it happens in our tabulation of Dynkin diagrams below.

Before tabulating cases, we make an observation about affine Dynkin diagrams in the irreducible case, analogous to a known feature of usual Dynkin diagrams. First we consider a class of examples:

Example 21.3 Let $\Psi = \Psi_\Phi$ for a reduced and irreducible root system Φ. Pick a basis $\Delta_0 = \{a_1, \dots, a_n\}$ of Φ, and let Φ^+ be the associated positive system of roots. One basis of Ψ is $\Delta = \{1 - a_0, a_1, \dots, a_n\}$ for a_0 the highest root in Φ^+. Then $\mathrm{Dyn}(\Psi)$ as a bare graph is obtained from $\mathrm{Dyn}(\Phi)$ (with vertex set Δ_0) by joining $1 - a_0$ to a_i for $1 \le i \le n$ exactly when $\langle a_0, a_i^\vee \rangle \neq 0$. Some such non-vanishing occurs since otherwise the nonzero a_0 would be orthogonal to $a_1, \dots, a_n$ that span the space, an absurdity.

Here is the general result subsuming the preceding example:

Lemma 21.4 *If Ψ is irreducible with $\nabla\Psi = \Phi$ then* $\mathrm{Dyn}(\Psi)$ *consists of* $\mathrm{Dyn}(\Phi)$ *joined along some weighted (possibly directed) edges to an additional vertex. In particular,* $\mathrm{Dyn}(\Psi)$ *inherits connectedness from* $\mathrm{Dyn}(\Phi)$.

Proof Since $W(\Psi)$ acts transitively on the set of bases, we can pick an extra special vertex x_0 such that Δ contains $\{\psi_1, \dots, \psi_\ell\}$ with all $\psi_i \in \Psi_{x_0}$ and with $\{\dot{\psi}_1, \dots, \dot{\psi}_\ell\}$ a basis of Φ. But $\#\Delta = 1 + \dim A = 1 + \dim \Phi$, so $\ell = \dim A$ and hence $\Delta = \{\psi_0, \psi_1, \dots, \psi_n\}$ with $\psi_0(x_0) > 0$ (as $\psi_i(x_0) = 0$ for all $i > 0$).

Thus, $\mathrm{Dyn}(\Psi) = \{\psi_0\} \coprod \mathrm{Dyn}(\Phi)$ in which $\mathrm{Dyn}(\Phi)$ is a subgraph with the same weighted edges and arrows, so it remains to show some edge joins ψ_0 to $\mathrm{Dyn}(\Phi)$. If there is no such edge then $\dot{\psi}_0$ is orthogonal to the vectors $\dot{\psi}_1, \dots, \dot{\psi}_\ell$ that span V^*, forcing $\dot{\psi}_0 = 0$, but ψ_0 is non-constant. □

In Fig. 7 we list affine diagrams for the reduced irreducible cases, as in [KP, Table 1.3.4]. We decorate diagram vertices ψ_i to record if the corresponding chamber vertex x_i (see Proposition 19.3) is extra special, not special, or special but not extra special. We use a filled-in circle for ψ_i when x_i is non-special, an empty circle when x_i is extra special, and a circle with $\times$ inside when x_i is special but not extra special. By Proposition 20.12(3) the latter can only happen if Φ is non-reduced, so for reduced Ψ (necessarily Ψ_Φ or its dual for an irreducible Φ) this can happen only for the self-dual Ψ_{BC_n}.

To efficiently determine what speciality each diagram vertex has, especially for high-rank exceptional types (such as E_6 and E_7) whose bases are "messy" to describe, it is useful to apply a combinatorial criterion for speciality in [KP, Proposition 1.3.47(3)]. Also, if Ψ is reduced then $\mathrm{Dyn}(\Psi^\vee)$ is obtained from $\mathrm{Dyn}(\Psi)$ by reversing arrows (and keeping the same decorations!), as we see from the definitions since $(\Psi^\vee)_x = (\Psi_x)^\vee$ and coroots of a basis for a *reduced* Ψ (resp. Φ) are a basis for $\Psi^\vee$ (resp. Φ), by considering walls of a chamber (remember that Ψ_{BC_n} is self-dual).

Remark 21.5 An observation that comes out of the exhaustive tabulation is that the *only* connected affine diagrams which contain a loop are Ψ_{A_n} for $n \ge 1$ (where "loop" means we can find vertices $a_1, \dots, a_m$ with $m \ge 2$ such that there is an edge from a_i to a_{i+1} for each i mod m).

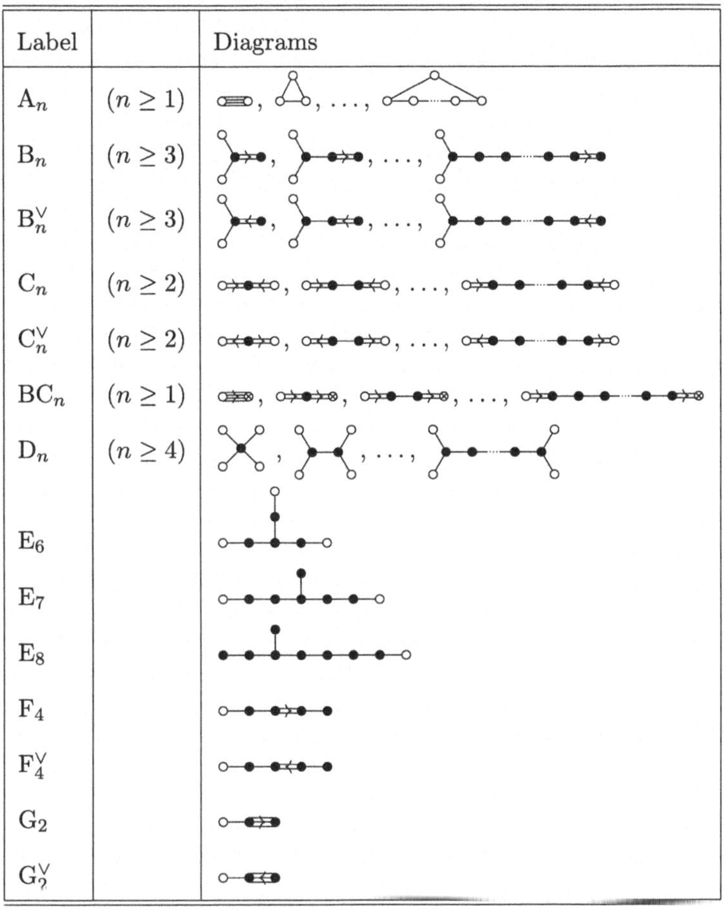

Fig. 7 The Dynkin diagrams of reduced irreducible affine root systems

This may seem like a niche fact of no utility, but it is a key ingredient in the proof of a striking result to come later via Bruhat-Tits theory (Theorem 30.13): if an absolutely simple and connected semisimple k-group G is anisotropic then it must be of type A and split over k^{un} (if the residue field is finite then G is even inner type A, which is to say G^{sc} is the algebraic group of units of reduced norm 1 in a finite-dimensional central division algebra over k: see Corollary 30.15).

To appreciate how striking this is, recall that over $\mathbf{R}$ there are anisotropic absolutely simple and connected semisimple groups of *every* Dynkin type ("compact forms"). Over non-Archimedean local fields k with finite residue field, we can see the non-existence of a k-anisotropic form for some Dynkin types via local class field theory. For example, consider G of type B_n, so $G^{\mathrm{ad}} \simeq \mathrm{SO}(q)$ for a non-degenerate quadratic space (V, q) with odd dimension $2n + 1$. Taking $n \geq 2$ (so B_n is not also type A), this quadratic space has dimension at least 5, so the quadratic form has a non-trivial zero by local class field theory and hence $G^{\mathrm{ad}} = \mathrm{SO}(q)$ is k-isotropic. Thus, G is k-isotropic too.

Remark 21.6 For reduced irreducible Φ, consider $\mathrm{Aut}(\mathrm{Dyn}(\Phi))$ as a subgroup of $\mathrm{Aut}(\mathrm{Dyn}(\Psi_\Phi))$ via automorphisms fixing the additional vertex (see Example 21.3). By [Bour, Chap. 6, Sect. 4.2, Proposition 1 and Sect. 4.3], the automorphism group of $\mathrm{Dyn}(\Psi_\Phi)$ is the semi-direct product of $\mathrm{Aut}(\mathrm{Dyn}(\Phi))$ acting on $P(\Phi^\vee)/Q(\Phi^\vee)$ (coweight lattice modulo coroot lattice).

This visualizes the center of split simply connected groups over general fields F as follows. Let G be a connected semisimple F-group that is simply connected and F-simple with a *split* maximal F-torus T, and suppose $\Phi(G, T) = \Phi$. The image T^{ad} of T in G^{ad} has character lattice $Q(\Phi)$ and $\mathrm{X}^*(T) = P(\Phi)$, so $\mathrm{X}^*(Z_G) = P(\Phi)/Q(\Phi)$ is naturally $\mathbf{Q}/\mathbf{Z}$-dual to $P(\Phi^\vee)/Q(\Phi^\vee)$ [Bour, Chap. 6, Sect. 1.9]. Thus, by looking at Fig. 7 and knowing where the additional vertex is located, we can read off Z_G from the automorphisms of the affine diagram; e.g., $Z_{\mathrm{E}_6^{\mathrm{sc}}}$ has order $|S_3|/|\mathbf{Z}/2\mathbf{Z}| = 3$.

In Fig. 8 we give a table of the affine diagrams for the non-reduced irreducible Ψ (which are determined up to isomorphism by the pair of isomorphism classes of reduced irreducible affine root systems Ψ^{nd} and Ψ^{nm}) as in [KP, Table 1.3.3]. The underlying diagram of Ψ as a weighted directed graph is that of the reduced irreducible Ψ^{nd}, and we have tabulated affine diagrams in the reduced irreducible cases, so rather than keep track of special and extra special properties we only keep track of multipliable vertices (by drawing an extra circle around such vertices).

Lemma 21.7 *For any irreducible Ψ with a basis $\Delta = \Delta_C$ and a facet $\mathcal{F}$ of $\overline{C}$, the usual root system $\Psi_{\mathcal{F}}$ has Dynkin diagram obtained from* $\mathrm{Dyn}(\Psi)$ *by removing the vertices of $\overline{\mathcal{F}}$ from $\Delta = \mathrm{vert}(\mathrm{Dyn}(\Psi))$ and removing the edges of* $\mathrm{Dyn}(\Psi)$ *touching such removed vertices.*

Proof This is seen by combining two facts: upon writing $\Delta = \{\psi_0, \ldots, \psi_n\}$, we can label $\mathrm{vert}(C)$ as $\{x_0, \ldots, x_n\}$ where $\psi_i(x_j) \in \mathbf{R}_{>0}\delta_{ij}$, and $\Psi_{\mathcal{F}}$ has $\Delta \cap \Psi_{\mathcal{F}}$ as a basis (Remark 20.7). □

Label		Diagrams
$(\mathrm{BC}_n, \mathrm{C}_n)$	$(n \geq 1)$	
$(\mathrm{C}_n^\vee, \mathrm{BC}_n)$	$(n \geq 1)$	
$(\mathrm{B}_n, \mathrm{B}_n^\vee)$	$(n \geq 2)$	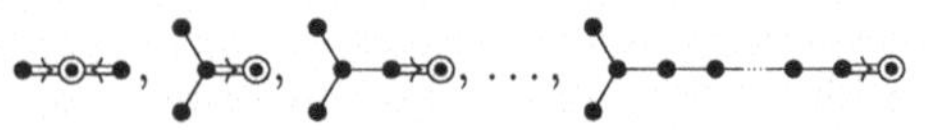
$(\mathrm{C}_n^\vee, \mathrm{C}_n)$	$(n \geq 1)$	

Fig. 8 The Dynkin diagrams of non-reduced irreducible affine root systems

Proposition 21.8 *The isomorphism class of an irreducible and reduced* Ψ *is determined by the weighted directed graph* $\mathrm{Dyn}(\Psi)$.

Since $\mathrm{Dyn}(\Psi) = \mathrm{Dyn}(\Psi^{\mathrm{nd}})$ for non-reduced irreducible Ψ (disregarding the decorations we add to the diagram record additional information), we cannot remove "reduced" from this result.

Proof The only rank-2 cases are Ψ_{A_1} and Ψ_{BC_1}, whose diagrams we compare directly. Hence, we can focus on rank ≥ 3, so these are Ψ_Φ or its dual for irreducible (possibly non-reduced) Φ with rank ≥ 2. To verify without case-checking of the tabulation that there is no repetition among the isomorphism class of the affine diagrams, one can harness the identification of $\mathrm{Dyn}(\Psi_x)$ as the subgraph of $\mathrm{Dyn}(\Psi)$ complementary to the vertex x for each $x \in \Delta$ (this is Lemma 21.7 with $\mathcal{F} = \{x\}$), and use that Ψ_x is determined up to isomorphism by $\mathrm{Dyn}(\Psi_x)$ (a fact in the theory of usual root systems). This is carried out in the proof of [KP, Proposition 1.3.67(2)]. □

22 Tits Systems (or *BN*-Pairs) I

The facets $\mathcal{F}$ of apartments of the building of a connected reductive group G over a Henselian discretely-valued field k with perfect residue field will correspond to bounded open subgroups $G(k)^0_{\mathcal{F}}$ called *parahoric* subgroups. To define such subgroups in general requires having the building (though for complete k with finite residue field there is a more direct characterization [KP, Proposition 13.5.2]), but some aspects of parahoric subgroups rest on the purely group-theoretic notion of a "Tits system" in $G(k)^0$. We now discuss this latter concept, and in Sect. 23 relate it to the notion of "abstract building".

Let G be an abstract group. A *Tits system* for G is a triple (B, N, S) consisting of subgroups $B, N \subset G$ and a subset $S \subset N/(B \cap N)$ such that:

(TS1) $T := B \cap N$ is normal in N, and $\langle B, N \rangle = G$;

(TS2) the subset $S \subset W := N/T$ consists of elements of order 2 that generate W (we call W the *Weyl group* of the Tits system);

(TS3) for all $s \in S$ and $w \in W$, $sBw \subset BwB \cup BswB$ (so in particular, taking $w = s$, we have $sBs \subset BsB \cup B$, so $BsB \cup B$ is a subgroup of G);

(TS4) for all $s \in S$ we have $sBs \neq B$ (equivalently $sBs \not\subset B$, since if $sBs \subset B$ then $B \subset s^{-1}Bs^{-1} = sBs$ because $s^2 \in B \cap N \subset B$).

The normality of T in N implies $T \subset \bigcap_{w \in W} wBw^{-1}$ for any Tits system, and when this containment is an equality (i.e., T is "as big as possible") then we say the Tits system is *saturated*.

If S is empty then $W = 1$ and so $N \subset B$, hence $G = \langle B, N \rangle = B$; conversely if $B = G$ then clearly $W = 1$ and so S is empty. In such cases, when moreover the Tits system is saturated then $B = T$ (equivalently, $B \subset N$), so in saturated cases S is empty if and only if $N = B = G$. Of course, it is primarily the case of non-empty S that is of interest.

Expanding on the parenthetical observation we made for (TS3), by [Bour, Chap. 4, §2.5, Cor.] the subset $S \subset W$ is uniquely determined as the set of those $w \in W$ for which $BwB \cup B$ is a subgroup of G. For this reason, the data of the Tits system is uniquely determined by the pair (B, N) and hence Tits systems are often called *BN-pairs*.

We are going to be interested in Tits systems for which the associated Weyl group W is infinite (in fact, it will be the Weyl group of an affine root system), but let's first record the classical source of examples with finite W (giving rise to "spherical buildings"):

Example 22.1 Let $G = \mathbf{G}(F)$ for an arbirary field F and connected reductive F-group $\mathbf{G}$. Let $B = \mathbf{P}(F)$ for a minimal parabolic F-subgroup $\mathbf{P}$ of $\mathbf{G}$, and $N = N_{\mathbf{G}}(\mathbf{S})(F)$ for a maximal split F-torus $\mathbf{S} \subset \mathbf{P}$. The structure theory of connected reductive groups over fields gives $T := B \cap N = Z_{\mathbf{G}}(\mathbf{S})(F)$, so $W = W(\mathbf{G}, \mathbf{S}) \simeq W(\Phi(\mathbf{G}, \mathbf{S}))$.

For the basis Δ of $\Phi(\mathbf{G}, \mathbf{S})$ corresponding to $\mathbf{P}$ under the bijection between the set of minimal parabolic F-subgroups $\mathbf{P}' \supset \mathbf{S}$ and the set of positive systems of roots in $\Phi = \Phi(\mathbf{G}, \mathbf{S})$ via

$$\mathbf{P}' \mapsto \{a \in \Phi : a \text{ is an } \mathbf{S}\text{-weight on } \mathrm{Lie}(\mathscr{R}_{u,F}(\mathbf{P}'))\},$$

if we let $S = \{r_a\}_{a \in \Delta} \subset W(\Phi) = W$ then (G, B, N, S) is a Tits system by [C3, Theorem V.2.7]. In such cases, S is empty precisely when $\mathscr{D}(\mathbf{G})$ is F-anisotropic.

By [C3, Remark V.2.8], such Tits systems not only have finite W but are saturated and clearly satisfy $B = T \ltimes U$ with $U = \mathbf{U}(F)$ for $\mathbf{U}$ the F-split smooth connected unipotent $\mathscr{R}_{u,F}(\mathbf{P})$. This U is a nilpotent group (since $\mathbf{U}$ is nilpotent as an algebraic group), and one says a BN-pair is *weakly split* when such a semi-direct product structure exists for B using a nilpotent subgroup $U \subset B$.

Remark 22.2 In the setting of Example 22.1, for $s = r_a$ the root group $\mathbf{U}_a$ has trivial schematic intersection with $s\mathbf{P}s^{-1}$ (since conjugation by r_a on $\mathrm{X}^*(\mathbf{S})_{\mathbf{Q}}$ negates a and carries $\Phi^+ - \mathbf{Q}a$ into itself), so $\mathbf{U}_a(F) \subset B/(B \cap s^{-1}Bs)$. It follows that when F is *infinite*, so $\mathbf{U}_a(F)$ is infinite for all $a \in \Phi$, then $B/(B \cap s^{-1}Bs)$ is infinite for all $s \in S$.

Remarkably there is a converse when F is infinite and $\mathscr{D}\mathbf{G}$ is F-simple: in such situations, any saturated and weakly split BN-pair (B, N) for $\mathbf{G}(F)$ such that its Weyl group is finite and the quotient $B/(B \cap sBs^{-1})$ infinite for all $s \in S$ *must* arise from a (uniquely determined) $(\mathbf{P}, \mathbf{S})$ by [Pra, Theorem B]. (See [Pra, Sect. 1, Remark 3] for counterexamples when F is finite.)

Definition 22.3 For a Tits system (G, B, N, S), a *parabolic subgroup* $P \subset G$ is a subgroup containing a conjugate of B.

Example 22.4 For a subset $X \subset S$, and $W_X := \langle X \rangle \subset W$, the subset $P_X := BW_XB \subset G$ is a subgroup by [Bour, Chap. 4, Sect. 2.5, Theorem 3(a)]. These subgroups clearly contain B. The subgroups containing B are called *standard* (parabolic). By [Bour, Chap. 4, Sect. 2.5, Theorem 3(b, d)], every standard parabolic subgroup has the form P_X for a unique $X \subset S$ (with $X = \emptyset$ for $P = B$).

We emphasize how remarkable it is that the *only* subgroups of G containing $B = P_\emptyset$ are the subgroups P_X for subsets $X \subset S$. (Even for $G = \mathrm{GL}_n(F)$ for a field F and B its subgroup of upper-triangular matrices, it is not obvious.) This is crucial for encoding group-theoretic information in terms of a geometric object (a "polysimplicial" building).

Theorem 22.5 *The following hold for any Tits system* (G, B, N, S).

(1) (Bruhat decomposition) $G = \coprod_{w \in W} BwB$.
(2) Each parabolic subgroup P *is conjugate to* P_X *for a unique* $X \subset S$ *(called the "type" of* P*).*
(3) Each P *is its own normalizer in* G.

Proof For (1), see [Bour, Chap. 4, Sect. 2.3, Theorem 1]. For (2), see [Bour, Chap. 4, Sect. 2.6, Proposition 4]. For (3), see [Bour, Chap. 4, Sect. 2.6, Theorem 4(iv)]. □

Tits was motivated to discover buildings in the search for a unified approach to geometrically interpreting groups through an action on a naturally associated non-positively curved space, much as some aspects of non-compact Lie groups G are understood through their action on the "symmetric space" G/K of non-compact type (for a maximal compact subgroup K in G).

For a connected reductive group $\mathbf{G}$ over a Henselian field k such that $\mathscr{D}\mathbf{G}$ has positive k-rank r, the relevant BN-pair in Bruhat-Tits theory for $\mathbf{G}$ will be $G = \mathbf{G}(k)^0$, $B = \mathbf{G}(k)^0_{\mathcal{C}}$ for a chamber $\mathcal{C}$ in the r-dimensional apartment $\mathcal{A}(\mathbf{S}) \subset \mathcal{B}(\mathbf{G})$ associated to a maximal split k-torus $\mathbf{S} \subset \mathbf{G}$, and $N = N_{\mathbf{G}}(\mathbf{S})(k) \cap G$. The group W will be identified with $W(\Psi)$ for an affine root system $\Psi(\mathbf{G}, \mathbf{S})$ (to be built!) on the affine space $\mathcal{A}(\mathbf{S})$ giving rise to a "polysimplicial structure" encoded in the data of the building, and $S \subset W = W(\Psi)$ will correspond to the set of reflections in walls of the chamber $\mathcal{C}$.

The *bounded* parabolic subgroups associated to this Tits system will be those of the form $P = G(k)^0_{\mathcal{F}}$ for facets $\mathcal{F}$ of the "polysimplicial" $\mathcal{B}(\mathbf{G})$, and the "type" of P will be encoded by using the $G(k)^0$-action on the building to move $\mathcal{F}$ over to a uniquely determined facet of the choice of $\mathcal{C}$. Boundedness for a parabolic subgroup associated to this Tits system will have a characterization in purely group-theoretic terms as we shall see (but bear in mind that the ambient group for the Tits system is $G(k)^0$, which is generally not $G(k)$).

23 Tits Systems (or BN-Pairs) II

For a Tits system (G, B, N, S) and parabolic subgroup $P \subset G$, we defined its *type* $X \subset S$ where P is conjugate to a unique parabolic subgroup $P_0 \supset B$ that in turn is necessarily equal to $P_X := BW_XB$ for a unique $X \subset S$ ("standard" parabolics). By [Bour, Chap. 4, Sect. 2.5, Theorem 3], $P_X \subset P_Y$ if and only if $X \subset Y$, so $P_X \cap P_{X'} = P_{X\cap X'}$.

In particular, the maximal *standard* proper parabolic subgroups of G are $P_{S-\{s\}} =: G_s$ for $s \in S$, so each standard proper parabolic subgroup P_X is contained in a maximal one and is equal to $\bigcap_{s\notin X} G_s$ where the G_s's are *exactly* the maximal proper parabolic subgroups containing P_X (if we write $X = S - S'$ with non-empty S' then $X = \bigcap_{s\in S'}(S - \{s\})$).

It follows that if S is finite and non-empty then each proper parabolic subgroup P is contained in only *finitely many* maximal proper parabolic subgroups, say $P_1, \ldots, P_m$, and $P = \bigcap_i P_i$ with the type(P_i)'s exactly the sets $S - \{s\}$ for those s *not* in type(P). In particular, the P_i's are distinguished from each other by their types inside S.

In Sect. 24 we will see that the set of *proper* parabolic subgroups P of G is in an inclusion-reversing bijection with the set of facets $\mathcal{F}$ of an "abstract building" for (G, B, N, S). In geometric terms, the parabolic $P_{\mathcal{F}}$ corresponding to a facet $\mathcal{F}$ is the pointwise-stabilizer of $\mathcal{F}$ for the G-action on that "abstract building" (this is the visual reason why passing to a subfacet enlarges the corresponding parabolic subgroup). In particular, the set vert$(\mathcal{F})$ of vertices of such a facet will be in bijection with the set of maximal proper parabolic subgroups $Q \supset P_{\mathcal{F}}$.

We emphasize again that for connected reductive $\mathbf{G}$ over a Henselian discretely-valued field k (with perfect residue field), the associated Tits system to be made will have ambient group G equal to $\mathbf{G}(k)^0$, which is generally *not* $\mathbf{G}(k)$ (though $\mathbf{G}(k)$ will act isometrically on an associated Bruhat-Tits building, a richer structure than a $\mathbf{G}(k)^0$-action that is essentially group-theoretic in nature). For instance, in Example 10.12 we saw that for $\mathbf{G} = \mathrm{PGL}_2$, the open normal subgroup $\mathrm{PGL}_2(k)^0 \subset \mathrm{PGL}_2(k)$ has index 2: the elements of $\mathrm{PGL}_2(k)$ whose "determinant" in $k^\times/(k^\times)^2$ is in the image of $\mathfrak{o}^\times/(\mathfrak{o}^\times)^2$.

Via the invariance of Bruhat-Tits buildings (which will be instances of "abstract buildings" associated to Tits systems) under central quotients for the connected reductive group, there will be an equivariant (and isometric) identification

$$\begin{array}{ccccccc} & & \mathcal{B}(\mathrm{PGL}_2) & = & \mathcal{B}(\mathrm{SL}_2) & & \\ & & \circlearrowleft & & \circlearrowleft & & \\ \mathrm{PGL}_2(k)^0 & \underset{2}{\hookrightarrow} & \mathrm{PGL}_2(k) & \longleftarrow & \mathrm{SL}_2(k) & = & \mathrm{SL}_2(k)^0. \end{array}$$

Under the relation of Tits systems (G, B, N, S) with abstract buildings to be discussed in Sect. 24, the pointwise G-stabilizer for a facet in an associated "abstract building" will always coincide with the G-stabilizer for the facet (i.e., if some $g \in G$ carries a facet onto itself then it does so by *fixing all vertices too*): see

Theorem 24.11(2). Thus, if the action of some $g \in \mathbf{G}(k)^0 \subset \mathbf{G}(k)$ on $\mathcal{B}(\mathbf{G}) = \mathcal{B}(\mathbf{G}^{\mathrm{ad}})$ preserves a facet then it fixes all vertices of that facet and is even *trivial* on the facet. (This applies to all $g \in \mathbf{G}(k)$ when $\mathbf{G}$ is simply connected since then $\mathbf{G}(k)^0 = \mathbf{G}(k)$ by Definition 10.7(1).) However, that property is generally *false* for the action of $g \in \mathbf{G}(k) - \mathbf{G}(k)^0$ on $\mathcal{B}(\mathbf{G})$. Here is a counterexample:

Example 23.1 Taking $\mathbf{G} = \mathrm{PGL}_2$, by Example 2.4 some $g \in \mathrm{PGL}_2(k) - \mathrm{PGL}_2(k)^0$ acts on the tree $\mathcal{B}(\mathrm{PGL}_2) = \mathcal{B}(\mathrm{SL}_2)$ preserving an edge $\mathcal{C}$ but swapping its two endpoints! This reinforces awareness that beyond the simply connected case the $\mathbf{G}(k)$-action on $\mathcal{B}(\mathbf{G})$ is richer structure than what is provided by Tits systems alone (though even just *defining* the Tits system structure on $\mathbf{G}(k)^0$ is highly non-trivial in general, particularly beyond the quasi-split case).

For a *central* quotient map $\mathbf{G} \to \mathbf{G}'$ of connected reductive groups, the reader can check from the definitions that $\mathbf{G}(k)^0$ is carried into $\mathbf{G}'(k)^0$. (The general functoriality of $\mathbf{G}(k)^0$ in $\mathbf{G}$ will not be known beyond the case $k = K$ from Proposition 10.13 until Theorem 30.7; the case of central quotient maps is rather special.) We'll have a natural equivariant isometry

$$\begin{array}{ccc} \mathcal{B}(\mathbf{G}) & \xrightarrow{\sim} & \mathcal{B}(\mathbf{G}') \\ \circlearrowleft & & \circlearrowleft \\ \mathbf{G}(k) & \longrightarrow & \mathbf{G}'(k) \end{array}$$

but the map of groups $\mathbf{G}(k)^0 \to \mathbf{G}'(k)^0$ corresponding to the Tits systems is usually *not* surjective!

Motivated by this situation, it is natural to seek a relationship between the Tits system structures for groups related by a "not quite surjective" homomorphism (such as $\mathbf{G}(k)^0 \to \mathbf{G}'(k)^0$ above). This is provided by the following result, for which some hard work [KP, Lemma 7.5.2] verifies its hypotheses in the context of $\mathbf{G}(k)^0 \to \mathbf{G}'(k)^0$ when $k = K$:

Proposition 23.2 *Let (G, B, N, S), (G', B', N', S') be two Tits systems, and suppose $f : G \to G'$ satisfies $f(B) \subset B'$, $f(N) \subset N'$, and*

(1) $\ker(f) \subset T := B \cap N$;
(2) $f(G) \lhd G'$, *$G'/f(G)$ is abelian;*
(3) *$T' := B' \cap N'$ normalizes $f(B)$, and $B' = f(B)T'$.*

Then

- $f : W \xrightarrow{\sim} W'$ *and* $S \xrightarrow{\sim} S'$;
- *there is a bijection* $\mathrm{Par}(G) \xrightarrow{\sim} \mathrm{Par}(G')$ *between sets of parabolic subgroups via $f^{-1}(P') \leftarrow\!\shortmid P'$, $P \mapsto N_{G'}(f(P))$, and this is both type-preserving and inclusion-preserving in both directions.*

Proof See [KP, Lemmas 1.4.13, 1.4.14]. □

For a connected reductive group $\mathbf{G}$ over k for which $\mathscr{D}\mathbf{G}$ is k-isotropic (equivalent to $\mathcal{B}(G)$ not being a point, as we'll see in Proposition 26.1), there will be an equivariant isometry of buildings

$$\mathcal{B}(\mathbf{G}) = \mathcal{B}(\mathbf{G}^{\mathrm{ad}}) = \prod_j \mathcal{B}(\mathbf{G}_j^{\mathrm{ad}})$$

where $\mathbf{G}^{\mathrm{ad}} = \prod_i \mathbf{G}_i^{\mathrm{ad}}$ for the k-simple "factors" $\mathbf{G}_i$ of $\mathscr{D}\mathbf{G}$ (so some of the $\mathbf{G}_i$'s are k-isotropic) and $\{\mathbf{G}_j^{\mathrm{ad}}\}$ is the set of k-*isotropic* simple factors of $\mathbf{G}^{\mathrm{ad}}$.

The chambers in $\mathcal{B}(\mathbf{G})$ will be exactly the subsets $\mathcal{C} = \prod \mathcal{C}_j$ for chambers $\mathcal{C}_j \subset \mathcal{B}(\mathbf{G}_j^{\mathrm{ad}})$, and the latter are all *simplices* (so $\mathcal{C}$ is a "polysimplex": product of simplices). Upon fixing such a chamber there will be an associated Tits system structure on $G = \mathbf{G}(k)^0$ via $B = \mathbf{G}(k)^0_{\mathcal{C}}$ and $N = N_{\mathbf{G}}(\mathbf{S})(k) \cap G$ (where $\mathbf{S}$ is a maximal split k-torus whose associated apartment $\mathcal{A}(\mathbf{S})$ contains $\mathcal{C}$), with the set S of involutions for the Tits system in natural bijection with the set

$$\mathrm{wall}(\mathcal{C}) = \coprod \mathrm{wall}(\mathcal{C}_j). \tag{23.1}$$

In addition to the decomposition $S = \coprod S_j$ encoding (23.1), the Tits system will have Weyl group $W = W(\Psi)$ for an affine root system $(\mathcal{A}(\mathbf{S}), \Psi)$ whose irreducible factors are the $(\mathcal{A}(\mathbf{S}_j), \Psi_j)$'s for the maximal split tori $\mathbf{S}_j \subset \mathbf{G}_j^{\mathrm{ad}}$ that are the images of $\mathbf{S}$. The resulting product decomposition $W(\Psi) = \prod W(\Psi_j)$ induces $S = \coprod S_j$ at the level of reflections, encoding bases for the affine root systems Ψ and $\{\Psi_j\}$ (a basis for an affine root system is labeled by the set of walls of a chamber).

Now we state a purely group-theoretic mechanism to construct the decomposition $S = \coprod S_j$ without reference to buildings. For a Tits system (G, B, N, S) with S finite and non-empty, by [Bour, Chap. 4, Sect. 1] the pair (W, S) is a "Coxeter system" (a group generated by a set S of involutions satisfying certain axioms that we don't state here), and there is a unique decomposition $S = \coprod_j S_j$ for non-empty *pairwise commuting* subsets $S_j \subset S$ that are themselves *irreducible*: $S_j \neq S'_j \coprod S''_j$ for non-empty commuting subsets $S'_j, S''_j \subset S_j$. Concretely, the sets S_j are in bijection with the sets of vertices of connected components of the Coxeter graph of (W, S) (an analogue of Dynkin diagrams, used for classifying Coxeter systems).

For what we need, each S_j will be the set of vertices of a connected *affine* Dynkin diagram, so $\#S_j \geq 2$ (this is important: remember that the size of a basis of an irreducible affine root system is 1 more than the dimension of the affine space, and the affine space for an affine root system always has dimension at least 1). Relative to this *canonical* decomposition $S = \coprod S_j$, a subset $X \subset S$ is called *admissible* if $X \cap S_j \subsetneq S_j$ for all j. Correspondingly, a parabolic subgroup $P \subset G$ is called *admissible* when its type is admissible. The Bruhat-Tits building $\mathcal{B}(\mathbf{G})$ will have facets corresponding to *admissible* (proper) parabolic subgroups of $\mathbf{G}(k)^0$.

The significance of admissibility is that it is a "proper parabolic" property relative to all k-isotropic k-simple "factors" of $\mathbf{G}$, as we now sketch informally. Even though $\mathbf{G}(k)^0$ doesn't generally decompose as a direct product corresponding to the k-simple factors of $\mathbf{G}^{\mathrm{ad}}$, much as $\mathbf{G}(k)$ doesn't, by the final part of Proposition 23.2 applied to $\mathbf{G}(k)^0 \to \mathbf{G}^{\mathrm{ad}}(k)^0$ the *set* of parabolic subgroups P of $\mathbf{G}(k)^0$ will decompose as a product of *sets* of parabolic subgroups $P_i \subset \mathbf{G}_i^{\mathrm{ad}}(k)^0$ over all i. When $P \mapsto \{P_i\}$ under this dictionary, the subset $X = \mathrm{type}(P) \subset S$ will satisfy $X \cap S_i = \mathrm{type}(P_i)$ for all i, so admissibility for P corresponds to the subgroup $P_i \subset \mathbf{G}_i^{\mathrm{ad}}(k)^0$ being proper for every k-isotropic $\mathbf{G}_i^{\mathrm{ad}}$.

Remark 23.3 It is natural to wonder why we make a fuss about different simple factors of $\mathbf{G}$ rather than only considering k-simple $\mathbf{G}$ from the outset when developing the general theory of buildings. The reason is that ultimately $\mathcal{B}(\mathbf{G})$ will be *constructed* as $\mathcal{B}(\mathbf{G}_K)^\Gamma$ for a natural isometric action by $\Gamma = \mathrm{Gal}(K/k)$ on $\mathcal{B}(\mathbf{G}_K)$, so we need a general theory of buildings that applies to the K-group $\mathbf{G}_K$ which is generally not K-simple when $\mathbf{G}$ is k-simple (e.g., $\mathbf{G} = \mathrm{R}_{k'/k}(\mathbf{G}')$ for a non-trivial finite unramified extension k'/k and a connected semisimple k'-group $\mathbf{G}'$ that is absolutely simple).

24 Abstract Buildings

Now we use Tits systems to create combinatorial "geometric" structures called (abstract) *buildings* (in which a non-empty finite set of "vertices" is a proxy for a simplex). In Sect. 25 we will proceed in reverse to show that an abstract building equipped with a sufficiently rich action by a group G arises from a Tits system structure on G via the construction discussed below. These two inverse prodedures—going from Tits systems to abstract buildings, and from abstract buildings with a suitable group action to a Tits system—both play crucial roles in the construction of Bruhat-Tits theory.

The discussion in this section is used over K to make a building from a Tits system structure for $\mathbf{G}(K)^0$ that can be constructed in the quasi-split case (as always holds over K, by Steinberg's Theorem 13.2). That construction over K ultimately rests on many commutator calculations and an extensive study of subgroups in $\mathrm{SL}_{2,K}$ and the *quasi-split* $\mathrm{SU}_3(K'/K)$ for a quadratic Galois extension K'/K (see [C3, Exs. Q.4.2, Q.4.4] for some basic facts concerning quasi-split SU_3 and its root system). The inverse procedure in the other direction (to be given in Sect. 25) is essential for the "unramified descent" that creates a Tits system structure for $\mathbf{G}(k)^0$ via a building made for $\mathbf{G}$ from a building for $\mathbf{G}_K$.

Definition 24.1 An (abstract) *simplicial complex* is a pair $(V, \mathcal{B})$ where V is a non-empty set (of "vertices") and $\mathcal{B}$ is a set of *non-empty* finite subsets $\mathcal{F} = \{x_1, \ldots, x_n\} \subset V$ (called *facets*) such that:

(i) $\{x\} \in \mathcal{B}$ for all $x \in V$,
(ii) for all $\mathcal{F} \in \mathcal{B}$ and a non-empty subset $\mathcal{F}' \subset \mathcal{F}$, we have $\mathcal{F}' \in \mathcal{B}$.

The motivation for Definition 24.1(ii) comes from the fact that in convex linear geometry each facet of a simplex is the open convex hull of its vertices, and any non-empty set of vertices of such a facet corresponds to a facet of the same simplex (via

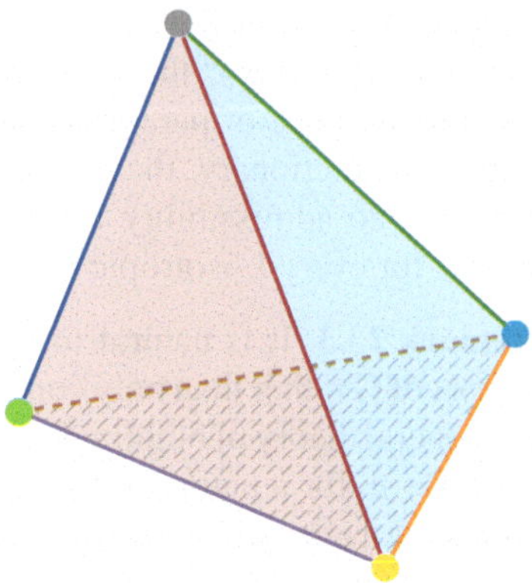

Fig. 9 A 3-dimensonal simplex and its facets (in different colors)

open convex hull). For instance, in Fig. 9 the "Euclidean realization" of facets of a 3-dimensional simplex are shown with different colors: the interior of a tetrahedron, the interior of each of the 4 triangular faces (avoiding the edges), the interior of each of the 6 edges (avoiding the endpoints), and the 4 vertices.

We define the *dimension* $\dim \mathcal{F}$ of a facet $\mathcal{F}$ to be $\#\mathcal{F} - 1$ (e.g., a vertex has dimension $1 - 1 = 0$), so the abstract facets corresponding to the colored regions in Figure 9 are assigned the "correct" dimension. There is also an evident notion of *isomorphism* between abstract simplicial complexes.

An abstract facet is the data of vertices for what determines a simplex in a Euclidean space, and for a simplicial complex we are retaining the information via vertices of how simplices are "glued" along facets (e.g., if $\mathcal{F}$ and $\mathcal{F}'$ are two simplices that have non-empty intersection, then $\mathcal{F} \cap \mathcal{F}' \in \mathcal{B}$ is the set of vertices for a facet along which we imagine gluing $\mathcal{F}$ and $\mathcal{F}'$). Note that V is the set of minimal facets; we often write $\mathcal{B}$ rather than $(V, \mathcal{B})$ to refer to the simplicial complex.

Recall for an affine root system (A, Ψ), chambers being simplices is equivalent to Ψ being irreducible, and for general Ψ with irreducible decomposition $\coprod \Psi_i$ the chambers are products of the simplex chambers of the Ψ_i's. This motivates us to introduce a broader notion than abstract simplicial complex in which we allow for direct products of simplices (again recorded via vertices, now as tuples of one vertex per factor):

Definition 24.2 An (abstract) *polysimplicial complex* is a tuple $\mathcal{B} = \{\mathcal{B}_i\}_{i \in I}$ for a non-empty finite index set I with the $\mathcal{B}_i$'s abstract simplicial complexes. The set V of *vertices* of $\mathcal{B}$ is defined to be $\prod V_i$ for $V_i = \text{vert}(\mathcal{B}_i)$, and the *facets* of $\mathcal{B}$ are the non-empty finite subsets of V having the form $\mathcal{F} = \prod \mathcal{F}_i$ for $\mathcal{F}_i \in \mathcal{B}_i$ (note that such an $\mathcal{F}$ is also an element of $\prod \mathcal{B}_i$).

The *dimension* $\dim \mathcal{F}$ of such an $\mathcal{F}$ is $\sum(\#\mathcal{F}_i - 1)$. A facet that is maximal with respect to inclusion (as subsets of V) is called a *chamber*.

It is easy to check that a chamber of $\mathcal{B}$ is exactly $C = \prod C_i$ for chambers $C_i \in \mathcal{B}_i$. An *isomorphism* $\varphi : \mathcal{B} \simeq \mathcal{B}'$ of polysimplicial complexes is (by definition) a collection of isomorphisms $\varphi_i : \mathcal{B}_i \simeq \mathcal{B}'_{\sigma(i)}$ for a bijection $\sigma : I \simeq I'$ of the index sets for the "simplicial factors". The next notion introduces a class of polysimplicial complexes relevant to us, and for these all chambers have the same dimension.

Definition 24.3 A polysimplicial complex $\mathcal{B}$ is called a *chamber complex* if

(1) each facet $\mathcal{F}$ is contained in a chamber, and
(2) for any two chambers $\mathcal{C}$ and $\mathcal{C}'$, there exists a sequence of chambers $\mathcal{C} = \mathcal{C}_0, \mathcal{C}_1, \ldots, \mathcal{C}_N = \mathcal{C}'$ such that for all $0 \le i < N$, $\mathcal{C}_i \cap \mathcal{C}_{i+1}$ is non-empty (hence a facet in $\mathcal{B}$) with codimension 1 in both $\mathcal{C}_i$ and $\mathcal{C}_{i+1}$ (sometimes called a common "panel" or "wall"). This datum is called a *gallery* joining $\mathcal{C}$ to $\mathcal{C}'$. In particular, all chambers have the *same dimension* (denoted $\dim \mathcal{B}$).

Remark 24.4 A polysimplicial complex $\mathcal{B}$ is a chamber complex if and only if all $\mathcal{B}_i$ are so; this is easy to check.

We will soon see that a large supply of chamber complexes (of a rather special type) arises from Tits systems. But first we give a basic and admittedly uninteresting example that is however worthwhile to include in the general theory since a k-isotropic connected semisimple group $\mathbf{G}$ over a Henselian discretely-valued field k can have a k-anisotropic k-simple "factor" (giving rise to a 1-point direct factor of the associated Bruhat-Tits building $\mathcal{B}(\mathbf{G})$):

Example 24.5 A 0-dimensional chamber complex is a point. Indeed, in the 0-dimensional case points are chambers and if C and C' are *distinct* chambers then $C \cap C'$ has to be empty. Thus, no gallery can join such C to C' since we require $\emptyset \notin \mathcal{B}_i$ for all i, so there is only one point.

Definition 24.6 A chamber complex $\mathcal{B} = \{\mathcal{B}_i\}_{i \in I}$ is *thick* if each codimension-1 facet $\mathcal{F}$ of $\mathcal{B}$ is contained in at least 3 chambers (equivalently, all $\mathcal{B}_i$'s are thick). A chamber complex is *thin* if every codimension-1 facet $\mathcal{F}$ of $\mathcal{B}$ is contained in exactly two chambers (equivalently, all $\mathcal{B}_i$'s are thin).

Remark 24.7 A 0-dimensional chamber complex is a point and so has *no* codimension-1 facets. Consequently, it is both thick and thin. We mention this only so that indeed 1-point chamber complexes will be (abstract) buildings in the sense to be defined shortly.

We need one more notion before we can define what an abstract building is. By a *subcomplex* of a chamber complex $\mathcal{B} = \{\mathcal{B}_i\}_{i \in I}$, we mean a polysimplicial complex $\mathcal{B}' = \{\mathcal{B}'_i\}_{i \in I}$ with $\mathcal{B}'_i \subset \mathcal{B}_i$ for all i such that if $\mathcal{F}'$ is a facet of $\mathcal{B}'$ and $\mathcal{F}'_0$ is a facet of $\mathcal{B}$ with $\mathcal{F}'_0 \subset \mathcal{F}'$ then $\mathcal{F}'_0$ is a facet of $\mathcal{B}'$.

Definition 24.8 An *(abstract) building* is a thick chamber complex $\mathcal{B}$ equipped with a collection of thin chamber subcomplexes $\mathcal{A} \subset \mathcal{B}$ (called *apartments*), such that

(BLI) any two chambers $\mathcal{C}, \mathcal{C}'$ of $\mathcal{B}$ lie in a common apartment,

(BLII) for any two facets $\mathcal{F}_1, \mathcal{F}_2$ and apartments $\mathcal{A}, \mathcal{A}' \ni \mathcal{F}_1, \mathcal{F}_2$ there exists an isomorphism $\mathcal{A} \simeq \mathcal{A}'$ of polysimplicial complexes that is the *identity* on $\mathcal{F}_1$ and on $\mathcal{F}_2$.

Example 24.9 We now illustrate with a picture the requirements (BLI) and (BLII) for a building that is tree with valence 3 at every vertex. We draw a "geometric realization", but logically it is the vertices and pairs of vertices from endpoints of edges that are the data for the abstract simplicial complex.

In Fig. 10 there are two chambers $\mathcal{F}$ and $\mathcal{F}'$ as dark black edges, and through them we show two (among the very many!) apartments in red and blue. (The unique collection of edges joining these edges is the unique purple gallery linking them together; the black and purples edges are in both apartments.) Such apartments illustrate (BLI).

Regarding both the red and blue apartments (each of which contains the black and purple edges too) as unions of unit intervals attached end-to-end, there is an evident simplicial isomorphism between the red and blue apartments that is the identity on each of the two darkened edges; this is (BLII).

Remark 24.10 In higher dimension than in the preceding example, there is no uniqueness for a gallery linking two chambers, not even for a gallery of minimal length. Indeed, by [KP, Cor. 1.5.27] a gallery of minimal length linking two chambers C and C' is contained in every apartment that contains both C and C', so by (BLII) in Definition 24.8 it suffices to analyze a single apartment. The tiling of $\mathbf{R}^2$ with equilateral triangles as in Fig. 3 is a simplicial apartment for $\mathcal{B}(\mathrm{SL}_3)$, so a pair of opposite triangles in a hexagon does the job.

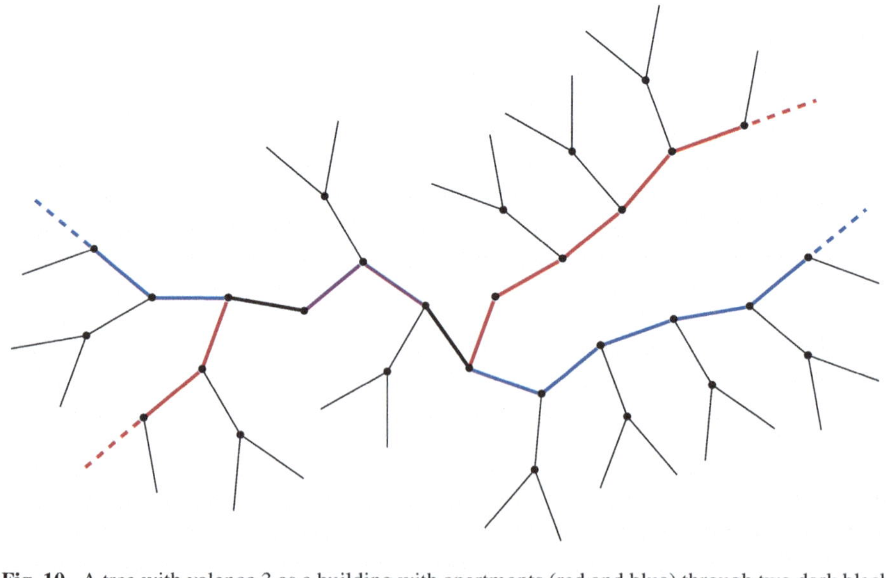

Fig. 10 A tree with valence 3 as a building with apartments (red and blue) through two dark black edges (as chambers) joined by a purple gallery

Another example is a pair of opposite triangles around an extra special vertex in the simplicial apartment for $\mathcal{B}(\mathrm{Sp}_4)$, shown in Fig. 2 (with x_0 and x_1 each extra special).

Now we have the language to express how the set of proper parabolic subgroups of a Tits system organizes itself into the set of facets of an abstract building:

Theorem 24.11 *Let (G, B, N, S) be a Tits system, with S finite and $\#S \geq 2$ (in particular, $S \neq \emptyset$ and hence $B \neq G$, so there* **are** *proper parabolic subgroups of G).*

Let V be the set of maximal **proper** *parabolic subgroups $P \subset G$, and let $\mathcal{B}$ be the set of non-empty subsets $\mathcal{F} = \{P_1, \dots, P_n\} \subset V$ for which $P_{\mathcal{F}} := \bigcap P_i$ is parabolic. Let G act on $\mathcal{B}$ by $g.\mathcal{F} = \{gP_1g^{-1}, \dots, gP_ng^{-1}\}$ (this is induced by the action on V via $P \mapsto gPg^{-1}$). Moreover, define*

- $\mathcal{C}_0 = \{Q \in V : Q \supset B\}$ *(so $P_{\mathcal{C}_0} = B$).*
- $\mathcal{A}_0 = \{n.\mathcal{C}_0 : n \in N\} \subset \mathcal{B}$.

The following hold:

(1) The pair $(V, \mathcal{B})$ is a simplicial complex with the G-action on it via (poly)simplicial automorphisms, and it is an abstract building with apartments $\{g.\mathcal{A}_0 : g \in G\}$. Its chambers are $\{g.\mathcal{C}_0 : g \in G\}$ (so there is a natural bijection between the set of minimal parabolic subgroups and the set of chambers), and Axiom (BLII) holds using an isomorphism of apartments arising from the action of some $g \in G$.

(2) The subgroup $P_{\mathcal{F}} \subset G$ is equal to the stabilizer $\{g \in G : g.\mathcal{F} = \mathcal{F}\}$ of $\mathcal{F}$ inside $\mathcal{B}$. The effect of $P_{\mathcal{F}}$ on the vertices of $\mathcal{F}$ is **trivial**, *and more generally $\mathcal{B}^{P_{\mathcal{F}}}$ is the set of facets of $\mathcal{F}$ (in particular, $V^{P_{\mathcal{F}}} = \mathcal{F}$ as subsets of V).*

(3) The group G acts transitively on the set of pairs $\{(\mathcal{A}, \mathcal{C}) : \mathcal{C} \in \mathcal{A}\}$, and $P_{\mathcal{F}}$ acts transitively on the set $\{\mathcal{A} \ni \mathcal{F}\}$ of apartments containing $\mathcal{F}$

(4) Each facet $\mathcal{F} \in \mathcal{B}$ is G-conjugate to a unique *facet $\mathcal{F}_0 \le \mathcal{C}_0$ (corresponding to the* **complement** *in S of the "type" of $P_{\mathcal{F}}$, via the natural bijection of S with the set of walls of $\mathcal{C}_0$), and if $\mathcal{F}, \mathcal{F}' \in \mathcal{A}_0$ are G-conjugate then they are N-conjugate.*

Proof See [KP, Propositions 1.5.6 and 1.5.13]. The reason (2) holds is because for a facet $\mathcal{F}'$ of $\mathcal{B}$ to be fixed by $g \in G$ amounts to requiring g to normalize $P_{\mathcal{F}'}$ and hence belong to $P_{\mathcal{F}'}$ (since every parabolic subgroup of G is its own normalizer). This implies a facet $\mathcal{F}'$ is carried to itself by $P_{\mathcal{F}}$ exactly when $P_{\mathcal{F}} \subset P_{\mathcal{F}'}$, which in turn says exactly $\mathcal{F}'$ is a facet of $\mathcal{F}$. □

The building $\mathcal{B}$ is called the *Tits building* for (G, N, B, S). We emphasize that the triviality of the $P_{\mathcal{F}}$-action on the set of vertices of $\mathcal{F}$ underlies why the Bruhat-Tits building for $\mathbf{G}$ arises from a Tits system structure for the ambient group $\mathbf{G}(k)^0$ rather than for $\mathbf{G}(k)$. This is illustrated by the fact (as in Example 23.1) that $\mathrm{PGL}_2(k)$ acting on $\mathcal{B}(\mathrm{PGL}_2) = \mathcal{B}(\mathrm{SL}_2)$ has elements that flip an edge (i.e., preserve its unordered pair of vertices by swapping them) whereas for the index-2 subgroup $\mathrm{PGL}_2(k)^0$ any $g \in \mathrm{PGL}_2(k)^0$ preserving an edge fixes its vertices (it never swaps them).

25 From Buildings to Tits Systems

Given a Tits system (G, B, N, S) with non-empty finite S having size at least 2, in Sect. 24 we constructed an abstract simplicial building $\mathcal{B}$ equipped with a G-action such that the G-stabilizers of the facets are exactly the proper parabolic subgroups. The properness requirement on the parabolic subgroups is too weak for our needs when the associated Coxeter system (W, S) is reducible, as we now illustrate with a product situation.

Suppose $G = G_1 \times G_2$ is a product of two Tits systems (with $B = B_1 \times B_2$, $N = N_1 \times N_2$, $S = S_1 \coprod S_2$). The properness condition $P = P_1 \times P_2 \subsetneq G$ does not force $P_i \subsetneq G_i$ for *both* $i = 1, 2$. This is relevant to the Bruhat-Tits building that will be related to a Tits system structure for $G = \mathbf{G}(k)^0$ when $\mathbf{G}^{\mathrm{ad}} = \prod_i \mathbf{G}_i$ is not k-simple, since we want the stabilizers of facets in the Bruhat-Tits building to be *bounded* open subgroups (not only "bounded in some factor").

Thus, with a general Tits system for which S is finite of size at least 2 but (W, S) might be reducible, we can focus on those P for which the type $X = \mathrm{type}(P) \subset S = \coprod S_i$ has $X_i = X \cap S_i \subsetneq S_i$ for all i. This properness condition for all i is a proxy for a "factorwise proper parabolic" property even when the given Tits system isn't a product but is closely related to a product through a map that is nearly a central quotient: see Proposition 23.2 and the discussion both preceding and following it.

We are going to organize the parabolic subgroups with an admissible type into an abstract polysimplicial building, but before stating how that goes we should explore how imposing admissibility affects what should be regarded as a vertex. Rather than have maximal proper parabolic subgroups correspond to vertices, and then using the *unique* expression of a general proper parabolic subgroup P as the intersection of the finitely many maximal proper parabolic subgroups P_i containing P, we shall now impose "admissibility" throughout.

More specifically, let's consider a Tits system (G, B, N, S) for which S is non-empty and finite with (W, S) having irreducible decomposition $(\prod W_i, \coprod S_i)$ satisfying $\#S_i \geq 2$ for all i. Now consider a parabolic subgroup $P \subset G$ whose type $X = \mathrm{type}(P) \subset S = \coprod S_i$ is *admissible*: $X_i := X \cap S_i \subsetneq S_i$ for all i. (Note that *minimal* parabolic subgroups have admissible type, since their type is empty!) The *maximal admissible* parabolic subgroups Q are those whose type $X \subset S$ is maximal among those that are admissible; i.e., $X = \coprod X_i$ for $X_i = X \cap S_i$ of the form $S_i - \{s_i\}$ for some $s_i \in S_i$ for each i.

The key point is this: an *admissible* $X = \coprod X_i \subset \coprod S_i = S$ is contained in exactly the maximal admissible subsets $Y = \coprod Y_i$ with $Y_i = S_i - \{s_i\}$ for some point $s_i \in S_i - X_i$ for each i (such s_i exist since $X_i \subsetneq S_i$ for all i), and

$$X = \bigcap_{\text{admissible } Y \supset X} Y.$$

Concretely, all we're saying is that a proper subset X_0 of a non-empty finite set S_0 is the intersection of the sets $S_0 - \{s_0\}$ as s_0 varies through $S_0 - X_0$, and such sets $S_0 - \{s_0\}$ are precisely the maximal proper subsets of S_0 containing X_0.

Example 25.1 If $P \subset G$ is a parabolic subgroup with an admissible type $X = \coprod X_i$ for proper subsets $X_i \subset S_i$ for all i then

$$P = \bigcap_j P_j \tag{25.1}$$

where P_j varies through the maximal *admissible* parabolic subgroups of G containing P. The types of such P_j vary through the *maximal* admissible subsets of S that contain X.

We want to organize the maximal *admissible* parabolic subgroups into the vertices of a (polysimplicial) abstract building, to be called the *restricted Tits building* of (G, B, N, S), for which (25.1) replaces the role of the description $P_{\mathcal{F}} = \bigcap_{v \in \mathcal{F}} P_v$ using maximal proper parabolic subgroups $P_v \subset G$ with the Tits building discussed in Sect. 24. Here is the result:

Proposition 25.2 *Let (G, B, N, S) be a Tits system with finite $S \neq \emptyset$ and assume the irreducible components $S_i \subset S$ satisfy $\#S_i > 2$ for all i. Let $\mathcal{B}$ be the associated (simplicial) Tits building, and let $\mathcal{B}_i$ be the simplicial Tits building for (G, P_{S-S_i}, N, S_i) (by* [KP, Lemma 1.4.8], *this is a Tits system with Weyl group W_i; the facets of $\mathcal{B}_i$ are the proper parabolic subgroups of (G, B, N, S) that contain a conjugate of P_{S-S_i}).*

For $(P_i) \in \prod_{i \in I} \mathcal{B}_i$ the subgroup $P := \bigcap P_i \subsetneq G$ belongs to $\mathcal{B}$ (i.e., contains a conjugate of B) and has **admissible** *type $\coprod \mathrm{type}(P_i) \subset \coprod S_i = S$. Conversely, every $P \in \mathcal{B}$ with admissible type arises uniquely in this way, using the unique $P_i \in \mathcal{B}_i$ that contains P and has type $(S - S_i) \coprod (S_i \cap \mathrm{type}(P)) \subsetneq S$.*

The polysimplicial complex $\mathcal{B}' = \{\mathcal{B}_i\}_{i \in I}$ has its set of facets in bijection with the set $F'(\mathcal{B})$ of facets $\mathcal{F}$ of $\mathcal{B}$ such that $\mathrm{type}(P_{\mathcal{F}}) \subset S = \coprod S_i$ is admissible. In particular, its set of vertices is identified with

$$\begin{aligned} V' &:= \{P \subset G : \text{maximal among parabolics with } \textbf{admissible } \mathrm{type}(P) \subset S = \coprod S_i\} \\ &= \{P \subset G : \mathrm{type}(P) = \coprod_i (S_i - \{s_i\}) \text{ for some } s_i \in S_i\} \end{aligned}$$

and each P with admissible type is equal to the intersection of all $P' \in V'$ containing P. The polysimplicial complex $\mathcal{B}'$ is a building when equipped with apartments $\mathcal{A} \cap F'(\mathcal{B})$ for apartments $\mathcal{A}$ of $\mathcal{B}$.

Proof See [KP, Props. 1.4.10, 1.5.18]. □

We call $\mathcal{B}'$ the *restricted Tits building* for (G, B, N, S). When (W, S) is irreducible, it is exactly the simplicial Tits building for (G, B, N, S). From its construction, the building $\mathcal{B}'$ inherits a G-action from that on $\mathcal{B}$ and inherits properties (1)–(4) of Theorem 24.11 from $\mathcal{B}$ apart from being polysimplicial rather than simplicial; e.g., G acts transitively on the set of pairs $(\mathcal{A}', \mathcal{C}')$ for $\mathcal{B}'$.

Remark 25.3 Although the notions of *vertex* for $\mathcal{B}$ and $\mathcal{B}'$ don't generally coincide, for each facet F' of $\mathcal{B}'$ and the corresponding facet F of $\mathcal{B}$ ("corresponding" means equality of the associated proper parabolic subgroups of G) every subfacet (e.g., vertex) of F' is also a subfacet of F. This is compatible with the actions of G on both buildings, so the end of Theorem 24.11(1) (i.e., Axiom BLII using an isomorphism of apartments arising from some $g \in G$) indeed carries over to $\mathcal{B}'$.

The sets of apartments in $\mathcal{B}$ and in $\mathcal{B}'$ are in bijection since the sets of chambers in the buildings coincide. Both $\mathcal{B}$ and $\mathcal{B}'$ satisfy the following further property:

Corollary 25.4 *For any two pairs $(\mathcal{A}_1, \mathcal{C}_1)$, $(\mathcal{A}_2, \mathcal{C}_2)$, there exists $g \in G$ such that $g.(\mathcal{A}_1, \mathcal{C}_1) = (\mathcal{A}_2, \mathcal{C}_2)$ with g the* identity *on the vertices of $\mathcal{C}_1 \cap \mathcal{C}_2$.*

Proof Pick any $g \in G$ such that $g.(\mathcal{A}_1, \mathcal{C}_1) = (\mathcal{A}_2, \mathcal{C}_2)$. If $\mathcal{C}_1 \cap \mathcal{C}_2 = \emptyset$ there is nothing to do, so we may suppose that $\mathcal{C}_1 \cap \mathcal{C}_2 \neq \emptyset$. Hence, $\mathcal{F} := \mathcal{C}_1 \cap \mathcal{C}_2$ is a facet, and g carries $\mathcal{F} \subset \mathcal{C}_1$ into $g.\mathcal{C}_1 = \mathcal{C}_2$. But G carries each facet to only *one* facet of a chosen chamber, yet $\mathcal{F} \subset \mathcal{C}_2$ too, so $g.\mathcal{F} = \mathcal{F}$. It follows that $g \in P_{\mathcal{F}}$, so g is the identity on the vertices of $\mathcal{F}$ by Theorem 24.11(2). □

As a key application of the restricted building, we shall soon give a reverse construction: for a (polysimplicial) building $\mathcal{B}$ with an action by a group G such that the action satisfies a couple of axioms (including a property encoding the conclusion of Corollary 25.4), we shall create a Tits system structure (G, B, N, S) for G with a finite non-empty S such that if (W, S) has all irreducible components S_i satisfying $\#S_i \geq 2$ then the restricted building associated to (G, B, N, S) is naturally G-equivariantly isomorphic to the given $\mathcal{B}$!

Before we carry this out, we emphasize the importance of having procedures in both directions: from Tits systems to (restricted) buildings and from polysimplicial buildings with a suitable group action to Tits systems such that the latter is inverse to the former under some mild hypotheses. For a connected reductive group $\mathbf{G}$ over a discretely-valied Henselian field k with a perfect residue field and maximal unramified extension K/k, the construction of $\mathcal{B}(G)$ will proceed in two steps, each using one of those two procedures, as follows:

(i) Since $\mathbf{G}_K$ is quasi-split (Steinberg's Theorem 13.2), the group $\mathbf{G}(K)$ can be studied using work with $\mathrm{SL}_{2,K}$ and $\mathrm{SU}_3(K'/K)$ (for quadratic Galois K'/K) due to Proposition 25.7 and Remark 25.8 below. Intricate commutator calculations in $\mathbf{G}(K)$ involving pieces of relative root groups cut out by congruence conditions give rise to a Tits system structure on $\mathbf{G}(K)^0$, from which one gets an abstract building.
That abstract building yields (via hard work with a "geometric realization") the Bruhat-Tits building $\mathcal{B}(\mathbf{G}_K)$ as a complete metric space with an isometric action by $\mathbf{G}(K) \rtimes \Gamma$ for $\Gamma = \mathrm{Gal}(K/k)$. A list of "Bruhat-Tits theory axioms" is satisfied, and this uniquely characterizes $\mathcal{B}(\mathbf{G}_K)$ up to unique $\mathbf{G}(k)$-equivariant isomorphism (Proposition 30.1).

(ii) A detailed study of the geometry of $\mathcal{B}(\mathbf{G}_K)$ and its relation to the group theory of $\mathbf{G}(K)$ ultimately shows that the complete metric space

$$\mathcal{B}(\mathbf{G}) := \mathcal{B}(\mathbf{G}_K)^{\Gamma}$$

inherits a natural structure of polysimplicial building respected by the isometric $\mathbf{G}(k)$-action.
The $\mathbf{G}(k)^0$-action on $\mathcal{B}(\mathbf{G})$ satisfies the hypotheses in Theorem 25.5 below, which gives a Tits system structure on $\mathbf{G}(k)^0$ whose associated abstract (restricted) building is $\mathcal{B}(\mathbf{G})$ (i.e., the "simplicial factors" of $\mathcal{B}(\mathbf{G})$ encode the Tits system structure)! Moreover, $\mathcal{B}(\mathbf{G})$ satisfies "Bruhat-Tits theory axioms" characterizing it up to unique $\mathbf{G}(k)$-equivariant isomorphism.

Next we give the "reverse construction" of a Tits system from a building.

Theorem 25.5 *Let $\mathcal{B} = \{\mathcal{B}_i\}_{i\in I}$ be a polysimplicial building, where each $\mathcal{B}_i$ is simplicial, and assume it is equipped with an action by a group G such that:*

(i) for all $(\mathcal{A}, \mathcal{C})$ and $(\mathcal{A}', \mathcal{C}')$, there exists $g \in G$ such that $g.(\mathcal{A}, \mathcal{C}) = (\mathcal{A}', \mathcal{C}')$ with $g|_{\mathcal{C}\cap\mathcal{C}'} = \mathrm{id}$;

(ii) for all chambers $\mathcal{C}$ of $\mathcal{B}$, the stabilizer $\mathrm{Stab}_G(\mathcal{C})$ acts trivially on $\mathrm{vert}(\mathcal{C})$.

Choose an apartment $\mathcal{A}_0$ and a chamber $\mathcal{C}_0 \in \mathcal{A}_0$. Define $B = \mathrm{Stab}_G(\mathcal{C}_0)$, $N = \mathrm{Stab}_G(\mathcal{A}_0)$, so the set $W := N/(B\cap N)$ is in bijection with the set of chambers in $\mathcal{A}_0$ via $w \mapsto w\mathcal{C}_0$. For each wall $\mathcal{F}$ of $\mathcal{C}_0$, let $r_{\mathcal{F}} \in W$ correspond to the unique chamber $\mathcal{C} \neq \mathcal{C}_0$ in $\mathcal{A}_0$ sharing the wall $\mathcal{F}$ with $\mathcal{C}_0$ (recall $\mathcal{A}_0$ is a thin chamber complex). In particular, the subset $S \subset W$ consisting of such $r_{\mathcal{F}}$'s is in bijection with the set of walls of $\mathcal{C}_0$, so S is finite (and non-empty if $\dim\mathcal{B} > 0$). Let $P_{\mathcal{F}} := \mathrm{Stab}_G(\mathcal{F})$ for facets $\mathcal{F}$ of $\mathcal{B}$.

(1) The 4-tuple (G, B, N, S) is a saturated Tits system (so $B\cap N \lhd N$ and (W, S) is a Coxeter system).

(2) If the irreducible decomposition $(W, S) = (\prod_{j\in J} W_j, \coprod_{j\in J} S_j)$ satisfies $\#\{i \in I : \dim\mathcal{B}_i > 0\} \geq \#J > 0$ and $\#S_j \geq 2$ for all j then $\mathcal{F} \mapsto P_{\mathcal{F}} \subset G$ is a G-equivariant isomorphism of $\mathcal{B}$ onto the restricted building of (G, B, N, S).

The hypothesis $\#S_j \geq 2$ for all j in (2) is required to define the "restricted building" for (G, B, N, S) (see $\mathcal{B}'$ in Proposition 25.2). The hypothesis in (2) relating $\#J$ to the number of positive-dimensional $\mathcal{B}_i$'s prevents $\mathcal{B}$ from being the *simplicial* ("unrestricted") Tits building (Theorem 24.11) associated to a *reducible* Tits system, and requiring $\#J > 0$ avoids $S = \emptyset$ (in which case $B = N = G$, so $\dim \mathcal{B} = 0$).

Proof Part (1) is [KP, Proposition 1.5.28]. For part (2), we need to show the parabolic subgroups P of (G, B, N, S) with *admissible* $\text{type}(P) \subset S$ (i.e., $\text{type}(P) \cap S_j \subsetneq S_j$ for all j) are the $P_{\mathcal{F}}$'s for facets $\mathcal{F}$ of $\mathcal{B}$, and that $P_{\mathcal{F}} \subset P_{\mathcal{F}'}$ if and only if $\mathcal{F}' \subset \mathcal{F}$. This is proved as Theorem A.11 in Appendix A. □

We have finally (!) assembled all notions needed to list 11 axioms for Bruhat-Tits theory, from which its results can be deduced and be shown to uniquely characterize the Bruhat-Tits building equipped with its various structures. These axioms are called Axiom 4.1.x in [KP, Sect. 4.1] for $x = 1, 2, 4, 6, 8, 9, 16, 17, 20, 22, 27$, so we will call them "Axiom x" for such x for ease of cross-referencing.

We don't need to refer to general Tits systems anymore, so we resume writing G rather than $\mathbf{G}$ for a connected reductive group over a Henselian discretely-valued field k with perfect residue field $\mathfrak{f}$.

Axiom 1. There exists an abstract polysimplicial building $\mathcal{B}(G)$ equipped with a $G(k)$-action and a bijective indexing of its set of apartments by the set of maximal split k-tori $S \subset G$, say $\mathcal{A}(S)$ indexed by S. This indexing is $G(k)$-equivariant in the sense that $g.\mathcal{A}(S) = \mathcal{A}(gSg^{-1})$ for all $g \in G(k)$, so $\text{Stab}_{G(k)}(\mathcal{A}(S)) = N_G(S)(k)$.

Also, for any central quotient $G \twoheadrightarrow \overline{G}$ there is a specified isomorphism $\mathcal{B}(G) \simeq \mathcal{B}(\overline{G})$ equivariant with respect to $G(k) \to \overline{G}(k)$ and compatible with composition of central quotients.

Axiom 2. For all facets $\mathcal{F}$ of $\mathcal{B}(G)$, the stabilizer $G(k)^0_{\mathcal{F}} := \text{Stab}_{G(k)^0}(\mathcal{F})$ of $\mathcal{F}$ under $G(k)^0$ is a bounded open subgroup (bounded in G). Call these *parahoric* subgroups of $G(k)$. When $\mathcal{F}$ is a chamber, call it an *Iwahori subgroup*.

Remark 25.6 A subsequent axiom (Axiom 9) will ensure that the subgroup $G(k)^0_{\mathcal{F}} \subset G(k)^0$ acts trivially on the set of vertices of $\mathcal{F}$ and determines $\mathcal{F}$, and that $G(k)^0$ acts transitively on the set of chambers. Hence, $G(k)^0_{\mathcal{F}} \subseteq G(k)^0_{\mathcal{F}'}$ when $\mathcal{F}'$ is a facet of $\mathcal{F}$, so the Iwahori subgroups are the minimal parahoric subgroups and constitute a single conjugacy class of bounded open subgroups of $G(k)$.

The following result explains the importance of SL_2 and SU_3 for work with quasi-split groups:

Proposition 25.7 *Let H be a* **quasi-split** *connected semisimple group over a field F. For a maximal split F-torus $S \subset H$ and each root a in the relative root system $\Phi = \Phi(H, S)$, let $S_a \subset S$ be the codimension-1 subtorus killed by a and let $H_a = \mathscr{D}(Z_H(S_a))$ (a quasi-split rank-1 connected semisimple F-group).*

Assume H is absolutely simple and simply connected (so Φ is irreducible). If a is longest in Φ then $H_a \simeq \text{SL}_2$ except when a is divisible, in which case H_a is the quasi-split $\text{SU}_3(F'/F)$ for a quadratic Galois extension F'/F. In all cases, for longest a the S-root group U_a is 1-dimensional.

Remark 25.8 Before proving Proposition 25.7, we make some observations. Firstly, H_a is quasi-split because $Z_H(S_a)$ is quasi-split, as the S-centralizer in $Z_H(S_a)$ coincides with $Z_H(S)$ that is a torus due to H being quasi-split. Furthermore, since H is simply connected the subgroup H_a is too since it is the derived group of a torus centralizer and so inherits simple connectedness from H [C2, Cor. 9.5.11].

The "absolutely simple" hypothesis in Proposition 25.7 is needed to rule out the intervention of non-trivial finite separable Weil restriction. Indeed, by [C3, Cor. M.1.2], if we remove "absolutely simple" then $H \simeq \prod \mathrm{R}_{F_i/F}(H_i')$ for finite separable F_i/F and simply connected F_i-groups H_i' that are absolutely simple. These H_i''s inherit the quasi-split property from H [C3, Lemma P.6.1], and S corresponds exactly to a choice of maximal split F_i-torus $S_i' \subset H_i'$ for each i (with the non-empty $\Phi(H_i', S_i')$'s identified with the irreducible components of $\Phi(H, S)$); see [C3, Proposition P.6.2, Sect. Q.2].

The "longest" hypothesis on a is also essential: for quasi-split $\mathrm{SU}_n(F'/F)$ with $n \geq 3$, there are non-longest a (such as roots corresponding to all but one node in the relative diagram) for which $H_a \simeq \mathrm{R}_{F'/F}(\mathrm{SL}_2)$ and so U_a is 2-dimensional. See [C3, Sect. Q.4] (especially [C3, Exs. Q.4.2–Q.4.4]) for explicit descriptions with such special unitary groups, also going beyond the quasi-split case (see Remark 25.9 as well).

Proof It will first be proved that U_a is 1-dimensional. This implies the simply connected H_a cannot be a Weil restriction through a non-trivial finite separable extension of F and so it must be *absolutely simple* (see [C2, Cor. M.1.2]). That will open the door to showing the possibilities for H_a are as claimed.

G. Prasad provided the following uniform argument for the 1-dimensionality of U_a, avoiding casework. Let $T = Z_H(S)$; this is a maximal F-torus since H is quasi-split. Let F'/F be a splitting field of T and let $\Phi = \Phi(H, S)$, $\Phi' = \Phi(H_{F'}, T_{F'})$. Since H is quasi-split, the restriction map $\Phi' \to \Phi \cup \{0\}$ induced by $\mathrm{X}^*(T_{F'}) \twoheadrightarrow \mathrm{X}^*(S_{F'}) = \mathrm{X}^*(S)$ has image Φ (in other words, Δ_0 in the Tits index is empty; see [C3, Remark 12.1.2]).

For a linear hyperplane in $\mathrm{X}^*(S)_{\mathbf{R}}$ disjoint from Φ, its hyperplane preimage in $\mathrm{X}^*(T_{F'})_{\mathbf{R}}$ is disjoint from Φ'. Thus, if we pick a positive system of roots $\Phi^+ \subset \Phi$ then its preimage $\Phi'^+ \subset \Phi'$ is also a positive system of roots. (We are not claiming that all positive systems of roots for Φ' arise in this way.) Let $a \in \Phi$ be a longest root, and pick $a' \in \Phi'$ such that $a'|_{S_{F'}} = a \in \Phi$. We claim a' is also a longest root. To see this, we first claim that the longest roots in an irreducible root system are the same as the highest roots for positive systems of roots. Since the Weyl group of an *irreducible* root system acts transitively on all roots with a given length and acts transitively on the set of positive systems of roots, it suffices to show that a highest root for a positive system of roots in an irreducible root system is a longest root; this fact is [Bour, Chap. 4, Sect. 1.8, Proposition 25(iii)].

Returning to our longest $a \in \Phi$ and a preimage $a' \in \Phi'$ of a, let Φ^+ be a positive system of roots for which a is highest, and let Φ'^+ be the preimage positive system for Φ', so $a' \in \Phi'^+$. We claim a' is the highest positive root, so a' is then both a longest root in Φ' and is *unique* (the choice of Φ^+ which gave rise to Φ'^+ has nothing to do with the choice of a'). If a' is not highest then the highest positive root $b' \in \Phi'^+$ is $a' + \sum c'_i$ for a non-empty set of simple positive roots c'_i with respect to Φ'^+. Then restricting to $S_{F'}$ yields the positive root $b := b'|_{S_{F'}} \in \Phi^+$ that strictly dominates a, contradicting that a is the highest positive root in Φ^+.

For the split group $H_{F'}$, all $T_{F'}$-root groups are 1-dimensional. For each $c' \in \Phi'$, the $S_{F'}$-action on the F'-line $\text{Lie}(U_{c'})$ is via $c = c'|_{S_{F'}} \in \Phi$. But a' is the *unique* $c' \in \Phi'$ restricting to a, and $\text{Lie}(H_{F'})$ is the direct sum of $T_{F'}$-root lines and $\text{Lie}(T_{F'}) = \text{Lie}(T)_{F'}$ on which the adjoint action by $S_{F'}$ is trivial, so the $S_{F'}$-representation space $\text{Lie}(H_{F'}) = \text{Lie}(H)_{F'}$ over F' has a 1-dimensional weight space for $a_{F'}$. Hence, the S-weight space decomposition of $\text{Lie}(H)$ must have a 1-dimensional a-weight space. This coincides with $\text{Lie}(U_a)$ except possibly if a is multipliable (due to how root groups are defined in terms of triviality for the action of the group scheme $\ker(a) \subset S$). But a is a longest root, so it cannot be multipliable. This completes the proof that $\dim U_a = 1$.

The quasi-split and simply connected H_a with F-rank 1 is now known to be *absolutely simple* over F. If H_a is split then it is SL_2. It remains to show the non-split quasi-split *absolutely simple* and simply connected F-groups are exactly the quasi-split groups $\text{SU}_3(F'/F)$ for quadratic Galois F'/F.

To prove that a simply connected quasi-split group G that is non-split and *absolutely simple* is $\text{SU}_3(F'/F)$ for some quadratic Galois F'/F, we have to do a small amount of casework. Using the relationship between absolute and relative root systems in [C2, Sect. 12.1], the property $\Delta_0 = \emptyset$ in the quasi-split case [C2, Remark 12.1.2] implies that the non-trivial Galois action on the *connected* absolute diagram must have *only one* Galois orbit. But looking at the diagrams for reduced irreducible root systems, A_2 is the only one admitting a non-trivial transitive group action. Hence, G must be a non-split quasi-split form of SL_3!

The diagram automorphism group is $\mathbf{Z}/2\mathbf{Z}$, so such G are classified by classes in $\text{H}^1(F, \mathbf{Z}/2\mathbf{Z}) - \{0\}$ [C2, Theorem 7.1.9(3), Propositions 7.2.2, 7.2.12]. The groups $\text{SU}_3(F'/F)$ account for all such classes, so these exhaust the non-split quasi-split forms of SL_3 up to isomorphism. □

Remark 25.9 For an alternative proof of Proposition 25.7, we can use casework with the possibilities for the irreducible (and reduced) absolute root system and the elegant algorithm in [KP, Sect. 2.6(c)] for computing the (possibly non-reduced but irreducible) relative root system.

The set of isomorphism classes of quasi-split forms of a split connected reductive group is in natural bijection with $\text{H}^1(F, \Gamma)$ for the automorphism group Γ of the root datum equipped with a choice of basis (with Γ given the trivial Galois action), by [C2, Theorem 7.1.9(3), Proposition 7.2.2]. The group Γ is the automorphism group of the Dynkin diagram when the group is simply connected (or of adjoint type), so the simply connected and absolutely simple quasi-split H is split except when the corresponding connected Dynkin diagram for H_{F_s} has a non-trivial automporphism: types A_n for $n \geq 2$, D_n for $n \geq 4$, and E_6.

In split cases everything is clear, so we can focus on the *non-split* quasi-split cases (which must be one of the absolute types just listed). The Tits index as in

[C3, Sect. 12.1] for such a quasi-split group involves a non-trivial (continuous) Galois action on the absolute diagram for which the subset Δ_0 is empty (see [C3, Remark 12.1.2]). An algorithm in [KP, Sect. 2.6(c)] (using a tabulation [KP, Table 2.6.1] of highest root coefficients for the irreducible root systems) describes the diagram of the relative root system $\Phi(H, S)$ in terms of the diagram for the absolute root system $\Phi(H_{F_s}, T_{F_s})$. The algorithm also detects when a root in the relative diagram is multipliable or not.

When this algorithm is run on types A, D, and E_6 for *non-split* quasi-split forms, it works out as follows: absolute type A_{2m} ($m \geq 1$) yields relative diagram BC_m, absolute type A_{2m+1} ($m \geq 1$) yields relative diagram C_{m+1}, absolute type E_6 yield relative diagram F_4, absolute type D_n ($n \geq 4$) equipped with an involution yields relative diagram C_{n-1}, and the triality forms ${}^3D_4$ and ${}^6D_4$ each have relative diagram G_2. The miracle is that for the natural surjection from the set of nodes in the absolute diagram onto the set of nodes in the relative diagram [C2, Proposition 12.1.3], in *all* cases with a reduced relative diagram (i.e., away from absolute type A_{2m}) the longest nodes a in the relative diagram arise from roots on the absolute diagram with *trivial* Galois action. The corresponding simply connected group H_a of relative rank 1 is therefore split in such cases, so it is SL_2.

Finally, for absolute type A_{2m} the non-split relative roots are the divisible ones and arise already from the Galois-twisting applied to the A_2 subgroup corresponding to the pair of central nodes in the A_{2m} diagram. In other words, $H_a = SU_3(F'/F)$ where F'/F is the quadratic Galois extension cut out by the order-2 Galois action on the absolute diagram.

26 More Affine Spaces

In Sect. 25 we stated the first two axioms for Bruhat-Tits theory. In Sect. 27 we'll state three more axioms, giving enough to make the non-positively curved metric on the ("geometric realization" of the) building. This will rest on a result concerning affine spaces that we build up to in Lemma 26.2. But first we record a basic consequence of Axiom 1 from Sect. 25.

Proposition 26.1 *The building $\mathcal{B}(G)$ is a point (equivalently, there is exactly one vertex) if and only if $\mathscr{D}(G)$ is k-anisotropic.*

We give a proof using only Axiom 1 since that is simplest. One can argue in other ways by also using Axiom 2, thanks to $G(k)^0$ having finite index in $G(k)^1$ (by Theorem 12.1).

Proof First assume the building is a point x. In this case there is only one apartment, but the set of apartments is indexed bijectively by the set of maximal split k-tori in G by Axiom 1, so there is only one maximal split k-torus $S \subset G$. This forces $N_G(S)(k) = G(k)$. That equality for a connected reductive group over any field forces anisotropicity of the derived group over the field by using the Bruhat decompostion for the group of rational points (since the unipotent radical of a minimal parabolic subgroup is an affine space, which has at least 2 points if its dimension is positive).

Conversely, assume $\mathscr{D}(G)$ is k-anisotropic, so there is only one maximal split k-torus S in G (namely the maximal split k-torus A_G in the maximal central torus of G). By Axiom 1 it follows that $\mathcal{B}(G)$ has only one apartment A. Every chamber lies in some apartment, and every facet lies in some chamber (as the building is a chamber complex), so all facets lie in A and hence $\mathcal{B}(G) = A$.

Apartments are thin chamber complexes and buildings are thick chamber complexes, so the only way $\mathcal{B}(G)$ can be equal to A is if there are no codimension-1 facets at all (so the thick and thin properties are vacuous and hence can be simultaneously satisfied). This happens precisely when $\dim \mathcal{B}(G) = 0$. By Example 24.5, this says $\mathcal{B}(G)$ is a point. □

Let S be a maximal split k-torus in G, $N = N_G(S)$, and $Z = Z_G(S)$. Let $S' = (S \cap \mathscr{D}(G))^0_{\mathrm{red}}$ be the corresponding maximal split k-torus in $\mathscr{D}(G)$. The next two axioms for Bruhat-Tits theory, Axioms 4 and 6 to be stated in Sect. 27, involve two constructions for which the second one *will only make sense* when $\mathscr{D}(G)$ is k-isotropic (equivalently $\Phi(G, S)$ is non-empty, or equivalently $S' \neq 1$):

- a *canonical* affine space $A = A(S)$ for the $\mathbf{R}$-vector space $V(S') := \mathrm{X}_*(S')_{\mathbf{R}}$ (which we will sometimes denote as V'; note that $\dim A = \dim S'$),
- when $S' \neq 1$, an affine root system $\Psi = \Psi(G, S)$ for that affine space (actually two such affine root systems, which however coincide when $\Phi(G, S)$ is reduced),

required to satisfy an array of properties. Among these properties (assuming $S' \neq 1$) will be:

(i) the abstract polysimplicial complex underlying (A, Ψ) is identified with the (thin chamber complex) apartment $\mathcal{A}(S)$ in the abstract polysimplicial building $\mathcal{B}(G)$,
(ii) $\nabla\Psi = \Phi(G, S)$ in $V'^* = \mathrm{X}^*(S')_{\mathbf{R}} \subset \mathrm{X}^*(S)_{\mathbf{R}}$ (for the inclusion: recall S is an isogenous quotient of $A_G \times S'$, so $\mathrm{X}^*(S')_{\mathbf{Q}}$ is naturally a direct summand of $\mathrm{X}^*(S)_{\mathbf{Q}}$).

Axiom 4 will be concerned with the affine space A over V', and Axiom 6 will be concerned with the affine root system (A, Ψ) (assuming $\dim A > 0$). In case $\mathscr{D}(G)$ is k-anisotropic, or equivalently $\mathcal{B}(G)$ is a point, we have $V' = 0$ and so the unique apartment in $\mathcal{B}(G)$ has a unique structure of affine space over V' anyway. In this sense, Axiom 4 will be meaningful and trivially satisfied when $S' = 1$.

There is no way to speak of an affine root system $\Psi(G, S)$ when $S' = 1$ since a 1-point affine space A satisfies $A^* = \mathbf{R}$ yet an affine root system (whose members are non-constant) is always non-empty (see Remark 16.2). Thus, Axiom 6 will *require assuming* $\mathscr{D}(G)$ is k-isotropic (equivalently $V' \neq 0$). Here is a key uniqueness result for an affine space which removes any ambiguity from later constructions.

Lemma 26.2 *Up to unique isomorphism, there exists an affine space* $A = A(S)$ *for* $V' = \mathrm{X}_*(S')_{\mathbf{R}}$ *equipped with an* $N(k)$*-action* $f \colon N(k) \to \mathrm{Aff}(A)$ *through affine automorphisms of* A *such that*

(1) for all $n \in N(k)$, the linear map $\nabla(f(n)) \in \mathrm{Aut}(V')$ is the natural action on V' (arising from the quotient map $N(k) \twoheadrightarrow N(k)/Z(k) = W(\Phi(G, S))$ using that the $N(k)$-conjugation action on G preserves S and hence preserves S'),

(2) for $z \in Z(k)$ (so $\nabla(f(z)) = \mathrm{id}$, ensuring the affine automorphism $f(z)$ of A is translation by some vector in V'), the map $f(z)$ is translation by vector $\nu(z) \in V'$ given by the image of z under

$$Z(k) \xrightarrow{-\omega_Z} \mathrm{X}^*(Z)^\vee_{\mathbf{R}} = \mathrm{X}_*(S)_{\mathbf{R}} \simeq \mathrm{X}_*(S')_{\mathbf{R}} \oplus \mathrm{X}_*(A_G)_{\mathbf{R}} \overset{\mathrm{pr}_1}{\twoheadrightarrow} \mathrm{X}_*(S')_{\mathbf{R}} = V',$$

where the isomorphism $\mathrm{X}^*(Z)_{\mathbf{R}} \simeq \mathrm{X}^*(S)_{\mathbf{R}}$ arises from the map of k-tori $S \to Z/\mathscr{D}(Z)$ carrying S isogenously onto the maximal split *k-torus in $Z/\mathscr{D}(Z)$. Here, ω_Z sends $g \in Z(k)$ to $\chi \mapsto \mathrm{ord}(\chi(g))$.*

The construction of A goes as follows (see [KP, Proposition 4.4.3] for the details). For $G' = \mathscr{D}(G)$ and its maximal split k-torus S', the normalizer $N' = N_{G'}(S')$ and centralizer $Z' = Z_{G'}(S')$ satisfy $N'(k)/Z'(k) \simeq W := W(\Phi(G, S))$ since $\Phi(G', S') = \Phi(G, S)$. Since $\mathscr{D}(Z')$ is k-anisotropic (as S' is central in Z'), there is a unique maximal bounded subgroup $Z'(k)_b$ in $Z'(k)$ (namely the preimage under $Z'(k) \to (Z'/\mathscr{D}(Z'))(k)$ of the maximal bounded subgroup of the group of k-points of the k-torus $Z'/\mathscr{D}(Z')$, due to Proposition 4.1(4) applied to the isogeny $Z' \to (Z'/C) \times (Z'/\mathscr{D}(Z'))$ with C the maximal central k-torus of Z' since Z'/C is k-anisotropic). By uniqueness $Z'(k)_b$ is normal in $N'(k)$, so we get a short exact sequence of groups

$$1 \to Z'(k)/Z'(k)_b \to N'(k)/Z'(k)_b \to W \to 1. \tag{26.1}$$

Since $\nu : Z(k) \to V' := \mathrm{X}_*(S')_{\mathbf{R}}$ kills all bounded subgroups (as ω_Z does), $\nu|_{Z'(k)}$ kills $Z'(k)_b$. One can then form a "non-commutative pushout" of (26.1) along $\nu : Z'(k)/Z'(k)_b \to V'$ to get a short exact sequence of groups

$$1 \to V' \to E \to W \to 1. \tag{26.2}$$

The $N'(k)$-equivariance of $\nu|_{Z'(k)}$ (or even the $N(k)$-equivariance of ν) implies that this extension of W by the commutative V' makes E-conjugation on V' induce the natural W-action on V' (and E has a natural structure of Lie group, as it contains V' with finite index). The set of splittings of (26.2) is then an affine space for V', and it is the sought-after A.

Remark 26.3 The preceding construction of A is "too abstract" for many purposes. It can be described in more concrete terms using "valuations on the root datum," a way of putting compatible "norms" on the root subgroups $U_a(k) \subset G(k)$ to cut out bounded subgroups (beware $U_a(k)$ is often non-commutative when a is multipliable). That interpretation is most practical for quasi-split G, as then (by considering G^{sc}) up to an intervention of Weil restriction as in Remark 25.8 the root groups for G are found inside the rank-1 simply connected semisimple quasi-split k-groups SL_2 and $\mathrm{SU}_3(k'/k)$ for quadratic Galois k'/k by Proposition 25.7.

Remark 26.4 By Lemma 26.2(2) the group $Z(k)^1 = \ker \omega_Z$ acts trivially on A, so all bounded subgroups of $Z(k)$ act trivially on A.

Remark 26.5 For $z = s \in S(k) \subset Z(k)$ and $a \in \Phi(G, S) \subset \mathrm{X}^*(S')_{\mathbf{R}} = V'^*$, we claim

$$a(\nu(s)) = -\operatorname{ord}(a(s)) \in \mathbf{Z}. \tag{26.3}$$

This claim is clear from the definition of ν in two cases: $s \in S'(k)$ and $s \in A_G(k)$. Hence, it holds for $s \in S'(k)A_G(k)$ by the homomorphism property in s for both sides of (26.3). The quotient $S(k)/(S'(k)A_G(k))$ is killed by some positive integer N since the multiplication map $S' \times A_G \to S$ is an isogeny, so by multiplying both sides of (26.3) by N (as we may do) we can replace s with s^N on both sides and thereby reduce (26.3) in general to the settled case $s \in S'(k)A_G(k)$.

For $\Phi = \Phi(G, S)$, the reflection $r_a \in N(k)/Z(k) = W(\Phi)$ for $a \in \Phi$ arises from $G(k)$ as follows.

Proposition 26.6 (Tits) *For $a \in \Phi = \Phi(G, S)$ and $u \in U_a(k) - \{1\}$, there exist unique $u', u'' \in U_{-a}(k)$ such that $m(u) := u'uu'' \in N(k)$. Moreover, $u', u'' \neq 1$ and $m(u)$ acts on $\mathrm{X}^*(S)_{\mathbf{R}}$ by the reflection r_a. Furthermore, if $a \in \Phi^{\mathrm{nm}}$ then $u' = u'' = m(u)^{-1}um(u)$ and $m(u)^2 \in S(k)$.*

Proof See [CGP, Proposition C.2.24]. (In the split case $m(u)^2 = a^\vee(-1)$, but beyond the split case there is no analogous explicit formula in $S(k)$ for $m(u)^2$; see Example 26.8.) □

Example 26.7 When $G = \mathrm{SL}_2$, for the diagonal split maximal torus S and root $a : \left(\begin{smallmatrix} t & 0 \\ 0 & 1/t \end{smallmatrix}\right) \mapsto t^2$ we have $U_a = \{\left(\begin{smallmatrix} 1 & * \\ 0 & 1 \end{smallmatrix}\right)\}$ and $U_{-a} = \{\left(\begin{smallmatrix} 1 & 0 \\ * & 1 \end{smallmatrix}\right)\}$. For any *non-trivial* $u = \left(\begin{smallmatrix} 1 & x \\ 0 & 1 \end{smallmatrix}\right) \in U_a(k)$ (so $x \neq 0$) one can check directly that

$$u' = u'' = \begin{pmatrix} 1 & 0 \\ -1/x & 1 \end{pmatrix}, \quad m(u) = \begin{pmatrix} 0 & x \\ -1/x & 0 \end{pmatrix}$$

(and that these are the only possibilities for u', u'').

The preceding example for SL_2 yields Proposition 26.6 for *split* G, whereas beyond the split case one has to do more work to prove Proposition 26.6. In particular, to prove Proposition 26.6 for quasi-split G it suffices (after passing to G^{sc} and invoking Proposition 25.7 up to the intervention of Weil restriction as in Remark 25.8) to check the result for the quasi-split $\mathrm{SU}_3(k'/k)$ for quadratic Galois k'/k. The following example, whose details we leave to the reader, nicely illustrates how Proposition 26.6 beyond the split case is genuinely different from the split case (e.g., $u' \neq u''$ can occur, and the point $m(u)^2$ may be outside $S(k)$ when $a \notin \Phi^{\mathrm{nm}}$; i.e., when a is multipliable).

Example 26.8 For a quadratic Galois extension k'/k with non-trivial automorphism denoted $z \mapsto \overline{z}$, consider the non-degenerate Hermitian form $h : k'^3 \times k'^3 \to k'$ defined by

$$h(\mathbf{x}, \mathbf{y}) = x_1\overline{y}_3 - x_2\overline{y}_2 + x_3\overline{y}_1.$$

This is the same as the case $c = -1$ considered in the quasi-split calculations in [C3, Ex. Q.4.4] except we have swapped the roles of the 2nd and 3rd standard basis vectors in k'^3 (this makes root groups work out a bit more cleanly below as upper-triangular 3×3 matrices).

A maximal split k-torus S in $G = \mathrm{SU}(h)$ is given by

$$S = \left\{ \begin{pmatrix} t & & \\ & 1 & \\ & & 1/t \end{pmatrix} \right\} \simeq \mathbf{G}_m$$

with 2-dimensional maximal k-torus

$$Z = Z_G(S) = \left\{ \begin{pmatrix} t' & & \\ & \overline{t}'/t' & \\ & & 1/\overline{t}' \end{pmatrix} : t' \in \mathrm{R}_{k'/k}(\mathbf{G}_m) \right\} \simeq \mathrm{R}_{k'/k}(\mathbf{G}_m).$$

The root $a : \mathrm{diag}(t, 1, 1/t) \mapsto t$ is multipliable with root group

$$U_a = \left\{ \begin{pmatrix} 1 & u & v \\ & 1 & \overline{u} \\ & & 1 \end{pmatrix} : v + \overline{v} = u\overline{u} \right\}$$

for $u, v \in \mathrm{R}_{k'/k}(\mathbf{G}_a)$, and $u = 0$ defines U_{2a}. Likewise U_{-a} has the analogous lower-triangular recipe.

For a point $x \in U_a(k)$ given by the above recipe in terms of $u, v \in k'$ with $v \in k'^{\times}$, we have $x', x'' \in U_{-a}(k)$ given by

$$x' = \begin{pmatrix} 1 & & \\ -\overline{u}/v & 1 & \\ 1/\overline{v} & -u/\overline{v} & 1 \end{pmatrix}, \quad x'' = \begin{pmatrix} 1 & & \\ -\overline{u}/\overline{v} & 1 & \\ 1/\overline{v} & -u/v & 1 \end{pmatrix}$$

(so $x'' \neq x'$ when $u \in k'^{\times}$ and $v \in k' - k$). By direct calculation,

$$m(x) = x'xx'' = \begin{pmatrix} & & v \\ & -\overline{v}/v & \\ 1/\overline{v} & & \end{pmatrix}$$

so $m(x)^2 = \mathrm{diag}(v/\overline{v}, (\overline{v}/v)^2, v/\overline{v}) \in Z(k)$ is *outside* $S(k)$ when $\overline{v}/v \neq \pm 1$ (note that $a \notin \Phi^{\mathrm{nm}}$).

In the general case, via Axiom 6 to be discussed in Sect. 27, the action by $m(u) \in N(k)$ on $A = A(S)$ is an affine reflection, so $A^{m(u)=1}$ is a hyperplane. From such hyperplanes we shall obtain the sought-after affine roots. Note that the existence of $u \in U_a(k) - \{1\}$ requires the existence of a, so Φ is non-empty and hence $\dim A = \dim S' > 0$.

27 Apartment Axioms

Let G be a connected reductive k-group, $S \subset G$ a maximal split k-torus, $N = N_G(S)$, $Z = Z_G(S)$, $\Phi = \Phi(G, S)$, $S' = (S \cap \mathscr{D}(G))^0_{\mathrm{red}}$. In Sect. 26 we discussed the existence and uniqueness up to unique $N(k)$-equivariant isomorphism of an affine space $A = A(S)$ over $\mathrm{X}_*(S')_{\mathbf{R}}$ that satisfies a couple of specific properties (see Lemma 26.2). This uniqueness up to *unique* isomorphism says in particular that A admits no non-trivial $N(k)$-equivariant automorphisms as an *affine space*.

We shall soon impose a polysimplicial structure on A, or more precisely build an $N(k)$-equivariant locally finite hyperplane arrangement in A whose underlying set of facets is an abstract polysimplicial complex A_0. It will turn out that A_0 also admits no non-trivial $N(k)$-equivariant automorphisms; this is rather different from the corresponding statement for A, as it doesn't involve the affine space structure.

Proposition 26.6 provides $m(u) = u'uu'' \in N(k)$ for every non-trivial $u \in U_a(k)$ for $a \in \Phi$, with unique $u', u'' \in U_{-a}(k)$. This uniqueness implies $nm(u)n^{-1} = m(nun^{-1})$ for $n \in N(k)$ (where $nun^{-1} \in U_{n.a}(k) - \{1\}$). This provides ingredients used in the following Bruhat-Tits theory axiom.

Axiom 6. Assume $S' \neq 1$ (equivalently $\Phi \neq \emptyset$, or equivalently $\mathscr{D}(G)$ is k-isotropic, or equivalently $\dim A > 0$).

(AS1) For all $a \in \Phi$ and $u \in U_a(k) - \{1\}$, the action on A by $m(u) \in N(k)$ (whose derivative r_a is a reflection of V') has fixed-point locus $A^{m(u)=1}$ a hyperplane H_u in A that is an affine space for the codimension-1 subspace $\ker(a) \subset \mathrm{X}_*(S')_{\mathbf{R}} = V'$ (equivalently, H_u is defined by a non-constant affine functional $\psi \in A^*$ whose derivative is a nonzero scalar multiple of a, hence ψ can be uniquely scaled to satisfy $\dot{\psi} = a$). In particular, we can define $\psi_a^u \in A^* - \mathbf{R}$ to be the unique non-constant affine functional ψ on A that satisfies $H_\psi = H_u$ and $\dot{\psi} = a$.

(AS2) The set $\Psi' := \{\psi_a^u : a \in \Phi, u \in U_a(k) - \{1\}\} \subset A^* - \mathbf{R}$ is an affine root system and its derivative $\nabla(\Psi')$ coincides with Φ inside $V'^* = \mathrm{X}^*(S')_{\mathbf{R}}$.

(AS3) Note that if non-constant ψ_1, ψ_2 satisfy $\psi_1 \geq \psi_2$ then $\psi_1 = c + \psi_2$ for some $c \in \mathbf{R}_{\geq 0}$ since the affine functional $\psi_1 - \psi_2$ doesn't take values with both signs and so must be constant. In particular $\dot{\psi}_1 = \dot{\psi}_2$. For $\psi \in A^*$ with $\dot{\psi} = a \in \Phi$, we require that the subset

$$U_\psi = \{u \in U_a(k) : u = 1 \text{ or } \psi_a^u \geq \psi\}$$

of $U_a(k)$ is a subgroup (equivalently $H_u \subset A^{\psi \leq 0}$, since the property $\psi(H_u) \leq 0$ forces H_u not to touch both connected components of $A - H_\psi$ and hence H_ψ is *parallel* to H_u; in algebraic terms, such parallelism says that the defining affine functionals ψ and ψ_a^u with the *same* derivative differ by a constant). If $\psi_1 \geq \psi_2$ and their common derivative is a then clearly $U_{\psi_1} \subset U_{\psi_2}$. Thus, for affine $\psi' > \psi$ we have $\dot{\psi}' = \dot{\psi} = a$ and $U_{\psi'}$ grows as ψ' decreases toward ψ from above, so the union

$$U_{\psi+} = \bigcup_{\psi' > \psi} U_{\psi'}$$

is directed (as ψ' decreases) and hence is also a subgroup. We further require that

$$\Psi := \{\psi \in \Psi' : U_\psi \not\subset U_{\psi+} U_{2a}(k)\}$$

(where by convention $U_{2a}(k) = 1$ when $2a \notin \Phi$) is an affine root system with derivative root system $\nabla \Psi = \Phi$ as well.

(AS4) $\Psi' \subset \Psi \cup (1/2)\Psi$ (so since $\Psi \subset \Psi'$ too, the affine root systems Ψ and Ψ' define the *same* set of vanishing hyperplanes in A and hence have the same affine Weyl group inside $\mathrm{Aff}(A)$).

(AS5) The image $\mathrm{im}(f)$ of the map $f : N(k) \to \mathrm{Aff}(A)$ from Lemma 26.2 satisfies

$$W(\Psi) \subset \mathrm{im}(f) \subset W(\Psi)^{\mathrm{ext}}$$

where the *extended affine Weyl group* $W(\Psi)^{\mathrm{ext}}$ is defined by

$$W(\Psi)^{\mathrm{ext}} = \{\alpha \in \mathrm{Aff}(A) : \alpha(\Psi) = \Psi, \dot{\alpha} \in W(\Phi)\}.$$

We also require that if $G = G^{\mathrm{sc}}$ then $\mathrm{im}(f) = W(\Psi)$ and if G is quasi-split of adjoint type then $\mathrm{im}(f) = W(\Psi)^{\mathrm{ext}}$ (e.g., for $G = \mathrm{PGL}_2$ the element $\left(\begin{smallmatrix} 0 & 1 \\ \pi & 0 \end{smallmatrix}\right) \in N(k)$ reflects an apartment in the tree $\mathcal{B}(\mathrm{PGL}_2) = \mathcal{B}(\mathrm{SL}_2)$ around the midpoint of an edge; it lies outside $W(\Psi)$).

[Note that $W(\Psi)$ is normal in $W(\Psi)^{\mathrm{ext}}$: any $\alpha \in \mathrm{Aff}(A)$ with $\dot{\alpha} \in W(\Phi)$ preserves distance on A defined by a $W(\Phi)$-invariant inner product on $\mathrm{X}_*(S')_{\mathbf{R}}$ and so if it preserves Ψ then it must permute the set of reflections r_ψ in the corresponding vanishing hyperplanes. The index of $W(\Psi)$ in $W(\Psi)^{\mathrm{ext}}$ is *finite*. Indeed, any $\alpha \in W(\Psi)^{\mathrm{ext}}$ preserves the polysimplicial strucutre of A and hence carries a choice of chamber $\mathcal{C}$ to another chamber $\mathcal{C}'$, yet $W(\Psi)$ acts transitively on the set of chambers, so we can focus on those $\alpha \in W(\Psi)^{\mathrm{ext}}$ preserving $\mathcal{C}$. But there are only finitely many automorphisms of $\mathcal{C}$ (look at the vertices), so we can further arrange that α fixes all vertices of $\mathcal{C}$ and thus $\alpha = 1$ by affineness.]

(AS6) The $N(k)$-equivariant polysimplicial complex A_0 underlying both (A, Ψ) and (A, Ψ') is the apartment $\mathcal{A}(S)$ equipped with its natural $N(k)$-action via polysimplicial automorphisms.

In (AS6) there is a subtle point: to paraphrase Bill Clinton, what is the meaning of "is"? All joking aside, the issue is that (AS6) only asserts the existence of *some* $N(k)$-equivariant polysimplicial isomorphism $A_0 \simeq \mathcal{A}(S)$, so for well-posedness we have to make sure such an isomorphism is unique if it exists. Put another way, if we *ignore* the affine space structure on A and instead remember only its underlying $N(k)$-equivariant polysimplicial structure A_0 (from both Ψ and Ψ') then is there no non-trivial $N(k)$-equivariant automorphism? Fortunately the answer is affirmative:

Lemma 27.1 *The $N(k)$-equivariant polysimplicial complex A_0 underlying (A, Ψ) and (A, Ψ') has no non-trivial $N(k)$-equivariant automorphism.*

Proof Choose a vertex v of A_0. This is an intersection of finitely many of the hyperplanes H_u in A, so it is a vertex of finitely many abstract codimension-1 facets $\mathcal{F}_j$ of A_0 and $\bigcap_j \mathcal{F}_j = \{v\}$. Each $\mathcal{F}_j$ belongs to a unique hyperplane H_{u_j} in A since it has codimension 1.

Since $H_u = A^{m(u)=1}$, we see that $A_0^{m(u)=1}$ is the union of all facets of A_0 corresponding to facets in H_u, so $\bigcap_j A_0^{m(u_j)=1} = \{v\}$. This description of v is *unaffected* by any polysimplicial automorphism of A_0 that *commutes* with the $N(k)$-action (e.g., commutes with the $m(u)$-actions for all $u \in U_a(k) - \{1\}$). Thus, all such automorphisms must fix v. But v was an arbitrary vertex of A_0, so we are done. □

Axiom 4. Via the unique $N(k)$-equivariant identification of the polysimplicial complex $\mathcal{A}(S)$ with $A(S)_0$ from Axiom 6, for each $g \in G(k)$ the polysimplicial isomorphism $g : \mathcal{A}(S) \simeq \mathcal{A}(gSg^{-1})$ arising from the g-action on $\mathcal{B}(G)$ arises from a (visibly necessarily unique) *affine* automorphism $g : A(S) \simeq A(gSg^{-1})$. Moreover, the *linear* derivative isomorphism $\nabla g : \mathrm{X}_*(S')_{\mathbf{R}} \simeq \mathrm{X}_*(gS'g^{-1})_{\mathbf{R}}$ is the canonical one (arising from post-composing cocharacters with g-conjugation on G).

The preceding Axioms 6 and 4 concern apartments as polysimplical structures and affine spaces. We will use this in Sect. 28 to *enrich* the combinatorial $\mathcal{B}(G)$ to a more substantive "space" that we'll endow with a metric (to be proved complete and non-positively curved). This will require a further axiom that relates $\mathcal{B}(G)$ to a Tits system structure on $G(k)^0$.

28 Metric and Enrichment of the Building

We now state an axiom of group-theoretic nature. Using this and the apartment axioms from Sect. 27, we will enrich the combinatorial restricted building to a more interesting kind of "space" for which a useful metric structure can be defined.

Axiom 9. Assume $\mathscr{D}G$ is k-isotropic (so $\dim \mathcal{B}(G) > 0$). Choose a chamber $\mathcal{C} \in \mathcal{A}(S)$. Define $\mathcal{G} = G(k)^0$, $\mathcal{I} = G(k)^0_{\mathcal{C}}$, and $\mathcal{N} = N(k) \cap \mathcal{G}$. We require that the

$\mathcal{G}$-action on $\mathcal{B}(G)$ satisfies hypotheses (i) and (ii) in Theorem 25.5, so $(\mathcal{I}, \mathcal{N})$ is a saturated BN-pair in $\mathcal{G}$. (Hypothesis (i) also follows from: Axiom 1, $G^{\mathrm{sc}}(k)$-conjugacy of all S's, and $N(k)$-conjugacy of all chambers in $\mathcal{A}(S)$ by (AS5) and (AS6) in Axiom 6.) We further require that for $W := \mathcal{N}/(\mathcal{I} \cap \mathcal{N})$ and its associated non-empty finite subset $\mathcal{S}$ of elements of order 2, all irreducible components $(W_i, \mathcal{S}_i)$ of $(W, \mathcal{S})$ have $\#\mathcal{S}_i \geq 2$. (The k-anisotropic k-simple factors of $\mathscr{D}G$ do not contribute to $(W, \mathcal{S})$.)

The subgroup $\mathcal{I}$ is called an *Iwahori subgroup* of $G(k)$; it varies through a $G(k)$-conjugacy class as we vary the choice of $\mathcal{C}$ (and when $\mathscr{D}G$ is k-anisotropic we define $\mathcal{I} = \mathcal{N} = G(k)^0$ since $\mathcal{B}(G)$ is then a point). Consquences of Axiom 9 (applied to maximal split k-tori $S \subset G$) arise from Theorem 25.5(2):

- $\mathcal{B}(G)$ is naturally identified with the restricted building of the BN-pair $(\mathcal{I}, \mathcal{N})$ (this is trivial when $\mathscr{D}G$ is k-anisotropic, as then $\mathcal{B}(G)$ is a point and the BN-pair satisfies $\mathcal{I} = \mathcal{N} = G(k)^0$);
- $G(k)^0_{\mathcal{F}} := \mathrm{Stab}_{\mathcal{G}}(\mathcal{F})$ acts trivially on $\mathrm{vert}(\mathcal{F})$ for every facet $\mathcal{F}$ of $\mathcal{B}(G)$;
- the subgroups $G(k)^0_{\mathcal{F}} \subset G(k)^0$ vary (without repetition) through precisely the parabolic subgroups of the BN-pair $(\mathcal{I}, \mathcal{N})$ as $\mathcal{F}$ varies through the facets of $\mathcal{B}(G)$.

To prepare for the metric space structure, we now *enhance* the abstract polysimplicial building. Since $\mathcal{B}(G)$ is covered by the thin chamber complexes $\mathcal{A}(S)$ that meet along *subcomplexes*, we can create a larger "space" (for now just a set) $B(G)$ by gluing the affine spaces $A(S)$ along the subsets corresponding to the union of facets where the $\mathcal{A}(S)$'s meet each other (the triple overlap condition is readily verified). The affineness aspect of Axiom 4 thereby allows us to define the $G(k)$-action on $B(G)$ by gluing the *affine* isomorphisms $g : A(S) \simeq A(gSg^{-1})$; the triple overlap compatibility holds since the restriction of an affine isomorphism to a facet *closure* is determined by its effect on the vertices (although we haven't put a topology on $B(G)$, the notion of *closure* $\overline{\mathcal{F}}$ of a facet $\mathcal{F}$ makes sense since it may be computed in any apartment containing the facet as the union of its subfacets).

By design, $B(G)$ is covered by *disjoint* facets for which the "closure" relation

$$\mathcal{F}' \prec \mathcal{F}$$

defined by $\mathcal{F}' \subset \overline{\mathcal{F}}$ corresponds to the combinatorial property $\mathrm{vert}(\mathcal{F}') \subset \mathrm{vert}(\mathcal{F})$ (i.e., subset inclusion for facets in the abstract polysimplicial complex).

For the $G(k)$-action on $B(G)$, the stabilizer of a facet is the *same* as the stabilizer of the corresponding abstract facet in $\mathcal{B}(G)$. Let $\mathcal{F}_0$ be an abstract facet in $\mathcal{B}(G)$ corresponding to a facet $\mathcal{F}$ in $B(G)$. The fixed-point locus of $G(k)^0_{\mathcal{F}_0}$ in the abstract polysimplicial complex $\mathcal{B}(G)$ consists of the abstract subfacets of $\mathcal{F}_0$, and $G(k)^0_{\mathcal{F}_0}$ acts *trivially* on $\mathrm{vert}(\mathcal{F}_0) = \mathrm{vert}(\mathcal{F})$ (see Theorem 24.11), so the fixed-point locus in $B(G)$ for the action by $G(k)^0_{\mathcal{F}}$ is $\overline{\mathcal{F}}$. Note that distinct facets in $B(G)$ are *disjoint* whereas in $\mathcal{B}(G)$ facets admit inclusion relations since they are combinatorial data of vertex sets.

Now we *may rename* $B(G)$ *as* $\mathcal{B}(G)$ *without causing confusion* (provided we are attentive to working with the relation $F' \prec F$ in place of $\mathrm{vert}(F') \subset \mathrm{vert}(F)$ from the combinatorial setting). In particular, the apartments $\mathcal{A}(S)$ are now really affine spaces and it makes sense to form the *pointwise stabilizers* $G(k)^0_x$ and $G(k)^1_x$ for all $x \in \mathcal{B}(G)$. Since the action of $G(k)^0$ carries facets to facets and we have seen that the $\mathcal{F}$-stabilizer $G(k)^0_{\mathcal{F}}$ coincides with $\bigcap_{x\in\mathcal{F}} G(k)^0_x$, the following definition makes sense:

Definition 28.1 For a non-empty bounded subset Ω of an apartment, define $G(k)^*_{\Omega} = \bigcap_{x\in\Omega} G(k)^*_x$ for $* = 0, 1$.

This is certainly bounded since each $G(k)^1_x$ is bounded (as it contains $G(k)^0_x$ with finite index, and $G(k)^0_x$ is bounded by Axiom 2), and it is also *open* because Ω meets only finitely many facets $\mathcal{F}_1, \ldots, \mathcal{F}_n$ and $G(k)^0_{\Omega} = \bigcap_j G(k)^0_{\mathcal{F}_j}$. The analogue of this latter equality for $G(k)^1$ is false in general, such as $G = \mathrm{PGL}_2$ and Ω a midpoint of an edge of the tree $\mathcal{B}(G)$. If Ω is contained in a facet $\mathcal{F}$ then $G(k)^0_{\Omega} = G(k)^0_{\mathcal{F}}$ coincides with the $G(k)^0$-stabilizer of Ω.

Definition 28.2 For a general non-empty bounded subset Ω of an apartment, define the *stabilizer*

$$G(k)^{\dagger}_{\Omega} = \{g \in G(k)^1 : g(\Omega) = \Omega\}$$

of Ω in $G(k)^1$.

Note that $G(k)^{\dagger}_{\Omega}$ is generally bigger than the pointwise-stabilizer $G(k)^1_{\Omega}$ (e.g., $G = \mathrm{PGL}_2$ and Ω a chamber). However, it contains the open $G(k)^1_{\Omega}$ with *finite index* and hence is also bounded (and of course open). Indeed, Ω meets only finitely many facets $\mathcal{F}_1, \ldots, \mathcal{F}_n$, so $G(k)^{\dagger}_{\Omega}$ permutes the *finite* set of vertices of the $\mathcal{F}_j$'s. Thus, there is a finite-index subgroup which fixes all of those vertices pointwise and hence acts as the identity on all $\mathcal{F}_j$'s, so that finite-index subgroup is contained in $G(k)^1_{\Omega}$.

Remark 28.3 The preceding discussion shows $G(k)^0_x = G(k)^0_y$ inside $G(k)^0$ if and only if x, y lie in the same facet. This is generally *false* for $G(k)^1$, as happens for PGL_2 and x the midpoint of an edge (where "midpoint" is defined using the affine space structure of apartments; this has no counterpart for the abstract building!).

Now we can finally make the function $d : \mathcal{B}(G) \times \mathcal{B}(G) \to \mathbf{R}_{\geq 0}$ that will be a metric with good properties. (As with affine root systems, ultimately this metric will allow for rescalings corresponding to each k-simple "factor" of G.) To get started, we pick a maximal split k-torus $S_0 \subset G$, let $\Phi_0 = \Phi(G, S_0)$, and pick a $W(\Phi_0)$-invariant inner product on $\mathrm{X}_*(S_0')_{\mathbf{R}}$. This defines a notion of distance on that vector space, and so also a notion of distance on the affine space $\mathcal{A}(S_0)$ since the "difference" between two points in this affine space is a vector in $\mathrm{X}_*(S_0')_{\mathbf{R}}$.

We transport this notion of distance to every apartment $\mathcal{A}(S)$ by picking $g \in G(k)$ such that $S = gS_0g^{-1}$ and using the *affine isomorphism* $g : \mathcal{A}(S_0) \simeq \mathcal{A}(S)$ to transport the distance function. To see this is independent of the choice of g, we observe that all the choices are related through right multiplication on g by $N_G(S_0)(k)$, so we just need to make sure that the $N_G(S_0)(k)$-action on $\mathcal{A}(S_0)$ through affine automorphisms is an isometry. Such an affine action commutes with "differences" in the affine space, so this becomes the claim that the natural derivative $N_G(S_0)(k)$-action on $X_*(S_0')_{\mathbf{R}}$ is an isometry. That in turn holds since we chose distance on this vector space to be $W(\Phi_0)$-invariant.

Let $d_{\mathcal{A}}$ denote the distance function we have just made on each apartment $\mathcal{A}$.

Proposition 28.4 *There is a unique function $d : \mathcal{B}(G) \times \mathcal{B}(G) \to \mathbf{R}_{\geq 0}$ restricting to $d_{\mathcal{A}}$ on each apartment, and it is $G(k)$-invariant.*

Proof Any two points $x, y \in \mathcal{B}(G)$ lie in respective facets $\mathcal{F}$ and $\mathcal{F}'$, and hence in chamber closures: $x \in \overline{\mathcal{C}}$ and $y \in \overline{\mathcal{C}}'$. But any two chambers lie in a common apartment (from the axioms for an abstract building), so there exists an apartment $\mathcal{A}$ that contains both chambers and hence their chamber closures. Thus, $x, y \in \mathcal{A}$. This proves the uniqueness of d. For existence, it suffices to show that if $\mathcal{A}, \mathcal{A}'$ are two apartments each containing x and y then

$$d_{\mathcal{A}}(x, y) = d_{\mathcal{A}'}(x, y).$$

Once such existence is settled, $G(k)$-invariance is immediate since for any $g \in G(k)$ the affine isomorphism $g : \mathcal{A} \simeq g\mathcal{A}$ is an isometry.

For our enriched notion of $\mathcal{B}(G)$, apartments are unions of facets and distinct facets are *disjoint*. Thus, if we pick the unique respective facets $\mathcal{F}, F$ in $\mathcal{A}$ containing respective points $x, y \in \mathcal{A} \cap \mathcal{A}'$ then these two facets also lie in $\mathcal{A}'$. The corresponding apartments $\mathcal{A}_0, \mathcal{A}_0'$ in the abstract polysimplicial restricted building associated to the Tits system structure on $G(k)^0$ each contain the corresponding abstract facets $\mathcal{F}_0 = \mathrm{vert}(\mathcal{F})$ and $F_0 = \mathrm{vert}(F)$. By Remark 25.3 (here we're using the consequence of Axiom 9 that the polysimplicial complex underlying $\mathcal{B}(G)$ is the reduced building for a Tits system structure on $G(k)^0$), there exists $g \in G(k)^0$ such that $\mathcal{A}_0' = g\mathcal{A}_0$ and g acts as the identity on the vertex sets $\mathcal{F}_0$ and F_0.

The affine isomorphism $g : \mathcal{A} \simeq \mathcal{A}'$ acts as the identity on the *vertices* of the facets $\mathcal{F}, F \subset \mathcal{A} \cap \mathcal{A}'$, but the effect of an *affine* isomorphism on a facet is determined by what it does to the vertices of the facet, so g is the identity on $\mathcal{F}$ and on F. Hence, $g(x) = x$ and $g(y) = y$. Since $g : \mathcal{A} \simeq \mathcal{A}'$ is an isometry (!), we obtain $d_{\mathcal{A}}(x, y) = d_{\mathcal{A}'}(g(x), g(y)) = d_{\mathcal{A}'}(x, y)$ as desired. □

Remark 28.5 Let's discuss how d changes if we change the initial choice of Weyl-invariant inner product on $X_*(S_0')_{\mathbf{R}}$. Let $\{G_i^{\mathrm{ad}}\}$ be the set of k-simple factors of the central quotient G^{ad}, and $\prod S_i^{\mathrm{ad}}$ be the corresponding decomposition of the maximal split torus image S^{ad} of S in G^{ad}, so the irreducible decomposition of the root system $(X_*(S_0')_{\mathbf{R}}, \Phi(G, S_0))$ is $\prod (X_*(S_i')_{\mathbf{R}}, \Phi_i)$ for $\Phi_i = \Phi(G_i^{\mathrm{ad}}, S_i^{\mathrm{ad}})$ provided we drop those i for which G_i^{ad} is k-anisotropic (equivalently, Φ_i is empty). Via the natural bijection $\mathcal{B}(G) \simeq \prod \mathcal{B}(G_i^{\mathrm{ad}})$ respecting polysimplicial structures and affine space structures on apartments (the factor $\mathcal{B}(G_i^{\mathrm{ad}})$ is a point exactly when the k-simple G_i^{ad} is k-anisotropic), the decomposition $W(\Phi) = \prod W(\Phi_i)$ has $W(\Phi_i)$ absolutely irreducible on $X_*(S'_i)_{\mathbf{R}}$ (when $\Phi_i \neq \emptyset$) and $d := \sqrt{\sum d_i^2}$ where d_i on $\mathcal{B}(G_i^{\text{ad}})$ arises from a choice of $W(\Phi_i)$-invariant inner product on $X_*(S'_i)_{\mathbf{R}}$.

In other words, d on $\mathcal{B}(G)$ is assembled as a "Euclidean sum" of corresponding d_i's on the $\mathcal{B}(G_i^{\text{ad}})$'s. The choice of each d_i is unique up to a positive scaling factor since $X_*(S'_i)_{\mathbf{R}}$ is absolutely irreducible under $W(\Phi_i)$, so such scalings on those factors (only interesting for G_i^{ad} that are k-isotropic) account for the possible variation of d. In particular, any two such d's are bounded by constant multiples of each other. Thus, for the way d is used in Bruhat-Tits theory such variation in d is harmless.

In Appendix A we give the proof that d is a metric that makes $\mathcal{B}(G)$ a complete and contractible uniquely geodesic metric space which is moreover non-positively curved. The Bruhat-Tits fixed point lemma then applies to this metric space, but to use that for the study of bounded subgroups of $G(k)$ we need to relate boundedness in the metric space (the focus of the fixed-point lemma) to boundedness in $G(k)$. That relationship between notions of boundedness is discussed in Sect. 29.

29 Boundedness and Applications

Assume $\mathscr{D}G$ is k-isotropic. Let $S \subset G$ be a maximal split k-torus (so $\Phi := \Phi(G, S)$ is non-empty), $N = N_G(S)$, $Z = Z_G(S)$, $\mathcal{A} = \mathcal{A}(S)$ (so $\dim \mathcal{A} > 0$), and let $\Psi, \Psi' \subset \mathcal{A}^*$ be the affine root systems from Axiom 6 in Sect. 27.

Axiom 8. The following properties hold:

(i) For all $\psi \in \Psi$ with $\dot{\psi} = a \in \Phi$, the half-space $\mathcal{A}^{\psi \geq 0}$ is preserved by $U_\psi \subset U_a(k)$.

(ii) Let $\Omega \subset \mathcal{A}$ be a non-empty bounded subset. For $a \in \Phi$, let ψ_a^Ω be the least $\psi \in \Psi$ with $\dot{\psi} = a$ such that $\Omega \subset \mathcal{A}^{\psi \geq 0}$.
The group $G(k)_\Omega^0$ is generated by its subgroups $Z(k)^0$ (which acts trivially on $\mathcal{A}$, as does $Z(k)^1$) and the $U_{\psi_a^\Omega}$'s for $a \in \Phi(G, S)$. (Since Z is k-anisotropic modulo its center, $Z(k)^0 = Z(K)^0 \cap Z(k)$ by definition.)

(iii) For $a \in \Phi$, $G(k)_\Omega^0 \cap U_a(k) = U_{\psi'^\Omega_a}$ where ψ'^Ω_a is the least $\psi' \in \Psi'$ with derivative a such that $\Omega \subset \mathcal{A}^{\psi' \geq 0}$.

Remark 29.1 The containment $Z(k)^0 \subset G(k)_\Omega^0 \cap Z(k)$ is an equality. This is easy to prove granting the existence of the building satisfying the preceding axioms, since any $z \in Z(k)$ acts on $\mathcal{A}$ as a translation encoded by $\omega_Z(z)$, so fixing even one point of Ω forces $z \in Z(k)^1$ (and hence if also $z \in G(k)^0$ then $z \in Z(k)^0$ by Proposition 11.4). However, such reasoning is misleading because the *construction* of Bruhat-Tits theory over k includes a direct verification of that equality, as is done in [KP, Sect. 9.4].

Axiom 16. (Iwahori decomposition) Let $\mathcal{C} \subset \mathcal{A}$ be a chamber, $\mathcal{I} = G(k)^0_{\mathcal{C}}$ the associated Iwahori subgroup, $\Phi^+ \subset \Phi$ a positive system of roots, and $\Phi^- = -\Phi^+$. Let $U^{\pm} = \mathscr{R}_{u,k}(P^{\pm})$ for the minimal parabolic k-subgroups $P^{\pm} \subset G$ containing S that correspond to $\Phi^{\pm}$.

The multiplication map

$$(U^+(k) \cap \mathcal{I}) \times Z(k)^0 \times (U^-(k) \cap \mathcal{I}) \to \mathcal{I}$$

is bijective, as is the multiplication map (for each sign)

$$\prod_{a \in \Phi^{\pm,\mathrm{nd}}} U_{\psi_a^{\mathcal{C}}} \to U^{\pm}(k) \cap \mathcal{I}$$

(using any choice of enumeration of $\Phi^{\pm,\mathrm{nd}}$ for each sign).

Remark 29.2 There is an axiom [KP, Axiom 4.1.17] which we don't precisely state here that encodes the behavior of $\mathcal{B}(G)$ with respect to isomorphisms in the pair (k, G); it is also a formal consequence of Proposition 30.1 below. The purpose of this axiom is to provide an *action* by $\mathrm{Gal}(K/k)$ on $\mathcal{B}(G_K)$ through isometric polysimplicial automorphisms when Bruhat-Tits theory is available over K.

The existence of the Galois action amounts to (unique) polysimplicial isomorphisms of affine spaces (even affine root systems) $[\gamma]_T : \mathcal{A}(T) \simeq \mathcal{A}(\gamma^*(T))$ for maximal split K-tori $T \subset G_K$ and $\gamma \in \mathrm{Gal}(K/k)$ such that $[\gamma'\gamma]_T = [\gamma']_{\gamma^*(T)} \circ [\gamma]_T$, and showing they glue (as we vary T) to an action on $\mathcal{B}(G_K)$. This action is a key ingredient in "unramified descent" for buildings to be stated in Sect. 30.

Proposition 29.3 *We have* $G(k) = \mathcal{I}N(k)\mathcal{I}$.

Remark 29.4 Since $Z(k)^0 \subset G(k)^0_{\mathcal{C}} = \mathcal{I}$, this proposition implies that a set of representatives for $N(k)/Z(k)^0$ in $N(k)$ represents all $\mathcal{I}$ double-cosets for $G(k)$. In fact, it represents them *uniquely*: this follows from a more refined argument given in Appendix D (see Proposition D.1), where we additionally use Bruhat-Tits theory to prove the *Cartan decomposition* that gives double-coset representatives (also unique modulo $Z(k)^0$) for $G(k)$ relative to certain *maximal* parahoric subgroups.

Proof Let $\mathcal{G} = G(k)^0$, and define $\mathcal{N} = N(k) \cap \mathcal{G} = \mathrm{Stab}_{\mathcal{A}}(\mathcal{G})$ (the final equality uses that $N(k) = \mathrm{Stab}_{\mathcal{A}}(G(k))$). By Axiom 6 in Sect. 27 the pair $(\mathcal{I}, \mathcal{N})$ is a (saturated) BN-pair for $\mathcal{G}$, so

$$\mathcal{G} = \mathcal{I}\mathcal{N}\mathcal{I}$$

by the Bruhat decomposition for abstract Tits systems.

Since $\mathcal{G}$ acts transitively on the set of apartments in $\mathcal{B}(G)$ (as this building is the restricted building associated to a Tits system structure for $\mathcal{G} = G(k)^0$), we have

$G(k) = N(k)\mathcal{G}$. The group $\mathcal{G}$ acts transitively on the set of pairs $(\mathcal{A}', \mathcal{C}')$ (Remark 25.3), so $\mathcal{N}$ acts transitively on the set of chambers in $\mathcal{A}$. But $N(k)$ also acts on that set of chambers, so $N(k) = N(k)_{\mathcal{C}}\mathcal{N}$. Hence,

$$G(k) = N(k)\mathcal{G} = N(k)_{\mathcal{C}}\mathcal{N}\mathcal{G} = N(k)_{\mathcal{C}}\mathcal{G}.$$

Plugging in the Bruhat decomposition for $\mathcal{G}$, we get

$$G(k) = N(k)_{\mathcal{C}}\mathcal{I}\mathcal{N}\mathcal{I}.$$

Conjugation by elements of $N(k)_{\mathcal{C}}$ carries $\mathcal{I} = G(k)^0_{\mathcal{C}}$ back onto itself, so

$$G(k) = \mathcal{I}N(k)_{\mathcal{C}}\mathcal{N}\mathcal{I} = \mathcal{I}N(k)\mathcal{I}.$$

□

We have

$$N(k) \supset Z(k) \supset Z(k)^1 \supset Z(k)^0,$$

where the first and third inclusions have finite index, and the second inclusion comes from the injective homomorphism

$$\omega_Z : Z(k)/Z(k)^1 \hookrightarrow \mathrm{Hom}_{\mathbf{Z}}(\mathrm{X}^*_k(Z), \mathbf{Z})$$

so $Z(k)/Z(k)^1$ is a $\mathbf{Z}$-lattice. Since Iwahori subgroups are bounded, this motivates the hope that a subset $\Sigma \subset G(k)$ is bounded if and only if its Iwahori representatives in $N(k)$ have finitely many images modulo $Z(k)^0$. But $Z(k)^0 \subset \mathcal{I}$, so the hope is that Σ is bounded in $G(k)$ if and only if it meets only finitely many $\mathcal{I}$ double cosets in $G(k)$; this would be a link between the valuation topology of $G(k)$ and the group theory of $G(k)$.

The next result makes this hope precise and uses it as a bridge giving a crucial link between boundedness inside $G(k)$ (for its valuation topology) and boundedness in the metric topology on $\mathcal{B}(G)$. This opens the door to using the Bruhat-Tits fixed point lemma (Theorem 14.1(2)) and other aspects of the building to analyze bounded subgroups of $G(k)$.

Theorem 29.5 *For a subset $\Sigma \subset G(k)^1$, the following are equivalent:*

(1) Σ is bounded in $G(k)$;
(2) the subset $\Sigma \subset G(k) = \mathcal{I}N(k)\mathcal{I}$ meets only finitely many $\mathcal{I}$ double cosets;
(3) for all $x \in \mathcal{B}(G)$, the "orbit" $\Sigma.x := \{\sigma.x : \sigma \in \Sigma\} \subset \mathcal{B}(G)$ is bounded;
(3′) for some $x \in \mathcal{B}(G)$, the "orbit" $\Sigma.x$ is bounded.

In practice we care about the relationship of (1) to (3) (or (3′)), but we need (2) to make that bridge.

Proof Since $\mathcal{I}$ is bounded in $G(k)$, (1) is equivalent to boundedness of $\mathcal{I}\Sigma\mathcal{I}$ in $G(k)$. This in turn is equivalent to $(\mathcal{I}\Sigma\mathcal{I}) \cap N(k)$ being bounded (since every $\mathcal{I}$-double coset in $G(k)$ has a representative in $N(k)$), so (1) is equivalent to the property $\Sigma \subset \mathcal{I}\mathcal{Y}\mathcal{I}$ for a bounded subset $\mathcal{Y} \subset N(k)$.

Since $N(k)/Z(k)$ is finite, we can pick a finite subset $\mathcal{S} \subset N(k)$ representing $N(k)/Z(k)$, so a bounded subset of $N(k)$ has bounded intersection with each of the finitely many cosets $\sigma Z(k)$ for $\sigma \in \mathcal{S}$. Thus, (1) is equivalent to the property $\Sigma \subset \mathcal{I}\mathcal{S}\mathcal{Y}'\mathcal{I}$ for a bounded subset $\mathcal{Y}' \subset Z(k)$.

Since $Z(k)^1$ is *bounded* (because Z is k-anisotropic modulo its center), a subset $\mathcal{Y}' \subset Z(k)$ is bounded if and only if $\mathcal{Y}'Z(k)^1$ is bounded. The injectivity of

$$\omega_Z : Z(k)/Z(k)^1 \hookrightarrow \mathrm{Hom}(\mathrm{X}_k^*(Z), \mathbf{Z})$$

implies that $Z(k)/Z(k)^1$ is a finite free $\mathbf{Z}$-module. Hence, it is reasonable to expect that a subset of $Z(k)$ is bounded if and only if its image in $Z(k)/Z(k)^1$ is *finite*. This intuitive expectation requires a proof:

Lemma 29.6 *A subset $Y \subset Z(k)$ is bounded if and only if its image in $Z(k)/Z(k)^1$ is finite.*

Proof The "if" implication is clear, since $Z(k)^1$ is bounded in $Z(k)$. Now assume Y is bounded; we want to show Y has finite image in $Z(k)/Z(k)^1$.

For the maximal *split* torus quotient A of Z, a bounded subset of $Z(k)$ has bounded image in $A(k)$. Since A is a k-split torus, a subset of $A(k)$ is bounded if and only if its image in $A(k)/A(k)^1$ is finite. But the preimage of $A(k)^1$ under $Z(k) \to A(k)$ is $Z(k)^1$ (!), so $Z(k)/Z(k)^1 \to A(k)/A(k)^1$ is injective. Hence, Y has finite image in $Z(k)/Z(k)^1$ as desired. □

Since $Z(k)^1/Z(k)^0$ is finite (Theorem 12.1), finiteness for the image of a subset $\mathcal{Y}' \subset Z(k)$ in $Z(k)/Z(k)^1$ is the same as finiteness of its image in $Z(k)/Z(k)^0$. But $Z(k)^0 \subset \mathcal{I}$, so if $\Sigma \subset G(k)$ is bounded then $\Sigma \subset \mathcal{I}\mathcal{S}\mathcal{Y}'\mathcal{I}$ for a subset $\mathcal{Y}' \subset Z(k)$ contained in finitely many $Z(k)^0$-cosets and hence $\mathcal{I}\mathcal{S}\mathcal{Y}'\mathcal{I}$ consists of only finitely many $\mathcal{I}$ double cosets. This shows (1) implies (2). The converse is obvious (as $\mathcal{I}$ is bounded in $G(k)$).

We will now show (2) $\Rightarrow$ (3) and (3′) $\Rightarrow$ (2); since (3) $\Rightarrow$ (3′) is obvious, the equivalence of all four conditions in Theorem 29.5 would then be established. Since (1) doesn't mention a specific choice of chamber or apartment and (1) is equivalent to (2), to deduce (3) for a given $x \in \mathcal{B}(G)$ from (2) we can pick a chamber $\mathcal{C}$ whose closure contains x and work with $\mathcal{I} = G(k)_{\mathcal{C}}^0$ in (2). For each $n \in N(k)$ we just need to show $\mathcal{I}n\mathcal{I}.x$ is bounded in $\mathcal{B}(G)$.

The Iwahori subgroup $\mathcal{I}$ acts trivially on $\overline{\mathcal{C}}$, so $\mathcal{I}n\mathcal{I}.x = \mathcal{I}n.x$. To show this is bounded in the metric of the building, it suffices to show that as g varies through $\mathcal{I}$, $d(gn.x, x)$ is bounded. But d is $G(k)$-invariant, so $d(gn.x, x) = d(n.x, g^{-1}.x) = d(n.x, x)$ since $\mathcal{I}$ acts trivially on x. This completes the proof that (2) implies (3).

Finally, we show (3′) for some $x_0 \in \mathcal{B}(G)$ implies (2). Since (2) is equivalent to (1) which doesn't mention a specific chamber or apartment, it suffices to show (3′) implies (2) for $\mathcal{I} = G(k)^0_{\mathcal{C}}$ for a chamber $\mathcal{C}$ whose closure contains x_0 and an apartment $\mathcal{A} \supset \mathcal{C}$.

Arguing by the contrapositive, it suffices to show that if $\{n_i\}$ is a subset of $N(k)$ representing infinitely many $\mathcal{I}$ double cosets meeting a subset $\Sigma \subset G(k)$ then $\Sigma.x_0$ is *not* bounded in $\mathcal{B}(G)$, as that would contradict the hypothesis (3′). Since $N(k)/Z(k)$ is finite, by passing to a suitable subset of $\{n_i\}$ we can assume all n_i's lie in the same $Z(k)$-coset of $N(k)$: $n_i = n_0 z_i$ for some fixed $n_0 \in N(k)$ and various $z_i \in Z(k)$.

We shall prove the collection of chambers $n_i\mathcal{C}$ in the apartment $\mathcal{A}$ is *infinite*. Granting that for a moment, by local finiteness of the polysimplicial structure in $\mathcal{A}$ and the containment $x_0 \in \overline{\mathcal{C}}$ we could then conclude that the collection of points $n_i.x_0$ has unbounded diameter. Since $\mathcal{I}$ acts on the building through isometries and fixes x_0, by the metric calculation at the end of the proof that (2) $\Rightarrow$ (3) it follows that *all* points in the "orbit" $\mathcal{I}n_i.x_0 = \mathcal{I}n_i\mathcal{I}.x_0$ have the same distance $d(n_i.x_0, x_0)$ to x_0. Hence, unboundedness of $d(n_i.x_0, x_0)$ as i varies forces points in $\Sigma.x_0$ to have unbounded distance from x_0, as desired to get a contradiction.

It remains to show that the collection of chambers $n_i\mathcal{C}$ in $\mathcal{A}$ is infinite. Suppose for some i, j that $n_i\mathcal{C} = n_j\mathcal{C}$, so $n_j^{-1}n_i \in N(k)_{\mathcal{C}}$. Recall we passed to the case $n_i = n_0 z_i$ for some fixed $n_0 \in N(k)$ and $z_i \in Z(k)$. Hence, $n_j^{-1}n_i = z_j^{-1}z_i \in Z(k)$. But $Z(k)$ acts on $\mathcal{A}$ through translations via $-\omega_Z$ and $n_j^{-1}n_i$ *preserves* $\mathcal{C}$, which no nonzero translation can do, so $z_j^{-1}z_i \in Z(k)^1$ and hence $n_i Z(k)^1 = n_j Z(k)^1$ for such i, j.

Since $Z(k)^1/Z(k)^0$ is finite, it follows that if there are only finitely many distinct chambers $n_i\mathcal{C}$ as we vary i then there are only finitely many possibilities for $n_i Z(k)^0$ as we vary i. But $Z(k)^0 \subset \mathcal{I}$, so there are only *finitely many* distinct double cosets $\mathcal{I}n_i\mathcal{I}$ as we vary i. This contradicts what we assumed about the n_i's, so the collection of chambers $n_i\mathcal{C}$ is infinite as desired. □

We can combine Theorem 29.5 with the Bruhat-Tits fixed point lemma to finally obtain some long-promised relationships between maximal bounded subgroups of $G(k)$ and point-stabilizers in the building:

Corollary 29.7 *The following are true.*

(1) For $x \in \mathcal{B}(G)$ a vertex, *$G(k)^1_x$ is a maximal bounded subgroup of $G(k)$.*
(2) Each maximal bounded (open) subgroup $\mathcal{K} \subset G(k)$ is $G(k)^1_x$ for some $x \in \mathcal{B}(G)$. If $G(k)^0 = G(k)^1$ (as happens when G is semisimple and simply connected, for example) then such an x must be a vertex.
(3) Maximal bounded subgroups of $G(k)^0$ are exactly $G(k)^0_x$ for vertices x (maximal parahorics!).

Before proving this corollary, we make some comments.

Remark 29.8 In (2), x is not necessarily a vertex (when G isn't simply connected). For example, with $G = \mathrm{PGL}_2$ maximality happens with x the midpoint of an edge of the tree $\mathcal{B}(\mathrm{PGL}_2) = \mathcal{B}(\mathrm{SL}_2)$. This illustrates the importance of going beyond the purely combinatorial theory of abstract buildings (which knows only about subfacets of facets, and particularly vertices of facets).

In connection with (3), first note that every bounded subgroup $\mathcal{K}$ of $G(k)^0$ lies in an open bounded subgroup (e.g., $G(k)^0 \cap \mathcal{K}'$ for an open bounded subgroup $\mathcal{K}'$ of $G(k)$ containing $\mathcal{K}$), and more importantly every bounded open subgroup of $G(k)^0$ lies in a maximal bounded subgroup of $G(k)^0$ (since the Zorn's Lemma method of proof of Proposition 1.3 that bounded open subgroups of the open subgroup $G(k)^0 \subset G(k)$ lie in maximal bounded subgroups of $G(k)$ adapts to using only subgroups of $G(k)^0$). Although (3) focuses on such subgroups, we emphasize that any maximal bounded subgroup $\mathcal{K}$ of $G(k)^0$ has finite index in a maximal bounded subgroup of $G(k)$. This holds because bounded subgroups of $G(k)$ lie in $G(k)^1$ and $G(k)^1/G(k)^0$ is finite. (Indeed, if $\mathcal{K}' \subset G(k)^1$ is a bounded subgroup of $G(k)$ containing $\mathcal{K}$ then $\mathcal{K}' \cap G(k)^0$ is a bounded subgroup containing $\mathcal{K}$ and hence *equal* to $\mathcal{K}$ by maximality of $\mathcal{K}$ in $G(k)^0$. Thus, $\mathcal{K}'/\mathcal{K} \to G(k)^1/G(k)^0$ is injective, so $[\mathcal{K}' : \mathcal{K}]$ is bounded independently of $\mathcal{K}'$).

Proof Suppose $\mathcal{K} \subset G(k)$ is a *bounded* subgroup, so $\mathcal{K} \subset G(k)^1$. The non-empty subset $\mathcal{K}.x_0 \subset \mathcal{B}(G)$ is bounded by Theorem 29.5 and preserved by $\mathcal{K}$! Thus there exists $y \in \mathcal{B}(G)$ such that $\mathcal{K} \subset G(k)^1_y$ by the Bruhat-Tits fixed point lemma (e.g., y can be taken to be the metric center of the convex hull of $\mathcal{K}.x_0$). But $G(k)^1_y$ is *bounded* since it contains $G(k)^0_y$ with finite index (as $G(k)^1/G(k)^0$ is finite) and all point stabilizers for $G(k)^0$ are bounded by Axiom 2 in Sect. 25. Therefore $\mathcal{K} = G(k)^1_y$ when $\mathcal{K}$ is maximal as a bounded subgroup of $G(k)$. This shows the first assertion in (2).

For the second assertion in (2), suppose $G(k)^0 = G(k)^1$, and assume $G(k)^1_x = G(k)^0_x$ is maximal as a bounded open subgroup of $G(k)$ for some $x \in \mathcal{B}(G)$. We want to show that x must be a vertex. Let $\mathcal{F}$ be the facet of $\mathcal{B}(G)$ containing x and let y be a vertex of $\mathcal{F}$, so $G(k)^0_x \subset G(k)^0_{\mathcal{F}} \subset G(k)^0_y$ (recall $G(k)^0_{\mathcal{F}}$ acts trivially on $\mathcal{F}$). The assumed maximality of $G(k)^0_x$ as a bounded subgroup of $G(k)$ then makes the containment $G(k)^0_x \subset G(k)^0_y$ of bounded subgroups of $G(k)$ an *equality*. But for any $z \in \mathcal{B}(G)$ the subgroup $G(k)^0_z \subset G(k)$ coincides with the parahoric subgroup $G(k)^0_{\mathcal{F}'}$ for the unique facet $\mathcal{F}'$ containing z (see Remark 28.3), so taking $z = x$ (so $\mathcal{F}' = \mathcal{F}$) yields $G(k)^0_{\mathcal{F}} = G(k)^0_y$. This is an equality of parahoric subgroups of $G(k)^0$, so by the theory of Tits systems it forces the corresponding facets $\mathcal{F}$ and $\{y\}$ to *coincide*. But $x \in \mathcal{F}$, so $x = y$ is a vertex as desired. This completes the proof of (2).

Turning to the proof of (1), consider a vertex $x \in \mathcal{B}(G)$ and the bounded subgroup $G(k)^1_x$. This is contained in a maximal bounded subgroup, which by (2) is $G(k)^1_y$ for some $y \in \mathcal{B}(G)$. Thus, y is fixed by $G(k)^1_x$, and in particular by $G(k)^0_x$. But since x is a vertex, hence a (0-dimensional) facet, $\mathcal{B}(G)^{G(k)^0_x} = \overline{\{x\}} = \{x\}$. This forces $y = x$, so the maximal bounded $G(k)^1_y$ equals the initial $G(k)^1_x$.

Finally, to prove (3) we consider a maximal bounded subgroup $\mathcal{K}$ of $G(k)^0$. By Remark 29.8, $\mathcal{K} = G(k)^0 \cap \mathcal{K}'$ for a maximal bounded subgroup $\mathcal{K}'$ of $G(k)$. By (2), we have $\mathcal{K}' = G(k)^1_y$ for some $y \in \mathcal{B}(G)$. Hence, $G(k)^0 \cap \mathcal{K}' = G(k)^0_y$, and this in turn is equal to $G(k)^0_{\mathcal{F}}$ for the unique facet $\mathcal{F}$ containing y (see Remark 28.3). The containment $\mathcal{K} \subset G(k)^0_{\mathcal{F}}$ of bounded subgroups of $G(k)^0$ must be an equality since we assumed $\mathcal{K}$ is maximal as such a subgroup of $G(k)^0$, so then $G(k)^0_{\mathcal{F}}$ is in particular maximal as a parahoric subgroup of $G(k)^0$. This latter maximality forces $\mathcal{F}$ to be a vertex $\{x\}$, yet $y \in \mathcal{F}$, so $y = x$. Hence, $\mathcal{K} = G(k)^0_x$ for a vertex x, as desired. □

Corollary 29.9 *There are finitely many $G(k)$-conjugacy classes of maximal bounded subgroups of $G(k)$.*

Proof Let $\mathcal{C}$ be a chamber in $\mathcal{B}(G)$. By Corollary 29.7, any maximal bounded subgroup $\mathcal{K}$ of $G(k)$ has the form $G(k)^1_x$ for some $x \in \mathcal{B}(G)$. The $G(k)$-orbit of x meets the closure $\overline{\mathcal{C}}$. Since $g(G(k)^1_x)g^{-1} = G(k)^1_{g.x}$, considering $\mathcal{K}$ up to $G(k)$-conjugacy allows us to reduce to considering only $x \in \overline{\mathcal{C}}$.

The chamber closure $\overline{\mathcal{C}}$ is the union of the finitely many facets $\mathcal{F} \prec \mathcal{C}$. There is a unique such $\mathcal{F}$ that contains x, and this is the only facet in $\mathcal{B}(G)$ containing x, so

$$\mathcal{K} = G(k)^1_x \subset \mathrm{Stab}_{G(k)^1}(\mathcal{F}) \subset \mathrm{Stab}_{G(k)^1}(\overline{\mathcal{F}}).$$

Since $\overline{\mathcal{F}}$ is a non-empty closed bounded convex subset of $\mathcal{B}(G)$, by Theorem 14.1(1) there is a unique metric center $c \in \overline{\mathcal{F}}$. Thus, $\mathrm{Stab}_{G(k)^1}(\overline{\mathcal{F}}) \subset G(k)^1_c$. But $G(k)^1_c$ is *bounded* (as $G(k)^1_y$ is for every $y \in \mathcal{B}(G)$), so the inclusions

$$\mathcal{K} \subset \mathrm{Stab}_{G(k)^1}(\mathcal{F}) \subset G(k)^1_c$$

are equalities due to the assumed maximality of $\mathcal{K}$ as a bounded subgroup of $G(k)$. Thus $\mathcal{K} = \mathrm{Stab}_{G(k)^1}(\mathcal{F})$, so the number of $G(k)$-conjugacy classes of maximal bounded subgroups of $G(k)$ is at most the (finite) number of facets $\mathcal{F} \prec \mathcal{C}$. □

In general it is difficult to explicitly count the finite number in the preceding result. However, in an important special case we can make the count (and even conjugacy class representatives) explicit:

Corollary 29.10 *If G is semisimple and simply connected with k-simple factors G_i having respective k-ranks r_i then the number of conjugacy classes of maximal bounded subgroups of $G(k)$ is $\prod(r_i + 1)$. These classes are uniquely represented by the point-stabilizers $G(k)_x$ for vertices x of a single chamber $\mathcal{C}$ in $\mathcal{B}(G)$.*

Proof Since G is semisimple we have $G(k) = G(k)^1$, and since G is simply connected we have $G(k)^1 = G(k)^0$. By Corollary 29.7 the maximal bounded subgroups of $G(k)$ are precisely $G(k)_x$ for the vertices x of $\mathcal{B}(G)$. Since $gG(k)_x g^{-1} = G(k)_{g.x}$, and the $G(k)$-action on $\mathcal{B}(G)$ moves each point into the closure of a fixed choice of chamber $\mathcal{C}$ (i.e., into a subfacet of $\mathcal{C}$), we can focus on vertices of $\mathcal{C}$.

Next we check that for distinct vertices x, y of $\mathcal{C}$, if the corresponding stabilizers $G(k)_x$ and $G(k)_y$ are $G(k)$-conjugate, or equivalently x and y are in the same $G(k)$-orbit, then $x = y$. More generally, if $\mathcal{F}$ and $\mathcal{F}'$ are subfacets of $\mathcal{C}$ in the same $G(k)$-orbit then we'll show they are the same facet.

Since $G(k) = G(k)^0$, by Axiom 9 in Sect. 28 we have a BN-pair for $G(k)$ whose associated restricted building is the polysimplicial complex underlying $\mathcal{B}(G)$ (whose chambers coincide with those of the "unrestricted" abstract building). Thus, if two facets of a common chamber of $\mathcal{B}(G)$ are in the same $G(k)$-orbit then they must be the *same* facet by Theorem 24.11(4) (facets of $\mathcal{B}(G)$ are admissible facets of the simplicial building associated to a Tits system structure for $G(k)^0$). The equality $g\mathcal{F}' = \mathcal{F}$ for some $g \in G(k)$ then forces $\mathcal{F} = \mathcal{F}'$ as desired.

To count the number of vertices of the polysimplex $\mathcal{C}$ as $\prod(r_i + 1)$, let $\{G_i^{\mathrm{ad}}\}$ be the k-simple factors of G^{ad}, so we have a polysimplicial isomorphism $\mathcal{B}(G) = \mathcal{B}(G^{\mathrm{ad}}) = \prod \mathcal{B}(G_i^{\mathrm{ad}})$ under which a chamber $\mathcal{C} \subset \mathcal{B}(G)$ is a product $\prod \mathcal{C}_i$ of chambers $\mathcal{C}_i \subset \mathcal{B}(G_i^{\mathrm{ad}})$, and a vertex v of $\mathcal{C}$ is a tuple (v_i) of vertices v_i of $\mathcal{C}_i$. We always use the convention that a 1-point building has one chamber and one vertex (the unique point).

We thereby easily reduce to the k-simple case, so $\mathcal{C}$ is a *simplex* in an r-dimensional affine space for the k-rank r of G (this works even when $r = 0$, which is to say G is k-anisotropic, in which case $\mathcal{B}(G)$ is a point). Such a simplex has $r + 1$ vertices (even when $r = 0$). □

30 Bruhat-Tits Group Schemes and Applications

The Bruhat-Tits theory axioms stated so far determine the building uniquely:

Proposition 30.1 *If $\mathcal{B}_1$ and $\mathcal{B}_2$ covered by the canonical affine spaces $\mathcal{A}(S)$ for maximal k-split tori S in G satisfy Axioms 1, 4, 8, and 9 then there is a unique $G(k)$-equivariant isomorphism $\mathcal{B}_1 \simeq \mathcal{B}_2$ inducing the unique $N_G(S)(k)$-equivariant isomorphism of affine spaces $\mathcal{A}_1(S) \simeq \mathcal{A}_2(S)$ for some S (and then for all S).*

Proof See [KP, Sect. 4.4], particularly [KP, Proposition 4.4.6, Cor. 4.4.7] (which rests on most of [KP, Sect. 4.4]). □

Axiom 27. (Unramified descent) Let $\Gamma = \mathrm{Gal}(K/k)$, equipped with its natural action on $\mathcal{B}(G_K)$ that commutes with the action of the subgroup $G(k) \subset G(K)$.

(i) For all maximal k-split tori $S \subset G$, there exists a k-torus $T \subset G$ containing S (so $T \subset Z_G(S)$) for which T_K is a maximal split K-torus in G_K (such T are called *special*), and the subset $\mathcal{A}(T_K)^\Gamma \subset \mathcal{B}(G_K)^\Gamma$ is independent of such T (hence it is stable under $N_G(S)(k)$).

(ii) Via the inclusion $X_*(S')_{\mathbf{R}} \subset X_*(T'_K)_{\mathbf{R}}$, the subset $\mathcal{A}(T_K)^{\Gamma} \subset \mathcal{A}(T_K)$ is an affine space over $X_*(S')_{\mathbf{R}}$ and is $N_G(S)(k)$-equivariantly isomorphic (uniquely, by Lemma 26.2) to $\mathcal{A}(S)$.
(iii) The maps $\mathcal{A}(S) \simeq \mathcal{A}(T_K)^{\Gamma} \hookrightarrow \mathcal{B}(G_K)$ glue to a $G(k)$-equivariant bijection $\mathcal{B}(G) \simeq \mathcal{B}(G_K)^{\Gamma}$.
(iv) When $\mathcal{B}(G)$ is not a point (equivalently, $S' \neq 1$), the surjections $\mathcal{A}(T_K)^* \twoheadrightarrow \mathcal{A}(S)^*$ carry $\Psi(G_K, T_K)$ into $\Psi(G, S) \cup \mathbf{R}$ and hit the entirety of $\Psi(G, S)$.

Remark 30.2 In [KP, Axiom 4.1.27], the corresponding axiom is stated with a slightly different ordering of parts, and it includes a further condition that we incorporate into Axiom 20 below.

Note that if T is a special k-torus in G containing a maximal split k-torus $S \subset G$ then the inclusion $X_*(S')_{\mathbf{R}} = X_*(S'_K)_{\mathbf{R}} \hookrightarrow X_*(T'_K)_{\mathbf{R}}$ is equivariant for the natural map $N_G(T)(k)/Z_G(T)(k) \to W(G, S)$.

Remark 30.3 The proof of existence of special k-tori is rather non-trivial, relying on a lot of Bruhat-Tits theory over K, particularly Bruhat-Tits integral models discussed in Axiom 20 below; see [KP, Proposition 9.3.4].

For non-empty bounded $\Omega \subset \mathcal{A}(S)$, the inclusion $\mathcal{A}(S) \hookrightarrow \mathcal{A}(T_K)$ via Axiom 27 allows us to view Ω as a non-empty bounded subset of an apartment of $\mathcal{B}(G_K)$ too, so it makes sense to form $G(K)^*_{\Omega}$ for $* = 0, 1$ and clearly $G(K)^*_{\Omega} \cap G(k)^* = G(k)^*_{\Omega}$. Here we are using notations for pointwise stabilizer groups from Definitions 28.1 and 28.2.

One reason for interest in general Ω beyond facets is that applications of the fixed-point lemma often gives points that aren't vertices (e.g., midpoint of an edge for the tree associated to PGL_2 and SL_2), and another is that when we want to prove a result with a facet $\mathcal{F}$ of $\mathcal{B}(G)$ by viewing it as a subset of $\mathcal{B}(G_K)$ then in there it is generally not a facet (it is merely a non-empty bounded subset of an apartment).

Remark 30.4 Beware that it's subtle to describe exactly where the facets of $\mathcal{B}(G)$ occur inside $\mathcal{B}(G_K)$ with respect to the polysimplicial structure of the latter. For instance, when G is k-anisotropic (so $\mathcal{B}(G)$ is a point) and $\dim(\mathfrak{f}) \leq 1$ then the unique point in $\mathcal{B}(G)$ winds up occurring inside a chamber of $\mathcal{B}(G_K)$ (in particular it is *not* a vertex of $\mathcal{B}(G_K)$, assuming G isn't a torus), as we'll see in the proof of Theorem 30.13.

Proposition 30.5 *The orbits for* $\Gamma = \mathrm{Gal}(K/k)$ *on* $\mathcal{B}(G_K)$ *are finite. For any special k-torus $T \subset G$ (so T_K is split), if k'/k is a finite subextension of K/k that splits T then the action by* $\mathrm{Gal}(K/k')$ *on* $\mathcal{B}(G_K)$ *is trivial on* $\mathcal{A}(T_K)$.

Proof The inclusion $\mathcal{B}(G_{k'}) \subset \mathcal{B}(G_K)$ identifies $\mathcal{A}(T_{k'})$ with $\mathcal{A}(T_K)$ as affine spaces for $X_*(T'_{k'})_{\mathbf{R}} = X_*(T'_K)_{\mathbf{R}}$. The polysimplicial action by $\mathrm{Gal}(K/k')$ on $\mathcal{A}(T_K)$ is an $N_{G_{k'}}(T_{k'})(k')$-equivariant automorphism of the affine space $\mathcal{A}(T_{k'})$ (this includes commuting with the action of the vector space $X_*(T'_{k'})_{\mathbf{R}}$). But that affine space admits no non-trivial equivariant automorphism (as an affine space) by Lemma 26.2, so this $\mathrm{Gal}(K/k')$-action is trivial.

We conclude that the Γ-action on $\mathcal{A}(T_K)$ has finite orbits (of size at most $[k' : k]$). A general point $x \in \mathcal{B}(G_K)$ lies in an apartment $\mathcal{A}(\mathbb{T})$ for a maximal split K-torus $\mathbb{T} \subset G_K$. There exists $g \in G(K)$ such that $\mathbb{T} = gT_Kg^{-1}$. We can pick a finite extension k''/k' inside K so that $g \in G(k'')$, so $\mathcal{A}(\mathbb{T}) = g.\mathcal{A}(T_K)$. The action of $\mathrm{Gal}(K/k'')$ on $\mathcal{A}(\mathbb{T})$ is therefore trivial, so $\Gamma.x$ is finite (of size at most $[k'' : k]$). □

To use the metric on $\mathcal{B}(G_K)$ in the study of metric questions for $\mathcal{B}(G)$, we record:

Proposition 30.6 *Let d_K on $\mathcal{B}(G_K) \times \mathcal{B}(G_K)$ be as in Proposition 28.4 for G_K.*

(i) The restriction $d := d_K|_{\mathcal{B}(G)\times\mathcal{B}(G)}$ is as in Proposition 28.4 for G.
(ii) Every choice of d as in Proposition 28.4 for G arises in this way from some choice of d_K.
(iii) The function d_K can be chosen to be invariant under the action by $\Gamma = \mathrm{Gal}(K/k)$ on $\mathcal{B}(G_K)$.

In (iii), the Γ-invariance *can* fail for some choice of d_K. For example, consider $G = \mathrm{R}_{k'/k}(G')$ for a non-trivial finite unramified extension k'/k split by K/k and a k'-simple connected semisimple k'-group G'. We have $G_K = \prod_{\sigma:k'\to K} G'_{\sigma,K}$ for $G'_{\sigma,K} := G' \otimes_{k',\sigma} K$. Thus, $\mathcal{B}(G_K) = \prod \mathcal{B}(G'_{\sigma,K})$ has its factors permuted transitively by the Γ-action via $\mathcal{B}(G_{\sigma,K}) \simeq \mathcal{B}(G_{\gamma\circ\sigma,K})$ for $\gamma \in \Gamma$. The function d_{G_K} is a Euclidean sum of a choice of the $d_{G_{\sigma,K}}$'s, so Γ-invariance of d requires $d_{G_{\gamma^{-1}\circ\sigma,K}} = d_{G_{\sigma,K}}(\gamma(\cdot), \gamma(\cdot))$ for all $\gamma \in \Gamma$. Since k'/k is non-trivial unramified, we can make such an equality fail for some γ by rescaling $d_{G_{\sigma,K}}$ for some σ (since the buildings $\mathcal{B}(G_{\sigma,K})$ have positive dimension).

Proof Pick a maximal split k-torus $S \subset G$ and let $T \subset G$ be a special k-torus containing S. Since d_K is $G(K)$-equivariant, so its restriction d on $\mathcal{B}(G)$ is $G(k)$-equivariant, for (i) it suffices to show that the metric $d_{\mathcal{A}(T_K)}$ on $\mathcal{A}(T_K)$ restricts to a metric on $\mathcal{A}(S)$ arising from a $W(G, S)$-equivariant inner product on $\mathrm{X}_*(S')_{\mathbf{R}}$.

Letting $\langle\cdot,\cdot\rangle_{T_K}$ be a $W(G_K, T_K)$-invariant inner product on $\mathrm{X}_*(T'_K)_{\mathbf{R}}$, since the inclusion $\mathcal{A}(S) \hookrightarrow \mathcal{A}(T_K)$ is equivariant with respect to $j : \mathrm{X}_*(S')_{\mathbf{R}} \hookrightarrow \mathrm{X}_*(T'_K)_{\mathbf{R}}$ it is equivalent to show $\langle\cdot,\cdot\rangle_{T_K}$ restricts to a $W(G, S)$-invariant inner product on $\mathrm{X}_*(S')_{\mathbf{R}}$. In other words, we want $\langle\cdot,\cdot\rangle_{T_K}$ to be $N_G(S)(k)$-invariant on the subspace $\mathrm{X}_*(S')_{\mathbf{R}}$.

For $n \in N_G(S)(k)$ the k-torus $\widetilde{T} = nTn^{-1}$ containing S is also special, and n-conjugation induces an isomorphism $[n] : \mathrm{X}_*(T'_K)_{\mathbf{R}} \simeq \mathrm{X}_*(\widetilde{T}'_K)_{\mathbf{R}}$ extending the natural n-action on $\mathrm{X}_*(S')_{\mathbf{R}}$. We need that the transport of $\langle\cdot,\cdot\rangle_{T_K}$ to $\mathrm{X}_*(\widetilde{T}'_K)_{\mathbf{R}}$ by $[n]$ has the same restriction to $\mathrm{X}_*(S')_{\mathbf{R}}$. This in turn says exactly that the isomorphism $\mathcal{A}(T_K) \simeq \mathcal{A}(\widetilde{T}_K)$ of affine spaces arising from the n-action on $\mathcal{B}(G_K)$ preserves the restriction of $d_{\mathcal{B}(G_K)}$ to the subset $\mathcal{A}(S) \subset \mathcal{A}(T_K) \cap \mathcal{A}(\widetilde{T}_K)$. But $d_{\mathcal{B}(G_K)}$ is $G(K)$-equivariant, so (i) is proved.

For (ii), we can use the central quotient map $G \to G^{\mathrm{ad}}$ to reduce to the case when G is adjoint type. Then $G = \prod G_i$ for k-simple G_i, so we can pass to the G_i's separately to reduce to the case that G is k-simple. Hence, by Remark 28.5, d is unique up to a positive scaling factor and so it is enough to prove (ii) for *some* choice of d. But (i) provides some d that arises from a choice of d_K, so (ii) is settled.

For (iii), recall that d_K is built from a choice of $W(G_K, T_K)$-invariant inner product on $X_*(T'_K)_{\mathbf{R}}$ for a special k-torus $T \subset G$. Thus, it suffices to show that this inner product can be chosen to be invariant under the natural Γ-action on $X_*(T'_K)_{\mathbf{R}}$. (Indeed, if $\mathbb{T}$ is a maximal split K-torus in G_K and $g \in G(K)$ satisfies $g\mathbb{T}g^{-1} = T_K$ then for $x, y \in \mathcal{A}(\mathbb{T})$ and $\gamma \in \Gamma$ we have $d(x, y) = d_{\mathcal{A}(T_K)}(g.x, g.y)$ and $d(\gamma(x), \gamma(y)) = d_{\mathcal{A}(T_K)}(\gamma(g).\gamma(x), \gamma(g).\gamma(y)) = d_{\mathcal{A}(T_K)}(\gamma(g.x), \gamma(g.y))$ since $\gamma(g)(\gamma^*\mathbb{T})\gamma(g)^{-1} = T_K$. Hence, the desired equality $d(x, y) = d(\gamma(x), \gamma(y))$ follows from Γ-invariance of $d_{\mathcal{A}(T_K)}$.)

If k'/k is a finite Galois subextension of K/k such that $T_{k'}$ is split then $W(G_K, T_K) = W(G_{k'}, T_{k'}) = N_G(T)(k')/T(k')$ and $X_*(T'_K) = X_*(T'_{k'})$, so the Γ-action on $X_*(T'_K)_{\mathbf{R}}$ factors through the action by $\mathrm{Gal}(k'/k)$ on $V = X_*(T'_{k'})_{\mathbf{R}}$. The finite group $H := W(G_{k'}, T_{k'}) \rtimes \mathrm{Gal}(k'/k)$ thereby makes sense and acts on V, so an H-invariant inner product on V does the job. □

As an interesting application of unramified descent, we can bootstrap the functoriality of $G(K)^0$ in G (see Proposition 10.13) to that of $G(k)^0$ in G:

Theorem 30.7 *The inclusion $G(k)^0 \subset G(K)^0 \cap G(k)$ from Proposition 11.1 is an equality.*

Our deduction of this result below from the axioms of Bruhat-Tits theory (particularly Axiom 27) is a bit misleading, even though logically correct, because the *construction* of Bruhat-Tits theory over k from the theory over K involves directly proving this result along the way: see [KP, Proposition 9.3.24] and its use in the proof of [KP, Proposition 9.3.25]. So although not circular, it's a bit unsettling (somewhat like the distinction between the Greek and Babylonian approaches to mathematics). Of course, this is always an issue that can arise when using something as a black box. Nonetheless, the geometric ideas in the proof of [KP, Proposition 9.3.24] are very close to what we do in the proof of Theorem 30.7.

Proof Pick $g \in G(K)^0 \cap G(k)$. We will adjust it through multiplication by an element of $G(k)^0$ to arrive at the case when also $g \in Z(k)^1$. Then the equality $Z(k)^1 \cap G(K)^0 = Z(k)^0$ from Proposition 11.4 and the containment $Z(k)^0 \subset G(k)^0$ from Definition 10.7(3) would conclude the proof.

Since $G(k)$ acts on $\mathcal{B}(G)$ as a polysimplicial complex, yet the underlying abstract building coincides with the restricted one for a Tits system structure on $G(k)^0$, the subgroup $G(k)^0$ acts transitively on the set of pairs (A, C) consisting of an apartment A in $\mathcal{B}(G)$ and a chamber C in A. Hence, upon choosing such a pair (A, C) and applying our choice of $g \in G(K)^0 \cap G(k)$ to it, we can find $g' \in G(k)^0$ such that $(g'A, g'C) = (gA, gC)$. By replacing g with ${g'}^{-1}g$ (as we may do), it can then be arranged that g preserves both A and C. Writing $A = \mathcal{A}(S)$ for a maximal split k-torus $S \subset G$, we have $\mathrm{Stab}_{G(k)}(A) = N(k)$ for $N := N_G(S)$. Thus, $g \in N(k)$.

We are going to prove the affine transformation $g : A \simeq A$ is the identity on C. Granting this, let's see how to finish. First, we claim g acts as the identity on A. This is clear if A is a point (i.e., $\mathcal{B}(G)$ is a point, which is to say $\mathscr{D}G$ is k-anisotropic). Otherwise A supports an affine root system $\Psi(G, S)$ whose irreducible decomposition gives rise to decompositions $A = \prod A_i$ and $C = \prod C_i$ where each C_i is a simplex in A_i. An affine transformation of A that is the identity on C and hence on its vertices is then readily seen to be the identity on A. Thus, g indeed acts on A as the identity. In particular, the derivative action of g on $X_*(S')_{\mathbf{R}}$ is trivial. The latter is the natural action of the image of g under $N(k) \to N(k)/Z(k) = W(\Phi)$ for $Z = Z_G(S)$ and $\Phi = \Phi(G, S)$, so the faithfullness of the $W(\Phi)$-action on $X_*(S')_{\mathbf{R}}$ then forces g to have trivial image in $N(k)/Z(k)$; i.e., $g \in Z(k)$. But then the (trivial!) action of g on A is through a translation given in terms of $\omega_Z(g)$, and more specifically the vanishing of that translation vector forces $g \in Z(k)^1$. As we saw near the start of the proof, this yields $g \in G(k)^0$ as desired.

It remains to show that the action of g on $\mathcal{B}(G)$ is trivial on C. If $\mathcal{B}(G)$ is a point then this is clear, so we may assume $\dim \mathcal{B}(G) > 0$. In particular, the apartments have positive dimension and support affine root systems. Let $S \subset G$ be the maximal split k-torus corresponding to $A \supset C$, and let $T \subset G$ be a special k-torus containing S. Axiom 27(iv) tells us that the hyperplanes in A for its polysimplicial structure arise from those of $\mathcal{A}(T_K)$ that meet A in a non-empty proper subspace. In particular, the *connected* C is contained in a facet $\mathcal{F}$ of $\mathcal{A}(T_K)$. But $\mathcal{F}$ is certainly the unique such facet, so g viewed in $G(K)$ acting on $\mathcal{B}(G_K)$ must preserve $\mathcal{F}$ since it preserves C. Since $g \in G(K)^0$, we have that $g \in G(K)^0_{\mathcal{F}}$, yet such a parahoric subgroup of $G(K)$ acts as the *identity* on $\mathcal{F}$. Thus, g is the identity on the subset $C \subset \mathcal{F}$. □

Axiom 20. (Bruhat-Tits integral models) For non-empty bounded $\Omega \subset \mathcal{A}(S)$, there exists a (necessarily unique!) $\mathfrak{o}$-structure $\mathscr{G}^1_\Omega$ of G such that $\mathscr{G}^1_\Omega(\mathcal{O}) = G(K)^1_\Omega$. In particular:

$$\mathscr{G}^1_\Omega(\mathfrak{o}) = \mathscr{G}^1_\Omega(\mathcal{O}) \cap G(k) = G(K)^1_\Omega \cap G(k) = G(k)^1_\Omega$$

(by Corollary 7.4, $G(K)^1 \cap G(k) = G(k)^1$). Its (affine) relative identity component $\mathscr{G}^0_\Omega$ satisfies:

- $\mathscr{G}^0_\Omega(\mathfrak{o}) = G(k)^0_\Omega$ (applying this over K by unramified descent, also $\mathscr{G}^0_\Omega(\mathcal{O}) = G(K)^0_\Omega$),
- the closed k-torus $S \subset G$ is the generic fiber of a closed $\mathfrak{o}$-torus $\mathscr{S} \subset \mathscr{G}^0_\Omega$.

If Ω is contained in a facet $\mathcal{F} \subset \mathcal{B}(G)$ then $\mathscr{G}^0_\Omega = \mathscr{G}^0_{\mathcal{F}}$ (these $\mathfrak{o}$-structures on G are called *parahoric*). To see the equality, compare $\mathcal{O}$-points since $\mathcal{F}$ is *contained in* a facet in $\mathcal{B}(G_K)$ (by Axiom 27(iv)).

Remark 30.8 By the deformation theory of tori in smooth affine groups [C2, Proposition 3.1.6] (this uses that $\mathfrak{o}$ is Henselian), the split $\mathfrak{f}$-torus $\overline{\mathscr{S}} \subset \overline{\mathscr{G}}^0_\Omega$ is *maximal* split. Indeed, otherwise the *smooth* affine quotient $Z_{\mathscr{G}^0_\Omega}(\mathscr{S})/\mathscr{S}$ would have a non-trivial split torus in its special fiber and hence have a non-trivial split closed $\mathfrak{o}$-subtorus. Its k-fiber $Z_G(S)/S$ would then have a non-trivial split k-torus, contradicting that S is a *maximal* split k-torus in $Z_G(S)$. This equality of the k-rank of G and $\mathfrak{f}$-rank of $\overline{\mathscr{G}}^0_\Omega$ (equivalently the $\mathfrak{f}$-rank of its maximal reductive quotient G^0_Ω) applied over K

implies maximal (usually non-split) $\mathfrak{f}$-tori of G^0_Ω have the same dimension as maximal *split* K-tori of G_K, which in turn is generally *smaller* than the dimension of maximal k-tori of G (and maximal K-tori of G_K).

If S is a maximal split k-torus of G then the split $\mathfrak{o}$-torus relative identity component $\mathscr{S}$ of the Néron model of S is the only split $\mathfrak{o}$-torus model of S. Indeed, for any two split $\mathfrak{o}$-tori, any homomorphism between their generic fibers extends uniquely over $\mathfrak{o}$ since $\mathrm{End}_{\mathfrak{o}}(\mathbf{G}_m) = \mathbf{Z} = \mathrm{End}_k(\mathbf{G}_m)$.

If Ω is a non-empty bounded subset of an apartment of $\mathcal{B}(G)$ then Axiom 20 tells us that $S \to G$ extends to a closed immersion $j : \mathscr{S} \to \mathscr{G}^0_\Omega$ when $\Omega \subset \mathcal{A}(S)$. Interestingly, the converse is true. Here is the result, with a proof provided by G. Prasad. (This will be included in the 2nd edition of [KP]; a proof over K is given in [KP, Proposition 9.3.5].)

Proposition 30.9 *If $S \to G$ extends to a closed immersion $\iota : \mathscr{S} \to \mathscr{G}^0_\Omega$ then $\Omega \subset \mathcal{A}(S)$.*

If an $\mathfrak{o}$-homomorphism $f : \mathscr{S} \to \mathscr{G}^0_\Omega$ extending $S \hookrightarrow G$ exists then it *must* be a closed immersion. Indeed $\ker f$ is multiplicative type over $\mathfrak{o}$ by [C3, Exer. 2.4.3] and hence trivial since $(\ker f)_k = \ker(f_k) = 1$, so f is monic and thus a closed immersion by [C3, Lemma B.1.3].

Proof Pick an apartment $\mathcal{A}' = \mathcal{A}(S')$ containing Ω, and let $T' \subset G$ be a special k-torus containing S'. In particular, T'_K is maximal K-split in G_K, so T' is unramified and hence the relative identity component $\mathscr{T}'$ of its Néron model is an $\mathfrak{o}$-torus (since the formation of the Néron model commutes with the unramified scalar extension to $\mathcal{O}$, and over $\mathcal{O}$ that identity component is clearly a split torus).

By Axiom 20 and Remark 30.8 over $\mathcal{O}$, the containment $\Omega \subset \mathcal{A}(S') \subset \mathcal{A}(T'_K)$ gives a closed $\mathcal{O}$-torus inclusion $j : \mathscr{T}'_{\mathcal{O}} \subset (\mathscr{G}^0_\Omega)_{\mathcal{O}}$ extending $T'_K \hookrightarrow G_K$. But j is Γ-invariant as this can be seen over K, so by étale descent for maps (applicable due to the unramifiedness of K/k) we see that $T' \hookrightarrow G$ extends to a closed immersion $\mathscr{T}' \hookrightarrow \mathscr{G}^0_\Omega$ over $\mathfrak{o}$ descending j.

The inclusion $S' \subset T'$ of unramified k-tori inside G extends to a closed immersion of $\mathfrak{o}$-tori $\mathscr{S}' \subset \mathscr{T}'$ due to the anti-equivalence between $\mathfrak{o}$-tori and discrete Γ-lattices (as $\mathfrak{o}$ is Henselian), so $\mathscr{S}'$ is thereby a closed split $\mathfrak{o}$-torus in $\mathscr{G}^0_\Omega$ whose special fiber is maximal $\mathfrak{f}$-split in $\overline{\mathscr{G}}^0_\Omega$ by Remark 30.8. It is a general fact due to Borel and Tits [CGP, Theorem C.2.3] that maximal split tori in smooth connected affine groups over any field are conjugate by a rational point, so for the two maximal split $\mathfrak{f}$-tori $\overline{\mathscr{S}}, \overline{\mathscr{S}}' \subset \overline{\mathscr{G}}^0_\Omega$ there exists $g_0 \in \overline{\mathscr{G}}^0_\Omega(\mathfrak{f}) = \mathscr{G}^0_\Omega(\mathfrak{f})$ such that $\overline{\mathscr{S}} = g_0 \overline{\mathscr{S}}' g_0^{-1}$.

Since $\mathfrak{o}$ is Henselian, the deformation theory of tori in smooth affine groups (see [KP, Proposition 8.2.1(2)]) provides $g \in \mathscr{G}^0_\Omega(\mathfrak{o})$ lifting g_0 such that $\mathscr{S} = g\mathscr{S}'g^{-1}$. Thus, the closed $\mathfrak{o}$-torus $\mathscr{T} := g\mathscr{T}'g^{-1} \subset \mathscr{G}^0_\Omega$ contains $\mathscr{S}$ and its generic fiber $T \supset S$ is special since $T_K = gT'_K g^{-1}$ inside G_K. The g-action on $\mathcal{B}(G_K)$ carries $\mathcal{A}(T'_K)$ onto $\mathcal{A}(T_K)$ yet is the identity on Ω since $g \in \mathscr{G}^0_\Omega(\mathfrak{o}) = G(k)^0_\Omega$, so $\Omega \subset \mathcal{A}(T_K)$. But $\Omega \subset \mathcal{B}(G) = \mathcal{B}(G_K)^\Gamma$, so $\Omega \subset \mathcal{A}(T_K)^\Gamma = \mathcal{A}(S)$. □

Since $G(K)^1_{\overline{\Omega}} = G(K)^1_{\Omega}$, we have $\mathscr{G}^1_{\overline{\Omega}}(\mathcal{O}) = \mathscr{G}^1_{\Omega}(\mathcal{O})$ and hence $\mathscr{G}^1_{\overline{\Omega}} = \mathscr{G}^1_{\Omega}$ as $\mathfrak{o}$-models of G (for such equality it is sufficient to check as $\mathcal{O}$-models of G_K, where it may be checked on $\mathcal{O}$-points by smoothness). Hence, for non-empty bounded $\Omega, \Omega' \subset \mathcal{B}(G)$ in an apartment that satisfy $\Omega \prec \Omega'$ (i.e., $\Omega \subset \overline{\Omega}'$),

$$\mathscr{G}^1_{\Omega'}(\mathcal{O}) = \mathscr{G}^1_{\overline{\Omega}'}(\mathcal{O}) = G(K)^1_{\overline{\Omega}'} \subset G(K)^1_{\Omega} = \mathscr{G}^1_{\Omega}(\mathcal{O}),$$

so there exists a unique map of $\mathfrak{o}$-models

$$\rho_{\Omega,\Omega'} : \mathscr{G}^1_{\Omega'} \to \mathscr{G}^1_{\Omega}$$

(first over $\mathcal{O}$, and then over $\mathfrak{o}$ by $\mathrm{Gal}(K/k)$-equivariance due to K/k being unramified) and thus of relative identity components $\mathscr{G}^0_{\Omega'} \to \mathscr{G}^0_{\Omega}$ inducing a map on special fibers

$$\overline{\rho}_{\Omega,\Omega} : \overline{\mathscr{G}}^0_{\Omega'} \to \overline{\mathscr{G}}^0_{\Omega}.$$

Axiom 22 (Structure of special fiber) The kernel of $\overline{\rho}_{\Omega,\Omega'}$ is smooth connected unipotent. For facets $\mathcal{F} \prec \mathcal{F}'$, the image $\mathfrak{p}_{\mathcal{F},\mathcal{F}'} := \overline{\rho}_{\mathcal{F},\mathcal{F}'}(\overline{\mathscr{G}}^0_{\mathcal{F}'}) \subset \overline{\mathscr{G}}^0_{\mathcal{F}}$ is a parabolic $\mathfrak{f}$-subgroup, and the preimage under $\mathscr{G}^0_{\mathcal{F}}(\mathfrak{o}) \to \overline{\mathscr{G}}^0_{\mathcal{F}}(\mathfrak{f})$ of $\mathfrak{p}_{\mathcal{F},\mathcal{F}'}(\mathfrak{f})$ is $\mathscr{G}^0_{\mathcal{F}'}(\mathfrak{o})$. For fixed $\mathcal{F}$, as we vary through all $\mathcal{F}' \succ \mathcal{F}$ the subgroups $\mathfrak{p}_{\mathcal{F},\mathcal{F}'} \subset \overline{\mathscr{G}}^0_{\mathcal{F}}$ vary without repetition through the parabolic $\mathfrak{f}$-subgroups of $\overline{\mathscr{G}}^0_{\mathcal{F}}$ (or equivalently of its maximal reductive quotient $\mathsf{G}_{\mathcal{F}}$ over the perfect field $\mathfrak{f}$); the labeling by such facets $\mathcal{F}'$ is inclusion-reversing, so the set of chambers $\mathcal{C} \succ \mathcal{F}$ labels the set of minimal parabolic $\mathfrak{f}$-subgroups of $\mathsf{G}_{\mathcal{F}}$.

If $\mathcal{A}(S)$ is an apartment containing a facet $\mathcal{F}$ then via the natural identification of (co)character lattices of the split tori $\overline{\mathscr{S}}$ and S, the relative root datum of $\mathsf{G}_{\mathcal{F}}$ with respect to its maximal split $\mathfrak{f}$ torus $\overline{\mathscr{S}}$ is $(\mathrm{X}^*(S), \Phi_{\mathcal{F}}, \mathrm{X}_*(S), \Phi^\vee_{\mathcal{F}})$ where $\Phi_{\mathcal{F}}$ is the image under $\nabla : \Psi(G,S) \twoheadrightarrow \Phi(G,S)$ of the finite set of $\psi \in \Psi(G,S)$ vanishing on $\mathcal{F}$ and $\Phi^\vee_{\mathcal{F}}$ is the set of associated coroots in $\Phi(G,S)^\vee$.

Example 30.10 Suppose $\mathfrak{f}$ is finite or $\mathfrak{f} = \overline{\mathfrak{f}}$ (or more generally $\mathfrak{f}$ is perfect with $\dim \mathfrak{f} \le 1$), so the minimal parabolic $\mathfrak{f}$-subgroups of $\mathsf{G}_{\mathcal{F}}$ are Borel subgroups. Thus, the Iwahori subgroups $G(k)^0_{\mathcal{C}}$ for chambers $\mathcal{C} \succ \mathcal{F}$ are the preimages in $\mathscr{G}^0_{\mathcal{F}}(\mathfrak{o})$ of the groups of $\mathfrak{f}$-points of Borel $\mathfrak{f}$-subgroups of $\mathsf{G}_{\mathcal{F}}$.

Example 30.11 Assume G is semisimple, so $S' = S$. Suppose $x \in \mathcal{B}(G)$ is a vertex in $\mathcal{B}(G_K)$ (which implies x is a vertex in $\mathcal{B}(G)$ by unramified descent; the converse often fails). Let's show the maximal reductive quotient G_x of $\overline{\mathscr{G}}^0_x$ is semisimple. We may assume $k = K$. Pick an apartment $\mathcal{A}(S)$ containing x. Axiom 22 tells us the root datum for G_x over $\mathfrak{f}$ that we have arranged to be algebraically closed, so it suffices to show the subset $\Phi_x \subset \mathrm{X}^*(S)_{\mathbf{R}}$ is a spanning set. This set consists of the affine roots whose vanishing hyperplanes pass through x, and equations of those hyperplanes span the dual $\mathrm{X}^*(S)_{\mathbf{R}}$ of $\mathrm{X}_*(S)_{\mathbf{R}} = \mathrm{X}_*(S')_{\mathbf{R}}$ because x is a vertex.

It is shown in Appendix B that reductive $\mathfrak{o}$-models $\mathscr{G}$ of G are *exactly* the parahoric group schemes $\mathscr{G}_x^0$ for $x \in \mathcal{B}(G)$ that are special (always a vertex) in $\mathcal{B}(G_K)$, and that such x are special in $\mathcal{B}(G_{k'})$ for all unramified k'/k. (These x are called *hyperspecial*, subject to the requirement that G_K splits, as is automatic when G admits a reductive $\mathfrak{o}$-model. When G is split, so apartments of $\mathcal{B}(G)$ are apartments of $\mathcal{B}(G_K)$, special vertices of $\mathcal{B}(G)$ are hyperspecial since the relative root systems over k and K coincide.)

In particular, if $\mathscr{G}$ is a reductive $\mathfrak{o}$-model whose special fiber is quasi-split (automatic when $\dim \mathfrak{f} \leq 1$) then the parahoric subgroups $G(k)^0_{\mathcal{F}}$ for $\mathcal{F}$ whose closure contains the corresponding hyperspecial vertex x are exactly the preimages under $\mathscr{G}(\mathfrak{o}) \to \mathscr{G}(\mathfrak{f})$ of $P_0(\mathfrak{f})$ for the parabolic $\mathfrak{f}$-subgroups $P_0 \subset \mathscr{G}_{\mathfrak{f}}$. When there is a non-special vertex y, the corresponding parahoric subgroup $\mathscr{G}_y^0(\mathfrak{o})$ is not obtained (even up to $G(k)$-conjugacy) from the constructions inside $\mathfrak{o}$-points of reductive $\mathfrak{o}$-models since the $G(k)$-action on $\mathcal{B}(G)$ preserves the property of being special (and hence of being not special).

Example 30.12 Non-special vertices for split G arise already for the *simply connected* $G = \mathrm{Sp}_4$, whose chambers $\mathcal{C}$ have 2 special vertices x_0, x_1 and one non-special vertex $\widetilde{x}$ (at the right angle of the triangle). The corresponding parahoric subgroups $\mathscr{G}_x^0(\mathfrak{o}) = G(k)_x^0 = G(k)_x$ for $x = x_0, x_1, \widetilde{x}$ can be computed using an *integral theory* of open cell decomposition for smooth affine groups with connected fibers, such as [KP, Theorem 8.2.5, Lemma 9.4.20, Lemma 9.4.21]. For the (hyper)special x_0 and x_1 the reductive $\mathfrak{o}$-models $\mathscr{G}_{x_j}^0$ are *not* related through $G(k)$-conjugacy, but when $\mathfrak{f}$ has dimension at most 1 they are related through the action of $G^{\mathrm{ad}}(k)$ on G [C2, Theorem 7.2.16].

We end with an application of the existence of Bruhat-Tits integral models. Applications of such models to reductive $\mathfrak{o}$-structures, the construction of k-anisotropic maximal k-tori, computing Galois cohomology $\mathrm{H}^1(k, G)$ for semisimple G, and the Cartan decomposition are given in Appendices B–D.

Theorem 30.13 *Assume* $\dim(\mathfrak{f}) \leq 1$. *If G is absolutely simple and k-anisotropic then G_K is split of type* A.

Proof Consider the embedding of the 1-point space $\mathcal{B}(G)$ into $\mathcal{B}(G_K)$. Letting x be the image point, it lies in a unique facet $\mathcal{F}$ of $\mathcal{B}(G_K)$. Since the unique point of $\mathcal{B}(G)$ is a chamber, by the self-contained Lemma C.4 (which uses that $\dim \mathfrak{f} \leq 1$) necessarily $\mathcal{F}$ is a chamber of $\mathcal{B}(G_K)$; we now denote $\mathcal{F}$ as $\mathcal{C}$. Note that $\mathcal{B}(G_K)$ is not a point since G_K is non-trivial, semisimple, and quasi-split.

For $\Gamma = \mathrm{Gal}(K/k)$, the Γ-action on $\mathcal{B}(G_K)$ preserves $\mathcal{C}$ since it fixes $x \in \mathcal{C}$. Let T be a special k-torus in G such that the *positive-dimensional* $\mathcal{A}(T_K)$ contains x and hence contains $\mathcal{C}$. The set Δ of vertices of $\mathcal{C}$ is permuted by the polysimplicial action by Γ. We will show this action is transitive by using a Γ-invariant metric on $\mathcal{B}(G_K)$, as we can find by Proposition 30.6. If there are at least two Γ-orbits in Δ then the metric centers of the convex hulls of both orbits would be a pair of distinct points in

$\mathcal{B}(G_K)^\Gamma = \mathcal{B}(G)$. This contradicts that $\mathcal{B}(G)$ is a single point, so the Γ-action on Δ has one orbit: it is *transitive*.

The affine root system $\Psi = \Psi(G_K, T_K)$ is *irreducible* because its derivative root system $\Phi(G_K, T_K)$ is irreducible (as G is absolutely simple, hence the quasi-split G_K is K-simple), so its affine diagram is connected. The vertices of the diagram are (by definition) the non-divisible affine roots defining the walls of the chamber $\mathcal{C}$, and the weights on the edges and direction (if any) for higher-weight edges are defined using pairings between the associated derivative roots and coroots for $\nabla\Psi = \Phi$ and $\Phi^\vee$. Hence, Γ acts on the connected affine diagram preserving the weights and directions on edges. Since $\mathcal{C}$ is a simplex (by irreducibility), the set of diagram vertices is in Γ-equivariant bijection with Δ by associating to each basis root the vertex of $\mathcal{C}$ opposite the wall in the corresponding affine root hyperplane. Thus, the Γ-action on the diagram is *transitive* on the vertices.

An inspection of the classification of all such diagrams in Figs. 7 and 8, viewing them as graphs equipped with edge weights and directions, shows that only type A admits a transitive action by a group (due to the "cyclic" nature of that diagram)! Hence, $\Phi(G_K, T_K) = \nabla\Psi$ is type A.

We have proved that G_K has *relative root system* of type A, and it remains to show G_K is split. It is quasi-split (by Steinberg's theorem), so Tits' $*$-action on the absolute diagram is the natural Galois action on the geometric character lattice of a maximal K-torus in a Borel K-subgroup of G_K. We want to show this action is trivial (so G_K has a split maximal K-torus).

We now invoke a general recipe that computes the relative diagram in terms of two ingredients: the $*$-action on the absolute diagram D, and the Galois-stable proper subset D_0 of absolute roots restricting trivially on T_K. This recipe is described and proved in [KP, Sect. 2.6(c)], and it involves the positive integer coefficients with which the roots in a basis of an irreducible root system occur in the expression for the corresponding unique highest positive root. A table of such coefficients is given in [KP, Table 2.6.1].

In our setting, $Z_{G_K}(T_K) = Z_G(T)_K$ is a torus (as G_K is quasi-split with T_K a maximal split K-torus) and so no absolute root group can be contained in $Z_G(T)_{\overline{k}}$. Hence, D_0 is empty. Since the highest root for any type-A root system has all coefficients equal to 1 when written in terms of a basis, step (3) of the algorithm in [KP, Sect. 2.6(c)] with empty D_0 forces all Galois-orbits in D to be singletons. The Galois action on the absolute diagram is thus trivial, so G_K is split as desired. □

Example 30.14 Assuming the perfect residue field $\mathfrak{f}$ satisfies $\dim \mathfrak{f} \leq 1$ (e.g., finite $\mathfrak{f}$), let's deduce that any non-degenerate quadratic space (V, q) over k with $n = \dim V \geq 5$ has an isotropic vector: there exists $v \in V - \{0\}$ such that $q(v) = 0$. (Note we allow $\mathrm{char}(k) = 2$.)

The existence of an isotropic vector is equivalent to k-isotropicity of the connected semisimple group $\mathrm{SO}(q)$, so we just have to rule out the possibility that $\mathrm{SO}(q)$ is k-anisotropic. For this we use Theorem 30.13: the root system for $\mathrm{SO}(q)_{k_s}$ is B_m when $n = 2m+1$ is odd, and B_m isn't type A when $m \geq 2$ (i.e., $n = 2m+1 \geq 5$). This settles the case of odd n.

For even n the associated symmetric bilinear form $B_q : V \times V \to k$ is non-degenerate. If $\mathrm{char}(k) \neq 2$ then we can algebraically diagonalize q, so for even $n \geq 6$ we can find a 5-dimensional subspace $V' \subset V$ such that (V', q) is non-degenerate and hence isotropic. If $\mathrm{char}(k) = 2$ then B_q is symplectic, so (V, q) is an orthogonal sum of 2-dimensonal non-degenerate quadratic spaces. Thus, for even $n \geq 6$ we can again build a 5-dimensional $V' \subset V$ such that (V', q) is non-degenerate and hence isotropic.

Corollary 30.15 *If $\mathfrak{f}$ is finite then any absolutely simple k-anisotropic G must be of* inner *type A; i.e., $G^{\mathrm{sc}} = \mathrm{SL}_1(D)$ for a finite-dimensional central division algebra D over k.*

Finiteness of $\mathfrak{f}$ is crucial: see [KP, Ex. 10.3.3] for some G of non-inner type A with $\mathfrak{f} = \mathbf{C}(x)$.

Proof By Theorem 30.13, G^{sc} has absolute type A. Assume it is not inner type A, so its absolute diagram A_n has a non-trivial automorphism and hence $n \geq 2$. We seek a contradiction.

We first reduce to the case of complete k as follows. By Theorem 4.3, $G(k)$ is bounded in G. But $G(k)$ is dense in $G(\widehat{k})$, so $G(\widehat{k})$ is likewise bounded in $G_{\widehat{k}}$. Thus, $G_{\widehat{k}}$ is $\widehat{k}$-anisotropic. If G_0 is the split form of G over k then the short exact sequence of k-groups

$$1 \to G_0^{\mathrm{ad}} \to \mathrm{Aut}_{G_0/k} \to \mathbf{Z}/2\mathbf{Z} \to 1$$

yields an exact sequence of pointed sets

$$\mathrm{H}^1(k, G_0^{\mathrm{ad}}) \to \mathrm{H}^1(k, \mathrm{Aut}_{G_0/k}) \to \mathrm{H}^1(k, \mathbf{Z}/2\mathbf{Z})$$

whose formation is compatible with extension on k. The assumed non-inner property for G says exacty that the class $[G] \in \mathrm{H}^1(k, \mathrm{Aut}_{G_0/k})$ has non-trivial image in $\mathrm{H}^1(k, \mathbf{Z}/2\mathbf{Z})$. The equivalence of Galois theories for k and $\widehat{k}$ (Proposition 3.5) implies that the natural map $\mathrm{H}^1(k, \mathbf{Z}/2\mathbf{Z}) \to \mathrm{H}^1(\widehat{k}, \mathbf{Z}/2\mathbf{Z})$ is bijective, so $G_{\widehat{k}}$ is non-inner too. Hence, to get a contradiction it suffices to work over $\widehat{k}$, so now we may and do assume k is complete.

By [Mil2, Theorem 24.44], the simply connected G^{sc} of non-inner type A is the algebraic group of units u of reduced norm 1 satisfying $u^*u = 1$ for a central simple algebra A of dimension $(n+1)^2$ over a quadraic Galois extension k'/k equipped with an anti-multiplicative involution $*$ restricting to the non-trivial k-automorphism $z \mapsto \overline{z}$ of k' ("involution of the second kind"). We have $A \simeq \mathrm{Mat}_r(D)$ for a finite-dimensional central division algebra D over k', and by [Sch, Cor. 8.8.3] there is an involution of the second kind on D. (At the end of the proof of [Sch, Lemma 8.1] there is a mathematical typo: $(aa^J b)^I$ should be $(aa^J b)^{J_a}$.)

The involution of the second kind of D implies $D^{\mathrm{opp}} \simeq \overline{D}$ (the twist of D by the non-trivial k-automorphism $z \mapsto \overline{z}$ of k'). But the isomorphism $\mathrm{Br}(k') \simeq \mathbf{Q}/\mathbf{Z}$ from local class field theory for the complete k' is functorial with respect to isomorphisms in k', so $[\overline{D}] = [D]$ in $\mathrm{Br}(k')$. Hence, $[D] = [D^{\mathrm{opp}}] = [D]^{-1}$, so $[D]$ is 2-torsion in $\mathrm{Br}(k')$.

In other words, $D = k'$ or D is a quaternion division algebra over k'. The involution of the second kind on D is a Galois descent datum relative to k'/k, so $D \simeq k' \otimes_k D_0$ for the k-subalgebra $D_0 = D^{*=1}$ and hence D_0 is a quaternion division algebra over k when D is over k'. But the natural map $\mathrm{Br}(k) \to \mathrm{Br}(k')$ kills the classes of quaternion algebras since $[k' : k] = 2$, so D and D_0 cannot both be quaternion division algebras. We conclude $D = k'$, so A is unconditionally a matrix algebra over k': $A \simeq \mathrm{Mat}_{n+1}(k') = \mathrm{End}(V')$ for $V' = k'^{n+1}$.

The involution $a \mapsto a^*$ on $A = \mathrm{End}(V')$ is anti-multiplicative and Galois semi-linear, so by the Skolem-Noether theorem we have $a^* = b\overline{a}^{\top} b^{-1}$ for some $b \in \mathrm{GL}(V')$ unique up to $k'^{\times}$-scaling. The property $a^{**} = a$ is encoded by an identity $\overline{b}^{\top} = c'b$ for some $c' \in k'^{\times}$. Applying $M \mapsto \overline{M}^{\top}$ to that identity yields $b = \overline{c}'\overline{b}^{\top} = (\overline{c}'c')b$, so $\mathrm{N}_{k'/k}(c') = 1$ and hence $c' = z/\overline{z}$ for some $z \in k'^{\times}$. Replacing b with zb brings us to the case $\overline{b}^{\top} = b$.

Non-degenerate Hermitian forms on V' are given by $h(v', v'') = h_0(v', Mv'')$ for $h_0(v', v'') = v'^{\top}\overline{v}''$ and $M \in \mathrm{GL}(V')$ that satisfies $\overline{M}^{\top} = M$. Using $M = b^{-1}$, a calculation shows $h(av', av'') = h(v', a^*av'')$ for $a \in A$, so the k-group G^{sc} coincides with $\mathrm{SU}(h)$. The quadratic form $q_h : V' \to k$ defined by $q_h(v') = h(v', v')$ is non-degenerate with $\dim_k V' = 2(n+1) \geq 6 > 5$, so by Example 30.14 there is an isotropic vector: $v' \in V' - \{0\}$ such that $q_h(v') = 0$. Thus, by [C3, Remark Q.4.1] the k-group $\mathrm{SU}(h)$ contains $\mathbf{G}_m$. But $\mathrm{SU}(h) \simeq G^{\mathrm{sc}}$ is k-anisotropic, so we have a contradiction. Hence, such G of non-inner type A over k don't exist. □

Appendix A. Some Results on Buildings

This appendix studies the metric structure of the Bruhat-Tits building $\mathcal{B}(G)$ for a connected reductive group G and shows that an abstract building with a suitable action by an abstract group *must* be the restricted building arising from a Tits system structure on that group.

To verify that d in Proposition 28.4 is a metric, we need a couple of lemmas. The first introduces a very useful retraction of the building onto an apartment ("centered" at a chamber); this often allows global questions in the building to be reduced to more tractable questions in an apartment.

Lemma A.1 *Let $\mathcal{C} \subset \mathcal{A} \subset \mathcal{B}(G)$ be a chamber in an apartment. There is a unique retraction $\rho = \rho_{\mathcal{A},\mathcal{C}} : \mathcal{B}(G) \to \mathcal{A}$ such that for every apartment $\mathcal{A}'$ containing $\mathcal{C}$ the restriction $\rho|_{\mathcal{A}'} : \mathcal{A}' \to \mathcal{A}$ is an affine isomorphism. Moreover, $\rho|_{\mathcal{A}'}$ is an isomorphism of polysimplicial complexes (i.e., it carries facets to facets), and $\rho^{-1}(x) = \{x\}$ for all $x \in \overline{\mathcal{C}}$.*

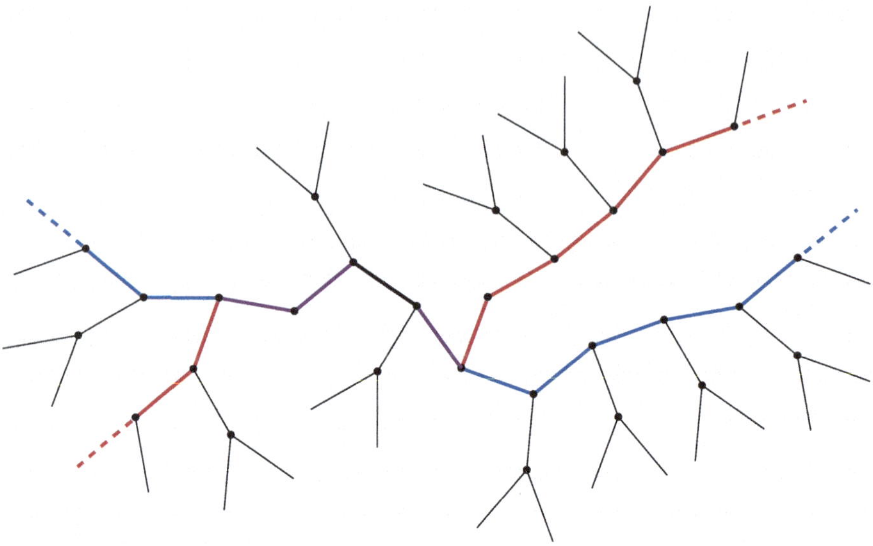

Fig. 11 A tree with valence 3 as a building with a red apartment $\mathcal{A}$ containing a dark black chamber $\mathcal{C}$. The blue apartment $\mathcal{A}'$ through $\mathcal{C}$ overlaps with $\mathcal{A}$ in the purple region and visibly uniquely "retracts" onto $\mathcal{A}$ keeping $\mathcal{C}$ fixed in place

We illustrate the idea with a 3-valent tree in Fig. 11, where $\mathcal{A}$ is the red apartment and $\mathcal{C}$ is the dark black chamber inside it (an open edge). For any other apartment $\mathcal{A}'$ through $\mathcal{C}$, such as the blue one, the unique simplicial isomorphism $\rho|_{\mathcal{A}'} : \mathcal{A}' \to \mathcal{A}$ extending the identity on $\mathcal{C}$ is readily seen.

Proof Any two chambers, and hence chamber closures, lie in a common apartment by the axioms for a building. Every point in the building lies in a chamber closure, so $\mathcal{B}(G)$ is covered by apartments $\mathcal{A}'$ through $\mathcal{C}$. Hence, for uniqueness of ρ it suffices to show uniqueness of an affine isomorphism $\mathcal{A}' \simeq \mathcal{A}$ restricting to the identity on $\mathcal{C}$.

This latter uniqueness amounts to checking that $\mathcal{A}$ has no non-trivial affine automorphism that is the identity on $\mathcal{C}$. To show an affine automorphism of $\mathcal{A}$ is determined by its effect on a single chamber $\mathcal{C}$, we can pass to the k-simple case (because $\mathcal{A}$ is naturally identified with the product of its quotients $\mathcal{A}_j$ modulo $V_j = \prod_{i \neq j} \mathrm{X}_*(S_i')_{\mathbf{R}}$ for each j, under which $\mathcal{C}$ is carried onto the product $\prod \mathcal{C}_j$ of its images $\mathcal{C}_j \subset \mathcal{A}_j$ over all j). This makes the chamber closure a simplex rather than just a polysimplex. Then use one vertex of $\mathcal{C}$ as a base point to reduce to considering a linear automorphism with a given effect on a basis (the other vertices of the chamber simplex).

With uniqueness settled, now we treat existence. By Remark 25.3, for each apartment $\mathcal{A}'$ through $\mathcal{C}$ there exists $g \in G(k)^0_{\mathcal{C}}$ such that $g\mathcal{A}' = \mathcal{A}$. The isomorphism $g : \mathcal{A}' \simeq \mathcal{A}$ of affine spaces extending the identity on $\mathcal{C}$ is also polysimplicial. This provides the existence of an affine polysimplicial isomorphism $\rho_{\mathcal{A}'} : \mathcal{A}' \simeq \mathcal{A}$ extending the identity on $\mathcal{C}$, and we know this is unique. To make ρ by gluing these isomorphisms, it suffices to show that $\rho_{\mathcal{A}'}$ and $\rho_{\mathcal{A}''}$ coincide on $\mathcal{A}' \cap \mathcal{A}''$ for any two apartments $\mathcal{A}'$ and $\mathcal{A}''$ through $\mathcal{C}$. Using $\mathcal{A}'' = \mathcal{A}$ would then ensure the gluing ρ is a retraction. Moreover, it would then be automatic that $\rho^{-1}(x) = \{x\}$ for $x \in \overline{\mathcal{C}} \subset \mathcal{A}$: we have $\rho(x) = x$ by the retraction property, and if $y \in \rho^{-1}(x)$ then we can pick an apartment $\mathcal{A}'$ through $\mathcal{C}$ and y, so also $x \in \overline{\mathcal{C}} \subset \mathcal{A}'$ and the affine isomorphism $\rho_{\mathcal{A}'} : \mathcal{A}' \simeq \mathcal{A}$ carries both x and y to x, forcing $y = x$.

For the agreement of $\rho_{\mathcal{A}'}$ and $\rho_{\mathcal{A}''}$ on $\mathcal{A}' \cap \mathcal{A}''$, observe that this overlap is a union of facets, so it suffices to show for *every* facet $\mathcal{F}$ of $\mathcal{B}(G)$ and apartment $\mathcal{A}'$ through $\mathcal{F}$ and $\mathcal{C}$ (such apartments exist through any pair of facets, by the building axioms) that $\rho_{\mathcal{A}'}|_{\mathcal{F}} : \mathcal{F} \to \mathcal{A}$ is independent of $\mathcal{A}'$. If $\mathcal{A}''$ is a second apartment through $\mathcal{F}$ and $\mathcal{C}$, by Remark 25.3 there exists $h \in G(k)^0$ satisfying $h\mathcal{A}'' = \mathcal{A}'$ and inducing the identity on $\mathcal{F}$ *and* $\mathcal{C}$. Thus, if $g \in G(k)^0_{\mathcal{C}}$ computes $\rho_{\mathcal{A}'}$ then $gh \in G(k)^0_{\mathcal{C}}$ computes $\rho_{\mathcal{A}''}$; i.e., $\rho_{\mathcal{A}''} = \rho_{\mathcal{A}'} \circ h$. But $h|_{\mathcal{F}}$ is the identity, so $\rho_{\mathcal{A}'}$ and $\rho_{\mathcal{A}''}$ coincide on $\mathcal{F}$ as desired. □

Lemma A.2 *Letting* $\rho = \rho_{\mathcal{A},\mathcal{C}} : \mathcal{B}(G) \to \mathcal{A}$ *for a pair* $\mathcal{C} \subset \mathcal{A}$, $d(\rho(x), \rho(y)) \le d(x, y)$ *for all* $x, y \in \mathcal{B}(G)$. *Equality holds if either* $x \in \overline{\mathcal{C}}$ *or if* $x, y \in \overline{\mathcal{F}}$ *for some facet* $\mathcal{F} \subset \mathcal{B}(G)$.

Proof First we verify equality in the asserted cases, and then we bootstrap from those cases to the general inequality. Each restriction $\rho : \mathcal{A}' \to \mathcal{A}$ for an apartment $\mathcal{A}'$ through $\mathcal{C}$ is an *isometry*, so for the equality cases it suffices to find such an $\mathcal{A}'$ containing both x and y. If $x \in \overline{\mathcal{C}}$ then we can pick a facet $\mathcal{F}$ containing y, and an apartment $\mathcal{A}'$ containing $\mathcal{C}$ and $\mathcal{F}$ does the job. If instead $x, y \in \overline{\mathcal{F}}$ then an apartment $\mathcal{A}'$ through $\mathcal{F}$ and $\mathcal{C}$ (hence containing $\overline{\mathcal{F}}$) does the job.

With the equality cases settled, let's now show the general inequality. For $x, y \in \mathcal{B}(G)$, pick an apartment $\mathcal{A}'$ through x and y. Let $[x, y] \subset \mathcal{A}'$ be the line segment joining x and y in that affine space. The locally finite collection of facets in $\mathcal{A}'$ (for its underlying thin chamber complex structure) then partitions $[x, y]$ into finitely many successive segments $[x_j, x_{j+1}]$ with $x = x_0 \le x_1 \le \dots \le x_n = y$ such that $[x_j, x_{j+1}] \subset \overline{\mathcal{F}}_j$ for a facet $\mathcal{F}_j \subset \mathcal{A}'$. Hence, the settled equality case ensures $d(\rho(x_j), \rho(x_{j+1})) = d(x_j, x_{j+1})$ for all j, so $d(\rho(x), \rho(y)) \le \sum_j d(\rho(x_j), \rho(x_{j+1})) = \sum_j d(x_j, x_{j+1}) = d(x, y)$ (the final equality uses that the x_j's are a finite partition of the segment $[x, y] \subset \mathcal{A}'$, and that d restricts to a Euclidean distance function on the affine space $\mathcal{A}'$). □

Proposition A.3 *The function d is a metric on $\mathcal{B}(G)$ and for all $x, y \in \mathcal{B}(G)$ the set*

$$L = \{z \in \mathcal{B}(G) : d(x, y) = d(x, z) + d(z, y)\}$$

coincides with the segment $[x, y]$ for any apartment $\mathcal{A}$ containing x and y (in particular, this segment is independent of the choice of such $\mathcal{A}$).

From now on, we'll write $[x, y]$ for the set L, and for $0 \leq t \leq 1$ we write $tx + (1-t)y$ for the unique point $z \in [x, y]$ satisfying $d(x, z) = (1-t)d(x, y)$ (and then also $d(z, y) = td(x, y)$).

Proof The symmetry and vanishing properties for d are immediate from the case of a Euclidean metric on apartments. For the triangle inequality, given $x, y, z \in \mathcal{B}(G)$ we pick an apartment $\mathcal{A}$ through x and y and a chamber $\mathcal{C} \subset \mathcal{A}$, so for $\rho = \rho_{\mathcal{A},\mathcal{C}}$ we have $\rho(x) = x$ and $\rho(y) = y$, yielding

$$\begin{aligned} d(x, y) &= d(\rho(x), \rho(y)) = d_{\mathcal{A}}(\rho(x), \rho(y)) \\ &\leq d_{\mathcal{A}}(\rho(x), \rho(z)) + d_{\mathcal{A}}(\rho(z), \rho(y)) \\ &= d(\rho(x), \rho(z)) + d(\rho(z), \rho(y)) \\ &\leq d(x, z) + d(z, y) \end{aligned} \tag{A.1}$$

(using Lemma A.2 at the end).

Now we pick $x, y \in \mathcal{B}(G)$ and define L as in the statement. It's clear that L contains $[x, y] \subset \mathcal{A}$ for any apartment $\mathcal{A}$ through x and y. For the reverse containment, we pick $z \in L$ and an apartment $\mathcal{A}$ through x and y (to define $[x, y]$), and want to show $z \in [x, y]$. Since $0 \leq d(x, z) \leq d(x, y) = d_{\mathcal{A}}(x, y)$ with $d(z, y) = d(x, y) - d(x, z)$, clearly there is a unique $z' \in [x, y]$ such that $d(x, z') = d(x, z)$, so $d(z', y) = d_{\mathcal{A}}(z', y) = d_{\mathcal{A}}(x, y) - d_{\mathcal{A}}(x, z') = d(x, y) - d(x, z') = d(x, y) - d(x, z) = d(z, y)$. Our aim is to show $z = z'$.

Pick a chamber $\mathcal{C} \subset \mathcal{A}$ such that $z' \in \overline{\mathcal{C}}$, and let $\rho = \rho_{\mathcal{A},\mathcal{C}}$, so $\rho(x) = x$ and $\rho(y) = y$. In (A.1) we have equality of the outer terms because $z \in L$, so all intermediate inequalities are equalities. In particular, $d(x, z) = d(\rho(x), \rho(z)) = d(x, \rho(z))$ and likewise $d(z, y) = d(\rho(z), y)$. In other words, $\rho(z)$ is a point in the affine space $\mathcal{A}$ satisfying

$$d(x, \rho(z)) = d(x, z) = d(x, z'), \quad d(\rho(z), y) = d(z, y) = d(z', y).$$

The point $\rho(z)$ lies on the sphere in $\mathcal{A}$ centered at x with radius $d(x, z')$ and on the sphere in $\mathcal{A}$ centered at y with radius $d(z', y) = d(x, y) - d(x, z')$, but those spheres have only the point z' in common, so $\rho(z) = z'$. Hence, $z \in \rho^{-1}(z')$, but this ρ-fiber is equal to $\{z'\}$ since $z' \in \overline{\mathcal{C}}$, so $z = z' \in L$. □

Remark A.4 There is no obvious way to express the metric topology on $\mathcal{B}(G)$ as a "gluing" of the topologies on the apartments for general k since we have to allow the case of an infinite residue field (such as for k^{un} in place of k) and in such cases there is no "local finiteness" for the covering of $\mathcal{B}(G)$ by apartments.

Note also that each apartment $\mathcal{A}$ is complete for the induced metric $d_{\mathcal{A}}$ from d, so $\mathcal{A}$ is *closed* in the metric space $\mathcal{B}(G)$. In particular, the notion of "facet closure" $\overline{\mathcal{F}}$ defined in terms of subfacets coincides with metric closure in $\mathcal{B}(G)$.

Theorem A.5 *The metric space $(\mathcal{B}(G), d)$ is uniquely geodesic and non-positively curved.*

Although non-positive curvature implies uniquely geodesic (Lemma 13.12), we give a short direct proof of the uniquely geodesic property below to be more self-contained.

Proof Since $d(x, y)$ can be computed using any apartment through both x and y, the segment $[x, y]$ parameterized by arc length is a geodesic joining x and y. If $c : [0, 1] \to \mathcal{B}(G)$ is a geodesic from x to y then $d(x, c(t)) = d(c(0), c(t)) = td(x, y)$ and $d(c(t), y) = d(c(t), c(1)) = (1 - t)d(x, y)$, so $d(x, c(t)) + d(c(t), y) = d(x, y)$. Hence, $c(t) \in [x, y]$ by Proposition A.3, and the parameterization by arc length then forces $c(t)$ to be the affine combination $tx + (1 - t)y \in [x, y]$ in the affine space $\mathcal{A}$. This completes the proof that $\mathcal{B}(G)$ is uniquely geodesic.

To show $\mathcal{B}(G)$ is non-positively curved, for $x, y, z \in \mathcal{B}(G)$ and the midpoint $m \in [x, y]$ we want to prove

$$d(x, z)^2 + d(z, y)^2 \geq 2d(m, z)^2 + (1/2)d(x, y)^2.$$

Let $\mathcal{A}$ be an apartment through x and y, so $[x, y] \subset \mathcal{A}$ and hence $m \in \mathcal{A}$. Let $\mathcal{C} \subset \mathcal{A}$ be a chamber with $m \in \overline{\mathcal{C}}$, and let $\rho = \rho_{\mathcal{A},\mathcal{C}} : \mathcal{B}(G) \to \mathcal{A}$ be the retraction along $\mathcal{A}$ centered at $\mathcal{C}$. Let $z' = \rho(z) \in \mathcal{A}$.

Since $\mathcal{A}$ is an affine space for a Euclidean space, so its metric $d_{\mathcal{A}}$ satisfies the parallelogram law, for $x, y, z' \in \mathcal{A}$ we have

$$d(x, z')^2 + d(z', y)^2 = 2d(m, z')^2 + (1/2)d(x, y)^2.$$

We shall use the interaction of ρ with d as in Lemma A.2 to get the desired inequality variant of that equality upon replacing z' with z.

By Lemma A.2, since $m \in \overline{\mathcal{C}}$ (so $\rho(m) = m$) we have $d(m, z') = d(\rho(m), \rho(z)) = d(m, z)$. On the other hand $\rho(x) = x$ and $\rho(y) = y$, so $d(x, z') = d(\rho(x), \rho(z)) \leq d(x, z)$ by Lemma A.2 and likewise $d(z', y) \leq d(z, y)$. Thus,

$$d(x, z)^2 + d(z, y)^2 \geq d(x, z')^2 + d(z', y)^2 = 2d(m, z')^2 + (1/2)d(x, y)^2 = 2d(m, z)^2 + (1/2)d(x, y)^2$$

as desired. □

Next we give a few more properties of the metric space structure on the Bruhat-Tits building.

Proposition A.6 *The action map $G(k) \times \mathcal{B}(G) \to \mathcal{B}(G)$ is continuous.*

Proof Since $G(k)$ and $\mathcal{B}(G)$ are both metric spaces, we can use the sequential continuity criterion: if $g_n \to g$ in $G(k)$ and $x_n \to x$ in $\mathcal{B}(G)$ then we'll show $g_n x_n \to gx$ in $\mathcal{B}(G)$. Since $G(k)$ acts through isometries, we have

$$d(g_n x_n, gx) \leq d(g_n x_n, g_n x) + d(g_n x, gx) = d(x_n, x) + d(g^{-1} g_n x, x).$$

As $n \to \infty$ we have $d(x_n, x) \to 0$ and $h_n := g^{-1} g_n \to 1$, so we just need $d(h_n x, x) \to 0$ as $n \to \infty$ for $x \in \mathcal{B}(G)$. By Axiom 2 in Sect. 25 the stabilizer $G(k)^0_x$ is open in $G(k)$, so $h_n \in G(k)^0_x$ for all large n and hence $d(h_n x, x) = 0$ for all large n. □

Proposition A.7 *The mapping*

$$[0, 1] \times \mathcal{B}(G) \times \mathcal{B}(G) \to \mathcal{B}(G)$$

defined by $(t, x, y) \mapsto tx + (1 - t)y$ *is continuous. In particular, for* $x_0 \in \mathcal{B}(G)$ *the mapping* $[0, 1] \times \mathcal{B}(G) \to \mathcal{B}(G)$ *defined by* $(t, y) \mapsto tx_0 + (1 - t)y$ *is a homotopy from the identity map to the constant map onto* x_0*, so* $\mathcal{B}(G)$ *is contractible.*

Proof We prove continuity at each (t_0, x, y), first treating the case $x = y$. For continuity at (t_0, x_0, x_0), we pick $\varepsilon > 0$ and want to show that if $y, y' \in \mathcal{B}(G)$ are sufficiently near x_0 and t is sufficiently near t_0 then $ty + (1 - t)y'$ is at distance $< \varepsilon$ from $t_0 x_0 + (1 - t_0)x_0 = x_0$. The point $ty + (1 - t)y'$ belongs to the segment $[y, y']$ of length $d(y, y')$, so all points on it are at most $d(y, y')$ from y' (for example). Hence,

$$\begin{aligned} d(ty + (1 - t)y', x_0) &\leq d(ty + (1 - t)y', y') + d(y', x_0) \\ &\leq d(y, y') + d(y', x_0) \leq d(y, x_0) + 2d(y', x_0), \end{aligned}$$

so if $d(y, x_0), d(y', x_0) < \varepsilon/3$ then $d(ty + (1 - t)y', x_0) < \varepsilon$ for all $t \in [0, 1]$.

Now assuming $x \neq y$, so $d(x, y) > 0$, we will prove continuity at any (t, x, y) granting the following lemma that we'll prove at the end:

Lemma A.8 *Pick* $\delta > 0$*. For distinct* $x, y \in \mathcal{B}(G)$ *and any* $z' \in \mathcal{B}(G)$ *and* $z \in [x, y]$*, if*

$$d(x, z') \leq d(x, z) + \delta d(x, y), \quad d(y, z') \leq d(y, z) + \delta d(x, y) \tag{A.2}$$

then

$$d(z, z')^2 \leq 4\delta d(x, z) d(y, z) + \delta^2 d(x, y)^2. \tag{A.3}$$

Granting Lemma A.8, to prove continuity at (t, x, y) with $x \neq y$ we pick $\varepsilon > 0$. We shall apply Lemma A.8 to $z = tx + (1 - t)y \in [x, y]$ and $z' = t'x' + (1 - t')y'$ for $t, t' \in [0, 1]$ assuming

$$|t' - t| \leq \eta, \ \ d(y', y), d(x', x) \leq \min((\delta/2), (\eta/2))d(x, y)$$

where $0 < \eta \leq \sqrt{1 + (\delta/2)} - 1$ for $\delta > 0$ to be determined in terms of ε later.

First note that

$$d(x', y') \leq d(x', x) + d(x, y) + d(y, y') \leq (1 + \eta)d(x, y) \tag{A.4}$$

and (since $[x, y]$ is parameterized by arc length)

$$d(x, z) = (1 - t)d(x, y), \ \ d(y, z) = td(x, y).$$

To use our temporary assumption that Lemma A.8 is true, we now verify (A.2) holds: this amounts to verifying

$$d(x, z') \leq ((1 - t) + \delta)d(x, y), \ \ d(y, z') \leq (t + \delta)d(x, y). \tag{A.5}$$

Since

$$d(x, z') \leq d(x, x') + d(x', z') = d(x, x') + (1 - t')d(x', y') \leq ((\delta/2) + (1 - t')(1 + \eta))d(x, y)$$

(using (A.4) at the end) and likewise

$$d(y, z') \leq ((\delta/2) + t'(1 + \eta))d(x, y),$$

to verify (A.5) it suffices to check

$$(\delta/2) + (1 - t')(1 + \eta) \leq (1 - t) + \delta, \ \ (\delta/2) + t'(1 + \eta) \leq t + \delta.$$

In other words, we want

$$(1 - t')(1 + \eta) \leq (1 - t) + \delta/2, \ \ t'(1 + \eta) \leq t + \delta/2.$$

This will be proved using only that $|t' - t| \leq \eta$; since $|(1 - t') - (1 - t)| = |t' - t|$ it's enough to check the second of those two desired inequalities. We have

$$t'(1 + \eta) \leq (t + \eta)(1 + \eta) = t + t\eta + \eta(1 + \eta) \leq t + \eta(2 + \eta),$$

so it is sufficient that $\eta(2 + \eta) \leq \delta/2$. This is equivalent to $(1 + \eta)^2 \leq 1 + (\delta/2)$, which holds since $\eta \leq \sqrt{1 + (\delta/2)} - 1$. Thus, the hypothesis we're temporarily assuming implies (A.3) has been verified, so

$$\begin{aligned} d(z, z')^2 &\leq 4\delta d(x, z)d(y, z) + \delta^2 d(x, y)^2 \\ &= (4\delta(1 - t)t + \delta^2)d(x, y)^2 \leq (4\delta + \delta^2)d(x, y)^2. \end{aligned}$$

To get $d(z, z') < \varepsilon$ it suffices to find $\delta > 0$ small enough that $\sqrt{4\delta + \delta^2}\, d(x, y) < \varepsilon$. This can be done since $d(x, y) > 0$, so Proposition A.7 is proved granting Lemma A.8. □

It remains to prove Lemma A.8. This is [KP, Lemma 4.2.8] (where our δ is denoted as ε), whose proof we now give along with some additional details on the relevant Euclidean geometry.

Proof For any points $x, y, z, z' \in \mathcal{B}(G)$ with $x \neq y$ and $z \in [x, y]$, we aim to show (A.2) implies (A.3). The cases $z = x$ and $z = y$ are easy, so we may and do assume $z \neq x, y$.

Let $\mathcal{A}$ be an apartment through x and y, so $[x, y] \subset \mathcal{A}$ and in particular $z \in \mathcal{A}$. We are going to reduce our task to the case when also z' lies in the affine space $\mathcal{A}$, so then it will become Euclidean geometry. Let $\mathcal{C} \subset \mathcal{A}$ be a chamber whose closure contains z, and let $\rho = \rho_{\mathcal{A},\mathcal{C}} : \mathcal{B}(G) \to \mathcal{A}$ be the associated retraction to $\mathcal{A}$ centered at $\mathcal{C}$. In particular, $\rho(x) = x$, $\rho(y) = y$, and $\rho(z) = z$. The equality case in Lemma A.2 ensures $d(z, z') = d(\rho(z), \rho(z')) = d(z, \rho(z'))$, and the fact that ρ causes distance to monotonically decrease in general gives

$$d(x, \rho(z')) = d(\rho(x), \rho(z')) \leq d(x, z')$$

and likewise $d(y, \rho(z')) \leq d(y, z')$. Hence, our hypotheses on x, y, z, z' are also satisfied by $x, y, z, \rho(z')$, and the desired conclusion for x, y, z, z' is *the same* as for $x, y, z, \rho(z')$. We may therefore replace z' with $\rho(z')$, so now all 4 points lie in the affine space $\mathcal{A}$ (on which $d = d_{\mathcal{A}}$) and the desired claim becomes one in Euclidean geometry. More specifically, if we replace $\mathcal{A}$ with the affine span of $[x, y]$ and z' then our task takes place in an affine Euclidean plane P (or line, which we can put into a plane).

The hypotheses and conclusion are unaffected by rescaling the distance function, so we may take x to be the origin of P and set up coordinates with y at 1 on a coordinate line L through 0 (i.e., rescale d so $d(x, y) = 1$). Taking L to be the horizontal coordinate axis, the hypotheses say z is a point $t \in (0, 1) \subset L$ and z' is a point inside both the closed disk $D = D((0, 0), t + \delta)$ and the closed disk $D' = D((1, 0), (1 - t) + \delta)$ with $\delta > 0$.

These disks meet at two distinct points (for each $\delta > 0$) since $-(t + \delta) < 1 - ((1 - t) + \delta) < t + \delta < 1 + ((1 - t) + \delta)$. The desired conclusion is that $d(z, z')^2 \leq 4\delta t(1 - t) + \delta^2$. In other words, we want to show that the distance r from a point $z' \in D \cap D'$ to the point $(t, 0)$ for $0 < t < 1$ satisfies

$$r^2 \leq 4\delta t(1 - t) + \delta^2.$$

In Fig. 12, we see that z is the midpoint of a shaded interval of length 2δ in the horizontal axis, and the farthest distance R from it in the overlap $D \cap D'$ of the closed disks is attained at the two indicated points P and Q where the circle boundaries of D and D' meet. (If $\delta > 0$ is taken large enough then the picture isn't accurate since $t + \delta > 1$ may occur, but that shaded interval of length 2δ doesn't matter in what

Fig. 12 Two closed disks D (blue) and D' (red) respectively centered at $(0, 0)$ and $(1, 0)$ with respective radii $t + \delta$ and $(1 - t) + \delta$

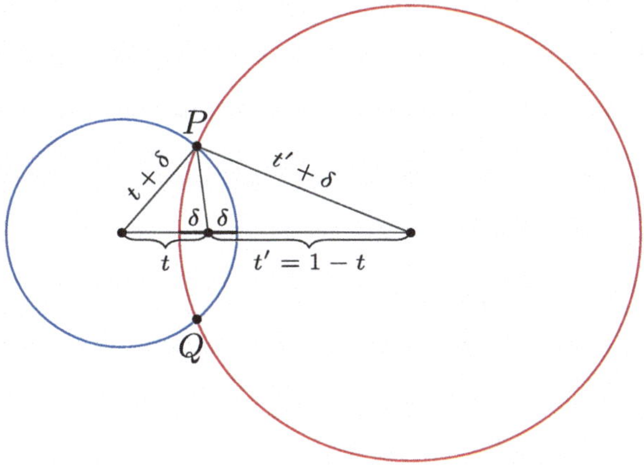

follows. It is the geometry of transversally crossing circles which matters, and we've noted that does always happen and we'll work with them via coordinate geometry.) Our aim is to show $R^2 = 4\delta tt' + \delta^2$ for $t' = 1 - t$.

Writing $P = (a, b)$ and $Q = (a, -b)$, we have

$$a^2 + b^2 = (t + \delta)^2, \quad (a - 1)^2 + b^2 = (t' + \delta)^2, \tag{A.6}$$

and

$$\begin{aligned} R^2 = \|(a, \pm b) - (t, 0)\|^2 = (a - t)^2 + b^2 &= (a^2 + b^2) - 2ta + t^2 \\ &= (t + \delta)^2 - 2ta + t^2 \\ &= 2t^2 + 2t(\delta - a) + \delta^2. \end{aligned}$$

Subtracting the first equation in (A.6) from the second gives

$$1 - 2a = (t' + \delta)^2 - (t + \delta)^2 = (t' - t)(1 + 2\delta). \tag{A.7}$$

We want $R^2 = 4\delta tt' + \delta^2$, or equivalently

$$2t^2 + 2t(\delta - a) = 4\delta tt',$$

which is the same as checking $t + (\delta - a) = 2\delta t'$. Multiplying both sides by 2 and adding 1 allows us to plug in (A.7) to rewrite this as

$$2t + 2\delta + (t' - t)(1 + 2\delta) \stackrel{?}{=} 1 + 4\delta t'.$$

Plugging in $t' = 1 - t$ and doing some algebra, the equality of both sides is obtained. □

Theorem A.9 *The metric space $(\mathcal{B}(G), d)$ is complete and for any set Σ of facets in $\mathcal{B}(G)$ the union $\bigcup_{\mathcal{F}\in\Sigma} \overline{\mathcal{F}}$ is closed in $\mathcal{B}(G)$.*

Completeness of the metric is a key ingredient (along with non-positive curvature) in the Bruhat-Tits fixed-point lemma (Theorem 14.1(2)). The closedness result for the union in Theorem A.9 is of interest since it produces closed subsets of $\mathcal{B}(G)$ in a combinatorial manner.

Proof We'll prove both assertions at the same time by showing that any such union $M = \bigcup_{\mathcal{F}\in\Sigma} \overline{\mathcal{F}}$ is complete for the metric d. The first key step is to show some features of the polysimplicial structure and $G(k)^0$-action that are plausible when the collection of apartments is locally finite in $\mathcal{B}(G)$ but are otherwise not so apparent:

Lemma A.10 *For any $x \in \mathcal{B}(G)$, any facet near enough to x contains x in its closure and the only point in the orbit $G(k)^0 x$ that is near enough to x is x.*

More precisely, there is an open metric ball B in $\mathcal{B}(G)$ centered at x that meets exacly those facets $\mathcal{F}$ for which $x \in \overline{\mathcal{F}}$, and it meets the orbit $G(k)^0 x$ in $\{x\}$.

Proof Let's first see that an open ball B centered at x that meets only those facets $\mathcal{F}$ whose closure contains x also meets $G(k)^0 x$ in exactly $\{x\}$. Consider any $y \in (G(k)^0 x) \cap B$. Let $\mathcal{F}$ be the facet of $\mathcal{B}(G)$ containing y, so $\mathcal{F}$ meets B and hence $x \in \overline{\mathcal{F}}$. But $y = gx$ for some $g \in G(k)^0$, so $y \in g\overline{\mathcal{F}} = \overline{g\mathcal{F}}$. Hence, the facet $\mathcal{F}$ containing y meets $\overline{g\mathcal{F}}$.

Distinct facets in our "geometric realization" $\mathcal{B}(G)$ are disjoint, so $\mathcal{F}$ must *be* one of the subfacets of $g\mathcal{F}$. However, these two facets have the *same* dimension, so such a subfacet relation forces equality: $g\mathcal{F} = \mathcal{F}$. This says $g \in G(k)^0_{\mathcal{F}}$, yet the effect of $G(k)^0_{\mathcal{F}}$ on $\overline{\mathcal{F}}$ is *trivial*! Thus, $y = gx = x$ as desired.

Now we turn to finding a sufficiently small open ball B in $\mathcal{B}(G)$ centered at x for which the only facets meeting B are those whose closure contains x. To figure out the radius for this ball, first we'll find such a radius that does the job for those facets within a single apartment $\mathcal{A}$ through x and then harness the isometric property of the $G(k)$-action on the building to show the same radius works at the level of the entire building.

The case when $\mathcal{B}(G)$ is a point is trivial, so we may assume $\dim \mathcal{B}(G) > 0$ (i.e., $\mathscr{D}G$ is k-isotropic). Let $\mathcal{A}$ be an apartment containing x. The polysimplicial structure of $\mathcal{A}$ is that of an affine root system (Axiom 6 in Sect. 27, which applies since now $\mathscr{D}G$ is isotropic), so it is "locally finite" in $\mathcal{A}$. Hence, there are only finitely many facets of $\mathcal{A}$ meeting any bounded neighborhood of x in $\mathcal{A}$. Among those facets, if we remove the finitely many facet closures not containing x then what remains is an *open* neighborhood of x in $\mathcal{A}$. Hence, for sufficiently small $\varepsilon > 0$, the ε-ball in $\mathcal{A}$ around x meets only those facets of $\mathcal{A}$ whose closure contains x.

Let B be the ε-ball centered at x in $\mathcal{B}(G)$ and consider a facet $\mathcal{F}$ of $\mathcal{B}(G)$ meeting B. We aim to show $x \in \overline{\mathcal{F}}$. Let $\mathcal{A}'$ be an apartment of $\mathcal{B}(G)$ containing two specific facets of the building: $\mathcal{F}$ and a choice of chamber $\mathcal{C} \subset \mathcal{A}$ with $x \in \overline{\mathcal{C}}$ (it is part of the definition of buildings that any two chambers lie in a common apartment and hence any two facets do too, as every facet is a subfacet of a chamber). Since $\mathcal{A}$ and $\mathcal{A}'$ are apartments containing the facet $\mathcal{C}$, by Remark 25.3 and the identification of $\mathcal{B}(G)$ (really its underlying polysimplicial complex) with the abstract restricted building for a Tits system structure on $G(k)^0$ in Axiom 9 in Sect. 28, there exists $g \in G(k)^0_{\mathcal{C}}$ such that $g\mathcal{A}' = \mathcal{A}$.

The effect of $G(k)^0_{\mathcal{C}}$ on $\overline{\mathcal{C}}$ is trivial, so $gx = x$. Now we bring in the isometric aspect of the $G(k)$-action on the building: g acts on $\mathcal{B}(G)$ through isometries and fixes x, so g carries each open metric ball centered at x into itself (even onto itself). Hence, $gB \subset B$ (even equality), so the assumption that the facet $\mathcal{F}$ meets B implies the facet $g\mathcal{F}$ meets $gB \subset B$. But $\mathcal{F}$ is a facet in $\mathcal{A}'$, so $g\mathcal{F}$ is a facet of the apartment $g\mathcal{A}' = \mathcal{A}$ (recall how we chose g). Clearly $B \cap \mathcal{A}$ is the open ε-ball of $\mathcal{A}$ centered at x, and we chose ε so small that this ball in $\mathcal{A}$ meets only those facets of $\mathcal{A}$ whose closure contains x. But $g\mathcal{F}$ is a facet of $\mathcal{A}$ meeting that ball, so $x \in \overline{g\mathcal{F}} = g\overline{\mathcal{F}}$. Since $x = gx$, we conclude that $x \in \overline{\mathcal{F}}$, as desired. □

Now we use Lemma A.10 to show the subset $M = \bigcup_{\mathcal{F}\in\Sigma} \overline{\mathcal{F}}$ of $\mathcal{B}(G)$ is complete. Choose a Cauchy sequence $\{x_n\}$ in M. To prove it has a limit in M, it suffices to show some subsequence has a limit in M. Pick a chamber $\mathcal{C}$ in the building, so $\overline{\mathcal{C}}$ is a compact subset of the building. Every $G(k)^0$-orbit meets the closure $\overline{\mathcal{C}}$ since: every point is in a facet, every facet is in a chamber closure, and $G(k)^0$ acts transitively on the set of chambers. Thus, for each n there is a point $y_n := g_n^{-1}x_n \in \overline{\mathcal{C}}$ with $g_n \in G(k)^0$, and by compactness of this chamber closure we can pass to a subsequence so that $y_n \to y \in \overline{\mathcal{C}}$.

Since $x_n = g_n y_n$ and $y_n \to y$, we're motivated to look at the points $g_n y$. We'll show this sequence is eventually constant, with terminal value that is the sought-after limit of the x_n's (first in the ambient building, and then in M as desired).

The key point is that $G(k)$-invariance of the metric ensures the $g_n y$'s are also Cauchy sequence:

$$\begin{aligned} d(g_m y, g_n y) &\le d(g_m y, g_m y_m) + d(g_m y_m, g_n y_n) + d(g_n y_n, g_n y) \\ &= d(y, y_m) + d(x_m, x_n) + d(y_n, y) \to 0 \end{aligned}$$

as $\min(m, n) \to \infty$. Hence, by taking m and n to be sufficiently large, the distance $d(y, g_m^{-1}g_n y) = d(g_m y, g_n y)$ can be made as small as we wish; i.e., the points $g_m^{-1}g_n y \in G(k)^0 y$ get arbitrarily close to y. Applying Lemma A.10 then ensures such points in $G(k)^0 y$ sufficiently near y must be *equal* to y, so $g_m y = g_n y$ for all large m and n.

We have shown that the sequence $\{g_n y\}$ *stabilizes*, say at $x \in \mathcal{B}(G)$. By passing to only large enough n, we can arrange that $g_n y = x$ for all n, so

$$d(x_n, x) = d(g_n y_n, g_n y) = d(y_n, y) \to 0$$

as $n \to \infty$. This produces a limit x in $\mathcal{B}(G)$, and it remains to show $x \in M$. But again using Lemma A.10, the only facets touching a small enough open ball B centered at x are those whose closure contains x. Take n big enough so $x_n \in B$. By assumption $x_n \in M$, so for some $\mathcal{F} \in \Sigma$ we have $x_n \in \overline{\mathcal{F}}$. This tells us $\overline{\mathcal{F}}$ meets B, so $\mathcal{F}$ meets B too (as B is open). But then $x \in \overline{\mathcal{F}}$ due to the smallness of B, so $x \in M$ as desired. □

To finish this appendix, we prove Theorem 25.5(2) (whose hypotheses force the given building to have positive dimension, so we state that explictly as an assumption from the outset below):

Theorem A.11 *Let $\mathcal{B} = \{\mathcal{B}_i\}_{i \in I}$ be a positive-dimensional polysimplicial building, with each $\mathcal{B}_i$ simplicial, and assume it is equipped with an action by an abstract group G such that:*

(i) for all $(\mathcal{A}, \mathcal{C})$ and $(\mathcal{A}', \mathcal{C}')$, there exists $g \in G$ such that $g.(\mathcal{A}, \mathcal{C}) = (\mathcal{A}', \mathcal{C}')$ with $g|_{\mathcal{C} \cap \mathcal{C}'} = \mathrm{id}$;
(ii) for all chambers $\mathcal{C}$ of $\mathcal{B}$, the stabilizer $\mathrm{Stab}_G(\mathcal{C})$ acts trivially on $\mathrm{vert}(\mathcal{C})$.

Choose an apartment $\mathcal{A}_0$ and a chamber $\mathcal{C}_0 \in \mathcal{A}_0$. Define $B = \mathrm{Stab}_G(\mathcal{C}_0)$, $N = \mathrm{Stab}_G(\mathcal{A}_0)$, so the set $W := N/(B \cap N)$ is in bijection with the set of chambers in $\mathcal{A}_0$ via $w \mapsto w\mathcal{C}_0$. For each wall $\mathcal{F}$ of $\mathcal{C}_0$, let $r_{\mathcal{F}} \in W$ correspond to the unique chamber $\mathcal{C} \neq \mathcal{C}_0$ in $\mathcal{A}_0$ sharing the wall $\mathcal{F}$ with $\mathcal{C}_0$ (recall $\mathcal{A}_0$ is a thin chamber complex). In particular, the subset $S \subset W$ consisting of such $r_{\mathcal{F}}$'s is in bijection with the set of walls of $\mathcal{C}_0$, so S is finite (and non-empty since $\dim \mathcal{B} > 0$).

For the saturated Tits system (G, B, N, S), if the irreducible decomposition $(W, S) = (\prod_{j \in J} W_j, \coprod_{j \in J} S_j)$ satisfies $\#\{i \in I : \dim \mathcal{B}_i > 0\} \geq \#J$ and $\#S_j \geq 2$ for all j then $\mathcal{F} \mapsto P_{\mathcal{F}} := \mathrm{Stab}_G(\mathcal{F}) \subset G$ is a G-equivariant isomorphism of $\mathcal{B}$ onto the restricted building of (G, B, N, S).

That (G, B, N, S) is a saturated Tits system is Theorem 25.5(1) (which is [KP, Proposition 1.5.28]).

Proof **Step 1**: We first verify that for each facet $\mathcal{F}$ of $\mathcal{B}$, every $g \in P_{\mathcal{F}}$ fixes each of the finitely many vertices v of $\mathcal{F}$; i.e., $P_{\mathcal{F}} = \bigcap_{v \in \mathrm{vert}(\mathcal{F})} P_{\{v\}}$. Letting $\mathcal{C}$ be a chamber with $\mathcal{F}$ as a face, the chamber $\mathcal{C}' := g\mathcal{C}$ also has $\mathcal{F}$ as a face since G acts on $\mathcal{B}$ through polysimplicial automorphisms and $g\mathcal{F} = \mathcal{F}$. By hypothesis (i) on the G-action, we can find $h \in G$ such that $h\mathcal{C}' = \mathcal{C}$ and $h|_{\mathcal{C} \cap \mathcal{C}'} = \mathrm{id}$. In particular, h fixes the vertices of $\mathcal{F}$ $(\subset \mathcal{C} \cap \mathcal{C}')$ and $hg \in P_{\mathcal{C}}$. Hypothesis (ii) on the G-action implies hg acts trivially on $\mathrm{vert}(\mathcal{C}) \supset \mathrm{vert}(\mathcal{F})$, so $g = h^{-1}(hg)$ is trivial on $\mathrm{vert}(\mathcal{F})$, as desired.

Step 2: Now we pass to the case where $\dim \mathcal{B}_i > 0$ for all i (recall we assume $\dim \mathcal{B} > 0$). Suppose there is a 0-dimensional simplicial factor, and let $\mathcal{B}_{i_1}, \ldots, \mathcal{B}_{i_r}$ be the non-empty set of such factors. Let $\{x_j\}$ be the unique point of $\mathcal{B}_{i_j}$, so for the non-empty *proper* subet $I' = \{i_1, \ldots, i_r\} \subset I$ (recall we assume $\dim \mathcal{B} > 0$), the facets of $\mathcal{B}$ have the form

$$\left(\prod_{i \notin I'} \mathcal{F}_i\right) \times (x_{i_1}, \ldots, x_{i_r}).$$

The effect of $g \in G$ on $\mathcal{B}$ permutes the $\mathcal{B}_i$'s preserving their dimension, so g must permute the x_{i_j}'s. In particular, g takes a vertex

$$((x_i)_{i\in I-I'}, (x_{i_1}, \ldots, x_{i_r}))$$

to

$$((y_i)_{i\in I-I'}, (x_{i_1}, \ldots, x_{i_r})).$$

We conclude that omitting mention of the x_{i_j}'s has no effect on how the facets of $\mathcal{B}$ are permuted by any $g \in G$, and in particular no effect on the Tits system (G, B, N, S). Thus, it is no loss of generality to assume $\dim \mathcal{B}_i > 0$ for all i, as we now do. In particular, $\#I \geq \#J$. Our hypothesis $\dim \mathcal{B} > 0$ ensures I is non-empty.

Step 3: Next we show the action by any $g \in G$ on $\mathcal{B}$ carries each $\mathcal{B}_i$ into itself (I am grateful to Sean Cotner for providing the following argument). As a preliminary step, we show that if an automorphism f of $\mathcal{B}$ is the identity on some codimension-1 facet $\mathcal{F}$ (such facets exist since $\dim \mathcal{B} > 0$) then f carries each $\mathcal{B}_i$ into itself. Let $\sigma : I \to I$ be the bijection for which $f(\mathcal{B}_i) = \mathcal{B}_{\sigma(i)}$, and let $f_i : \mathcal{B}_i \simeq \mathcal{B}_{\sigma(i)}$ be the resulting isomorphism from f for each i. If $\#I = 1$ then there is nothing to do, so we can assume $\#I \geq 2$.

Writing $\mathcal{F} = \mathcal{F}_{i_0} \times \prod_{i\neq i_0} \mathcal{C}_i$ for a codimension-1 facet $\mathcal{F}_{i_0}$ in $\mathcal{B}_{i_0}$ and chambers $\mathcal{C}_i$ in $\mathcal{B}_i$ for $i \neq i_0$, pick a vertex x of $\mathcal{F}_{i_0}$ and *distinct* vertices x_i, x'_i of $\mathcal{C}_i$ for all $i \neq i_0$ (this can be done since $\dim \mathcal{C}_i = \dim \mathcal{B}_i > 0$ for all i). By hypothesis f is the identity on $\mathcal{F}$, so for the vertices

$$\xi = (x, (x_i)_{i\neq i_0}), \quad \xi' = (x, (x'_i)_{i\neq i_0})$$

of $\mathcal{F}$ we have $f(\xi) = \xi$ and $f(\xi') = \xi'$. But $f(\xi)$ and $f(\xi')$ *each* have $\sigma(i_0)$th component $f_{i_0}(x)$ yet ξ and ξ' have *unequal* ith components $x_i \neq x'_i$ for all $i \neq i_0$. This rules out that $\sigma(i_0) \neq i_0$, so $\sigma(i_0) = i_0$.

Bijectivity of $\sigma : I \to I$ implies $\sigma(i) \neq i_0$ for $i \neq i_0$. If $\#I = 2$ we are done, so we can assume $\#I \geq 3$. Since $\sigma(i_0) = i_0$, for each $i_1 \neq i_0$ the argument we did for i_0 can be carried out for i_1 (and the behavior of σ on $I - \{i_0\}$) upon using x_{i_1} rather than x'_{i_1} in the i_1th component of ξ'; this rules out that $\sigma(i_1) \in I - \{i_0, i_1\}$, so $\sigma(i_1) = i_1$. Since i_1 was arbitrary in $I - \{i_0\}$, we have shown $\sigma = \mathrm{id}_I$.

With the preliminary step done, now consider the action on $\mathcal{B}$ by a general $g \in G$. Let $\mathcal{C}$ be a chamber of $\mathcal{B}$. If $g(\mathcal{C}) = \mathcal{C}$ then g acts trivially on all vertices of $\mathcal{C}$ by hypothesis (ii) on the G-action, so in particular it is the identity on each codimension-1 facet of $\mathcal{C}$. Thus, in such cases we can apply the preceding argument with the g-action as f (upon choosing a codimension-1 facet $\mathcal{F}$ of $\mathcal{C}$, as we may do since $\dim \mathcal{B} > 0$) to conclude that g carries each $\mathcal{B}_i$ into itself.

Suppose instead that the chamber $g(\mathcal{C})$ is different from $\mathcal{C}$. Since $\mathcal{B}$ is a chamber complex, we can find a gallery linking $\mathcal{C}$ to $g(\mathcal{C})$: a sequence of chambers $\mathcal{C} = \mathcal{C}_1, \ldots, \mathcal{C}_m = g\mathcal{C}$ ($m \geq 2$) such that $\mathcal{F}_j := \mathcal{C}_j \cap \mathcal{C}_{j+1}$ is a common wall of $\mathcal{C}_j$ and $\mathcal{C}_{j+1}$ for all $1 \leq j < m$. By hypothesis (i) for the G-action, there are $g_1, \ldots, g_{m-1} \in G$ satisfying $g_j\mathcal{C}_j = \mathcal{C}_{j+1}$ and $g_j|_{\mathcal{F}_j} = \mathrm{id}_{\mathcal{F}_j}$ for all j. Our preliminary step at the start ensures each g_j carries every $\mathcal{B}_i$ into itself. On the other hand $g_{m-1}\cdots g_1\mathcal{C}_1 = \mathcal{C}_m$, which is to say $g_{m-1}\cdots g_1$ carries $\mathcal{C}$ to $g\mathcal{C}$. Thus, $h := (g_{m-1}\cdots g_1)^{-1}g \in \mathrm{Stab}_G(\mathcal{C})$. But we've already seen that any element of G which preserves $\mathcal{C}$ has to carry each $\mathcal{B}_i$ into itself, so h carries each $\mathcal{B}_i$ into itself. Since the same holds for all g_j's, we conclude the same for $g_{m-1}\cdots g_1 h = g$ as desired.

Step 4: For any facet $\mathcal{F}$ of $\mathcal{B}$, next we use the combinatorial geometry of $\mathcal{B}$ to describe $P_{\mathcal{F}}$ in terms of the Tits system structure on G, upon choosing a chamber having $\mathcal{F}$ as a face. To do this, first we'll treat $\mathcal{F}$ of codimension 1, and then the general case.

Let $\mathcal{F}$ be a codimension-1 facet in the building $\mathcal{B}$. Pick a chamber $\mathcal{C}$ of $\mathcal{B}$ with $\mathcal{F}$ as a wall, and an apartment $\mathcal{A} \supset \mathcal{C}$. The stabilizers $B' := \mathrm{Stab}_G(\mathcal{C})$ and $N' := \mathrm{Stab}_G(\mathcal{A})$ are respectively conjugated onto B and N by a common $g \in G$ due to hypothesis (i). Thus, there is a unique order-2 element $r' \in W' := N'/(B' \cap N')$ whose effect on $\mathcal{A}$ carries $\mathcal{C}$ onto the unique other chamber in $\mathcal{A}$ with $\mathcal{F}$ as a wall. For $\nu' \in N'$ representing r', the subset $G_{r'} := B' \cup B'\nu' B'$ is independent of the choice of ν'.

We claim $G_{r'} = P_{\mathcal{F}}$. Via g-conjugation we may assume $\mathcal{C} = \mathcal{C}_0$ and $\mathcal{A} = \mathcal{A}_0$, so $B' = B$, $N' = N$, and $r' = s := r_{\mathcal{F}} \in S$. The theory of Tits systems (see [Bour, Chap. 4, Sect. 1.8, Cor. 2 to Proposition 7], [Bour, Chap. 4, Sect. 2.5, Theorem 3]) applied to (G, B, N, S) gives that the set of subgroups of G containing B is in inclusion-preserving bijection with the set of subsets $X \subset S$ via $G_X := BW_XB$ where $W_X \subset W$ is the subgroup generated by X and $W_X \cap S = X$. Since $P_{\mathcal{F}}$ is a subgroup of G containing B (as B acts trivially on $\mathrm{vert}(\mathcal{C}_0) \supset \mathrm{vert}(\mathcal{F})$), we have $P_{\mathcal{F}} = G_X = BW_XB$ for a unique subset $X \subset S$. But $s \in W$ has order 2 and carries $\mathcal{C}_0$ to the unique other chamber $\mathcal{C}_1 \neq \mathcal{C}_0$ in $\mathcal{A}_0$ with $\mathcal{F}$ as a wall, so $s(\mathcal{C}_1) = \mathcal{C}_0$. Thus, s preserves $\mathcal{C}_1 \cap \mathcal{C}_0 = \mathcal{F}$. This says $s \in P_{\mathcal{F}}$, so $s \in W_X \cap S = X$ and hence $G_s = G_{\{s\}} \subset G_X = P_{\mathcal{F}}$.

To prove G_s exhausts $P_{\mathcal{F}}$, we'll show $X = \{s\}$. If not, say X contains some $s' \in S$ distinct from s. We have $s' = r_{\mathcal{F}'}$ for a wall $\mathcal{F}' \neq \mathcal{F}$ of $\mathcal{C}_0$. Let $n, n' \in N$ represent $s, s' \in W$, so $n' \in P_{\mathcal{F}'}$. The apartments of $\mathcal{B} = \{\mathcal{B}_i\}_{i \in I}$ are precisely the direct products $\prod \mathcal{A}_i$ for apartments $\mathcal{A}_i$ of $\mathcal{B}_i$, and the facets of $\mathcal{B}$ are precisely the direct products $\prod \mathcal{F}_i$ for facets $\mathcal{F}_i$ of $\mathcal{B}_i$. Writing $\mathcal{C}_0 = \prod \mathcal{C}_i$ for a chamber $\mathcal{C}_i$ in $\mathcal{B}_i$ for each $i \in I$, clearly

$$\mathcal{F} = \mathcal{F}_{i_0} \times \prod_{i \neq i_0} \mathcal{C}_i, \quad \mathcal{F}' = \mathcal{F}'_{i_1} \times \prod_{i \neq i_1} \mathcal{C}_i$$

for walls $\mathcal{F}_{i_0} \subset \mathcal{C}_{i_0}$, $\mathcal{F}'_{i_1} \subset \mathcal{C}_{i_1}$ for some $i_0, i_1 \in I$.

Since n' is the identity on $\mathrm{vert}(\mathcal{F}')$ (Step 1) and the G-action carries each $\mathcal{B}_i$ into itself, the action of n' on $\mathcal{F}'_{i_1}$ and on $\mathcal{C}_i$ for $i \neq i_1$ is the identity. Further, since n' preserves $\mathcal{F}$, its action preserves $\mathcal{F}_{i_0}$ and is the identity on every $\mathcal{C}_i$ for $i \neq i_0$. If $i_0 \neq i_1$, then n' is the identity on all $\mathcal{C}_i$, forcing $n' \in \mathrm{Stab}_G(\mathcal{C}_0) = B$, an absurdity (since $s' \neq 1$ in W). Thus $i_0 = i_1$, so $\mathcal{F}_{i_0} \neq \mathcal{F}'_{i_0}$ are distinct walls of the *simplex* $\mathcal{C}_{i_0}$. This implies $\mathrm{vert}(\mathcal{C}_{i_0}) = \mathrm{vert}(\mathcal{F}_{i_0}) \cup \mathrm{vert}(\mathcal{F}_{i_1})$, so yet again n' preserves $\mathcal{C}_0$, a contradiction.

Now consider a facet $\mathcal{F}$ of $\mathcal{B}$ with any codimension (possibly 0). Once again letting $\mathcal{C}$ be a chamber of $\mathcal{B}$ with $\mathcal{F}$ as a face and letting $\mathcal{A}$ be an apartment containing $\mathcal{C}$, define the minimal parabolic subgroup $B' = \mathrm{Stab}_G(\mathcal{C})$ and define $N' = \mathrm{Stab}_G(\mathcal{A})$ as before. For $W' := N'/(B' \cap N')$ we have the Bruhat decomposition $G = B'W'B'$. For the subset $S' \subset W'$ of order-2 elements corresponding to the walls of $\mathcal{C}$, via conjugation by g carrying $(\mathcal{A}_0, \mathcal{C}_0)$ to $(\mathcal{A}, \mathcal{C})$ we see that the set of subgroups of G containing B' is in inclusion-preserving bijection with the set of subsets $X' \subset S'$ via $G_{X'} := B'W'_{X'}B'$ where $W'_{X'} \subset W'$ is the subgroup generated by X'.

Since $B' = P_{\mathcal{C}}$ fixes all vertices of $\mathcal{C}$, so in particular fixes the vertices of $\mathcal{F}$, we have $B' \subset P_{\mathcal{F}}$ and so $P_{\mathcal{F}} = G_{X'}$ for a unique subset $X' \subset S'$. The elements of S' correspond to the walls of $\mathcal{C}$, and we claim X' is exactly the set of walls F of $\mathcal{C}$ that contain the face $\mathcal{F}$ of $\mathcal{C}$ (e.g., X' is empty when $\mathcal{F} = \mathcal{C}$; i.e., when $\mathcal{F}$ is a chamber of $\mathcal{B}$).

The elements of X' are exactly the reflections $r_F \in W'$ that lie in $P_{\mathcal{F}}$, or more precisely for which some—or equivalently every—representative $n'_F \in N' = \mathrm{Stab}_G(\mathcal{A})$ of r_F is contained in $P_{\mathcal{F}}$. Such membership in $P_{\mathcal{F}}$ says exactly that the effect of $n'_F : \mathcal{A} \simeq \mathcal{A}$ on $\mathrm{vert}(\mathcal{C})$ (which is independent of the choice of representative of r_F) is trivial on $\mathrm{vert}(\mathcal{F})$. We have $\mathcal{C} = \prod \mathcal{C}_i$ for a chamber $\mathcal{C}_i$ of $\mathcal{B}_i$ for each i, and F is the direct product of a wall F_{i_0} of some $\mathcal{C}_{i_0}$ and the chambers $\mathcal{C}_i$ for all $i \neq i_0$. It follows that $n'_F : \mathcal{A} \simeq \mathcal{A}$ (carrying $\mathcal{C}$ to the unique other chamber of $\mathcal{A}$ with F as a wall) is the direct product of some $f_{i_0} : \mathcal{A}_{i_0} \simeq \mathcal{A}_{i_0}$ (carrying $\mathcal{C}_{i_0}$ to the unique other chamber of $\mathcal{A}_{i_0}$ with F_{i_0} as a wall) and the identity on $\mathcal{A}_i$ for all $i \neq i_0$.

Since $\mathcal{C}_{i_0}$ is a *simplex* (not just a polysimplex), so $\mathcal{C}_{i_0}$ has exactly one vertex not in the wall F_{i_0}, it follows that the set $\mathrm{vert}(F) = \mathrm{vert}(F_{i_0}) \times \prod_{i \neq i_0} \mathrm{vert}(\mathcal{C}_i)$ of vertices of F is exactly the subset of $\mathrm{vert}(\mathcal{C}) = \prod_i \mathrm{vert}(\mathcal{C}_i)$ fixed pointwise by r_F. Hence, for walls F of $\mathcal{C}$, $\mathcal{F} \subset F$ if and only if $n'_F \in P_{\mathcal{F}}$. This proves the desired description of X'.

Step 5: We have shown each $P_{\mathcal{F}} \subset G$ is parabolic relative to the Tits system structure (G, B, N, S) and they have been described via the combinatorial geometry of the given polysimplicial building $\mathcal{B}$ (as subgroups $G_{X'}$ for X' as described above in terms of a choice of chamber $\mathcal{C}$ with face $\mathcal{F}$ and apartment $\mathcal{A} \supset \mathcal{C}$). It remains to show:

(a) the $P_{\mathcal{F}}$'s are the parabolic subgroups of G with an *admissible* type (as a subset of $S = \coprod S_j$),
(b) for facets $\mathcal{F}$ and $\mathcal{F}'$ of $\mathcal{B}$ we have $P_{\mathcal{F}} \subset P_{\mathcal{F}'}$ inside G if and only if $\mathcal{F}' \subset \mathcal{F}$ inside $\mathcal{B}$.

We first prove (b). If $\mathcal{F}' \subset \mathcal{F}$ then $P_{\mathcal{F}} := \mathrm{Stab}_G(\mathcal{F})$ is contained in $P_{\mathcal{F}'}$ because $P_{\mathcal{F}}$ acts trivially on $\mathrm{vert}(\mathcal{F}) \supset \mathrm{vert}(\mathcal{F}')$ (Step 1). Conversely, if $P_{\mathcal{F}} \subset P_{\mathcal{F}'}$ then for each $x' \in \mathrm{vert}(\mathcal{F}')$ we have $P_{\mathcal{F}} \subset P_{\{x'\}}$. Thus, to deduce $\mathcal{F}' \subset \mathcal{F}$ (and thereby settle (b)) it suffices to show:

Lemma A.12 *For any facet $\mathcal{F}$ of $\mathcal{B}$, the vertices of $\mathcal{B}$ fixed by $P_{\mathcal{F}}$ are the vertices of $\mathcal{F}$.*

Proof Let's see that it suffices to treat the case when $\mathcal{F}$ is a chamber. Granting such cases, next consider facets $\mathcal{F}$ with codimension 1, so $\mathcal{F}$ is a wall of two chambers $\mathcal{C}$ and $\mathcal{C}'$ in an apartment $\mathcal{A}$ of the given building $\mathcal{B}$. By working inside the thin chamber complex $\mathcal{A}$ we see that $\mathcal{F} = \mathcal{C} \cap \mathcal{C}'$ as sets of vertices. Hence, $P_{\mathcal{C}}, P_{\mathcal{C}'} \subset P_{\mathcal{F}}$, so assuming the case of chambers is settled yields

$$\mathcal{B}^{P_{\mathcal{F}}} \subset \mathcal{B}^{P_{\mathcal{C}}} \cap \mathcal{B}^{P_{\mathcal{C}'}} = \mathcal{C} \cap \mathcal{C}' = \mathcal{F}$$

at the level of vertices. This (conditionally) settles the case of codimension-1 facets. The case of a general facet $\mathcal{F}$ of positive codimension is similarly deduced from the case of codimension-1 facets by picking a chamber $\mathcal{C}$ containing $\mathcal{F}$ and expressing $\mathcal{F}$ as the intersection of the finitely many walls of $\mathcal{C}$ that contain $\mathcal{F}$.

It remains to show for a general chamber $\mathcal{C}$ of $\mathcal{B}$ that a vertex x of $\mathcal{B}$ *outside* $\mathcal{C}$ is moved by some element of $P_{\mathcal{C}}$. Some apartment contains $\mathcal{C}$ and x, so by hypothesis (ii) for the G-action we can reduce to the case $\mathcal{C} = \mathcal{C}_0$ and $x \in \mathcal{A}_0$. Since $\mathcal{B}$ is a chamber complex, upon choosing a chamber $\mathcal{C}'$ containing x there is a finite sequence of chambers $\mathcal{C}_0, \mathcal{C}_1, \ldots, \mathcal{C}_M = \mathcal{C}'$ (so $M > 0$ since $x \notin \mathcal{C}_0$) with each pair $\mathcal{C}_m, \mathcal{C}_{m+1}$ sharing a common wall for $0 \leq m < M$ (see the "gallery axiom" in Definition 24.3). Pick $\mathcal{C}'$ making $M > 0$ as small as possible. Since $x \in \mathcal{A}_0$, by [KP, Cor. 1.5.27] all $\mathcal{C}_j$'s are contained in $\mathcal{A}_0$.

Consider the case $M = 1$: $\mathcal{C}' \neq \mathcal{C}_0$ is a chamber in $\mathcal{A}_0$ sharing a common wall $\mathcal{F}$ with $\mathcal{C}_0$ and $x \in \mathcal{C}'$ is a chosen vertex outside $\mathcal{F} = \mathcal{C}_0 \cap \mathcal{C}'$. We seek $b \in B$ moving x. Since $\mathcal{C}' = s\mathcal{C}_0$ for $s \in S \subset W$ the reflection through $\mathcal{F}$, the G-stabilizer of $\mathcal{C}'$ is $sBs^{-1} = sBs$. But $B \not\subset sBs^{-1}$ (otherwise $B \subset sBs$, which would force the reverse inclusion since $s^2 = 1$ in W and then contradict that B is its own normalizer in G), so we can pick $b \in B$ that does *not* preserve $\mathcal{C}'$. We claim this does the job.

The chamber $b\mathcal{C}'$ is distinct from $\mathcal{C}'$ yet has $\mathcal{F}$ as a wall and is also different from $b\mathcal{C}_0 = \mathcal{C}_0 \neq \mathcal{C}'$. Thus, $b\mathcal{C}'$ is a third distinct chamber in $\mathcal{B}$ with $\mathcal{F}$ as a wall, so it is *not* contained in $\mathcal{A}_0$ (because apartments are thin chamber complexes). The "adjacency" of $b\mathcal{C}'$ to the chamber $\mathcal{C}_0 \subset \mathcal{A}_0$ (i.e., having a common wall) implies that *none* of the vertices of $b\mathcal{C}'$ outside the common wall $\mathcal{F}$ with $\mathcal{C}_0$ can lie in $\mathcal{A}_0$ (otherwise by [KP, Cor. 1.5.27] the entire chamber $b\mathcal{C}'$ would have to be contained in $\mathcal{A}_0$, which is not the case). In particular $bx \notin \mathcal{A}_0$, so $bx \neq x$ as desired. This settles the case $M = 1$.

The case $M > 1$ is now treated as follows. Let $\mathcal{F}$ be the common wall for $\mathcal{C}_0$ and $\mathcal{C}_1$, and $x' \in \mathcal{C}_1$ a vertex not in $\mathcal{F}$, so $x' \notin \mathcal{C}_0$. The case "$M = 1$" applies to x', so there exists $b \in B$ that moves x'. Hence, $b\mathcal{C}_1 \neq \mathcal{C}_1$, yet $b\mathcal{C}_1$ is a chamber with $\mathcal{F}$ as a wall and it is distinct from $b\mathcal{C}_0 = \mathcal{C}_0$ too. Since $\mathcal{A}_0$ is a thin chamber complex, this third chamber $b\mathcal{C}_1$ is *not* contained in $\mathcal{A}_0$. But $\{b\mathcal{C}_m\}_{0 \leq m \leq M}$ is a *minimal-length* gallery from $\mathcal{C}_0$ to a chamber containing bx. Since the gallery is not entirely inside $\mathcal{A}_0$ (as $b\mathcal{C}_1$ is not), it follows from [KP, Cor. 1.5.27] that $bx \notin \mathcal{A}_0$. In particular, $bx \neq x$! □

Step 6: Let's turn to proving (a) from Step 5: each parabolic subgroup $P_{\mathcal{F}} \subset G$ has type in $S = \coprod S_j$ that is admissible, and every parabolic subgroup with an admissible type arises in this way. For each $g \in G$, let $g_i : \mathcal{B}_i \simeq \mathcal{B}_i$ be the visibly simplicial automorphism arising from g. This G-action on $\mathcal{B}_i$ satisfies analogues of (i) and (ii). For instance, if the pairs $(\mathcal{A}_i, \mathcal{C}_i)$ and $(\mathcal{A}'_i, \mathcal{C}'_i)$ each consist of an apartment in $\mathcal{B}_i$ and a chamber contained in that apartment then by choosing such pairs $(\mathcal{A}_{i'}, \mathcal{C}_{i'})$ for all $i' \in I - \{i\}$ and applying (i) to

$$\mathcal{A} := \mathcal{A}_i \times (\prod_{i' \neq i} \mathcal{A}_{i'}) \supset \mathcal{C}_i \times (\prod_{i' \neq i} \mathcal{C}_{i'}) =: \mathcal{C}, \ \ \mathcal{A}' = \mathcal{A}'_i \times (\prod_{i' \neq i} \mathcal{A}_{i'}) \supset \mathcal{C}'_i \times (\prod_{i' \neq i} \mathcal{C}_{i'}) =: \mathcal{C}'$$

yields $g \in G$ such that its effect $g_i : \mathcal{B}_i \simeq \mathcal{B}_i$ on the ith simplicial factor satisfies $g_i.(\mathcal{A}_i, \mathcal{C}_i) = (\mathcal{A}'_i, \mathcal{C}'_i)$ and $g_i|_{\mathcal{C}_i \cap \mathcal{C}'_i} = \text{id}$ because $\mathcal{C} \cap \mathcal{C}' = (\mathcal{C}_i \cap \mathcal{C}'_i) \times (\prod_{i' \neq i} \mathcal{C}_{i'})$. The verification of (ii) for the G-action on $\mathcal{B}_i$ is similar. Defining $B_i := \text{Stab}_G(\mathcal{C}_i)$ and $N_i := \text{Stab}_G(\mathcal{A}_i)$, by applying the settled Theorem 25.5(1) to the G-action on the positive-dimensional $\mathcal{B}_i$ we see that (G, B_i, N_i, S_i) is a saturated Tits system where $S_i \subset W_i := N_i/(B_i \cap N_i)$ is the set of involutions corresponding to the walls of $\mathcal{C}_i \subset \mathcal{A}_i$.

Write $\mathcal{C}_0 = \prod_{i \in I} \mathcal{C}_i$ and $\mathcal{A}_0 = \prod_{i \in I} \mathcal{A}_i$, so the walls of $\mathcal{C}_0$ are $\mathcal{F} = \mathcal{F}_{i_0} \times (\prod_{i \neq i_0} \mathcal{C}_i)$ for a wall $\mathcal{F}_{i_0}$ of $\mathcal{C}_{i_0}$ for some i_0. Hence, $s = r_{\mathcal{F}} \in W$ carries $\mathcal{C}_0$ to the unique chamber $\mathcal{C}'_{i_0} \times (\prod_{i \neq i_0} \mathcal{C}_i)$ in $\mathcal{A}_0$ distinct from $\mathcal{C}_0$ and having $\mathcal{F}$ as a wall. In particular, $\mathcal{C}'_{i_0}$ is the unique chamber inside $\mathcal{A}_{i_0}$ that is distinct from $\mathcal{C}_{i_0}$ and has $\mathcal{F}_{i_0}$ as a wall. For $i \neq i_0$, the effect $n_i : \mathcal{A}_i \simeq \mathcal{A}_i$ of a representative $n \in N$ of $r_{\mathcal{F}}$ is the identity on the chamber $\mathcal{C}_i$ since n acts trivially on vert$(\mathcal{F})$ by Step 3, so $n_i = \text{id}_{\mathcal{A}_i}$ for such i by [KP, Lemma 1.5.23].

Elements of W are determined by where they carry $\mathcal{C}_0$ in the set of chambers in $\mathcal{A}$. Thus, if $\mathcal{F}'$ is a wall of $\mathcal{C}_0$ whose i_1th factor is a wall $\mathcal{F}'_{i_1}$ of $\mathcal{C}_{i_1}$ with $i_1 \neq i_0$ then the elements $r_{\mathcal{F}}, r_{\mathcal{F}'} \in S$ *commute* since their composites in either order preserve $\mathcal{A}$ and carry $\mathcal{C}_0 = \prod_i \mathcal{C}_i$ to the *same* chamber $\mathcal{C}'_{i_0} \times \mathcal{C}'_{i_1} \times (\prod_{i \neq i_0, i_1} \mathcal{C}_i)$. Thus, for $S_i \subset S$ the non-empty subset corresponding to the walls of the simplex factor $\mathcal{C}_i$ of $\mathcal{C}_0$, we have $\#S_i = 1 + \dim \mathcal{B}_i \geq 2$ for all i and $S = \coprod S_i$ with the S_i's *pairwise commuting*.

Step 7: The construction of the irreducible decomposition $\{(W_j, S_j)\}_{j \in J}$ of a Coxeter system (with $\#S_j \geq 2$ for all j by hypothesis) implies that each S_i built above is a union of some S_j's for j in a non-empty subset $J_i \subset J$. This forces $\#I \leq \#J$, but we arranged in Step 2 that $\#I \geq \#J$, so $\#I = \#J$ and hence the S_i's are exactly the S_j's up to a change of index set.

The description of facets as I-indexed products thereby implies (via our proof of (b)) that the subset $X = \text{type}(P_{\mathcal{F}}) \subset S$ is in bijection with the set of walls of $\mathcal{C}_0$ that contain $\mathcal{F}$ and that $G_X = P_{\mathcal{F}}$. Hence, X is *admissible* for every facet of $\mathcal{C}_0$ and all admissible types arise! This completes the proof of (a). □

Appendix B. Reductive Models and Hyperspecial Points

This appendix uses unramified descent and Bruhat-Tits group schemes from Sect. 30 (particularly Axioms 27, 20, and 22 therein) to prove maximality of the group of integral points of a reductive model, and characterizes such models in terms of specific points in the building. We emphasize that the real need for Bruhat-Tits theory is to prove maximality when the residue field is not separably closed (especially finite, so the Zariski-density hypothesis in Proposition 6.8 fails); we'll give a direct proof of maximality when $\mathfrak{f} = \mathfrak{f}_s$, and use Bruhat-Tits theory to bootstrap from that case to general perfect $\mathfrak{f}$.

In order to state the maximality for the group of integral points of a reductive model in a form that is well-suited to proof using unramified descent, we need to introduce the notion of a *hyperspecial* vertex. To get started:

Definition B.1 A point $x \in \mathcal{B}(G)$ is *special* if either $\mathcal{B}(G)$ is a point (equivalently $\mathscr{D}(G)$ is k-anisotropic) or $\dim \mathcal{B}(G) > 0$ and x is special in some apartment $\mathcal{A}(S)$ that contains it; i.e., x is special for the affine root system $(\mathcal{A}(S), \Psi(G, S))$ (in the sense analyzed in Sect. 20, where it is shown that such points are necessarily vertices).

This definition is *independent* of the choice of such apartment containing x since for the unique facet $\mathcal{F}$ containing x the group $G(k)^0_x = G(k)^0_{\mathcal{F}}$ acts transitively on the set of apartments containing x (or equivalenly containing $\mathcal{F}$) due to the axioms of a building (see (BLII) in Definition 24.8).

Lemma B.2 *A point $x \in \mathcal{B}(G)$ is special in $\mathcal{B}(G_K)$ if and only if x is special in $\mathcal{B}(G_{k'})$ for every finite unramified extension k'/k.*

We say a point $x \in \mathcal{B}(G)$ is *hyperspecial* when the following two properties *both* hold: (i) it satisfies the equivalent properties in this lemma, (ii) G_K is split.

Proof We may and do assume G is not a torus, so the quasi-split $\mathscr{D}(G)_K$ is non-trivial and so has positive K-rank; i.e., $\mathcal{B}(G_K)$ is not a point. Let $S \subset G$ be a maximal split k-torus such that $x \in \mathcal{A}(S)$, and let $T \supset S$ be a special k-torus in G. There is a finite unramified extension k'/k inside K/k such that $T_{k'}$ is split, so (as seen in the proof of Theorem 30.13) $\mathcal{A}(T_{k'}) = \mathcal{A}(T_K)$ and $\Psi(G_K, T_K) = \Psi(G_{k'}, T_{k'})$. Likewise $\Phi(G_K, T_K) = \Phi(G_{k'}, T_{k'})$. Hence, it is immediate from Definition B.1 that x is special in $\mathcal{B}(G_{k'})$ if and only if it is special in $\mathcal{B}(G_K)$. In particular, if x is special in $\mathcal{B}(G_F)$ for all finite unramified extensions F of k then it is special in $\mathcal{B}(G_K)$.

It remains to show that if x is special in $\mathcal{B}(G_K)$ then it is special in $\mathcal{B}(G)$. By hypothesis, every affine root hyperplane in $\mathcal{A}(T_K)$ is parallel to one passing through x. Provided $\mathcal{B}(G)$ isn't a point, Axiom 27(iv) tells us that the affine root hyperplanes in $\mathcal{A}(S)$ arise exactly from the affine root hyperplanes in $\mathcal{A}(T_K)$ that meet $\mathcal{A}(S)$ in a non-empty proper subspace, so x is then special for $(\mathcal{A}(S), \Psi(G, S))$. On the other hand, when $\mathcal{B}(G)$ is a point then x is special by definition. □

Theorem B.3 *If $\mathscr{G}$ is a reductive $\mathfrak{o}$-model of G (i.e., $\mathscr{G}_{\mathfrak{f}}$ is also connected reductive) then $\mathscr{G}(\mathfrak{o})$ is a maximal bounded subgroup $G(k)$. More specifically there is a unique $x \in \mathcal{B}(G)$ such that $\mathscr{G} = \mathscr{G}^0_x$, and x is hyperspecial.*

Proof We first settle the maximality result over K (as $\mathscr{G}_{\mathcal{O}}$ is a reductive $\mathcal{O}$-model of G_K). Let $\mathcal{K}$ be a maximal bounded subgroup of $G(K)$ containing $\mathscr{G}(\mathcal{O})$. By Proposition 5.4, there is an $\mathcal{O}$-structure $\mathscr{H}$ on G_K such that $\mathscr{H}(\mathcal{O}) = \mathcal{K}$. Hence, by Proposition 6.8 we have an $\mathcal{O}$-homomorphism $f : \mathscr{G}_{\mathcal{O}} \to \mathscr{H}$ as $\mathcal{O}$-structures on G_K. The following lemma provided by S. Cotner shows reductivity of $\mathscr{G}$ implies f is an isomorphism, so $\mathscr{G}(\mathcal{O}) = \mathcal{K}$ and hence the desired maximality is proved over K.

Lemma B.4 *If $f : H \to H'$ is a map between smooth affine group schemes over a discrete valuation ring R with fraction field F such that H is reductive over R (i.e., connected reductive fibers) and f_F is an isomorphism then f is an isomorphism.*

Proof We may and do make a local extension on R to arrange that R is complete with algebraically closed residue field. The key point is to show f is flat. To do this, by the fibral flatness criterion [EGA, IV$_3$, 11.3.10] it suffices to show f is flat between fibers over Spec R. The case of the generic fibers is clear (as f_F is assumed to be an isomorphism), and for the map $f_0 : H_0 \to H_0'$ between special fibers we first note that the constant fiber dimensions for each of H and H' over Spec(R) are the same, so by smoothness for the special fibers and the Miracle Flatness Theorem [Mat, Theorem 23.1] we see f_0 is flat provided that it is quasi-finite, or equivalently has finite kernel.

Our task is equivalent to showing the normal smooth connected closed subgroup $\ker(f_0)^0_{\mathrm{red}}$ in the connected reductive H_0 over the algebraically closed residue field is trivial. By normality it is reductive, so it is trivial if its tori T_0 are trivial. Since R is complete, by the deformation theory of tori in smooth affine group schemes [C2, Cor. B.3.5] there is a closed R-torus $T \subset H$ with special fiber T_0. The composite map $T \to H \xrightarrow{f} H'$ is trivial on special fibers, so by [C2, Cor. B.3.5] this composite map is trivial; i.e., $T \subset \ker f$. But f_F is an isomorphism, so $T_F = 1$ and hence $T = 1$, so $T_0 = 1$. This shows $\ker(f_0)$ is indeed finite, so f is flat.

Since $\ker(f) \to \mathrm{Spec}(R)$ is quasi-finite, flat, and separated (even affine), by Zariski's Main Theorem [EGA, IV$_4$, Theorem 18.5.11(a)$\Leftrightarrow$(c)] it is a disjoint union of a finite flat R-scheme (the "finite part") and some F-scheme. The triviality of $\ker(f_F)$ and existence of the identity section thereby forces $\ker(f) = 1$, so f is a monomorphism. As a flat monomorphism of finite type, it is an open immersion [EGA, IV$_4$, Theorem 17.9.1]. Hence, H is the open union H'^0 of fibral identity components in H'. But H has connected reductive fibers, so connectedness of H_F' forces connectedness of H_0' by [C3, Proposition 3.1.12]. This says f is an isomorphism. □

By Corollary 29.7, $\mathscr{G}(\mathcal{O}) = \mathcal{K} = G(K)^1_x = \mathscr{G}^1_x(\mathcal{O})$ for some $x \in \mathcal{B}(G_K)$, so $\mathscr{G}_{\mathcal{O}} = \mathscr{G}^1_x$ and hence this is also equal to $\mathscr{G}^0_x$. To settle our task over K, it suffices to show that if $\mathscr{G}^0_x$ is reductive (over $\mathcal{O}$) then x is special in $\mathcal{B}(G_K)$ and G_K is split. Indeed, the special property forces x to be a vertex and we claim that yields the uniqueness of x (over K). If $y \in \mathcal{B}(G)$ satisfies $\mathscr{G} = \mathscr{G}^0_y$ then $\mathscr{G}^0_x = \mathscr{G}^0_y$, so passing to $\mathcal{O}$-valued points gives $G(K)^0_x = G(K)^0_y$. But x is a facet since it is a vertex, so that forces $y = x$, giving the desired uniqueness of x over K.

To show that reductivity of $\mathscr{G}_x^0$ (over $\mathcal{O}$) forces x to be special (over K), we pick an apartment $\mathcal{A}(S) \subset \mathcal{B}(G_K)$ containing x (here $S \subset G_K$ is a maximal split K-torus). Axiom 20 applied to $\Omega = \{x\}$ over K gives that S extends to a split closed $\mathcal{O}$-torus $\mathscr{S} \subset \mathscr{G}_x^0$. By Axiom 22 applied to the facet $\mathcal{F}$ containing x, the root system $\Phi(\overline{\mathscr{G}}_x^0, \overline{\mathscr{S}})$ of the special fiber of $\mathscr{G}_x^0 = \mathscr{G}_{\mathcal{F}}^0$ is the image under

$$\nabla : \Psi = \Psi(G_K, S) \to \Phi(G_K, S)$$

of the set $\Psi_{\mathcal{F}}$ of affine roots vanishing on $\mathcal{F}$. The restriction of ∇ to $\Psi_x \supset \Psi_{\mathcal{F}}$ is injective, so it suffices to show (i) $\#\Psi_{\mathcal{F}} = \#\Phi(G_K, S)$ (as then x is extra special and hence special) and (ii) G_K is split.

We know $\overline{\mathscr{S}}$ is a maximal split torus in $\overline{\mathscr{G}}_x^0$ (Remark 30.8), hence a maximal torus since the residue field of $\mathcal{O}$ is separably closed. Since $\mathscr{G}_x^0$ is reductive, so its special and generic fibers have maximal tori with the same dimension, S must be a maximal K-torus in G_K and so G_K is *split*. In particular $\Phi(G_K, S)$ is the root system of the geometric generic fiber of $\mathscr{G}_x^0$. The geometric special fiber has been seen to have root system $\Psi_{\mathcal{F}}$. These two root systems have the same size because the root system of each geometric fiber of a reductive group over a connected base is the same for all such fibers [C2, Lemma 5.1.3]. This completes the proof that x is hyperspecial over K.

The desired result has been established over K, so by working over k we note that the unique x in $\mathcal{B}(G_K)$ for which $\mathscr{G}(\mathcal{O}) = G(K)_x^1$ must be fixed by $\Gamma = \mathrm{Gal}(K/k)$. Hence, $x \in \mathcal{B}(G_K)^\Gamma = \mathcal{B}(G)$. This provides an $\mathfrak{o}$-group $\mathscr{G}_x^0$ descending the $\mathcal{O}$-group above with the same notation that was shown to equal $\mathscr{G}_{\mathcal{O}}$ as $\mathcal{O}$-structures on G_K. The point x is also hyperspecial (over k), by Lemma B.2 and the settled result over K.

Passing to Γ-invariants on the equality $\mathscr{G}(\mathcal{O}) = \mathscr{G}_x^1(\mathcal{O})$ gives $\mathscr{G}(\mathfrak{o}) = \mathscr{G}_x^1(\mathfrak{o})$, and we have seen that $\mathscr{G}_x^1 = \mathscr{G}_x^0$ as $\mathfrak{o}$-structures on G (as it suffices to check this over $\mathcal{O}$). If $y \in \mathcal{B}(G)$ satisfies $\mathscr{G} = \mathscr{G}_y^0$ then $\mathscr{G}_x^0 = \mathscr{G}_y^0$, so passing to $\mathfrak{o}$-valued points gives $G(k)_x^0 = G(k)_y^0$. But x is a facet since it is a vertex, so

$$y \in \mathcal{B}(G)^{G(k)_y^0} = \mathcal{B}(G)^{G(k)_x^0} = \{x\},$$

forcing $y = x$. That establishes the desired uniqueness of x in general (not just over K). □

We finish this appendix with a characterization of hyperspecial points in terms of reductivity of the stabilizer group scheme:

Corollary B.5 *A point $x \in \mathcal{B}(G)$ is hyperspecial if and only if $\mathscr{G}_x^0$ is reductive (i.e., has reductive special fiber too), in which case $\mathscr{G}_x^1 = \mathscr{G}_x^0$.*

In view of this result and Theorem B.3, for a reductive $\mathfrak{o}$-structure $\mathscr{G}$ of G the maximal bounded subgroup $\mathscr{G}(\mathfrak{o}) \subset G(k)$ is often called *hyperspecial*.

Proof We may and do assume $k = K$ since (i) the hyperspecial propery amounts to the special property for the corresponding point in $\mathcal{B}(G_K)$ and the split property for G_K, (ii) the formation of $\mathscr{G}_x^*$ for $* = 0, 1$ commutes with the scalar extension $\mathfrak{o} \to \mathcal{O}$.

If $\mathscr{G}_x^0$ is reductive then by Theorem B.3 with $\mathscr{G} = \mathscr{G}_x^0$ we see that x is hyperspecial. It remains to show conversely that if $k = K$ and x is hyperspecial (in particular, G is split) then $\mathscr{G}_x^1 = \mathscr{G}_x^0$ and its special fiber is reductive.

Let S be a maximal split k-torus in G such that $x \in \mathcal{A}(S)$. First suppose $G = S$, so the equality $G(k)^1 = G(k)^0$ (as the maximal bounded subgroup of $\mathfrak{o}$-points in the Néron model) is clear and hence $\mathscr{G}_x^1 = \mathscr{G}_x^0$ (recall that we have put ourselves in the case $k = K$). By dimension reasons, Axiom 20 implies $\mathscr{G}_x^0$ coincides with the $\mathfrak{o}$-torus relative identity component $\mathscr{S}$ of the Néron model of S. This settles the case $G = S$.

Now we may and do assume the split G is not a torus, so $\mathcal{B}(G)$ is not a point and hence the apartments have positive dimension. Thus, there is an affine root system for each apartment. Since the special point x is a vertex, hence a facet, by Axiom 22 the root system with respect to $\overline{\mathscr{S}}$ for the maximal reductive quotient G_x of the special fiber $\overline{\mathscr{G}}_x^0$ is identified with Ψ_x via $\nabla : \Psi \to \Phi(G, S) \subset \mathrm{X}^*(S) = \mathrm{X}^*(\overline{\mathscr{S}})$. Since $\nabla(\Psi_x) = \Phi(\mathsf{G}_x, \overline{\mathscr{S}}) \subset \Phi(G, S)$ with x assumed to be special, the vanishing hyperplanes for $\Psi_x \simeq \nabla(\Psi_x)$ in $\mathrm{X}_*(S')_{\mathbf{R}}$ coincide with those of $\Phi(G, S)$. But $\Phi(G, S)$ is a *reduced* root system since G is split, so the containment $\nabla(\Psi_x) \subset \Phi(G, S)$ is an equality.

The equality of root systems gives

$$\dim G = \#\Phi(G, S) + \dim S = \#\Psi_x + \dim \overline{\mathscr{S}} = \#\Phi(\mathsf{G}_x, \overline{\mathscr{S}}) = \dim \mathsf{G}_x.$$

But $\dim G = \dim \overline{\mathscr{G}}_x^0 = \dim \mathsf{G}_x + \dim \mathscr{R}_u(\overline{\mathscr{G}}_x^0)$, so $\mathscr{R}_u(\overline{\mathscr{G}}_x^0) = 1$; i.e., $\mathscr{G}_x^0$ has reductive special fiber. Thus, the smooth affine $\mathfrak{o}$-group $\mathscr{G}_x^1$ with connected reductive generic fiber G has special fiber with reductive identity component, so the special fiber is also connected [C2, Proposition 3.1.12] and hence $\mathscr{G}_x^0 = \mathscr{G}_x^1$. □

We have characterized how reductive $\mathfrak{o}$-groups relate to Bruhat-Tits integral models, so we can use that to describe the parahoric subgroups of $G(k)$ whose associated facet admits a hyperspecial vertex:

Proposition B.6 *Assume G admits a reductive $\mathfrak{o}$-model $\mathscr{G}$ and let $x \in \mathcal{B}(G)$ be the corresponding hyperspecial vertex x. Let G be the connected reductive special fiber of $\mathscr{G}$. The parahoric subgroups of $G(k)$ associated to facets of $\mathcal{B}(G)$ with x as a vertex are precisely the preimages under $\mathscr{G}(\mathfrak{o}) \to \mathscr{G}(\mathfrak{f}) = \mathsf{G}(\mathfrak{f})$ of the subgroups $\mathsf{P}(\mathfrak{f})$ for parabolic $\mathfrak{f}$-subgroups $\mathsf{P} \subset \mathsf{G}$.*

In particular, if P_0 is a minimal parabolic $\mathfrak{f}$-subgroup of G then the preimage of $\mathsf{P}_0(\mathfrak{f})$ under $\mathscr{G}(\mathfrak{o}) \to \mathsf{G}(\mathfrak{f})$ is an Iwahori subgroup of $G(k)$.

Proof Since $\mathscr{G} = \mathscr{G}_x^0$, we apply Axiom 22 with the facet $\mathcal{F} = \{x\}$. □

Every chamber for an affine root system has at least one special (even extra special) vertex, but to apply Proposition B.6 to such a vertex we need it to be hyperspecial. If G is split then all special vertices are hyperspecial, so for a split reductive $\mathfrak{o}$-model $\mathscr{G}$ of G the Iwahori subgroups of $G(k)$ are the conjugates of the preimage under $\mathscr{G}(\mathfrak{o}) \to \mathscr{G}(\mathfrak{f})$ of $\mathsf{B}(\mathfrak{f})$ for a Borel $\mathfrak{f}$-subgroup $\mathsf{B} \subset \mathscr{G}_{\mathfrak{f}}$.

Appendix C. Anisotropic Maximal Tori and Galois Cohomology

The aim of this appendix is to use Bruhat-Tits (parahoric) group schemes from Sect. 30 (particularly Axioms 20 and 22 for them) to prove two results about connected semisimple k-groups G when $\dim \mathfrak{f} \leq 1$. First, we prove the existence of anisotropic maximal k-tori in G when the residue field $\mathfrak{f}$ admits a sufficiently rich Galois theory [KP, Theorem 10.5.1]. For this, Bruhat-Tits group schemes make contact with connected semisimple groups over the residue field, for which such an anisotropicity result is proved (and used) as a separate lemma. Notably, this allows for a uniform treatment of all G, without any case-by-case considerations.

The second main result is the trivality of $\mathrm{H}^1(k, G)$ when G is simply connected [KP, Theorem 10.6.4]. This was originally proved by Kneser for non-Archimedean local fields k of characteristic 0, using a lot of case-work. Bruhat-Tits theory allows for a uniform argument that also works in positive characteristic. As a corollary one can describe $\mathrm{H}^1(k, G)$ when G isn't assumed to be simply connected but k is complete with finite residue field [KP, Theorem 10.6.8]; the proof is purely "Galois cohomological" and uses Bruhat-Tits theory via the existence of a maximal k-torus T that is *anisotropic* (so $\mathrm{H}^2(k, T) = 1$ by Tate local duality since $\mathrm{X}_k^*(T) = 1$; see [KP, Lemma 10.6.7]).

Theorem C.1 *Assume the perfect residue field $\mathfrak{f}$ of k satisfies* $\dim \mathfrak{f} \leq 1$ *and admits a cyclic Galois extension of every degree (e.g., a finite residue field). Every connected semisimple k-group G admits a k-anisotropic maximal k-torus that contains a k-torus T for which $T_K \subset G_K$ is a maximal split K-torus.*

When k is a non-Archimedean local field (complete with finite residue field), this implies G admits a maximal k-torus whose group of rational points is compact. To see how this contrasts with the situation over $\mathbf{R}$, consider the Lie group $\mathrm{SL}_n(\mathbf{R})$ that has its unique conjugacy class of maximal compact subgroups represented by $\mathrm{SO}(n) = \mathrm{SO}(q)(\mathbf{R})$ for $q = \sum x_i^2$. It follows that the anisotropic $\mathbf{R}$-tori in $\mathrm{SL}_{n,\mathbf{R}}$ have dimension at most the rank $[n/2]$ of $\mathrm{SO}(n)$, which for $n \geq 3$ is less than the dimension $n-1$ of maximal $\mathbf{R}$-tori of $\mathrm{SL}_{n,\mathbf{R}}$.

Proof The idea is to show a suitable Bruhat-Tits group scheme model $\mathscr{G} = \mathscr{G}_{\mathcal{F}}^0$ of G has special fiber $\overline{\mathscr{G}}$ containing an anisotropic maximal $\mathfrak{f}$-torus T_0, and lift T_0 to a closed $\mathfrak{o}$-torus $\mathscr{T} \subset \mathscr{G}$ whose generic fiber $\mathscr{T}_k \subset G$ is anisotropic and unramified (i.e., split over K) with centralizer in which we can find the desired *maximal* k-torus of G. To motivate a choice of parahoric group scheme $\mathscr{G}^0_{\mathcal{F}}$ to try for $\mathscr{G}$, we first note that (by perfectness of $\mathfrak{f}$, which ensures $\mathsf{G}_{\mathcal{F}} := \overline{\mathscr{G}}/\mathscr{R}_{u,\mathfrak{f}}(\overline{\mathscr{G}})$ is reductive) the maximal $\mathfrak{f}$-tori found in $\overline{\mathscr{G}}$ are "the same" as those found in its maximal connected reductive quotient $\mathsf{G} := \mathsf{G}_{\mathcal{F}}$. This leads us to want G to have anisotropic $\mathfrak{f}$-tori as "big" as possible.

Provided the connected semisimple $\mathsf{D} := \mathscr{D}(\mathsf{G})$ has an anisotropic maximal $\mathfrak{f}$-torus (a general property of connected semisimple $\mathfrak{f}$-groups shown in Lemma C.2), we want the maximal central torus Z in G to be anisotropic (as then $\mathsf{G} = \mathsf{Z} \cdot \mathsf{D}$ has an anisotropic maximal $\mathfrak{f}$-torus). Since Z maps isogenously onto the torus quotient G/D, for Z to be anisotropic we want the $\mathfrak{f}$-ranks of G and D to coincide.

The $\mathfrak{f}$-rank of G equals that of $\overline{\mathscr{G}}$, which is equal to the k-rank of G by Remark 30.8. Hence, we want equality of relative ranks for the connected semisimple D over $\mathfrak{f}$ and G over k. The relative rank of the derived group of a connected reductive group equals the dimension of its relative root system, so we seek $\mathcal{F} \subset \mathcal{B}(G)$ for which the relative root system of $\mathsf{G}_{\mathcal{F}}$ has the same dimension as that of G.

If the semisimple G has k-rank equal to 0 (i.e., $\mathcal{B}(G)$ is a point, so there's only one facet) then G has $\mathfrak{f}$-rank 0, so its maximal central torus is $\mathfrak{f}$-anisotropic as desired. Next suppose a maximal split k-torus S in G is *not* central, so the apartment $\mathcal{A}(S)$ in $\mathcal{B}(G)$ has positive dimension and hence has an associated affine root system $\Psi := \Psi(G, S)$ (Axiom 6 in Sect. 27). The relative root system $\Phi(\mathsf{G}_{\mathcal{F}}, \overline{\mathscr{S}})$ inside $\mathrm{X}^*(\overline{\mathscr{S}}) = \mathrm{X}^*(S)$ is identified (by Axiom 22) with $\Psi_{\mathcal{F}} \overset{\nabla}{\simeq} \nabla(\Psi_{\mathcal{F}}) \subset \Phi(G, S) \subset \mathrm{X}^*(S)$, where $\Psi_{\mathcal{F}}$ is the set of affine roots in Ψ vanishing on the facet $\mathcal{F} \subset \mathcal{A}(S)$. For this to be as big as possible, we want $\Psi_{\mathcal{F}}$ to span as big a subspace as possible. That brings us back to the notion of a special vertex: this is a point (necesarily a vertex) $x \in \mathcal{A}(S)$ for which Ψ_x agrees with $\Phi(G, S)$ up to scalar multiples. That is, every affine root hyperplane is parallel to one through x.

Putting it all together, we want to consider $\mathscr{G} := \mathscr{G}^0_x$ for a special point $x \in \mathcal{A}(S)$: these always exist and are always vertices, by Proposition 20.12 and the discussion after its proof. The discussion above shows that such $\overline{\mathscr{G}}$ has an anisotropic maximal $\mathfrak{f}$-torus $\overline{T}$ since connected semisimple $\mathfrak{f}$-groups have anisotropic maximal $\mathfrak{f}$-tori (Lemma C.2). Note that dim $\overline{T}$ is the $\overline{\mathfrak{f}}$-rank of the group $\overline{\mathscr{G}}_{\overline{\mathfrak{f}}}$ that is the special fiber of $\mathscr{G}_{\mathcal{O}}$, and this rank is in turn the K-rank of the generic fiber G_K by Remark 30.8.

Let $\mathscr{T} \subset \mathscr{G}$ be a closed $\mathfrak{o}$-torus lifting $\overline{T}$, as exists by the Henselian property of $\mathfrak{o}$ and the deformation theory of tori in smooth affine groups (see [C2, Proposition 3.1.6]). The k-torus $T := \mathscr{T}_k \subset \mathscr{G}_k = G$ is unramified since T_K is split: indeed, T_K is the generic fiber of the $\mathcal{O}$-torus $\mathscr{T}_{\mathcal{O}}$ whose special fiber $\overline{T}_{\overline{\mathfrak{f}}}$ is split. Next, we show T is k-anisotropic using the $\mathfrak{f}$-anisotropicity of $\overline{T}$: the geometric character lattice $\mathrm{X}^*(T_K)$ equipped with its action by $\mathrm{Gal}(K/k) = \mathrm{Gal}(\overline{\mathfrak{f}}/\mathfrak{f})$ is naturally identified with $\mathrm{X}^*(\overline{T}_{\overline{\mathfrak{f}}})$. The subspace of Galois-invariants in this latter lattice is 0 since $\overline{T}$ is $\mathfrak{f}$-anisotropic by design, so T is k-anisotropic as claimed.

The split K-torus T_K in G_K is *maximal* as a split K-torus in G_K since dim $T_K =$ dim T has been seen to be equal to the K-rank of G_K. Thus, $Z_G(T)_K/T_K$ is K-anisotropic (i.e., has K-rank 0), so all k-tori in $Z_G(T)/T$ are k-anisotropic. It follows from k-anisotropicity of T that every maximal k-torus in $Z_G(T)$ is k-anisotropic. But such k-tori are also maximal in G, so we are done. □

The following lemma was used in the preceding proof.

Lemma C.2 *Let* $\mathfrak{f}$ *be a field with* $\dim \mathfrak{f} \leq 1$ *such that it admits a cyclic Galois extension of every degree (e.g., a finite field). If* H *is a connected semisimple* $\mathfrak{f}$*-group then it has an anisotropic maximal* $\mathfrak{f}$*-torus.*

Proof By Steinberg's Theorem 13.2, H is quasi-split. We first treat the case when H is split, and then reduce the general quasi-split case over $\mathfrak{f}$ to the split case. Let $S \subset H$ be a maximal split $\mathfrak{f}$-torus that we assume is maximal as an $\mathfrak{f}$-torus (i.e., H is split). For $\Phi := \Phi(H, S)$, by semisimplicity of H the pair $(\Phi, \mathrm{X}^*(S)_{\mathbf{Q}})$ is a root system; we may and do assume the split semisimple H is non-trivial, so $\Phi \neq \emptyset$.

Choose a positive system of roots $\Phi^+ \subset \Phi$ and let $n \in N_H(S)(\mathfrak{f})$ represent the corresponding Coxeter element in

$$N_H(S)(\mathfrak{f})/S(\mathfrak{f}) \simeq W(\Phi).$$

By [Bour, Chap. 5, Sect. 6.2, Definition. 2, Theorem 1(i)] applied to each irreducible component of Φ, the action by n on $V = \mathrm{X}^*(S)_{\mathbf{Q}}$ satisfies $V^{n=1} = 0$. We claim the representative n can be chosen to have finite order in $N_H(S)(\mathfrak{f})$. Consider a split $\mathbf{Z}$-descent (H_0, S_0) of (H, S) (or of H^{sc} equipped with the split maximal torus preimage of S); this uses the Existence Theorem over $\mathbf{Z}$ and the Isomorphism Theorem over $\mathfrak{f}$ [C2, Proposition 6.1.16]. Since $\mathrm{H}^1_{\text{ét}}(\mathbf{Z}, \mathrm{GL}_1) = \mathrm{Pic}(\mathbf{Z}) = 1$, we have

$$N_{H_0}(S_0)(\mathbf{Z})/S_0(\mathbf{Z}) = (N_{H_0}(S_0)/S_0)(\mathbf{Z}) = W(\Phi)$$

(second equality by [C2, Proposition 5.1.6]). But $S_0(\mathbf{Z}) = (\mathbf{Z}^\times)^r$ is finite, so the finite-order representative is obtained. (If $\mathrm{char}(\mathfrak{f}) = p > 0$ we could instead use the Existence theorem over $\mathbf{F}_p$ rather than over $\mathbf{Z}$, but split descent to the prime field is inadequate if $\mathrm{char}(\mathfrak{f}) = 0$.)

Let $F/\mathfrak{f}$ be a finite cyclic extension with degree equal to the order of n, so upon choosing a generator γ of $\Gamma = \mathrm{Gal}(F/\mathfrak{f})$ we get a homomorphism $c : \Gamma \to G(\mathfrak{f})$ defined by $\gamma \mapsto n$. The homomorphism c can be regarded as a 1-cocycle in $Z^1(\Gamma, G(F))$, so it represents a class in the pointed set $\mathrm{H}^1(\mathfrak{f}, G)$ that is trivial by Steinberg's Theorem 13.2. The inflation map of pointed sets $\mathrm{H}^1(\Gamma, G(F)) \to \mathrm{H}^1(\mathfrak{f}, G)$ has trivial kernel [Ser2, Chap. 1, Sect. 5.8(a)], so $n = c(\gamma) = g^{-1}\gamma(g)$ for some $g \in G(F)$.

Since $\gamma(g) = gn$ with n valued in $N_H(S)$, so

$$gS_Fg^{-1} = \gamma(g)S_F\gamma(g)^{-1} = \gamma^*(gS_Fg^{-1}),$$

it follows that gS_Fg^{-1} descends to a maximal $\mathfrak{f}$-torus $T \subset G$. By design $T_F \simeq S_F$ is split, so to show T is $\mathfrak{f}$-anisotropic it suffices to show the γ-action on $\mathrm{X}^*(T_F)$ has fixed-locus 0. Every character of T_F has the form $\chi' : t \mapsto \chi_F(g^{-1}tg)$ for $\chi \in X^*(S)$ and points t of T_F, so

$$(\gamma^*(\chi'))(t) = \chi_F(\gamma(g)^{-1}t\gamma(g)) = \chi_F(n^{-1}g^{-1}tgn) = (n.\chi)_F(g^{-1}tg) = (n.\chi)'(t).$$

Hence, χ' is defined over $\mathfrak{f}$ (i.e., it is γ-fixed) if and only if $n.\chi = \chi$, but we've already noted that $X^*(S)_{\mathbf{Q}}^{n=1} = 0$, so this forces χ to be trivial and hence $\chi' = 1$. That establishes the $\mathfrak{f}$-anisotropicity of T, and so settles the case of split H.

Now consider a general H. It is harmless to replace H with its simply connected central cover, so $H = \prod H_i$ for $\mathfrak{f}$-simple and simply connected groups H_i. It suffices to treat the H_i's separately, so we can assume H is $\mathfrak{f}$-simple and simply connected. Then by [C3, Cor. M.1.2], $H \simeq \mathrm{R}_{\mathfrak{f}'/\mathfrak{f}}(H')$ for a finite separable $\mathfrak{f}'/\mathfrak{f}$ and an *absolutely simple* and simply connected $\mathfrak{f}'$-group H'. If we can find an anisotropic maximal $\mathfrak{f}'$-torus $T' \subset H'$ then $T := \mathrm{R}_{\mathfrak{f}'/\mathfrak{f}}(T')$ is a maximal $\mathfrak{f}$-torus of H that is anisotropic since $\mathrm{Hom}_{\mathfrak{f}}(\mathrm{GL}_1, T) = \mathrm{Hom}_{\mathfrak{f}'}(\mathrm{GL}_1, T') = 1$. Hence, we now may and do assume H is absolutely simple.

Since H is also quasi-split, hence not anisotropic, for a maximal split $\mathfrak{f}$-torus $S \subset H$ the (possibly non-reduced) relative root system $\Phi = \Phi(H, S)$ is non-empty and thus (by $\mathfrak{f}$-simplicity) is *irreducible* with $\mathbf{Q}\Phi = X^*(S)_{\mathbf{Q}}$.

The centralizer $Z = Z_H(S)$ is a maximal torus since H is quasi-split, so for its maximal anisotropic subtorus $T \subset Z$ the multiplication map $S \times T \to Z$ is an isogeny. Let $H' = \mathscr{D}(Z_H(T))$, so its maximal tori have dimension at most $\dim Z - \dim T = \dim S$. We'll show S is contained in H', so then it is a maximal $\mathfrak{f}$-torus in H' and hence the connected semisimple H' is thereby split (perhaps not $\mathfrak{f}$-simple). Once that is shown, the settled general split case provides an anisotropic maximal $\mathfrak{f}$-torus $T' \subset H'$. But then $\dim T' = \dim S = \dim Z - \dim T$ where $\dim Z$ is the dimension of the maximal $\mathfrak{f}$-tori of H. Since $T' \subset Z_H(T)$ and $T' \cap T \subset H' \cap T$ is finite, it then would follow that the anisotropic torus $T \cdot T'$ in H is maximal, as desired.

Our final task is to show $S \subset H'$. Since S normalizes $H' := \mathscr{D}(Z_H(T))$, it makes sense to form the group $S \cdot H'$ in which S is a maximal split torus and H' is the derived group. The subtorus $S' := (S \cap H')^0_{\mathrm{red}} \subset S$ is a maximal $\mathfrak{f}$-split torus in H' for which $\dim S'$ is equal to the span of $\Phi(S \cdot H', S)$ inside $V = X^*(S)_{\mathbf{Q}}$. Hence, to show $S' = S$ (yielding $S \subset H'$) it suffices to show that the subset $\Phi' := \Phi(S \cdot H', S) \subset \Phi(H, S) = \Phi$ spans V.

By irreducibility of the root system (V, Φ), the action of $W(\Phi)$ on V is absolutely irreducible and the longest roots constitute a single $W(\Phi)$-orbit. That orbit spans V by irreducibility, so it suffices to show all such roots belong to Φ'. Each S-root group $U_a \subset H$ for $a \in \Phi$ is normalized by T since T commutes with S. If T *centralizes* U_a, we'd get $U_a \subset Z_H(T)$ and hence $U_a \subset \mathscr{D}(Z_H(T)) = H'$, so $a \in \Phi(S \cdot H', S) = \Phi'$. Thus, it remains to prove that for the longest roots $a \in \Phi$ the T-conjugation action on the split unipotent U_a is trivial. We have $\dim U_a = 1$ by Proposition 25.7 (this uses that H is *absolutely simple* and *quasi-split*). Thus, the functor of S-equivariant automorphisms of the split unipotent U_a is GL_1, so the T-action on U_a is encoded in an $\mathfrak{f}$-homomorphism $T \to \mathrm{GL}_1$. This map must be trivial since T is anisotropic, so T centralizes U_a as desired. □

Theorem C.3 (Kneser) *If* $\dim \mathfrak{f} \leq 1$ *and* G *is a simply connected semisimple* k*-group then* $\mathrm{H}^1(k, G) = 1$.

Proof By Lemma 12.2, the map of pointed sets $\mathrm{H}^1(k, G) \to \mathrm{H}^1(\widehat{k}, G)$ is injective. Hence, we may and do assume k is complete (this will only be used for a successive approximation argument near the end of the proof). Since $\mathrm{H}^1(K, G_K) = 1$ by Steinberg's Theorem 13.2, by inflation-restriction we have $\mathrm{H}^1(k, G) = \mathrm{H}^1(K/k, G(K))$. Since G is simply connected, we have $G(K) = G(K)^0$. We'll prove $\mathrm{H}^1(K/k, G(K)^0) = 1$ by using a well-chosen Iwahori subgroup of $G(K)$.

To motivate how Iwahori subgroups are used, we recall a classical fact [Ser2, Chap. 3, Sect. 4.3, Lemma 6] in the Galois cohomology of smooth affine groups $\mathbf{G}$ over fields F: the map $\mathrm{H}^1(F, N_{\mathbf{G}}(\mathbf{T})) \to \mathrm{H}^1(F, \mathbf{G})$ is surjective for any maximal F-torus $\mathbf{T}$ in $\mathbf{G}$. This rests on a general result about Galois cohomology of normalizers [Ser2, Chap. 3, Sect. 2.2, Lemma 1] and the fact that in smooth affine F-groups (such as a Galois-twisted form of $\mathbf{G}$) the maximal F-tori remain maximal over F_s and are all conjugate over F_s.

Those features of maximal tori will be adapted with K/k in place of F_s/F and certain Iwahori subgroups of $G(K)^0$ in place of tori. Since Iwahori subgroups of $G(K)^0$ (like parahoric subgroups for Tits systems more generally) are their own normalizers, we will thereby arrive at the surjectivity of

$$\mathrm{H}^1(K/k, \mathcal{I}) \to \mathrm{H}^1(K/k, G(K))$$

for $\mathcal{I} = \mathscr{G}^0_{\mathcal{C}}(\mathcal{O})$ with a $\mathrm{Gal}(K/k)$-stable chamber $\mathcal{C} \subset \mathcal{B}(G_K)$ (the existence of such a chamber will use the hypothesis $\dim \mathfrak{f} \leq 1$). The triviality of $\mathrm{H}^1(K/k, \mathcal{I})$ will be proved by successive approximation based on work with infinitesimal fibers of the smooth affine group $\mathscr{G}^0_{\mathcal{C}} \to \mathrm{Spec}(\mathcal{O})$ (and the base case for successive approximation will rest on $\dim \mathfrak{f} \leq 1$ via Steinberg's Theorem 13.2).

Having laid out the motivation for the strategy of proof, we record a lemma whose proof uses crucially that $\dim \mathfrak{f} \leq 1$:

Lemma C.4 *If* $C \subset \mathcal{B}(G)$ *is a chamber then it is contained in a unique facet in* $\mathcal{B}(G_K)$ *and that facet is a chamber* $\mathcal{C}$.

Proof To show C lies in a unique facet of $\mathcal{B}(G_K)$, we pick an apartment $A = \mathcal{A}(S)$ of $\mathcal{B}(G)$ containing C, so $A = \mathcal{B}(G) \cap \mathcal{A}(T_K)$ for a special k-torus $T \supset S$ in G by Axiom 27. If A is a point then there is nothing to do, and when $\dim A > 0$ we need to show the connected C meets each hyperplane in $\mathcal{A}(T_K)$ in either C or $\emptyset$. This in turn holds because by Axiom 22, the affine root hyperplanes in A (when $\dim A > 0$, which is to say $\dim \mathcal{B}(G) > 0$) are $A \cap H$ for affine root hyperplanes H in $\mathcal{A}(T_K)$ meeting A in a non-empty proper subspace.

Let $\mathcal{F} \subset \mathcal{B}(G_K)$ be the unique facet containing a chamber C in $\mathcal{A}(S) \subset \mathcal{B}(G)$. In particular, for a point $x \in C$ we have

$$(\mathscr{G}^0_C)_{\mathcal{O}} = (\mathscr{G}^0_x)_{\mathcal{O}} = \mathscr{G}^0_{\mathcal{F}}$$

By Axiom 22, the chamber property for $\mathcal{F}$ (which we want to prove) corresponds to $\overline{\mathscr{G}}^0_{\mathcal{F}}$ having no proper parabolic $\bar{\mathfrak{f}}$-subgroup. That in turn says exactly that the maximal reductive quotient $\mathsf{G}_{\mathcal{F}}$ over $\bar{\mathfrak{f}}$ has no proper parabolic subgroup. The absence of such proper subgroups says exactly that the connected reductive $\mathsf{G}_{\mathcal{F}}$ over $\bar{\mathfrak{f}}$ is a torus, or equivalently $\overline{\mathscr{G}}^0_{\mathcal{F}}$ is solvable.

Since $\mathscr{G}^0_{\mathcal{F}} = (\mathscr{G}^0_{\mathcal{C}})_{\mathcal{O}}$, we see that $\overline{\mathscr{G}}^0_{\mathcal{F}} = (\overline{\mathscr{G}}^0_{\mathcal{C}})_{\bar{\mathfrak{f}}}$. Hence, solvability of $\overline{\mathscr{G}}^0_{\mathcal{F}}$ is equivalent to solvability of $\overline{\mathscr{G}}^0_{C}$. But since $\dim \mathfrak{f} \leq 1$, so every smooth connected affine $\mathfrak{f}$-group is quasi-split, solvability for such an $\mathfrak{f}$-group is equivalent to having no proper parabolic $\mathfrak{f}$-subgroup. The latter does hold for $\overline{\mathscr{G}}^0_{C}$ since C is a chamber, so running the equivalences in reverse shows that $\mathcal{F}$ is a chamber. □

Now pick a chamber $C \subset \mathcal{B}(G)$, so the unique chamber $\mathcal{C} \subset \mathcal{B}(G_K)$ containing C is stable under the action of $\Gamma = \mathrm{Gal}(K/k)$ due to its uniqueness. The Iwahori subgroup $\mathcal{I} = G(K)^0_{\mathcal{C}} \subset G(K)^0 = G(K)$ is then likewise Γ-stable. We shall prove $\mathrm{H}^1(K/k, \mathcal{I}) \to \mathrm{H}^1(K/k, G(K))$ is surjective.

Let $c : \Gamma \to G(K)$ be a continuous 1-cocycle representing a class in $\mathrm{H}^1(K/k, G(K))$, so c factors through $\mathrm{Gal}(k'/k) \to G(k')$ for some finite Galois unramified k'/k inside K/k. The associated k'/k-form ${}_cG$ of G has $({}_cG)(K) = G(K)$ with Γ-action $\gamma * g = c(\gamma)\gamma(g)c(\gamma)^{-1}$. For a chamber $C' \subset \mathcal{B}({}_cG)$ the chamber $\mathcal{C}' \subset \mathcal{B}(({}_cG)_K) = \mathcal{B}(G_K)$ containing C' (Lemma C.4) gives rise to an Iwahori subgroup $\mathcal{I}' = G(K)_{\mathcal{C}'}$ that is Γ-stable under the $*$-action (as $\mathcal{C}'$ is stable under the $*$-action of Γ on $G(K) = ({}_cG)(K)$). That is,

$$c(\gamma)\gamma(\mathcal{I}')c(\gamma)^{-1} = \mathcal{I}'. \tag{C.1}$$

We have $\mathcal{I}' = g\mathcal{I}g^{-1}$ for some $g \in G(K)$ since all Iwahori subgroups are conjugate. Plugging that into (C.1) yields that $g^{-1}c(\gamma)\gamma(g)$ normalizes $\mathcal{I}$. But $\mathcal{I}$ is its own normalizer in $G(K) = G(K)^0$ since it is a parabolic subgroup for a Tits system (see Axiom 9 in Sect. 28). Hence, $c'(\gamma) := g^{-1}c(\gamma)\gamma(g)$ is a continuous function $\Gamma \to \mathcal{I}$. By inspection, c' is 1-cohomologous to c [Ser2, Chap. 1, Sect. 5.1], so it is a 1-cocycle representing the same class as c in $\mathrm{H}^1(k, G)$. Since c' is valued in $\mathcal{I}$, we have shown $\mathrm{H}^1(K/k, \mathcal{I}) \to \mathrm{H}^1(K/k, G(K))$ is surjective.

It remains to prove the triviality of $\mathrm{H}^1(K/k, \mathcal{I})$, where $\mathcal{I} = \mathscr{G}^0_{\mathcal{C}}(\mathcal{O}) = (\mathscr{G}^0_{C})_{\mathcal{O}}(\mathcal{O}) = \mathscr{G}^0_{C}(\mathcal{O})$ (switching from $\mathcal{C} \subset \mathcal{B}(G_K)$ to $C \subset \mathcal{B}(G)$). More generally, if $\mathscr{G}$ is *any* smooth affine $\mathfrak{o}$-group with a connected special fiber (such as $\mathscr{G}^0_C$) then we claim $\mathrm{H}^1(K/k, \mathscr{G}(\mathcal{O})) = 1$. Anything in $\mathrm{H}^1(K/k, \mathscr{G}(\mathcal{O}))$ is inflated from $\mathrm{H}^1(k'/k, \mathscr{G}(\mathfrak{o}'))$ for a finite Galois unramified k'/k, so it suffices to show $\mathrm{H}^1(k'/k, \mathscr{G}(\mathfrak{o}')) = 1$ for each finite Galois unramified k'/k. This will use completeness of k, hence of k'.

Fix such k'/k, let $\Gamma_0 = \mathrm{Gal}(k'/k) = \mathrm{Gal}(\mathfrak{f}'/\mathfrak{f})$ (for the corresponding Galois extension $\mathfrak{f}'/\mathfrak{f}$ of residue fields), and let $c : \Gamma_0 \to \mathscr{G}(\mathfrak{o}')$ be a 1-cocycle. We seek $g \in \mathscr{G}(\mathfrak{o}')$ such that $c(\gamma) = g^{-1}\gamma(g)$ for all $\gamma \in \Gamma_0$. The inverse system of groups $G_n = \mathscr{G}(\mathfrak{o}'/\mathfrak{m}'^{n+1})$ for $n \geq 0$ has inverse limit $\mathscr{G}(\mathfrak{o}')$ by completeness of k', compatibly with the Γ_0-action. The transition maps $G_{n+1} \to G_n$ are surjective for all $n \geq 0$ by $\mathfrak{o}$-smoothness of $\mathscr{G}$, so $\mathscr{G}(\mathfrak{o}') \to G_n$ is surjective for all $n \geq 0$.

Let $K_n = \ker(\mathscr{G}(\mathfrak{o}') \to G_n)$. Since $\bigcap K_n = 1$ and any infinite product $\prod k_n$ with $k_n \in K_n$ converges in $\mathscr{G}(\mathfrak{o}')$ (by completeness of k'), a standard successive approximation argument reduces the task of proving $\mathrm{H}^1(\Gamma_0, \mathscr{G}(\mathfrak{o}')) = 1$ to proving $\mathrm{H}^1(\Gamma_0, G_0) = 1$ and $\mathrm{H}^1(\Gamma_0, K_n/K_{n-1}) = 1$ for all $n \geq 1$.

First we prove triviality of $\mathrm{H}^1(\Gamma_0, G_0) = \mathrm{H}^1(\mathfrak{f}'/\mathfrak{f}, \overline{\mathscr{G}}(\mathfrak{f}'))$. Inflation maps as we increase $\mathfrak{f}'/\mathfrak{f}$ have trivial kernel (a basic fact about general degree-1 non-abelian group cohomology), so it suffices to prove $\mathrm{H}^1(\mathfrak{f}, \overline{\mathscr{G}}) = 1$. That vanishing holds by Steinberg's Theorem 13.2 since $\dim \mathfrak{f} \leq 1$.

By smoothness of $\mathscr{G}$, it is standard that $K_n/K_{n-1} \simeq \mathrm{Lie}(\overline{\mathscr{G}}_{\mathfrak{f}'}) \otimes_{\mathfrak{f}'} (\mathfrak{m}'^n/\mathfrak{m}'^{n+1})$ (as groups equipped with an action by $\mathrm{Gal}(k'/k) = \mathrm{Gal}(\mathfrak{f}'/\mathfrak{f})$). But the right side is Galois-equivariantly identified with $V \otimes_{\mathfrak{f}} \mathfrak{f}'$ for the $\mathfrak{f}$-vector space

$$V = \mathrm{Lie}(\overline{\mathscr{G}}) \otimes_{\mathfrak{f}} (\mathfrak{m}^n/\mathfrak{m}^{n+1}),$$

so $\mathrm{H}^1(\Gamma_0, K_n/K_{n-1}) = V \otimes_{\mathfrak{f}} \mathrm{H}^1(\mathfrak{f}'/\mathfrak{f}, \mathfrak{f}') = 0$. □

When k is a local field (rather than merely being Henselian with finite residue field, for example), so Tate local duality for tori is available, Theorems C.1 and C.3 can be combined to prove:

Corollary C.5 *Let G be a connected semisimple group over a non-Archimedean local field k, and let μ be the kernel of the central isogeny $q : G^{\mathrm{sc}} \to G$. The connecting map $\delta : \mathrm{H}^1(k, G) \to \mathrm{H}^2(k, \mu)$ is bijective.*

Beware that in this result, the kernel μ may be non-étale when $\mathrm{char}(k) = p > 0$, so the construction of δ then requires a description of $\mathrm{H}^2(k, \mu)$ in terms of gerbes, on par with the description of $\mathrm{H}^1(k, G)$ in terms of torsors. See [C1, Appendix B] (especially [C1, Sect. B.3]) for a discussion of this. The proof of Corollary C.5 requires no Bruhat-Tits theory beyond its role in the proofs of Theorems C.1 and C.3.

Proof By Theorem C.1 we can find a k-anisotropic maximal k-torus $T \subset G$, so by centrality of $q : G^{\mathrm{sc}} \twoheadrightarrow G$ the preimage $T' = q^{-1}(T) \subset G^{\mathrm{sc}}$ is also a maximal k-torus and clearly is k-anistropic. By Tate local duality for tori [Mil1, Chap. 3, Theorem 6.9], $\mathrm{H}^2(k, T')$ is $\mathbf{Q}/\mathbf{Z}$-dual to the profinite completion of $\mathrm{H}^0(k, \mathrm{X}^*(T'_{k_s})) = \mathrm{X}^*_k(T') = \mathrm{Hom}_k(T', \mathrm{GL}_1) = 1$ (vanishing because T' is k-anisotropic). Thus, $1 \to \mu \to T' \to T \to 1$ gives rise to a long exact sequence

$$\mathrm{H}^1(k, T') \to \mathrm{H}^1(k, T) \to \mathrm{H}^2(k, \mu) \to \mathrm{H}^2(k, T') = 1,$$

so $\delta' : \mathrm{H}^1(k, T) \to \mathrm{H}^2(k, \mu)$ is surjective.

The commutative diagram of central extensions

$$\begin{array}{ccccccccc} 1 & \longrightarrow & \mu & \longrightarrow & T' & \longrightarrow & T & \longrightarrow & 1 \\ & & \| & & \downarrow & & \downarrow & & \\ 1 & \longrightarrow & \mu & \longrightarrow & G^{\mathrm{sc}} & \longrightarrow & G & \longrightarrow & 1 \end{array}$$

yields a commutative diagram of connecting maps

$$\begin{array}{ccc} \mathrm{H}^1(k, T) & \xrightarrow{\delta'} & \mathrm{H}^2(k, \mu) \\ \downarrow & & \| \\ \mathrm{H}^1(k, G) & \xrightarrow[\delta]{} & \mathrm{H}^2(k, \mu) \end{array}$$

from which we see $\delta : \mathrm{H}^1(k, G) \to \mathrm{H}^2(k, \mu)$ is surjective.

To show δ is injective, we first note that the diagram of pointed sets

$$\mathrm{H}^1(k, G^{\mathrm{sc}}) \to \mathrm{H}^1(k, G) \xrightarrow{\delta} \mathrm{H}^2(k, \mu)$$

is exact. Hence, $\delta^{-1}(1)$ is the image of $\mathrm{H}^1(k, G^{\mathrm{sc}})$. But this latter cohomology set is a point, by Theorem C.3. To handle the other fibers of δ (all of which are non-empty, as δ is surjective), we pick a right G-torsor Y over k representing an element of $\mathrm{H}^1(k, G)$ and aim to show the fiber over $\delta([Y])$ has only one point in it.

The Galois-cohomological technique of cocycle-twisting has a formulation in the language of torsors: if we denote by ${}_YG$ the k-group $\underline{\mathrm{Aut}}_G(Y)$ then by [C1, Sect. B.2] there is a natural bijection of pointed sets $t_Y : \mathrm{H}^1(k, G) \to \mathrm{H}^1(k, {}_YG)$ carrying $[Y]$ to the trivial class (and corresponding to subtracting $[Y]$ when G is commutative). Moreover, by [C1, Proposition B.3.3] there is an associated "Y-twisted" central extension

$$1 \to \mu \to {}_Y(G^{\mathrm{sc}}) \to {}_YG \to 1$$

whose connecting map fits into the top of a commutative diagram of sets

$$\begin{array}{ccc} \mathrm{H}^1(k, G) & \xrightarrow{\delta} & \mathrm{H}^2(k, \mu) \\ {\scriptstyle t_Y}\downarrow & & \downarrow{\scriptstyle -\delta([Y])} \\ \mathrm{H}^1(k, {}_YG) & \xrightarrow[{}_Y\delta]{} & \mathrm{H}^2(k, \mu) \end{array}$$

(In [C1, Proposition B.3.3(ii)] the role of subtraction of $\delta([Y])$ on the right side is accidentally not mentioned but is implicit in the proof and in how the statement is used in the proof of [C1, Proposition 7.1.3]. An alternative reference for this commutative square in vaster generality is [Gir, Chap. 4, Proposition 4.3.4].)

In this way, t_Y carries the δ-fiber through $[Y]$ over to the ${}_Y\delta$-fiber over 1. But the latter is a single point since it is the image of the pointed set $\mathrm{H}^1(k, {}_Y(G^{\mathrm{sc}}))$ that is trivial by Theorem C.3 since ${}_Y(G^{\mathrm{sc}})$ is simply connected (indeed, for a finite separable extension k'/k over which the smooth non-empty Y acquires a rational point, we have $({}_Y(G^{\mathrm{sc}}))_{k'} \simeq (G^{\mathrm{sc}})_{k'}$). □

Appendix D. Cartan Decomposition

In this appendix, we discuss double coset representatives for $G(k)$ relative to two types of parahoric subgroups: a minimal one (i.e., an Iwahori subgroup) and certain maximal ones (corresponding to special vertices). The former amounts to a refinement of Proposition 29.3, and the latter is called the *Cartan decomposition*. Here is the Iwahori case:

Proposition D.1 *Let $\mathcal{I} = G(k)^0_{\mathcal{C}}$ be the Iwahori subgroup associated to a chamber $\mathcal{C} \subset \mathcal{B}(G)$ and let $\mathcal{A} := \mathcal{A}(S)$ be an apartment containing $\mathcal{C}$, with $S \subset G$ a maximal split k-torus.*

For $N = N_G(S)$ and $Z = Z_G(S)$, in the double coset decomposition $G(k) = \mathcal{I}N(k)\mathcal{I}$ from Proposition 29.3 a pair of elements $n, n' \in N(k)$ have the same $\mathcal{I}$ double coset if and only if $n'n^{-1} \in Z(k)^0$.

Note that $Z(k)^0$ is a normal subgroup of $N(k)$ since $Z(k)^0 = Z(k)^1 \cap G(k)^0$ (Proposition 11.4).

Proof Since $Z(k)^0 \subset \mathcal{I}$, the implication "$\Leftarrow$" is clear. For the converse, consider $n, n' \in N(k)$ that are in the same $\mathcal{I}$ double coset. We aim to show $n \equiv n' \bmod Z(k)^0$.

Let $N(k)_{\mathcal{C}} = \mathrm{Stab}_{N(k)}(\mathcal{C})$ and $\mathcal{N} = \mathrm{Stab}_{G(k)^0}(\mathcal{A}) = N(k) \cap G(k)^0$. In the proof of Proposition 29.3 we saw that $N(k)_{\mathcal{C}}$ normalizes $\mathcal{I}$ and $N(k) = N(k)_{\mathcal{C}}\mathcal{N}$ (set of pairwise products), so likewise $N(k) = \mathcal{N}N(k)_{\mathcal{C}}$. Hence, we can write $n = \widetilde{n}n_0$ and $n' = \widetilde{n}'n_0'$ for $\widetilde{n}, \widetilde{n}' \in \mathcal{N}$ and $n_0, n_0' \in N(k)_{\mathcal{C}}$. We will show $n_0 \equiv n_0' \bmod Z(k)^0$ and $\widetilde{n} \equiv \widetilde{n}' \bmod Z(k)^0$.

Since $N(k)_{\mathcal{C}}$ normalizes $\mathcal{I}$, we have

$$\mathcal{I}n\mathcal{I} = \mathcal{I}\widetilde{n}\mathcal{I}n_0, \quad \mathcal{I}n'\mathcal{I} = \mathcal{I}\widetilde{n}'\mathcal{I}n_0'. \tag{D.1}$$

The *equality* of these double cosets yields $n_0'n_0^{-1} \in \mathcal{I}(\widetilde{n}')^{-1}\mathcal{I}\widetilde{n}\mathcal{I} \subset G(k)^0$, so $n_0'n_0^{-1} \in N(k)_{\mathcal{C}} \cap G(k)^0$. This final intersection is in $Z(k)^0$. Indeed, any g in this intersection belongs to $N(k)$, so to prove $g \in Z(k)^0$ it suffices to show triviality of the g-action on $\mathcal{A}$ since the kernel of that action is $Z(k)^1$ (and $Z(k)^1 \cap G(k)^0 = Z(k)^0$ by Proposition 11.4). But $g \in G(k)^0$ preserves the facet $\mathcal{C}$ and so must act on it as the identity (as for any facet), yet the membership $g \in N(k)$ implies g acts on $\mathcal{A}$ through an *affine* automorphism. Such an automorphism of $\mathcal{A}$ that is the identity on the polysimplex $\mathcal{C}$ must be the identity.

Writing $\widetilde{n}n_0 = n = h_1 n' h_2 = h_1 \widetilde{n}' n_0' h_2$ for some $h_1, h_2 \in \mathcal{I}$, applying this common element of $G(k)$ to $\mathcal{C}$ implies $\widetilde{n}\mathcal{C} = h_1\widetilde{n}'\mathcal{C}$, so $\widetilde{n}^{-1}h_1\widetilde{n}' \in G(k)^0_{\mathcal{C}} = \mathcal{I}$. Hence, $\widetilde{n}'$ and $\widetilde{n}$ lie in the same $\mathcal{I}$ double coset. But $(\mathcal{I}, \mathcal{N})$ is a BN-pair for $G(k)^0$, so $\widetilde{n} \equiv \widetilde{n}' \bmod \mathcal{N} \cap \mathcal{I}$ by the Bruhat decomposition for Tits systems. Thus, to deduce $\widetilde{n} \equiv \widetilde{n}' \bmod Z(k)^0$ it suffices to show the containment $Z(k)^0 \subset \mathcal{N} \cap \mathcal{I}$ is an equality. This intersection is contained in $N(k)_{\mathcal{C}} \cap G(k)^0$ that we have seen is equal to $Z(k)^0$, so we are done. □

Next we turn to the Cartan decomposition. As a warm-up, we first discuss the case $G = \mathrm{GL}_n$. It is a classical fact, deduced from the structure theorem for modules over a principal ideal domain, that if $\pi \in \mathfrak{o}$ is a uniformizer then the $\mathrm{GL}_n(\mathfrak{o})$-double cosets in $\mathrm{GL}_n(k)$ are uniquely represented by diagonal matrices

$$\mathrm{diag}(\pi^{e_1}, \ldots, \pi^{e_n}) \tag{D.2}$$

with $e_1 \geq \cdots \geq e_n$. The same statement is true for SL_n (subject to the evident necessary condition $\sum e_j = 0$), though it isn't quite obvious that this follows from the result for GL_n.

The Cartan decomposition generalizes this to a description of double coset representatives relative to a certain type of maximal parahoric subgroup for a general G. To state it, we need to define an analogue of the special type of diagonal matrices used for GL_n. For the split diagonal maximal torus $D_n \subset \mathrm{GL}_n$ and positive system of roots $\Phi^+ \subset \Phi(\mathrm{GL}_n, D_n)$ corresponding to the upper-triangular Borel subgroup, the positive roots are the characters

$$a_{ij} : \mathrm{diag}(t_1, \ldots, t_n) \mapsto t_i/t_j$$

for $i < j$. Since $D_n(k)^0$ consists of the diagonal matrices whose entries t_i all belong to $\mathfrak{o}^\times$, each coset in $D_n(k)/D_n(k)^0$ is represented by a unique diagonal matrix

$$\mathrm{diag}(\pi^{e_1}, \ldots, \pi^{e_n})$$

for $e_1, \ldots, e_n \in \mathbf{Z}$. The inequalities $e_j \geq e_{j+1}$ can be expressed in terms of the simple positive roots $a_{i,i+1}$ since $\mathrm{ord}(a_{i,i+1}(t)) = \mathrm{ord}(t_i/t_{i+1})$. Requiring these to all be non-negative is of course the same as requiring $\mathrm{ord}(a(t)) \geq 0$ for all $a \in \Phi^+$. To summarize, each coset in

$$\{t \in D_n(k) : \mathrm{ord}(a(t)) \geq 0 \text{ for all } a \in \Phi^+\}/D_n(k)^0$$

is uniquely represented by a diagonal matrix

$$\mathrm{diag}(\pi^{e_1}, \ldots, \pi^{e_n})$$

with $e_1 \geq \cdots \geq e_n$.

For a general connected reductive k-group G, maximal split k-torus $S \subset G$, and $Z = Z_G(S)$, the central S in the connected reductive Z is an isogeny-factor of Z. Hence, by anisotropicity of Z/S, composing $S \hookrightarrow Z$ with the maximal split torus quotient $q : Z \to \overline{S}$ yields an isogeny $S \to \overline{S}$. In particular, $\mathrm{X}^*(\overline{S}) \to \mathrm{X}^*(S)$ is a finite-index lattice inclusion, so naturally $\mathrm{X}^*(S)_{\mathbf{Q}} = \mathrm{X}^*(\overline{S})_{\mathbf{Q}}$. Thus, each $a \in \Phi(G, S) \subset \mathrm{X}^*(S)_{\mathbf{Q}}$ can be viewed as an element $a_{\mathbf{Q}} \in \mathrm{X}^*(\overline{S})_{\mathbf{Q}}$.

For each $\overline{s} \in \overline{S}(k)$ we have a homomorphism $\mathrm{X}^*(\overline{S}) \to \mathbf{Z}$ via $\chi \mapsto \mathrm{ord}(\chi(\overline{s}))$, and hence a homomorphism $\mathrm{Hom}_k(Z, \mathrm{GL}_1)_{\mathbf{Q}} = \mathrm{X}^*(\overline{S})_{\mathbf{Q}} \to \mathbf{Q}$ that we'll denote with the same notation. Thus, for each $\chi \in \mathrm{X}^*(\overline{S})_{\mathbf{Q}}$ we get a homomorphism $Z(k) \to \mathbf{Q}$ by $z \mapsto \mathrm{ord}(\chi(z))$. Upon choosing a positive system of roots $\Phi^+ \subset \Phi := \Phi(G, S)$, we can thereby define the subset

$$\mathcal{Z} := \{z \in Z(k) : \mathrm{ord}(a_{\mathbf{Q}}(z)) \geq 0 \text{ for all } a \in \Phi^+\} \subset Z(k);$$

note that this contains the kernel $Z(k)^1$ of $\omega_Z : Z(k) \to \mathrm{X}_*(S)_{\mathbf{R}}$ (see (7.1) applied to Z). For $G = \mathrm{GL}_n$, $S = D_n$, and the standard choice of Φ^+, the cosets in $\mathcal{Z}/Z(k)^0$ are uniquely represented by exactly the special diagonal matrices in (D.2).

Theorem D.2 (Cartan decomposition) *If $x \in \mathcal{B}(G)$ is a special vertex then*

$$G(k) = \bigcup_{z \in \mathcal{Z}} G(k)_x^0 z G(k)_x^0$$

where the $G(k)_x^0$-double cosets represented by $z, z' \in \mathcal{Z}$ coincide precisely when z, z' have the same image in $Z(k)/Z(k)^0$.

The parahoric subgroups associated to vertices (the minimal facets) are maximal among parahoric subgroups, and *there is always at least one special vertex on each chamber*: this is a tautology when $\mathcal{B}(G)$ is a point, and otherwise follows from Proposition 20.12.

Note in particular that if $z, z' \in Z(k)^1 \subset \mathcal{Z}$ are distinct modulo $Z(k)^0$ then $G(k)_x^0 z G(k)_x^0 \neq G(k)_x^0 z' G(k)_x^0$. This distinctness of double cosets is also immediate from the fact that $G(k)/G(k)^0$ contains $Z(k)^1/Z(k)^0$ as a subgroup (Proposition 11.4).

Proof Let $\mathcal{C}$ be a chamber in $\mathcal{B}(G)$ with x as a vertex, and let $\mathcal{A} = \mathcal{A}(S)$ be an apartment containing $\mathcal{C}$. Let $N = N_G(S)$, $Z = Z_G(S)$, and $\Phi = \Phi(G, S)$. The Iwahori subgroup $\mathcal{I} = G(k)_{\mathcal{C}}^0$ acts trivially on the closure $\overline{\mathcal{C}}$, so it is contained in $G(k)_x^0$.

The strategy for proving the Cartan decomposition is to refine the Bruhat decomposition $G(k) = \mathcal{I} N(k) \mathcal{I}$ from Proposition 29.3 in two steps: first we will massage it into $G(k)_x^0 Z(k) \mathcal{I}$ (increasing the left factor $\mathcal{I}$ to $G(k)_x^0$ and shrinking $N(k)$ in the middle to $Z(k)$), and then we will massage that into $G(k)_x^0 \mathcal{Z} G(k)_x^0$ (shrinking $Z(k)$ to $\mathcal{Z}$ and increasing the right factor $\mathcal{I}$ to $G(k)_x^0$). Identifying exactly when elements of $\mathcal{Z}$ represent the same $G(k)_x^0$ double coset in $G(k)$ then requires some careful group-theoretic work, as we shall see.

Let's first dispose of the degenerate situation in which $\mathcal{B}(G)$ is a point (so the relative root system Φ is empty); in such cases the unique point of the building is special by definition. In this case S is central and there are no relative roots, so $\mathcal{Z} = Z(k) = G(k)$ and $G(k)_x^0 = G(k)^0$. Thus, the Cartan decomposition is the obvious assertion $G(k) = G(k)^0 G(k) G(k)^0$. We have $G(k)^0 g G(k)^0 = g G(k)^0$ by normality of $G(k)^0$ in $G(k)$, so the proposed characterization of when two double cosets coincide is also obvious.

Now we assume $\mathcal{B}(G)$ is not a point, so $\mathcal{A}$ has positive dimension and Axiom 6 in Sect. 27 provides an affine root system $\Psi = \Psi(G, S)$ defining the polysimplicial structure on $\mathcal{A}$ and having derivative $\nabla\Psi = \Phi$. The affine Weyl group $W(\Psi) \subset \mathrm{Aff}(\mathcal{A})$ is generated by reflections in the walls of any single chamber, such as $\mathcal{C}$, and it coincides with the Weyl group $\mathcal{N}/(\mathcal{N} \cap \mathcal{I}) \subset \mathrm{Aff}(\mathcal{A})$ of the BN-pair $(\mathcal{I}, \mathcal{N})$ of $G(k)^0$ encoded in the restricted abstract building and the pair $(\mathcal{A}, \mathcal{C})$ (see Axiom 9 in Sect. 28 and the discussion immediately after it). Since $\mathcal{N} = \mathrm{Stab}_{G(k)^0}(\mathcal{A}) = N(k) \cap G(k)^0$, the subgroup $\mathcal{N}_x := \mathrm{Stab}_{\mathcal{N}}(x)$ is equal to $N(k) \cap G(k)_x^0$.

The x-stabilizer $W(\Psi)_x$ is clearly the image of $\mathcal{N}_x$, and (by Lemma 20.2) it is also carried isomorphically under $\nabla : W(\Psi) \to W(\Phi)$ onto the subgroup $W(\Psi_x)$. But x is special, so $W(\Psi_x) = W(\Phi)$ by Proposition 20.12(2). Hence, composing ∇ with the action of $N(k)$ on $\mathcal{A} \subset \mathcal{B}(G)$ carries $\mathcal{N}_x$ onto $W(\Phi)$. But that composite action is exactly the natural action of $N(k)$ on the root system $(\mathrm{X}^*(S')_{\mathbf{R}}, \Phi)$ by Axiom 4 in Sect. 27, yet this natural action identifies $N(k)/Z(k)$ with $W(\Phi)$, so we have shown that the natural map

$$N(k) \cap G(k)_x^0 = \mathcal{N}_x \to N(k)/Z(k)$$

is surjective. Hence, $N(k) \subset G(k)_x^0\, Z(k)$, so

$$G(k) = \mathcal{I} N(k) \mathcal{I} \subset \mathcal{I} G(k)_x^0\, Z(k) \mathcal{I} = G(k)_x^0\, Z(k) \mathcal{I}$$

(recall $\mathcal{I} := G(k)_{\mathcal{C}}^0 \subset G(k)_x^0$ since x is a vertex of $\mathcal{C}$). This completes the first step of our two-step strategy.

Since $\mathcal{I} \subset G(k)_x^0$, to prove the Cartan decomposition (without worrying yet about how to identify equality of double cosets for $z, z' \in \mathcal{Z}$), it suffices to show $Z(k) \subset G(k)_x^0\, \mathcal{Z} G(k)_x^0$. We'll prove the more precise containment $Z(k) \subset \mathcal{N}_x\, \mathcal{Z} \mathcal{N}_x$, and even more specifically we'll prove

$$\bigcup_{n \in \mathcal{N}_x} n \mathcal{Z} n^{-1} = Z(k)$$

(the left side is clearly contained in the right side since $Z(k)$ is normal in $N(k)$).

Our goal is to show that for each $z \in Z(k)$ there is some $n \in \mathcal{N}_x$ such that the element $n^{-1}zn \in Z(k)$ satisfies the non-negativity requirements which define $\mathcal{Z}$. For each $a \in \Phi^+$ and $n \in \mathcal{N}_x$ with image $w \in W(\Phi)$,

$$a_{\mathbf{Q}}(n^{-1}zn) = (w.a)_{\mathbf{Q}}(z).$$

As we vary through all such n, the corresponding w's vary through the entirety of $W(\Phi)$, and so the sets $w\Phi^+$ vary through all positive systems of roots in Φ (since $W(\Phi)$ acts transitively on the set of Weyl chambers, and hence on the set of positive systems of roots). Hence, it suffices to show that for each $z \in Z(k)$ there is *some* positive system of roots $R^+ \subset \Phi$ such that $\mathrm{ord}(a_{\mathbf{Q}}(z)) \geq 0$ for all $a \in R^+$.

By definition, $a_{\mathbf{Q}}(z)$ only depends on z through the image $q(z) \in \overline{S}(k)$, and we may replace z with a power z^N for $N \geq 1$, so we can arrange that $q(z)$ is in the image of $S(k) \to \overline{S}(k)$. Hence, we can forget about Z and even G, remembering only the split k-torus S and the root datum $(\mathrm{X}^*(S), \Phi, \mathrm{X}_*(S), \Phi^\vee)$ (generally not a root system since Φ may not span $\mathrm{X}^*(S)_{\mathbf{Q}}$): it suffices to show for every $s \in S(k)$ that there exists a positive system of roots $R^+ \subset \Phi$ such that $\mathrm{ord}(a(s)) \geq 0$ for all $a \in R^+$.

The pairing $\mathrm{X}^*(S) \times S(k) \to \mathbf{Z}$ defined by $(a, s) \mapsto \mathrm{ord}(a(s))$ makes $S(k)^0$ annihilate $\mathrm{X}^*(S)$. Using $k^\times \otimes_{\mathbf{Z}} \mathrm{X}_*(S) \simeq S(k)$ via $(t, \lambda) \mapsto \lambda(t)$, the induced pairing

$$\mathrm{X}^*(S) \times (S(k)/S(k)^0) \to \mathbf{Z}$$

becomes the evaluation pairing between $\mathrm{X}^*(S)$ and $\mathrm{X}_*(S)$. Thus, we are reduced to showing that for every $v \in \mathrm{X}_*(S)$ there exists a positive system of roots $R^+ \subset \Phi \subset \mathrm{X}^*(S)$ such that $a(v) \geq 0$ for all $a \in R^+$. Letting $V = \mathrm{X}_*(S)_{\mathbf{R}}$, it suffices to show that as R^+ varies through all positive systems of roots in Φ, the cones

$$C(R^+) = \{v \in V : a(v) \geq 0 \text{ for all } a \in R^+\}$$

cover V. Observe that this claim is insensitive to replacing V with its quotient modulo the subspace $V_0 \subset \mathrm{X}_*(S)_{\mathbf{R}}$ orthogonal to the span $V' = \mathbf{R}\Phi \subset \mathrm{X}^*(S)_{\mathbf{R}}$.

Since V/V_0 is naturally dual to V', we are brought to a statement about the *root system* (V', Φ): the dual space is covered by the cones defined by non-negativity of evaluation against positive systems of roots in Φ. A choice of $W(\Phi)$-invariant inner product turns this into the standard fact that the dual root system has underlying vector space covered by closures of Weyl chambers. This completes the proof that $G(k) = G(k)_x^0 \, \mathcal{Z} G(k)_x^0$, the second step of our two-step strategy.

Now we turn to showing that elements $z, z' \in \mathcal{Z}$ have the same $G(k)_x^0$ double coset if and only if z and z' have the same image in $Z(k)/Z(k)^0$ (from the definition of $\mathcal{Z}$, it is a union of $Z(k)^1$-cosets, hence also $Z(k)^0$-cosets). In what follows, all we will use about x is that it is a point in $\mathcal{A}$, not that it is a special vertex (or lies in the closure of $\mathcal{C}$).

The implication "$\Leftarrow$" is clear since $Z(k)^0 \subset G(k)_x^0$. For the converse, assume $z' \in G(k)_x^0 z G(k)_x^0$. The images of z and z' in the quotient group $G(k)/G(k)^0$ are the same, so $z^{-1}z' \in G(k)^0$. Since $Z(k)^0 = G(k)^0 \cap Z(k)^1$ (Proposition 11.4), to deduce $z^{-1}z' \in Z(k)^0$ it is sufficient to prove $z^{-1}z' \in Z(k)^1$; i.e., we want z and z' to have the same image in $Z(k)/Z(k)^1$. As a first step, we'll show $z' \in \mathcal{N}_x z\mathcal{N}_x$ for $\mathcal{N}_x := \mathcal{N} \cap G(k)_x = N(k) \cap G(k)^0_x$.

We're given that $z' = h_1 z h_2$ for some $h_1, h_2 \in G(k)^0_x$, so $z'h_2^{-1} = h_1 z$. The point x lies in the apartments $\mathcal{A}$ and $h_1\mathcal{A}$ (since $h_1.x = x$), and the point $z'.x$ also lies in both $\mathcal{A}$ and $h_1\mathcal{A}$: the former because $z' \in N(k)$ and the latter because

$$z'.x = z'.(h_2^{-1}.x) = h_1.(z.x) \in h_1\mathcal{A}.$$

Letting $\mathcal{F}$ and $\mathcal{F}'$ be the unique facets of the building respectively containing x and $z'.x$, the verification of building axiom (BLII) from Definition 24.8 for the (restricted) abstract building associated to a Tits system provides $g \in G(k)^0$ carrying $\mathcal{A}$ to $h_1\mathcal{A}$ and restricting to the identity on both $\mathcal{F}$ and $\mathcal{F}'$.

We have $g\mathcal{A} = h_1\mathcal{A}$ and

$$g \in G(k)^0_{\mathcal{F}} \cap G(k)^0_{\mathcal{F}'} \subset G(k)^0_x \cap G(k)^0_{z'.x} = G(k)^0_x \cap (z'G(k)^0_x z'^{-1}).$$

Hence, $g^{-1}h_1 \in G(k)^0_x \cap \mathrm{Stab}_{G(k)}(\mathcal{A}) = G(k)^0_x \cap N(k) = \mathcal{N}_x$ and $z'^{-1}gz' \in G(k)^0_x$, so

$$g^{-1}h_1 z = g^{-1}z'h_2^{-1} = z'(z'^{-1}g^{-1}z')h_2^{-1} \in \mathcal{N}_x\, z \cap z'G(k)^0_x.$$

We can control the factor from $G(k)^0_x$ on the right side: by writing $nz = z'\gamma$ for some $n \in \mathcal{N}_x$ and some $\gamma \in G(k)^0_x$, we have $\gamma = z'^{-1}nz \in N(k)$ and hence $\gamma \in G(k)^0_x \cap N(k) = \mathcal{N}_x$. This implies $g^{-1}h_1 z \in z'\mathcal{N}_x$, so

$$z' \in (g^{-1}h_1)z\mathcal{N}_x \subset \mathcal{N}_x z\mathcal{N}_x.$$

We have deduced the refined property $z' = \widetilde{n}_1 z\widetilde{n}_2$ for some $\widetilde{n}_i \in \mathcal{N}_x$, and we want z and z' to have the same image in $Z(k)/Z(k)^1$. Since $z' = \widetilde{n}_1 z\widetilde{n}_2 = (\widetilde{n}_1\widetilde{n}_2)(\widetilde{n}_2^{-1}z\widetilde{n}_2)$ with $\widetilde{n}_2^{-1}z\widetilde{n}_2 \in Z(k)$, we see

$$\widetilde{n}_1\widetilde{n}_2 \in Z(k) \cap \mathcal{N}_x.$$

This latter intersection is a bounded subgroup of $Z(k)$ since $\mathcal{N}_x \subset G(k)^0_x$ is bounded. But all bounded subgroups of $Z(k)$ are contained in $Z(k)^1$, so $\widetilde{n}_1\widetilde{n}_2 \in Z(k)^1 \cap \mathcal{N}_x$. Since it is harmless to multiply z' on either side by $Z(k)^1 \cap G(k)^0_x$, we can replace z' with $(\widetilde{n}_1\widetilde{n}_2)^{-1}z'$ to reduce to the situation

$$z' = \widetilde{n}_2^{-1}z\widetilde{n}_2$$

(i.e., the special case $\widetilde{n}_1 = \widetilde{n}_2^{-1}$).

If $w \in W(\Phi)$ is the image of $\widetilde{n}_2$ then $\mathrm{ord}(a_{\mathbf{Q}}(z')) = \mathrm{ord}((w.a)_{\mathbf{Q}}(z))$ for all $a \in \Phi^+$. Letting $\overline{s}, \overline{s}' \in \overline{S}(k)$ be the respective images $q(z), q(z')$ of z and z', these in turn have respective images in $\overline{S}(k)/\overline{S}(k)^0 = \mathrm{X}_*(\overline{S})$ denoted as v, v' that satisfy $v' = w^{-1}.v$ via the natural $W(\Phi)$-action on the space $\mathrm{X}_*(S)_{\mathbf{R}} = \mathrm{X}_*(\overline{S})_{\mathbf{R}}$ dual to $\mathrm{X}^*(S)_{\mathbf{R}}$ (this is the action of the dual Weyl group $W(\Phi)^* = W(\Phi^\vee)$). The property $z, z' \in \mathcal{Z}$ says $a(v), a(v') \geq 0$ for all $a \in \Phi^+$ (positive system of coroots of the dual root system), and we want to show $v = v'$.

Under the natural decomposition of $\mathrm{X}_*(S)_{\mathbf{R}}$ as $\mathbf{R}\Phi^\vee \oplus (\mathbf{R}\Phi)^\perp$ (encoding the isogeny decomposition of S into an almost direct product of S' and the maximal split *central* k-torus in G), $W(\Phi^\vee)$ acts trivially on the second factor and so the components of v and v' along that factor are the same due to the given equality $v' = w^{-1}.v$. Thus, we can project into the summands corresponding to the spans of the roots and coroots to bring ourselves to an assertion entirely about the dual root system $(V, \Phi^\vee)$ (with $V = \mathbf{R}\Phi^\vee$): if $v' = w^{-1}.v$ in V and both v and v' lie in the closed Weyl chamber $\overline{C}$ defined by $a \geq 0$ for all $a \in \Phi^+$ then $v' = v$. In other words, if v and v' lie in the same closed Weyl chamber for the dual root system and are in the same orbit under its Weyl group then we claim $v' = v$. This is immediate from the fact that a closed Weyl chamber for a root system over $\mathbf{R}$ is a fundamental domain for the action by the Weyl group on the ambient $\mathbf{R}$-vector space [Bour, Chap. 6, Sect. 1.5, Theorem 2(ii)]. □

Contents

References

[Ber] Berkovich, V.: Étale cohomology of non-archimedean analytic spaces. Publ. Math IHES **78**, 5–161 (1993)

[BH] Bridson, M.R., Haefliger, A.: Metric Spaces of Non-positive Curvature. Springer-Verlag, New York (1999)

[BLR] Bosch, S., Lütkebohmert, W., Raynaud, M.: Néron Models. Springer-Verlag, New York (1986)

[Bor] Borel, A.: Semi-simple Groups and Riemannian Symmetric Spaces. American Mathematical Society (1998)

[Bour] Bourbaki, N.: Lie Groups and Lie Algebras (Chaps. IV–VI). Springer-Verlag, New York (2002)

[BS] Borel, A., Springer, T.: Rationality properties of linear algebraic groups II. Tôhoku Math. J. **20**, 443–497 (1968)

[BT1] Bruhat, F., Tits, J.: Groupes réductifs sur un corps local. Publ. Math IHES 41 (1972)

[BT2] Bruhat, F., Tits, J.: Groupes r éductifs sur un corps local II. Schémas en groupes. Existence d'une donnée radicielle valuée. Publ. Math IHES 60 (1984)

[BT3] Bruhat, F., Tits, J.: Groupes algébriques sur un corps local. Chapitre III. Compléments et applications à cohomologie galoisiennne. J. Fac. Sci. Univ. Tokyo Sect. 1A Math. 34(3) (1987)

[C1] Conrad, B.: Finiteness theorems for algebraic groups over function fields. Compos. Math. **148**, 555–639 (2012)

[C2] Conrad, B.: Reductive Group Schemes. Panoramas et Synthèses **42–43**, 93–444 (2014)

[C3] Conrad, B.: Algebraic groups II. https://www.ams.org/open-math-notes/omn-view-listing?listingId=110663 (2024)

[CGP] Conrad, B., Gabber, O., Prasad, G.: Pseudo-Reductive Groups, 2nd edn. Cambridge University Press, Cambridge (2016)

[EGA] Grothendieck, A., Dieudonné, J.: Éléments de géométrie algébrique. Publ. Math IHES 4, 8, 11, 17, 20, 24, 28, 32 (1960–1967)

[Gir] Giraud, J.: Cohomologie Non Abélienne. Springer-Verlag, New York (1971)

[Hoch] Hochschild, G.: The Structure of Lie Groups. Holden-Day, San Francisco (1965)

[KP] Kaletha, T., Prasad, G.: Bruhat-Tits Theory, 1st edn. Cambridge University Press, Cambridge (2023)

[KPerr] Kaletha, T., Prasad, G.: Corrections and Additions in the Book: "Bruhat-Tits Theory: A New Approach. https://websites.umich.edu/~kaletha/v1errata.pdf
[Mac] MacDonald, I.: Affine root systems and Dedekind's η-function. Inventiones Mathematicae 15 (1972)
[Mat] Matsumura, H.: Commutative Ring Theory. Cambridge University Press (1986)
[Mil1] Milne, J.: Arithmetic Duality Theorems, 2nd edn. BookSurge, LLC (2006). https://www.jmilne.org/math/Books/ADTnot.pdf
[Mil2] Milne, J.: Algebraic Groups. Cambridge University Press (2017)
[Pra] Prasad, G.: Weakly-split spherical tits systems in quasi-reductive groups. Am. J. Math. 136(3)
[PY] Prasad, G., Yu, J.-K.: On quasi-reductive group schemes. J. Algebraic Geom. **15**, 507–549 (2006)
[Sch] Scharlau, W.: Quadratic and Hermitian Forms. Springer-Verlag, New York (1985)
[Ser1] Serre, J.-P.: Local Fields. Graduate Texts in Mathematics, vol. 67. Springer-Verlag, New York (1979)
[Ser2] Serre, J.-P.: Galois Cohomology. Springer Monographs in Mathematics. Springer-Verlag, New York (1997)
[SGA3] Demazure, M., Grothendieck, A.: Schemas en groupes I, II, III. Lecture Notes in Mathematics 151, 152, 153 (1970)
[Stacks] The Stacks Project Authors: Stacks Project. https://stacks.math.columbia.edu/
[Ste] Steinberg, R.: Regular elements of semisimple groups. Publ. Math IHES 25 (1965)
[Tits] Tits, J.: Reductive groups over local fields. In: Proceedings of Symposia in Pure Mathematics, vol. 33(I), pp. 29–69 (1979)

Bounded Generation in Linear Groups and Diophantine Approximation

Pietro Corvaja

Abstract A group is said to be boundedly generated if it is decomposable as a product of finitely many cyclic groups. In this survey, we explore the link between bounded generation in matrix groups, linear recurrence sequences, exponential Diophantine equations and Diophantine approximation. In particular, we present recent results obtained by the author in collaboration with J. Demeio, A. Rapinchuk, J. Ren and U. Zannier, which show that bounded generation cannot hold for a large class of groups of matrices.

Keywords Groups with bounded generation · Diophantine approximation · Heights

1 Introduction

An abstract group is said to be boundedly generated (BG) if it is decomposable as a product of cyclic groups, namely there exist finitely many elements $\gamma_1, \ldots, \gamma_r$ such that each element of the group can be written in the form $\gamma_1^{a_1} \cdots \gamma_r^{a_r}$, for integers $a_1, \ldots, a_r$. This is a stronger condition than the finite generation; the two properties clearly coincide in the abelian case.

We shall present the ideas and techniques, coming from Diophantine approximations and Diophantine geometry, behind the recent proofs by Demeio, Rapinchuk, Ren, Zannier and the author of these notes of the fact that (BG) cannot hold for large classes of discrete matrix groups. For instance, in [6, 8] we proved the following:

Theorem *Let* $\Gamma \subset \mathrm{GL}_d(\mathbb{C})$ *be a matrix group consisting only of semi-simple elements. Then* Γ *has (BG) if and only if it is finitely generated and virtually abelian.*

We shall provide a sketch of the proofs of this statement, and some of its generalizations, enlightening the arithmetic tools at the base of such proofs. The connection

---

P. Corvaja (✉)
Dipartimento di Scienze Matematiche, Informatiche e Fisiche, Udine, Italy
e-mail: pietro.corvaja@uniud.it

G. Prasad et al. (eds.), *Zariski Dense Subgroups, Number Theory and Geometric Applications*, Infosys Science Foundation Series,
https://doi.org/10.1007/978-981-95-3437-1_2

between matrix groups and exponential Dophantine equations arises via the theory of linear recurrence sequences; hence we devote a section (Sect. 3) to this theory.

Basically, the aim of this text is to introduce Diophantine approximation techniques to researchers in group theory, especially those interested in bounded generation and subgroup growth, but not vice-versa. The readers wanting to learn more on the latter subjects are addressed e.g. to the recent papers [10, 14, 17].

2 Bounded Generation

Let Γ be an abstract group[1]; we say it is *finitely generated* if it is generated by a finite set, i.e. there exists a finite set $F \subset \Gamma$ such that the minimal subgroup of Γ containing F is the whole of Γ. Finitely generated groups are quotients of the free groups $\mathcal{F}_n$, where, for $n \in \mathbb{N}$, $\mathcal{F}_n$ is defined to be the free group on n generators.

We say that a group Γ is *boundedly generated* (BG) if there exists a finite string of elements $(\gamma_1, \dots, \gamma_k) \in \Gamma^k$, where $k \in \mathbb{N}$ is a natural number, such that

$$\Gamma = \{\gamma_1^{n_1} \cdots \gamma_k^{n_k} \, : \, (n_1, \dots, n_k) \in \mathbb{Z}^k\}. \tag{1}$$

The following properties are easy to verify:

(i) Every quotient of a (BG) group satisfies (BG).
(ii) Extensions of (BG) groups by (BG) groups are (BG) groups.
(iii) If a group G admits a finite index (BG) subgroup, then G itself has (BG).

Clearly, every boundedly generated group is also finitely generated; for (virtually) abelian groups the two notions coincide. However, as we now show, there are examples of finitely generated groups which are not boundedly generated. Note that if the free groups $\mathcal{F}_n$ were boundedly generated, then the two notions of finite and bounded generation would coincide.

Hence, the first result to prove, before starting any investigation of the notion of bounded generation, is the following:

Theorem 2.1 *For $n \geq 2$ the free group $\mathcal{F}_n$ is not boundedly generated.*

Since $\mathcal{F}_2$ is a quotient of $\mathcal{F}_n$ for every $n \geq 2$, it suffices to prove that $\mathcal{F}_2$ has no (BG). The proof relies on the following:

Lemma 2.2 *For every positive integer m, let $P(m)$ be the maximum of the products $x_1 \cdots x_h$, where h is an arbitrary natural number and $x_1, \dots, x_h$ are positive real numbers such that $m = x_1 + \cdots + x_h$. Then*

$$P(m) < e^{m/e}. \tag{2}$$

---

[1] This simply means a group.

Proof By the arithmetic mean / geometric mean inequality, for each k the maximum is attained for $(x_1, \ldots, x_k) = (m/k, \ldots, m/k)$ and so equals $(m/k)^k$. Let us write this quantity as

$$\left(\frac{m}{k}\right)^k = \left(\left(\frac{m}{k}\right)^{k/m}\right)^m = \left(u^{1/u}\right)^m,$$

where $u = m/k$, so we are reduced to maximizing the quantity $u^{1/u}$ for $u > 0$ (forgetting that u must be a sub-multiple of m, in particular a rational number). The minimum over the positive real line of the function $u \mapsto u^{1/u}$ is attained at $u = e$, hence the estimate (2). □

Corollary 2.3 *For $m \geq 1$, the maximal order of an element in the group $\mathfrak{S}_m$ is $\leq e^{m/e}$.*

Proof Each element of $\mathfrak{S}_m$ can be factored as a product of disjoint cycles of lengths $x_1, \ldots, x_k$ where $x_1, \ldots, x_k$ are positive integers with $x_1 + \cdots + x_k \leq m$. The order of such an element is the least common multiple of the numbers $x_1, \ldots, x_k$, which is at most their product. Applying the lemma, the result follows. □

Proof of Theorem 2.1. As remarked, it suffices to treat the case $n = 2$. Since, for every $m \geq 1$, the symmetric group $\mathfrak{S}_m$ can be generated by two elements, it is a quotient of the free group $\mathcal{F}_2$. Suppose by contradiction that $\mathcal{F}_2$ is boundedly generated, so (1) holds for $\Gamma = \mathcal{F}_2$. Letting k be the number appearing in (1) we would obtain that the order of the finite group $\mathfrak{S}_m$, which equals $m!$, is bounded by the products of the orders of $g_1, \ldots, g_k$. By the above corollary, each element g_i has order $\leq e^{m/e}$, so the total number of elements of $\mathfrak{S}_m$ is bounded from above by $e^{mk/e}$. Then we obtain $m! \leq e^{mk/e}$, in particular $\log m! = O(m)$, which is clearly false. □

Another interesting and relevant example of a finitely generated non (BG) group is represented by the group $SL_2(\mathbb{Z})$, indeed, it contains the free group $\Gamma_2 \simeq \mathcal{F}_2$ as a finite index subgroup, where Γ_2 is the subgroup generated by matrices $\begin{pmatrix} 1 & 2 \\ 0 & 1 \end{pmatrix}, \begin{pmatrix} 1 & 0 \\ 2 & 1 \end{pmatrix}$.

Unlike $SL_2(\mathbb{Z})$, the groups $SL_n(\mathbb{Z})$, for $n \geq 3$, are known to be boundedly generated: D. Carter and G. Keller [3] proved in 1983 that these groups are products of finitely many unipotent subgroups generated by elementary matrices; for $SL_3(\mathbb{Z})$ it was recently proved [18] that 37 such matrices suffice.

For the ring of integers $\mathcal{O}$ of a number field containing infinitely many units, the group $SL_2(\mathcal{O})$ too has (BG).

`Example` There exist solvable finitely generated matrix groups which are not (BG). One example is provided by the so-called Lamplighter group, i.e. the group of matrices of the form

$$\begin{pmatrix} X^n & P(X) \\ 0 & 1 \end{pmatrix}, \qquad n \in \mathbb{Z},\ P(X) \in \mathbb{Z}[X, X^{-1}].$$

Although the whole group is finitely generated, its normal subgroup corresponding to $n = 0$ in the matrix above is not finitely generated. By what we stated at (ii) above, whenever a solvable group Γ is obtained by successive extensions of finitely generated abelian groups, Γ has (BG).

It is also to be noted that a result of Malcev asserts that every solvable finitely generated subgroup of $\mathrm{GL}_n(\mathbb{Z})$ is (BG).

In the sequel we shall be interested only in linear groups, i.e. subgroups of a group $G(\kappa)$ where κ is a field (especially a number field) and G is a linear algebraic group defined over κ.

In joint work with A. Rapinchuk, J. Ren and U. Zannier, we proved the following:

Theorem 2.4 (Main Theorem) *Let $\Gamma \subset \mathrm{GL}_d(\mathbb{C})$ be a linear group consisting only of semi-simple elements. The following are equivalent:*

(i) Γ has (BG).
(ii) Γ is virtually abelian.

In [6] we proved that for a linear group Γ which is not virtually solvable, for every (BG) representation of G as in (1) at least two of the elements g_i are not semi-simple. This result gives in particular a counterpart to Carter-Keller's theorem stating that $\mathrm{SL}_n(\mathbb{Z})$, for $n \geq 3$, is boundedly generated by elementary (in particular, non-semisimple) matrices.

We shall present a sketch of the proof of Theorem 2.4 based on the theory of exponential Diophantine equations, following [6].

In [7] we proved sparseness results for subsets of a matrix group which are boundedly generated by semi-simple elements. For instance, an easy corollary to Theorem 1.5 in [7] reads as follows:

Theorem 2.5 *Let $G \subset \mathrm{GL}_d$ be an algebraic group defined over $\mathbb{Q}$; let $\gamma_1, \ldots, \gamma_k$ be semi-simple matrices in $G(\mathbb{Z})$; set $\Sigma = \langle\gamma_1\rangle \cdots \langle\gamma_k\rangle \subset G(\mathbb{Z})$. Then the number of matrices in Σ whose entries have absolute value $\leq T$ grows asymptotically as $c(\log T)^r$ when $T \to \infty$, for some $c > 0$ and $0 \leq r \leq k$.*

Suppose now a subgroup $\Gamma \subset G(\mathbb{Z})$ contains a non-trivial unipotent element g. Then the growth of the powers g^n, $n = 1, 2, \ldots$, is polynomial in n. So "most" of them cannot be contained in the set Σ. In particular, if a group Γ contains a unipotent matrix then it cannot be boundedly generated by semi-simple matrices.

3 Linear Recurrence Sequences and Powers of Matrices

`Definition` Let $n \mapsto u_n \in \mathbb{C}$ be a sequence of complex numbers and $d \geq 1$ a natural number. We say that such a sequence is a linear recurrence sequence (l.r.s.) of order $\leq d$ if there exist complex numbers $a_1, \ldots, a_d$ with $a_d \neq 0$ such that identically

$$u_{n+d} = a_1 u_{n+d-1} + \cdots + a_d u_n. \tag{3}$$

Remark If such a relation holds for a string of coefficients $(a_1, \dots, a_d)$ with $a_d = 0$ we say that the sequence is eventually linearly recurrent. An example of a sequence which is eventually linearly recurrent but not a linear recurrence sequence is the sequence $1, 0, 0, \dots$, i.e. $u_0 = 1$ and $u_n = 0$ for $n > 0$. It satisfies for all $n \geq 0$ the order two recurrence relation $u_{n+2} = u_{n+1} = 1 \cdot u_{n+1} + 0 \cdot u_n$, while for $n > 0$ it also satisfies the relation of order one: $u_{n+1} = u_n$.

To a linear recurrence relation of the type (3) we associate the characteristic polynomial

$$P(X) = X^d - a_1 X^{d-1} - \dots - a_d \in \mathbb{C}[X]. \tag{4}$$

Given a sequence $\underline{u} : \mathbb{N} \to \mathbb{C}$, sending $n \mapsto u_n$, one can consider the set of linear recurrences satisfied by $\underline{u}$. Namely, the set of pairs $(d, (a_1, \dots, a_d))$, where $d \in \mathbb{N}$ and $(a_1, \dots, a_d) \in \mathbb{C}^d$ such that the relation (3) holds identically.

The set of polynomials of the form (4) associated to the linear recurrences satisfied by $\underline{u}$ generates an ideal $\mathcal{I}_{\underline{u}}$ in the ring $\mathbb{C}[X]$. If such an ideal $\mathcal{I}_{\underline{u}}$ contains a polynomial $P(X)$ with $P(0) \neq 0$ then the given sequence is a l.r.s.. If the ideal $\mathcal{I}_{\underline{u}}$ contains a non-zero polynomial then the sequence is eventually linearly recurrent.

Since the ring $\mathbb{C}[X]$ of polynomials is a principal ideal domain, the ideal $\mathcal{I}_{\underline{u}}$ can be generated by a single polynomial; this polynomial is the characteristic polynomial of the *minimal* recurrence relation satisfied by the sequence $\underline{u}$.

Examples Let us see the most relevant examples of linear recurrence sequences:

- The sequence $0, 1, 2, 3, \dots$ has the property that each term is the arithmetic mean between the previous term and the subsequent one. Hence it satisfies the linear relation $u_{n+1} = (u_n + u_{n+2})/2$, which can be written in the form of a binary linear relation as follows:

$$u_{n+2} = 2u_{n+1} - u_n.$$

  Every arithmetic progression satisfies the above linear recurrence relation.
- The geometric progressions $n \mapsto ab^n =: u_n$, where a, b are non-zero complex numbers, satisfy the linear recurrence relation:

$$u_{n+1} = b \cdot u_n.$$

- The sequence of Fibonacci is defined by setting $u_0 = 0$, $u_1 = 1$ and

$$u_{n+2} = u_{n+1} + u_n.$$

  It satisfies Binet's formula

$$u_n = \frac{\alpha^n - \beta^n}{\alpha - \beta},$$

  where α is the golden ratio and β its Galois conjugate over $\mathbb{Q}$; so α, β both satisfy the quadratic equation $x^2 - x - 1 = 0$.

- Given an invertible square matrix $T \in \mathrm{GL}_d(\mathbb{C})$, each entry of its power T^n is expressed by a linear recurrence sequence. This link between powers of matrices and linear recurrence sequences will turn out to be crucial (see Theorem 3.8 for a more complete statement).

We recall the following two fundamental characterizations of the set of linear recurrence sequences, which, in view of the following results, turns out to be an algebra of complex sequences.

Theorem 3.6 *Let $\mathbb{N} \ni n \mapsto u_n \in \mathbb{C}$ be a sequence of complex numbers. The following are equivalent:*

(i) The series $\sum_{n=0}^{\infty} u_n t^n$ represents a rational function vanishing at infinity;
(ii) the sequence is a linear recurrence sequence.

Also the following are equivalent:

(ia) The series $\sum_{n=0}^{\infty} u_n t^n$ represents a rational function;
(iia) the sequence is eventually linearly recurrent.

□

Theorem 3.7 *Let $\mathbb{N} \ni n \mapsto u_n \in \mathbb{C}$ be a sequence of complex numbers and $d \in \mathbb{N}$. The following are equivalent:*

(i) There exist an integer k with $1 \leq k \leq d$, polynomials $p_1(x), \dots, p_k(x)$ with

$$\sum_{i=1}^{k} (\deg(p_i) + 1) = d$$

and non-zero pairwise distinct complex numbers $\alpha_1, \dots, \alpha_k$ such that identically

$$u_n = \sum_{i=1}^{k} p_i(n) \alpha_i^n; \tag{5}$$

(ii) the sequence is linearly recurrent of order d.

□

The complex numbers $\alpha_1, \dots, \alpha_k$ appearing in the formula (5) are called *roots* of the linear recurrence. They are algebraic numbers whenever the sequence takes values in the field of algebraic numbers.

The reason why linear recurrence sequences are so relevant in the theory of linear groups is due to the following:

Theorem 3.8 *Let $d \in \mathbb{N}$ be a positive integer and let $T \in \mathrm{M}_d(\mathbb{C})$ be a $d \times d$ matrix with complex coefficients. Writing the entries of the powers T^n as $(a_{i,j}(n))$, for $(i,j) \in \{1,\dots,d\}^2$, each sequence $n \mapsto a_{i,j}(n)$ is eventually linearly recurrent. If the matrix T is invertible, then each such sequence is linearly recurrent of order $\leq d$. The characteristic polynomial of the recurrence is a divisor of the characteristic polynomial of the matrix.*

□

(We normalize the characteristic polynomial by setting, for a $d \times d$ matrix, $P_T(X) = (-1)^d(\det(T - X \cdot I_d))$.

Definition Expressions like (5) are also called *exponential polynomials*. Whenever the polynomial coefficients $p_i(x) \in \mathbb{C}[x]$ are constants, we shall speak of *power sum*.

Clearly, given a sequence $\underline{u} : n \mapsto u_n$ the following are equivalent:

(i) The sequence is a l.r.s. and the ideal $\mathcal{I}_{\underline{u}}$ contains a square-free polynomial;
(ii) the sequence $\underline{u}$ can be expressed by a power sum.

Note that if a matrix $T \in \mathrm{GL}_d(\mathbb{C})$ is semi-simple, then the entries $a_{i,j}(n)$ of T^n (for all $(i,j) \in \{1,\dots,d\}^2$) are expressed by power sums. On the contrary, if T is not semi-simple then at least one of the l.r. sequences $n \mapsto a_{i,j}(n)$ is not a power sum.

Examples

(1) Let $d = 2$, $T = \begin{pmatrix} 0 & 1 \\ 1 & 1 \end{pmatrix}$. Then $T^n = \begin{pmatrix} F_{n-1} & F_n \\ F_n & F_{n+1} \end{pmatrix}$, where $n \mapsto F_n$ is the Fibonacci sequence. Note that the characteristic polynomial of the matrix T is $\det(T - X \cdot I_2) = X^2 - X - 1$.
(2) Choose again $d = 2$ and let T be the unipotent matrix $T = \begin{pmatrix} 1 & 1 \\ 0 & 1 \end{pmatrix}$. Then $T^n = \begin{pmatrix} 1 & n \\ 0 & 1 \end{pmatrix}$ and each of the entries satisfies the linear recurrence $u_{n+2} = 2u_{n+1} - u_n$, which characterizes the arithmetic progression. Indeed, the characteristic polynomial of the matrix T is $\det(T - X \cdot I_2) = X^2 - 2X + 1 = (X - 1)^2$.
(3) The simplest case is represented by the 1×1 matrix $T = (a)$, where a is any non-zero complex number. Then $T^n = (a^n)$ and the recurrence reads $u_{n+1} = a \cdot u_n$ while the characteristic polynomial is $X - a$.

Exponential polynomials in several variables. Let us consider the analogue of exponential polynomials (5) in several variables. Restricting our attention to those with constant coefficients—what we named power sums—the general expression for a k-variable function is

$$f(\mathbf{x}) = \sum_{j=1}^{e} a_j \alpha_1^{l_{1,j}(\mathbf{x})} \cdots \alpha_r^{l_{r,j}(\mathbf{x})}, \tag{6}$$

where $\alpha_1, \ldots, \alpha_r$ are non-zero complex numbers, the constants $a_1, \ldots, a_e$ are complex numbers, the variable $\mathbf{x}$ is a vector in $\mathbb{Z}^k$ and the $l_{i,j} : \mathbb{Z}^k \to \mathbb{Z}$ are linear forms with integral coefficients.

We shall call such expressions *purely exponential polynomials* (PEP) in k variables. The relevance of these expressions in our context derives from the following fact: Given k semi-simple matrices $\gamma_1, \ldots, \gamma_k$, each entry of the product

$$\gamma_1^{x_1} \cdots \gamma_k^{x_k}$$

is a function of $\mathbf{x} = (x_1, \ldots, x_k)$ expressible in the form (6).

In the general case of arbitrary invertible matrices $\gamma_1, \ldots, \gamma_k$, we must allow also (ordinary) polynomials as coefficients in the expression (6) in place of the constants a_j.

4 Diophantine Approximation and *S*-Unit Equations

A fundamental tool in the proof of our results is represented by Diophantine approximation; in the first place, Diophantine approximation concerns the approximation of irrational numbers by rational numbers.

A celebrated theorem of Dirichlet states that

Theorem 4.1 *Let α be a real irrational number. Then the inequality*

$$\left| \alpha - \frac{p}{q} \right| < \frac{1}{q^2}$$

admits infinitely many solutions in rational numbers p/q, where p, q are coprime integers, $q > 0$.

The proof is elementary, being based on the so-called Dirichlet's box principle (or "pigeonhole principle").

It is not hard to prove that the exponent 2 in the denominator is almost always best possible, in the following sense: The set of real numbers α for which there exists a number ϵ such that the inequality $|\alpha - p/q| < q^{-2-\epsilon}$ has infinitely many solutions (in coprime integers p, q with $q > 0$) has measure zero. This result has been proved by Khintchin in 1927, and constitutes the base of the so-called metric theory of Diophantine approximation.

A fundamental theorem of Roth [23] asserts that every (irrational) algebraic number behaves as do almost all real numbers:

Theorem 4.2 (Roth's Theorem) *Let $\alpha \in \mathbb{R}$ be an irrational algebraic number and $\epsilon > 0$ a positive real number. Then the inequality*

$$\left|\alpha - \frac{p}{q}\right| < \frac{1}{q^{2+\epsilon}} \tag{7}$$

has only finitely many solutions in rational numbers p/q.

This is equivalent to saying that for every α and ϵ as above there exists a number $c = c(\alpha, \epsilon) > 0$ such that for all rational numbers p/q,

$$\left|\alpha - \frac{p}{q}\right| > \frac{c}{q^{2+\epsilon}}. \tag{8}$$

Homogenizing the left-hand side in (7) (or in (8)) we are led to consider the linear form

$$|q\alpha - p|.$$

Roth's Theorem states that

$$|q\alpha - p| \gg q^{-1-\epsilon}.$$

Alternatively, we can consider the product of linear forms

$$|q| \cdot |q\alpha - p|,$$

and rephrase Roth's Theorem by asserting that the inequality

$$|q| \cdot |q\alpha - p| < \max(|p|, |q|)^{-\epsilon} \tag{9}$$

has only finitely many solutions $(q, p) \in \mathbb{Z}^2, q \neq 0$. Note, that, on the opposite direction, the inequality $|q| \cdot |q\alpha - p| < 1 = q^{-0}$ does admit infinitely many solutions in view of Dirichlet's Theorem 4.1.

Our aim is to extend the present theory into three different directions: First of all we can replace the field of approximants $\mathbb{Q}$ by any other number field; also, we can replace the usual archimedean absolute value by any other absolute value, either archimedean or ultrametric; thirdly, we can consider simultaneous approximations to several different targets, necessarily with respect to different metrics.

Not surprisingly, if the field at our disposal has large degree over $\mathbb{Q}$, approximating by elements of that field will be easier. However, when working in an arbitrary number field, we have to find an analogue of the quantity $\max(|q|, |p|)$ appearing on the right-hand side of (9). This is the reason to introduce the notion of height.

Absolute values and heights.
Let κ be a number field. A place of κ is an equivalence class of absolute values, i.e. the set of absolute values inducing a same topology. For each place ν of κ we choose

a suitably normalized absolute value $|\cdot|_\nu$; our choice will guarantee that the product formula holds "without weights" and that the product of the archimedean absolute values of a rational number $a > 1$ coincides with a. There is just one choice with these properties.

We then define, for every element $\alpha \in \kappa$, its *height* $H(\alpha)$ to be

$$H(\alpha) = \prod_\nu \max(|\alpha|_\nu, 1),$$

where the product is taken over all the places of κ. Also, we set $h(\alpha) = \log H(\alpha)$ and call it the *logarithmic height*. It turns out that $H(\alpha)$, so also $h(\alpha)$, does not depend on the number field containing α.

However, we remark that the absolute values $|a|_\nu$, even when a is rational, will depend on κ: For instance, if κ is real quadratic, it has two archimedean absolute values $|\cdot|_{\infty,1}$, $|\cdot|_{\infty,2}$, coinciding on the subfield $\mathbb{Q}$ and satisfying $|2|_{\infty,1} = |2|_{\infty,2} = \sqrt{2}$.

More generally, for a vector $P = (a_1, \ldots, a_n) \in \kappa^n$ its affine height is defined to be

$$H_{\text{aff}}(P) = \prod_\nu \max(1, |a_1|_\nu, \ldots, |a_n|_\nu)$$

and consequently its affine logarithmic height $h_{\text{aff}}(P) = \log H_{\text{aff}}(P)$. The projective height $H(P)$, which only depends on the class of P in $\mathbb{P}_{n-1}(\kappa)$, is

$$H(P) = \prod_\nu \max(|a_1|_\nu, \ldots, |a_n|_\nu).$$

The fact that $H(P)$ is independent of the chosen homogeneous coordinates is a consequence of the product formula.

The mentioned generalized version of Roth's Theorem, first formulated by Lang in 1960 [15] (see also [9], Theorem 1.39), reads

Theorem 4.3 *Let κ be a number field, and let S be a finite set of places of κ. For each place $\nu \in S$, let α_ν be an element of κ. Let $\epsilon > 0$. Then the inequality*

$$\prod_{\nu \in S} |\alpha_\nu - \beta|_\nu < H(\beta)^{-2-\epsilon} \tag{10}$$

has only finitely many solutions $\beta \in \kappa$.

Note that in the above statement we do not distinguish between the field of approximants and the number field containing the target points.

The classical Roth's theorem where the approximants β must lie in $\mathbb{Q}$ and there is just one target α to be approximated in the usual real absolute value can be deduced from Theorem 4.3 via the following principle: If a sequence $n \mapsto \beta_n$ tends to an irrational real algebraic number α in the usual metric, then it also converges to each of its conjugates with respect to other archimedean metrics; this is due to the fact that all archimedean metrics are Galois conjugate. Letting κ be a Galois closure of $\mathbb{Q}(\alpha)/\mathbb{Q}$ and normalizing all the absolute values with respect to κ one can see that the rational approximations β to the irrational number α satisfying the inequality (7) provide solutions to the inequality (10).

The Subspace Theorem. The crucial extension of Roth's theorem to higher dimensions is the famous Schmidt's Subspace Theorem, proved by W. Schmidt in the archimedean case and generalized by H.-P. Schlickewei to include arbitrary valuations. We provide two equivalent formulations:

Theorem 4.4 (Subspace Theorem—Projective version) *Let κ be a number field and let S be a finite set of normalized absolute values of κ. Let $N \geq 2$ be an integer and, for each place $\nu \in S$, let $L_{\nu,1}, \ldots, L_{\nu,N}$ be a basis for the space of linear forms defined over κ in N variables. Finally, let $\epsilon > 0$ be a real number. Then the Zariski-closure of the set of solutions $\mathbf{x} = (x_1, \ldots, x_N) \in \kappa^N$ to the inequality*

$$\prod_{\nu\in S}\prod_{i=1}^{N} \frac{\left|L_{\nu,i}(\mathbf{x})\right|_\nu}{\|\mathbf{x}\|_\nu} < H(\mathbf{x})^{-N-\epsilon} \tag{11}$$

is a finite union of proper subspaces of $\mathbb{P}_{N-1}$.

□

To state the affine version, we need to introduce the last notation. Given a finite set S of places of a number field κ containing all the archimedean ones, the ring of S-integers of κ is the set

$$\mathcal{O}_S = \{x \in \kappa : |x|_\nu \leq 1, \quad \forall \nu \notin S\}.$$

Translating Roth's theorem into higher dimensions, using Roth's inequality in the form (9), we obtain the following affine version of the Subspace Theorem:

Theorem 4.5 (Subspace Theorem—affine version) *Let κ be a number field and let S be a finite set of normalized absolute values including the archimedean ones. Let $N \geq 2$ be an integer and, for each place $\nu \in S$, let $L_{\nu,1}, \ldots, L_{\nu,N}$ be a basis for the space of linear forms defined over κ in N variables. Finally, let $\epsilon > 0$ be a real number. Then the set of solutions $\mathbf{x} = (x_1, \ldots, x_N) \in \mathcal{O}_S^N$ to the inequality*

$$\prod_{\nu\in S}\prod_{i=1}^{N} \left|L_{\nu,i}(\mathbf{x})\right|_\nu < H_{\text{aff}}(\mathbf{x})^{-\epsilon} \tag{12}$$

lies in a finite union of hyperplanes of κ^N.

□

This second version could be stated by using the projective height instead of the affine one.

A simple application of the Subspace Theorem concerns the growth of linear recurrence sequences. Suppose $n \mapsto u_n$ is a linear recurrence sequence of algebraic numbers. Then it can be represented in the form (5). If some ratio α_i/α_j, for $1 \leq i < j \leq k$ between two roots α_i, α_j is a root of unity, let m be a common multiple of all the order of these roots of unity. Then the m sequences $n \mapsto u_{mn+r}$, for $r = 0, \ldots, m-1$, can be re-written using fewer terms in the form (5), where now no ratio of two distinct roots is a root of unity.

Now two cases can arise: (1) The resulting exponential polynomial is indeed a standard polynomial ($k = 1$ and $\alpha_1 = 1$); or (2) there is at least one root $\neq 1$.

In the first case, the growth of u_n is clearly of polynomial type, and $h(u_n) = O(\log n)$.

In the second case, there exists at least one place ν (of a number field κ containing all the roots α_i and the coefficients of the polynomials p_i in the expression (5)) and an index i such that $|\alpha_i|_\nu > 1$, so the sequence $n \mapsto p_i(n)\alpha_i^n$ grows exponentially with respect to the place ν. However, if there exists another root $\alpha_j \neq 1$, for some index $j \neq i$, then there might be some cancellation between $p_i(n)\alpha_i^n$ and $p_j(n)\alpha_j^n$. An application of the Subspace Theorem shows that this in fact does not happen, "up to an exponent ϵ":

Theorem 4.6 *Let $n \mapsto u_n = \sum_{i=1}^k p_i(n)\alpha_i^n$ be an exponential polynomial which is not a polynomial and such that no quotient of its roots is a root of unity. Then for every $\epsilon > 0$ there exists a number n_0 such that for $n > n_0$*

$$H(u_n) > \max(H(\alpha_1), \ldots, H(\alpha_k))^{n(1-\epsilon)}.$$

□

In particular, for every constant $c \in \kappa$, the Diophantine equation $u_n = c$, i.e.

$$\sum_{i=1}^{k} p_i(n)\alpha_i^n = c,$$

has only finitely many solutions n. This last fact, known as the Skolem-Mahler-Lech theorem, admits an easier proof, provided by Th. Skolem, K. Mahler and C. Lech using p-adic methods (for a sketch of the proof, see [9], Theorem 4.3).

The S-unit equation. One can generalize equations of the above form to several variables. In the case the polynomials factors $p_i(n)$ of the "main terms" α_i^n are constant, the equation in several variables takes the form

$$a_1\alpha_1^{n_1} + \cdots + a_k\alpha_k^{n_k} = c$$

to be solved in $(n_1, \ldots, n_k) \in \mathbb{N}^k$. To state the finiteness result we shall need in its most suitable form, we introduce the notion of S-unit. An S-unit is a unit of the ring of S-integers already defined. Hence, given a finite set of places S of a number field κ containing all the archimedean ones, an element $u \in \kappa$ is an S-unit if $|u|_\nu = 1$ for all places $\nu \notin S$.

Once $\alpha_1, \ldots, \alpha_k$ are fixed, one can always enlarge the set of places to a finite set S such that each element α_i becomes an S-unit. Then the above displayed equation is a particular case of a so-called S-unit equation, i.e. an equation of the form

$$a_1 u_1 + \ldots + a_k u_k = 1, \tag{13}$$

where $a_1, \ldots, a_k$ are fixed non-zero coefficients in κ and the unknown $u_1, \ldots, u_k$ vary in the group $\mathcal{O}_S^*$ of S-units.

The mentioned finiteness result reads:

Theorem 4.7 (S-unit equation Theorem) *Given an integer $k \geq 2$, a group of S-units $\mathcal{O}_S^*$ in a number field κ and non-zero elements $a_1, \ldots, a_k \in \kappa$, the Eq.* (13) *admits only finitely many solutions $(u_1, \ldots, u_k) \in (\mathcal{O}_S^*)^k$ such that no subsum on the left-hand side vanishes.*

□

Clearly, there can be infinite "trivial families" of solutions as in the following example: Assume $\mathcal{O}_S^*$ is infinite and consider the equation

$$u + v + w = 1. \tag{14}$$

Then letting $w = 1$ and $v = -u$, with varying $u \in \mathcal{O}_S^*$, one obtains one such infinite family, corresponding to the vanishing of $u + v$.

Geometrically, the solutions to the above equation correspond to the integral points on the linear torus $\mathbb{G}_m^3$, with coordinates u, v, w, lying on the subvariety $u + v + w = 1$. Indeed, the integral points on a torus $\mathbb{G}_m$, over any ring, correspond to the unit of that ring. This is due to the fact that the variety $\mathbb{G}_m$ can be viewed as the closed subset of the plane $\mathbb{A}^2$ given by the equation $xy = 1$; then its integral points correspond to the integral values of x such that $y = x^{-1}$ is also integral.

Let us draw the following useful corollary from Theorem 4.7:

Corollary 4.8 *Given an integer $k \geq 2$, a group of S-units $\mathcal{O}_S^*$ in a number field κ and non-zero elements $a_1, \ldots, a_k \in \kappa$, then for all but finitely many projective solutions $(u, u_1, \ldots, u_k) \in (\mathcal{O}_S^*)^{k+1}$ of the equation*

$$a_1 u_1 + \ldots + a_k u_k = u$$

there is an index $i \in \{1, \ldots, k\}$ such that $u = a_i u_i$.

While speaking of projective solutions, we identify two vectors of the form $(u, u_1, \ldots, u_k) \in (\mathcal{O}_S^*)^{k+1}$ whenever they differ multiplicatively by a unit.

Proof Dividing out by u we can reduce to the case $u = 1$, where now the adjective "projective" is omitted. From Theorem 4.7 we get that for all but finitely many solutions a subsum on the left vanishes. Then the complementary subsum takes the value 1. Applying again the theorem, we obtain an equation with only one term $a_i u_i$ on the left, and this term must be equal to 1. □

Coming back to Eq. (14), the trivial families of solutions, as the one given by $w = 1$, $v = -u$, correspond to algebraic curves which are images of $\mathbb{G}_m$ inside the surface defined in $\mathbb{G}_m^3$ by Eq. (14).

A generalization of the S-unit equation theorem, whose proof is based on the same Diophantine tools, i.e. the Subspace Theorem, is the following result, stated in terms of integral points on sub-varieties of tori:

Theorem 4.9 (Laurent's Theorem) *Let $V \subset \mathbb{G}_m^k$ be an algebraic subvariety of a torus defined over a number field κ. Let S be a finite set of places of κ containing the archimedean ones and let $\mathcal{O}_S$ be the corresponding ring of S-integers. Then the Zariski-closure of the set $V(\mathcal{O}_S)$ of integral points on V is a finite union of translates of algebraic subtori contained in V.*

□

We remark once again that a point on a torus is S-integral if and only if all of its coordinates are S-units. Hence the set $V(\mathcal{O}_S)$ consists of the intersection of V with the group $(\mathcal{O}_S^*)^k = \mathbb{G}_m^k(\mathcal{O}_S)$.

Also, this finitely generated group can be replaced by any finitely generated multiplicative subgroup of $\mathbb{G}_m^k(\mathbb{C}) = (\mathbb{C}^*)^k$. This generalization is obtained via standard specialization arguments to reduce to the algebraic case, and then to a power of an S-unit group.

The compact analogue of Theorem 4.9, formerly a conjecture by Lang, is a theorem of Faltings [11]; the torus $\mathbb{G}_m^k$ is replaced by an abelian variety. In view of the Mordell-Weil theorem, the result for abelian varieties contains Mordell's conjecture on the finiteness of rational points on curves of genus ≥ 2.

The proofs of the results presented in this section are completely outside the scope of this exposition. The reader is referred to a standard text of Diophantine geometry, e.g. the one by Bombieri and Gubler [2] or the older one by Hindry and Silverman [13], or the most recent one by the author and U. Zannier [9].

We will only give a proof of the S-unit equation theorem in two variables, due to Siegel and Mahler. Actually, we content to work out the following special case, whose proof shows the main ideas behind the proof of Theorem 4.9.

Theorem 4.10 *For every non-zero integer $c \in \mathbb{Z}$ the Diophantine equation*

$$2^m - 3^n = c \tag{15}$$

admits only finitely many solutions in natural numbers $m, n \in \mathbb{N}$.

Proof We can treat just the case $c > 0$, since $c < 0$ is symmetric and $c = 0$ is a trivial case. We shall apply Roth's theorem in the version given by the case $N = 2$ of the Subspace Theorem. We set $\kappa = \mathbb{Q}$, S consisting of the archimedean, the 2-adic and the 3-adic places of $\mathbb{Q}$. For the archimedean place $\nu = \infty$ we set $L_{\nu,1}(x_1, x_2) = x_1 - x_2$ and $L_{\nu,2}(x_1, x_2) = x_2$, while for the two p-adic places ν we set $L_{\nu,i}(x_1, x_2) = x_i$, $i = 1, 2$. To each solution $(m, n) \in \mathbb{N}^2$ to Eq. (15) we associate the point $\mathbf{x} = (x_1, x_2) := (2^m, 3^n) \in \mathbb{Q}^2$. Note that $\|\mathbf{x}\|_\infty = \max(2^m, 3^n) = 2^m$, while $\|\mathbf{x}\|_2 = \|\mathbf{x}\|_3 = 1$. The double product in (11) reads

$$\prod_{\nu\in S}\prod_{i=1}^{N}\frac{|L_{\nu,i}(\mathbf{x})|_\nu}{\|\mathbf{x}\|_\nu} = \frac{c}{2^m}\cdot\frac{3^n}{2^m}\cdot 2^{-m}\cdot 3^{-n} = \frac{c}{2^{3m}}.$$

Since the height of $\mathbf{x}$ equals 2^m, the double product is $\ll H(\mathbf{x})^{-3}$, so for $\epsilon = 1/2$ (say) the inequality (11) is satisfied for all large values of m, n. By the conclusion of the Subspace Theorem, the ratio $2^m/3^n$ has only finitely many possibilities, and clearly this gives the finiteness of the set of solutions to (15). □

Formally, the finiteness of the set of solutions to Eq. (15) can be stated in the following form:

$$\lim_{(m,n)\in\,\mathbb{N}^2} |2^m - 3^n| = +\infty,$$

where the limit is taken with respect to any ordering of the set $\mathbb{N}^2$.

An inspection of the proof of Theorem 4.10 easily shows that in fact the result admits a quantitative formulation: For every $\epsilon > 0$, for all but finitely many pairs $(m, n) \in \mathbb{N}^2$ such that $2^m > 3^n$,

$$2^m - 3^n > 2^{(1-\epsilon)m}, \tag{16}$$

Clearly, it is an optimal estimate "up to ϵ".

Still another formulation: Consider the set $\mathcal{B} = \{1, 3, 5, 7, 13, \ldots\}$ made of the positive integers representable in the form $2^m - 3^n$. Then for every ϵ and all large $N \in \mathbb{N}$,

$$\sharp(\mathcal{B}\cap\{1, \ldots, N\}) \le \frac{1+\epsilon}{\log 2}\cdot\log N. \tag{17}$$

Note that using the higher dimensional Subspace Theorem (precisely, Theorem 4.4 with $N = 4$), one can also deduce that the sequence of consecutive gaps in the set $\mathcal{B}$ tends to infinity exponentially.

Let us apply these Diophantine approximation results to purely exponential polynomials. Consider an expression of the form (6), where now all the involved terms are algebraic numbers. Namely, there exists a number field κ and a finite set of places S of κ, containing the archimedean ones, such that the constants $a_1, \ldots, a_e \in \kappa$ and $\alpha_1, \ldots, \alpha_r \in \mathcal{O}_S^*$.

One might be tempted to search for lower bounds for the height $h(f(\mathbf{x}))$ in terms of the norm of the vector $\mathbf{x}$, analogous to the lower bound (16) (note that the left-hand side in (16) in indeed a purely exponential polynomial in two variables). However, such general lower bounds cannot hold, since a given value $f(\mathbf{x})$ can be attained for infinitely many different vectors $\mathbf{x}$. By the S-unit equation theorem, this fact can hold only if some fixed subsum defining f vanishes infinitely often. In other terms, for every fixed number c, if the equation $f(\mathbf{x}) = c$ has infinitely many solutions $\mathbf{x} \in \mathbb{Z}^k$ there is a translate of a positive dimensional subgroup of $\mathbb{Z}^k$ on which identically $f(\mathbf{x}) = c$; also, there exist only finitely many maximal subgroups with this property.

What can be proved, and was actually proved in [8] (it is a particular case of Theorem 1.8 therein), is the following:

Theorem 4.11 *Let $f : \mathbb{Z}^k \to \kappa$ be a purely exponential polynomial whose coefficients and roots are defined over a number field κ. There exists an effectively computable constant $C = C(f) > 0$ such that for all but finitely many elements $b \in f(\mathbb{Z}^k) \subset \kappa$ there exists an integral vector $\mathbf{x} \in f^{-1}(b)$ such that*

$$h(f(\mathbf{x})) \geq C \cdot \|\mathbf{x}\|.$$

Here the norm $\|\mathbf{x}\|$ denotes the usual Euclidean norm of the vector $\mathbf{x} \in \mathbb{Z}^k \subset \mathbb{R}^k$.

Using the lower bound of Theorem 4.11 one can deduce the following result, which extends the bound (17) (note that the function $\mathbb{Z}^2 \to \mathbb{C}$ sending $(m, n) \mapsto 2^m - 3^n$ is indeed a purely exponential polynomial):

Corollary 4.12 *Let $f : \mathbb{Z}^k \to \kappa$ be a purely exponential polynomial in k variables whose coefficients and roots are algebraic numbers. There exist a number r with $0 \leq r \leq k$ and a computable constant c such that*

$$\sharp\{b \in f(\mathbb{Z}^k) : H_{\text{aff}}(b) \leq T\} \sim c(\log T)^r \qquad \text{for } T \to \infty. \tag{18}$$

Sparseness results of this kind will be crucial in our proofs.

5 Generic Elements in Linear Groups and Hilbert Irreducibility Theorem

Hilbert Irreducibility Theorem. The celebrated Hilbert Irreducibility Theorem, proved by Hilbert in 1892, states the following:

Theorem 5.13 *Let $P(X, Y) \in \mathbb{Z}[X, Y]$ be an irreducible polynomial, of degree ≥ 1 in Y. Then for infinitely many specializations $X \mapsto n \in \mathbb{Z}$ the specialized polynomial $P(n, Y)$ is irreducible in the ring $\mathbb{Q}[Y]$.*

The interesting case is when $\deg_Y P \geq 2$; in that case the Diophantine equation $P(n, y) = 0$ has no solution $y \in \mathbb{Q}$. We can rephrase this last fact in the following way: Consider the irreducible algebraic curve $\mathcal{C} \subset \mathbb{A}^2$ defined by the equation $P(x, y) = 0$ and the projection $\pi : \mathcal{C} \to \mathbb{A}^1$ provided by the first coordinate. Then, if the morphism π has no rational section, there exist infinitely many points $n \in \mathbb{Q}$ (actually even in $\mathbb{Z}$) such that the fiber $\pi^{-1}(n)$ contains no rational point.

With little effort, one can prove that for infinitely many specializations, the fiber remains irreducible. Actually, up to allowing the curve to be reducible, the weak form of the theorem asserting the existence of infinitely many rational points in $\mathbb{A}^1$ having no rational pre-image is equivalent to the apparently stronger assertion that infinitely many rational points in $\mathbb{A}^1$ have an irreducible fiber.

For a presentation of Hilbert Irreducibility Theorem, and especially a discussion on the above reduction, we refer to Chapter 9 of the book by Serre [25] or to section 3.8 of the book by Zannier and the author [9]. Moreover, if the cover $\pi : \mathcal{C} \to \mathbb{A}^1$ is Galois with Galois group Δ, one can find specializations conserving the Galois group, by this meaning that the splitting field of the fiber has Galois group isomorphic to Δ.

The Hilbert Irreducibility Theorem can be generalized to arbitrary number fields and dominant rational maps $\pi : X \to \mathbb{P}_n$. The general statement can be formulated as follows (see for instance Proposition 1 in Sect. 9.2 of [25]):

Theorem 5.14 *Let V be an algebraic variety defined over a number field κ and let $V \to \mathbb{P}_n$ be a dominant rational map admitting no rational section. Then the set of points $p \in \mathbb{P}_n(\kappa)$ such that the fiber $\pi^{-1}(p)$ admits no rational point is Zariski-dense in $\mathbb{P}_n$. If moreover V is irreducible, and the extension of function fields $\kappa(V)/\pi^*(\kappa(\mathbb{P}_n))$ is Galois with group Δ, the set of points $p \in \mathbb{P}_n(\kappa)$ such that the splitting field of the fiber $\pi^{-1}(p)$ has Galois group Δ is Zariski-dense in $\mathbb{P}_n$.*

An important problem in Diophantine geometry consists in understanding for which varieties other than projective spaces the conclusions of the above theorem hold. It is easy to extend Theorem 5.14 to rational varieties, while already its validity for unirational varieties is not known, and is linked to the inverse Galois problem.

Conserning linear algebraic groups, a result of Colliot-Thélène and Sansuc (Corollary 7.15 in [4]) states that:

Theorem 5.15 *Let G be a reductive algebraic group over a number field κ and let $\pi : V \to G$ be a cover defined over κ, without rational sections. Then the set of elements $g \in G(\kappa)$ such that $\pi^{-1}(g)$ has no rational point is Zariski-dense.*

Generic elements. Let $G \subset \mathrm{GL}_d$ be a a linear algebraic group defined over κ, where $\kappa \subset \mathbb{C}$ is a finitely generated field, say a number field to fix ideas. Then one can define the generic characteristic polynomial of G, to be the polynomial $P_G(X) \in \kappa[G][X]$

$$P_G(g, X) = (-1)^d \det(g - X I_d) = X^d - \mathrm{Tr}(g) X^{d-1} + \cdots \pm \det(g).$$

It is a polynomial in one indeterminate X with coefficients in the ring of regular functions on G. For every $g \in G(\kappa)$, it specializes to a polynomial in $\kappa[X]$. It has a generic Galois group over the field $\kappa(G)$ of rational functions on G, and, for every $g \in G(\kappa)$ a specialized Galois group over κ. Of course, both Galois groups depend on the representation $G \hookrightarrow \mathrm{GL}_d$.

Examples
Take $G = \mathrm{SL}_2$. Then $P_G(g, X) = X^2 - \mathrm{Tr}(g)X + 1$; its generic Galois group is the group with two elements $\mathfrak{S}_2$, since the discriminant function $(\mathrm{Tr}(g))^2 - 4$ is not a square in the field $\kappa(G)$.

Take $G = \mathrm{O}(3)$, the group of orthogonal matrices. Then every matrix $g \in \mathrm{O}(3, \kappa)$ has an eigenvalue equal to 1. The generic characteristic polynomial is then divisible by the polynomial $(X - 1)$ in the ring $\kappa[G][X]$. Its generic Galois group is the group with 2 elements $\mathfrak{S}_2$. The same holds for the orthogonal group associated to any non-degenerate quadratic form of rank three.

Let us have a closer look at this example. Suppose $q(x, y, z)$ is a non-degenerate quadratic form in three variables, and consider the special orthogonal group $G = \mathrm{SO}(q)$. The vanishing of q defines a conic over κ, possibly without κ-rational points. Every projective transformation associated to an element of $G(\kappa)$ leaves the conic invariant; we shall show that it admits a rational fixed point on the projective plane.

The transformation must fix one or two algebraic points on the conic, either rational or quadratic conjugates. If it fixes a rational point, then we are done; otherwise it fixes a pair of quadratic conjugate points, and so it leaves invariant the corresponding tangent lines. These tangent lines, which are defined over the same quadratic field and are conjugate one of the other, intersect on a rational point, which must be fixed by the transformation. Hence, every rational element has a rational fixed point, and correspondingly a rational eigenvalue. Actually, this eigenvalue is always equal to 1.

In general, if every rational element has a rational eigenvalue then there is a character $\xi : G \to \mathbb{G}_m$ providing the value of such an eigenvalue, i.e. such that $(X - \xi(g)) | P_G(g, X)$ in the ring $\kappa[G][X]$. An important consequence of the result of Prasad-Rapinchuk below implies the same conclusion (the existence of a character $\xi : G \to \mathbb{G}_m$ whose values are always eigenvalues) under the condition that each element of a Zariski-dense subgroup has a rational eigenvalue. (See [1, 5] for discussions on this topic).

One expects that in $G(\kappa)$ one can find elements whose characteristic polynomial has a Galois group coinciding with the Galois group of the generic characteristic polynomial.

Let us rephrase this problem. Consider the subvariety $V \subset \mathbb{A}^d \times G$:

$$V = \{(\lambda_1, \ldots, \lambda_d, g) \in \mathbb{A}^d \times G \mid P_G(g, X) = (X - \lambda_1) \cdots (X - \lambda_d)\}. \quad (19)$$

The affine variety V is endowed with a finite map $V \to G$, projecting $(\lambda_1, \ldots, \lambda_d, g) \mapsto g$.

Suppose V is irreducible. Then the projection $V \to G$ corresponds to a field extension $\kappa(V)/\kappa(G)$ which is Galois, whose Galois group coincides with the Galois group of the generic characteristic polynomial. If the pre-image of a rational element $g \in \mathbb{G}(\kappa)$ is irreducible, then the Galois group of the characteristic polynomial of g coincides with the Galois group of the generic characteristic polynomial. We shall say that g is a *generic* element.

Now, the variant of the Hilbert Irreducibility Theorem provided by Theorem 5.15 states that the set of rational elements $g \in G(\kappa)$ with this property is Zariski-dense. Let now Γ be a Zariski-dense subgroup of $G(\kappa)$. The mentioned theorem of Prasad and Rapinchuk states that we can find rational elements with the above property inside Γ:

Theorem 5.16 *Let κ be a finitely generated field of characteristic zero, let G be a semi-simple algebraic group over κ and let $\Gamma \subset G(\kappa)$ be a Zariski-dense subgroup. Then Γ contains generic elements.*

Let us provide an approach to the Prasad-Rapinchuk theorem based on Diophantine approximation and Diophantine equations. We shall treat the basic case where $G = \mathrm{SL}_2$ and κ is a number field. This analysis will also showa a link between Hilbert Irreducibility and Diophantine equations.

Let then $\Gamma \subset \mathrm{SL}_2(\kappa)$. If the conclusion were false, every element of $g \in \Gamma$ would have rational eigenvalues, so $(\mathrm{Tr}(g))^2 - 4$ is a square in κ, where as before $\mathrm{Tr}(g)$ denotes the trace of the matrix g.

Let us take two matrices of infinite order $g_1, g_2 \in \Gamma$ and consider all the traces of the products $g_1 g_2^n$. Suppose first that g_2 is semi-simple: After applying a conjugation we can write

$$\mathrm{Tr}(g_1 g_2^n) = \mathrm{Tr}\left(\begin{pmatrix} a & b \\ c & d \end{pmatrix} \cdot \begin{pmatrix} \alpha^n & 0 \\ 0 & \alpha^{-n} \end{pmatrix}\right) = a\alpha^n + d\alpha^{-n}.$$

We can also suppose that g_1 is NOT triangular (after performing that specific conjugation): Indeed, if all the matrices of Γ were triangular in a certain given base, then Γ would not be Zariski-dense. Also note that this condition amounts to the inequality

$$ad \neq 1.$$

Morevoer, since g_1 has infinite order $(a, d) \neq (0, 0)$. By symmetry, we can then assume without loss of generality that

$$a \neq 0.$$

Now, the condition that $(\mathrm{Tr}(g_1 g_2^n))^2 - 4$ is a square becomes

$$a^2\alpha^{2n} + d^2\alpha^{-2n} + 2ad - 4 = u^2 \tag{20}$$

for some $u \in \kappa$. Note that the left-hand side is a linear recurrence sequence in the variable n. After multiplying both sides in the above equation by α^{2n} and setting $y = u\alpha^n$, the equation reads

$$a^2\alpha^{4n} + (2ad - 4)\alpha^{2n} + d^2 - y^2 = 0.$$

Putting $x = \alpha^{2n}$ we obtain the equation of a smooth conic:

$$a^2x^2 + (2ad - 4)x + d^2 - y^2 = 0, \tag{21}$$

to be solved in $x, y \in \kappa$ with x in the multiplicative group generated by α^2. Note that y is automatically integral over the ring generated by the coefficients a, d and α, α^{-1}. Geometrically, this Diophantine problem corresponds to finding S-integral points on $\mathbb{P}_1 - \{4 \text{ points}\}$, where S is a set of valuations of κ, including all the archimedean ones, such that a, d are S-integers and $\alpha, 2a, ad - 1$ are S-units.

Let us explain this point. The projective completion of the conic is isomorphic to the projective line $\mathbb{P}_1$ and the points to remove are the two points at infinity plus the zero set of the rational function x. This last requirement is due to the fact that we look for solutions x with $x = \alpha^{2n}$ for some integer n, so x is an S-unit, which means it is not congruent to 0 modulo any prime ideal of $\mathcal{O}_S$.

Let us see how the S-unit equation theorem can play a role to assure the finiteness of such solutions.

Define the two rational functions on the conic

$$\xi := \frac{y + ax - d}{2ax}, \quad \eta = \frac{y - ax - d}{2ax}.$$

We claim that every solution $(x, y) \in \mathcal{O}_S^* \times \mathcal{O}_S$ of the equation of the conic (21) gives rise to S-unit values of the above rational functions. To prove this, note that

$$\begin{aligned}(y + ax - d)(y - ax + d) &= y^2 - (ax - d)^2 = y^2 - a^2x^2 - d^2 + 2adx \\ &= (2ad - 4)x + 2adx = (4ad - 4)x,\end{aligned}$$

where we used the equation of the conic to justify the third equality. Now, under our assumptions $(4ad - 4)$ is an S-unit; so the numerator of ξ, η take S-unit values and since their denominator $2ax$ is also an S-unit, we can conclude that ξ, η take S-unit values at all the points (x, y) on the conic with x an S-unit.

From the definition of ξ, η we get the identity

$$\xi - \eta = 1,$$

which constitutes a simple case of the S-unit equation. In this case Theorem 4.7 provides at once finiteness of the set of solutions, from which we deduce that the set of integers n such that $g_1g_2^n$ has rational eigenvalues is finite. In particular, there are infinitely many matrices in Γ whose eigenvalues are not rational.

Remark The left-hand side in (20) is a linear recurrence sequence which is a square in the ring of linear recurrence sequences precisely if $ad = 1$, which we have excluded. On the other hand, it is clear that if $ad = 1$, so g_1 is triangular, then each product $g_1 g_2^n$ would have rational eigenvalues.

Suppose now that g_2 is not semisimple. We can suppose it is the upper triangular matrix $g_2 = \begin{pmatrix} 1 & 1 \\ 0 & 1 \end{pmatrix}$. Again, we can suppose that g_1 is not upper triangular. Then, with the same notation for g_1,

$$\mathrm{Tr}(g_1 g_2^n) = a + cn + d,$$

where $c \neq 0$. We have to exclude that a degree one polynomial always takes square values in a given number field. This is a very particular case of Hilbert Irreducibility Theorem (formulated by Hilbert only for the rational number field): The polynomial in two variables $a + cx + d - y^2$ is irreducible in the ring $\kappa[x, y]$, and then it admits infinitely many specializations $x \mapsto n \in \mathbb{N}$ such that the specialized polynomial is still irreducible in $\kappa[y]$.

Let us prove this very special case.

Consider the equation $y^2 = (a + d) + cn$; take a ring of S-integers $\mathcal{O}_S \subset \kappa$ containing $a + d$ and c and select prime $\mathfrak{p} \subset \mathcal{O}_S$, with $c \notin \mathfrak{p}$, which splits completely, so that its residue field is $\mathbb{Z}/p\mathbb{Z}$ for some prime p, which we suppose to be $\neq 2$. Then we can choose a non-square w in $\mathbb{Z}/p\mathbb{Z}$ and the integer n such that $cn + (a + d) \equiv w$ modulo $\mathfrak{p}$. Then $cn + a + d$ cannot be a square in κ, and this happens for all other numbers n in the same class modulo p.

In the first situation, that of a semi-simple matrix, the Diophantine results we applied is the finiteness of the set of values of a certain linear recurrence sequence which are perfect squares; in the second case, such a finiteness does not hold but it remains true that infinitely many values are not squares, and actually that all the values in a suitable arithmetic progression are not squares.

A general result of Zannier [28] asserts precisely that if a linear recurrence sequence is not a perfect d-th power in the ring of linear recurrence sequences (for a certain integer $d > 1$), then its values cannot be all perfect d-th powers on any given number field.

Generalizing this result, Ferretti and Zannier [12] proved a general result for solutions of equations of the form

$$x^d + a_1(n)x^{d-1} + \cdots + a_d(n) = 0,$$

where $a_1, \ldots, a_d$ are linear recurrence sequences.

Basically, their result states that if for every $n \in \mathbb{N}$ the above equation has a solution in $x \in \kappa$, then the equation admits a functional solution $n \mapsto x(n)$ which is a linear recurrence sequence.

Using this Diophantine result, Liu [16], after preliminary work by the author [5], proved the following general result: Given an algebraic group G over a number field κ, a generically finite map $V \to G$ and a Zariski-dense subgroup $\Gamma \subset \mathbb{G}(\kappa)$, if every point of Γ lifts to a rational point in $V(\kappa)$, then there exists an isogeny $G' \to G$ which factors through V, namely a diagram $G' \dashrightarrow V \to G$, for a rational map $G' \dashrightarrow V$.

This result is used in the following way: If G is simply connected (to simplify matters) and $\pi : V \to G$ is a non-trivial ramified cover, then an isogeny $G' \to G$ factoring π does not exist; one can conclude that in every Zariski-dense subgroup $\Gamma \subset G(\kappa)$ there exist infinitely many points having κ-irreducible pre-image in V.

To obtain Theorem 5.16, we consider the cover $V \to G$, where G is embedded into GL_d, and V, defined again by (19), is endowed with its natural projection to G. From Liu's Theorem we deduce that Γ contains an element whose fiber is irreducible over κ; then the Galois group of its characteristic polynomial is necessarily the generic one.

The proof of Prasad and Rapinchuk, which will be outlined in a second, was however simpler, and made no use of Diophantine approximation. We wanted to digress in this direction to further enlighten possible applications of Diophantine approximation to linear group theory.

The method of Prasad-Rapinchuk (see [20, 21]) makes use of p-adic methods. The starting point is that given a p-adic valuation of a number field κ, a Zariski-dense subgroup $\Gamma \subset G(\kappa)$ of a linear algebraic group G, the topological closure of Γ in $G(\kappa_p)$, κ_p being the completion of κ at the p-adic valuation, contains an open subgroup. So, every matrix of $G(\kappa_p)$ sufficiently close to the identity in the p-adic metric can be approximated by elements of Γ.

Suppose first $G = \mathrm{SL}_2$, as in the previous case. We have to prove that infinitely many matrices $\gamma \in \Gamma$ have no rational eigenvalues. This corresponds, as we already remarked, to $(\mathrm{Tr}(\gamma))^2 - 4$ not being a square in κ. Since the regular function $(\mathrm{Tr}(g))^2 - 4$ is not identically a square, one can choose a prime $\mathfrak{p}$ of κ and a matrix $g \in \mathrm{SL}_2(\kappa)$ such that the eigenvalues of g do not even belong to the completion $\kappa_{\mathfrak{p}}$. Moreover, one can choose such a prime so that its residue field is $\mathbb{Z}/p\mathbb{Z}$, as in the last proof (in other words, κ embeds into $\mathbb{Q}_p$).

We can p-adically approximate this matrix g by an element $\gamma \in \Gamma$ of the subgroup. Since this matrix γ has no eigenvalues in $\mathbb{Q}_p$, *a fortiori* it has no eigenvalues in κ, proving our content.

Let us now show how to modify this argument in the case of the algebraic group SL_3. The generic Galois group of the characteristic polynomial of SL_3 is the full symmetric group $\mathfrak{S}_3$. We then want to find a matrix γ whose characteristic polynomial has Galois group $\mathfrak{S}_3$ over κ; this is a stronger condition than having no rational eigenvalues.

We need to work with two suitable primes $p \neq q$. We can find a matrix $g_p \in \mathrm{SL}_3(\mathbb{Q}_p)$ and a matrix $g_q \in \mathrm{SL}_3(\mathbb{Q}_q)$ with the following property: The matrix g_p has no eigenvalues in $\mathbb{Q}_p$ while the matrix g_q has exactly one rational eigenvalue in $\mathbb{Q}_q$, the other two being quadratic over $\mathbb{Q}_q$.

Approximating by a single matrix $\gamma \in \Gamma$ the two matrices g_p (in the p-adic sense) and g_q (in the q-adic sense), we obtain a characteristic polynomial $P(\gamma, X)$ whose splitting field over κ has the following property: There exists a prime $\mathfrak{p}$ of κ over which the polynomial remains irreducible and a prime $\mathfrak{q}$ over which the polynomial splits into the product of a linear and an irreducible quadratic factors. This implies that the Galois group of the splitting field of the characteristic polynomial $P(\gamma, X)$ over κ contains both a 3-cycle and a 2-cycle, hence it must be the full symmetric group $\mathfrak{S}_3$.

6 Proof of the Main Theorems

We shall start with the proof of Theorem 2.4 in the simplest case of the group $G = \mathrm{SL}_d$, following an approach based on Diophantine equations (namely the S-unit equation theorem). At the end of this section, we shall discuss an alternative proof based on sparseness results (Diophantine approximation).

We stress once again that although we state and prove our results for matrices defined over number fields, this restriction is not crucial. Indeed, from one hand the S-unit equation theorem that we shall use holds true over every finitely generated field; from the other hand, the validity of Theorem 2.4 for groups of matrices with algebraic entries easily implies its extension to any group of matrices: By its nature, one can reduce to finitely generated fields, and then the descent to number fields is obtained via standard specialization techniques (see, e.g. R. Rumely paper [24], especially Theorem 7 therein).

The result to prove is then the following:

Theorem 6.17 *Let κ be a number field, $\Gamma \subset \mathrm{SL}_d(\kappa)$, $d \geq 2$, a group formed only by semi-simple elements. Suppose Γ is Zariski-dense in SL_d. Then Γ has no (BG).*

The proof actually shows that one can drop the hypothesis that Γ is formed only by semi-simple elements, obtaining the conclusion that Γ cannot be boundedly generated by semisimple elements.

Proof Let $\gamma_1, \ldots, \gamma_k$ be (semi-simple) matrices in Γ. We shall prove the existence of a matrix $g \in \Gamma$ such that $\langle g \rangle \cap \langle \gamma_1 \rangle \cdots \langle \gamma_k \rangle$ is finite.

Let κ' be the number field generated over κ by all the eigenvalues of the $\gamma_1, \ldots, \gamma_k$. By the mentioned theorem of Prasad-Rapinchuk the group Γ contains a matrix g whose characteristic polynomial is irreducible over κ'. Let $\lambda_1, \ldots, \lambda_d$ be the eigenvalues of g. Then they satisfy the multiplicative dependence relation $\lambda_1 \cdots \lambda_d = 1$. The Galois group $\mathrm{Gal}(\kappa'(\lambda_1, \ldots, \lambda_d)/\kappa')$ acts on the set $\{\lambda_1, \ldots, \lambda_d\}$ as the full symmetric group $\mathfrak{S}_d$. From this fact it easily follows that $\lambda_1, \ldots, \lambda_d$ cannot satisfy a second multiplicative dependence relation modulo κ'^*. In particular:

`Claim`: *No power of the eigenvalue $\lambda_1 =: \lambda$ of g belongs to the multiplicative group generated by the eigenvalues of $\gamma_1, \ldots, \gamma_k$.*

Let us now conclude our proof, namely let us prove that only finitely many powers of g belong to the set $\langle \gamma_1 \rangle \cdots \langle \gamma_k \rangle$. Since g is diagonalizable, we can conjugate all the matrices over the splitting field κ'' of g over κ' and suppose that g is diagonal, so λ^n is one of the entries of g^n. From a relation of the form

$$g^n = \gamma_1^{a_{1,n}} \cdots \gamma_k^{a_{k,n}}$$

where $\mathbf{a} = (\mathbf{a}_n = a_{1,n}, \ldots, a_{k,n}) \in \mathbb{Z}^k$, we obtain that the entry λ_1^n of g^n can be expressed in terms of an exponential polynomial in k variables $(a_1, \ldots, a_k)$, calculated at the point $\mathbf{a} \in \mathbb{Z}^k$.

Writing the values of such an exponential polynomial as a linear combination of elements of the multiplicative group generated by the eigenvalues of $\gamma_1, \ldots, \gamma_k$, we obtain a relation of the form

$$\lambda^n = c_1 u_1 + \cdots + c_r u_r,$$

where $c_1, \ldots, c_r$ are fixed non-zero algebraic numbers in κ'' and $u_1, \ldots, u_r$ belong to the mentioned finitely generated multiplicative subgroup of κ'^*.

By the corollary to the S-unit equation theorem (Corollary 4.8), for all but finitely many solutions $(n, u_1, \ldots, u_r)$ there exists an index $i \in \{1, \ldots, d\}$ such that

$$\lambda^n = c_i u_i.$$

We claim that for each given c_i, the above relation cannot hold for more than one exponent n. Indeed, suppose by the contrary that

$$\lambda^n = c_i u, \qquad \lambda^{n'} = c_i u'$$

for certain u, u' belonging to the subgroup of κ' generated by the eigenvalues of $\gamma_1, \ldots, \gamma_k$. Then $\lambda^{n-n'} = u/u'$ would belong to such a subgroup, against our previous claim. This contradiction concludes the proof. □

We presented the detailed proof of Theorem 2.4 only in the case of the group SL_d. The general case follows the same lines. The details are given in the paper [6]. The key Diophantine statement (Theorem 4.1 of [6]) can be formulated (in a slightly weaker version) as follows:

Lemma 6.18 *Let κ be a number field and let $\gamma_1, \ldots, \gamma_k$ be semi-simple matrices in $\mathrm{GL}_d(\kappa)$. Let Λ denote the multiplicative group generated by the eigenvalues of $\gamma_1, \ldots, \gamma_k$. Then, for every matrix $\gamma \in \mathrm{GL}_d(\kappa)$ having an eigenvalue λ for which no power λ^n, $n \in \{1, 2, \ldots\}$, belongs to Λ, the set of exponents n such that γ^n can be written in the form $\gamma_1^{n_1} \cdots \gamma_k^{n_k}$ is finite.*

The lemma is proved in the following way. One can suppose that γ is upper-triangular, with its first entry on the diagonal equal to λ. Then if γ^n is of the form $\gamma_1^{n_1} \cdots \gamma_k^{n_k}$ in particular the algebraic number λ^n is expressed as the value of a purely exponential polynomial whose roots can be taken to be generators $u_1, \ldots, u_h$ of the finitely generated multiplicative group Λ. Applying Laurent's theorem (or the S-unit equation theorem) one concludes that if there are infinitely many n such that γ^n is expressible as a product $\gamma_1^{n_1} \cdots \gamma_k^{n_k}$ then a multiplicative dependence relation of the form $\lambda^n = c \cdot u_1^{a_1} \cdots u_h^{a_h}$, for a suitable fixed constant $c \in \bar{\mathbb{Q}}^*$ and infinitely many $(n, a_1, \ldots, a_h) \in \mathbb{Z}^{h+1}$, would hold. This, as in our proof above, would imply that some power of λ belongs to Λ, which we have excluded.

`Sparseness and the Proof of Theorem` 2.5. A slightly different approach to the proof of Theorem 2.4 consists in replacing Laurent's Theorem 4.9, or the S-unit equation Theorem 4.7, which concerns solutions of exponential Diophantine equations, by Theorem 4.11 or its Corollary 4.12.

This is the content of the works [7, 8], which consider more generally subsets of the affine space which are parametrized by (PEP) polynomial maps. See also the recent preprint by A. Puch and D. Smertnig [19]

Corollary 4.12 implies quite easily Theorem 2.5. To see how this is related to (BG) questions, recall the so-called Tits Alternative, which is the following statement proved by Tits in 1972 [27]:

Theorem (Tits Alternative). *Let κ be any field. A finitely generated subgroup of $\mathrm{GL}_d(\kappa)$ either is virtually solvable or contains a non-abelian free subgroup.*

Suppose now that $\Gamma \subset \mathrm{GL}_d(\kappa)$, κ being a number field, is not virtually solvable. Then it contains two elements α, β generating a free non-abelian group. Writing any element of this free group as a word in α, β as

$$g = \alpha^{a_1}\beta^{b_1}\cdots\alpha^{a_h}\beta^{b_h}, \tag{22}$$

where $a_1, \ldots, a_h, b_1, \ldots, b_h$ are integers, we have the easy estimate

$$h(g) \leq c \cdot (|a_1| + |b_1| + \cdots + |a_h| + |b_h|),$$

for a suitable constant c. In other words,

$$H_{\mathrm{aff}}(g) \leq \exp(c \cdot (|a_1| + |b_1| + \cdots + |a_h| + |b_h|)).$$

On the other hand, by the freeness of the group $\langle\alpha, \beta\rangle$, we claim that

`Claim`: *The number of elements which can be written as in* (22) *with* $|a_1| + |b_1| + \ldots + |a_h| + |b_h| \leq H$ *is* $\gg 2^H$.

We can argue as follows: Take $h = H/2$, if H is even, otherwise $h = (H-1)/2$. Each choice of exponents $a_i, b_i \in \{\pm 1\}$ in (22) gives rise to pairwise different elements with $|a_1| + |b_1| + \ldots + |a_h| + |b_h| \leq H$. The number of choices is 2^{2h}, which equals 2^H or $2^H/2$.

It follows in particular that the number of elements of affine height $\leq T$ in Γ grows (at least) polynomially in T. In view of Corollary 4.12, this proves that Γ cannot be parametrized by a PEP map, in particular it cannot be boundedly generated by semi-simple matrices.

With a slightly extra effort, one can prove that only virtually abelian finitely generated groups can be parametrized by PEP maps, and actually these are precisely the groups that admit (BG) by semi-simple elements.

Acknowledgements The author is grateful to the International Centre for Theoretical Sciences in Bangalore and to the organizers of the program "Zariski-dense subgroups, number theory and geometric applications" for inviting him to deliver a mini-course during the event.

The author is grateful to an anonymous referee for his/her careful reading of a previous version of this text and the many suggested improvements.

The visit of the author to the ICTS at Bangalore was partially supported by the PRIN 2017 *Geometric, Algebraic and Analytic methods in Arithmetic* financed by the Italian governement.

References

1. Bernik, J., Okninski, J.: On semigroups of matrices with eigenvalue 1 in small dimensions. Linear Algebra Appl. **405**, 67–73 (2005)
2. Bombieri, E., Gubler, W.: Heights in Diophantine Geometry. Cambridge University Press (2006)
3. Carter, D., Keller, G.: Bounded elementary generation in $\mathrm{SL}_n(\mathcal{O})$. Am. J. Math. **105**(3), 673–687 (1983)
4. Colliot-Thélène, J.-L., Sansuc, J.-J.: Principal homogeneous spaces under flasque tori: applications. J. Algebra **106**(1), 148–205 (1987)
5. Corvaja, P.: Rational fixed points for linear group actions. Annali Scuola Normale di Pisa **6** (2007)
6. Corvaja, P., Rapinchuk, A., Ren, J., Zannier, U.: Non virtually abelian anisotropic linear groups are not boundedly generated. Invent. Math. **227**, 1–26 (2022)
7. Corvaja, P., Demeio, J., Rapinchuk, A., Ren, J., Zannier, U.: Bounded generation by semi-simple elements: quantitative results. Comptes Rendus Math. Acad. Sci. Paris **360**, 1249–1255 (2022)
8. Corvaja, P., Demeio, J., Rapinchuk, A., Ren, J., Zannier, U.: Purely exponential parametrization and their group theoretic applications. arXiv preprint arXiv:2308.14013
9. Corvaja, P., Zannier, U.: Applications of Diophantine Approximation to Integral Points and Transcendence. Cambridge Tracts in Mathematics, vol. 212. Cambridge University Press (2018)
10. Erovenko, I.V., Rapinchuk, A.S.: Bounded generation of S-arithmetic subgroups of isotropic orthogonal groups over number fields. J. Number Theory **119**, 28–48 (2006)
11. Faltings, G.: Diophantine approximation on abelian varieties. Ann. Math. **133**, 549–576 (1991)
12. Ferretti, A., Zannier, U.: Equations in the Hadamard ring of rational functions. Ann. Scuola Norm. Sup. **6**, 457–475 (2007)
13. Hindry, M., Silverman, J.: Diophantine Geometry: An Introduction. Graduate Texts in Mathematics, vol. 201. Springer Verlag (2000)
14. Lubotzky, A., Segal, D.: Subgroup Growth. Progress in Math, vol. 212. Birkhäuser Verlag, Basel (2003)
15. Lang, S.: On integral points on curves. Publ. Math. I.H.E.S. **6**, 27–43 (1960)
16. Liu, F.: On Hilbert irreducibility theorem for linear algebraic groups. Ann. Scuola Norm. Sup. (2022)
17. Morgan, A.V., Rapinchuk, A.S., Sury, B.: Bounded generation of SL_2 over rings of S-integers with infinitely many units. Algebra Number Theory **12**, 1949–1974 (2018)
18. Nica, B.: On bounded elementary generation for SL_n over polynomial rings. Israel J. Math. **225**, 1949–1974 (2018)
19. Puch, A., Smertnig, D.: Factoring through monomial representations: arithmetic characterizations and ambiguity of weighted automata (2024). arXiv preprint arXiv:2410.03444
20. Prasad, G., Rapinchuk, A.: Existence of irreducible $\mathbb{R}$-regular elements in Zariski-dense subgroups. Math. Res. Lett. **10**, 21–32 (2003)

21. Prasad, G., Rapinchuk, A.: Generic elements in Zariski-dense subgroups and isospectral locally symmetric domains. Thin Subgroups Superstrong Approx. **61**, 211–252, Math. Sci. Res. Inst. Publ.**61**, Cambridge University Press (2014)
22. Prasad, G., Rapinchuk, A.: Generic elements in Zariski-dense subgroups form an open set. Trans. Mosc. Math. Soc. **78**, 299–314 (2017)
23. Roth, K.F.: Rational approximations to algebraic numbers. Mathematika **2**, 1–20 (1955)
24. Rumely, R.: Note on van der Poorten proof of Hadamard quotient theorem, I, II, Séminaire de Théorie des nombres de Paris, 1986–87, 349–382, 383–409, Progress in Math. 75, Birkhäuser, Boston (1988)
25. Serre, J.P.: Lectures on the Mordell-Weil Theorem, Aspects of Mathematics, Vieweg, 4rd edn (1997)
26. Vojta, P.: Integral points on subvarieties of semi-abelian varieties I. Inv. Math. **126**, 133–181 (1996)
27. Tits, J.: Free subgroups in linear groups. J. Algebra **20**, 250–270 (1972)
28. Zannier, U.: A proof of Pisot's d-th root conjecture. Ann. Math. **151**, 375–383 (2000)

Zariski-Dense Deformations of Standard Discontinuous Groups for Pseudo-Riemannian Homogeneous Spaces

Kazuki Kannaka and Toshiyuki Kobayashi

Abstract Let $X = G/H$ be a homogeneous space of a Lie group G. When the isotropy subgroup H is non-compact, a discrete subgroup Γ may fail to act properly discontinuously on X. In this article, we address the following question: in the setting where G and H are reductive Lie groups and $\Gamma\backslash X$ is a standard quotient, to what extent can one deform the discrete subgroup Γ while preserving the proper discontinuity of the action on X? We provide several classification results, including conditions under which local rigidity holds for compact standard quotients $\Gamma\backslash X$, when a standard quotient can be deformed into a non-standard quotient, a characterization of the largest Zariski-closure of discontinuous groups under small deformations, and conditions under which Zariski-dense deformations occur.

Keywords Discontinuous group · Proper action · Zariski dense subgroup · Rigidity · Locally symmetric space · Pseudo-Riemannian manifold · Clifford–Klein form · Spin group

2020 Mathematics Subject Classification Primary: 57S30, 58H15. Secondary: 22D50, 22E40, 22E46, 53C30, 58J50

K. Kannaka
Faculty of Mathematics and Physics, Institute of Science and Engineering, Kanazawa University, Kanazawa, Ishikawa, Japan
e-mail: kannaka@se.kanazawa-u.ac.jp

RIKEN Interdisciplinary Theoretical and Mathematical Sciences (iTHEMS), Wako, Saitama, Japan

T. Kobayashi (✉)
Graduate School of Mathematical Sciences, The University of Tokyo, Meguro, Tokyo, Japan
e-mail: toshi@ms.u-tokyo.ac.jp

French-Japanese Laboratory in Mathematics and its Interactions, FJ-LMI CNRS IRL2025, Tokyo, Japan

G. Prasad et al. (eds.), *Zariski Dense Subgroups, Number Theory and Geometric Applications*, Infosys Science Foundation Series,
https://doi.org/10.1007/978-981-95-3437-1_3

1 Introduction

Let X be a homogeneous space G/H, where G and H are real reductive algebraic groups. In this article, we focus on the case where the isotropy subgroup H is non-compact, and we consider the following two problems:

Problem 1.1 To what extent can we deform cocompact discontinuous groups for the homogeneous space X?

Problem 1.2 Does there exist a Zariski-dense subgroup Γ with cohomological dimension greater than 2 such that Γ acts properly discontinuously on X?

In this context, the term "discontinuous group for X" is used distinctively from "discrete subgroup of G". In fact, even when Γ is a discrete subgroup of G, the quotient space $X_\Gamma := \Gamma\backslash X$ is not always Hausdorff when H is non-compact. We say that a discrete subgroup Γ of a Lie group G is a *discontinuous group for* the homogeneous space X if the Γ-action on X is properly discontinuous and free. In this case, X_Γ carries a unique C^∞-manifold structure such that the quotient map $X \to X_\Gamma$ is a smooth covering of X_Γ, through which the quotient space X_Γ inherits any local G-invariant geometric structure on X. The resulting quotient manifold X_Γ is also referred to as a *Clifford–Klein form* of X, which is a typical example of (G, X)-manifolds in the sense of Ehresmann and Thurston. See, e.g., [44] for a detailed survey.

We now briefly explain our motivations for the aforementioned problems, the existing knowledge, and the contributions presented in this article. The necessary notations and basic concepts will be reviewed in Sect. 2.

The classical Selberg–Weil local rigidity theorem asserts that the compact quotient X_Γ of the irreducible *Riemannian* symmetric space $X = G/K$, where K is a compact subgroup, cannot be continuously deformed unless $\dim X = 2$ [70]. In contrast, a notable feature when H is non-compact, observed by Kobayashi in the early 1990s [39], is that cocompact discontinuous groups for *pseudo-Riemannian* symmetric spaces exhibit greater "flexibility". Specifically, there exist arbitrarily high-dimensional compact quotients X_Γ of irreducible symmetric spaces $X = G/H$ that do allow continuous deformation [43, Thms. A and B]. On the other hand, it has also been proven that there exist compact quotients X_Γ with local rigidity in pseudo-Riemannian symmetric spaces of the form $X = G/H$ [43, Prop. 1.8].

The difficulty regarding Problem 1.1 is that when H is not compact, small deformations of a discrete subgroup can easily destroy the proper discontinuity. Goldman [15] conjectured, in the context of the 3-dimensional compact anti-de Sitter space, that any small deformation of any standard cocompact discontinuous group preserves the proper discontinuity. This conjecture was proved by Kobayashi [43], by extending the properness criterion [37] to the general setting, as shown in [2, 41]. He also demonstrated the existence of compact "standard" quotients that allow "non-standard" deformation when X is locally isomorphic to the group manifold $SO(n, 1)$ and $SU(n, 1)$ for all $n \geq 2$, which are regarded as symmetric spaces of the form $(G \times G)/\operatorname{diag} G$. Deformation in the context of a discontinuous group involves the study of deformation spaces of the corresponding geometric structures modeled on (G, X). See also Ghys [14] for geometric interpretations in the case of $SL(2, \mathbb{C})$. These results have been extended to the symmetric space $G/H = SO(2, 2n)/U(1, n)$ by Kassel [27], where she constructed a small deformation into a Zariski-dense subgroup (Definition A.6 in Appendix A) while preserving the proper discontinuity of the action on G/H.

The deformation of the discrete subgroup described above has the following properties:

(1) (standard quotient) it starts with a discrete subgroup whose Zariski-closure acts properly on $X = G/H$, or more precisely, it corresponds to a "standard" discontinuous group in the sense of [29, Def. 1.4], see Definition 2.4;
(2) (deformation as a discontinuous group) the deformation of the discrete subgroup preserves its proper discontinuity;
(3) (non-standard deformation) after the deformation, the Zariski-closure of the new discrete subgroup no longer acts properly on X.

Keeping the assumption that a cocompact discontinuous group is standard, we examine Problem 1.1 in the general setting where G/H is a homogeneous space of *reductive type*, i.e., where $G \supset H$ are real reductive algebraic groups. We begin by providing a rigorous formulation in Question 2.5, which includes conditions for when local rigidity holds, when non-standard deformations are possible, and when Zariski-dense deformations occur.

We provide answers to these questions by dividing them into the following cases of compact standard quotients, denoted symbolically as $\Gamma\backslash G/H$ and $\Gamma_L\backslash G/\Gamma_H$, which are modeled on the homogeneous space G/H and the group manifold $(G \times G)/\operatorname{diag} G$, respectively. In the first case, $\Gamma\backslash G/H$, we provide answers to Question 2.5 in Table 3 for a classification, where G is a simple Lie group (Theorem 2.9). In the second case, $\Gamma_L\backslash G/\Gamma_H$, the answer is given in Theorem 2.21.

We observe from our classification that, in roughly half of the cases of $\Gamma\backslash G/H$, local rigidity holds, while in the case $\Gamma_L\backslash G/\Gamma_H$, more deformable cocompact discontinuous groups exist. We also observe that in the case $\Gamma_L\backslash G/\Gamma_H$ with G simple, there are no Zariski-dense deformations. In contrast, in the $\Gamma\backslash G/H$ case with G simple, we show that when deformation to a non-standard form is possible, the same homogeneous space G/H also admits a cocompact discontinuous group that can be deformed into a Zariski-dense subgroup. Such homogeneous spaces G/H include the indefinite-Kähler symmetric space $G/H = SO(2, 2n)/U(1, n)$ [27] and the following 7-dimensional space form, with pseudo-Riemannian metric of signature $(4, 3)$ and constant sectional curvature -1:

Theorem 1.3 (Answer to (Q3) in Table 3) *There exists a standard cocompact discontinuous group for $SO(4, 4)/SO(3, 4)$ that admits a small deformation which is Zariski-dense in $SO(4, 4)$.*

So far, we have outlined our results for the case of *cocompact* discontinuous groups in Problem 1.1. Now, in Problem 1.2, we investigate the possibility of Zariski-dense deformations. Keeping in mind that such deformations in G are more likely to occur for discrete groups with lower cohomological dimension, we allow discontinuous groups that are not necessarily cocompact for X in addressing Problem 1.2.

As one of numerical invariants for an abstract group Γ, we recall the projective dimension of the group ring $\mathbb{R}[\Gamma]$ is called the *cohomological dimension* of Γ, denoted as $\mathrm{cd}_{\mathbb{R}}(\Gamma)$. Free groups and surface groups (the fundamental groups of closed hyperbolic surfaces) have cohomological dimensions of 1 and 2, respectively.

The general upper bound on the cohomological dimension of a subgroup Γ acting properly discontinuously on a homogeneous space $X = G/H$ was established in [37, Cor. 5.5] as the inequality

$$\mathrm{cd}_{\mathbb{R}}(\Gamma) \leq d(X), \tag{1.1}$$

where we define the "non-compact dimension" of X by

$$d(X) := d(G) - d(H), \tag{1.2}$$

and define $d(G)$ for the group G to be the dimension of the Riemannian symmetric space G/K. The homogeneous space X is diffeomorphic to a vector bundle over a compact manifold with fiber $\mathbb{R}^{d(G)}$. The equality in (1.1) holds if and only if the Γ-action on X is cocompact [37, Cor. 5.5].

We have already discussed Problem 1.2 in the case where X_Γ is compact as a part of Problems 1.1. This is the case where the equality $\mathrm{cd}_{\mathbb{R}}(\Gamma) = d(X)$ holds.

In contrast, for cases where the cohomological dimension is lower, some solutions to Problem 1.2 are already known (see also Sect. 5.5 for some open questions):

- ($\mathrm{cd}_{\mathbb{R}}(\Gamma) = 1$) when Γ is isomorphic to a free group, see Benoist [2];
- ($\mathrm{cd}_{\mathbb{R}}(\Gamma) = 2$) when Γ is isomorphic to a surface group and G/H is a symmetric space, see a recent paper [24].

This article highlights Problem 1.2 in the setting where

$$2 < \mathrm{cd}_{\mathbb{R}}(\Gamma) < d(X)$$

by considering deformations of a discrete subgroup that is isomorphic to a cocompact discrete subgroup Γ of $Spin(n, 1)$, in particular, where $\mathrm{cd}_{\mathbb{R}}(\Gamma) = n$. Our answers to Problem 1.2 include the following examples, see Theorem 5.28: there exist Zariski-dense discontinuous groups with cohomological dimension 6 in the cases where $d(X) = 7$, 8, or 16 such as

$$X = \begin{cases} SO(8,\mathbb{C})/SO(7,\mathbb{C}) & d(X) = 7 \quad \text{(complex sphere)}, \\ SO(8,8)/SO(7,8) & d(X) = 8 \quad \text{(space form)}, \\ SU(8,8)/U(7,8) & d(X) = 16 \quad \text{(indefinite Kähler)}, \\ SL(16,\mathbb{R})/SL(15,\mathbb{R}) & d(X) = 16. \end{cases}$$

In proving the aforementioned results, particularly Theorems 2.9, 2.21, and 5.28, we focus on a special case of the following problem regarding deformations of cocompact discrete subgroups of $Spin(n, 1)$.

Problem 1.4 Let G be a real algebraic group, Γ a torsion-free cocompact discrete subgroup of $Spin(n, 1)$, and $\varphi\colon \Gamma \to G$ a group homomorphism. Find a small deformation $\varphi' \in \mathrm{Hom}(\Gamma, G)$ of φ that maximizes the Zariski-closure of $\varphi'(\Gamma)$.

The reader may wonder why Problem 1.4 and Theorem 1.6 below are formulated in terms of $Spin(n, 1)$ rather than $SO(n, 1)$, despite the technical difficulties involved. The main reasons for this choice are summarized in the following remark.

Remark 1.5 (1) (Proper actions). Some homogeneous spaces X, such as the 15-dimensional space form $SO(8, 8)/SO(7, 8)$ mentioned above, admit a proper action by reductive subgroups only if they are globally isomorphic to $Spin(n, 1)$. In particular, $SO(n, 1)$ cannot act properly on such space X, as stated in Theorem 5.25.

(2) (Direct implications). The deformation theory of discrete subgroups of $Spin(n, 1)$ will imply similar results for quotient groups such as $SO(n, 1)$, but not vice versa, as discussed below.

(3) (Counterexample to the lifting for $n \geq 4$). We consider whether the following claim (P_n) holds:
(P_n). Any torsion-free cocompact discrete subgroup of $SO_0(n, 1)$ can be lifted to $Spin(n, 1)$.

- (P_2) and (P_3) holds (Culler–Shalen [11] and Thurston [67]).
- (P_n) fails when $n \geq 4$ (Martelli–Riolo–Slavich [53]).

Since a torsion-free cocompact discrete subgroup of $Spin(2, 1)$ is a surface group, Problem 1.4 for the case $n = 2$ already presents non-trivial challenges, as it asks for the largest Zariski-closure of a surface subgroup in G among its small deformations. This problem has been studied in previous work, such as Burger–Iozzi–Wienhard [8] and Kim–Pansu [34], for example.

For general $n \geq 2$, and particularly when $n \geq 3$, we explore Problem 1.4 in the special case where the deformation starts from a standard quotient X_Γ. To be more explicit, we consider the setting where $\varphi\colon \Gamma \to G$ can be extended to a homomorphism $Spin(n, 1) \to G$, that is, a homomorphism of real algebraic groups. (There are several equivalent definitions for *homomorphisms* between real algebraic groups, as discussed in Lemma A.7 in Appendix A.)

In this case, we introduce a real algebraic subgroup of G, denoted by G^φ (Definition 3.7), and obtain the following theorem.

Theorem 1.6 (see Theorem 3.9) *Let $n \geq 2$, G a Zariski-connected real algebraic group, and $\varphi\colon Spin(n, 1) \to G$ a homomorphism.*

(1) There exists a cocompact arithmetic subgroup Γ of $Spin(n, 1)$ such that $\varphi(\Gamma)$ can be deformed so that its Zariski-closure is G^φ.

(2) *Conversely, the group G^φ gives the upper bound for the Zariski-closure of $\varphi'(\Gamma)$, up to G-conjugacy, for any small deformation φ' of $\varphi|_\Gamma$, provided that $n \geq 3$.*

We provide a description of the largest Zariski-closure G^φ purely in terms of finite-dimensional representations of orthogonal Lie algebras. We then carry out explicit computations of G^φ in certain cases, including those that arise naturally from the spin representations associated with indefinite quadratic forms. See Theorem 3.18 for a sufficient condition for G^φ to coincide with G, for instance.

The geometric idea behind our proof of Theorem 1.6(1) in Sect. 4 is largely parallel to that of Johnson–Millson [21]. This geometric approach is also used in Kassel [27] and Beyrer–Kassel [4].

However, these results highlight the case where φ is a natural embedding from $SO(n, 1)$ into a classical Lie group G such as $SO(p, q)$ or $PSL(n + 1, \mathbb{R})$ (see also Remark 4.8). In contrast, Theorem 1.6 addresses the situation where $\varphi\colon Spin(n, 1) \to G$ is a homomorphism to an arbitrary real algebraic group G, thus working in full generality.

This requires considerable preparation for the proof, not only in extending the bending construction [21] for $SO(n, 1)$ to $Spin(n, 1)$, but also in providing a detailed framework, within the context of Clifford algebras, for iterating the appropriate bending constructions to reach the largest Zariski-closure.

In Sects. 2–5, we have discussed the geometric questions of the quotient space X_Γ, including the Zariski-dense deformation addressed in Theorems 2.9 and 2.21. Now, turning our attention to the analytic aspects, in Sect. 6, we address analytic problems related to $L^2(X_\Gamma)$ for the quotient space X_Γ. This is a new area of research [29–32, 45], with a list of basic open problems and related topics discussed in a recent article [47].

We observe that the quotient space X_Γ inherits any G-invariant differential operators from X, such as the pseudo-Riemannian Laplacian, which are referred to as *intrinsic differential operators* in [29]. In contrast to the classical Riemannian setting, it is noteworthy that, when H is non-compact, there may be discrete spectra for such intrinsic differential operators, some of which are stable under deformations of X_Γ, while others are not.

In Sect. 6, we highlight the analytic aspects of this study from the perspective of deformations of discontinuous groups for $X = G/H$ with H non-compact. We briefly review the current state of the art regarding the stable discrete spectrum and its multiplicities, and pose several problems concerning the distribution of the stable discrete spectrum and its multiplicities.

At the end of this article, for the convenience of a broader audience, several technical details and concepts that may not be immediately obvious to the reader are collected in Appendices A–F, on the following topics:

- Appendix A: Basic concepts regarding real algebraic groups, particularly topologies;
- Appendix B: Basic notations for Clifford algebras and spin groups over a field of characteristic $\neq 2$;

- Appendix C: A sufficient condition for arithmetic groups to be torsion-free;
- Appendix D: Describing the fundamental group of a manifold endowed with a hypersurface in terms of an HNN extension;
- Appendix E: An upper bound for small deformations of discrete subgroups in terms of the first cohomology;
- Appendix F: An optimal bound on the number of generators for a continuous family of Zariski-dense subgroups.

Notation and Conventions

- $\mathbb{N} = \{0, 1, 2, \dots\}$, and $\mathbb{N}_+ = \{1, 2, \dots\}$.
- For a unital associative algebra A, we denote by $A^\times$ the group of invertible elements in A.
- We denote by id_S the identity map on a set S.
- We denote by $\#I$ the number of elements in a finite set I.
- We assume that algebraic groups are linear.
- For an algebraic group $\mathbf{G}$ over a field $\mathbb{F}$, and for an extension field $\mathbb{E}$ of $\mathbb{F}$, $\mathbf{G}(\mathbb{E})$ denotes the group of $\mathbb{E}$-points of $\mathbf{G}$. When $\mathbb{E} = \mathbb{R}$ or $\mathbb{C}$, we write $G = \mathbf{G}(\mathbb{R})$ and $G_\mathbb{C} = \mathbf{G}(\mathbb{C})$.
- The Lie algebras of Lie groups $G, H, L, \dots$ are denoted by the corresponding lower-case German letters $\mathfrak{g}, \mathfrak{h}, \mathfrak{l}, \dots$, respectively.
- For a Lie group homomorphism $\varphi\colon L \to G$, we denote by $d\varphi\colon \mathfrak{l} \to \mathfrak{g}$ the differential homomorphism of φ.

2 Deformation of Locally Homogeneous Spaces

In this section, we discuss Problem 1.1, which addresses the extent to which we can deform cocompact discontinuous groups for homogeneous spaces. To this end, we first review basic terminologies related to the deformation of discontinuous groups for homogeneous spaces in Sect. 2.1. We then proceed to provide a rigorous formulation in Question 2.5 in Sect. 2.2, which includes conditions for when local rigidity holds, when non-standard deformations are possible, and when Zariski-dense deformations occur.

We will provide answers to this question by dividing it into the following cases of compact standard quotients, denoted symbolically as $\Gamma\backslash G/H$ and $\Gamma_L\backslash G/\Gamma_H$, where G is a simple Lie group.

In the first case, $\Gamma\backslash G/H$, which is modeled on the homogeneous space G/H, we provide answers to Question 2.5 in Table 3 for a classification (Theorem 2.9). In the second case, $\Gamma_L\backslash G/\Gamma_H$, which is modeled on the group manifold $(G \times G)/\operatorname{diag} G$, the answer is given in Theorem 2.21.

We observe from our classification that, in roughly half of the cases of $\Gamma\backslash G/H$, local rigidity holds, while in the case $\Gamma_L\backslash G/\Gamma_H$, more deformable cocompact discontinuous groups exist. We also observe that in the case $\Gamma_L\backslash G/\Gamma_H$, there are no Zariski-dense deformations, whereas in the $\Gamma\backslash G/H$ case, some spaces of G/H admit Zariski-dense deformations of cocompact discontinuous groups.

The proof of the main results of this section, Theorems 2.9 and 2.21, will be given in Sects. 5.2 and 5.4, respectively.

2.1 Deformation Space $\mathcal{R}(\Gamma, G; X)$

As mentioned in Introduction, not every discrete subgroup Γ of a Lie group G acts properly discontinuously on the homogeneous space $X = G/H$ when H is non-compact. Moreover, a small deformation of a discrete subgroup Γ in G may destroy the proper discontinuity of the action on X. In light of these considerations, we provide a precise definition and notation here to formalize Problems 1.1 and 1.2 in the next section (see Question 2.5).

First, we set aside the space X, and consider the classical setting in which the Γ-action on $X = G/H$ is not relevant.

Let Γ be a finitely-generated discrete subgroup of a Lie group G. By deformation, we mean fixing an abstract group Γ and varying the homomorphisms from Γ into the group G. We set:

$$\mathcal{R}(\Gamma, G) := \{\varphi \in \mathrm{Hom}(\Gamma, G) \mid \varphi \text{ is faithful and discrete}\}.$$

We topologize $\mathrm{Hom}(\Gamma, G)$ by pointwise convergence and equip the subspace $\mathcal{R}(\Gamma, G)$ by the relative topology.

Remark 2.1 (*Goldman–Millson* [16, Thm. 1.1]) When G is a linear group, $\mathcal{R}(\Gamma, G)$ is a closed subset of $\mathrm{Hom}(\Gamma, G)$ for any torsion-free group Γ that can be realized as a Zariski-dense subgroup of a real semisimple algebraic group.

The group G acts on itself through inner automorphisms, that is, the subset $\mathcal{R}(\Gamma, G)$ invariant. We recall from Weil [71] the following definition:

Definition 2.2 A discrete and faithful representation $\varphi \in \mathcal{R}(\Gamma, G)$ is *locally rigid* if the G-orbit of φ is open in $\mathcal{R}(\Gamma, G)$.

Next, we return to the main theme of this article: the deformation of *discontinuous groups* for homogeneous spaces $X = G/H$. To formulate this properly, it is important to note that when H is non-compact, a discrete subgroup of G does not necessarily act properly discontinuously on G/H.

We recall from [39] (see also [43]) the definition of the following subset of $\mathcal{R}(\Gamma, G)$:

$$\mathcal{R}(\Gamma, G; X) := \{\varphi \in \mathcal{R}(\Gamma, G) \mid \varphi(\Gamma) \text{ is a discontinuous group for } G/H\},$$

which plays a basic role in the deformation theory of discontinuous groups for X.

There are two natural actions: those of the automorphism group $\mathrm{Aut}(\Gamma)$ and the inner automorphism group of G on $\mathrm{Hom}(\Gamma, G)$. These actions commute with each other and both leave the subset $\mathcal{R}(\Gamma, G; X)$ invariant.

We recall further from [44, Sect. 5.3] the definitions of $\mathcal{M}(\Gamma, G; X)$ and $\mathcal{T}(\Gamma, G; X)$ as follows:

$$\mathcal{M}(\Gamma, G; X) := \mathrm{Aut}(\Gamma)\backslash\mathcal{R}(\Gamma, G; X)/G.$$

In the case where G acts faithfully on X, we can think of $\mathcal{M}(\Gamma, G; X)$ as the *moduli space* of quotients of X by discontinuous groups isomorphic to Γ. Geometrically, this means that, for $\varphi_1, \varphi_2 \in \mathcal{R}(\Gamma, G; X)$, the locally homogeneous spaces $\varphi_1(\Gamma)\backslash X$ and $\varphi_2(\Gamma)\backslash X$ are isomorphic to each other if and only if $[\varphi_1] = [\varphi_2]$ in $\mathcal{M}(\Gamma, G; X)$. The "higher-dimensional Teichmüller space", which represents the local properties of the "moduli space", is defined as follows.

$$\mathcal{T}(\Gamma, G; X) := \mathcal{R}(\Gamma, G; X)/G.$$

Definition 2.3 (local rigidity as discontinuous groups, [43, Sect. 1]). We say that $\varphi \in \mathcal{R}(\Gamma, G; X)$ is *locally rigid as a discontinuous group for* X, or that $\varphi(\Gamma)\backslash X$ is *locally rigid*, if the G-orbit of φ is open in $\mathcal{R}(\Gamma, G; X)$; equivalently, $[\varphi]$ is an isolated point in $\mathcal{T}(\Gamma, G; X)$.

If this is not the case, we say that $\varphi \in \mathcal{R}(\Gamma, G; X)$ is *deformable as a discontinuous group for* X, or that $\varphi(\Gamma)\backslash X$ is *deformable*.

In the group case where $X = G/\{e\}$, we have $\mathcal{R}(\Gamma, G; X) = \mathcal{R}(\Gamma, G)$, and $\varphi \in \mathcal{R}(\Gamma, G)$ is locally rigid if and only if it is locally rigid as a discontinuous group for $X = G/\{e\}$. We summarize in Sect. 3.1 some known results regarding local rigidity of discrete subgroups of G, without considering their actions on $X = G/H$. In Sect. 5.1, we discuss the difference between the concepts of local rigidity as *discrete groups* and local rigidity as *discontinuous groups*.

2.2 *Question About Deformations of Standard Cocompact Discontinuous Groups*

In this section, we provide a rigorous formulation of Problem 1.1 concerning deformations of standard cocompact *discontinuous groups* for homogeneous spaces G/H of reductive type. This is one of the central questions addressed in this article, and it includes conditions for when local rigidity holds (Q1), when non-standard deformations are possible (Q2), and when Zariski-dense deformations occur (Q3).

The basic setup involves standard quotients X_Γ, which we now recall.

Definition 2.4 ([29, Def. 1.4]). Let $X = G/H$ be a homogeneous space of reductive type. A discontinuous group Γ for X is called *standard* if there exists a closed reductive subgroup L of G which contains Γ and acts properly on X.

We address the following question:

Question 2.5 Let $L \subset G \supset H$ be a triple of real reductive algebraic groups such that

$$L \text{ acts properly and cocompactly on } X = G/H. \qquad (2.1)$$

Classify the triples (G, H, L) for which L admits a (torsion-free), cocompact discrete subgroup Γ satisfying each of the following properties: (Q1), (Q2), or (Q3), where the conditions become progressively stronger, when H is non-compact.

(Q1). Γ is deformable, i.e., it is not locally rigid as a discontinuous group for X (Definition 2.3);
(Q2). Γ can be deformed into a non-standard discontinuous group for X;
(Q3). Γ can be deformed into a Zariski-dense discrete subgroup of G (Definition A.6 in Appendix A), while preserving the proper discontinuity of the action on X.

Clearly, the difference in the compact factors of the subgroups H does not affect the answer to Question 2.5. We will provide the classification in Theorem 2.9 for Question 2.5, based on the list of triples (G, H, L) satisfying (2.1), as presented in Tables 1 and 2, see Notation 2.6.

If a triple (G, H, L) satisfies the condition (2.1), that is, if L acts properly and cocompactly on $X = G/H$, then the double coset $\Gamma_H \backslash G / \Gamma_L$ becomes a compact manifold, referred to as an *exotic* quotient in [29], for any discrete subgroups Γ_H and Γ_L that have no torsion of H and L, respectively. (The term *exotic quotient* is also used to mean something different in other literature.) This quotient manifold $\Gamma_H \backslash G / \Gamma_L$ is a special case of the standard quotient of the group manifold G, considered as the symmetric space $(G \times G)/\operatorname{diag} G$. We will examine the deformation of these (exotic) quotients $\Gamma_H \backslash G / \Gamma_L$ and provide the classification results in Theorem 2.21, which answer Question 2.5, based on the same list of triples (G, H, L).

2.3 List of Triples (G, H, L) with Proper and Cocompact Actions

In this section, we provide a list of the triples of reductive Lie groups (G, H, L) for which Question 2.5 will be considered. The basic assumption is that the triple

(G, H, L) satisfies the condition (2.1), equivalently, one of the following equivalent conditions is satisfied:

(i) L acts properly and cocompactly on G/H;
(ii) H acts properly and cocompactly on G/L;
(iii) $H \times L$ acts properly and cocompactly on $(G \times G)/\operatorname{diag} G$.

These conditions are obviously satisfied if G is compact. Furthermore, these conditions are unchanged under taking a finite covering or replacing the connected component of the groups. The following list, Tables 1 and 2, exhibits such triples when G is a simple Lie group and when both H and L are non-compact. There are no new results in this section, but we provide a few comments.

A criterion for a triple (G, H, L) of reductive Lie groups to ensure that L acts properly on $X = G/H$ was established in [37, Thm. 4.1] in the late 1980s, along with a criterion for cocompactness in [37, Thm. 4.7]. These criteria are computable. Table 1 is taken from [50], by Kobayashi and Yoshino, and lists irreducible symmetric spaces G/H that admit proper and cocompact actions by reductive subgroups L. A recent work by Tojo [68] asserts that this list is exhaustive, as mentioned in Remark 2.8 below.

We have listed the triples (G, H, L) satisfying the condition (2.1) in Table 1 where G/H is a symmetric space and G is simple. Since the condition (2.1) is symmetric with respect to H and L, additional triples can be obtained by switching H and L. The following table is derived in this way, with cases where both G/H and G/L are symmetric spaces omitted.

The numbering i $(1 \leq i \leq 12)$ of Case i' in Table 2 corresponds to the numbering of Case i in Table 1, with H and L switched.

Table 1 Symmetric spaces G/H with G simple that admit a proper and cocompact action by a reductive subgroup L of G with $L_{ss} \subset L \subset L_{max}$

Case	G/H	L_{ss}	L_{max}
1	$SU(2n, 2)/Sp(n, 1)$	$SU(2n, 1)$	$U(2n, 1)$
2	$SU(2n, 2)/U(2n, 1)$	$Sp(n, 1)$	–
3	$SO(2n, 2)/U(n, 1)$	$SO(2n, 1)$	–
4	$SO(2n, 2)/SO(2n, 1)$	$SU(n, 1)$	$U(n, 1)$
5	$SO(4n, 4)/SO(4n, 3)$	$Sp(n, 1)$	$Sp(n, 1) \times Sp(1)$
6	$SO(8, 8)/SO(8, 7)$	$Spin(8, 1)$	–
7	$SO(4, 4)/(SO(4, 1) \times SO(3))$	$Spin(4, 3)$	–
8	$SO(8, \mathbb{C})/SO(7, \mathbb{C})$	$Spin(7, 1)$	–
9	$SO(8, \mathbb{C})/SO(7, 1)$	$Spin(7, \mathbb{C})$	–
10	$SO^*(8)/U(3, 1)$	$Spin(6, 1)$	–
11	$SO^*(8)/(SO^*(6) \times SO^*(2))$	$Spin(6, 1)$	–
12	$SO(4, 3)/(SO(4, 1) \times SO(2))$	$G_{2(2)}$	–

Table 2 Non-symmetric spaces G/H with G simple that admit a proper and cocompact action by a reductive subgroup L of G with $L_{ss} \subset L \subset L_{max}$

Case	G/H	L_{ss}	L_{max}
1'	$SU(2n, 2)/SU(2n, 1)$	$Sp(n, 1)$	–
4'	$SO(2n, 2)/SU(n, 1)$	$SO(2n, 1)$	–
5'	$SO(4n, 4)/Sp(n, 1)$	$SO(4n, 3)$	–
6'	$SO(8, 8)/Spin(8, 1)$	$SO(8, 7)$	–
7'	$SO(4, 4)/Spin(4, 3)$	$SO(4, 1)$	$SO(4, 1) \times SO(3)$
8'	$SO(8, \mathbb{C})/Spin(7, 1)$	$SO(7, \mathbb{C})$	–
9'	$SO(8, \mathbb{C})/Spin(7, \mathbb{C})$	$SO(7, 1)$	–
10'	$SO^*(8)/Spin(6, 1)$	$SU(3, 1)$	$U(3, 1)$
11'	$SO^*(8)/Spin(6, 1)$	$SO^*(6)$	$SO^*(6) \times SO^*(2)$
12'	$SO(4, 3)/G_{2(2)}$	$SO(4, 1)$	$SO(4, 1) \times SO(2)$

Regarding the subgroup L, there are several possible choices. While we will see eventually that the answers to Question 2.5 remain unaffected by these choices, we introduce some notation for clarity.

Notation 2.6 • Let L_{ss} be as given in Tables 1 or 2.

- Let L_{max} be the identity component of the normalizer of L_{ss} in G in the sense of Definition A.6.
- For a subgroup L satisfying $L_{ss} \subset L \subset L_{max}$, we denote by L_c the compact factor of L.

Remark 2.7 (1) We have included $n = 1$ in Cases 3 and 4 in Table 1 and in Case 4' in Table 2, although G is not a simple Lie group.

(2) The table includes triples that are locally isomorphic in the case of low dimensions; however, we have not excluded overlaps in the table. See Remark 2.16 for such examples.

Remark 2.8 Tables 1 and 2 complete the earlier list presented in Kulkarni [51] and Kobayashi [37, 38, 42]. In particular, Cases 4 and 5 are from [51], and Cases 1, 1', 2, 3, 4', and 5' are from [37].

It is plausible that Tables 1 and 2 list all the homogeneous spaces G/H with G simple, H non-compact and reductive, that admit a proper and cocompact action by a reductive subgroup L.

Based on the (infinitesimal) classification of irreducible symmetric spaces by Berger [3], Tojo [68] proved that irreducible symmetric spaces G/H that admit a reductive subgroup L satisfying the properness and cocompactness condition (2.1) are either listed in Table 1, Riemannian symmetric spaces G/K, or group manifolds $(G \times G)/\operatorname{diag} G$.

A more recent preprint of Bocheński–Tralle [6] shows, under the assumption that G is absolutely simple, that Tables 1 and 2 contain all the homogeneous spaces of the form G/H with H non-compact and reductive that admit a proper and cocompact action by a reductive subgroup L. We note that our list also includes the cases, such as Cases 8 and 9, where G is a complex simple Lie group.

2.4 Answers to Question 2.5: Classification of the Triples (G, H, L)

In this section, we provide classification results, as stated in Theorem 2.9, for (Q1), (Q2), and (Q3) of Question 2.5. This is based on the list of triples (G, H, L) that satisfy (2.1), as presented in Tables 1 and 2.

Theorem 2.9 *Let (G, H, L) be one of the triples in Tables 1 and 2, where $L_{ss} \subset L \subset L_{max}$. Then, the answer to (Q1), (Q2), and (Q3) in Question 2.5 is provided in Table 3.*

Remark 2.10 (1) Theorem 2.9 includes the claim that the choice of L, where $L_{ss} \subset L \subset L_{max}$, does not affect the answer to any of (Q1), (Q2), or (Q3) in Question 2.5.
(2) Theorem 2.9 remain unchanged if we replace (G, H, L) with locally isomorphic triples (Definition 2.12), as can be seen from the proof and Remark 2.13.

Theorem 2.9 will be proved in Sect. 5.2, using Theorem 3.9 from Sect. 3.

The triples (G, H, L) listed in Table 3 appear in either Table 1 or Table 2. The interpretation of Theorem 2.9 is illustrated using Case 1 as an example:

Example 2.11 In the first row of the table, Case 1 shows that for the symmetric space $G/H = SU(2n, 2)/Sp(n, 1)$ and $L = SU(2n, 1)$, (Q1) is "yes", (Q2) is "no" (and thus (Q3) is also "no"). This means that there exists a torsion-free cocompact discrete subgroup Γ of L such that Γ is deformable as a discontinuous group for G/H (Q1). However, Γ cannot be deformed into a non-standard discontinuous group (Q2), and, in particular, it cannot be deformed into a Zariski-dense subgroup while preserving the proper discontinuity of the action on G/H (Q3).

Regarding the concept of local isomorphisms of triples, we adopt the definition from [46], which allows for individual applications of *inner automorphisms* at the infinitesimal level. This definition fits well with (Q1)–(Q3) of Question 2.5 and is given as follows.

Definition 2.12 ([46, Def. 7.1]) The triples $L_1 \subset G_1 \supset H_1$ and $L_2 \subset G_2 \supset H_2$ are said to be *locally isomorphic* if there exists an isomorphism $\mathfrak{g}_1 \simeq \mathfrak{g}_2$ between the Lie algebras of the Lie groups G_1 and G_2, such that under this isomorphism, $\mathfrak{h}_1$ is conjugate to $\mathfrak{h}_2$ by an inner automorphism, and $\mathfrak{l}_1$ is conjugate to $\mathfrak{l}_2$ by another inner automorphism.

We say that the homogeneous spaces G_1/H_1 and G_2/H_2 are *locally isomorphic* if the triples $\{e\} \subset G_1 \supset H_1$ and $\{e\} \subset G_2 \supset H_2$ are locally isomorphic.

Table 3 Complete answers to Question 2.5 for the triples (G, H, L) with non-compact semisimple factors L_{ss} in Tables 1 and 2

Case	G/H	L_{ss}	Q1	Q2	Q3
1	$SU(2n, 2)/Sp(n, 1)$	$SU(2n, 1)$	Yes	No	No
1'-1	$SU(2n, 2)/SU(2n, 1)$ $(n \geq 2)$	$Sp(n, 1)$	No	No	No
1'-2	$SU(2, 2)/SU(2, 1)$	$Spin(4, 1)$	Yes	Yes	Yes
2–1	$SU(2n, 2)/U(2n, 1)$ $(n \geq 2)$	$Sp(n, 1)$	No	No	No
2–2	$SU(2, 2)/U(2, 1)$	$Spin(4, 1)$	Yes	Yes	Yes
3	$SO(2n, 2)/U(n, 1)$	$SO(2n, 1)$	Yes	Yes	Yes
4–1	$SO(2n, 2)/SO(2n, 1)$ $(n \geq 2)$	$SU(n, 1)$	Yes	No	No
4–2	$SO(2, 2)/SO(2, 1)$	$SU(1, 1)$	Yes	Yes	Yes
4'	$SO(2n, 2)/SU(n, 1)$	$SO(2n, 1)$	Yes	Yes	Yes
5–1	$SO(4n, 4)/SO(4n, 3)$ $(n \geq 2)$	$Sp(n, 1)$	No	No	No
5–2	$SO(4, 4)/SO(4, 3)$	$Spin(4, 1)$	Yes	Yes	Yes
5'	$SO(4n, 4)/Sp(n, 1)$	$SO(4n, 3)$	No	No	No
6	$SO(8, 8)/SO(8, 7)$	$Spin(8, 1)$	No	No	No
6'	$SO(8, 8)/Spin(8, 1)$	$SO(8, 7)$	No	No	No
7	$SO(4, 4)/(SO(4, 1) \times SO(3))$	$Spin(4, 3)$	No	No	No
7'	$SO(4, 4)/Spin(4, 3)$	$SO(4, 1)$	Yes	Yes	Yes
8	$SO(8, \mathbb{C})/SO(7, \mathbb{C})$	$Spin(7, 1)$	No	No	No
8'	$SO(8, \mathbb{C})/Spin(7, 1)$	$SO(7, \mathbb{C})$	No	No	No
9	$SO(8, \mathbb{C})/SO(7, 1)$	$Spin(7, \mathbb{C})$	No	No	No
9'	$SO(8, \mathbb{C})/Spin(7, \mathbb{C})$	$SO(7, 1)$	No	No	No
10	$SO^*(8)/U(3, 1)$	$Spin(6, 1)$	Yes	Yes	Yes
10'	$SO^*(8)/Spin(6, 1)$	$SU(3, 1)$	Yes	No	No
11	$SO^*(8)/(SO^*(6) \times SO^*(2))$	$Spin(6, 1)$	Yes	Yes	Yes
11'	$SO^*(8)/Spin(6, 1)$	$SO^*(6)$	No	No	No
12	$SO(4, 3)/(SO(4, 1) \times SO(2))$	$G_{2(2)}$	No	No	No
12'	$SO(4, 3)/G_{2(2)}$	$SO(4, 1)$	Yes	Yes	Yes

Remark 2.13 In Definition 2.12, we do not require the existence of a morphism between G_1 and G_2. However, if there is another triple $\tilde{L} \subset \tilde{G} \supset \tilde{H}$ with the following properties:

(1) all the three triples of linear Lie groups are locally isomorphic;
(2) there exists a morphism $\phi_j : \tilde{G} \to G_j$ for each $j = 1, 2$;
(3) $\tilde{\Gamma}$ is a discrete subgroup of $\tilde{L}$, and $\Gamma_j = \psi_j(\tilde{\Gamma})$,

then we have the following easy consequences:

(1) the following three conditions are equivalent:

   (i) the Γ_1-action on $X_1 = G_1/H_1$ is properly discontinuous (respectively, cocompact),

(ii) the Γ_2-action on $X_2 = G_2/H_2$ is properly discontinuous (respectively, cocompact),
(iii) the $\tilde{\Gamma}$-action on $\tilde{X} = \tilde{G}/\tilde{H}$ is properly discontinuous (respectively, cocompact).

(2) if $\tilde{\Gamma}$ is deformable as a discontinuous group for $\tilde{X}$, then Γ_j is deformable as a discontinuous group for X_j $(j = 1, 2)$.

Remark 2.14 For Cases 1', 2, and 5, there is an isomorphism of Lie groups L, $Sp(1, 1) \simeq Spin(4, 1)$. As can be seen in the proof, it is this isomorphism that causes the differences in the answers to Question 2.5 (for each of Cases 1', 2, and 5).

As a direct consequence of Theorem 2.9, we obtain the following list of homogeneous spaces G/H that have the property that (Q2) holds (and (Q3) holds as well). Since this property is unaffected by replacing H with a connected cocompact subgroup, the following classification result in (iii) is stated with this consideration in mind.

Theorem 2.15 *Let (G, H, L) be one of the triples in Tables 1 and 2, or let H be replaced by a connected cocompact subgroup of H. Then the following are equivalent:*

(i) ((Q2) in Question 2.5 for G/H) There exists a torsion-free, cocompact discrete subgroup Γ of L such that the quotient $\Gamma\backslash G/H$ can be deformed into a non-standard quotient.
(ii) ((Q3) in Question 2.5 for G/H) There exists a torsion-free, cocompact discrete subgroup Γ of L such that Γ can be deformed into a Zariski-dense subgroup in G, keeping the proper discontinuity of the action on G/H.
(iii) (Classification). G/H is one of the following homogeneous spaces, modulo compact factors of the subgroup H:

- $SU(2, 2)/U(2, 1)$;
- $SO(2, 2n)/U(1, n)$ $(n \geq 2)$;
- $SO(4, 4)/SO(4, 3)$;
- $SO^*(8)/U(3, 1)$;
- $SO^*(8)/(SO^*(6) \times SO^*(2))$;
- $SU(2, 2)/SU(2, 1)$;
- $SO(2, 2n)/SU(1, n)$ $(n \geq 2)$;
- $SO(4, 4)/Spin(4, 3)$;
- $SO(4, 3)/G_{2(2)}$;
- $SO^*(8)/SU(3, 1)$;
- $SO^*(8)/SO^*(6)$.

Contrary to the equivalence (i) $\Leftrightarrow$ (ii) in Theorem 2.15 for $\Gamma\backslash G/H$, such an equivalence does not hold for the exotic quotients $\Gamma_H\backslash G/\Gamma_L$, as stated in Theorem 2.21 (2).

Remark 2.16 Kassel proved in [27] that the answer to (Q3) is affirmative when $G/H = SO(2,2n)/U(1,n)$. To clarify the relationship between this space and the other homogeneous spaces in the list in Theorem 2.15(iii), we introduce the following symbols: $\dashrightarrow$, $\approx$, and $\xrightarrow{S^1}$.

(i) $G'/H' \approx G/H$: two homogeneous spaces G'/H' and G/H are locally isomorphic to each other;
(ii) $G/H' \xrightarrow{S^1} G/H$: a G-equivariant S^1 fiber bundle when H is a subgroup of H' such that H'/H is isomorphic to S^1;
(iii) $G'/H' \dashrightarrow G/H$: an injective homomorphism $\psi\colon G' \to G$ induces the diffeomorphism G'/H' onto G/H.

Obviously, the relation $\xrightarrow{S^1}$ does not affect the answer to (Q3). A typical example of this relation is

$$SO(2,2n)/SU(1,n) \xrightarrow{S^1} SO(2,2n)/U(1,n). \tag{2.2}$$

Using the binary relation symbols $\dashrightarrow$, $\approx$, and $\xrightarrow{S^1}$, the homogeneous spaces of lower dimensions in the list in Theorem 2.15(iii) can be grouped into two categories, which are connected to (2.2) as follows:

(1) ($n = 2$ in (2.2))

$$\begin{array}{ccc}
SU(2,2)/SU(1,2) & \xrightarrow{S^1} & SU(2,2)/U(1,2) \\
\wr\wr & & \wr\wr \\
SO(2,4)/SU(1,2) & \xrightarrow{S^1} & SO(2,4)/U(1,2) \\
\downarrow & & \\
SO(4,3)/G_{2(2)} & & \\
\downarrow & & \\
SO(4,4)/Spin(4,3) & & \\
\wr\wr & & \\
SO(4,4)/SO(4,3) & &
\end{array}$$

(2) ($n = 3$ in (2.2))

$$\begin{array}{ccc}
SO(2,6)/SU(1,3) & \xrightarrow{S^1} & SO(2,6)/U(1,3) \\
\wr\wr & & \wr\wr \\
SO^*(8)/SO^*(6) & \xrightarrow{S^1} & SO^*(8)/(SO^*(6) \times SO^*(2)) \\
\wr\wr & & \wr\wr \\
SO^*(8)/SU(1,3) & \xrightarrow{S^1} & SO^*(8)/U(1,3)
\end{array}$$

To be precise, we recall that $G_{2(2)} = \mathrm{Aut}(\mathbb{O}')$ (where $\mathbb{O}'$ is the 8-dimensional nonassociative algebra of split-octonions) has a maximal compact subgroup $(Sp(1) \times Sp(1))/\operatorname{diag}\{\pm 1\}$.

Second, We examine whether the binary relation $\approx$ affects the answer to (Q3).

As a typical example of the relation $G'/H' \approx G/H$, we consider the case where $\psi \colon G' \to G$ is a covering map. In this case, given a torsion-free cocompact discrete subgroup Γ of G, it may happen that there does not exist a torsion-free discrete subgroup Γ' of G' such that $\Gamma = \psi(\Gamma')$, as discussed in Remark 1.5(3) for $\psi \colon Spin(n,1) \to SO_0(n,1)$ with $n \geq 4$. Thus, the relation $\approx$ may, a priori, affect the answer to (Q3). This is why we formulate the deformation theory in $Spin(n,1)$ in Sect. 3, rather than $SO(n,1)$. From this formulation, we get the conclusion that our results remain unaffected by the binary relation $\approx$.

Third, regarding the relation $G'/H' \dashrightarrow G/H$, where the dimension of G is larger than that of G', the answers to (Q1)–(Q3) generally differ between the two (see Remark 2.17 below). However, somewhat surprisingly, it does not affect the results concerning the series of homogeneous spaces discussed above.

Remark 2.17 Consider the relation $G'/H' \dashrightarrow G/H$ in Remark 2.16, which is given by a group homomorphism $\psi \colon G' \to G$ that induces a diffeomorphism

$$G'/H' \stackrel{\sim}{\to} G/H$$

between the two homogeneous spaces G'/H' and G/H. For a discrete subgroup Γ' of G', we set

$$\Gamma := \psi(\Gamma'), \; L := \psi(L'). \tag{2.3}$$

Regarding (Q1)–(Q3) in Question 2.5, we may compare the Γ'-action on G'/H' with the Γ-action on G/H. Since the bijection $G'/H' \stackrel{\sim}{\to} G/H$ is a homeomorphism, any topological properties like proper discontinuity or cocompactness are the same for G'/H' and G/H, and some of obvious relationships follow: for instance, the existence of a cocompact discontinuous group for G'/H' implies that there exists a cocompact discontinuous group for G/H. However, there are delicate differences regarding deformations as follows:

On (Q1) Even if Γ' is locally rigid as a discontinuous group for G'/H', it is possible that $\Gamma = \psi(\Gamma')$ is deformable as a discontinuous group for G/H.

On (Q2) Even if Γ' cannot be deformed into a non-standard discontinuous group for G'/H', it is possible that $\Gamma = \psi(\Gamma')$ is deformed into a non-standard discontinuous group for G/H.

On (Q3) Even if Γ' is deformable into a Zariski-dense discrete subgroup of G' preserving the proper discontinuity of the action on G'/H', Γ is not necessarily deformable into a Zariski-dense discrete subgroup of G.

2.5 Group Manifold Case

In contrast to the Selberg–Weil rigidity theorem for the Riemannian symmetric space G/K, an irreducible pseudo-Riemannian symmetric space may admit a cocompact discontinuous group that is not locally rigid, even in higher dimensions. This was first observed in the early 90s (see [39]) for the group manifold G, viewed as a homogeneous space $(G \times G)/ \operatorname{diag} G$. The classification for such simple Lie groups G is obtained by Kobayashi as follows:

Fact 2.18 ([43, Thm. A]) *Let G be a non-compact linear semisimple Lie group. Then the following three conditions on G are equivalent:*

(i) the Lie algebra $\mathfrak{g}$ of G is isomorphic to either $\mathfrak{so}(n, 1)$ or $\mathfrak{su}(n, 1)$;
(ii) (affirmative answer to (Q1) in Question 2.5) there exists a cocompact discrete subgroup Γ of G such that $\Gamma \times \{1\}$ is not locally rigid as a discontinuous group for $(G \times G)/ \operatorname{diag} G$;
(iii) (affirmative answer to (Q2) in Question 2.5) there exists a cocompact discrete subgroup Γ of G such that the compact manifold $\Gamma\backslash G$ admits a non-standard deformation of the quotient of $(G \times G)/ \operatorname{diag} G$ ($\simeq G$).

The implication (i) $\Rightarrow$ (iii) in Fact 2.18, in the case where $\mathfrak{g} = \mathfrak{so}(2, 1) \simeq \mathfrak{su}(1, 1)$, is due to Goldman [15].

A natural extension of this problem is to strengthen the "non-standard deformation" in (iii) of Fact 2.18 and ask (Q3): Is it possible to deform it into a Zariski-dense discrete subgroup while preserving the proper discontinuity of the action on the group manifold $(G \times G)/ \operatorname{diag} G$? The answer is affirmative for $\mathfrak{g} = \mathfrak{so}(n, 1)$, as follows.

Theorem 2.19 *For any Zariski-connected real algebraic group G with Lie algebra $\mathfrak{so}(n, 1)$, there exist a torsion-free, cocompact discrete subgroup Γ of G and a small deformation Γ' of $\Gamma \times \{e\}$ such that Γ' is Zariski-dense in the direct product group $G \times G$, while preserving the proper discontinuity of the action on the group manifold $(G \times G)/ \operatorname{diag} G$.*

Theorem 2.19 is more or less well-known to experts, at least for $G = SO(n, 1)$ or its quotient groups. Nevertheless, we provide a proof for $G = Spin(n, 1)$ in Sect. 5.3, as an immediate application of Theorem 5.9.

Remark 2.20 In the case where G is a linear Lie group locally isomorphic to $SU(n, 1)$ with $n \geq 3$, the authors are unsure whether the implication (i) $\Rightarrow$ (iii) in Fact 2.18 can be strengthened, as in Theorem 2.19 for $\mathfrak{g} = \mathfrak{so}(n, 1)$. However, the answer is affirmative when $n = 2$, as noted in Tholozan [66, Sect. 2.3.2].

Theorem 2.21 *Let (G, H, L) be a triple of reductive groups in Table 3.*

(1) The following three conditions on the triple (G, H, L) are equivalent:

(i) ((Q1) in Question 2.5 for G/H). Up to switching H and L if necessary, there exists a torsion-free, cocompact discrete subgroup Γ of L such that Γ is deformable as a discontinuous group for G/H.

(ii) ((Q1) in Question 2.5 for $(G \times G)/\operatorname{diag} G$*). There exist torsion-free, cocompact discrete subgroups* Γ_H *and* Γ_L *of* H *and* L*, respectively, such that the discrete subgroup* $\Gamma_H \times \Gamma_L$ *is deformable as a discontinuous group for the group manifold* $(G \times G)/\operatorname{diag} G$.

(iii) (Classification) Up to switching H *and* L*, and up to compact factors, the triple* (G, H, L) *does not belong to one of the following lists:*

- *(Case 5,* $n \geq 2$*)* $(SO(4n, 4), SO(4n, 3), Sp(n, 1))$;
- *(Case 6)* $(SO(8, 8), SO(8, 7), Spin(8, 1))$;
- *(Case 8)* $(SO(8, \mathbb{C}), SO(7, \mathbb{C}), Spin(7, 1))$;
- *(Case 9)* $(SO(8, \mathbb{C}), SO(7, 1), Spin(7, \mathbb{C}))$.

(2) The following three conditions on the triple (G, H, L) *are equivalent:*

(i) ((Q2) in Question 2.5 for G/H*). Up to switching* H *and* L *if necessary, there exists a torsion-free, cocompact discrete subgroup* Γ *of* L *such that* Γ *can be deformed into a non-standard, cocompact, discontinuous group for* G/H.

(ii) ((Q2) in Question 2.5 for $(G \times G)/\operatorname{diag} G$*). There exist torsion-free, cocompact discrete subgroups* Γ_H *and* Γ_L *of* H *and* L*, respectively, such that the discrete subgroup* $\Gamma_H \times \Gamma_L$ *can be deformed into a non-standard, cocompact, discontinuous group for* $(G \times G)/\operatorname{diag} G$.

(iii) (Classification). Up to switching H *and* L *and up to compact factors, the triple* (G, H, L) *belongs to one of the following lists:*

- *(Cases 1 and 2,* $n = 1$*)* $(SU(2, 2), Sp(1, 1), SU(2, 1))$;
- *(Cases 3 and 4)* $(SO(2n, 2), SO(2n, 1), SU(n, 1))$;
- *(Case 5,* $n = 1$*)* $(SO(4, 4), SO(4, 3), Spin(4, 1))$;
- *(Case 7)* $(SO(4, 4), SO(4, 1), Spin(4, 3))$;
- *(Case 10)* $(SO^*(8), SU(3, 1), Spin(6, 1))$;
- *(Case 11)* $(SO^*(8), SO^*(6), Spin(6, 1))$;
- *(Case 12)* $(SO(4, 3), SO(4, 1), G_{2(2)})$.

(3) ((Q3) in Question 2.5 for $(G \times G)/\operatorname{diag} G$*). Assume* G *is simple. Then there does not exist a cocompact, direct product discrete subgroup* $\Gamma_H \times \Gamma_L$ *in* $H \times L$*, which can be deformed into a Zariski-dense subgroup of* $G \times G$ *while keeping the proper discontinuity of the action on* $(G \times G)/\operatorname{diag} G$ *for any triple* (G, H, L) *in Table 3.*

As stated in Remark 2.10 for Theorems 2.9, 2.21 also includes the claim that the choice of H and L, where $H_{ss} \subset H \subset H_{max}$ and $L_{ss} \subset L \subset L_{max}$, respectively, does not affect the answer to any of (Q1), (Q2), or (Q3) in Question 2.5, and that the answer remains unchanged if we replace (G, H, L) with locally isomorphic triples (Definition 2.12).

Theorem 2.21 will be proved in Sect. 5.4.

In the last statement (3) of Theorem 2.21, we assumed that G is simple. An analogous statement fails when G is not simple, as observed in the following proposition, where $G = SO(2, 2)$ in Cases 3 and 4 with $n = 1$.

Proposition 2.22 *The six-dimensional group manifold $G = SO(2, 2)$ has an exotic compact quotient of the form $\Gamma_1\backslash G/\Gamma_2$, where both Γ_1 and Γ_2 are torsion-free, Zariski-dense discrete subgroups of G.*

This proposition will be proven also in Sect. 5.4.

3 Deformation of Discrete Subgroups

Let Γ be a discrete subgroup of G. In this section, in order to study the deformation of Γ as a *discontinuous group* for a G-manifold X, we will momentarily set aside the space X and focus on the deformation of Γ within G. To put it succinctly, the focus will be solely on the pair (Γ, G) rather than the triplet (Γ, G, X), which simplifies the argument.

In Sect. 3.1, we recall results related to the local rigidity of representations of discrete groups, such as Raghunathan's vanishing theorem and a theorem of Klingler [35].

Owing to these local rigidity theorems, to examine the extent to which the standard quotient can be deformed (Problems 1.1 and 1.2), it suffices to focus on the case where Γ is a cocompact discrete subgroup of $Spin(n, 1)$. This will be the topic of the latter part of this section.

In Sect. 3.2, we will consider a general setup where G is a Zariski-connected real algebraic group and $\varphi\colon Spin(n, 1) \to G$ is a homomorphism, and introduce a real algebraic subgroup G^{φ} (Definition 3.7). We prove in Theorem 3.9(3), for any cocompact discrete subgroup Γ of $Spin(n, 1)$ and any small deformation φ' of $\varphi|_{\Gamma}$ in G, that $\varphi'(\Gamma)$ is contained in G^{φ} up to G-conjugacy. This upper bound is optimal: we shall prove in Theorem 3.9(1) that there exist Γ and φ' such that the Zariski-closure of $\varphi'(\Gamma)$ coincides with G^{φ}.

In Sect. 3.3, we will discuss in more detail the special cases where $n = 2$ or φ is a spin representation. In the case where $n = 2$, we explain how Theorem 3.9 connects the Zariski-closure of deformations of specific surface group representations to the combinatorics of nilpotent orbits in complex reductive Lie algebras. In the case where φ is a spin representation, we compute G^{φ} in terms of the Clifford algebra associated with the indefinite quadratic form of signature (p, q) and provide sufficient conditions on the triple (p, q, n) for the classical group $G(p, q)$ to admit a Zariski-dense subgroup as a deformation of a cocompact discrete subgroup of $Spin(n, 1)$ (see Theorem 3.18). For notation related to $G(p, q)$ and Clifford algebras, see Appendix B.

3.1 Local Rigidity Theorems After Weil, Matsushima–Murakami, Raghunathan, Goldman–Millson, and Klingler

A finitely-generated discrete subgroup Γ of G is locally rigid if it is *infinitesimally rigid*, that is, the first cohomology $H^1(\Gamma, \mathfrak{g})$ vanishes (Weil [71]).

We begin with the classical vanishing theorems for the first cohomology, notably those by Matsushima–Murakami [55] and Raghunathan:

Fact 3.1 (Raghunathan [60]) *Let L be a non-compact, simple Lie group, π be an irreducible finite-dimensional representation of L on a real vector space V, and Γ be a cocompact discrete subgroup without torsion. If $H^1(\Gamma, V) \neq 0$, then one of the following holds:*

(1) The Lie algebra $\mathfrak{l}$ is isomorphic to $\mathfrak{so}(n, 1)$ with $n \geq 3$, and V contains a non-zero $\mathfrak{so}(n-1, 1)$-invariant vector.
(2) The Lie algebra $\mathfrak{l}$ is isomorphic to $\mathfrak{su}(n, 1)$ with $n \geq 1$, and (π, V) is isomorphic to the symmetric tensor of the standard representation or its dual.

Remark 3.2 (1) For an irreducible representation π of $\mathfrak{so}(n, 1)$ on a finite-dimensional real vector space V, the following two conditions are equivalent:

(i) V contains a non-zero $\mathfrak{so}(n-1, 1)$-invariant vector;
(ii) V contains a non-zero $\mathfrak{so}(n)$-invariant vector.

For $n \geq 2$, the complexification of such a representation π can be realized in the space of spherical harmonics of degree k, defined by

$$\{f \in C^\infty(S^n) \mid \Delta_{S^n} f = -k(k+n-1)f\},$$

where Δ_{S^n} is the Laplacian of the unit sphere S^n. We refer to π as the *spherical harmonics representation* of degree k.

(2) There is an isomorphism $\mathfrak{so}(2, 1) \simeq \mathfrak{su}(1, 1)$ of Lie algebras. For this Lie algebra, we refer to the second statement for $\mathfrak{su}(1, 1)$.

Thanks to Fact 3.1, a cocompact discrete subgroup of L can potentially be deformed within a larger Lie group G only if it is locally isomorphic to $SO(n, 1)$ or $SU(n, 1)$.

Infinitesimal rigidity is a necessary condition for local rigidity, it is not sufficient. For complex hyperbolic lattices, Goldman–Millson [17] discovered a striking example where local rigidity holds and infinitesimal rigidity fails. This example has been further generalized by a theorem of Klingler [35, Thm. 1.3.7], whose special cases we recall here:

Fact 3.3 (Klingler [35, Thm. 1.3.8]) *Let Γ be a cocompact discrete subgroup of $SU(n, 1)$, and $\varphi\colon SU(n, 1) \hookrightarrow SU(p, q)$ $(p \geq n, q \geq 1)$ be the standard embedding, Then any morphism φ' suffiently close to $\varphi|_\Gamma$ is conjugate to a representation of* *the form $\varphi \cdot \chi$, where $\chi\colon \Gamma \to SU(p,q)$ is a deformation of the trivial representation in the centralizer of $SU(n,1)$ in $SU(p,q)$.*

Remark 3.4 The above results hold for locally isomorphic groups. That is, let $\phi\colon L \to G$ be a Lie group homomorphism, and Γ be a cocompact discrete subgroup of L. Suppose the differentials of ϕ and φ (from the theorem above) agree. In this case, any homomorphism sufficiently close to $\phi|_\Gamma$ maps into $\phi(L) \cdot Z_G(L)$ up to conjugation by G, where $Z_G(L)$ denotes the centralizer of $\phi(L)$ in G.

In fact, the proof of Fact 3.3 is based on the obstruction of the integrability of infinitesimal deformation, which can be traced back to Goldman–Millson [17]: the conclusion of Fact 3.3 holds if

$$\text{for any } c \in H^1(\Gamma, \mathfrak{g}) \setminus H^1(\Gamma, \mathfrak{z}_\mathfrak{g}(\mathfrak{l})),\ [c,c] \neq 0 \text{ in } H^2(\Gamma, \mathfrak{g}), \tag{3.1}$$

and this statement holds under the same assumption at the Lie algebra level.

We will use Fact 3.3 along with Remark 3.4 in the proof (Step 4) of Theorem 2.9 in Sect. 5.2.

In the sequel, we discuss cocompact discrete subgroup in a group which is locally isomorphic to $SO(n,1)$.

3.2 Deformation of Cocompact Discrete Subgroups of $Spin(n,1)$

Let $\mathbf{G}$ be a Zariski-connected real algebraic group, G the Lie group of real points $\mathbf{G}(\mathbb{R})$, and $\varphi\colon Spin(n,1) \to G$ a homomorphism ($n \geq 2$) in the sense of Lemma A.7. For basic notions of algebraic groups, we refer to Appendix A. In this section, we discuss to what extent the cocompact discrete subgroup Γ of $Spin(n,1)$ can be deformed in G through φ.

Definition 3.5 (*Lie algebra $\mathfrak{g}^\varphi$*) We regard the Lie algebra $\mathfrak{g}$ of G as a $\mathfrak{spin}(n,1)$-module via φ. Let $\mathfrak{g}^\varphi$ denote the Lie subalgebra of $\mathfrak{g}$, generated by $d\varphi(\mathfrak{spin}(n,1))$ and all the $\mathfrak{spin}(n,1)$-submodules that are isomorphic to irreducible spherical harmonics modules of $\mathfrak{spin}(n,1)$ (see Remark 3.2). In other words, $\mathfrak{g}^\varphi$ is the smallest Lie subalgebra of $\mathfrak{g}$ that contains

$$d\varphi(\mathfrak{spin}(n,1)) + \mathfrak{z}_\mathfrak{g}(d\varphi(\mathfrak{spin}(n-1,1))), \tag{3.2}$$

where $\mathfrak{z}_\mathfrak{g}(d\varphi(\mathfrak{spin}(n-1,1)))$ denotes the centralizer of $d\varphi(\mathfrak{spin}(n-1,1))$ in $\mathfrak{g}$.

We define a Zariski-connected real algebraic group $\mathbf{G}^\varphi$ corresponding to the Lie algebra $\mathfrak{g}^\varphi$ (we refer to Definition A.4 in Appendix A for the definition of Zariski-connectedness of real algebraic groups). For this purpose, we use the following lemma, the proof of which will be given in Sect. 4.1.

Lemma 3.6 *Let $G_{\mathbb{C}}$ be the complex Lie group of complex points $\mathbf{G}(\mathbb{C})$, and $G^{\varphi}_{\mathbb{C}}$ the analytic subgroup of $G_{\mathbb{C}}$ corresponding to $\mathfrak{g}^{\varphi} \otimes_{\mathbb{R}} \mathbb{C}$. Then $G^{\varphi}_{\mathbb{C}}$ is a Zariski-closed subset defined over $\mathbb{R}$.*

Since $G^{\varphi}_{\mathbb{C}}$ is connected in the usual topology, it is also connected in the Zariski-topology.

Definition 3.7 (*Real algebraic group G^{φ}*) Let $\mathbf{G}^{\varphi}$ be the Zariski-connected real algebraic group whose set of $\mathbb{C}$-points is the Zariski-closed subset $G^{\varphi}_{\mathbb{C}}$ given in Lemma 3.6. Furthermore, define $G^{\varphi} := \mathbf{G}^{\varphi}(\mathbb{R})$.

Remark 3.8 If G is reductive, then G^{φ} is also reductive.

Here is the main result of this section.

Theorem 3.9 *Let G be a Zariski-connected real algebraic group, and $\varphi\colon Spin(n, 1) \to G$ a homomorphism, where $n \geq 2$.*

(1) There exist a torsion-free cocompact discrete subgroup Γ of $Spin(n, 1)$ and a small deformation φ' of $\varphi|_{\Gamma}$ in $\mathrm{Hom}(\Gamma, G)$ such that the Zariski-closure of $\varphi'(\Gamma)$ coincides with G^{φ}, where G^{φ} is the real algebraic group, as given in Definition 3.7.
(2) If G is reductive and if $\varphi\colon Spin(n, 1) \to G$ is non-trivial, then we can choose $\varphi' \in \mathrm{Hom}(\Gamma, G)$ as in (1) such that φ' is discrete and faithful.
(3) Assume $n \geq 3$. For any torsion-free cocompact discrete subgroup Γ of $Spin(n, 1)$, there exists a neighborhood U of $\varphi|_{\Gamma}$ in $\mathrm{Hom}(\Gamma, G)$ such that $\varphi'(\Gamma) \subset G^{\varphi}$ up to G-conjugacy for any $\varphi' \in U$.

Remark 3.10 Analogous results to Theorem 3.9 hold as well if we replace $Spin(n, 1)$ with $SO(n, 1)$. The results are likewise new even in that context, and can be directly deduced from Theorem 3.9 via the double covering map $Spin(n, 1) \to SO_0(n, 1)$.

In general, we have

$$d\varphi(\mathfrak{spin}(n, 1)) + \mathfrak{z}_{\mathfrak{g}}(d\varphi(\mathfrak{spin}(n, 1))) \subset \mathfrak{g}^{\varphi} \subset \mathfrak{g}.$$

As special cases of Theorem 3.9, corresponding to the minimal case

$$\mathfrak{g}^{\varphi} = d\varphi\,(\mathfrak{spin}(n, 1)) + \mathfrak{z}_{\mathfrak{g}}(d\varphi(\mathfrak{spin}(n, 1)))$$

and the maximal case $\mathfrak{g}^{\varphi} = \mathfrak{g}$, we obtain Corollaries 3.11 and 3.13, respectively.

We begin with the particular case where $\mathfrak{g}^{\varphi}$ is minimal.

Corollary 3.11 *Let G be a Zariski-connected real algebraic group and let $n \geq 3$. Suppose that $\varphi\colon Spin(n, 1) \to G$ is a non-trivial homomorphism. Then the following two conditions are equivalent:*

(*i*) $\mathfrak{z}_{\mathfrak{g}}(d\varphi(\mathfrak{spin}(n,1))) = \mathfrak{z}_{\mathfrak{g}}(d\varphi(\mathfrak{spin}(n-1,1)))$.
(*ii*) *For any cocompact discrete subgroup* Γ *of* $Spin(n,1)$, *any morphism* $\varphi' \colon \Gamma \to G$ *sufficiently close to* $\varphi|_{\Gamma}$ *is conjugate to a representation of the form* $\varphi \cdot \chi$, *where* $\chi \colon \Gamma \to G$ *is a deformation of the trivial representation with image in the centralizer of* $\varphi(Spin(n,1))$ *in* G.

Remark 3.12 Suppose that $\rho \colon L \to G$ is a homomorphism. In the case where $L = SU(n,1)$, Klingler [35, Thm.1.3.7] established a sufficient condition under which every torsion-free cocompact discrete subgroup $\Gamma \subset L$ admits no non-trivial deformations beyond the centralizer of L. In contrast, Corollary 3.11 considers the analogous problem for $L = Spin(n,1)$, and provides a necessary and sufficient condition formulated in terms of finite-dimensional representations of Lie algebras.

Proof of Corollary 3.11 By Theorem 3.9 (3), it suffices to show that the condition (i) is equivalent to $\mathfrak{g}^{\varphi} = d\varphi(\mathfrak{spin}(n,1)) + \mathfrak{z}_{\mathfrak{g}}(d\varphi(\mathfrak{spin}(n,1)))$, but this is clear from the alternative definition of $\mathfrak{g}^{\varphi}$, as stated in Remark 3.5. □

By contrast, the following is an immediate consequence of Theorem 3.9(1) in the particular case where $\mathfrak{g}^{\varphi}$ is maximal.

Corollary 3.13 *Let* G *be a Zariski-connected real algebraic group. If* $\mathfrak{g}^{\varphi} = \mathfrak{g}$ *holds for a homomorphism* $\varphi \colon Spin(n,1) \to G$, *then* $\varphi(\Gamma)$ *can be deformed into a Zariski-dense subgroup. The converse also holds, provided that* $n \geq 3$.

We end this section with the proof of Theorem 3.9(2) and (3). The proof of (1) is differed to Sect. 4.

Proof of Theorem 3.9 (2) If $\varphi \colon Spin(n,1) \to G$ is non-trivial, then $\varphi|_{\Gamma}$ is discrete. Since Γ is torsion-free, $\varphi|_{\Gamma}$ is faithful. Therefore, the claim that a small deformation of $\varphi|_{\Gamma}$ preserves faithfulness and discreteness is a special case of the stability of discontinuous groups, as will be explained in Sect. 5.1. (We can apply Fact 5.6 by Kassel to $X = G/\{e\}$.) □

Remark 3.14 The proof of the stability of the discontinuous group for homogeneous spaces $X = G/H$ of reductive groups G, using a quantitative estimate, may be traced back to Kobayashi [43, Thm. 2.6]. In the above proof of (2) with $H = \{e\}$, one can apply Guichard [20, Thm. 2] if G is semisimple.

Proof of Theorem 3.9 (3) We apply a general result concerning an upper bound for the local deformation of discrete subgroups, which will be presented in Proposition E.2 in Appendix E.

Let $n \geq 3$, and we consider the $\mathfrak{spin}(n,1)$-module $\mathfrak{g}/\mathfrak{g}^{\varphi}$. By the definition of $\mathfrak{g}^{\varphi}$, this module does not contain any spherical harmonics representation as an irreducible component. Furthermore, since $n \geq 3$, the Lie group $Spin(n,1)$ is not locally isomorphic to $SU(m,1)$ for any $m \in \mathbb{N}_{+}$. Then, it follows from Raghunathan's vanishing theorem (Fact 3.1) that $H^1(\Gamma, \mathfrak{g}/\mathfrak{g}^{\varphi}) = 0$. By applying Proposition E.2 to $L = G^{\varphi}$, we obtain the desired conclusion. □

Remark 3.15 The assumption $n \geq 3$ in Theorem 3.9(3) is necessary and cannot be dropped for this upper bound result. In fact, an analogous statement fails when $n = 2$. For instance, let $G = SL(3, \mathbb{R})$, and let $L = SL(2, \mathbb{R})$ be the standard subgroup. Since $Spin(2, 1)$ is isomorphic to $SL(2, \mathbb{R})$, we may consider the embedding

$$\varphi\colon Spin(2, 1) \simeq SL(2, \mathbb{R}) \hookrightarrow SL(3, \mathbb{R}).$$

Applying Kim–Pansu [34, Thm. 1] to this setting, one sees that a discrete surface subgroup of $SL(2, \mathbb{R})$ of genus ≥ 128 can be deformed into a Zariski-dense subgroup of $SL(3, \mathbb{R})$. On the other hand, G^{φ} is given by

$$G^{\varphi} = S(GL(2, \mathbb{R}) \times GL(1, \mathbb{R})),$$

which is a proper subgroup of G.

3.3 *Examples of Theorem 3.9*

In this section, we illustrate Theorem 3.9 with some simple examples.

Theorem 3.9, applied to the deformation of discontinuous groups for $X = G/H$, will be discussed in Theorem 5.9, which is used in the proof of the main theorems in Sect. 2 for the compact standard quotients X_Γ as well as in the deformation theory of non-compact standard quotients X_Γ in Sect. 5.5.

We begin with the case $n = 2$. Theorem 3.9 in the $n = 2$ case and Corollary 3.17 were recently proved in [24], where it is noted that $Spin(2, 1)$ is isomorphic to $SL(2, \mathbb{R})$.

Let G be a Zariski-connected real reductive algebraic group and $\varphi\colon SL(2, \mathbb{R}) \to G$ a homomorphism. We define

$$\sigma^{\varphi} := \varphi\begin{pmatrix} -1 & 0 \\ 0 & -1 \end{pmatrix} = \exp\left(\pi\sqrt{-1}d\varphi\begin{pmatrix} 1 & 0 \\ 0 & -1 \end{pmatrix}\right) \in G.$$

Then it follows from [24, Lem. 3.3] that $\mathfrak{g}^{\varphi}$ is the centralizer of σ^{φ} in $\mathfrak{g}$ and that G^{φ} is the identity component of the centralizer of σ^{φ} in G in the Zariski-topology.

The following equivalence makes the condition $G = G^{\varphi}$ transparent:

Definition-Lemma 3.16 Let G be a Zariski-connected real reductive algebraic group and $\varphi\colon SL(2, \mathbb{R}) \to G$ a homomorphism. The following four conditions on a homomorphism $\varphi\colon SL(2, \mathbb{R}) \to G$ are equivalent:

(i) $G^{\varphi} = G$;
(ii) $\mathfrak{g}^{\varphi} = \mathfrak{g}$;

(iii) The complex nilpotent orbit in $\mathfrak{g}_{\mathbb{C}} := \mathfrak{g} \otimes_{\mathbb{R}} \mathbb{C}$ given by

$$\mathrm{Int}(\mathfrak{g}_{\mathbb{C}}) \cdot d\varphi \begin{pmatrix} 0 & 1 \\ 0 & 0 \end{pmatrix}$$

is even in the sense that the weights in the weighted Dynkin diagram associated to this nilpotent orbit are even, see [10, Chap. 3.8], for example;

(iv) We regard $\mathfrak{g}$ as an $SL(2, \mathbb{R})$-module via $\mathrm{Ad} \circ \varphi$. Then, every irreducible component of $\mathfrak{g}$ has an odd dimension, or equivalently, its highest weight is even.

We say that the homomorphism φ is *even* if it satisfies one of these equivalent conditions.

Proof (i)$\Leftrightarrow$(ii): The implication (i) $\Rightarrow$ (ii) is clear. Conversely, assume $\mathfrak{g}^{\varphi} = \mathfrak{g}$. The complex algebraic group $G_{\mathbb{C}}$ is Zariski-connected and, by Lemma A.1, also connected in the usual topology. Now, recall from Definition 3.7 that the analytic subgroup of $G_{\mathbb{C}}$ corresponding to $\mathfrak{g}^{\varphi} \otimes_{\mathbb{R}} \mathbb{C}$ is $G_{\mathbb{C}}^{\varphi}$. Thus, we conclude that $G_{\mathbb{C}} = G_{\mathbb{C}}^{\varphi}$. By taking real points of these two algebraic groups, it follows that $G = G^{\varphi}$.

(ii)$\Leftrightarrow$(iv): For any irreducible representation (τ, V) of $SL(2, \mathbb{R})$, the element

$$\tau \begin{pmatrix} -1 & 0 \\ 0 & -1 \end{pmatrix}$$

acts trivially on V if and only if V has odd dimension. The equivalence (ii) $\Leftrightarrow$ (iv) follows immediately from this fact.

(iii)$\Leftrightarrow$(iv): This is an elementary representation theory of $\mathfrak{sl}(2, \mathbb{R})$, see [10, Lem. 3.8.7], for instance. □

As a corollary to Theorem 3.9(1) applied to the $n = 2$ case, we obtain the following:

Corollary 3.17 ([24, Cor. 1.7]) *Let G be a Zariski-connected real reductive algebraic group and $\varphi: SL(2, \mathbb{R}) \to G$ an even homomorphism. Then there exists a discrete surface subgroup Γ of $SL(2, \mathbb{R})$ such that $\varphi(\Gamma)$ can be deformed into a Zariski-dense subgroup of G.*

Next, we consider the case where φ is the spin representation of $Spin(n, 1)$ for general n. As an example of Theorem 3.9(1), we present classical groups that contain a Zariski-dense discrete subgroup isomorphic to a cocompact discrete subgroup of $Spin(n, 1)$.

To be more precise, let $C(p, q)$ be the Clifford algebra associated with a real quadratic form of signature (p, q), and let G be the group $G(p, q)$, introduced in [50, Def. 4.3.1]. We refer to Appendix B for the notation of $G(p, q)$, and to Proposition B.2 for the identifications of $G(p, q)$ with classical groups that exhibit certain periodicities of $p + q$ and $p - q$.

When $p \geq n$ and $q \geq 1$, there exists a natural injective homomorphism $\varphi: Spin(n, 1) \to G(p, q)$.

Theorem 3.18 *Let $p \geq n \geq 2$ and $q \geq 1$. We set $m = p + q$. Assume that one of the following conditions holds:*

(1) When $m \equiv i \mod 4$ (where $i = 1, 2, 3, 4$), it holds that $m - n \geq i + 2$.
(2) $m < 10$ and $n \leq 6$.

Then, there exist a torsion-free, cocompact discrete subgroup Γ of $Spin(n, 1)$ and a small deformation $\varphi' \in \mathcal{R}(\Gamma, G(p, q))$ of the map $\varphi|_\Gamma : \Gamma \to G(p, q)$ such that the Zariski-closure of $\varphi'(\Gamma)$ coincides with the identity component of the group $G(p, q)$ in the Zariski-topology.

Owing to Theorem 3.9(1), the proof of Theorem 3.18 reduces to the following proposition on the finite-dimensional representation theory of Lie algebras.

Proposition 3.19 *In the setting of Theorem 3.18, we have $\mathfrak{g}^\varphi = \mathfrak{g}$.*

Before entering the proof, let us provide some examples of Theorem 3.18 in the special case $n = 2$.

Example 3.20 Let us consider the following inclusion maps:

(1) $Spin(4, 1) \to G(4, 3) \simeq O(4, 4)$;
(2) $Spin(6, 1) \to G(6, 1) \simeq O^*(8)$;
(3) $Spin(6, 1) \to G(7, 1) \simeq O(8, \mathbb{C})$;
(4) $Spin(6, 1) \to G(7, 2) \simeq O^*(16)$;
(5) $Spin(6, 1) \to G(7, 3) \simeq GL(8, \mathbb{H})$;
(6) $Spin(6, 1) \to G(8, 1) \simeq O(8, 8)$;
(7) $Spin(6, 1) \to G(8, 2) \simeq U(8, 8)$;
(8) $Spin(6, 1) \to G(8, 3) \simeq Sp(8, 8)$;
(9) $Spin(6, 1) \to G(9, 1) \simeq GL(16, \mathbb{R})$;
(10) $Spin(6, 1) \to G(10, 1) \simeq Sp(16, \mathbb{R})$;
(11) $Spin(6, 1) \to G(11, 1) \simeq Sp(16, \mathbb{C})$.

Then, we have $\mathfrak{g}^\varphi = \mathfrak{g}$. In particular, there exists a cocompact discrete subgroup Γ of $Spin(6, 1)$ that can be deformed into a Zariski-dense subgroup of the identity component of these classical groups in the Zariski-topology.

Proof of Example 3.20 The fact that each $G(p, q)$ is isomorphic to the classical groups described above follows from Proposition B.2. □

To prove Proposition 3.19, let us introduce some notation and lemmas. Let $C(p, q)$ denote the Clifford algebra associated with the standard quadratic form $x_1^2 + \cdots + x_p^2 - x_{p+1}^2 - \cdots - x_{p+q}^2$. Let $\{e_1, \ldots, e_{p+q}\}$ be the standard basis of $\mathbb{R}^{p+q}$. We define the basis of $C(p, q)$ by $\{e_I\}$, where

$$e_I := e_{i_1} \cdots e_{i_k} \in C(p, q),$$

for $I = \{i_1, \ldots, i_k\}$ with $1 \leq i_1 < \cdots < i_k \leq p + q$. In the following lemma, we regard $C(p, q)$ as a Lie algebra over $\mathbb{R}$ with the Lie bracket defined by $[x, y] = xy - yx$.

Lemma 3.21 *Given two subsets $I, J \subset \{1, \ldots, p+q\}$, where $\#I$ and $\#J$ are even, we have $[e_I, e_J] \neq 0$ if and only if $\#(I \cap J)$ is odd. Furthermore, $[e_I, e_J]$ is a scalar multiple of $e_{I \triangle J}$, where $I \triangle J = (I \setminus J) \cup (J \setminus I)$ denotes the symmetric difference.*

Proof For each $j \in J$,

$$e_j e_I = e_I e_j \times \begin{cases} -1 & (j \in I), \\ 1 & (j \notin I). \end{cases}$$

Hence, we have

$$e_J e_I = (-1)^{\#(I \cap J)} \, e_I e_J.$$

Thus, the assertions follow immediately. □

For each $k = 0, \ldots, m = p + q$, we write $\wedge^k$ for the $\mathbb{R}$-vector subspace of $C(p, q)$ spanned by the elements e_I for all subsets $I \subset \{1, \ldots, p+q\}$ with $\#I = k$.

With this notation, the module structure of the Lie algebras $\mathfrak{g}(p, q)$ and $\mathfrak{spin}(p, q)$ is given as follows

$$\mathfrak{g}(p, q) = \bigoplus_{\substack{0 \le k \le m \\ k \equiv 2 \bmod 4}} \wedge^k \quad \text{and} \quad \mathfrak{spin}(p, q) = \wedge^2, \tag{3.3}$$

see Lemma B.1 in Appendix B. Furthermore, each k, the subspace $\wedge^k$ is a $\mathfrak{spin}(p, q)$-submodule of $\mathfrak{g}(p, q)$ and is isomorphic to the k-th exterior tensor representation $\wedge^k V$, where V is the standard representation of $\mathfrak{spin}(p, q)$.

Lemma 3.22 *Let ℓ be an even number. If $\min(\ell, m - \ell) \ge 3$, then the Lie algebra $\mathfrak{g}(p, q)$ is generated by $\mathfrak{spin}(p, q)$ and $\wedge^\ell$.*

Proof The key step in the proof is to show the following claim: $\wedge^6 \subset [\wedge^\ell, \wedge^\ell]$ if $\min(\ell, m - \ell) \ge 3$. To prove this claim, let I' be an arbitrary subset of $\{1, \ldots, m\}$ with six elements. Since $\min(\ell, m - \ell) \ge 3$ by assumption, we can find subsets $I, J \subset \{1, \ldots, m\}$, each containing ℓ elements such that $I \triangle J = I'$. Since ℓ is even, $\#(I \cap J)$ is odd, and consequently, by Lemma 3.21, we have $e_{I'} \in [\wedge^\ell, \wedge^\ell]$. This completes the proof of the claim.

A similar argument shows that for an even number k, we have

$$\wedge^{k+4} \subset [\wedge^6, \wedge^k] \qquad \text{if } k \le m - 4,$$
$$\wedge^{k-4} \subset [\wedge^6, \wedge^k] \qquad \text{if } k \ge 4.$$

By (3.3), our assertion follows inductively. □

Let $V = \mathbb{R}^{p,q}$ and $W = \mathbb{R}^{n,1}$ be the standard representations of $\mathfrak{spin}(p,q)$ and $\mathfrak{spin}(n,1)$, respectively. Then we have the following branching law of the $\mathfrak{spin}(p,q)$-module V when restricted to $\mathfrak{spin}(n,1)$ via φ:

$$V \simeq W \oplus \mathbf{1}^{\oplus(m-(n+1))},$$

where $\mathbf{1}$ denotes the trivial representation and $m = p + q$.

Lemma 3.23 *Suppose $m \geq n + 1 \geq 3$.*

(1) If $m - n \geq \ell$ or if $\ell \geq n$, then $\wedge^\ell$ contains a non-zero spherical harmonics representation of $\mathfrak{spin}(n,1)$.
(2) $\mathfrak{g}^\varphi \supset \mathfrak{spin}(p,q)$.

Proof (1): The assertion is clear if $\ell = 0$ or m. Suppose that $1 \leq \ell \leq m - 1$. We have the following decomposition as $\mathfrak{spin}(n,1)$-modules:

$$\wedge^\ell \simeq \bigoplus_{k=0}^{\ell} \wedge^k W \otimes \wedge^{\ell-k}(\mathbb{C}^{m-(n+1)}), \tag{3.4}$$

where $\mathfrak{spin}(n,1)$ acts on $\wedge^k W$ as the k-th exterior power of the standard representation W, and trivially on $\wedge^{\ell-k}(\mathbb{C}^{m-(n+1)})$. Therefore, $\wedge^\ell$ contains a non-zero spherical harmonics if $W \otimes \wedge^{\ell-1}(\mathbb{C}^{m-(n+1)}) \neq 0$, which occurs if $m - n \geq \ell$.

Furthermore, we have $\wedge^k V \simeq \wedge^{m-k} V$ for all $0 \leq k \leq m$ as $\mathfrak{spin}(p,q)$-modules when $m \geq 3$. Hence, the same conclusion holds if $m - n \geq m - \ell$, i.e., if $\ell \geq n$. Thus, the first statement is proved.

(2): The $\ell = 2$ case in (3.4) gives an isomorphism of $\mathfrak{spin}(n,1)$-modules:

$$\wedge^2 V \simeq \wedge^2 W \oplus W^{\oplus(m-(n+1))} \oplus \mathbf{1}^{\oplus\binom{m-(n+1)}{2}}. \tag{3.5}$$

As we have seen in (3.3), the Lie algebra $\mathfrak{g} = \mathfrak{g}(p,q)$ contains $\wedge^2 = \wedge^2 V$ as a Lie subalgebra, isomorphic to $\mathfrak{spin}(p,q)$. Furthermore, the first component $\wedge^2 W$ in (3.5) corresponds to the Lie subalgebra $d\varphi(\mathfrak{spin}(n,1))$, while all other irreducible components are spherical harmonics representations of $\mathfrak{spin}(n,1)$. Thus, by Definition 3.5, we conclude that $\mathfrak{g}^\varphi \supset \mathfrak{spin}(p,q)$. □

Before proving Proposition 3.19(1) and (2), we state a general principle used in the proof. Since $\mathfrak{g}^\varphi$ is a Lie algebra, it is also a $\mathfrak{spin}(p,q)$-submodule of $\mathfrak{g} = \mathfrak{g}(p,q)$ by Lemma 3.23. Thus, to show that an irreducible $\mathfrak{spin}(p,q)$-submodule U of $\mathfrak{g}$ is contained in $\mathfrak{g}^\varphi$, it suffices to verify that $\mathfrak{g}^\varphi \cap U \neq 0$.

We are ready to show Proposition 3.19.

Proof of Proposition 3.19 (1): First, we consider the case where $m \geq 10$ and $m \neq 12$. In this case, the $\mathfrak{spin}(p,q)$-module $\wedge^{m-2-i}$ is irreducible because $m - 2 - i > m/2$ and $i \in \{1, 2, 3, 4\}$. Moreover, since $m \equiv i \mod 4$, it follows from we have (3.3) that $\wedge^{m-2-i} \subset \mathfrak{g} = \mathfrak{g}(p,q)$. Since $m - 2 - i \geq n$ by the assumption of

(1), $\mathfrak{g}^{\varphi} \cap \wedge^{m-2-i} \neq 0$ by Lemma 3.23(1). By the principle above, it follows that $\mathfrak{g}^{\varphi} \supset \wedge^{m-2-i}$. Furthermore, since $m \geq 10$, we have $\min(m-2-i, i+2) \geq 3$, it follows from Lemma 3.22 that $\mathfrak{g}^{\varphi} = \mathfrak{g}(p, q)$.

Next, we consider the case $m = 12$. In this case, the assumption (1) implies that $m = 12$, $i = 4$, and $n \leq 6$. Then we have

$$\mathfrak{g}(p, q) = \wedge^2 + \wedge^6 + \wedge^{10}.$$

By Lemma 3.22, it suffices to show that $\wedge^6$ is contained in $\mathfrak{g}^{\varphi}$.

In contrast to the above cases, we need to be careful because the module $\wedge^6$ is not irreducible as a $\mathfrak{spin}(p, q)$-module and splits into the direct sum of two distinct irreducible submodules, which we denote by V_1 and V_2. Since $n + 1 < m$, the modules V_1 and V_2 are isomorphic to each other as $\mathfrak{spin}(n, 1)$-modules. Moreover, since $n \leq 6$, the module $\wedge^6 \simeq \wedge^6 V$ contains at least two irreducible components, each of which is isomorphic to the standard representation $W \simeq \wedge^n W$ as a $\mathfrak{spin}(n, 1)$-module. Hence, both V_1 and V_2 contain W as an irreducible component when regarded as $\mathfrak{spin}(n, 1)$-modules. Thus, for each $i = 1, 2$, we obtain $\mathfrak{g}^{\varphi} \cap V_i \neq 0$, which, by the principle above, implies $V_i \subset \mathfrak{g}^{\varphi}$. Consequently, we conclude that $\wedge^6 \subset \mathfrak{g}^{\varphi}$, which completes the proof in this case.

(2): When $m \leq 5$, we have $\mathfrak{g}(p, q) = \wedge^2 = \mathfrak{spin}(p, q)$, and thus it follows that $\mathfrak{g}^{\varphi} = \mathfrak{g}(p, q)$. Hence, the assertion is obvious.

Next, when $6 \leq m < 10$, we have $\mathfrak{g}(p, q) = \wedge^2 + \wedge^6$. Since $n \geq 6$ by the assumption of (2), $\mathfrak{g}^{\varphi} \cap \wedge^6 \neq 0$ by Lemma 3.23(1). Since $\wedge^6$ is an irreducible $\mathfrak{spin}(p, q)$-module, based on the above principle, we conclude that $\mathfrak{g}^{\varphi} = \mathfrak{g}$. Thus, the proof of (2) is complete. □

As mentioned, the proof of Theorem 3.18 is derived from Proposition 3.19 by Theorem 3.9(1). The proof of Theorem 3.9(1) is the main task of the next section.

4 Deformations of the Representations of Spin Hyperbolic Lattices that Maximize the Zariski-Closure

Let $\mathbf{G}$ be a Zariski-connected real algebraic group, $G = \mathbf{G}(\mathbb{R})$, $G_{\mathbb{C}} = \mathbf{G}(\mathbb{C})$, and let $\varphi\colon Spin(n, 1) \to G$ be a homomorphism (we refer to Lemma A.7 in Appendix A for equivalent definitions in different categories). Suppose that Γ is a torsion-free, cocompact discrete subgroup of $Spin(n, 1)$.

In this section, as in the previous one, we set aside the space $X = G/H$ and focus on the deformation of $\varphi|_{\Gamma} \in \mathrm{Hom}(\Gamma, G)$. We construct a pair (Γ, φ'), where Γ is a torsion-free, cocompact discrete subgroup and φ' is a small deformation of $\varphi|_{\Gamma}$, such that the pair achieves the maximal Zariski-closure of $\varphi'(\Gamma)$ up to G-conjugacy, when $n \geq 3$. The argument in this section completes the proofs of the statements postponed in Sect. 3.2, including Lemma 3.6 and Theorem 3.9(1).

Lemma 3.6 was necessary for the definition of the maximal Zariski-closure G^{φ} in Definition 3.7 and is proven in Sect. 4.1.

The proof of Theorem 3.9(1) consists of three steps. In Sect. 4.2, we explain the role of each step. The details of each step are provided in Sects. 4.3, 4.4, and 4.5, where we also complete the proof of the theorem.

4.1 Proof of Lemma 3.6

In this section, we prove Lemma 3.6.

Let $\mathbf{G} = \mathbf{S} \cdot \mathbf{U}$ be a Levi decomposition, where $S = \mathbf{S}(\mathbb{R})$ is a maximal real reductive algebraic subgroup containing $\varphi(Spin(n,1))$ and $\mathbf{U}$ is the unipotent radical of $\mathbf{G}$. Then the real Lie algebra $\mathfrak{s}^{\varphi}$ is defined in a manner similar to $\mathfrak{g}^{\varphi}$ in Definition 3.5, using $\varphi\colon Spin(n,1) \to S$. Lemma 3.6 asserts that the analytic subgroup $G^{\varphi}_{\mathbb{C}}$ of $G_{\mathbb{C}}$ corresponding to $\mathfrak{g}^{\varphi} \otimes_{\mathbb{R}} \mathbb{C}$ is Zariski-closed in $G_{\mathbb{C}}$ and is defined over $\mathbb{R}$.

First, let us show that the analytic subgroup $S^{\varphi}_{\mathbb{C}}$ of $S_{\mathbb{C}} = \mathbf{S}(\mathbb{C})$ corresponding to $\mathfrak{s}^{\varphi} \otimes_{\mathbb{R}} \mathbb{C}$ is Zariski-closed. Take a Cartan involution θ of $\mathbf{S}(\mathbb{R})$ which preserves $\varphi(Spin(n,1))$. By definition, θ also preserves $\mathfrak{s}^{\varphi}$, and thus we see that the Lie algebra $\mathfrak{s}^{\varphi}$ is reductive. Let $(S^{\varphi}_{\mathbb{C}})_{ss}$ be the analytic subgroup of $\mathbf{S}(\mathbb{C})$ corresponding to the semisimple part of $\mathfrak{s}^{\varphi} \otimes_{\mathbb{R}} \mathbb{C}$. Then, by Lemma A.2(1) in Appendix A, we see that $(S^{\varphi}_{\mathbb{C}})_{ss}$ is Zariski-closed. Furthermore, let us consider the identity component $\mathbf{Z}$ of the following real algebraic group in the Zariski-topology:

$$\{g \in \mathbf{S} \mid g \text{ centralizes } d\varphi(\mathfrak{spin}(n,1)) \text{ and all the submodules of } \mathfrak{s} \text{ isomorphic to some spherical harmonics of } \mathfrak{spin}(n,1)\}.$$

By the definition of $\mathfrak{s}^{\varphi}$, we see that the Lie algebra of $\mathbf{Z}(\mathbb{C})$ coincides with the center of $\mathfrak{s}^{\varphi} \otimes_{\mathbb{R}} \mathbb{C}$. Since $\mathbf{Z}(\mathbb{C})$ is also connected in the usual topology by Lemma A.1, $\mathbf{Z}(\mathbb{C})$ is the analytic subgroup corresponding to the center of $\mathfrak{s}^{\varphi} \otimes_{\mathbb{R}} \mathbb{C}$. Hence, we have $S^{\varphi}_{\mathbb{C}} = (S^{\varphi}_{\mathbb{C}})_{ss} \cdot \mathbf{Z}(\mathbb{C})$, and thus $S^{\varphi}_{\mathbb{C}}$ is Zariski-closed.

Next, let us show our assertion. Denote by $\mathfrak{u}$ the nilpotent radical of $\mathfrak{g}$. Since $\mathfrak{u}$ is stable under the adjoint action of $d\varphi(\mathfrak{spin}(n,1))$, we get a decomposition

$$\mathfrak{g}^{\varphi} = \mathfrak{s}^{\varphi} + \mathfrak{u}^{\varphi},$$

where $\mathfrak{u}^{\varphi}$ is a Lie subalgebra of $\mathfrak{u}$. Let $U^{\varphi}_{\mathbb{C}}$ be the analytic subgroup of $\mathbf{U}(\mathbb{C})$ corresponding to $\mathfrak{u}^{\varphi} \otimes_{\mathbb{R}} \mathbb{C}$. Here we note that the exponential map $\exp\colon \mathfrak{u} \otimes_{\mathbb{R}} \mathbb{C} \to \mathbf{U}(\mathbb{C})$ gives an isomorphism of algebraic varieties. Hence, $U^{\varphi}_{\mathbb{C}} = \exp(\mathfrak{u}^{\varphi} \otimes_{\mathbb{R}} \mathbb{C})$ is Zariski-closed, and thus $G^{\varphi}_{\mathbb{C}} = S^{\varphi}_{\mathbb{C}} \cdot U^{\varphi}_{\mathbb{C}}$ is also Zariski-closed. Since $G^{\varphi}_{\mathbb{C}}$ is connected in the usual topology and its Lie algebra is defined over $\mathbb{R}$, $G^{\varphi}_{\mathbb{C}}$ is stable under the complex conjugation of $G_{\mathbb{C}}$. Thus the Zariski-closed subset $G^{\varphi}_{\mathbb{C}}$ is defined over $\mathbb{R}$. Thus Lemma 3.6 is proved.

4.2 Outline of the Proof of Theorem 3.9(1)

This section provides an outline of Theorem 3.9(1). We recall the setting: G is a Zariski-connected real algebraic group, $L = Spin(n, 1)$, $\varphi: L \to G$ is a homomorphism, and G^{φ} is the algebraic group introduced in Definition 3.7. What we need to do is to find a torsion-free cocompact discrete subgroup Γ and a small deformation of $\varphi|_{\Gamma}$ such that the Zariski-closure of $\varphi'(\Gamma)$ coincides with G^{φ}.

The proof of Theorem 3.9(1) consists of the following three steps:

Step 1 Construction of a cocompact discrete subgroup Γ of $Spin(n, 1)$;
Step 2 An overview of Johnson–Millson's bending deformation;
Step 3 Finding the Zariski-closure of a specific deformation of $\varphi|_{\Gamma}$.

First, we will outline the summary of each step, and then proceed with the proofs of each step in the following sections.

Step 1 (Construction of Γ). Let $L = Spin(n, 1)$ $(n \geq 2)$, $L' = Spin(n-1, 1)$ a subgroup of L, $L_K = Spin(n)$ a maximal compact subgroup of L, and $L'_K = L' \cap L_K \simeq Spin(n-1)$. Then $X' = L'/L'_K$ is a totally geodesic hypersurface of the n-dimensional hyperbolic space $X = L/L_K$. For a torsion-free cocompact discrete subgroup Γ of L, the quotient space $\Gamma\backslash X$ is an orientable connected compact hyperbolic n-manifold.

We consider the following condition for Γ:

Condition 4.1 Let k be a positive integer. There exist k orientable connected totally geodesic closed hypersurfaces $N_1, \ldots, N_k$ of $M := \Gamma\backslash X$ such that

- $N_i \cap N_j = \emptyset$ for any $i \neq j$.
- $M \smallsetminus (N_1 \cup \cdots \cup N_k)$ is connected.

Step 1 is to prove the following:

Theorem 4.2 *For any positive integer k, there exists a torsion-free cocompact arithmetic subgroup $\Gamma \equiv \Gamma_k$ of $L = Spin(n, 1)$ satisfying Condition 4.1.*

Example 4.3 When $n = 2$, Theorem 4.2 clearly holds. In fact, let $M = \Sigma_k$ be an orientable compact hyperbolic Riemann surface of genus $k \geq 2$, and we express $\pi_1(\Sigma_k)$ in terms of generators and relations as follows:

$$\pi_1(\Sigma_k) = \langle a_1, b_1, \ldots, a_k, b_k \mid [a_1, b_1] \cdots [a_k, b_k] = 1 \rangle.$$

Let N_i be a simple closed geodesic representing the free homotopy class of a_i for each $i = 1, \ldots, k$. Since the holonomy representation $\pi_1(\Sigma_k) \to PSL(2, \mathbb{R})$ can be lifted to $SL(2, \mathbb{R}) \simeq Spin(2, 1)$, Theorem 4.2 holds when $n = 2$.

We shall prove Theorem 4.2 in Sect. 4.3. When $n \geq 3$, we will actually construct such a group (see Example 4.17).

Theorem 4.2 was originally proved by Millson [56] where he treated the case where $L = SO(n, 1)$.

Remark 4.4 (1) Theorem 4.2 in the case $L = Spin(n, 1)$ implies an analogous result in the case where $L = SO(n, 1)$.

(2) As seen in Remark 1.5(3), the converse implication holds for $n = 2, 3$, but fails for $n \geq 4$.

We shall need a torsion-free cocompact discrete subgroup Γ in $L = Spin(n, 1)$ rather than in $L = SO(n, 1)$, as stated in Theorem 4.2. As we will see in Sect. 5.5, for a certain class of homogeneous spaces $X = G/H$, any non-abelian and standard discontinuous groups are virtually contained in $Spin(n, 1)$ but not in $SO(n, 1)$ for some n. For example, deformations of compact standard quotients of $SO(4, 4)/SO(3, 4)$ are obtained via $Spin(4, 1)$ (see Sect. 5.2), and those of non-compact standard quotients of $SO(8, 8)/SO(7, 8)$ via $Spin(6, 1)$ (see Sect. 5.5).

Step 2 (An overview of bending construction). In Step 2, based on the geometric idea of bending construction by Johnson–Millson [21], we reformulate a general principle of the construction of small deformation, that we shall use (Lemma 4.7).

From now on, fix $k \in \mathbb{N}$. By Theorem 4.2, we take a torsion-free cocompact discrete subgroup $\Gamma \equiv \Gamma_k$ of $Spin(n, 1)$ satisfying Condition 4.1 for $k \in \mathbb{N}$. Let $X = Spin(n, 1)/Spin(n)$, $M = \Gamma \backslash X$ and $N_1, \dots, N_k$ as in Condition 4.1.

For each $i = 1, \dots, k$, by the tubular neighborhood theorem, we choose an open neighborhood $\tilde{N}_i^{(4)}$ of N_i in M and a diffeomorphism

$$f_i \colon N_i \times (-4, 4) \simeq \tilde{N}_i^{(4)}$$

such that $f_i(N_i \times \{0\})$ coincides with N_i. We define

$$\tilde{N}_i := f_i(N \times (-2, 2)),$$

fix $y_i \subset N_i$, and put

$$y_{i,+} := f_i(y_i, 1) \text{ and } y_{i,-} := f_i(y_i, -1).$$

Since $N_1, \dots, N_k$ are disjoint and since

$$S := M \setminus (N_1 \cup \dots \cup N_k)$$

is path-connected, we may and do assume

- $\tilde{N}_1, \dots, \tilde{N}_k$ are disjoint;
- $M \setminus (\tilde{N}_1^{(4)} \cup \dots \cup \tilde{N}_k^{(4)})$ is path-connected.

Definition 4.5 Fix a base point $x_0 \in M \setminus (\tilde{N}_1^{(4)} \cup \dots \cup \tilde{N}_k^{(4)})$. For each $i = 1, \dots, k$, we define the oriented loop ν_i in M, starting at x_0, as the composition of the following paths:

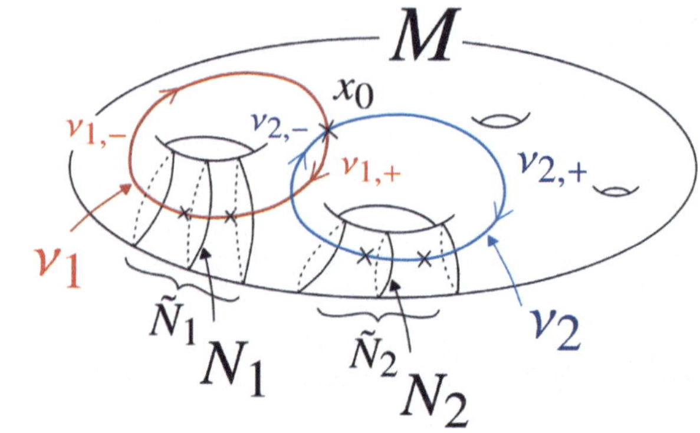

Fig. 1 The loops ν_1 and ν_2 in the case $k = 2$

- a path from x_0 to $f_i(y_i, 3)$ inside $M \setminus f_i(N_i \times (-3, 3))$;
- the path from $f_i(y_i, 3)$ to $f_i(y_i, -3)$, given by $f_i(\{y_i\} \times [-3, 3])$;
- a path from $f_i(y_i, -3)$ to x_0 inside $M \setminus f_i(N_i \times (-3, 3))$.

Furthermore, we can take the loop ν_i to be a closed *submanifold* of M. Let $\nu_{i,+}$ denote the segment of the loop ν from x_0 to $y_{i,+}$, and let $\nu_{i,-}$ denote the segment of ν from $y_{i,-}$ back to x_0.

Figure 1 summarizes some of the notation introduced so far.

Let $i = 1, \ldots, k$. For an oriented loop ℓ starting at y_i in N_i, $f_i(\ell, 1)$ and $f_i(\ell, -1)$ form oriented loops in M starting at $y_{i,+}$ and $y_{i,-}$, respectively. With the above paths $\nu_{i,+}$ and $\nu_{i,-}$, we define $j_{i,+}, j_{i,-} : \pi_1(N_i, y_i) \to \pi_1(S, x_0)$ as two group homomorphisms

$$j_{i,+}([\ell]) := [\nu_{i,+}^{-1} \circ f_i(\ell, 1) \circ \nu_{i,+}], \qquad (4.1)$$
$$j_{i,-}([\ell]) := [\nu_{i,-} \circ f_i(\ell, -1) \circ \nu_{i,-}^{-1}].$$

Here, b^{-1} denotes the path obtained by reversing the orientation of the path b, and $[c]$ denotes the homotopy class defined by the loop c.

For the deformation of a representation of Γ, we require the following two lemmas. As discussed later, via deck transformations, the discontinuous group Γ for the hyperbolic space X can be identified with the fundamental group $\pi_1(M)$ of $M = \Gamma\backslash X$. Lemma 4.6 states that the group structure of $\pi_1(M)$ is obtained as an iterated HNN extension of $\pi_1(S)$. Lemma 4.7 utilizes this group structure to explicitly construct a small deformation of a representation of $\Gamma \simeq \pi_1(M, x_0)$ into a Lie group. For the proofs, see Sect. 4.4.

Lemma 4.6 *Let F_k be the free group generated by the words $a_1, \ldots, a_k$. We define a homomorphism Ψ from the free product $\pi_1(S, x_0) * F_k$ to $\pi_1(M, x_0)$,*

$$\Psi : \pi_1(S, x_0) * F_k \to \pi_1(M, x_0), \qquad (4.2)$$

as the homomorphism induced by the natural map $\pi_1(S, x_0) \to \pi_1(M, x_0)$ *and by* $\Psi(a_i) = [\nu_i]$ *for each* $i = 1, \ldots, k$. *Then* Ψ *is surjective and its kernel is the normal subgroup* $\mathcal{N}$ *generated by*

$$a_i j_{i,+}([\ell])a_i^{-1} j_{i,-}([\ell])^{-1} \quad \textit{for } [\ell] \in \pi_1(N_i, y_i) \textit{ and } i = 1, \ldots, k.$$

In order to state the second lemma, we introduce some notation. Let $L' := Spin(n-1, 1)$ be regarded as a subgroup of $L := Spin(n, 1)$, and let $X' := Spin(n-1, 1)/Spin(n-1)$ be viewed as a totally geodesic hypersurface of the hyperbolic space $X = Spin(n, 1)/Spin(n)$. Recall that $N_1, \ldots, N_k$ are orientable, connected, totally geodesic hypersurfaces of the hyperbolic manifold $M = \Gamma\backslash X$. Hence, for each $i = 1, \ldots, k$, there exist an element $\alpha_i \in L$ and a diffeomorphism $N_i \simeq (\alpha_i L' \alpha_i^{-1} \cap \Gamma)\backslash\alpha_i X'$ such that the following diagram commutes:

$$\begin{array}{ccc} N_i & \xhookrightarrow{\text{inclusion}} & M \\ \downarrow\simeq & \circlearrowleft & \downarrow = \\ (\alpha_i L'\alpha_i^{-1} \cap \Gamma)\backslash\alpha_i X' & \xrightarrow{\text{natural}} & \Gamma\backslash X. \end{array} \tag{4.3}$$

Now we recall how the discontinuous group Γ for X is identified with the fundamental group $\pi_1(M, x_0)$ of $M = \Gamma\backslash X$. Since the hyperbolic space X is simply-connected, the quotient map $\pi_\Gamma : X \to \Gamma\backslash X = M$ is a universal covering. We fix a point $\tilde{x}_0 \in X$ in the discrete fiber $X_{x_0} := \pi_\Gamma^{-1}(x_0)$ of $x_0 \in M$. Then, we get a group isomorphism $D^M_{\tilde{x}_0} : \pi_1(M, x_0) \to \Gamma$, by the following relation of the deck transformation:

$$[\ell] \cdot \tilde{x}_0 = D^M_{\tilde{x}_0}([\ell]) \cdot \tilde{x}_0 \quad \text{for } [\ell] \in \pi_1(M, x_0). \tag{4.4}$$

Lemma 4.7 *Let* G *be a Lie group,* $\varphi \colon \Gamma \to G$ *a group homomorphism,* $L' = Spin(n-1, 1)$, *and* $\alpha_1, \ldots, \alpha_k \in L = Spin(n, 1)$ *as above. For each* $i = 1, \ldots, k$, *we take a possibly zero element* v_i *of* $\mathfrak{g}$ *such that*

$$v_i \textit{ is } \varphi(\alpha_i L' \alpha_i^{-1} \cap \Gamma)\textit{-invariant}. \tag{4.5}$$

Retain the notation as in Lemma 4.6. For $t \in \mathbb{R}$, *we put*

(a) $\varphi_t \equiv \varphi$ *on* $D^M_{\tilde{x}_0} \circ \Psi(\pi_1(S, x_0))$;
(b) $\varphi_t(D^M_{\tilde{x}_0}([\nu_i])) := \varphi(D^M_{\tilde{x}_0}([\nu_i])) \exp(t v_i)$ *for each* $i = 1, \ldots, k$.

Then φ_t *induces a group homomorphism* $\varphi_t \colon \Gamma \to G$.

Step 3 (Zariski-closure of specific deformation). Let G be a Zariski-connected real algebraic group, $L := Spin(n, 1)$ and $\varphi \colon L \to G$ a homomorphism. In Step 3, we construct a small deformation φ' of $\varphi|_\Gamma$ such that the Zariski-closure of $\varphi'(\Gamma)$ coincides with G^φ (Proposition 4.9), where Γ is a certain cocompact discrete subgroup of L. This is achieved by finding appropriate vectors $v_i \in \mathfrak{g}$ $(1 \le i \le k)$ in the setting of Lemma 4.7, considering the full generality of G and $\varphi\colon L \to G$. We shall also optimize the number k, which corresponds to the number of totally geodesic hypersurfaces in Condition 4.1 that we use for bending constructions.

Remark 4.8 When $\mathfrak{g}^{\varphi} = \mathfrak{g}$, the conclusion of Proposition 4.9 below implies that Γ can be deformed into a Zariski-dense subgroup in G. The previous results for the case $L = SO(n, 1)$, such as Johnson–Millson [21] (φ maps into $G = SO(n+1, 1)$ or $PSL(n+1, \mathbb{R})$), and Kassel [27] (φ maps into $SO(n, 2)$), are (implicitly) based on the following structure in their proofs: *the* $\mathfrak{l}$*-module* $\mathfrak{g}/\mathfrak{l}$ *is an irreducible spherical harmonics module.* Along the same lines of argument, this assumption can be relaxed to the existence of an increasing sequence of reductive Lie algebras

$$\mathfrak{l} = \mathfrak{l}^0 \subset \mathfrak{l}^1 \subset \cdots \subset \mathfrak{l}^m = \mathfrak{g}$$

such that $\mathfrak{l}^{k+1}/\mathfrak{l}^k$ is an irreducible $\mathfrak{l}^k$-module and that it contains a non-zero spherical harmonics representation of $\mathfrak{l}$. For example, this applies to the case $G = SO(p, q)$ $(p \geq n,\ q \geq 1)$, as discussed in a recent preprint [4, Appendix A] by Beyrer and Kassel.

However, in our general setting of φ and G, we cannot rely on such a restrictive structure for $\mathfrak{l}$ and $\mathfrak{g}$. For instance, in the context of Theorem 3.18, this is not the case for the natural homomorphism $\varphi\colon Spin(n, 1) \to G(p, q)$ when p and q take arbitrary values.

To construct a desired small deformation, we first introduce some notation.

Let V_i denote the irreducible spherical harmonics representation of $L := Spin(n, 1)$ of degree i (see Remark 3.2). Given a homomorphism $\varphi\colon L \to G$, we denote by $Z_G(\varphi(L))$ the identity component (in the Zariski topology) of the centralizer of $\varphi(L)$ in G. For $i \in \mathbb{N}$, let $\mathfrak{g}(V_i)$ denote the isotypic component of V_i in $\mathfrak{g}$. Then $\mathfrak{g}(V_0) = \mathfrak{z}_{\mathfrak{g}}(d\varphi(\mathfrak{l}))$, the Lie algebra of $Z_G(\varphi(L))$.

We consider a decomposition of $\mathfrak{g}$ as an L-module via $\varphi \circ \mathrm{Ad}$:

$$\mathfrak{g} = d\varphi(\mathfrak{l}) \oplus \mathfrak{z}_{\mathfrak{g}}(d\varphi(\mathfrak{l})) \oplus \bigoplus_{i=1}^{\infty} \mathfrak{g}(V_i) \oplus (\text{other representations}). \tag{4.6}$$

By Definition 3.5, $\mathfrak{g}^{\varphi}$ is the Lie algebra generated by

$$d\varphi(\mathfrak{l}) \oplus \mathfrak{z}_{\mathfrak{g}}(d\varphi(\mathfrak{l})) \oplus \bigoplus_{i=1}^{\infty} \mathfrak{g}(V_i).$$

Needless to say, the summation is finite because $\dim \mathfrak{g} < \infty$. The multiplicity of the irreducible L-module V_i in $\mathfrak{g}$ is denoted by

$$[\mathfrak{g} : V_i] := \dim_{\mathbb{R}} \mathrm{Hom}_L(V_i, \mathfrak{g}).$$

Then, we may decompose

$$\mathfrak{g}(V_i) = \bigoplus_{j=1}^{[\mathfrak{g}:V_i]} V_i^{(j)},$$

where each $V_i^{(j)}$ is isomorphic to V_i.

We set $m := \max_{i \in \mathbb{N}_+} [\mathfrak{g} : V_i]$. We define k, for example, by

$$k := \max(\eta(Z_G(\varphi(L))), m) \in \mathbb{N}, \tag{4.7}$$

with the notation as defined in Definition F.3 of Appendix F. We note that $\eta(Z_G(L)) \leq 2$ if G is reductive by Theorem F.5.

We take Γ satisfying Condition 4.1 for k by Theorem 4.2. Let $\alpha_1, \ldots, \alpha_k$ be elements of $L = Spin(n, 1)$ as in (4.3).

We recall $L' = Spin(n-1, 1)$. For each $i \in \mathbb{N}_+$, we take $v_i^{(1)}, \ldots, v_i^{(k)}$ as follows: $v_i^{(j)}$ is a non-zero $\varphi(\alpha_i L' \alpha_i^{-1})$-invariant element in $V_i^{(j)}$ for $1 \leq j \leq [\mathfrak{g} : V_i]$ and $v_i^{(j)} := 0$ for $[\mathfrak{g} : V_i] + 1 \leq j \leq k$. We note that $v_i^{(j)} = 0$ for $i \gg 0$.

We take $u_1, \ldots, u_k \in \mathfrak{z}_{\mathfrak{g}}(\mathfrak{l})$ as follows: With the notation as defined in Definition F.1, $G(t\{u_1, \ldots, u_{\eta(Z_G(\varphi(L)))}\}) = Z_G(L)$ and $u_j := 0$ if $\eta(Z_G(L)) < j \leq k$.

We now define $v_j \in \mathfrak{g}$, for $1 \leq j \leq k$, by a finite sum:

$$v_j := u_j + \sum_{i=1}^{\infty} v_i^{(j)}. \tag{4.8}$$

Proposition 4.9 *Let G be a Zariski-connected, real algebraic group, and let $\varphi \colon Spin(n, 1) \to G$ a homomorphism. Let $\varphi_t \colon \Gamma \to G$ be the homomorphism associated with $v_1, \ldots, v_k$, as defined in Lemma 4.7. Then, the Zariski-closure of $\varphi_t(\Gamma)$ coincides with G^{φ} for any sufficiently small real number $t \neq 0$.*

This proposition will be proved in Sect. 4.5, and thus the proof of Theorem 3.9(1) is complete.

4.3 Proof of Theorem 4.2 (Step 1)

In this section, we prove a $Spin(n, 1)$ analogue of Millson's theorem [56] for $SO(n, 1)$ by reformulating his argument [56] in terms of Clifford algebras (Theorem 4.2). The part of the proof in [56] that relies on the strong approximation theorem becomes slightly simpler when working with $Spin(n, 1)$ instead of $SO(n, 1)$ since the algebraic group $\mathbf{Spin}_{n,1}$ is simply-connected (Proposition 4.15).

To prove Theorem 4.2, we introduce some notation. Let $\mathbb{F}$ be a totally real number field of degree $r > 1$, $\{\sigma_1, \ldots, \sigma_r\}$ the set of embeddings of the field $\mathbb{F}$ into $\mathbb{R}$, $\mathcal{O}$ the ring of integers of $\mathbb{F}$, V an $(n + 1)$-dimensional $\mathbb{F}$-vector space equipped with a quadratic form $Q = Q(v)$ on V, and $C_{\text{even}}(V)$ the even Clifford algebra associated to (V, Q). We refer to Appendix B for the notation related to Clifford algebras and spin groups.

Take an orthogonal $\mathbb{F}$-basis $\{e_0, \dots, e_n\}$ of V with respect to Q, such that $Q(e_i) \in \mathcal{O}$ for any $i = 0, \dots, n$. Let Λ be the $\mathcal{O}$-submodule of V generated by $e_0, \dots, e_n$, and let $C_{\text{even}}(\Lambda)$ be the $\mathcal{O}$-subalgebra of $C_{\text{even}}(V)$ generated by $e_{i_1} \cdots e_{i_\ell}$ for any even number ℓ and any $0 \leq i_1 < \cdots < i_\ell \leq n$. We set

$$\Gamma_\Lambda := C_{\text{even}}(\Lambda) \cap Spin(V). \tag{4.9}$$

For an ideal I of $\mathcal{O}$, we also consider its congruence subgroup

$$\Gamma_\Lambda(I) := \{\gamma \in \Gamma_\Lambda \mid \gamma - 1 \in I C_{\text{even}}(\Lambda)\}. \tag{4.10}$$

This is a finite-index subgroup of Γ_Λ. We will prove that $\Gamma_\Lambda(I)$ satisfies Condition 4.1 under certain assumptions (Proposition 4.16).

First, we verify that Γ_Λ satisfies the following properties:

- (Lemma 4.10) Γ_Λ can be regarded as an $\mathcal{O}$-arithmetic group of the $\mathbb{F}$-group $\mathbf{Spin}_V$ (see, e.g., [59, p. 227]);
- (Proposition 4.11) Under certain conditions, Γ_Λ can be regarded as a cocompact discrete subgroup of $Spin(n, 1)$.

See (B.2) in Appendix B for the definition of the $\mathbb{F}$-group $\mathbf{Spin}_V$.

Using the $\mathbb{F}$-basis of $C_{\text{even}}(V)$

$$\{e_{i_1} \cdots e_{i_\ell} \mid \ell \text{ is even and } 0 \leq i_1 < \ldots < i_\ell \leq n\}, \tag{4.11}$$

we identify the $\mathbb{F}$-vector space $C_{\text{even}}(V)$ with $\mathbb{F}^{2^n}$. For any extension field $\mathbb{E}$ of $\mathbb{F}$, we identify the $\mathbb{E}$-algebra $C_{\text{even}}(V) \otimes_{\mathbb{F}} \mathbb{E}$ naturally with the Clifford algebra $C_{\text{even}}(V \otimes_{\mathbb{F}} \mathbb{E})$. For each $x \in C_{\text{even}}(V) \otimes_{\mathbb{F}} \mathbb{E}$, consider the left multiplication map by x. This yields an injective $\mathbb{E}$-algebra homomorphism

$$\iota_{\mathbb{E}} \colon C_{\text{even}}(V) \otimes_{\mathbb{F}} \mathbb{E} \to \operatorname{End}_{\mathbb{E}}(C_{\text{even}}(V) \otimes_{\mathbb{F}} \mathbb{E}) \simeq M(2^n, \mathbb{E}).$$

Using this embedding, we regard $\mathbf{Spin}_V$ as an $\mathbb{F}$-algebraic group and define the group of $\mathcal{O}$-points by

$$\mathbf{Spin}_V(\mathcal{O}) := \mathbf{Spin}_V(\mathbb{E}) \cap \iota_{\mathbb{E}}^{-1}(GL(2^n, \mathcal{O})).$$

This definition does not depend on the choice of the extension field $\mathbb{E}$.

Lemma 4.10 *The group Γ_Λ coincides with $\mathbf{Spin}_V(\mathcal{O})$. In particular, Γ_Λ is an $\mathcal{O}$-arithmetic subgroup of $\mathbf{Spin}_V$.*

Proof Note that (4.11) forms an $\mathcal{O}$-basis of $C_{\text{even}}(\Lambda)$. We have

$$\begin{aligned}
\mathbf{Spin}_V(\mathcal{O}) &= \mathbf{Spin}_V(\mathbb{F}) \cap \iota_{\mathbb{F}}^{-1}(GL(2^n, \mathcal{O})) \\
&= Spin(V) \cap \iota_{\mathbb{F}}^{-1}(\{x \in GL(2^n, \mathbb{F}) \mid x\mathcal{O}^{2^n} = \mathcal{O}^{2^n}\}) \\
&= \{x \in Spin(V) \mid xC_{\text{even}}(\Lambda) = C_{\text{even}}(\Lambda)\} \\
&= Spin(V) \cap C_{\text{even}}(\Lambda) \\
&= \Gamma_\Lambda .
\end{aligned}$$

Hence, the assertion is proved. □

Proposition 4.11 *For each $i = 1, \ldots, r$, let $\mathbb{F}_{\sigma_i}$ denote the field $\mathbb{R}$ regarded as an extension field of $\mathbb{F}$ via the embedding $\sigma_i : \mathbb{F} \to \mathbb{R}$. Assume the following:*

- *$\sigma_1(Q(e_i)) > 0$ for any $i = 0, \ldots, n-1$ and $\sigma_1(Q(e_n)) < 0$;*
- *$\sigma_j(Q(e_i)) > 0$ for any $i = 0, \ldots, n$ and $j = 2, \ldots, r$;*

Then we have

$$\mathbf{Spin}_V(\mathbb{F}_{\sigma_i}) \simeq \begin{cases} Spin(n, 1) & (i = 1), \\ Spin(n+1) & (i > 1). \end{cases}$$

Furthermore, the inclusion map

$$\mathbf{Spin}_V(\mathcal{O}) = \Gamma_\Lambda \to \mathbf{Spin}_V(\mathbb{F}_{\sigma_1}) \simeq Spin(n, 1)$$

is discrete and cocompact.

Proof The first statement follows immediately from the assumption on the quadratic form Q. Thus we focus on proving the second statement. Since $r > 1$, there exists an inclusion

$$\mathbf{Spin}_V(\mathbb{F}) = Spin(V) \to \mathbf{Spin}_V(\mathbb{F}_{\sigma_2}) \simeq Spin(n+1),$$

which implies that $Spin(V)$ has no non-trivial unipotent elements. Hence, $\mathbf{Spin}_V$ is anisotropic over $\mathbb{F}$. By a well-known criterion (e.g., [59, Thm. 4.17 (3)]), the diagonal embedding

$$\mathbf{Spin}_V(\mathcal{O}) = \Gamma_\Lambda \to \prod_{i=1}^{r} \mathbf{Spin}_V(\mathbb{F}_{\sigma_i}) \simeq Spin(n, 1) \times Spin(n+1)^{r-1}$$

is discrete and cocompact. Since $Spin(n+1)^{r-1}$ is compact, the second statement follows. □

Next, let I be an ideal of $\mathcal{O}$. We examine properties of $\Gamma_\Lambda(I)$. By an argument similar to the proof of Lemma 4.10, we obtain the identity

$$\Gamma_\Lambda(I) = \Gamma_\Lambda \cap \iota_{\mathbb{F}}^{-1}(\{g \in GL(2^n, \mathcal{O}) \mid g \equiv 1 \mod I\}). \tag{4.12}$$

Using Lemma C.1 in Appendix C, we have

Lemma 4.12 *Let P, Q be prime ideals of $\mathcal{O}$ such that $P \cap \mathbb{Z} \neq Q \cap \mathbb{Z}$. Then $\Gamma_\Lambda(PQ)$ is torsion-free.*

To prove Theorem 4.2, we analyze the finite group $\Gamma_\Lambda / \Gamma_\Lambda(I)$ for some ideal I (Proposition 4.15). To state the proposition, we first review some basic concepts of algebraic groups over number fields needed in our discussion. For further details, see, for example, [59].

For a maximal ideal P of the ring of integers $\mathcal{O}$, the completion of the global field $\mathbb{F}$ with respect to the P-adic topology defines a non-Archimedean local field $\mathbb{F}_P$. Let $\mathcal{O}_P$ denote the ring of integers of $\mathbb{F}_P$.

In the same manner as the definition of $\mathbf{Spin}_V(\mathcal{O})$, we define the $\mathcal{O}_P$-points $\mathbf{Spin}_V(\mathcal{O}_P)$ as

$$\mathbf{Spin}_V(\mathcal{O}_P) := \mathbf{Spin}_V(\mathbb{F}_P) \cap \iota_{\mathbb{F}_P}^{-1}(GL(2^n, \mathcal{O}_P)),$$

via the left regular representation $\iota_{\mathbb{F}_P} \colon C_{\mathrm{even}}(V \otimes_{\mathbb{F}} \mathbb{F}_P) \to M(2^n, \mathbb{F}_P)$ using the $\mathbb{F}$-basis from (4.11).

Furthermore, assume that the maximal ideal P does not contain $2Q(e_0) \cdots Q(e_n)$. In this case, the residue field $k_P := \mathcal{O}/P$ has characteristic greater than 2. We define a k_P-vector space by

$$\overline{\Lambda}_P := \Lambda / P\Lambda,$$

and denote by $\overline{Q}_P$ the quadratic form on $\overline{\Lambda}_P$ induced by Q. Thus the Clifford algebra $C_{\mathrm{even}}(\overline{\Lambda}_P)$ and the spin group $Spin(\overline{\Lambda}_P)$ are defined. By definition, it follows immediately that $Spin(\overline{\Lambda}_P)$ provides a *reduction* of $\mathbf{Spin}_V$ modulo P in the sense of [59, p. 142]. Hence, by [59, Prop. 3.20], we obtain the following:

Lemma 4.13 *There exists a finite set of maximal ideals E_V of $\mathcal{O}$, depending only on (V, Q, Λ) such that every maximal ideal $P \notin E_V$ does not contain $2Q(e_0) \cdots Q(e_n)$ and that the reduction map*

$$\mathbf{Spin}_V(\mathcal{O}_P) \to Spin(\overline{\Lambda}_P)$$

is surjective.

Remark 4.14 For the orthogonal group $\mathbf{SO}_V$, the reduction map

$$\mathbf{SO}_V(\mathcal{O}_P) \to SO(\overline{\Lambda}_P)$$

is surjective for every maximal ideal P that does not contain 2 or the discriminant $Q(e_0)\cdots Q(e_n)$. Millson [56] used this fact for the orthogonal group $\mathbf{SO}_V$. However, the authors do not know whether a similar statement holds for the spin group $\mathbf{Spin}_V$, as it is unclear whether it is possible to take E_V to be the set of all maximal ideals P that do not contain $2Q(e_0)\cdots Q(e_n)$ in Lemma 4.13. This statement would strengthen Example 4.17 below by removing the assumption of "sufficiently large" from it.

We are now ready to state the proposition on $\Gamma_\Lambda/\Gamma_\Lambda(I)$:

Proposition 4.15 *Suppose that* $\dim_{\mathbb{F}} V \geq 3$ *(recall that* $\dim_{\mathbb{F}} V = n+1$*) and that all the assumptions in Proposition 4.11 hold. Let* E_V *be the finite set of maximal ideals in Lemma 4.13, and let* $P_1, \ldots, P_m$ *be maximal ideals of* $\mathcal{O}$ *that do not belong to* E_V*. Then, the reduction map induces an isomorphism*

$$\Gamma_\Lambda/\Gamma_\Lambda(P_1\cdots P_m) \simeq \prod_{i=1}^{m} Spin(\overline{\Lambda}_{P_i}).$$

Proof Since $\dim_{\mathbb{F}} V \geq 3$, the $\mathbb{F}$-group $\mathbf{Spin}_V$ is simply-connected. Under the assumptions in Proposition 4.11, $\mathbf{Spin}_V(\mathbb{F}_{\sigma_1}) \simeq Spin(n,1)$, which is a non-compact simple Lie group for every $n \geq 2$. Thus, the *strong approximation theorem* (Kneser [36], see [59, Thm. 7.12]) applies to $\mathbf{Spin}_V$. Although the following proof relies on the standard argument using the strong approximation theorem, we provide a proof for readers who may be unfamiliar with it.

Let us recall the statement of the strong approximation theorem in the form required for our setting. We define the finite adelization of $\mathbf{Spin}_V$ as

$$\mathbf{Spin}_V(\mathbb{A}_{\mathbb{F}}^f) := \bigcup_S \left(\prod_{P\in S} \mathbf{Spin}_V(\mathbb{F}_P) \times \prod_{P\notin S} \mathbf{Spin}_V(\mathcal{O}_P) \right),$$

where S ranges over all finite sets of maximal ideals of $\mathcal{O}$. Furthermore, we endow $\mathbf{Spin}_V(\mathbb{A}_{\mathbb{F}}^f)$ with the inductive limit topology arising from the direct system of locally compact totally disconnected groups

$$\prod_{P\in S} \mathbf{Spin}_V(\mathbb{F}_P) \times \prod_{P\notin S} \mathbf{Spin}_V(\mathcal{O}_P).$$

With this topology, $\mathbf{Spin}_V(\mathbb{A}_{\mathbb{F}}^f)$ becomes a locally compact Hausdorff topological group. Moreover, the diagonal morphism $\mathbf{Spin}_V(\mathbb{F}) \to \mathbf{Spin}_V(\mathbb{A}_{\mathbb{F}}^f)$ is well-defined, and thus we can regard $\mathbf{Spin}_V(\mathbb{F})$ as a subgroup of $\mathbf{Spin}_V(\mathbb{A}_{\mathbb{F}}^f)$. The strong approximation theorem asserts that $\mathbf{Spin}_V(\mathbb{F}) = Spin(V)$ is dense in $\mathbf{Spin}_V(\mathbb{A}_{\mathbb{F}}^f)$.

We now prove our assertion. Since the kernel of the reduction map

$$\Gamma_\Lambda \to \prod_{i=1}^{m} Spin(\overline{\Lambda}_{P_i})$$

is clearly $\Gamma_\Lambda(P_1 \cdots P_m)$, it suffices to prove that the reduction map is surjective.

For each $i = 1, \ldots, m$, take an element $\bar{g}_i \in Spin(\overline{\Lambda}_{P_i})$, and consider the open subset of $\mathbf{Spin}_V(\mathbb{A}_{\mathbb{F}}^f)$ given by

$$U = \prod_{i=1}^{m} \{g \in \mathbf{Spin}_V(\mathcal{O}_{P_i}) \mid g \bmod P_i = \bar{g}_i\} \times \prod_{P \neq P_i} \mathbf{Spin}_V(\mathcal{O}_P).$$

By Lemma 4.13, we see that U is non-empty. Since $Spin(V)$ is dense in $\mathbf{Spin}_V(\mathbb{A}_{\mathbb{F}}^f)$, we obtain $U \cap Spin(V) \neq \emptyset$.

Here note

$$\mathbf{Spin}_V(\mathcal{O}) = \prod_P \mathbf{Spin}_V(\mathcal{O}_P) \cap \mathbf{Spin}_V(\mathbb{F}) \quad \text{in} \quad \mathbf{Spin}_V(\mathbb{A}_{\mathbb{F}}^f).$$

Thus it follows that any element in $U \cap Spin(V)$ must belong to $\Gamma_\Lambda = \mathbf{Spin}_V(\mathcal{O})$. Hence, we conclude that $U \cap \Gamma_\Lambda \neq \emptyset$, which implies that the reduction map is surjective. This completes the proof. □

Finally, we prove that for a suitable ideal I, the $\mathcal{O}$-arithmetic subgroup $\Gamma_\Lambda(I)$ satisfies Condition 4.1 (Proposition 4.16). To state the proposition, we introduce some notation.

Recall that $e_0, \ldots, e_n$ form an $\mathbb{F}$-basis of V. Let V' be the $\mathbb{F}$-subspace of V generated by $e_1, \ldots, e_n$, and define $\Lambda' := V' \cap \Lambda$. As a result, the following are defined: the Clifford algebra $C_{\mathrm{even}}(V') \equiv C_{\mathrm{even}}(V', Q|_{V'})$, its $\mathcal{O}$-subalgebra $C_{\mathrm{even}}(\Lambda')$, the spin group $\mathbf{Spin}_{V'}$, the $\mathcal{O}$-arithmetic subgroup $\Gamma_{\Lambda'}$, and its congruence subgroup $\Gamma_{\Lambda'}(I)$ for an ideal I.

The following proposition is the main goal of this section. The proof is achieved by reformulating the argument in Millson [56, p. 245] regarding the construction of a family of compact hyperbolic manifolds with arbitrarily large first Betti numbers in terms of Clifford algebras.

Proposition 4.16 *Assume that* $\dim_{\mathbb{F}} V \geq 4$ *(recall that* $\dim_{\mathbb{F}} V = n + 1$*). Take* E_V *and* $E_{V'}$ *be finite sets of maximal ideals of* $\mathcal{O}$ *as given by applying Lemma 4.13 to* (V, Q, Λ) *and* $(V', Q|_{V'}, \Lambda')$*, respectively.*

For $k \in \mathbb{N}$*, choose an integer* $m \geq 2$ *such that* $2^m - 1 \geq k$*. Suppose that there exist* $m + 2$ *distinct prime ideals* $P_{-1}, P_0, \ldots, P_m$ *satisfying the following conditions:*

- $P_i \cap \mathbb{Z} \neq P_j \cap \mathbb{Z}$ *for any* $i \neq j$*;*
- *Each* P_i *belongs to neither* E_V *nor* $E_{V'}$*.*

Moreover, in addition to the assumptions in Proposition 4.11, suppose that $Q(e_0) \in \mathcal{O}^\times$. *Then, the torsion-free cocompact discrete subgroup* $\Gamma_\Lambda(P_{-1} \cdots P_m)$ *of* $Spin(n, 1)$ *satisfies Condition 4.1 for k.*

Example 4.17 Let $n \geq 3$ and $\mathbb{F} = \mathbb{Q}(\sqrt{2})$. We consider the following quadratic form with coefficients in $\mathcal{O} = \mathbb{Z}[\sqrt{2}]$:

$$Q(x) = x_0^2 + \cdots + x_{n-1}^2 - \sqrt{2}x_n^2.$$

We regard this as a quadratic form on $V = \mathbb{F}^{n+1}$ with standard basis $e_0, \ldots, e_n$, and let $\Lambda = \mathcal{O}^{n+1}$. The triple (V, Q, Λ) satisfies all the assumptions in Proposition 4.11 and $Q(e_0) = 1$ is invertible in $\mathcal{O}$.

Primes of the form $8a \pm 3$ exist infinitely and remain primes in $\mathcal{O}$. For $k \in \mathbb{N}$, take a natural number $m \geq 2$ such that $2^m - 1 \geq k$, and choose $m + 2$ such sufficiently large primes $p_{-1}, p_0, \ldots, p_m$. By Proposition 4.16, $\Gamma_\Lambda(p_{-1} \cdots p_m\mathcal{O})$ is a torsion-free cocompact discrete subgroup of $Spin(n, 1)$ satisfying Condition 4.1 for $k \in \mathbb{N}$.

Thus, once Proposition 4.16 is established, the proof of Theorem 4.2 for $n \geq 3$ is complete (see Example 4.3 for the proof in the case $n = 2$). The rest of this section is devoted to the proof of Proposition 4.16. For this purpose, we use two geometric lemmas from Millson [56] (Lemmas 4.18 and 4.19).

The proof of the following lemma is essentially the same as that of [56, Lem. 2.1] (referred to as Jaffe's lemma). In [56], the statement was proved under the assumptions that X is a hyperbolic manifold and T is a group of order 2. Since the results hold in a more general setting, we provide a proof in this broader context for future reference.

Lemma 4.18 *Let* $X := G/H$ *be a homogeneous space,* Γ *a discontinuous group for* X, *and* $T \subset \mathrm{Aut}(G)$ *a finite subgroup such that* H *and* Γ *are both stable under* T.

Put $G' := \{g \in G \mid t(g) = g \text{ for any } t \in T\}$, $H' := G' \cap H$, $\Gamma' = G' \cap \Gamma$, *and* $X' := G'/H'$. *Then the map* $\pi\colon \Gamma'\backslash X' \to \Gamma\backslash X$ *defined by* $\Gamma' g' H' \mapsto \Gamma g' H$ *is a diffeomorphism into a regular submanifold of* $\Gamma\backslash X$.

Proof Define the action of $t \in T$ on $gH \in X$ by $t(gH) = t(g)H$. Under this action, T fixes every element of X'.

Take any point $x \in X'$. Then we choose a neighborhood U of x in X such that, if $\gamma \in \Gamma$ satisfies $\gamma U \cap U \neq \emptyset$, then $\gamma = 1$. By replacing U with $\bigcap_{t \in T} t(U)$, we may assume that U is stable under the T-action.

We claim that if $\gamma \in \Gamma$ satisfies $\gamma U \cap X' \neq \emptyset$, then $\gamma \in \Gamma'$. Indeed, suppose there exists $y \in U$ such that $\gamma y \in X'$. Then, for any $t \in T$, we have $t(\gamma)t(y) = \gamma y$. Since $t(y) \in U$, it follows that $t(\gamma) = \gamma$. Hence, $\gamma \in \Gamma'$. This proves the claim. In particular, we also see that the map $\pi\colon \Gamma'\backslash X' \to \Gamma\backslash X$ is injective.

Let $\pi_\Gamma\colon X \to \Gamma\backslash X$ and $\pi_{\Gamma'}\colon X' \to \Gamma\backslash X'$ be the quotient maps. Then the image of π is $\pi_\Gamma(X')$. It follows from the previous claim that $\pi_\Gamma(U) \cap \pi_\Gamma(X') = \pi_\Gamma(U \cap X')$. Noting that $\pi_\Gamma|_U\colon U \to \pi_\Gamma(U)$ is a diffeomorphism by the choice of U, we see that

$\pi_\Gamma(X')$ is a regular submanifold of $\Gamma\backslash X$. Since $\pi_\Gamma|_{X'} = \pi \circ \pi_{\Gamma'}$, the map π is locally diffeomorphic to the regular submanifold $\pi_\Gamma(X')$. From the above, our assertion is proved. □

The following lemma is proved in Millson [56, Sect. 4, p. 245], where $\tilde{M} \supset F$, $\gamma_1, \ldots, \gamma_r$, and σ therein correspond to our $M \supset N, \alpha_1, \ldots, \alpha_k$, and τ, respectively.

Lemma 4.19 *Let M be an orientable connected Riemannian manifold, τ an involutive isometry on M, and N an orientable connected hypersurface of M such that τ is the identity on N. Suppose that we are given isometries $\alpha_1, \ldots, \alpha_k$ on M satisfying the following conditions:*

(a) $\alpha_i(N) \cap N = \emptyset$ and $\alpha_i(N) \cap \alpha_j(N) = \emptyset$ for any $i \neq j$;
(b) $\tau(\alpha_i(N)) = \alpha_i(N)$ for any i;
(c) τ preserves the orientation of $\alpha_i(N)$ for any i.

Then $M \setminus (\alpha_1(N) \cup \cdots \cup \alpha_k(N))$ is connected.

We are ready to show Proposition 4.16.

Proof of Proposition 4.16 Recall Proposition 4.11. We shall write $\mathbb{F}_{\sigma_1}$ simply as $\mathbb{R}$ and let us define

$$L = \mathbf{Spin}_V(\mathbb{R}) \simeq Spin(n, 1).$$

Now we fix the embeddings of the subgroups

$$L' = Spin(n-1, 1), \quad L_K = Spin(n), \quad L'_K = L' \cap L_K = Spin(n-1)$$

into L. Recall that V' is the $\mathbb{F}$-subspace of V generated by $e_1, \ldots, e_n$, where $\{e_0, \ldots, e_n\}$ forms an $\mathbb{F}$-basis of V. Hence, the Clifford algebra $C_{\mathrm{even}}(V') \equiv C_{\mathrm{even}}(V', Q|_{V'})$ is naturally embedded in $C_{\mathrm{even}}(V)$. This allows us to regard $C_{\mathrm{even}}(\Lambda')$ as an $\mathcal{O}$-subalgebra of $C_{\mathrm{even}}(\Lambda)$ and $\mathbf{Spin}_{V'}$ as an $\mathbb{F}$-subgroup of $\mathbf{Spin}_V$. We then define

$$L' := \mathbf{Spin}_{V'}(\mathbb{R}) \simeq Spin(n-1, 1).$$

Similarly, we define L_K and L'_K as the Lie groups of $\mathbb{R}$-points of the spin groups associated with the $\mathbb{F}$-subspaces of V generated by $\{e_0, \ldots, e_{n-1}\}$ and $\{e_1, \ldots, e_{n-1}\}$, respectively. Then we have $L_K \simeq Spin(n)$ and $L'_K \simeq Spin(n-1)$.

We consider the ideal of $\mathcal{O}$:

$$I = P_{-1} P_0 \cdots P_m.$$

By Proposition 4.11 and Lemma 4.12, the quotient spaces

$$X_\Lambda(I) := \Gamma_\Lambda(I)\backslash X \text{ and } X'_{\Lambda'}(I) := \Gamma_{\Lambda'}(I)\backslash X'$$

are both orientable connected compact hyperbolic manifolds, where $X = L/L_K$ and $X' = L'/L'_K$.

To show that $X'_{\Lambda'}(I)$ is a hypersurface of $X_\Lambda(I)$, we apply Lemma 4.18 to $(G, H, \Gamma) = (L, L_K, \Gamma_\Lambda(I))$. For this purpose, now we verify that the assumptions of Lemma 4.18 are satisfied.

Recall that elements of V are regarded as elements of the Clifford algebra $C(V) = C(V, Q)$. Let $\tau\colon C_{\mathrm{even}}(V) \to C_{\mathrm{even}}(V)$ be the involutive $\mathbb{F}$-automorphism defined by

$$\tau(x) = e_0 x e_0^{-1}.$$

It is straightforward to verify that

$$C_{\mathrm{even}}(V') = \{x \in C_{\mathrm{even}}(V) \mid \tau(x) = x\}.$$

In particular, we have

$$L' = \{x \in L \mid \tau(x) = x\}. \tag{4.13}$$

Since $Q(e_0) \in \mathcal{O}^\times$, the map τ preserves $C_{\mathrm{even}}(\Lambda)$. Hence, we also obtain $C_{\mathrm{even}}(\Lambda') = \{x \in C_{\mathrm{even}}(\Lambda) \mid \tau(x) = x\}$, and thus

$$\Gamma_{\Lambda'}(I) = \Gamma_\Lambda(I) \cap L'. \tag{4.14}$$

Similarly, we have $L_K = \{x \in L \mid e_n x e_n^{-1} = x\}$ and $L'_K = L_K \cap L'$.

The map $\tau\colon C_{\mathrm{even}}(V) \to C_{\mathrm{even}}(V)$ defines an involutive automorphism of L, which we also denote by $\tau \in \mathrm{Aut}(L)$. It is clear that τ preserves both $\Gamma_\Lambda(I)$ and L_K. Thus, by (4.13) and (4.14), Lemma 4.18 implies that $X'_{\Lambda'}(I)$ can be naturally identified with a totally geodesic hypersurface of $X_\Lambda(I)$.

Next, to complete the proof of our assertion, we apply Lemma 4.19 to

$$(M, N) = (X_\Lambda(I), X'_{\Lambda'}(I)).$$

For this purpose, we find isometries $\alpha_1, \ldots, \alpha_k$ of $X_\Lambda(I)$ that satisfy conditions *(a)*–*(c)* in Lemma 4.19.

Before proceeding, we make the following observations. By Lemma 4.12, the group $\Gamma_\Lambda(P_{-1}P_0)$ is torsion-free. Moreover, by an argument similar, we have $\Gamma_{\Lambda'}(P_{-1}P_0) = \Gamma_\Lambda(P_{-1}P_0) \cap L'$ (see (4.14)) and Lemma 4.18 can also be applied to $(G, H, \Gamma) = (L, L_K, \Gamma_\Lambda(P_{-1}P_0))$. Thus, by the injectivity result of Lemma 4.18, for each $\alpha \in \Gamma_\Lambda(P_{-1}P_0)$, we see that

$$\alpha(X') \cap X' \neq \emptyset \text{ if and only if } \alpha \in \Gamma_{\Lambda'}(P_{-1}P_0). \tag{4.15}$$

We define finite groups Φ and Φ' as

$$\Phi := \Gamma_\Lambda(P_{-1}P_0)/\Gamma_\Lambda(I) \text{ and } \Phi' := \Gamma_{\Lambda'}(P_{-1}P_0)/\Gamma_{\Lambda'}(I).$$

The natural left actions of the finite group Φ and the involution $\tau \in \mathrm{Aut}(L)$ on the Riemannian manifold $X_\Lambda(I)$ are isometric. Moreover, the action of Φ preserves the orientation of $X_\Lambda(I)$, while the involution $\tau \in \mathrm{Aut}(L)$ reverses it.

In what follows, we seek isometries $\alpha_1, \ldots, \alpha_k$ satisfying conditions (a)–(c) in Lemma 4.19 among the elements of Φ. By (4.15), we observe that for each $\alpha \in \Gamma_\Lambda(P_{-1}P_0)$,

$$\alpha(X'_{\Lambda'}(I)) \cap X'_{\Lambda'}(I) \neq \emptyset \text{ if and only if } \alpha \in \Gamma_{\Lambda'}(P_{-1}P_0). \tag{4.16}$$

On condition (a): Let $\alpha, \beta \in \Gamma_\Lambda(P_{-1}P_0)$. By (4.16), we see that

$$\alpha(X'_{\Lambda'}(I)) \cap \beta(X'_{\Lambda'}(I)) \neq \emptyset$$

if and only if $\alpha \equiv \beta \mod \Gamma_{\Lambda'}(P_{-1}P_0)$.

On condition (b): Let $\alpha \in \Gamma_\Lambda(P_{-1}P_0)$. For $x' \in X'_{\Lambda'}(I)$, we have

$$\tau \cdot (\alpha \cdot x') = \tau(\alpha) \cdot (\tau \cdot x') = (e_0 \alpha e_0^{-1}) \cdot x'. \tag{4.17}$$

By (4.16), we see that $\tau(\alpha(X'_{\Lambda'}(I))) = \alpha(X'_{\Lambda'}(I))$ if and only if $e_0 \alpha e_0^{-1} \equiv \alpha \mod \Gamma_{\Lambda'}(P_{-1}P_0)$.

On condition (c): Let $\alpha \in \Gamma_\Lambda(P_{-1}P_0)$. In our setting, condition (c) holds automatically under condition (b). That is, we show that if $\alpha(X'_{\Lambda'}(I))$ is τ-stable, then τ preserves the orientation of $\alpha(X'_{\Lambda'}(I))$. By (4.17), the action of τ on $\alpha(X'_{\Lambda'}(I))$ coincides with the action of $e_0\alpha e_0^{-1}\alpha^{-1} \in \Gamma_{\Lambda'}(P_{-1}P_0)$. This preserves the orientation of $\alpha(X'_{\Lambda'}(I))$.

Therefore, in order to apply Lemma 4.19, it suffices to find $\alpha_1, \ldots, \alpha_k \in \Gamma_\Lambda(P_{-1}P_0)$ satisfying the following conditions:

(i) $[\alpha_i] \not\equiv 1 \mod \Phi'$ and $[\alpha_i] \not\equiv [\alpha_j] \mod \Phi'$ for any $i \neq j$;
(ii) $[e_0\alpha_i e_0^{-1}] \equiv [\alpha_i] \mod \Phi'$,

where $[\alpha]$ means the element of $\Phi = \Gamma_\Lambda(P_{-1}P_0)/\Gamma_\Lambda(I)$ represented by $\alpha \in \Gamma_\Lambda(P_{-1}P_0)$.

To this end, we analyze the finite set Φ/Φ'. Since $\dim_{\mathbb{F}} V' \geq 3$, Proposition 4.15 can be applied to both $\mathbf{Spin}_V$ and $\mathbf{Spin}_{V'}$. Hence, we have

$$\Phi \simeq \prod_{i=1}^m Spin(\overline{\Lambda}_{P_i}) \text{ and } \Phi' \simeq \prod_{i=1}^m Spin(\overline{\Lambda'}_{P_i}),$$

and thus obtain

$$\Phi/\Phi' \simeq \prod_{i=1}^m \left(Spin(\overline{\Lambda}_{P_i})/Spin(\overline{\Lambda'}_{P_i})\right).$$

Let us consider the natural action of each $Spin(\overline{\Lambda}_{P_i})$ on the vector space $\overline{\Lambda}_{P_i}$ over the finite field k_{P_i}, which is given by

$$\bar{v} \in \overline{\Lambda}_{P_i} \mapsto \bar{g}\bar{v}\bar{g}^{-1} \in \overline{\Lambda}_{P_i}$$

for $\bar{g} \in Spin(\overline{\Lambda}_{P_i})$. Recall that elements of $\overline{\Lambda}_{P_i}$ are regarded as elements of the Clifford algebra $C_{\text{even}}(\overline{\Lambda}_{P_i})$. This action factors through the adjoint representation $\text{Ad}\colon Spin(\overline{\Lambda}_{P_i}) \to SO(\overline{\Lambda}_{P_i})$.

Here, note that the image of Ad is the subgroup O' of $SO(\overline{\Lambda}_{P_i})$ consisting of elements with *spinor norm* 1 (see, e.g., [64, Sect. 24, the identity (24.7a)]). By Millson [56, Lem. 3.1], the O'-orbit of e_0 mod P_i, equivalently the $Spin(\overline{\Lambda}_{P_i})$-orbit, coincides with the sphere through e_0 mod P_i in $\overline{\Lambda}_{P_i}$. Moreover, the isotropy group of e_0 mod P_i is $Spin(\overline{\Lambda'}_{P_i})$. Thus, we obtain the bijection

$$\Phi/\Phi' \simeq \prod_{i=1}^{m} \{\bar{x} \in \overline{\Lambda}_{P_i} \mid Q(x) \equiv Q(e_0) \mod P_i\}, \tag{4.18}$$

where $\bar{x}$ means the element of $\overline{\Lambda}_{P_i}$ represented by $x \in \Lambda$. This map assigns to each $[g] \in \Phi$ (where $g \in \Gamma_\Lambda(P_{-1}P_0)$) the element on the right-hand side whose i-th component is given by ge_0g^{-1} mod P_i.

In the right-hand side of (4.18), choose k distinct elements from the set

$$\{(\pm\overline{e_0}, \dots, \pm\overline{e_0})\} \smallsetminus \{(\overline{e_0}, \dots, \overline{e_0})\},$$

which consists of $2^m - 1$ elements (recall $2^m - 1 > k$). Let $\alpha_1, \dots, \alpha_k$ be elements of Φ that represent the corresponding elements of Φ/Φ'. It is immediate that these satisfy conditions (i) and (ii).

Now we define k orientable connected totally geodesic closed hypersurfaces $N_1, \dots, N_k$ of $M = X_\Lambda(I)$ by

$$N_i := \alpha_i(X'_{\Lambda'}(I)) \quad (i = 1, \dots, k).$$

Applying Lemma 4.19 to $(M, N) = (X_\Lambda(I), X'_{\Lambda'}(I))$ and $\alpha_1, \dots, \alpha_k$, we conclude that $N_i \cap N_j = \emptyset$ for all $i \neq j$ and that $M \smallsetminus (N_1 \cup \dots \cup N_k)$ is connected. Thus, $\Gamma_\Lambda(I)$ satisfies Condition 4.1 for k, which completes the proof. □

4.4 Proof of Step 2 (An Overview of Bending Construction)

In Step 2, we provide a proof of general properties (Lemmas 4.6, 4.7, and 4.20) concerning iterated HNN extensions and bending construction. In Step 3, we will apply these results to construct the desired small deformation.

Fix $k \in \mathbb{N}_+$. Let Γ be a torsion-free cocompact discrete subgroup of $L = Spin(n, 1)$ satisfying Condition 4.1. Lemma 4.6 asserts that the fundamental group Γ of the hyperbolic manifold $M = \Gamma\backslash X$ can be expressed as an iterated HNN extension of length k. Lemma 4.7 provides a method for constructing a small deformation of a representation of Γ through k iterations of the bending construction. Furthermore, Lemma 4.20 provides a lower bound of the Zariski-closure of such a small deformation.

Proof of Lemma 4.6 The proof follows by repeatedly applying Proposition D.2.

We proceed by induction on k. For $k = 1$, the claim follows immediately by applying Proposition D.2 to our setting.

Suppose $k \geq 2$, and assume that the assertion holds for $k - 1$.

We set $S' := M \setminus (N_2 \cup \cdots \cup N_k)$. For each $i = 2, \ldots, k$, the two homomorphisms

$$j_{i,+}, j_{i,-} : \pi_1(N_i, y_i) \to \pi_1(S, x_0)$$

can be factored as $\pi_1(N_i, y_i) \to \pi_1(S', x_0) \to \pi_1(S, x_0)$. We also denote the first maps $\pi_1(N_i, y_i) \to \pi_1(S', x_0)$ by the same symbols $j_{i,+}$ and $j_{i,-}$.

Let F_{k-1} be the free group generated by $a_2, \ldots, a_k$, and let $\mathcal{N}'$ be the normal subgroup of $\pi_1(S', x_0) * F_{k-1}$ generated by

$$a_i j_{i,+}([\ell]) a_i^{-1} j_{i,-}([\ell])^{-1}, \quad ([\ell] \in \pi_1(N_i, y_i),\ i = 2, \ldots, k).$$

By the induction hypothesis, there is a natural isomorphism

$$\pi_1(M, x_0) \simeq (\pi_1(S', x_0) * F_{k-1})/\mathcal{N}'.$$

Next, let $S := S' \setminus N_1$, and let $\mathcal{N}_1$ be the normal subgroup of $\pi_1(S, x_0) * a_1^{\mathbb{Z}}$ generated by

$$a_1 j_{1,+}([\ell]) a_1^{-1} j_{1,-}([\ell])^{-1}, \quad ([\ell] \in \pi_1(N_1, y_1)).$$

Applying Proposition D.2 to the orientable connected manifold S', we obtain the isomorphism

$$\pi_1(S', x_0) \simeq (\pi_1(S, x_0) * a_1^{\mathbb{Z}})/\mathcal{N}_1.$$

Combining these isomorphisms, we have

$$\pi_1(M, x_0) \simeq (((\pi_1(S, x_0) * a_1^{\mathbb{Z}})/\mathcal{N}_1) * F_{k-1})/\mathcal{N}' \simeq (\pi_1(S, x_0) * F_k)/\mathcal{N}.$$

The second isomorphism is derived from the universal property of the free product. It is clear that the above isomorphism is induced by the surjective homomorphism Ψ given by (4.2), which completes the proof. □

Before proving Lemma 4.7, we fix some notation. We recall that

$$\pi_\Gamma : X \to M = \Gamma \backslash X$$

is the covering map, and we fix $\tilde{x}_0 \in X$ such that $\pi_\Gamma(\tilde{x}_0) = x_0$, as in the paragraph preceding Lemma 4.7. For every $i = 1, \ldots, k$, recall that the loop ν_i (see Definition 4.5) intersects N_i at exactly one point, namely, at y_i. Let $\nu'_{i,+}$ denote the segment of the loop ν_i from x_0 to y_i. Via the covering map π_Γ, we lift the path $\nu'_{i,+}$ to a path in X that starts at $\tilde{x}_0 \in X$, and we denote its endpoint by $\tilde{y}_i \in X$. Then, $\pi_\Gamma(\tilde{y}_i) = y_i \in N_i$. It follows from the commutative diagram (4.3) that $\tilde{y}_i$ lies in $\alpha_i X'$. Thus, just as the isomorphism $D^M_{\tilde{x}_0} : \pi_1(M, x_0) \simeq \Gamma$ (see (4.4)) is defined, we also define the isomorphism $D^N_{\tilde{y}_i} : \pi_1(N, y_i) \simeq \alpha_i L' \alpha_i^{-1} \cap \Gamma$.

We are ready to prove Lemma 4.7.

Proof of Lemma 4.7 First, we show that for each $i = 1, \dots, k$ and any

$$D^M_{\tilde{x}_0}(\Psi(j_{i,+}([\ell]))) \in \alpha_i L' \alpha_i^{-1} \cap \Gamma \quad \text{for any } [\ell] \in \pi_1(N_i, y_i). \tag{4.19}$$

To verify (4.19), we consider the following diagram.

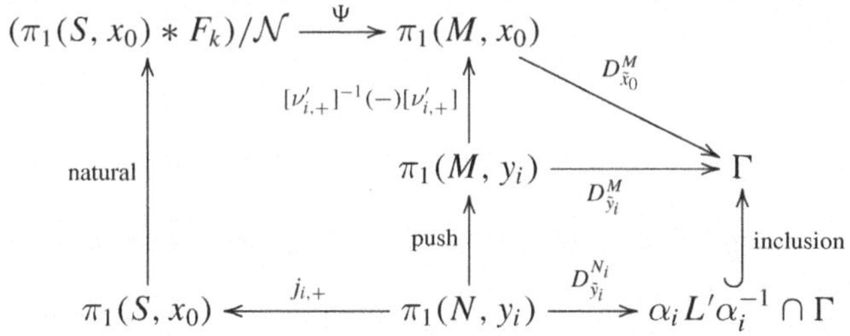

The commutativity of this diagram follows from the unique lifting property of covering maps with respect to paths. For the definition of $j_{i,+}$, see (4.1). Recall that $\nu'_{i,+}$ is the path from x_0 to y_i along the oriented loop ν_i. From the commutativity of this diagram, (4.19) follows immediately.

Now we prove our assertion. Put $\psi := D^M_{\tilde{x}_0} \circ \Psi$. By Lemma 4.6, it suffices to show that for each $i = 1, \dots, k$ and for any $[\ell] \subset \pi_1(N_i, y_i)$,

$$\varphi_t(\psi(a_i))\varphi_t(\psi(j_{i,+}([\ell])))\varphi_t(\psi(a_i))^{-1}\varphi_t(\psi(j_{i,-}([\ell]))^{-1} = 1.$$

Since $j_{i,+}([\ell]), j_{i,-}([\ell]) \in \pi_1(S, x_0)$, we have

$$\begin{aligned}(\text{LHS}) &= \varphi(\psi(a_i))e^{tv_i}\varphi(\psi(j_{i,+}([\ell])))(\varphi(\psi(a_i))e^{tv_i})^{-1}\varphi(\psi(j_{i,-}([\ell])))^{-1} \\ &= \varphi(\psi(a_i j_{i,+}([\ell]) a_i^{-1} j_{i,-}([\ell])^{-1})) = 1\end{aligned}$$

where the second equality follows from (4.19) and assumption (4.5). Thus our assertion is proved. □

For the proof of Step 3, we give the following lemma concerning the Zariski-closure of small deformations of Γ obtained via Lemma 4.7. The following proof is in the same line with Kassel [27, Sect. 6] for $\varphi : SO(n, 1) \to SO(n, 2)$ using Johnson–Millson [21, Lem. 5.9].

Lemma 4.20 (cf. the argument in [27, Lem. 6.4]) *Let G be a Zariski-connected real algebraic group. Under the assumptions and notation of Lemma 4.7, assume that the group homomorphism $\varphi\colon \Gamma \to G$ arises from a homomorphism $\varphi\colon Spin(n,1) \to G$. Then, for any $t \in \mathbb{R}$, the Zariski-closure of $\varphi_t(\Gamma)$ contains $\varphi(Spin(n,1))$ and e^{tv_i} for $i = 1, \ldots, k$.*

Proof Let L_t denote the Zariski-closure of $\varphi_t(\Gamma)$ in G.

Let $\varpi\colon Spin(n,1) \to SO_0(n,1)$ be the double cover, and consider the group isomorphism

$$\psi := D^M_{\tilde{x}_0} \circ \Psi\colon \pi_1(S, x_0) * F_k \to \Gamma \ (\subset Spin(n,1)).$$

By [21, Lem. 5.9], $\varpi(\psi(\pi_1(S, x_0)))$ is Zariski-dense in $SO(n,1)$. Hence, $\psi(\pi_1(S, x_0))$ is also Zariski-dense in $Spin(n,1)$.

We recall that $\varphi\colon Spin(n,1) \to G$ is a homomorphism in the sense of Lemma A.7. Since φ is continuous in the Zariski-topology, and since

$$\varphi_t(\psi(\pi_1(S, x_0))) = \varphi(\psi(\pi_1(S, x_0)))$$

by the construction of φ_t (see (a) in Lemma 4.7), the Zariski-closure of $\varphi_t(\psi(\pi_1(S, x_0)))$ contains $\varphi(Spin(n,1))$. Thus, L_t contains $\varphi(Spin(n,1))$. Then it follows that $e^{tv_i} \in L_t$. Indeed, the inclusive relation $\varphi(Spin(n,1)) \subset L_t$ implies that $\varphi(D^M_{\tilde{x}_0}(\nu_i)) \in L_t$ for every $i = 1, \ldots, k$. Thus, it follows that

$$e^{tv_i} = \varphi(D^M_{\tilde{x}_0}(\nu_i))^{-1} \varphi_t(D^M_{\tilde{x}_0}(\nu_i)) \in L_t.$$

Thus, the proof is completed. □

4.5 Proof of Step 3 (Zariski-Closure of Specific Deformation)

We recall from Theorem 3.9(3) that when $n \geq 3$, the algebraic subgroup G^φ (see Definition 3.7) achieves the upper bound of the Zariski-closure of any small deformation of $\varphi|_\Gamma$, up to G-conjugacy. In this section, we provide a proof of Proposition 4.9, which states that repeating the bending construction k times results in a small deformation $\varphi_t(\Gamma)$ of $\varphi|_\Gamma$, whose Zariski-closure in G, to be denoted by L_t, attains the upper bound G^φ. The proof requires a careful discussion to handle the general situation in which we cannot expect $v_i^{(j)} \in \mathfrak{l}_t$, nor the existence of towers of symmetric pairs $\mathfrak{l} = \mathfrak{l}^0 \subset \mathfrak{l}^1 \subset \cdots \subset \mathfrak{l}^k = \mathfrak{g}$, as mentioned in Remark 4.8.

Before proceeding with the proof of Proposition 4.9, we first establish a couple of auxiliary lemmas.

Lemma 4.21 *Let $\mathbb{F}$ be a field, and A be an associated $\mathbb{F}$-algebra, and let V be an A-module which is finite-dimensional over $\mathbb{F}$. For $n \geq 1$, let $V^{\oplus n}_{gen}$ denote the set of n-tuples $v_1, \ldots, v_n \in V$ such that*

$$Av_1 + \cdots + Av_n = V.$$

Then $V_{gen}^{\oplus n}$ is Zariski-open in $V^{\oplus n}$.

Example 4.22 In the case $n = 1$, an element in $V_{gen} (:= V_{gen}^{\oplus 1})$ is called a *cyclic vector*. We note that $V_{gen} = V \smallsetminus \{0\}$ if and only in V is irreducible.

Proof of Lemma 4.21 Suppose V is m-dimensional. Then $V_{gen}^{\oplus n}$ is the union of the Zariski-open set

$$\{(v_1, \ldots, v_n) | \dim_{\mathbb{F}}(\mathrm{span}_{\mathbb{F}}\{a_1 v_1, \ldots, a_n v_n\}) = m\}$$

where $(a_1, \ldots, a_n)$ runs over $A \times \cdots \times A$. □

For a Lie algebra $\mathfrak{l}$, we denote by $U(\mathfrak{l})_+$ the augmentation ideal of the universal enveloping algebra of $\mathfrak{l}$.

We extend the adjoint action of the Lie algebra $\mathfrak{g}$ to an action on its universal enveloping algebra $U(\mathfrak{g})$, which we continue to denote by ad.

We also use the following notation, which will be studied in Appendix F. For a finite subset $\mathcal{X} \subset \mathfrak{g}$ and a subgroup $L \subset G$, we denote by $G(L; \mathcal{X})$ the identity component of the Zariski closure of the subgroup of G generated by L and the elements e^X for $X \in \mathcal{X}$ (Definition F.1). Its Lie algebra is denoted by $\mathfrak{g}(L; \mathcal{X})$. Furthermore, for each $t \in \mathbb{R}$, we define $t\mathcal{X} := \{tX | X \in \mathcal{X}\}$.

Lemma 4.23 *Let $L \subset G$ be two Zariski-connected real algebraic groups. For any $X \in \mathfrak{l}$ and for any $u \in U(\mathfrak{l})_+$, there exists an analytic map*

$$a \equiv a_{X,u} : \mathbb{R} \to \mathfrak{g}$$

such that $a(t) \in \mathfrak{g}(L, t\{X\})$ for all $t \in \mathbb{R} \smallsetminus \{0\}$ and $a(0) = \mathrm{ad}(u)X$.

Proof We fix $X \in \mathfrak{l}$. The set of $u \in U(\mathfrak{l})_+$ for which there exists such an analytic map $a : \mathbb{R} \to \mathfrak{g}$ with $a(0) = \mathrm{ad}(u)X$ forms a vector space. Thus, it suffices to show the claim in the case $u = Y_1 \cdots Y_n$ for some $Y_1, \ldots, Y_n \in U(\mathfrak{l})$ with $n \geq 1$. We prove this by the induction on n.

First, consider the case $n = 1$. When $u = Y \in \mathfrak{l}$, we set

$$a(t) := \frac{1}{t}(\mathrm{Ad}(e^{-tX})Y - Y).$$

Clearly, $a : \mathbb{R} \to \mathfrak{g}$ is an analytic map with $a(0) = -[X, Y] = \mathrm{ad}(Y)X$. Moreover, since $Y \in \mathfrak{l}$ and $e^{tX} \in G(L; t\{X\})$, we conclude that $a(t) \in \mathfrak{g}(L; t\{X\})$ for any $t \neq 0$.

Next, suppose that $v = Y_2 \cdots Y_n \in U(\mathfrak{l})_+$ and that there exists an analytic map $b : \mathbb{R} \to \mathfrak{g}$ such that $b(t) \in \mathfrak{g}(L; t\{X\})$ for any $t \neq 0$ and $b(0) = \mathrm{ad}(v)X$. Let $u := Y_1 v$ where $Y_1 \in \mathfrak{l}$. We set

$$a(t) := \frac{1}{t}(\mathrm{Ad}(e^{tY_1})b(t) - b(t)).$$

Then $a\colon \mathbb{R} \to \mathfrak{g}$ is an analytic map and we have $a(0) = [Y_1, b(0)] = \mathrm{ad}(Y)(\mathrm{ad}(v)X) = \mathrm{ad}(Y_1 v)X$. Since $e^{tY} \in L$ and $b(t) \in \mathfrak{g}(L; t\{X\})$, we have $a(t) \in \mathfrak{g}(L; t\{X\})$ for any $t \neq 0$.

By induction, the lemma is proved. □

Proof of Proposition 4.9 Let $L = Spin(n, 1)$, and L_t denote the Zariski-closure of $\varphi_t(\Gamma)$ in G.

First, we claim that $L_t \subset G^{\varphi}$ for every $t \in \mathbb{R}$. It follows from the definition (4.8) of v_j that $v_j \in \mathfrak{g}^{\varphi}$, where we recall Definition 3.5 for $\mathfrak{g}^{\varphi}$. Moreover, $\varphi(L) \subset G^{\varphi}$ by the definition of G^{φ} in Definition 3.7. Therefore, the construction of φ_t in Lemma 4.7 implies that $\varphi_t(\Gamma) \subset G^{\varphi}$, which concludes that $L_t \subset G^{\varphi}$.

Second, we claim that $\mathfrak{l}_t \supset \mathfrak{g}^{\varphi}$ if t is sufficiently small, where we write $\mathfrak{l}_t$ for the (real) Lie algebra of L_t. This claim will imply that $L_t = G^{\varphi}$ because G^{φ} is Zariski-connected.

We recall from Lemma 4.20 the following two properties:

$$\varphi(L) \subset L_t, \tag{4.20}$$

$$\exp(t v_j) \in L_t \quad \text{for any } 1 \leq j \leq k. \tag{4.21}$$

We might expect from (4.21) that $v_j \in \mathfrak{l}_t$ for a sufficiently small t. However, such a statement is not true in our general setting. Therefore, in what follows, we will proceed with a more careful discussion.

The properties (4.20) and (4.21) imply that $G(\varphi(L); t\{v_1, \ldots, v_k\}) \subset L_t$ with the notation as in Definition F.1 of Appendix F, and hence, we have $\mathfrak{g}(\varphi(L), t\{v_1, \ldots, v_m\}) \subset \mathfrak{l}_t$.

Recall that $\mathfrak{g}(V_i)$ ($\simeq \mathrm{Hom}_L(V_i, \mathfrak{g}) \otimes V_i$) denote the isotypic component of V_i in $\mathfrak{g}$. We note that $\mathfrak{g}(V_i) = \mathfrak{g}^{\varphi}(V_i)$ by Definition 3.5 of $\mathfrak{g}^{\varphi}$ for any $i \in \mathbb{N}$. Let

$$\mathrm{pr}_i \colon \mathfrak{g} \to \mathfrak{g}(V_i)$$

denote the projection to $\mathfrak{g}(V_i)$, see (4.6). Then, pr_i is an L-homomorphism. Moreover, for $1 \leq i$, by the definition (4.8) of v_j, we have $\mathrm{pr}_i(v_j) = v_i^{(j)}$ for any j satisfying $1 \leq j \leq [\mathfrak{g} : V_i]$.

We note that there are only finitely many non-zero $v_i^{(j)}$ because $\dim \mathfrak{g} < \infty$. Since V_i is not the trivial representation for any $i \geq 1$, there exists $Y \in \mathfrak{l}$ such that $\mathrm{ad}(Y)v_i^{(j)} \neq \{0\}$ whenever $v_i^{(j)} \neq 0$. In turn, we have $U(\mathfrak{l})(\mathrm{ad}(Y)v_i^{(j)}) = V_i^{(j)}$ by the irreducibility of $V_i^{(j)} \simeq V_i$, and consequently

$$\mathrm{pr}_i\left(\sum_{j=1}^{m} U(\mathfrak{l})[Y, v_j]\right) = \sum_{j=1}^{m} U(\mathfrak{l})[Y, v_i^{(j)}] = \sum_{j=1}^{m} V_i^{(j)} = \mathfrak{g}(V_i).$$

For each $i \geq 1$, let $a_i \colon \mathbb{R} \to \mathfrak{g}$ be an analytic map, as given in Lemma 4.23 in Appendix F, satisfying $a_i(0) = [Y, v_i]$ and such that $a_i(t) \in \mathfrak{g}(\varphi(L); t\{v_i\})$ for any $t \neq 0$. Then, for any sufficiently small $t \neq 0$, we have

$$\mathrm{pr}_i\left(\sum_{j=1}^{m} U(\mathfrak{l})a_j(t)\right) = \sum_{j=1}^{m} U(\mathfrak{l})\,\mathrm{pr}_i(a_j(t)) = \mathfrak{g}(V_i).$$

by Lemma 4.21. Since the L-module $\sum_{j=1}^{m} U(\mathfrak{l})a_j(t)$ is contained in $\mathfrak{g}(\varphi(L); t\{v_1, \ldots, v_m\}) \subset \mathfrak{l}_t$, we have shown, for any $i \in \mathbb{N}_+$,

$$\mathfrak{l}_t \supset \mathfrak{g}(V_i). \tag{4.22}$$

Now we prove (4.22) for $i = 0$, that is, $\mathfrak{l}_t \supset \mathfrak{z}_{\mathfrak{g}}(d\varphi(\mathfrak{l}))$.

Let L' denote the Zariski closure of the group generated by $\varphi(L)$ and $\exp(\mathfrak{g}(V_i))$ for all $i \in \mathbb{N}_+$. It follows from (4.22) that $L_t \supset L'$. Moreover, the identity component $Z_G(\varphi(L))$ of the centralizer normalizes L'. Therefore, it follows from the Baker–Campbell–Hausdorff formula that

$$\exp(-tu_j)\exp(tv_j) = \exp(-tu_j)\exp(t(u_j + \sum_i v_i^{(j)})) \in L'\ (\subset L_t)$$

because $u_j \in \mathfrak{z}_{\mathfrak{g}}(d\varphi(\mathfrak{l}))$ and $\sum_i v_i^{(j)} \in \mathfrak{l}'$. Thus $\exp(tu_j)$ lies in the subgroup L_t because $\exp(tv_j) \in L_t$. By the choice of $u_1, \ldots, u_{\eta(Z_G(\varphi(L)))}$, L_t contains $Z_G(\varphi(L)) = G(t\{u_1, \ldots,$
$u_{\eta(Z_G(\varphi(L)))}\})$ for sufficiently small $t > 0$.

This proves (4.22) in the case $i = 0$.

Therefore, we obtain $\mathfrak{l}_t \supset \mathfrak{g}^{\varphi}$, which we wanted to prove. □

5 Proof of Main Results

In this section, we study deformations of discontinuous groups for $X = G/H$, and in particular, we complete the proofs of the main theorems of this article, Theorems 2.9 and 2.21.

This section is organized as follows:

In Sect. 5.1, we review preliminaries on deformations of discontinuous groups for $X = G/H$ with non-compact H, as well as results on local rigidity and the stability of proper discontinuity and cocompactness. In Sect. 5.2, we complete the proof of Theorem 2.9, which answers Question 2.5 (Q1)–(Q3). In Sect. 5.3, we prove Theorem 2.19. In Sect. 5.4, we complete the proof of Theorem 2.21, which answers Question 2.5 (Q1)–(Q3) for the exotic quotient $\Gamma_L\backslash G/\Gamma_H$. In Sect. 5.5, we construct discontinuous groups for certain homogeneous spaces $X = G/H$ which are Zariski-dense subgroups of G and have cohomological dimension between 2 and the non-compact dimension $d(X)$, in connection with Problem 1.2.

5.1 Preliminaries on Deformations of Quotients $\Gamma\backslash X$

Let $X = G/H$ be a homogeneous space of a Lie group G with H non-compact. In this section, we summarize some results of the local rigidity of discontinuous groups, and also discuss when local rigidity fails, along with how small deformations of discrete subgroups can preserve (or sometimes not preserve) the properties of the action, such as proper discontinuity and cocompactness.

Since proper discontinuity may change under small deformations in the general setting, the following concept introduced by Kobayashi and Nasrin will be useful in clarifying the subsequent discussion.

Definition 5.1 (*Stability of proper discontinuity,* [48, 49]) Let $\varphi \in \mathcal{R}(\Gamma, G; X)$, as defined in Sect. 2.1. The Γ-action on X via φ is called *stable as a discontinuous group for* X (under any small deformation) if $\mathcal{R}(\Gamma, G; X)$ is a neighborhood of φ in $\mathrm{Hom}(\Gamma, G)$, that is, if any small deformation φ' of φ in $\mathrm{Hom}(\Gamma, G)$ is faithful and discrete, and if the Γ-action on X through φ' is properly discontinuous.

Let us make a simple observation on the relationship between local rigidity for $\mathcal{R}(\Gamma, G; X)$ (see Definition 2.3) and local rigidity for $\mathcal{R}(\Gamma, G) \equiv \mathcal{R}(\Gamma, G; G/\{e\})$ (see Sect. 3.1):

Lemma 5.2 *Let Γ be a discontinuous group for G/H, and let $\iota \colon \Gamma \to G$ denote the inclusion map. The following claims hold:*

(1) If ι is locally rigid, then it is also locally rigid as a discontinuous group for X.
(2) If ι is stable as a discontinuous group, then the converse of (1) holds.

Proof (1) is straightforward, and (2) follows directly from the definition of stability. □

Example 5.3 In the group case where $X = G/\{e\}$, we have $\mathcal{R}(\Gamma, G; X) = \mathcal{R}(\Gamma, G)$. For $\varphi \in \mathcal{R}(\Gamma, G; X)$ to be stable in the sense of Definition 5.1, it is equivalent to the condition that any small deformation of φ in $\mathrm{Hom}(\Gamma, G)$ is faithful and discrete.

Remark 5.4 (*Weil* [70]) Let G be a (not necessarily reductive) Lie group, and Γ a cocompact discrete subgroup of G. Then, the Γ-action on $G/\{e\}$ is stable as a discontinuous group for $G/\{e\}$.

We now consider the case where H is non-compact.

Example 5.5 (1) (cocompact but unstable). Let $G := \mathrm{Aff}(\mathbb{R})$ be the affine transformation group of $X := \mathbb{R}$. For $\varepsilon \neq 1$, we define a homomorphism $\varphi_\varepsilon \colon \mathbb{Z} \to G$ by mapping the generator $1 \in \mathbb{Z}$ to the affine transformation $x \mapsto (1-\varepsilon)x + 1$. Clearly, $\varphi_0(\mathbb{Z})$ is a cocompact discontinuous group for $X = \mathbb{R}$, and $\varphi_\varepsilon \in \mathrm{Hom}(\Gamma, G)$ is faithful and discrete. However, the properly discontinuous action of $\Gamma = \mathbb{Z}$ on $X = \mathbb{R}$ is not stable at φ_0, even though $\varphi_\varepsilon(\mathbb{Z})$ is discrete in G. Indeed, $\varphi_\varepsilon(\mathbb{Z})$ has the fixed point $\varepsilon^{-1} \in X$ for any $\varepsilon \neq 0, 1$, and consequently, the action of $\varphi_\varepsilon(\mathbb{Z})$ on X cannot be properly discontinuous.

(2) (cocompact and stable). In the setting where $X = G/H$ is of reductive type with non-compact H, the first example of a *stable* cocompact discontinuous group Γ, in arbitrarily high dimensions, was proven by Kobayashi [43], as a solution to a conjecture by Goldman [15].
(3) (stability varies with a homomorphism φ). For some triples (Γ, G, X), it is possible that $\mathcal{R}(\Gamma, G; X)$ contains both stable and unstable elements. (For an explicit description, see [48] by Kobayashi and Nasrin, particularly when G is a nilpotent Lie group.)

In contrast to the failure of stability for cocompact properly discontinuous actions, as presented in Example 5.5(1) in the non-reductive case, we expect that proper discontinuity is an open condition in a reasonable setting where $X = G/H$ is of reductive type. After the solution to Goldman's conjecture [15] by Kobayashi [43] in 1990s, several sufficient conditions have been developed to ensure the stability property in the reductive setting (see, for example, Guéritaud–Kassel [19], Guéritaud–Guichard–Kassel–Wienhard [18], Kannaka [23], and Kassel–Tholozan [33] (preprint), for recent work). In this article, we need:

Fact 5.6 (Kassel [27, Thm. 1.3]) *Let G be a real reductive algebraic group, $X = G/H$ a homogeneous space of reductive type, L a real simple algebraic group of real rank 1, and $\varphi\colon L \to G$ a Lie group homomorphism such that the L-action on $X = G/H$ via φ is proper. If Γ is a torsion-free cocompact discrete subgroup of L, then the Γ-action on X via φ is stable as a discontinuous group for X.*

In the case where Γ is a discontinuous group for a group manifold $(G \times G)/\operatorname{diag} G$ of the form $\Gamma = \Gamma_H \times \Gamma_L$, Γ is not an irreducible lattice, and thus the assumption of Fact 5.6 is not satisfied. However, an argument in the same spirit can still be applied, and the stability of discontinuous groups holds in the following form:

Fact 5.7 (Kassel and Kobayashi [29]) *Let G be a real reductive algebraic group, H and L two closed reductive subgroups of G such that the action of $H \times L$ on $G \simeq (G \times G)/\operatorname{diag} G$ is proper, and let Γ_H and Γ_L be torsion-free irreducible cocompact discrete subgroups of H and L, respectively. Then the action of $\Gamma_H \times \Gamma_L$ on G is stable as a discontinuous group for $(G \times G)/\operatorname{diag} G$.*

Fact 5.7 is derived from [29, Lem. 4.23], which provides a quantitative estimate of the sharpness constant in this setting.

Finally, with respect to the cocompactness of the action, as shown by Kobayashi in [37], cocompactness is always preserved as long as the stability of the discontinuous group holds.

Fact 5.8 ([37, Cor. 5.5]) *Let Γ be a cocompact discontinuous group for $X = G/H$. Then for any $\varphi \in \mathcal{R}(\Gamma, G; X)$, the quotient $X_{\varphi(\Gamma)}$ is also compact.*

We end this section by providing a key framework for proving the answer to Question 2.5 (3), which follows from the main result of Sect. 3 along with the aforementioned stability theorem.

Theorem 5.9 *Let G be a Zariski-connected real reductive algebraic group, and $X = G/H$ a homogeneous space of reductive type. Let $\varphi\colon Spin(n,1) \to G$ be a homomorphism such that the action of $Spin(n,1)$ on X, via φ, is proper.*

(1) There exists a pair (Γ, φ') such that the Zariski-closure of $\varphi'(\Gamma)$ equals G^{φ} (Definition 3.7), where Γ is a torsion-free cocompact discrete subgroup of $Spin(n,1)$ and $\varphi' \in \mathcal{R}(\Gamma, G; X)$ is a small deformation of $\varphi|_{\Gamma}$ in Hom(Γ, G).
(2) If $\mathfrak{g}^{\varphi} = \mathfrak{g}$ (see Definition 3.5), then $\varphi'(\Gamma)$ is Zariski-dense in G, where (Γ, φ') is the pair in (1).

Proof We have shown in Theorem 3.9 (1) that there exists a pair (Γ, φ') where Γ is a cocompact discrete subgroup Γ of L without torsion and φ' is a small deformation of $\varphi|_{\Gamma}$ such that $\varphi'(\Gamma)$ is Zariski-dense in G^{φ}. Then the theorem follows from Fact 5.6 applied to $L = Spin(n,1)$. □

5.2 *Proof of Theorem 2.9*

In this section, we complete the proof of Theorem 2.9. In the first half of this section, we discuss the case where the discontinuous group is locally rigid, and in the second half, we deal with the case where it is deformable.

As shown in [43, Sect. 3.12] using Weil [71], the vanishing of the first cohomology of Γ serves as an effective first sufficient condition to the local rigidity of a cocompact discontinuous group Γ acting on a homogeneous space G/H. This condition can be addressed without considering the subgroup H, see Steps 1 and 3 below. We shall prove in Steps 4 and 5 that this approach covers all the local rigidity cases in Theorem 2.9. That is, in the setting where (G, H, L) is one of the triples in Table 3, there exists a cocompact discrete subgroup Γ of L for which $H^1(\Gamma, \mathfrak{g}) \neq 0$ if and only if there exists such a Γ that is deformable as a discontinuous group for $X = G/H$.

We recall that L_{ss} is the semisimple factor of L. The proof of Theorem 2.9 is organized in the following five steps.

Step 1: Proof of local rigidity in the cases where $\mathfrak{l}_{ss}$ is isomorphic to neither $\mathfrak{so}(n,1)$ nor $\mathfrak{su}(n,1)$.
Step 2: Irreducible decomposition of the $\mathfrak{l}_{ss}$-module $\mathfrak{g}/\mathfrak{l}_{ss}$ in the cases where $\mathfrak{l}_{ss}$ is isomorphic to either $\mathfrak{so}(n,1)$ or $\mathfrak{su}(n,1)$.
Step 3: Proof of local rigidity in the cases where $\mathfrak{l}_{ss}$ is isomorphic to either $\mathfrak{so}(n,1)$ or $\mathfrak{su}(n,1)$ and $H^1(\Gamma, \mathfrak{g}) = 0$ for any cocompact discrete subgroup Γ of L.
Step 4: We discuss Γ which is deformable but cannot be deformed into a non-standard discontinuous group for X. This happens when $\mathfrak{l}_{ss}$ is isomorphic to $\mathfrak{su}(n,1)$.
Step 5: We discuss Γ which can be deformed into a Zariski-dense subgroup. This happens when $\mathfrak{l}$ is isomorphic to $\mathfrak{so}(n,1)$.

The proof of the theorem is mainly carried out for $L = L_{ss}$. For each $j = 1, 2, 3$, if (Qj) is affirmative for L_{ss}, then (Qj) is also affirmative for all L such that $L_{ss} \subset L \subset L_{max}$, as in Notation 2.6. For the reverse direction, we use the following lemma.

Lemma 5.10 *Let L_c denote the compact factor of a real reductive group L without split center. Then we have*

$$L = L_{ss} L_c.$$

Let Γ be a torsion-free discrete subgroup of L. Then, the quotient map

$$L \to L/L_c \simeq L_{ss}/(L_{ss} \cap L_c)$$

is injective when restricted to Γ, and therefore, we may regard Γ as a discrete subgroup of the semisimple Lie group $L_{ss}/(L_{ss} \cap L_c)$.

Proof Since Γ is torsion-free, the finite subgroup $\Gamma \cap L_c$ must be the singleton $\{e\}$. Hence, the assertion is clear. □

Step 1. We prove the cases where $\mathfrak{l}_{ss}$ is not isomorphic to $\mathfrak{so}(n, 1)$ or $\mathfrak{su}(n, 1)$. The proof is parallel to [43, Sect. 3.12]. Let Γ be a torsion-free cocompact discrete subgroup of L. It follows from Lemma 5.10 that we can deduce that Γ is a torsion-free cocompact discrete subgroup of the semisimple Lie group $L_{ss}/(L_{ss} \cap L_c)$. Since $\mathfrak{l}_{ss}$ is not isomorphic to $\mathfrak{so}(n, 1)$ or $\mathfrak{su}(n, 1)$, we apply Raghunathan's vanishing theorem

Table 4 Irreducible decomposition of $\mathfrak{g}/\mathfrak{l}_{ss}$ as an $\mathfrak{l}_{ss}$-module

	$\mathfrak{g}$	$\mathfrak{l}_{ss}$	$\mathfrak{g}/\mathfrak{l}_{ss}$
Case 1	$\mathfrak{su}(2n, 2)$	$\mathfrak{su}(2n, 1)$	$\mathbb{C}^{2n,1} + \mathbf{1}$
Case 1'-2 Case 2–2	$\mathfrak{su}(2, 2)$	$\mathfrak{spin}(4, 1)$	$\mathbb{R}^{4,1}$
Case 3	$\mathfrak{so}(2n, 2)$	$\mathfrak{so}(2n, 1)$	$\mathbb{R}^{2n,1}$
Case 4–1-a	$\mathfrak{so}(2n, 2)$ $(n \geq 3)$	$\mathfrak{su}(n, 1)$	$\bigwedge^2 \mathbb{C}^{n,1} + \mathbf{1}$
Case 4–1-b	$\mathfrak{so}(4, 2)$	$\mathfrak{su}(2, 1)$	$\mathbb{C}^{2,1} + \mathbf{1}$
Case 4–2	$\mathfrak{so}(2, 2)$	$\mathfrak{so}(2, 1)$	$\mathbf{1}^{\oplus 3}$
Case 4'	$\mathfrak{so}(2n, 2)$	$\mathfrak{so}(2n, 1)$	$\mathbb{R}^{2n,1}$
Case 5–2	$\mathfrak{so}(4, 4)$	$\mathfrak{spin}(4, 1)$	$(\mathbb{R}^{4,1})^{\oplus 3} + \mathbf{1}^{\oplus 3}$
Case 6	$\mathfrak{so}(8, 8)$	$\mathfrak{spin}(8, 1)$	$\bigwedge^3 \mathbb{R}^{8,1}$
Case 7'	$\mathfrak{so}(4, 4)$	$\mathfrak{so}(4, 1)$	$(\mathbb{R}^{4,1})^{\oplus 3} + \mathbf{1}^{\oplus 3}$
Case 8	$\mathfrak{so}(8, \mathbb{C})$	$\mathfrak{spin}(7, 1)$	$\bigwedge^2 \mathbb{R}^{7,1}$
Case 9'	$\mathfrak{so}(8, \mathbb{C})$	$\mathfrak{so}(7, 1)$	$\bigwedge^2 \mathbb{R}^{7,1}$
Case 10	$\mathfrak{so}^*(8)$	$\mathfrak{spin}(6, 1)$	$\mathbb{R}^{6,1}$
Case 10'	$\mathfrak{so}^*(8)$	$\mathfrak{su}(3, 1)$	$\bigwedge^2 \mathbb{C}^{3,1} + \mathbf{1}$
Case 11	$\mathfrak{so}^*(8)$	$\mathfrak{spin}(6, 1)$	$\mathbb{R}^{6,1}$
Case 12'	$\mathfrak{so}(4, 3)$	$\mathfrak{so}(4, 1)$	$\mathbb{R}^{4,1} + \mathbf{1}^{\oplus 2}$

(Fact 3.1), which implies that we have $H^1(\Gamma, \mathfrak{g}) = 0$. This implies that the inclusion map $\iota\colon \Gamma \to G$ is locally rigid by a theorem of Weil [71] in Sect. 3.1. Therefore, by Lemma 5.2(1), ι is locally rigid as a discontinuous group for G/H. This gives a proof of Theorem 2.9 in Cases 1'–1, 2–1, 5–1, 5', 6', 7, 8', 9, 11', and 12 in Table 3.

The remaining cases are written out in Table 4 for the next steps.

Step 2. Suppose that $\mathfrak{l}_{ss}$ is isomorphic to $\mathfrak{so}(n, 1)$ or $\mathfrak{su}(n, 1)$. As preparation for the following steps, we will provide the irreducible decomposition of $\mathfrak{g}/\mathfrak{l}_{ss}$ as an $\mathfrak{l}_{ss}$-module in the following lemma, for the cases listed in Table 4:

Lemma 5.11 *In the cases listed in Table 4, the irreducible decomposition of the real vector space $\mathfrak{g}/\mathfrak{l}_{ss}$ as an $\mathfrak{l}_{ss}$-module is given. Here, $\mathbb{R}^{n,1}$ denotes the standard representation of $\mathfrak{l}_{ss} = \mathfrak{so}(n, 1)$ or $\mathfrak{spin}(n, 1)$, and $\mathbb{C}^{n,1}$ represents the standard representation of $\mathfrak{l}_{ss} = \mathfrak{su}(n, 1)$. The symbol **1** denotes the trivial representation.*

Step 3. We prove Cases 6, 8, and 9' in Table 4.

In these cases, $\mathfrak{l}_{ss}$ is isomorphic to $\mathfrak{so}(n, 1)$. In light of the irreducible decomposition of $\mathfrak{g}/\mathfrak{l}_{ss}$ in Table 4, as stated in Lemma 5.11, we again apply Raghunathan's vanishing theorem (Fact 3.1), and we have $H^1(\Gamma, \mathfrak{g}) = 0$ for any torsion-free cocompact discrete subgroup Γ of L. In these cases, as in Step 1, the inclusion map $\Gamma \to G$ is locally rigid as a discontinuous group for G/H.

Step 4. We consider Cases 1, 4–1, and 10'. In these cases, the following two conditions hold:

- $\mathfrak{l}_{ss}$ is isomorphic to $\mathfrak{su}(n, 1)$ $(n \geq 2)$;
- there exists a torsion-free cocompact discrete subgroup Γ of L such that $H^1(\Gamma, \mathfrak{g}) \neq 0$.

Proposition 5.12 *Suppose that (G, H, L) is one of Cases 1, 4–1, and 10'.*

(1) ((Q1) in Question 2.5 is yes.) There exists a torsion-free cocompact discrete subgroup Γ of L such that Γ is deformable as a discontinuous group for G/H.
(2) ((Q2) in Question 2.5 is no.) No torsion-free cocompact discrete subgroup Γ can be deformed into a non-standard discontinuous group for G/H.

Proof of Proposition 5.12 Before entering the proof, we observe that Case 4–1-b is locally isomorphic to Case 1 with $n = 1$, that is,

$$(G, H, L_{ss}) \approx (SU(2, 2), Sp(1, 1), SU(2, 1)),$$

whereas Case 10' is locally isomorphic to Case 4–1-a with $n = 3$, that is,

$$(G, H, L_{ss}) \approx (SO(6, 2), SO(6, 1), SU(3, 1)).$$

(1). This statement was proved in [43, Thm. B].

(2). Using the notation introduced in Notation 2.6, in all cases under consideration, L_{max} is isomorphic to $U(n, 1)$ up to finite covering. Accordingly, Γ is cocompact in L_{max}, and for the purpose of our proof, we may assume without loss of generality that $\mathfrak{l}$ is isomorphic to $\mathfrak{u}(n, 1)$.

We begin with the Cases 4–1-a and 10', in which the first cohomology group $H^1(\Gamma, \mathfrak{g}/\mathfrak{l}) = 0$ vanishes.

In Case 4–1-a, $\mathfrak{g}/\mathfrak{l}$ is isomorphic to $\bigwedge^2 \mathbb{C}^{n,1}$ as a $\mathfrak{u}(n, 1)$-module. Therefore, we have $H^1(\Gamma, \mathfrak{g}/\mathfrak{l}) = 0$ by the vanishing theorem of Raghunathan [60], see Fact 3.1. We apply Proposition E.2 in Appendix E to $\mathfrak{l} = \mathfrak{u}(n, 1)$ and conclude that any homomorphism $\Gamma \to G$ which is sufficiently close to the initial inclusion $\Gamma \to U(n, 1)$ remains inside $U(n, 1)$, up to conjugation by elements of G.

In Case 10', $\mathfrak{g}/\mathfrak{l}$ is isomorphic to $\bigwedge^2 \mathbb{C}^{3,1}$ as a $\mathfrak{u}(3, 1)$-module. Hence, the proof is similar to Case 4–1-a.

We next consider Cases 1 and 4–1-b. In these cases, the first cohomology group does not necessarily vanish, so the analysis is more involved. In these cases, the pair (L, G) is isomorphic finite covering of the pair

$$(U(n, 1), SU(n, 2)).$$

We prove the claim in the spirit of the original work of Goldman–Millson [16], which treated the pair $(SU(n, 1), SU(n + 1, 1))$. Applying their argument to the c-dual symmetric pair of $(U(n, 1), SU(n + 1, 1))$, namely to the symmetric pair $(L, G) = (U(n, 1), SU(n, 2))$, we confirm that the argument extends to pairs that are only locally isomorphic. This is carried out in the following two steps.

The first step consists of proving Proposition E.9. For future reference, we give a proof in Appendix E under more general assumptions, where we not only drop the requirement that $L \subset G$ form a reductive symmetric pair, but also eliminate the assumption that L and G are reductive altogether.

The second step is to show that the three assumptions of Proposition E.9 are satisfied for the specific pair $(L, G) = (U(n, 1), SU(n, 2))$. In the cases $(L, G) = (U(n, 1), SU(n + 1, 1))$ and $(U(n, 1), SU(n, 2))$, the only non-trivial assumption in Proposition E.9 is condition (3) therein. Condition (3), while formulated in terms of the cohomology of a Γ-module, can be reinterpreted—using the result of Matsushima—Murakami [55]—as a statement about harmonic forms with values in a local system over the compact complex hyperbolic manifold $\Gamma\backslash\mathbb{H}^n_{\mathbb{C}}$, which facilitates its verification. Moreover, the same argument remains valid in each step, provided the groups are locally isomorphic.

Since $U(n, 1)$ acts properly on G/H, this deformation remains standard.

Thus the proposition is shown. □

Remark 5.13 Alternatively, the second statement of Proposition 5.12 may be verified in Cases 1 and 4–1-b when L is locally isomorphic to $SU(n, 1)$, by applying Klingler's local rigidity (Fact 3.3) together with Remark 3.4, which asserts in this setting that any homomorphism $\Gamma \to G$ sufficiently close to the initial inclusion $\Gamma \hookrightarrow L$ is, up to conjugation, contained in $L \cdot Z_G(L) = U(n, 1)$.

Step 5. The remaining cases are Cases 1'-2, 2–2, 3, 4–2, 4', 5–2, 7', 10, 11, and 12'. In these cases, the following two conditions hold:

- $\mathfrak{l}_{ss}$ is isomorphic to $\mathfrak{so}(n, 1)$ $(n \geq 2)$;
- there exists a torsion-free cocompact discrete subgroup Γ of L such that $H^1(\Gamma, \mathfrak{g}) \neq 0$.

We take a closer look at the Lie algebra $\mathfrak{g}^{\varphi}$ in Definition 3.5, or the corresponding algebraic group G^{φ} in Definition 3.7:

Proposition 5.14 *Let (G, H, L) be one of Cases 1'-2, 2–2, 3, 4–2, 4', 5–2, 7', 10, 11, or 12'. We denote by φ the natural morphism $L_{ss} \to G$. Then we have $\mathfrak{g}^{\varphi} = \mathfrak{g}$.*

Note that Case 5–2 of Proposition 5.14 has already been proven in Example 3.20 (1) using $G(p, q)$, and Cases 10 and 11 have been shown in Example 3.20 (2) using $G(p, q)$.

Proof We recall from Definition 3.5 that $\mathfrak{g}^{\varphi}$ is the Lie algebra containing $\mathfrak{l}_{ss}$ and all the subrepresentations in the complementary subspace that contain non-zero $\mathfrak{l}_{ss}$-spherical harmonics. It then follows immediately from Lemma 5.11 that $\mathfrak{g}^{\varphi}$ coincides with $\mathfrak{g}$ in each of the above cases. □

Thus, from Theorem 5.9, we conclude that the answer to (Q3) is "yes" in these cases.

5.3 Proof of Theorem 2.19

In this section, we provide a proof of Theorem 2.19.

For any Zariski-connected real algebraic group G that is locally isomorphic to $SO(n, 1)$, there is a covering homomorphism $Spin(n, 1) \to G$. Hence, it suffices to prove the case where $G = Spin(n, 1)$. We apply Theorem 5.9 to the setting where $(G \times \{e\}, G \times G, \mathrm{diag}\, G)$ plays the role of (L, G, H).

Via the natural embedding $\varphi\colon G \times \{e\} \hookrightarrow G \times G$, the Lie algebra $\mathfrak{g} \oplus \mathfrak{g}$ is decomposed into irreducible representations $\mathfrak{g} + \mathbf{1}^{\oplus \dim \mathfrak{g}}$ as a $\mathfrak{g} \oplus \{0\}$-module.

Then it follows from Definition 3.5 that $(\mathfrak{g} \oplus \mathfrak{g})^{\varphi}$ coincides with $\mathfrak{g} \oplus \mathfrak{g}$. Thus, from Theorem 5.9, we conclude that there exist a torsion-free, cocompact discrete subgroup Γ of G and a small deformation Γ' of $\Gamma \times \{e\}$ such that Γ' is Zariski-dense in the direct product group $G \times G$, while preserving the proper discontinuity of the action on the group manifold $(G \times G)/\mathrm{diag}\, G$.

5.4 Proof Of Theorem 2.21

The majority of the proof of Theorem 2.21 in the $\Gamma_L \backslash G / \Gamma_H$ case reduces to the proof of Theorem 2.9 in the $\Gamma \backslash G / H$ case, given in Sect. 5.2. In addition, let us prepare the necessary lemmas and propositions in advance.

In the sequel, we assume that G is a Zariski-connected real reductive algebraic group, H and L are real reductive algebraic subgroups of G, Γ_H and Γ_L are torsion-free cocompact discrete subgroups of H and L, respectively, and that $\iota_H : \Gamma_H \hookrightarrow G$, $\iota_L : \Gamma_L \hookrightarrow G$ denote the natural embeddings.

The proof of the following lemma is straightforward and we omit it.

Lemma 5.15 *Suppose that $\varphi \in \mathrm{Hom}(\Gamma_H \times \Gamma_L, G \times G)$ is sufficiently close to the original embedding $\iota_H \times \iota_L : \Gamma_H \times \Gamma_L \to G \times G$. Then, for each group $J = H$ or L, there exist group homomorphisms*

- *$j_J : \Gamma_J \hookrightarrow G$, sufficiently close to the original ι_J,*
- *$\rho_J : \Gamma_J \to G$, sufficiently close to the trivial homomorphism,*

such that

$$\varphi(\gamma_H, \gamma_L) = (j_H(\gamma_H)\rho_L(\gamma_L), j_L(\gamma_L)\rho_H(\gamma_H)) \tag{5.1}$$

for all $\gamma_H \in \Gamma_H$ and $\gamma_L \in \Gamma_L$.

Moreover, we have

$$\rho_L(\Gamma_L) \subset Z_G(H'),$$
$$\mathrm{pr}_1 \circ \varphi(\Gamma_H \times \Gamma_L) \subset H' \cdot Z_G(H'),$$

where $\mathrm{pr}_1 : G \times G \to G$ denotes the first projection, H' is the Zariski-closure of $j_H(\Gamma_H)$, and $Z_G(H')$ denotes the centralizer of H' in G.

Proposition 5.16 *Suppose that the direct product group $H \times L$ acts properly on the group manifold $(G \times G)/\mathrm{diag}(G)$ ($\simeq G$). Furthermore, we assume the following three conditions:*

(a) Both the inclusion map $\iota_H : \Gamma_H \hookrightarrow G$ and the trivial representation $\mathbf{1}_H : \Gamma_H \to G$ are locally rigid;
(b) The image of any small deformation of $\iota_L : \Gamma_L \hookrightarrow G$ remains inside L, up to conjugation by elements of G;
(c) The centralizer $Z_G(H)$ of H in G is a compact group.

Under these assumptions, $\Gamma_H \times \Gamma_L$ cannot be deformed into a non-standard discontinuous group for the group manifold $(G \times G)/\mathrm{diag}(G) \simeq G$.

Proof Suppose that $\varphi \in \mathcal{R}(\Gamma_H \times \Gamma_L, G \times G; G)$ is a small deformation of $\iota_H \times \iota_L$. It follows from Lemma 5.15 that φ is of the form (5.1) for some group homomorphisms $j_H, \rho_H : \Gamma_H \to G$, $j_L, \rho_L : \Gamma_L \to G$.

By the assumptions (a) and (b), if necessary, we can replace φ by an appropriate conjugation in $G \times G$, and assume that $j_H = \iota_H$, $\rho_H = \mathbf{1}_H$, and $j_L(\Gamma_L) \subset L$. Furthermore, by Borel's density theorem (e.g., [61, Cor. 5.16 (ii)]), Γ_H is Zariski-dense in H. Applying again Lemma 5.15, we obtain

$$\varphi(\Gamma_H \times \Gamma_L) \subset (H \cdot Z_G(H)) \times L.$$

Since $Z_G(H)$ is compact by the assumption (c), the group $(H \cdot Z_G(H)) \times L$ acts properly on the group manifold G. Hence, $\varphi(\Gamma_H \times \Gamma_L)$ is a standard discontinuous group for G. □

Proposition 5.17 *Retain the setting and the assumption (a) in Proposition 5.16. Furthermore, suppose that the assumptions (b) and (c) are strengthened by the following stronger assumptions:*

(b') $\iota_L : \Gamma_L \hookrightarrow G$ is locally rigid;
(c') The centralizer $Z_G(H)$ of H in G is a finite group.

Then, $\iota_H \times \iota_L$ is not deformable as a discontinuous group for the group manifold $(G \times G)/\operatorname{diag}(G)$ $(\simeq G)$.

Proof As we already know from the proof of Proposition 5.16, $\rho_L(\Gamma_L)$ is contained in $Z_G(H)$, which is a finite group in our setting. Hence, the small deformation ρ_L must be the trivial representation. Thus, the local rigidity follows. □

Proposition 5.18 *Assume the following:*

(a) G is simple and Zariski-connected;
(b) The inclusion map $\iota_H : \Gamma_H \to G$ cannot be deformed into a Zariski-dense discrete subgroup of G.

Under these assumptions, the direct product group $\Gamma_H \times \Gamma_L$ cannot be deformed into a Zariski-dense discrete subgroup of $G \times G$.

Proof Take any $\varphi \in \mathcal{R}(\Gamma_H \times \Gamma_L, G \times G)$ that is suffiently close to the original embedding $\iota_H \times \iota_L$. Then, by Lemma 5.15, there exist group homomorphisms $j_H, \rho_H : \Gamma_H \to G$ and $j_L, \rho_L : \Gamma_L \to G$ satisfying (5.1). Let H' be the Zariski-closure of $j_H(\Gamma_H)$ in G. Since $j_H(\Gamma_H) \simeq \Gamma_H$ is an infinite group, H' has positive dimension. On the other hand, by the assumption (b), H' is a proper algebraic subgroup of G. Therefore, by the assumption (a), we can conclude that $H' \cdot Z_G(H')$ is a proper algebraic subgroup of the simple algebraic group G. On the other hand, from Lemma 5.15, we have

$$\mathrm{pr}_1 \circ \varphi(\Gamma_H \times \Gamma_L) \subset H' \cdot Z_{G'}(H').$$

Therefore, the image of φ cannot be Zariski-dense. □

We are ready to prove Theorem 2.21.

Proof of Theorem 2.21 (1). The equivalence of (i) and (iii): It follows from Theorem 2.15 that collecting the triples (G, H, L) for which the answer to (Q1) is "yes" for at least one of Case i or Case i' in Table 3 yields the classification in (iii).

The implication from (i) to (ii) is straightforward. In fact, if Γ_L is deformable as a discontinuous group for G/H, then it is also deformable for G/Γ_H. Consequently, the direct product group $\Gamma_L \times \Gamma_H$ is deformable as a discontinuous group for the group manifold $G \times G/\operatorname{diag}(G)$.

Let us prove the implication from (ii) to (iii) by contraposition. Suppose that the triple (G, H, L) belongs to one of the following cases, up to switching H and L, and up to compact factors:

- Case 5. $(SO(4n, 4), SO(4n, 3), Sp(n, 1))$ $(n \geq 2)$;
- Case 6. $(SO(8, 8), SO(8, 7), Spin(8, 1))$;
- Case 8. $(SO(8, \mathbb{C}), SO(7, \mathbb{C}), Spin(7, 1))$;
- Case 9. $(SO(8, \mathbb{C}), Spin(7, \mathbb{C}), SO(7, 1))$.

Let Γ_H and Γ_L be torsion-free cocompact discrete subgroups of H and L, respectively. Then the triples (G, H, L) satisfy all the assumptions of Propositions 5.16 and 5.17. In fact, the assumptions (a) in Proposition 5.16 and (b') in Proposition 5.17 follow from the vanishing of the first cohomologies, $H^1(\Gamma_H, \mathfrak{g}) = H^1(\Gamma_L, \mathfrak{g}) = 0$, which is verified using Lemma 5.11 and Raghunathan's vanishing theorem (Fact 3.1). Furthermore, it is clear that $Z_G(H)$ is a finite group in each case. Hence, by Proposition 5.17, $\Gamma_H \times \Gamma_L$ is locally rigid as a discontinuous group for the group manifold G.

(2). The equivalence of (i) and (iii) has been proven in Theorem 2.15. Let us verify the implications (i)⇒(ii) and (ii)⇒(iii).

(i)⇒(ii): Suppose that (i) holds. Then, by Theorem 2.15(i)⇒(ii), one can find a small deformation $j_L \in \mathcal{R}(\Gamma, G; G/H)$ of the original embedding $\iota_L : \Gamma_L \to G$ such that $j_L(\Gamma)$ is Zariski-dense in G, while preserving the proper discontinuity of the action on G/H. We take any cocompact discrete subgroup Γ_H of H, and denote by $\iota_H : \Gamma_H \hookrightarrow G$ the natural embedding map. Then, $\iota_H \times j_L$ gives an element of $\mathcal{R}(\Gamma_H \times \Gamma_L, G \times G; G)$, since the Zariski-closure of $\Gamma_H \times j_L(\Gamma_L)$ is equal to $H \times G$, which cannot act properly on the group manifold because H is non-compact.

Thus, $\Gamma_H \times j_L(\Gamma_L)$ is a non-standard discontinuous group for the group manifold $G \simeq (G \times G)/\operatorname{diag} G$.

(ii)⇒(iii):

We prove the implication from (ii) to (iii) by contraposition. The only triples that are not in the list of (iii), yet satisfy the condition of being deformable (i.e., fulfilling the classification in Theorem 2.21(1), are the following:
Cases 1 and 2: $(SU(2n, 2), Sp(n, 1), U(2n, 1))$ $(n \geq 2)$.

This triple (G, H, L) satisfies all the assumptions of Proposition 5.16. In fact, the assumption (a) follows from Raghunathan's vanishing theorem (Fact 3.1), and the assumption (c) can be easily verified. Moreover, the assumption (b) is verified using Klingler's local rigidity theorem, which holds for Γ as a torsion-free cocompact discrete subgroup of $SU(n, 1)$, and also holds for $U(n, 1)$. Thus, we conclude from Proposition 5.16 that (ii) is not true. Therefore, the implication (ii)⇒(iii) is established.

(3). We observe that the answer to (Q3) is "no" for at least one of Case i or Case i' in Table 3, except for Case 4.2, where $G = SO(2,2)$ is not simple. Hence, since G is a simple Lie group, the direct product group $\Gamma_H \times \Gamma_L$ cannot be deformed into a Zariski-dense subgroup of $G \times G$, as stated in Proposition 5.18.

Thus, the proof of Theorem 2.21 is complete. □

We have excluded the case $G = SO(2,2)$ in Theorem 2.21(3), which is not a simple Lie group, and the conclusion differs as stated in Proposition 2.22. We now provide a proof of Proposition 2.22. Specifically, this occurs in Cases 3 and 4 ($n = 1$), i.e., in the following case where

$$(G, H, L) = (SO(2,2), SO(2,1), SU(1,1))$$

up to the switching of H and L, and up to compact factors of H and L.

Proof of Proposition 2.22 Let $\iota_H\colon H \hookrightarrow G$ and $\iota_L\colon L \hookrightarrow G$ denote the natural embeddings. The Lie algebra $\mathfrak{g} = \mathfrak{so}(2,2)$ is decomposed into the sum of irreducible representations of $H = SO(2,1)$ and $L = SU(1,1)$ via the adjoint actions as follows:

$$\mathfrak{g} \simeq \mathfrak{h} + \mathbb{R}^{2,1}, \quad \mathfrak{g} \simeq \mathfrak{l} + \mathbf{1}^{\oplus 3}.$$

Then, we have $\mathfrak{g}^{\iota_H} = \mathfrak{g}^{\iota_L} = \mathfrak{g}$. Applying Theorem 3.9 to $H = SO(2,1)$ and $L = SU(1,1) \simeq Spin(2,1)$, there exist torsion-free cocompact discrete subgroups Γ_H and Γ_L of H and L, respectively, that can be deformed into Zariski-dense subgroups of G. Therefore, the direct product group $\Gamma_H \times \Gamma_L$ also has a small deformation $\phi\colon \Gamma_H \times \Gamma_L \to G \times G$ that is Zariski-dense in $G \times G$. By the stability of proper discontinuity (Fact 5.7), the action of $\phi(\Gamma_H \times \Gamma_L)$ on G is properly discontinuous. Hence, $\Gamma_H \times \Gamma_L$ can be deformed into a Zariski-dense subgroup of $G \times G$, preserving the proper discontinuity of the action on the group manifold $G = (G \times G)/\operatorname{diag}(G)$. □

5.5 Some Results for Problem 1.2

In this section, we present several results regarding Problem 1.2, namely, the existence problem of Zariski-dense discontinuous groups Γ for $X = G/H$, when the cohomological dimension $\mathrm{cd}_{\mathbb{R}}(\Gamma)$ is between 2 and $d(X)$.

First, we consider the case where Γ is isomorphic to the surface group $\pi_1(\Sigma_g)$, for which $\mathrm{cd}_{\mathbb{R}}(\Gamma) = 2$. In this case, by combining Theorem 5.9 (for $n = 2$) with Okuda's results, as discussed in Remark 5.20 below, we obtain the following theorem. For the proof, see [24].

Corollary 5.19 ([24, Thm. 5.4 (v)⇒(iii)]) *Let $X = G/H$ be a symmetric space, where G is a Zariski-connected real reductive algebraic group. Assume that there exists a discontinuous group for X which is isomophic to a surface group $\pi_1(\Sigma_g)$. Then, among such discontinuous groups, there exists a subgroup that is Zariski-dense in G.*

Remark 5.20 Several equivalent statements are known for the assumption in Corollary 5.19, which we recall from Okuda [57, 58]. For a semisimple symmetric space $X = G/H$, the following six conditions are equivalent:

(i) X admits a proper action of $SL(2, \mathbb{R})$ via a homomorphism $SL(2, \mathbb{R}) \to G$;
(ii) X admits a proper action of $SL(2, \mathbb{R})$ via an *even* homomorphism $SL(2, \mathbb{R}) \to G$ (Definition-Lemma 3.16);
(iii) X admits a properly discontinuous action of the free group F_k of rank k for *some* $k \geq 2$ via a group homomorphism $F_k \to G$;
(iv) X admits a properly discontinuous action of the free group F_k of rank k for *any* $k \geq 2$ via a group homomorphism $F_k \to G$;
(v) X admits a properly discontinuous action of the surface group $\pi_1(\Sigma_g)$ of genus g for *some* $g \geq 2$ via a group homomorphism $\pi_1(\Sigma_g) \to G$;
(vi) X admits a properly discontinuous action of the surface group $\pi_1(\Sigma_g)$ of genus g for *any* $g \geq 2$ via a group homomorphism $\pi_1(\Sigma_g) \to G$.

The above equivalence between (v) and (vi) suggests the following question:

Question 5.21 Let $\pi_1(\Sigma_g)$ be as in the assumption of Corollary 5.19. Does there exist a Zariski-dense subgroup in G isomorphic to $\pi_1(\Sigma_g)$?

We also do not know the answer to the following question:

Question 5.22 Can the assumption that G/H is symmetric be removed in Corollary 5.19?

The following result provides yet another perspective on deformations with discontinuous groups of lower cohomological dimension for reductive symmetric spaces.

Corollary 5.23 *Let $X = G/H$ be a symmetric space, where G is a Zarski-connected real semisimple algebraic group. If X admits a finitely-generated discontinuous group which is Zariski-dense in G, then there exists a discrete subgroup Γ of G with the following three properties: the comological dimension of Γ is greater than 1, Γ is Zariski-dense in G, and Γ acts properly discontinuously on X.*

Lemma 5.24 *Let G be a Zariski-connected real algebraic group, Γ a Zariski-dense subgroup of G, and Γ' a finite-index normal subgroup of Γ. Then Γ' is also Zariski-dense in G.*

Proof Let $G'_{\mathbb{C}}$ be the Zariski-closure of Γ' in the set $G_{\mathbb{C}}$ of complex points of G. Then $G'_{\mathbb{C}}$ is a Zariski-closed normal subgroup of $G_{\mathbb{C}}$. Hence, $G_{\mathbb{C}}/G'_{\mathbb{C}}$ is a complex algebraic group. The image of the finite group Γ/Γ' into $G_{\mathbb{C}}/G'_{\mathbb{C}}$ is Zariski-dense. Since a finite subgroup of an algebraic group is Zariski-closed, we see that $G_{\mathbb{C}}/G'_{\mathbb{C}}$ is a finite group. Hence, we have $\dim G'_{\mathbb{C}} = \dim G_{\mathbb{C}}$. Since $G_{\mathbb{C}}$ is Zariski-connected, $G'_{\mathbb{C}}$ coincides with $G_{\mathbb{C}}$, which means that Γ' is Zariski-dense in G. This proves our lemma. □

Proof of Corollary 5.23 Suppose that Λ is a finitely-generated, Zariski-dense discrete subgroup of G such that it acts properly on X. By Selberg's lemma [62, Lem. 8], there exists a torsion-free normal subgroup Λ' of Λ such that Λ' is of finite index in Λ. Since Λ' is also Zariski-dense in G by Lemma 5.24, and since the cohomological dimension $\mathrm{cd}_{\mathbb{R}}(\Lambda')$ is the same with $\mathrm{cd}_{\mathbb{R}}(\Lambda)$, we may replace Λ with Λ' and assume that Λ is torsion-free.

Since Λ is Zariski-dense in G, the cohomological dimension $\mathrm{cd}_{\mathbb{R}}(\Lambda)$ of Λ must be positive. If $\mathrm{cd}_{\mathbb{R}}(\Lambda) \geq 2$, we can take $\Gamma = \Lambda$. Suppose $\mathrm{cd}_{\mathbb{R}}(\Lambda) = 1$.

By the Stallings theorem [65], Λ must be a free group F_k. If $k = 1$, then Λ is an abelian group, which cannot be Zariski-dense in the non-abelian group G. If $k \geq 2$, then the proper discontinuity of the action of Λ on X implies that there exists an even homomorphism $\varphi\colon SL(2, \mathbb{R}) \to G$ (Definition–Lemma 3.16) such that $SL(2, \mathbb{R})$ acts properly on X via φ, by Remark 5.20. Then, there exist a discrete subgroup in $\varphi(SL(2, \mathbb{R}))$ that is isomorphic to a surface group, and a small deformation of it into a Zariski-dense subgroup of G, as shown in Corollary 3.17. This small deformation preserves the proper discontinuity of the action on $X = G/H$, as stated by the stability theorem (Fact 5.6). See also Theorem 5.9 in the case $n = 2$. □

Next, we consider the case where the cohomological dimension of the discontinuous group Γ for $X = G/H$ is greater than 2. As mentioned in Remark 1.5(1), there exist natural families of homogeneous spaces on which $SO(n, 1)$ does not act properly, while $Spin(n, 1)$ does act properly. These families yield a somewhat surprising connection between the existence of proper actions of $Spin(n, 1)$ and the Hurwitz–Radon number, as will be detailed in the forthcoming paper [25] by Kannaka and Tojo, for which we now provide a brief overview and applications of Theorem 5.9.

Let $\mathrm{ord}_2\colon \mathbb{Q}^\times \to \mathbb{Z}$ denote the 2-adic valuation, that is, for $t = 2^m(\mathrm{odd})/(\mathrm{odd})$ with $m \in \mathbb{Z}$, we define $\mathrm{ord}_2(t) := m$. Furthermore, when $\mathrm{ord}_2 t = 4a + b$ with $a \in \mathbb{Z}$ and $0 \leq b \leq 3$, we define $\rho(t) = 8a + 2^b$, which is called the *Hurwitz–Radon number* of $t \in \mathbb{Q}^\times$. Since $\rho(t)$ depends only on $\mathrm{ord}_2 t$, we summarize its values for powers of 2 in Table 5.

Table 5 Table of $\rho(t)$

t	...	1/16	1/8	1/4	1/2	1	2	4	8	16	...
$\rho(t)$	...	−7	−6	−4	0	1	2	4	8	9	...

Table 6 The parameters p and q range over all integers satisfying $0 \leq p < N/2$ and $0 \leq q < N$, respectively

G	H	$\rho(\mathfrak{g}, \iota)$
$SL(2N, \mathbb{R})$	$SO(N+1, N-1)$	$\rho(N)+1$
$SL(2N, \mathbb{R})$	$SL(2p+1, \mathbb{R}) \times SL(2N-(2p+1), \mathbb{R})$	$\rho(N)+1$
$SL(2N, \mathbb{C})$	$SU(N+1, N-1)$	$2\,\mathrm{ord}_2(N)+3$
$SL(2N, \mathbb{C})$	$SL(2p+1, \mathbb{C}) \times SL(2N-(2p+1), \mathbb{C})$	$2\,\mathrm{ord}_2(N)+3$
$SL(2N, \mathbb{H})$	$Sp(N+1, N-1)$	$\rho(N/2)+5$
$SL(2N, \mathbb{H})$	$SL(2p+1, \mathbb{H}) \times SL(2N-(2p+1), \mathbb{H})$	$\rho(N/2)+5$
$SO(4N, \mathbb{C})$	$SO(2p+1, \mathbb{C}) \times SO(4N-2p-1, \mathbb{C})$	$\rho(N/4)+7$
$SO(4N, \mathbb{C})$	$SO(2N+1, 2N-1)$	$\rho(N/4)+7$
$SU(N, N)$	$S(U(p, p+1) \times U(N-p, N-p-1))$	$2\,\mathrm{ord}_2(N)+2$
$SO(N, N)$	$SO(p, p+1) \times SO(N-q, N-q-1)$	$\rho(N)$
$SO^*(4N)$	$SO^*(4p+2) \times SO^*(4N-4p-2)$	$\rho(N/4)+6$
$SO^*(4N)$	$U(N+1, N-1)$	$\rho(N/4)+6$
$Sp(N, N)$	$Sp(p, p+1) \times Sp(N-p, N-p-1)$	$\rho(N/2)+4$
$Sp(N, \mathbb{R})$	$Sp(N-1, \mathbb{R})$	$\rho(N/2)+2$
$Sp(N, \mathbb{C})$	$Sp(N-1, \mathbb{C})$	$\rho(N/2)+3$

In [25], for a pair $(\mathfrak{g}, \iota)$ of a reductive Lie algebra $\mathfrak{g}$ and its completely reducible representation $\iota\colon \mathfrak{g} \to \mathfrak{gl}(N, \mathbb{C})$, a certain integer $\rho(\mathfrak{g}, \iota)$ is introduced. It is shown that for a classical Lie algebra $\mathfrak{g}$ and its natural representation ι, the value of $\rho(\mathfrak{g}, \iota)$ can be computed using either the Hurwitz–Radon number $\rho(N)$ or the 2-adic valuation $\mathrm{ord}_2(N)$. In particular, for the standard representation $\iota\colon \mathfrak{so}(N, N) \to \mathfrak{gl}(2N, \mathbb{C})$,

$$\rho(\mathfrak{so}(N, N), \iota) = \rho(N).$$

Furthermore, for certain families of homogeneous spaces G/H, the semisimple Lie groups that can act properly on G/H can be classified using $\rho(\mathfrak{g}, \iota)$ as follows:

Theorem 5.25 ([25]) *Let G/H be one of the classical homogeneous spaces in Table 6, ι the standard representation of the classical Lie algebra $\mathfrak{g}$, and L a connected semisimple Lie group with no compact factors. Then G/H admits a proper L-action via a homomorphism from L to G if and only if L is isomorphic to $Spin(n, 1)$ for some $2 \leq n \leq \rho(\mathfrak{g}, \iota)$.*

For example, the space form $SO(N, N)/SO(N-1, N)$ admits a proper action of a connected semisimple Lie group L with no compact factors if and only if L is isomorphic to $Spin(n, 1)$ for some $2 \leq n \leq \rho(N)$.

Furthermore, the following lemma provides a criterion to determine which actions of $Spin(n, 1)$ are proper on the homogeneous spaces G/H listed in Table 6. This lemma is proved based on the properness criterion [37, Thm. 4.1] established by Kobayashi:

Lemma 5.26 ([25]) *Let G/H be one of the classical homogeneous spaces in Table 6, and ι the standard representation of the classical group G. Then, for any non-trivial homomorphism $\varphi\colon Spin(n,1) \to G$ ($n \geq 2$), the $Spin(n,1)$-action on G/H via φ is proper if and only if $\iota(\varphi(-1)) = -I$, where 1 is the identity element in the Clifford algebra $C_{\text{even}}(n,1)$ and I is the identity matrix. In particular, any subgroup of G isomorphic to $SO_0(n,1)$ cannot acts properly on G/H.*

We now return to one of our main themes, Problem 1.2, and provide examples of discontinuous groups Γ for certain family homogeneous spaces $X = G/H$ that are Zariski-dense subgroups of G with cohomological dimension 6. To be more precise, we shall take G to be the identity component in the Zariski-topology, $G(p,q)_0$, of the group $G(p,q)$ (see Appendix B for the definition). We take Γ to be the discrete subgroups of $G(p,q)_0$ constructed by combining Theorem 3.9(1) with Example 3.20, which are of cohomological dimension 6 and Zariski-dense.

As a corollary of Lemma 5.26, the following result follows immediately:

Proposition 5.27 *Among the classical homogeneous spaces G/H listed in Table 6, we consider those for which there exist some $p, q \in \mathbb{N}_+$ such that G is isomorphic to $G(p,q)_0$. For any natural number $n \leq p$, the action of $Spin(n,1)$ on G/H, induced via the inclusion $Spin(n,1) \to G(p,q)_0 \simeq G$, is proper.*

By combining Proposition 5.27, Example 3.20, and Theorem 5.9, we obtain the following:

Theorem 5.28 *Every homogeneous space $X = G/H$ in Table 7 admits a discontinuous group for X which is a Zariski-dense subgroup of G with cohomological dimension 6.*

As seen in the proof of [37, Cor. 5.5], the quantity $d(X) - \mathrm{cd}_{\mathbb{R}}(\Gamma)$ represents the degree of deviation from compactness of X_Γ.

We extract examples of homogeneous spaces $X = G/H$ for which $d(X) \leq 16$ from Table 7.

$$X = \begin{cases} SO^*(8)/(SO^*(6) \times SO^*(2)) & (d(X) = 6), \\ SO^*(8)/U(3,1) & (d(X) = 6), \\ SO(8,\mathbb{C})/SO(7,\mathbb{C}) & (d(X) = 7), \\ SO(8,8)/SO(7,8) & (d(X) = 8), \\ SO^*(16)/(SO^*(14) \times SO^*(2)) & (d(X) = 14), \\ SL(16,\mathbb{R})/SL(15,\mathbb{R}) & (d(X) = 16), \\ SU(8,8)/U(7,8) & (d(X) = 16). \end{cases}$$

In the first two cases where $d(X) = 6$, the quotient space X_Γ is compact, and these cases have already been discussed in Theorem 2.15. In contrast, X_Γ is non-compact in the other cases where $d(X) > 6$.

To analyze the results obtained here, we introduce two numerical invariants related to discontinuous groups for homogeneous spaces and their deformations.

Definition 5.29 Let $X = G/H$ be a homogeneous space of reductive type.

(1) Let $\delta(X)$ denote the maximum cohomological dimension $\mathrm{cd}_{\mathbb{R}}(\Gamma)$ among all finitely-generated discontinuous groups Γ for X.
(2) Let $\delta_Z(X)$ denote the maximum cohomological dimension $\mathrm{cd}_{\mathbb{R}}(\Gamma)$ among all finitely-generated discontinuous groups Γ for X such that Γ is Zariski-dense in G. We set $\delta_Z(X) := 0$ if there is no such discontinuous group Γ.

By definition and from [37, Cor. 5.5], we have the following inequalities:

$$\delta_Z(X) \leq \delta(X) \leq d(X),$$

where we recall from (1.2) that $d(X)$ is the "non-compact dimension" of X, which can be explicitly computed. If the answer to Question (Q3) in Question 2.5 is affirmative, then we have the following equality:

$$\delta_Z(X) = \delta(X) = d(X).$$

We summarize some established results and new findings in terms of these numerical invariants:

(1) ([37, Cor. 4.4]) $\delta(X) = 0$ if and only if $\mathrm{rank}_{\mathbb{R}}(G) = \mathrm{rank}_{\mathbb{R}}(H)$ (the criterion for the Calabi-Markus phenomenon).
(2) If a reductive subgroup L of G acts properly on X, then $d(L) \leq \delta(X)$.
(3) ([37, Cor. 5.5]) $\delta(X) = d(X)$ if and only if X admits a cocompact discontinuous group.
(4) (cf. Benoist [2, Thm. 1.1]) A criterion for $\delta_Z(X) = 0$ is explicitly given in terms of the action of the Weyl group.

Table 7 Some homogeneous spaces admitting Zariski-dense discontinuous groups of cohomological dimension 6

G	H	p	$d(X)$
$SL(16, \mathbb{R})$	$SO(7, 9)$		72
$SL(16, \mathbb{R})$	$SL(2p+1, \mathbb{R}) \times SL(15-2p, \mathbb{R})$	$0 \leq p \leq 3$	$-4p^2 + 28p + 16$
$SL(8, \mathbb{H})$	$SL(2p+1, \mathbb{H}) \times SL(7-2p, \mathbb{H})$	$p = 0, 1$	$-16p^2 + 48p + 29$
$Sp(16, \mathbb{R})$	$Sp(p, \mathbb{R}) \times Sp(15-p, \mathbb{R})$	$0 \leq p \leq 7$	$2(p+1)(16-p)$
$Sp(16, \mathbb{C})$	$Sp(p, \mathbb{C}) \times Sp(15-p, \mathbb{C})$	$0 \leq p \leq 7$	$-4p^2 + 60p + 63$
$SO(8, 8)$	$SO(p, p+1) \times SO(8-p, 7-p)$	$0 \leq p \leq 3$	$-2p^2 + 14p + 8$
$SU(8, 8)$	$S(U(p, p+1) \times U(8-p, 7-p))$	$0 \leq p \leq 3$	$-4p^2 + 28p + 16$
$Sp(8, 8)$	$Sp(p, p+1) \times Sp(8-p, 7-p)$	$0 \leq p \leq 3$	$-8p^2 + 56p + 32$
$SO^*(8)$	$SO^*(6) \times SO^*(2)$		6
$SO^*(8)$	$U(3, 1)$		6
$SO^*(16)$	$SO^*(4p+2) \times SO^*(14-4p)$	$p = 0, 1$	$2(2p+1)(7-2p)$
$SO^*(16)$	$U(3, 5)$		26
$SO(8, \mathbb{C})$	$SO(2p+1, \mathbb{C}) \times SO(7-2p, \mathbb{C})$	$p = 0, 1$	$(2p+1)(8-2p)$

(5) (Theorem 2.15) $\delta_Z(X) = \delta(X) = d(X)$ holds when the homogeneous spaces G/H are among those listed in (iii) of Theorem 2.15.
(6) (Theorem 2.19) When $X = (G \times G)/\operatorname{diag} G$ with $\mathfrak{g} = \mathfrak{so}(n, 1)$, then $\delta_Z(X) = d(G) = n$.
(7) (Corollary 5.23) For any semisimple symmetric space $X = G/H$, $\delta_Z(X) \neq 1$.
(8) (Theorem 5.28) $6 \leq \delta_Z(X)$ for all $X = G/H$ in Table 7.

We do not know the explicit values of $\delta_Z(X)$ and $\delta(X)$ in the general setting. It may be of interest to provide effective lower or upper bounds for these numerical invariants.

6 Spectral Analysis of X_Γ Under Small Deformation of Discontinuous Groups Γ

In this section, we discuss analytic perspectives of the quotient space X_Γ. Spectral analysis on locally symmetric spaces, beyond the classical Riemannian setting, is a rapidly developing area of research [29–32, 45], with many open problems remaining to be addressed. We focus on the aspects of this analysis from the viewpoint of deformations of the discontinuous group Γ, which we have discussed in Sects. 2–5. At the end of this section, we will present a set of questions for further exploration.

6.1 Discrete Spectra on Compact Quotients

We begin by outlining some basic setup and notation. For more details, we refer to the forthcoming book [32], as well as the article introducing open problems and providing a brief survey [47].

Let $X = G/H$ be a homogeneous space of reductive type. The manifold X carries a pseudo-Riemannian structure with a transitive isometry group; namely, the reductive group G acts transitively and isometrically on X. Such a pseudo-Riemannian structure can be defined using the Killing form B if G is semisimple.

Let $\mathbb{D}_G(X)$ denote the algebra of G-invariant differential operators on X. Suppose that Γ is a discontinuous group for X. Then, any $D \in \mathbb{D}_G(X)$ induces a differential operator D_Γ on the quotient $X_\Gamma = \Gamma\backslash X$ via the covering map $\varpi \colon X \to X_\Gamma$. We consider the set

$$\mathbb{D}(X_\Gamma) := \{D_\Gamma \mid D \text{ is a } G\text{-invariant differential operator on } X\}$$

as the algebra of *intrinsic differential operators* on the (G, X)-manifold X_Γ. Clearly, the correspondence $D \mapsto D_\Gamma$ defines an isomorphism between the algebras $\mathbb{D}_G(X)$ and $\mathbb{D}(X_\Gamma)$. For example, X_Γ inherits a pseudo-Riemannian structure from X via ϖ, and the Laplacian Δ_{X_Γ} on X_Γ is locally the same operator Δ_X on X via the diffeomorphism ϖ; hence, Δ_{X_Γ} belongs to $\mathbb{D}(X_\Gamma)$. For global analysis on X_Γ, it is important to note that the Laplacian is no longer an elliptic differential operator in our setting, where the metric is not positive-definite.

From now on, we assume that $X = G/H$ is a symmetric space, where G is a real reductive algebraic group. Let $\mathfrak{q}$ be the orthogonal complementary subspace of $\mathfrak{h}$ in $\mathfrak{g}$ with respect to an $\mathrm{Ad}(H)$-invariant non-degenerate symmetric bilinear form (e.g., the Killing form if $\mathfrak{g}$ is semisimple) of the Lie algebra $\mathfrak{g}$. We fix a maximal semisimple abelian subspace $\mathfrak{j}$ of $\sqrt{-1}\mathfrak{q}$ and denote by W the Weyl group of the restricted root system of $(\mathfrak{g}_\mathbb{C}, \mathfrak{j}_\mathbb{C})$. The Harish-Chandra isomorphism $\mathbb{D}_G(X) \simeq S(\mathfrak{j}_\mathbb{C})^W$ gives the bijection

$$\mathrm{Hom}_{\mathbb{C}\text{-alg}}(\mathbb{D}_G(X), \mathbb{C}) \simeq \mathfrak{j}_\mathbb{C}^\vee/W \quad (\chi_\lambda \leftrightarrow \lambda).$$

In this section, we explore the following object related to locally symmetric spaces $X_\Gamma = \Gamma\backslash G/H$, which goes beyond the classical Riemannian case:

$$\begin{aligned} L^2(X_\Gamma, \mathcal{M}_\lambda) &:= \{f \in L^2(X_\Gamma) \mid D_\Gamma f = \chi_\lambda(D) f \text{ for any } D \in \mathbb{D}_G(X)\}, \\ \mathrm{Spec}_d(X_\Gamma) &:= \{\lambda \in \mathfrak{j}_\mathbb{C}^\vee/W \mid L^2(X_\Gamma, \mathcal{M}_\lambda) \neq 0\}, \end{aligned}$$

where $\chi_\lambda \colon \mathbb{D}_G(X) \to \mathbb{C}$ is the algebra homomorphism associated to $\lambda \in \mathfrak{j}_\mathbb{C}^\vee/W$. The differential equation

$$D_\Gamma f = \chi_\lambda(D) f$$

is interpreted in the weak sense, or in the sense of distributions. We refer to an element in $\mathrm{Spec}_d(X_\Gamma)$ as a *discrete spectrum* of X_Γ.

Example 6.1 If $X = G/H$ is a symmetric space of rank one, such as $SO(p, q+1)/SO(p, q)$, then the $\mathbb{C}$-algebra $\mathbb{D}_G(X)$ is generated by the Laplacian of the pseudo-Riemannian manifold X. In this case, $\mathrm{Spec}_d(X_\Gamma)$ is identified with the set of discrete spectrum of the Laplacian.

6.2 *Constructing Stable Spectra by Poincaré Series [29]*

We discuss the discrete spectrum, which is stable under small deformations of X_Γ, following its construction in Kassel and Kobayashi [29].

A general construction of the discrete spectrum for reductive symmetric spaces $X = G/H$ was established by Flensted-Jensen [13] under the rank condition

$$\mathrm{rank}(G/H) = \mathrm{rank}(K/(H \cap K)). \tag{6.1}$$

This is a generalization of the celebrated rank condition due to Harish-Chandra, and the necessity was proved by Matsuki–Oshima [54]. In our notation, Flensted-Jensen's theorem says that there exist countably many elements in $\mathrm{Spec}_d(X)$ in the case $\Gamma = \{e\}$.

Let us now consider a setting where Γ is a discontinuous group for a reductive symmetric space $X = G/H$. A general construction of the discrete spectrum for the locally symmetric space X_Γ, when X satisfies the rank condition (6.1) and Γ is a "sharp" discontinuous group for X, was provided in Kassel and Kobayashi [29] through the introduction of a generalized Poincaré series. It is noted that infinitely many discrete spectra remain stable under any small deformation of a discontinuous group Γ, as formulated in the following definition:

Definition 6.2 ([29, Def. 1.6]) Let Γ be a discontinuous group for G/H via the inclusion map $\iota \colon \Gamma \to G$. We say λ is a *stable (discrete) spectrum* if there exists a neighborhood U of $\iota \in \mathcal{R}(\Gamma, G; X)$ such that

$$\lambda \in \bigcap_{\varphi \in U} \operatorname{Spec}_d(X_{\varphi(\Gamma)}).$$

We recall from Definition 2.3 that $X_{\varphi(\Gamma)}$ is locally rigid if the G-orbit through φ is open in $\mathcal{R}(\Gamma, G; X)$. Since $\operatorname{Spec}_d(X_{\varphi(\Gamma)})$ remains unchanged when φ is replaced with its G-conjugate, all the discrete spectra are *stable* in the sense of Definition 6.2 if $X_{\varphi(\Gamma)}$ is locally rigid. Thus, we are interested in the case where $X_{\varphi(\Gamma)}$ is deformable. This is how affirmative answers to (Q1)–(Q3) in Question 2.5 regarding deformability relate to the theme of this section.

The following theorem for the existence of stable spectra is formulated for a subgroup L when it is of real-rank one, such as $Spin(n, 1)$.

Fact 6.3 ([29, Thm. 1.7]) *Let $X = G/H$ be a reductive symmetric space satisfying the rank condition (6.1). Let L be a real rank one semisimple Lie subgroup of G, which acts properly on X, and Γ be a torsion-free, cocompact discrete subgroup of L. Then there exist infinite stable discrete spectra for X_Γ.*

The idea of the proof in [29] is to construct periodic joint eigenfunctions of intrinsic differential operators by averaging eigenfunctions that decay rapidly at infinity over the Γ-orbits. To demonstrate the convergence of this infinite sum (the *generalized Poincaré series*) and, furthermore, to prove that the total sum is non-zero, we require both analytic estimates for the eigenfunctions and geometric estimates related to the action of the discontinuous group Γ. The geometric estimates include counting Γ-orbits $\Gamma \cdot x$ in pseudo-balls, which is in a different setting compared to the counting estimate by Eskin–McMullen [12], as well as a Kazhdan–Margulis type estimate for discontinuous groups. The existence of the stable discrete spectrum follows from the fact that these estimates can be provided *uniformly* with respect to the variables that appear in each of the estimates.

Combining Fact 6.3 with Theorem 5.9, we obtain the following:

Theorem 6.4 *Let G be a Zariski-connected real reductive algebraic group, let $X = G/H$ be a symmetric space satisfying the rank condition (6.1), and let $\varphi \colon Spin(n, 1) \to G$ be a homomorphism such that the action of $Spin(n, 1)$ on X is proper via φ. Let G^φ denote the algebraic subgroup of G, as introduced in Definition 3.7. Then there exists a pair (Γ, φ') with the following properties:*

- *Γ is a torsion-free cocompact arithmetic subgroup of $Spin(n, 1)$,*
- *φ' is a small deformation of $\varphi|_\Gamma$,*
- *the Zariski-closure of $\varphi'(\Gamma)$ is G^φ,*
- $\sharp(\mathrm{Spec}_d(X_{\varphi(\Gamma)}) \cap \mathrm{Spec}_d(X_{\varphi'(\Gamma)})) = \infty$.

As an illustration, the geometric results in Theorems 1.3 and 5.28 yield the following examples, respectively:

Example 6.5 There exists a 7-dimensional compact manifold M, with a pseudo-Riemannian metric of signature $(4, 3)$ and constant sectional curvature -1, satisfying the following properties:

- $\pi_1(M)$ is isomorphic to a torsion-free arithmetic cocompact discrete subgroup of $Spin(4, 1)$;
- the holonomy representation $\pi_1(M) \to SO(4, 4)$ has a Zariski-dense image;
- for a sufficiently large integer N, $m(6 - m)$ is a discrete spectrum of the pseudo-Riemannian Laplacian Δ_M for any integer $m \geq N$.

Example 6.6 There exists a 15-dimensional non-compact manifold M, with a pseudo-Riemannian metric of signature $(8, 7)$ and constant sectional curvature -1, satisfying the following properties:

- $\pi_1(M)$ is isomorphic to a torsion-free arithmetic cocompact discrete subgroup of $Spin(6, 1)$;
- the holonomy representation $\pi_1(M) \to SO(8, 8)$ has a Zariski-dense image;
- for a sufficiently large integer N, $m(14 - m)$ is a discrete spectrum of the pseudo-Riemannian Laplacian Δ_M for any integer $m \geq N$.

6.3 *Multiplicity of the Discrete Spectrum on Compact Quotients X_Γ*

In our setting, which goes beyond Riemannian geometry, the Laplacian is no longer an elliptic differential operator. As a result, the multiplicity of the discrete spectrum of the Laplacian can be infinite, even when the manifold X_Γ is compact. Thus, there is no guarantee that the multiplicity of the spectrum, $\dim L^2(X_\Gamma, \mathcal{M}_\lambda)$, is finite.

This section briefly reviews the latest findings regarding the multiplicity of the discrete spectrum λ, specifically, $\dim_{\mathbb{C}} L^2(X_\Gamma, \mathcal{M}_\lambda)$ given by Kassel and Kobayashi.

Fact 6.7 ([28, 32]) *Let $X = G/H$ be a reductive symmetric space with H non-compact. Assume that there exists a reductive subgroup L of G such that L acts properly on G/H, and that $L_{\mathbb{C}}$ acts spherically on the complexification $X_{\mathbb{C}} = G_{\mathbb{C}}/H_{\mathbb{C}}$.*

Table 8 The triples (L, G, H) that apply to Fact 6.7

Case	G/H	L
1	$SU(2n, 2)/Sp(n, 1)$	$U(2n, 1)$
2	$SU(2n, 2)/U(n, 1)$	$Sp(n, 1)$
3_{even}	$SO(2n, 2)/U(n, 1)$	$SO(2n, 1)$ (n is even)
4	$SO(2n, 2)/SO(2n, 1)$	$U(n, 1)$
5	$SO(4n, 4)/SO(4n, 3)$	$Sp(1) \cdot Sp(n, 1)$
6	$SO(8, 8)/SO(8, 7)$	$Spin(8, 1)$
7'	$SO(4, 4)/Spin(4, 3)$	$Spin(4, 1) \times SO(3)$
12'	$SO(4, 3)/G_{2(2)}$	$SO(4, 1) \times SO(2)$

Then, for any torsion-free discrete subgroup Γ *of* L*, we have*

$$\dim_{\mathbb{C}} L^2(X_\Gamma, \mathcal{M}_\lambda) = \infty$$

for every $\lambda \in \mathrm{Spec}_d(X)$ *sufficiently away from the walls.*

A key step in the proof is a geometric variant of the discrete decomposability theorem [40], which states that any H-distinguished irreducible unitary representation of G decomposes into irreducible unitary representations of L without continuous spectrum, under the assumptions of Fact 6.7. This is combined with the method of generalized Poincaré series introduced in the proof of Fact 6.3.

The triples (G, H, L), where G is simple and G/H is allowed to be non-symmetric, that satisfy the assumptions in Fact 6.7 were classified in [32, Table 1.1]. We extract the following cases from that table: G/H satisfies the rank condition (6.1) and L is maximal. These are summarized in Table 8. The homogeneous spaces G/H in Cases 7' and 12' are not symmetric spaces; however, the $\mathbb{C}$-algebra $\mathbb{D}_G(G/H)$ is generated by the Laplacian, and Fact 6.7 extends to these cases as well.

6.4 Multiplicity of Stable Discrete Spectra on Compact Quotients

Let Γ be a discontinuous group for $X = G/H$ via the inclusion map $\iota \colon \Gamma \to G$.

By Definition 6.2, λ is said to belong to the stable discrete spectrum if there exists a neighborhood U of ι in $\mathcal{R}(\Gamma, G, X)$ such that

$$\min_{\varphi \in U} \dim_{\mathbb{C}} L^2(X_{\varphi(\Gamma)}, \mathcal{M}_\lambda) \geq 1.$$

As the neighborhood U becomes smaller, the left-hand side of this inequality tends to increase. Thus, we consider the following definition:

Definition 6.8 Let $\mathcal{U}_\Gamma$ denote the set of all neighborhoods of ι in $\mathcal{R}(\Gamma, G, X)$. For $\lambda \in \mathrm{Spec}_d(X_\Gamma)$, we define the *multiplicity of the stable discrete spectrum* by

$$\widetilde{\mathcal{N}}_{X_\Gamma}(\lambda) := \sup_{U \in \mathcal{U}_\Gamma} \min_{\varphi \in U} \dim_{\mathbb{C}} L^2(X_{\varphi(\Gamma)}, \mathcal{M}_\lambda).$$

When the inclusion map $\iota\colon \Gamma \to G$ is locally rigid, the multiplicity of the stable discrete spectrum, $\widetilde{\mathcal{N}}_{X_\Gamma}(\lambda)$, reduces to the dimension of $L^2(X_\Gamma, \mathcal{M}_\lambda)$. In contrast, when ι is not locally rigid as a discontinuous group for G/H, we have $\widetilde{\mathcal{N}}_{X_\Gamma}(\lambda) \neq 0$ if and only if λ belongs to the stable discrete spectrum.

In the special case where X is the 3-dimensional anti-de Sitter space $SO(2,2)/SO(1,2)$, the $\mathbb{C}$-algebra $\mathbb{D}_G(X)$ is generated by the Laplacian. Therefore, it is sufficient to consider eigenvalues of the Laplacian, and we can regard $\mathrm{Spec}_d(X_\Gamma)$ as a subset of $\mathbb{C}$. We note that X_Γ has abundant non-standard deformations. In this case, Kannaka found a lower bound for the multiplicity of the stable discrete spectrum, $\widetilde{\mathcal{N}}_{X_\Gamma}(\lambda)$, as follows. We define $\lambda_m := 4m(m-1)$ for $m \in \mathbb{N}$.

Fact 6.9 ([22]) *Let Γ be a cocompact discontinuous group for the 3-dimensional anti-de Sitter space $X = SO(2,2)/SO(1,2)$. Then there exists a positive constant c_Γ such that*

$$\widetilde{\mathcal{N}}_{X_\Gamma}(\lambda_m) \geq \log_3 m - c_\Gamma$$

holds for all $m \in \mathbb{N}$.

We recall from Theorem 2.9 a list of triples (G, H, L) such that $X = G/H$ admits a compact standard quotient that is deformable. Combining Table 3 with Table 8, we obtain the following proposition.

Proposition 6.10 *Let (G, H, L) be one of Cases 1, 2 ($n = 1$), 3_{even}, 4, 5 ($n = 1$), 7', and 12' in Table 8. Then, Facts 6.3 and 6.7 apply to any such triple (G, H, L).*

Remark 6.11 (1) We have excluded Cases 2 ($n \geq 2$), 5 ($n \geq 2$), and 6 because any compact standard quotient X_Γ is locally rigid in these cases.
(2) By Theorem 2.15, there exists a cocompact standard quotient X_Γ that admits a Zariski-dense deformation in Cases 2 ($n = 1$), 3_{even}, 5 ($n = 1$), 7', and 12'.

Regarding the stable discrete spectrum, including its multiplicity and distribution, we are particularly interested in the setting where a compact standard quotient X_Γ of $X = G/H$ admits a non-standard deformation. We address the following questions.

Question 6.12 Is the set of stable discrete spectra $\{\lambda \in \mathrm{Spec}_d(X_\Gamma) \mid \widetilde{\mathcal{N}}_{X_\Gamma}(\lambda) \neq 0\}$ discrete in $\mathfrak{j}_{\mathbb{C}}^\vee / W$?

Question 6.13 Does there exist a small deformation $\varphi \in \mathcal{R}(\Gamma, G; X)$ such that $\dim_{\mathbb{C}} L^2(X_{\varphi(\Gamma)}, \mathcal{M}_\lambda) < \infty$ holds for any $\lambda \in \mathfrak{j}_{\mathbb{C}}^\vee$?

An affirmative answer to Question 6.13 would imply that $\widetilde{\mathcal{N}}_{X_\Gamma}(\lambda) < \infty$. Moreover, if Question 6.12 has an affirmative answer, the following question might become meaningful.

Question 6.14 Does there exist any geometric information about the global structure of the (G, X)-manifold X_Γ that we can extract from the asymptotic behavior of the multiplicity of the stable discrete spectrum, $\widetilde{\mathcal{N}}_{X_\Gamma}(\lambda)$?

We examine Question 6.12 in the classical Riemannian case:

Remark 6.15 Let $X_\Gamma = \Gamma\backslash X$ be a compact quotient of an irreducible Riemannian symmetric space $X = G/K$ of non-compact type. Then X_Γ is not locally rigid if and only if X is the two-dimensional hyperbolic space by the local rigidity theorem [70]. Moreover, since K is compact, the G-action on X is proper, and thus there does not exist a non-standard deformation of X_Γ.

(1) ($\dim X = 2$). When X is a two-dimensional hyperbolic space, one cannot obtain any information about X_Γ because $\widetilde{\mathcal{N}}_{X_\Gamma}(\lambda) = 0$ for $\lambda \geq 1/4$ (Wolpert [72, Thm. 5.14]).
(2) ($\dim X \geq 3$). In this case, X_Γ is locally rigid, and therefore, $\widetilde{\mathcal{N}}_{X_\Gamma}(\lambda) = \dim_{\mathbb{C}} L^2(X_\Gamma, \mathcal{M}_\lambda)$, which is finite because the Laplacian in the Riemannian case is an elliptic differential operator. Weyl's law in spectral theory tells us that one can extract geometric quantities, such as the dimension and volume, of X_Γ from the asymptotic behavior of $\widetilde{\mathcal{N}}_{X_\Gamma}(\lambda)$ as $\|\lambda\| \to \infty$.

Acknowledgements The first author was partially supported by the Special Postdoctoral Researcher Program at RIKEN and JSPS under Grant-in Aid for Scientific Research (JP24K16929).
The second author was partially supported by the JSPS under Grant-in Aid for Scientific Research (JP23H00084) and by the Institut Henri Poincaré (Paris) and the Institut des Hautes Études Scientifiques (Bures-sur-Yvette). He would like to express his gratitude to Gopal Prasad, Andrei Rapinchuk, B. Sury, and Aleksy Tralle for their warm hospitality during the Conference "Zariski Dense Subgroups, Number Theory, and Geometric Applications," held at the Ramanujan Lecture Hall, ICTS, Bangalore, India, from January 1 to 12, 2024, where he gave a series of three lectures on related topics.

Appendix A. Topologies of Real Algebraic Groups

In this section, we provide a brief review of basic concepts for complex algebraic groups as well as real algebraic groups that are used throughout this article. In particular, we highlight the relationship between the two topologies on complex (or real) algebraic groups: the usual topology and the Zariski-topology, which is weaker than the usual topology.

We begin with the connectedness of complex algebraic groups.

Lemma A.1 *Let $G_{\mathbb{C}}$ be a Zariski-connected complex algebraic group. Then it is also connected in the usual topology.*

Proof By [7, Chap. 1, Prop. 1.2], $G_{\mathbb{C}}$ is irreducible in the Zariski-topology. Hence, by [63, Chap. VII.2], $G_{\mathbb{C}}$ is connected in the usual topology. □

Every connected complex semisimple Lie group is linear. Moreover, the categories of connected complex semisimple Lie groups and connected complex semisimple algebraic groups are equivalent, as described below.

Lemma A.2 *(1) Let $G_{\mathbb{C}}$ be a complex algebraic group, and $\mathfrak{l}_{\mathbb{C}}$ a complex semisimple Lie subalgebra of $\mathfrak{g}_{\mathbb{C}}$. Then, the analytic subgroup $L_{\mathbb{C}}$ of $G_{\mathbb{C}}$ corresponding to $\mathfrak{l}_{\mathbb{C}}$ is Zariski-closed.*

(2) Let $L_{\mathbb{C}}$ be a connected complex semisimple algebraic group, $G_{\mathbb{C}}$ a complex algebraic group, $\varphi_{\mathbb{C}}\colon L_{\mathbb{C}} \to G_{\mathbb{C}}$ a holomorphic Lie group homomorphism. Then $\varphi_{\mathbb{C}}$ is a morphism of algebraic groups.

For the proof, we recall the following theorem from Chevalley [9, Chap. 2, Thm. 13]:

Fact A.3 *Let $G_{\mathbb{C}}$ be a closed complex Lie subgroup of $GL(n, \mathbb{C})$ with finitely many connected components, $\overline{G_{\mathbb{C}}}$ its Zariski-closure in $GL(n, \mathbb{C})$, and $\overline{\mathfrak{g}_{\mathbb{C}}}$ the complex Lie algebra of $\overline{G_{\mathbb{C}}}$. Then, their derived algebras are identical:* $[\mathfrak{g}_{\mathbb{C}}, \mathfrak{g}_{\mathbb{C}}] = [\overline{\mathfrak{g}}_{\mathbb{C}}, \overline{\mathfrak{g}}_{\mathbb{C}}]$.

Proof of Lemma A.2 (1). It suffices to show that the Zariski-closure $\overline{L_{\mathbb{C}}}$ of $L_{\mathbb{C}}$ coincides with $L_{\mathbb{C}}$. Since $L_{\mathbb{C}}$ is connected in the usual topology, $L_{\mathbb{C}}$ is Zariski-connected. Hence, $\overline{L_{\mathbb{C}}}$ is also Zariski-connected. By Lemma A.1, $\overline{L_{\mathbb{C}}}$ is connected in the usual topology.

Since $\mathfrak{l}_{\mathbb{C}}$ is semisimple, we have $[\mathfrak{l}_{\mathbb{C}}, \mathfrak{l}_{\mathbb{C}}] = \mathfrak{l}_{\mathbb{C}}$, which implies that $[L_{\mathbb{C}}, L_{\mathbb{C}}] = L_{\mathbb{C}}$ due to the connectedness of $L_{\mathbb{C}}$. Furthermore, by [7, Chap. I, Cor. 2.3], we also obtain $[\overline{L_{\mathbb{C}}}, \overline{L_{\mathbb{C}}}] = \overline{L_{\mathbb{C}}}$. Thus, applying Fact A.3, we conclude that

$$\overline{\mathfrak{l}_{\mathbb{C}}} = [\overline{\mathfrak{l}_{\mathbb{C}}}, \overline{\mathfrak{l}_{\mathbb{C}}}] = [\mathfrak{l}_{\mathbb{C}}, \mathfrak{l}_{\mathbb{C}}] = \mathfrak{l}_{\mathbb{C}}.$$

Since both $L_{\mathbb{C}}$ and $\overline{L_{\mathbb{C}}}$ are connected, it follows that $\overline{L_{\mathbb{C}}} = L_{\mathbb{C}}$. This proves the assertion.

(2). Consider the graph of $\varphi_{\mathbb{C}}\colon L_{\mathbb{C}} \to G_{\mathbb{C}}$ given by

$$G(\varphi_{\mathbb{C}}) := \{(\ell, \varphi_{\mathbb{C}}(\ell)) \mid \ell \in L_{\mathbb{C}}\}.$$

The connected complex Lie group $G(\varphi_{\mathbb{C}})$ is the analytic subgroup of $L_{\mathbb{C}} \times G_{\mathbb{C}}$ corresponding to the graph of the differential $d\varphi_{\mathbb{C}}$,

$$\mathfrak{g}(d\varphi_{\mathbb{C}}) := \{(X, d\varphi_{\mathbb{C}}(X)) \mid X \in \mathfrak{l}_{\mathbb{C}}\}.$$

Since the Lie algebra $\mathfrak{g}(d\varphi_{\mathbb{C}})$ is isomorphic to $\mathfrak{l}_{\mathbb{C}}$, it is semisimple. Thus, by (1), we conclude that $G(\varphi_{\mathbb{C}})$ is Zariski-closed in $L_{\mathbb{C}} \times G_{\mathbb{C}}$. It follows that $G(\varphi_{\mathbb{C}})$ is a Zariski-connected complex algebraic group.

Next, consider the first projection $\mathrm{pr}_1 : G(\varphi_{\mathbb{C}}) \to L_{\mathbb{C}}$, which is a bijective morphism of algebraic groups. It follows from Zariski's main theorem (see [7, Chap. AG,

Thm. 18.2]) that this map is an isomorphism of algebraic groups. Since $\varphi_{\mathbb{C}}: L_{\mathbb{C}} \to G_{\mathbb{C}}$ is the composition of the isomorphism $\mathrm{pr}_1^{-1}: L_{\mathbb{C}} \to G(\varphi_{\mathbb{C}})$ and the second projection $\mathrm{pr}_2: G(\varphi_{\mathbb{C}}) \to G_{\mathbb{C}}$, we see that $\varphi_{\mathbb{C}}$ is also a morphism of algebraic groups. □

Next, we focus on real algebraic groups. For a real algebraic group $\mathbf{G}$, we denote by G the Lie group of real points $\mathbf{G}(\mathbb{R})$ and by $G_{\mathbb{C}}$ the complex Lie group of complex points $\mathbf{G}(\mathbb{C})$. Note that $G_{\mathbb{C}}$ provides a complexification of G, that is, the Lie algebra $\mathfrak{g}_{\mathbb{C}}$ of $G_{\mathbb{C}}$ is naturally identified with the complexification $\mathfrak{g} \otimes_{\mathbb{R}} \mathbb{C}$ of the Lie algebra $\mathfrak{g}$ of G.

Definition A.4 A real algebraic group $\mathbf{G}$ is said to be *Zariski-connected* if $G_{\mathbb{C}}$ is connected in the Zariski-topology.

Remark A.5 In contrast to the case of complex algebraic groups, even if $\mathbf{G}$ is Zariski-connected, G is not necessarily connected in the usual topology. For instance, $SO(n, 1)$ is Zariski-connected in the sense of Definition A.4 but is not connected in the usual topology.

Definition A.6 Let $\mathbf{G}$ be a real algebraic group. The *Zariski-topology* on G means the relative topology induced by the Zariski-topology on $G_{\mathbb{C}}$. The *Zariski-closure* of a subset S of G is defined as its closure with respect to the Zariski-topology. We say that S is *Zariski-dense* in G if it is dense in G with respect to the Zariski-topology.

If $\mathbf{G}$ is Zariski-connected, then G is Zariski-dense in $G_{\mathbb{C}}$. Hence, if $\mathbf{G}$ is Zariski-connected, then a subset S of G is Zariski-dense in G if and only if S is Zariski-dense in $G_{\mathbb{C}}$.

In the case of semisimple real algebraic groups, the following lemma can be compared with Lemma A.2:

Lemma A.7 *Let* $\mathbf{L}$ *be a Zariski-connected real semisimple algebraic group,* $\mathbf{G}$ *a real algebraic group, and* $\varphi: L \to G$ *an abstract group homomorphism. Assume that* $L_{\mathbb{C}}$ *is simply-connected. Then the following claims about* φ *are equivalent:*

(i) φ *is continuous in the usual topology;*
(ii) φ *is of* C^∞*-class;*
(iii) φ *is real analytic;*
(iv) φ *can be extended to a holomorphic homomorphism* $\varphi_{\mathbb{C}}: L_{\mathbb{C}} \to G_{\mathbb{C}}$ *of complex Lie groups;*
(v) φ *is obtained from an* $\mathbb{R}$*-morphism* $\mathbf{L} \to \mathbf{G}$ *of algebraic groups.*

We simply refer to f *as a* homomorphism. *In particular, the image of* $f_{\mathbb{C}}$ *is Zariski-closed, and* f *is continuous in the Zariski-topology.*

Proof It is well known that (i), (ii), and (iii) are equivalent, and (v) ⇒ (i) is obvious. Thus, we focus on proving (iii) ⇒ (iv) ⇒ (v).

(iii)⇒(iv). Since $L_{\mathbb{C}}$ is simply-connected, the complexification of the differential $df: \mathfrak{l} \to \mathfrak{g}$ can be lifted to a holomorphic homomorphism $f_{\mathbb{C}}: L_{\mathbb{C}} \to G_{\mathbb{C}}$ of complex Lie groups. It is clear that $f_{\mathbb{C}}$ extends f.

(iv)⇒(v). By Lemma A.2(2), $f_{\mathbb{C}}$ is a morphism of algebraic groups. Thus, it suffices to show that $f_{\mathbb{C}}$ is defined over $\mathbb{R}$.

Let $\sigma_L : L_{\mathbb{C}} \to L_{\mathbb{C}}$ and $\sigma_G : G_{\mathbb{C}} \to G_{\mathbb{C}}$ denote the complex conjugations associated with the real algebraic groups **L** and **G**, respectively. Identifying $\mathfrak{l}_{\mathbb{C}} = \mathfrak{l} \otimes_{\mathbb{R}} \mathbb{C}$ and $\mathfrak{g}_{\mathbb{C}} = \mathfrak{g} \otimes_{\mathbb{R}} \mathbb{C}$, we have $df_{\mathbb{C}} = df \otimes_{\mathbb{R}} \mathrm{id}_{\mathbb{C}}$, $d\sigma_L = \mathrm{id}_{\mathfrak{l}} \otimes \sigma$, and $d\sigma_G = \mathrm{id}_{\mathfrak{g}} \otimes \sigma$, where σ is the complex conjugate of $\mathbb{C}$. Hence, it follows that $df_{\mathbb{C}} \circ d\sigma_L = d\sigma_G \circ df_{\mathbb{C}}$. Since $L_{\mathbb{C}}$ is connected, we see that $f_{\mathbb{C}} \circ \sigma_L = \sigma_G \circ f_{\mathbb{C}}$. Thus, the morphism of algebraic groups $f_{\mathbb{C}}$ is defined over $\mathbb{R}$. □

Appendix B. Clifford Algebras and Spin Groups

Throughout this article, we use Clifford algebras and spin groups extensively. In this appendix, we introduce some notation related to these concepts.

Let $\mathbb{F}$ be a field of characteristic $\neq 2$, and V be an n-dimensional $\mathbb{F}$-vector space equipped with a non-degenerate symmetric bilinear form $Q : V \times V \to \mathbb{F}$. As usual, we put $Q(v) := Q(v, v)$, which defines a quadratic form on the $\mathbb{F}$-vector space V. Let $TV := \bigoplus_{i \in \mathbb{N}} V^{\otimes i}$ denote the tensor algebra. Let let $I(Q)$ be the two-sided ideal generated by $v \otimes v - Q(v)$ for all $v \in V$. The quotient $\mathbb{F}$-algebra

$$C(V) \equiv C(V, Q) := TV/I(Q)$$

is called the *Clifford algebra* associated with the quadratic form Q.

When $\mathbb{E}$ is an extension field of $\mathbb{F}$, we consider the $\mathbb{E}$-linear extension of the $\mathbb{F}$-bilinear form Q on the $\mathbb{E}$-vector space $V \otimes_{\mathbb{F}} \mathbb{E}$. Then the $\mathbb{E}$-algebra $C(V) \otimes_{\mathbb{F}} \mathbb{E}$ is naturally isomorphic to $C(V \otimes_{\mathbb{F}} \mathbb{E})$.

Let $q : TV \to C(V)$ be the quotient homomorphism. If $\{e_1, \ldots, e_n\}$ is an $\mathbb{F}$-basis of V, then the elements $q(e_{i_1}) \cdots q(e_{i_k})$ $(1 \leq i_1 < \cdots < i_k \leq n,\ 0 \leq k \leq n)$ form an $\mathbb{F}$-basis of $C(V)$. In particular, $\dim_{\mathbb{F}} C(V) = 2^n$ and the restriction of q to the subspace V of TV is injective. Throughout this article, we think of V as a subspace of $C(V)$ via q and the image $q(v)$ of $v \in V$ is also denoted by the same symbol v.

The $\mathbb{F}$-linear map $V \to C(V)$ defined by $v \mapsto -v$ is uniquely extended to an involutive $\mathbb{F}$-algebra automorphism $(-)'$ of $C(V)$. We define an $\mathbb{F}$-subalgebra $C_{\mathrm{even}}(V)$ of $C(V)$ by

$$C_{\mathrm{even}}(V) \equiv C_{\mathrm{even}}(V, Q) := \{x \in C(V, Q) \mid x' = x\},$$

which is called the *even Clifford algebra* of Q. If $\{e_1, \ldots, e_n\}$ is an $\mathbb{F}$-basis of V, then we have $(e_{i_1} \cdots e_{i_k})' = (-1)^k e_{i_1} \cdots e_{i_k}$ for $1 \leq i_1 < \cdots < i_k \leq n$ and $0 \leq k \leq n$. Hence, $e_{i_1} \cdots e_{i_k} \in C_{\mathrm{even}}(V)$ if and only if k is even. In particular, $\dim_{\mathbb{F}} C_{\mathrm{even}}(V) = 2^{n-1}$. As is well-known, $C_{\mathrm{even}}(V)$ is a semisimple $\mathbb{F}$-algebra.

The $\mathbb{F}$-linear map $V \to C(V)$ defined by $v \mapsto v$ is uniquely extended to an $\mathbb{F}$-algebra involutive *anti*-automorphism $(-)^*$ of $C(V)$. The involution $*$ preserves $C_{\mathrm{even}}(V)$. For an extension field $\mathbb{E}$ of $\mathbb{F}$, we define

$$\mathbf{G}_V(\mathbb{E}) := \{x \in C_{\text{even}}(V \otimes_{\mathbb{F}} \mathbb{E})^\times \mid xx^* = 1\} \tag{B.1}$$
$$\mathbf{Spin}_V(\mathbb{E}) := \{x \in \mathbf{G}_V(\mathbb{E}) \mid x(V \otimes_{\mathbb{F}} \mathbb{E})x^{-1} \subset V \otimes_{\mathbb{F}} \mathbb{E}\}. \tag{B.2}$$

Here note $C_{\text{even}}(V \otimes_{\mathbb{F}} \mathbb{E})$ is naturally isomorphic to $C_{\text{even}}(V) \otimes_{\mathbb{F}} \mathbb{E}$ as an $\mathbb{E}$-algebra. Hence, we can regard both $\mathbf{G}_V$ and $\mathbf{Spin}_V$ as $\mathbb{F}$-algebraic groups by identifying $C_{\text{even}}(V)$ with $\mathbb{F}^{2^{n-1}}$. We put

$$G(V) := \mathbf{G}_V(\mathbb{F}) \text{ and } Spin(V) := \mathbf{Spin}_V(\mathbb{F}).$$

The group $Spin(V)$ is called the *spin group* of Q. Note that the elements $\pm 1 \in C_{\text{even}}(V)$ belong to $Spin(V)$. Since $C_{\text{even}}(V)$ is a semisimple $\mathbb{F}$-algebra, $G(V)$ is a classical group defined over $\mathbb{F}$.

For $x \in Spin(V)$ and $v \in V$, we have $xvx^{-1} \in V$ in $C(V)$ by definition. One can check that the $\mathbb{F}$-linear map $v \mapsto xvx^{-1}$ defines an element of the special orthogonal group $SO(V)$ with respect to Q. Thus we obtain an $\mathbb{F}$-homomorphism

$$\mathrm{Ad}\colon \mathbf{Spin}_V \to \mathbf{SO}_V, \quad x \mapsto (v \mapsto xvx^{-1}).$$

The kernel of this homomorphism is $\{\pm 1\}$. If $\mathbb{F}$ is algebraically closed, then the map Ad is surjective. However, this is not true in general. Indeed, when $\mathbb{F} = \mathbb{R}$, the image of Ad coincides with the identity component of the orthogonal Lie group $SO(V)$, which has two connected components if the quadratic form Q is not positive-definite.

From now on, we assume $\mathbb{F} = \mathbb{R}$. Then $G(V)$ and $Spin(V)$ are Lie groups. Let us recall that their Lie algebras $\mathfrak{g}(V)$ and $\mathfrak{spin}(V)$ are realized as Lie subalgebras of $C_{\text{even}}(V)$. Since $C_{\text{even}}(V)$ is an $\mathbb{F}$-algebra, we note that $C_{\text{even}}(V)$ is a Lie algebra over $\mathbb{F}$ with the bracket $[x, y] = xy - yx$. Further, via the exponential map $\exp\colon C_{\text{even}}(V) \to C_{\text{even}}(V)^\times$ $(x \mapsto \sum_{n=0}^{\infty} x^n/n!)$, we think of $C_{\text{even}}(V)$ as the Lie algebra of the Lie group $C_{\text{even}}(V)^\times$. In particular, we have

$$\mathfrak{g}(V) = \{x \in C_{\text{even}}(V, Q) \mid x + x^* = 0\}$$
$$\mathfrak{spin}(V) = \{x \in \mathfrak{g}(V) \mid \mathrm{ad}(x)(V) \subset V\},$$

where $\mathrm{ad}(x)(v) = [x, v]$. From this, we obtain the following lemma easily:

Lemma B.1 *Let $e_1, \dots, e_n$ be an orthogonal basis of the $\mathbb{F}$-vector space V with respect to Q. We write $\wedge^k$ $(0 \le k \le n)$ for the $\mathbb{F}$-vector subspace of $C_{\text{even}}(V)$ spanned by the elements $e_{i_1} \cdots e_{i_k}$ $(1 \le i_1 < \cdots < i_k \le n)$. Then we have*

$$\mathfrak{g}(V) = \bigoplus_{\substack{0 \le k \le n \\ k \equiv 2 \bmod 4}} \wedge^k \quad \textit{and} \quad \mathfrak{spin}(V) = \wedge^2.$$

Let $\mathbb{R}^{p,q}$ be the real vector space $\mathbb{R}^{p+q}$ equipped with the quadratic form $x_1^2 + \cdots + x_p^2 - x_{p+1}^2 - \cdots - x_{p+q}^2$. In this case, $C(\mathbb{R}^{p,q})$, $C_{\text{even}}(\mathbb{R}^{p,q})$, $G(\mathbb{R}^{p,q})$, $\mathfrak{g}(\mathbb{R}^{p,q})$, $Spin(\mathbb{R}^{p,q})$, and $\mathfrak{spin}(\mathbb{R}^{p,q})$ are denoted by $C(p, q)$, $C_{\text{even}}(p, q)$, $G(p, q)$,

$\mathfrak{g}(p,q)$, $Spin(p,q)$, and $\mathfrak{spin}(p,q)$, respectively. For the structure of the even Clifford algebra $C_{\text{even}}(p,q)$ as an $\mathbb{R}$-algebra, see, for example, [50, Prop. 4.4.1]. Moreover, Kobayashi and Yoshino provided an explicit identification of the groups $G(p,q)$ with classical groups, which is useful for a uniform treatment of sporadic cases in Table 1. We extract the classification table of $G(p,q)$ from [50]:

Proposition B.2 ([50, Prop. 4.4.4]) *For $p, q \geq 1$, the group $G(p,q)$ is isomorphic to one of the following classical Lie groups. Here, $n = 2^{(p+q-\alpha)/2}$ and $\alpha \in \{3,4,5,6\}$ are given according to $p - q \mod 8$ as in the table.*

$G(p,q)$	α	$p-q \mod 8$	$p+q \mod 8$
$O(n,n)^2$			0
$GL(2n,\mathbb{R})$	4	0	± 2
$Sp(n,\mathbb{R})^2$			4
$O(n,n)$	3	± 1	± 1
$Sp(n,\mathbb{R})$			± 3
$O(2n,\mathbb{C})$			0
$U(n,n)$	4	± 2	± 2
$Sp(n,\mathbb{C})$			4
$O^*(4n)$	5	± 3	± 1
$Sp(n,n)$			± 3
$O^*(4n)^2$			0
$GL(2n,\mathbb{H})$	6	4	± 2
$Sp(n,n)^2$			4

Lemma B.1 and Proposition B.2 are used in Sect. 3.3.

Appendix C. Torsion Freeness of Congruence Subgroups

In this appendix, we prove the following lemma concerning torsion of arithmetic groups, which is used in Sect. 4.3. It seems to be well-known to experts, but we give a proof for the convenience of the reader.

Lemma C.1 *Let $\mathbb{F}$ be a number field of finite degree and $\mathcal{O}$ the ring of integers of $\mathbb{F}$. Suppose that prime ideals P, Q of $\mathcal{O}$ satisfy $P \cap \mathbb{Z} \neq Q \cap \mathbb{Z}$. Then the congruence subgroup*

$$\{g \in GL(n,\mathcal{O}) \mid g \equiv 1 \mod PQ\}$$

is torsion-free.

We give an elementary lemma for the proof:

Lemma C.2 *Let $\mathbb{F}$, $\mathcal{O}$, P, and Q be as in Lemma C.1. For $m \in \mathbb{N}_+$, if a primitive m-th root ζ_m in $\mathbb{F}$ satisfies $\zeta_m \equiv 1 \mod PQ$, then $m = 1$.*

Proof Let $k = \mathbb{Q}(\zeta_m)$ and $\mathcal{O}_k = \mathbb{Z}[\zeta_m]$ the ring of integers of k. If m has two distinct prime factors, then $1 - \zeta_m$ is a unit in $\mathcal{O}_k$ (see, e.g., by [64, Lem. 17.2(2)]). Hence, $1 - \zeta_m$ is not contained in any prime ideal of $\mathcal{O}_k$. However, since $P \cap \mathcal{O}_k$ is a prime ideal of $\mathcal{O}_k$ and $1 - \zeta_m \in P$ by assumption, this leads to a contradiction. Hence, suppose $m = \ell^r$ for some prime number ℓ and non-negative integer r.

For contradiction, suppose $m > 1$, equivalently $r > 0$. Then $(1 - \zeta_m)\mathcal{O}_k$ is a prime ideal of $\mathcal{O}_k$ containing ℓ (see, e.g., by [64, Thm. 17.5]). Hence, the prime number ℓ is contained in both P and Q, which contradicts the assumption that $P \cap \mathbb{Z} \neq Q \cap \mathbb{Z}$. Therefore, we have $m = 1$. □

We are ready to show Lemma C.1.

Proof of Lemma C.1 Let g be an element of $GL(n, \mathcal{O})$ such that $g \equiv 1 \mod PQ$, and let $f(t) \in \mathcal{O}[t]$ be the characteristic polynomial of g.

Since $g \equiv 1 \mod PQ$, we have $f(t) \equiv (t-1)^n \mod PQ$. Let $\mathbb{E}$ be a finite extension field of $\mathbb{F}$ containing all the eigenvalues of g, and let P' and Q' be prime ideals of the integers $\mathcal{O}_{\mathbb{E}}$ in $\mathbb{E}$ containing P and Q, respectively. Then, it follows that $f(t) \equiv (t-1)^n \mod P'Q'$.

Now suppose that g has finite order. In this case, g is diagonalizable over $\mathbb{E}$, and its eigenvalues are all roots of unity. To prove our assertion, it suffices to show that these eigenvalues are equal to 1.

Let ζ_m be an eigenvalue of g, which is a primitive m-th root of unity. Since $f(\zeta_m) = 0$, we have $(\zeta_m - 1)^n \equiv 0 \mod P'Q'$. Since both P' and Q' are prime ideals of $\mathcal{O}_{\mathbb{E}}$, it follows that $\zeta_m - 1 \equiv 0 \mod P'Q'$. By Lemma C.2, we conclude that $m = 1$. This completes the proof. □

Appendix D. A Lemma on the HNN Extension

Let M be an orientable connected manifold and N an orientable connected compact hypersurface such that $S := M \setminus N$ is connected. In this appendix, we present Proposition D.2, which explains how the fundamental group $\pi_1(M)$ of M is described by $\pi_1(S)$ and $\pi_1(N)$. We see that $\pi_1(M)$ is an HNN extension of $\pi_1(S)$. Although the argument using the van Kampen theorem is standard to experts, we provide a proof for the convenience of readers who may not be familiar with it. This proposition is used in Sect. 4.4, where M and N are hyperbolic manifolds.

Let us introduce some notations. By the tubular neighborhood theorem, one can and do take an open neighborhood $\tilde{N}^{(4)}$ of N in M and a diffeomorphism

$$f : N \times (-4, 4) \simeq \tilde{N}^{(4)}$$

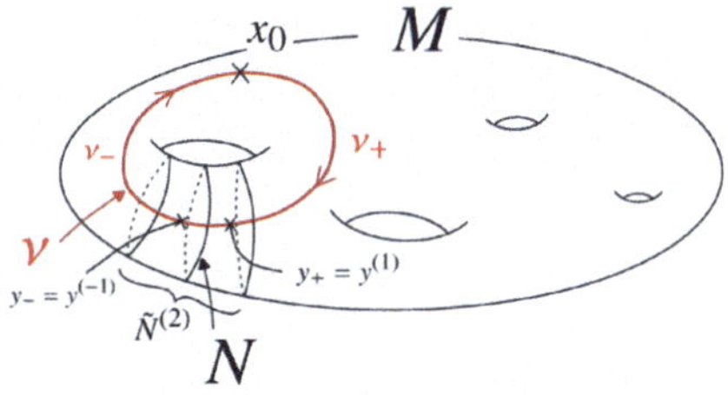

Fig. 2 Notation in Definition D.1 for the geometry away from the hypersurface N

such that $f(N \times \{0\})$ coincides with N. Fix $y \in N$ and put $\tilde{N}^{(t)} := f(N \times (-t, t))$ and $y^{(t)} := f(y, t)$ for $-4 < t < 4$. Since $S = M \smallsetminus N$ is connected, we may and do assume that $M \smallsetminus \tilde{N}^{(4)}$ is path-connected.

Definition D.1 Fix a base point $x_0 \in M \smallsetminus \tilde{N}^{(4)}$. We define the oriented loop ν in M, starting at x_0, as the composition of the following paths:

- a path from x_0 to $y^{(3)}$ inside $M \smallsetminus \tilde{N}^{(3)}$;
- the path from $y^{(3)}$ to $y^{(-3)}$ given by $f(\{y\} \times [-3, 3])$;
- a path from $y^{(-3)}$ to x_0 inside $M \smallsetminus \tilde{N}^{(3)}$.

Furthermore, we can take the loop ν to be a closed *submanifold* of M. Additionally, let ν_+ be the path along the loop ν from x_0 to $y^{(1)}$, and let ν_- be the path along ν from $y^{(-1)}$ to x_0.

Figure 2 summarizes some of the notation introduced so far.

For an oriented loop ℓ starting at y in N and a fixed $-4 < t < 4$, $f(\ell, t)$ forms an oriented loop in M starting at $y^{(t)}$. With the above paths ν_+ and ν_-, we define $j_+, j_- \colon \pi_1(N, y) \to \pi_1(S, x_0)$ as two group homomorphisms

$$j_+([\ell]) := [\nu_+^{-1} \circ f(\ell, 1) \circ \nu_+], \; j_-([\ell]) := [\nu_- \circ f(\ell, -1) \circ \nu_-^{-1}].$$

Here, b^{-1} denotes the path obtained by reversing the orientation of the path b, and $[c]$ denotes the homotopy class defined by the loop c.

Proposition D.2 *Let $a^{\mathbb{Z}}$ be the infinite cyclic group generated by a. Under the above setting, we define a group homomorphism*

$$\Psi \colon \pi_1(S, x_0) * a^{\mathbb{Z}} \to \pi_1(M, x_0)$$

by the natural map $\pi_1(S, x_0) \to \pi_1(M, x_0)$, and by sending the generator a to $[\nu]$. Then, Ψ is surjective, and its kernel is the normal subgroup $\mathcal{N}$, generated by the elements

$$a j_+([\ell]) a^{-1} j_-([\ell])^{-1}, \quad \textit{for every } [\ell] \in \pi_1(N, y).$$

Proof Since M is orientable, one can take a tubular neighborhood U_ν of ν in M and a diffeomorphism $g\colon \nu \times U \to U_\nu$, where U is a contractible open neighborhood U of y in N and $\nu = g(\nu \times \{y\})$. By taking a sufficiently small U_ν, we may and do assume the following:

- $g(y^{(t)}, u) = f(u, t)$ holds for $-2 \leq t \leq 2$ and $u \in U$.
- $U_\nu \setminus f(N \times [-2, 2]) = g((\nu \setminus \{y^{(t)} \mid -2 \leq t \leq 2\}) \times U)$.

In particular, $U_\nu \cap \tilde{N}^{(2)}$ and $U_\nu \setminus f(N \times [-2, 2])$ are both contractible.

We put

$$V := \tilde{N}^{(2)} \cup U_\nu.$$

Since $M = V \cup S$ and $V \cap S$ is path-connected, by the van Kampen theorem, we obtain the following pushout diagram:

$$\begin{array}{ccc} \pi_1(V \cap S, x_0) & \longrightarrow & \pi_1(S, x_0) \\ \downarrow & \circlearrowleft & \downarrow \\ \pi_1(V, x_0) & \longrightarrow & \pi_1(M, x_0). \end{array} \tag{D.1}$$

Now we compute $\pi_1(V, x_0)$ and $\pi_1(V \cap S, x_0)$ by the van Kampen theorem. Since $\tilde{N}^{(2)} \cap U_\nu$ is contractible, we have

$$\pi_1(V, y^{(1)}) \simeq \pi_1(\tilde{N}^{(2)}, y^{(1)}) * \pi_1(U_\nu, y^{(1)}).$$

Recall that U_ν and $\tilde{N}^{(2)}$ are homotopic to ν and N, respectively. Hence, we have $\pi_1(\tilde{N}^{(2)}, y^{(1)}) \simeq \pi_1(N, y)$ and $\pi_1(U_\nu, y^{(1)}) \simeq a^{\mathbb{Z}}$. Furthermore, we use ν_+ to change the base point $y^{(1)}$ to x_0. The resulting isomorphism

$$\pi_1(V, x_0) \underset{\simeq}{\overset{\lambda}{\longleftarrow}} \pi_1(N, y) * a^{\mathbb{Z}}, \tag{D.2}$$

is obtained by $\lambda([\ell]) = [\nu_+^{-1} \circ f(\ell, 1) \circ \nu_+]$ for $[\ell] \in \pi_1(N, y)$ and $\lambda(a) = [\nu]$.

We define

$$\tilde{N}_+^{(2)} := f(N \times (0, 2)), \quad \tilde{N}_-^{(2)} := f(N \times (-2, 0))$$

(see Fig. 3). Then, $V \cap S$ is the union of the two open sets

$$A_+ := \tilde{N}_+^{(2)} \cup (U_\nu \setminus f(N \times [-2, 2))) \text{ and } A_- := \tilde{N}_-^{(2)} \cup (U_\nu \setminus f(N \times (-2, 2])).$$

The intersection $A_+ \cap A_-$ is given by $U_\nu \setminus f(N \times [-2, 2])$, which is contractible. Thus, by the van Kampen theorem, we obtain

$$\pi_1(V \cap S, x_0) \simeq \pi_1(A_+, x_0) * \pi_1(A_-, x_0).$$

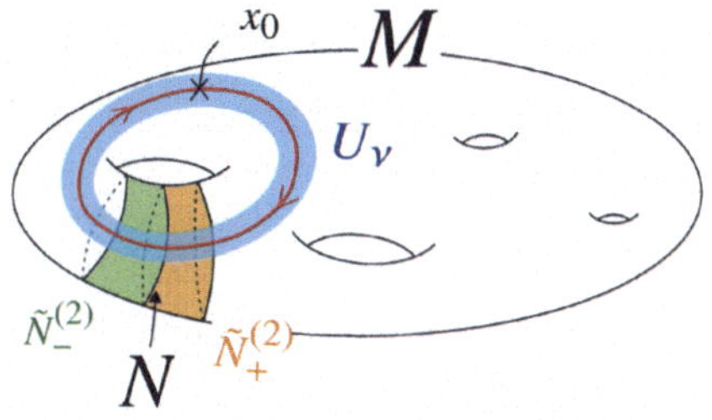

Fig. 3 Tubular neighborhoods of the hypersurface N and the loop ν

Furthermore, by the choice of U_ν, both A_+ and A_- are homotopy equivalent to N. Hence, we obtain the isomorphism

$$\pi_1(V \cap S, x_0) \underset{\simeq}{\overset{\mu}{\leftarrow}} \pi_1(N, y) * \pi_1(N, y), \tag{D.3}$$

which is induced by the two homomorphisms $\pi_1(N, y) \to \pi_1(V \cap S, x_0)$ mapping $[\ell] \in \pi_1(N, y)$ to $[\nu_+^{-1} \circ f(\ell, 1) \circ \nu_+]$ and $[\nu_- \circ f(\ell, -1) \circ \nu_-^{-1}]$ in $\pi_1(V \cap S, x_0)$, respectively.

Here we note

$$[\nu_- \circ f(\ell, -1) \circ \nu_-^{-1}] = [\nu \circ \nu_+^{-1} \circ f(\ell, 1) \circ \nu_+ \circ \nu^{-1}] \quad \text{in} \quad \pi_1(V, x_0).$$

Hence, from (D.2) and (D.3), the following diagram commutes:

$$\begin{array}{ccc} \pi_1(N, y) * \pi_1(N, y) & \underset{\simeq}{\overset{\mu}{\longrightarrow}} & \pi_1(V \cap S, x_0) \\ {\scriptstyle \mathrm{id} * (a(-)a^{-1})} \downarrow & \circlearrowright & \downarrow {\scriptstyle \text{natural}} \\ \pi_1(N, y) * a^{\mathbb{Z}} & \underset{\simeq}{\overset{\lambda}{\longrightarrow}} & \pi_1(V, x_0). \end{array}$$

Therefore, by (D.1), we obtain the following pushout diagram:

$$\begin{array}{ccc} \pi_1(N, y) * \pi_1(N, y) & \xrightarrow{j_+ * j_-} & \pi_1(S, x_0) \\ {\scriptstyle \mathrm{id} * (a(-)a^{-1})} \downarrow & \circlearrowright & \downarrow {\scriptstyle \text{natural}} \\ \pi_1(N, y) * a^{\mathbb{Z}} & \xrightarrow{j_+ * (a \mapsto [\nu])} & \pi_1(M, x_0). \end{array} \tag{D.4}$$

On the other hand, by the definition of the normal subgroup $\mathcal{N}$, we have the following pushout diagram:

$$\begin{array}{ccc} \pi_1(N, y) * \pi_1(N, y) & \xrightarrow{j_+ * j_-} & \pi_1(S, x_0) \\ {\scriptstyle \mathrm{id} * (a(-)a^{-1})} \downarrow & \circlearrowleft & \downarrow \\ \pi_1(N, y) * a^{\mathbb{Z}} & \xrightarrow{j_+ * \mathrm{id}} & (\pi_1(S, x_0) * a^{\mathbb{Z}})/\mathcal{N}. \end{array} \tag{D.5}$$

By comparing the two pushout diagrams (D.4) and (D.5), we obtain the desired conclusion of the theorem. □

Appendix E. An Upper Bound for Local Deformation of Discrete Subgroups

The main result of this appendix is Proposition E.2, which provides an upper bound for the Zariski-closure of potential deformations of discrete subgroups in the general setting. This proposition is used in the proof of Theorem 3.9(3), as well as in part of the proof of Proposition 5.12(2). While the results of this section are likely known to experts, we formulate them in a form suitable for our purposes and prove them using the *curve selection lemma* in the subanalytic geometry. See Lemma E.7(3). We include a proof for the benefit of readers who may not be familiar with them.

Setting E.1 *Let G be a Lie group, L a closed subgroup of G, and Γ a finitely-presented group. By choosing N generators of Γ satisfying a finite set of relations, we identify* $\mathrm{Hom}(\Gamma, G)$ *with a subset of $X := G^N$.*

We do not require L or G to be reductive here. In this setting, we provide a general local rigidity theorem. In other words, we give a necessary condition for a representation of Γ into L to be deformed in G, beyond L, up to G-conjugacy. Recall that G acts on $\mathrm{Hom}(\Gamma, G)$ by inner automorphisms, where $(g \cdot \varphi)(\gamma) := g\varphi(\gamma)g^{-1}$.

Proposition E.2 *In Setting E.1, let $\varphi \colon \Gamma \to G$ be a group homomorphism such that $\varphi(\Gamma) \subset L$. We regard the Lie algebras $\mathfrak{g}$ and $\mathfrak{l}$ as Γ-modules via φ. If $H^1(\Gamma, \mathfrak{g}/\mathfrak{l}) = 0$, then there exists a neighborhood U of φ in* $\mathrm{Hom}(\Gamma, G)$ *such that for any $\varphi' \in U$, we have $\varphi'(\Gamma) \subset L$, up to G-conjugacy.*

Our proof relies on properties of real analytic subsets. A subset A of a real analytic manifold X is called *real analytic* in X if, for each point $x \in X$, there exist an open neighborhood U of x in X and analytic functions $f_1, \dots, f_n$ on U such that

$$U \cap A = \{u \in U \mid f_1(u) = \cdots = f_n(u) = 0\}.$$

In Setting E.1, $\mathrm{Hom}(\Gamma, G)$ is a real analytic subset of $X = G^N$. Since L is real analytic in G, it follows that $\mathrm{Hom}(\Gamma, L)$ is also real analytic in X.

Step 1. As a first step, we prove Proposition E.2, assuming that the small deformation of φ is described by an analytic curve φ_t. The following discussion relying on Artin's approximation theorem is also seen in Goldman–Millson [16, Sect. 2]. To be precise, we show the following lemma:

Lemma E.3 *In the setting of Proposition E.2, for any one-parameter family of group homomorphisms $\varphi_t \colon \Gamma \to G$, depending analytically on $t \in \mathbb{R}$, with $\varphi_0 = \varphi$, there exists an analytic curve g_t in G with $g_0 = e$ such that $g_t \cdot \varphi_t \in \mathrm{Hom}(\Gamma, L)$ for any sufficiently small t.*

Proof of Lemma E.3 For each $\gamma \in \Gamma$, the real analytic curve $\varphi_t(\gamma)\varphi(\gamma)^{-1}$ in G, starts at e when $t = 0$. Its Taylor expansion gives rise to a sequence of maps $u_\ell \equiv u_\ell(\varphi_t) \colon \Gamma \to \mathfrak{g}$ ($\ell = 1, 2, \dots$), such that

$$\varphi_t(\gamma) = \exp(u_1(\gamma)t + u_2(\gamma)t^2 + \cdots)\varphi(\gamma),$$

where the domain of the convergence may depend on γ.

We fix $k \geq 1$, and prove the following claim: *assume that $u_\ell(\Gamma) \subset \mathfrak{l}$ for $\ell = 1, \dots, k-1$. Then, there exists an analytic curve in G, to be denoted by $g_t^{(k)}$, with the following two properties:*

(1) $u_\ell(\varphi_t) = u_\ell(g_t^{(k)} \cdot \varphi_t)$ for $\ell = 1, \dots, k-1$,
(2) $u_k(g_t^{(k)} \cdot \varphi_t)(\Gamma) \subset \mathfrak{l}$,

where $g_t^{(k)} \cdot \varphi_t$ denotes a homomorphism $\Gamma \to \mathfrak{g}$ that is obtained by conjugating φ_t with $g_t^{(k)}$.

To prove the claim, we compute $\varphi_t(\gamma\eta) = \varphi_t(\gamma)\varphi_t(\eta)$ using the Baker–Campbell–Hausdorff formula:

$$\begin{aligned}\varphi_t(\gamma\eta) &= \exp\left(\sum_{n=1}^{\infty} u_n(\gamma)t^n\right)\varphi(\gamma)\exp\left(\sum_{n=1}^{\infty} u_n(\eta)t^n\right)\varphi(\eta)\\ &= \exp\left(\sum_{n=1}^{\infty} u_n(\gamma)t^n\right)\exp\left(\mathrm{Ad}(\varphi(\gamma))\left(\sum_{n=1}^{\infty} u_n(\eta)t^n\right)\right)\varphi(\gamma\eta)\\ &= \exp(A(t) + B(t) + \frac{1}{2}[A(t), B(t)] + \cdots)\varphi(\gamma\eta),\end{aligned}$$

where $A(t) = \sum_{n=1}^{\infty} u_n(\gamma)t^n$ and $B(t) = \mathrm{Ad}(\varphi(\gamma))(\sum_{n=1}^{\infty} u_n(\eta)t^n)$. Hence, we have

$$\sum_{n=1}^{\infty} u_n(\gamma\eta)t^n = A(t) + B(t) + \frac{1}{2}[A(t), B(t)] + \cdots.$$

By assumption, the images of $u_1, \dots, u_{k-1}$ are contained in $\mathfrak{l}$. Comparing the coefficients of t^k in the above identity modulo $\mathfrak{l}$, we obtain

$$u_k(\gamma\eta) \equiv u_k(\gamma) + \mathrm{Ad}(\varphi(\gamma))u_k(\eta) \mod \mathfrak{l}.$$

Thus, u_k mod $\mathfrak{l}$ defines a 1-cocycle. By the assumption $H^1(\Gamma, \mathfrak{g}/\mathfrak{l}) = 0$, there exists an element $C_k \in \mathfrak{g}$ such that,

$$u_k(\gamma) \equiv \mathrm{Ad}(\varphi(\gamma))C_k - C_k \mod \mathfrak{l},$$

for any $\gamma \in \Gamma$. Now, we set $g_t^{(k)} := \exp(t^k C_k)$. Clearly, $u_1, \ldots, u_{k-1}$ remain unchanged, when we take the conjugate $g_t^{(k)} \cdot \varphi_t$. On the other hand, the coefficient of t^k in

$$\log((g_t^{(k)} \cdot \varphi_t)\varphi(\gamma)^{-1}) = \log(\exp(t^k C_k) \exp(A(t)) \exp(-t^k \mathrm{Ad}(\varphi(\gamma))C_k))$$

amounts to $C_k + u_k(\gamma) - \mathrm{Ad}(\varphi(\gamma))C_k \in \mathfrak{l}$. Hence, our claim is proved.

We now use the claim inductively on $k \geq 1$. We start with φ_t, and proceed with $g_t^{(1)} \cdot \varphi_t$, $g_t^{(2)} g_t^{(1)} \cdot \varphi_t$, and so on. Correspondingly, the family of maps from Γ to $\mathfrak{g}$,

$$u_\ell^{(k)} := u_\ell(g_t^{(k)} \cdots g_t^{(1)} \cdot \varphi_t)$$

has the following properties:

$$u_\ell^{(k)} = u_\ell^{(\ell)} \text{ and } u_\ell^{(k)}(\Gamma) \subset \mathfrak{l} \text{ for all } k \geq \ell.$$

Consequently, we obtain the formal power series

$$g_t = \cdots g_t^{(2)} g_t^{(1)}.$$

Then, g_t provides a formal solution to the equation

$$g_t \cdot \varphi_t \in \mathrm{Hom}(\Gamma, L). \tag{E.1}$$

Since $\mathrm{Hom}(\Gamma, L)$ is a real analytic subset of $X = G^N$ and φ_t is an analytic curve, the condition (E.1) defines an analytic equation for g_t. By Artin's approximation theorem [1, Thm. 1.2], there exists an analytic solution $g_t \in G$ satisfying $g_0 = 1$. Thus, our assertion is proved. □

Step 2. We complete the proof of Proposition E.2 by reducing it to the analytic curve case (Step 1). The following discussion uses the curve selection lemma as in Goldman–Millson [16, Sect. 1], where they use the fact from complex hyperbolic geometry that $G \cdot \mathrm{Hom}(\Gamma, L)$ is a locally algebraic subset of $\mathrm{Hom}(\Gamma, G)$ in the setting that $G = SU(n+1, 1)$, $L = U(n, 1)$, and Γ is a cocompact discrete subgroup of $U(n, 1)$. In the following, we simplify this discussion using the concept of subanalytic subsets, as stated in Lemma E.8.

Now we review briefly the definition and basic properties of subanalytic subsets (Definition E.4 to Lemma E.7), as presented in Kashiwara–Schapira [26, Sect. 8.2].

Definition E.4 Let X be a real analytic manifold. A subset Z of X is called *subanalytic* if, for each point x of X, there exists an open neighborhood U of x in X such that

$$Z \cap U = \bigcup_{i=1}^{p} (f_{i1}(Y_{i1}) \smallsetminus f_{i2}(Y_{i2})), \tag{E.2}$$

where, for each $i = 1, \ldots, p$ and $j = 1, 2$, Y_{ij} is a compact real analytic manifold and $f_{ij} \colon Y_{ij} \to U$ is a real analytic map.

Remark E.5 The above definition is equivalent to the following definition given in Bierstone–Milman [5, Prop. 3.13]: for each point x of X, there exists a neighborhood U of x in X, such that

$$Z \cap U = \bigcup_{i=1}^{p} (f_{i1}(A_{i1}) \smallsetminus f_{i2}(A_{i2})),$$

where, for each $i = 1, \ldots, p$ and $j = 1, 2$, A_{ij} is a closed real analytic subset of a real analytic manifold Y_{ij}, $f_{ij} \colon Y_{ij} \to U$ is a real analytic map, and $f_{ij}|_{A_{ij}}$ is proper. The non-trivial implication from the definition in Remark E.5 to Definition E.4 can be verified using the following Hironaka's uniformization theorem:

Fact E.6 (Hironaka's uniformization theorem [5, Thm. 0.1]) *Let Z be a closed subanalytic subset of a real analytic manifold X in the sense of Remark E.5. Then there exist a real analytic manifold Y and a proper real analytic map $f \colon Y \to X$ such that $f(Y) = Z$.*

We now explain immediate consequences of Definition E.4 and Remark E.5. By Definition E.4, a closed ball in Euclidean space $\mathbb{R}^n$ is subanalytic, as it can be expressed as the projection of the unit sphere S^n. Thus, every point in a real analytic manifold admits a compact subanalytic neighborhood. Furthermore, by Remark E.5, any real analytic subset of a real analytic manifold X is subanalytic.

We also need the following:

Lemma E.7 *Let X be a real analytic manifold.*

(1) If Z_1 and Z_2 are subanalytic subsets in X, then $Z_1 \smallsetminus Z_2$ is subanalytic in X.
(2) Let $f \colon X \to Y$ be an analytic map of real analytic manifolds. If Z is a subanalytic subset of X and f is proper on $\overline{Z}$, then $f(Z)$ is a subanalytic subset of Y.
(3) (Curve Selection Lemma) Let Z be a subanalytic subset of X and $z \in \overline{Z}$. Then there exists an analytic curve $c \colon (-1, 1) \to X$ such that $c(0) = z$ and $c(t) \in Z$ for any $t \neq 0$.

Under the preparation above, we provide a lemma for the proof of Proposition E.2:

Lemma E.8 *Suppose that we are in Setting E.1. Let W be a compact subanalytic neighborhood of the identity element in G. Then the subset*

$$Z := \mathrm{Hom}(\Gamma, G) \smallsetminus (W \cdot \mathrm{Hom}(\Gamma, L))$$

is subanalytic in $X = G^N$.

Proof The map $G \times \mathrm{Hom}(\Gamma, G) \to \mathrm{Hom}(\Gamma, G)$ defined by $(g, \varphi) \mapsto g \cdot \varphi$ is proper on $W \times \mathrm{Hom}(\Gamma, L)$. Hence, the image $W \cdot \mathrm{Hom}(\Gamma, L)$ is a subanalytic subset of X by Lemma E.7(2). By Lemma E.7(1), Z is subanalytic in X. □

We are ready to prove Proposition E.2.

Proof of Proposition E.2 Retain the notation from Lemma E.8.

To prove Proposition E.2, it suffices to show that $G \cdot \mathrm{Hom}(\Gamma, L)$ is a neighborhood of φ in $\mathrm{Hom}(\Gamma, G)$. Suppose, for the sake of contradiction, that this is not the case. Then $W \cdot \mathrm{Hom}(\Gamma, L)$ is also not a neighborhood of φ in $\mathrm{Hom}(\Gamma, G)$, which implies that $\varphi \in \overline{Z}$.

By Lemma E.8, the subset Z is subanalytic in X. Applying the curve selection lemma (Lemma E.7(3)), we obtain an analytic curve $\{\varphi_t\}_{-1<t<1}$ in X such that $\varphi_0 = \varphi$ and $\varphi_t \in Z$ for all $t \neq 0$. In particular, $\varphi_t \notin W \cdot \mathrm{Hom}(\Gamma, L)$ for all $t \neq 0$.

On the other hand, since $\varphi_t \in \mathrm{Hom}(\Gamma, G)$, Lemma E.3 (Step 1) ensures the existence of an analytic curve g_t in G such that $g_t \cdot \varphi_t \in \mathrm{Hom}(\Gamma, L)$. Hence, $\varphi_t \in W \cdot \mathrm{Hom}(\Gamma, L)$ for sufficiently small t, which is a contradiction. Thus, $G \cdot \mathrm{Hom}(\Gamma, L)$ must be a neighborhood of φ in $\mathrm{Hom}(\Gamma, G)$. This completes the proof of Proposition E.2. □

Although the following result is not used in the main body of this article, we include it for future reference, as it can be derived using the same idea originally due to Goldman–Millson [16]. The proof outlines only the necessary modifications to Proposition E.2.

In the following proposition, we recall that a 1-cochain $c \colon \Gamma \to \mathfrak{g}$ is a 1-cocycle if it satisfies

$$c(\gamma\eta) = c(\gamma) + \mathrm{Ad}(\varphi(\gamma))c(\eta),$$

for all $\gamma, \eta \in \Gamma$, and that the associated 2-cochain

$$[c, c] \colon \Gamma \times \Gamma \to \mathfrak{g}, \quad [c, c](\gamma, \eta) := [c(\gamma), \mathrm{Ad}(\varphi(\gamma))(c(\eta))]$$

is a 2-cocycle whenever c is a 1-cocycle.

Proposition E.9 *In Setting E.1, let $\varphi \colon \Gamma \to G$ be a group homomorphism such that $\varphi(\Gamma) \subset L$. We regard the Lie algebras $\mathfrak{g}$ and $\mathfrak{l}$ as Γ-modules via φ. Assume the following:*

(1) There exists an L*-invariant complementary subspace* $\mathfrak{m}$ *of* $\mathfrak{l}$ *in* $\mathfrak{g}$*;*
(2) $H^1(\Gamma, \mathfrak{l}/\mathfrak{z}) = 0$*, where* $\mathfrak{z}$ *denotes the center of* $\mathfrak{l}$*;*
(3) For any non-zero element $c \in H^1(\Gamma, \mathfrak{m})$*, we have*

$$\mathrm{pr}_{\mathfrak{l}}([c, c]) \neq 0 \quad \textit{in } H^2(\Gamma, \mathfrak{l}),$$

where $\mathrm{pr}_{\mathfrak{l}}$ *denotes the first projection onto* $H^2(\Gamma, \mathfrak{l})$ *with respect to the direct sum decomposition*

$$H^2(\Gamma, \mathfrak{g}) = H^2(\Gamma, \mathfrak{l}) \oplus H^2(\Gamma, \mathfrak{m}).$$

Then there exists a neighborhood U *of* φ *in* $\mathrm{Hom}(\Gamma, G)$ *such that for any* $\varphi' \in U$*, the image* $\varphi'(\Gamma)$ *is conjugate into* L *by an element of* G*.*

Proof By applying a similar argument using the *curve selection lemma* (see Lemma E.7 (3)) as in Step 2 of the proof of Proposition E.2, our assertion reduces to the case of an analytic deformation. Namely, for any family of group homomorphisms $\varphi_t : \Gamma \to G$, depending analytically on $t \in \mathbb{R}$, with $\varphi_0 = \varphi$, it suffices to show that there exists an analytic curve g_t in G with $g_0 = e$ such that $g_t \cdot \varphi_t \in \mathrm{Hom}(\Gamma, L)$ for all sufficiently small t.

Suppose that $\varphi_t : \Gamma \to G$ is a family of group homomorphisms that depends analytically on $t \in \mathbb{R}$, with $\varphi_0 = \varphi$. For each $\gamma \in \Gamma$, the real analytic curve $\varphi_t(\gamma)\varphi(\gamma)^{-1}$ in G, starts at e when $t = 0$. The Taylor expansion of this curve defines a sequence of maps $u_\ell \equiv u_\ell(\varphi_t) : \Gamma \to \mathfrak{g}$ $(\ell = 1, 2, \dots)$, such that

$$\varphi_t(\gamma) = \exp(u_1(\gamma)t + u_2(\gamma)t^2 + \cdots)\varphi(\gamma),$$

where the domain of convergence may depend on γ.

We fix $k \geq 1$, and prove the following claim: *assume that* $u_\ell(\Gamma) \subset \mathfrak{z}$ *for* $\ell = 1, \dots, k-1$*. Then, there exists an analytic curve in* G*, to be denoted by* $g_t^{(k)}$*, with the following two properties:*

(1) $u_\ell(\varphi_t) = u_\ell(g_t^{(k)} \cdot \varphi_t)$ for all $\ell = 1, \dots, k-1$,
(2) $u_k(g_t^{(k)} \cdot \varphi_t)(\Gamma) \subset \mathfrak{z}$,

where $g_t^{(k)} \cdot \varphi_t$ denotes a homomorphism $\Gamma \to \mathfrak{g}$ that is obtained by conjugating φ_t with $g_t^{(k)}$.

Once this claim has been proved, we may apply Artin's approximation theorem [1, Thm. 1.2] to the formal power series $g_t = \cdots g_t^{(2)} g_t^{(1)}$ to conclude the proposition, just as in Step 1 of the proof of Proposition E.2.

The remainder of the argument is devoted to proving the claim. To this end, we compute $\varphi_t(\gamma\eta) = \varphi_t(\gamma)\varphi_t(\eta)$ using the Baker–Campbell–Hausdorff formula:

$$\begin{aligned}\varphi_t(\gamma\eta) &= \exp(\sum_{n=1}^{\infty} u_n(\gamma)t^n)\varphi(\gamma)\exp(\sum_{n=1}^{\infty} u_n(\eta)t^n)\varphi(\eta)\\ &= \exp(\sum_{n=1}^{\infty} u_n(\gamma)t^n)\exp(\mathrm{Ad}(\varphi(\gamma))(\sum_{n=1}^{\infty} u_n(\eta)t^n))\varphi(\gamma\eta)\\ &= \exp(A(t)+B(t)+\tfrac{1}{2}[A(t),B(t)]+\cdots)\varphi(\gamma\eta),\end{aligned}$$

where $A(t) = \sum_{n=1}^{\infty} u_n(\gamma)t^n$ and $B(t) = \mathrm{Ad}(\varphi(\gamma))(\sum_{n=1}^{\infty} u_n(\eta)t^n)$. Hence, we have

$$\sum_{n=1}^{\infty} u_n(\gamma\eta)t^n = A(t) + B(t) + \frac{1}{2}[A(t), B(t)] + \cdots . \tag{E.3}$$

Let us consider the decomposition $u_n = v_n + w_n$ $(n = 1, 2, \ldots)$ of the 1-cochains corresponding to the decomposition $\mathfrak{g} = \mathfrak{l} + \mathfrak{m}$ as L-modules. By assumption, the images of $v_1, \ldots, v_{k-1}$ are contained in $\mathfrak{z}$ and $w_1, \ldots, w_{k-1}$ vanish. Comparing the $\mathfrak{l}$-components of the coefficient of t^k in the identity (E.3), we see that both v_k and w_k define 1-cocycles. Since $H^1(\Gamma, \mathfrak{l}/\mathfrak{z}) = 0$ by assumption, there exists an element $C'_k \in \mathfrak{l}$ such that,

$$v_k(\gamma) \equiv \mathrm{Ad}(\varphi(\gamma))C'_k - C'_k \mod \mathfrak{z},$$

for all $\gamma \in \Gamma$.

By comparing the $\mathfrak{l}$-component of the coefficient of t^{2k} in the identity (E.3), we obtain

$$v_{2k}(\gamma\eta) = v_{2k}(\gamma) + \mathrm{Ad}(\varphi(\gamma))v_{2k}(\eta) + \frac{1}{2}\big(\text{the } \mathfrak{l}\text{-component of } [w_k, w_k](\gamma, \eta)\big).$$

This shows that the $\mathfrak{l}$-component of the 2-cocycle $[w_k, w_k]$ is a 2-coboundary. By assumption (3), $w_k \colon \Gamma \to \mathfrak{m}$ is a 1-coboundary. Hence, there exists an element $C''_k \in \mathfrak{m}$ such that, for all $\gamma \in \Gamma$,

$$w_k(\gamma) = \mathrm{Ad}(\varphi(\gamma))C''_k - C''_k.$$

Now, we set $C_k := C'_k + C''_k \in \mathfrak{g}$ and define $g_t^{(k)} := \exp(t^k C_k) \in G$. Clearly, $u_1, \ldots, u_{k-1}$ remain unchanged, when we take replace φ_t with its conjugate $g_t^{(k)} \cdot \varphi_t \in G$. On the other hand, the coefficient of t^k in

$$\log((g_t^{(k)} \cdot \varphi_t)\varphi(\gamma)^{-1}) = \log(\exp(t^k C_k)\exp(A(t))\exp(-t^k \mathrm{Ad}(\varphi(\gamma))C_k))$$

is equal to $C_k + u_k(\gamma) - \mathrm{Ad}(\varphi(\gamma))C_k \in \mathfrak{z}$. Hence, the claim is proved, and the proposition follows. □

Appendix F. Continuous Family of Zariski-Dense Subgroups

This section discusses a *continuous family* of Zariski-dense subgroups in Zariski-connected, algebraic groups G. In Theorem F.5, we determine the optimal number of generators when G is reductive.

Let G be a Zariski-connected real algebraic group, and let $\mathfrak{g}$ be the real Lie algebra of G.

Definition F.1 For a finite subset $\mathcal{X} = \{X_1, \ldots, X_n\}$, we denote by $G(\mathcal{X})$ the identity component of the Zariski-closure of the subgroup of G generated by $e^{X_1}, \ldots, e^{X_n}$. It is also useful to introduce the notation $G(L; \mathcal{X})$, which denotes the identity component of the Zariski-closure of the subgroup of G generated by L and $e^{X_1}, \ldots, e^{X_n}$, when L is a subgroup of G.

For each fixed $t \in \mathbb{R}$, we define $t\mathcal{X} := \{tX \mid X \in \mathcal{X}\}$, and accordingly define the Zariski-connected subgroups $G(t\mathcal{X})$ and $G(L; t\mathcal{X})$, separately for each t.

Example F.2 Let G be a unipotent real algebraic group, $X \in \mathfrak{g}$, and set $\mathcal{X} = \{X\}$. Then the Lie algebra $\mathfrak{g}(\mathcal{X})$ of $G(\mathcal{X})$—the Zariski-closure of the subgroup generated by $\exp(X)$—is $\mathbb{R}X$, because the exponential map $\exp\colon \mathfrak{g} \to G$ is an isomorphism of algebraic groups.

In Definition F.1, it is important to note that the subgroups $G(t\mathcal{X})$ do not necessarily vary continuously with t, as will already be evident in the abelian case (see Proposition F.13 below).

Definition F.3 Let G be a Zariski-connected real algebraic group. We define $\underline{\eta}(G), \eta(G), \overline{\eta}(G)$ as the minimal cardinalities of finite subset $\mathcal{X}$ satisfying the following conditions, respectively:

$$\begin{aligned}
\eta(G) &: \quad G(t\mathcal{X}) = G \text{ for all } t > 0;\\
\eta(G) &: \quad G(t\mathcal{X}) = G \text{ for any } \delta > t > 0 \text{ with some } \delta > 0;\\
\underline{\eta}(G) &: \quad G(\mathcal{X}) = G.
\end{aligned}$$

We observe that $\eta(G)$ and $\overline{\eta}(G)$ depend on the *family* of Zariski-connected subgroups $G(t\mathcal{X})$, indexed by the continuous parameter t, whereas $\underline{\eta}(G)$ does not involve any continuous parameter. It is clear that

$$\underline{\eta}(G) \leq \eta(G) \leq \overline{\eta}(G).$$

Although our primary focus is on $\eta(G)$ for its application in Sects. 4.2 and 4.5, we frequently discuss $\underline{\eta}(G)$ and $\overline{\eta}(G)$ as well. First of all, we show that the value of $\underline{\eta}(G)$ can be arbitrarily large:

Example F.4 For $G = \mathbb{R}^n$ regarded as a unipotent commutative algebraic group, we have $\underline{\eta}(G) = \eta(G) = \overline{\eta}(G) = n$.

In contrast to the unipotent commutative case discussed in Example F.4, there exists an upper bound for $\overline{\eta}(G)$ in the reductive algebraic group.

Theorem F.5 *Let G be a Zariski-connected real reductive algebraic group. Then we have*

$$\underline{\eta}(G) = \eta(G) = \overline{\eta}(G) = 2,$$

except for the following cases:

- $(\underline{\eta}, \eta, \overline{\eta}) = (1, 1, 1)$ *when G is a split $\mathbb{R}$-torus ($G \simeq (\mathbb{R}^\times)^b$);*
- $(\underline{\eta}, \eta, \overline{\eta}) = (1, 2, 2)$ *when G is a non-split $\mathbb{R}$-torus ($G \simeq \mathbb{T}^a \times (\mathbb{R}^\times)^b$);*
- $(\underline{\eta}, \eta, \overline{\eta}) = (2, 2, 3)$ *when $\mathfrak{g}$ contains $\mathfrak{su}(2)$ as an ideal.*

The upper bounds for $\eta(G)$, $\underline{\eta}(G)$, and $\overline{\eta}(G)$ when G is *non-reductive* will be given in Lemma F.10.

Remark F.6 It is known that $\underline{\eta}(G) = 2$ for semisimple G; see, for example, Labourie [52, Lem. 5.3.13] for a proof in the case where G is the split group $SL(n, \mathbb{R})$. Our approach to the invariants η and $\overline{\eta}$ as well as $\underline{\eta}$, however, offers an alternative method of proof that extends to the family $G(t\mathcal{X})$ as the parameter $t \in \mathbb{R}$ varies.

Before turning to the proof of Theorem F.5, we first summarize some basic properties of $\eta(G)$, $\underline{\eta}(G)$, and $\overline{\eta}(G)$ in Lemmas F.7 and F.10 without assuming that G is reductive.

Lemma F.7 *For any homomorphism of real algebraic groups $\varphi \colon G \to G'$ and any finite subset $\mathcal{X} \subset \mathfrak{g}$, we have the identity*

$$d\varphi(\mathfrak{g}(\mathcal{X})) = \mathfrak{g}'(d\varphi(\mathcal{X})).$$

Proof Consider the extension of φ to a morphism of complex algebraic groups $\varphi_{\mathbb{C}} \colon G_{\mathbb{C}} \to G'_{\mathbb{C}}$. Then $\varphi_{\mathbb{C}}(G_{\mathbb{C}}(\mathcal{X}))$ is Zariski-closed and contains $\exp(d\varphi(X))$ for each $X \in \mathcal{X}$. Hence $\varphi_{\mathbb{C}}(G_{\mathbb{C}}(\mathcal{X})) \supset G'_{\mathbb{C}}(d\varphi(\mathcal{X}))$. The reverse inclusion follows from the Zariski-continuity of $\varphi_{\mathbb{C}}$. Therefore, we obtain the equality

$$\varphi_{\mathbb{C}}(G_{\mathbb{C}}(\mathcal{X})) = G'_{\mathbb{C}}(d\varphi(\mathcal{X})).$$

Taking the corresponding real Lie algebras then yields the desired identity. □

Definition F.8 Suppose that A is an associated $\mathbb{R}$-algebra, and that V is a finitely generated A-module. Let $\kappa(A, V)$ denote the minimal number of generators of V as an A-module. When V is a G-module for a group G, we define

$$\kappa(G, V) := \kappa(\mathbb{R}[G], V),$$

where $\mathbb{R}[G]$ is the group ring.

For the reader's convenience, we provide an explicit formula for $\kappa(S, \mathfrak{u})$ in terms of the multiplicities of irreducible representations. In the following lemma, the formula applies with $A = U(\mathfrak{s})$ and $V = \mathfrak{u}$.

Lemma F.9 *Let A be an associative $\mathbb{R}$-algebra, and V an A-module which is finite-dimensional over $\mathbb{R}$. Assume V decomposes $\bigoplus_i V_i^{n_i}$, where $V_1, V_2, \ldots$ are mutually non-isomorphic irreducible A-module and $n_i \in \mathbb{N}$ is the multiplicity of V_i in V. Put $D_i := \mathrm{End}_A(V_i)$, which is isomorphic to either $\mathbb{R}$, $\mathbb{C}$, or $\mathbb{H}$. We have*

$$\kappa(A, V) = \max_i \left(\left\lceil \frac{n_i}{\dim_{D_i} V_i} \right\rceil \right).$$

Proof By Jacobson's density theorem, the natural homomorphism of $\mathbb{R}$-algebras $A \to \prod_i \mathrm{End}_{D_i}(V_i)$ induced by the action of A on V_i is surjective. Therefore, we see that the A-module $\bigoplus_i V_i^{m_i}$ ($m_i \in \mathbb{N}$) is cyclic if and only if $m_i \leq \dim_{D_i} V_i$. Our conclusion follows from this fact. □

Lemma F.10 *Let G be a Zariski-connected real algebraic group. In what follows, ξ denotes any one of $\underline{\eta}$, η, or $\overline{\eta}$.*

(1) (Quotient). Let $\varphi \colon G \to G'$ be a homomorphism of Zariski-connected real algebraic groups such that $d\varphi$ is surjective. Then

$$\xi(G) \geq \xi(G').$$

Furthermore, if $d\varphi$ is injective, then equality holds.

(2) (Almost direct group). Let S and V be Zariski-connected normal subgroups such that $G = S \cdot V$ with $S \cap V$ finite, and with the Lie algebra $\mathfrak{s}$ being semisimple. Viewed as $\mathfrak{s}$-modules, we assume that the irreducible constituents of $\mathfrak{v}$ and $\mathfrak{s}$ are distinct. Then we have

$$\xi(G) = \max(\xi(S), \xi(V)).$$

(3) (Levi decomposition) Let $G = S \cdot U$ be a Levi decomposition of a Zariski-connected real algebraic group G, where S is a maximal real reductive algebraic subgroup and U is the unipotent radical of G. Then, we have

$$\xi(G) \leq \xi(S) + \kappa(S, \mathfrak{u}).$$

Proof In the proof below, since the argument remains unchanged when the parameter t is included, we treat $\underline{\eta}$, η, and $\overline{\eta}$ simultaneously, without explicitly introducing the parameter t.

(1). Let $\mathcal{X}$ be a finite subset of the Lie algebra $\mathfrak{g}$ such that $G(\mathcal{X}) = G$. By Lemma F.7, we have

$$\mathfrak{g}'(d\varphi(\mathcal{X})) = d\varphi(\mathfrak{g}(\mathcal{X})) = \mathfrak{g}'.$$

Since G' is Zariski-connected, it follows that $G'(d\varphi(\mathcal{X})) = G'$. Hence, the desired inequality follows.

Now suppose that $d\varphi$ is injective (and hence bijective). Let $\mathcal{X}'$ be a finite subset of the Lie algebra $\mathfrak{g}'$ such that $G'(\mathcal{X}') = G'$. Set $\mathcal{X} := (d\varphi)^{-1}(\mathcal{X}')$. Again, by Lemma F.7,

$$d\varphi(\mathfrak{g}(\mathcal{X})) = \mathfrak{g}'(\mathcal{X}') = \mathfrak{g}'.$$

This implies that $\mathfrak{g}(\mathcal{X}) = \mathfrak{g}$. Since G is Zariski-connected, we obtain $G(\mathcal{X}) = G$, which proves the reverse inequality.

(2). We know that $\xi(G) \geq \max(\xi(S), \xi(V))$ by (1). We now prove the reverse inequality.

Let $\mathrm{pr}_1 : \mathfrak{g} \to \mathfrak{s}$ and $\mathrm{pr}_2 : \mathfrak{g} \to \mathfrak{v}$ be the projection maps associated with the direct sum decomposition $\mathfrak{g} = \mathfrak{s} + \mathfrak{v}$. Choose a finite subset $\mathcal{X} = \{X_1, \ldots, X_m\} \subset \mathfrak{s}$ such that $S(\mathcal{X}) = S$, a finite subset $\mathcal{Y} = \{Y_1, \ldots, Y_n\} \subset \mathfrak{v}$ such that $V(\mathcal{Y}) = V$. Define

$$\mathcal{Z} = \{Z_1, \ldots, Z_{\max(m,n)}\} \text{ with } Z_i := X_i + Y_i,$$

where we take $X_i = 0$ for $i > m$ and $Y_i = 0$ for $i > n$. Let $\mathfrak{g}'$ denote the Lie algebra of $G(\mathcal{Z})$.

By Lemma F.7, we have $\mathrm{pr}_1(\mathfrak{g}') = \mathfrak{s}$ and $\mathrm{pr}_2(\mathfrak{g}') = \mathfrak{v}$. Since none of the $\mathfrak{s}$-irreducible constituents of $\mathfrak{v}$ appear in $\mathfrak{s}$, it follows from $\mathrm{pr}_1(\mathfrak{g}') = \mathfrak{s}$ that $\mathfrak{g}'$ contains the direct sum $\mathfrak{s} \oplus 0$. Therefore, from $\mathrm{pr}_2(\mathfrak{g}') = \mathfrak{v}$, we conclude that $\mathfrak{g}' = \mathfrak{g}$, and consequently, $G(\mathcal{Z}) = G$. Hence, we have established the reverse inequality $\xi(G) \leq \max(\xi(S), \xi(V))$.

(3). Let $\mathcal{X}$ be a finite subset of $\mathfrak{s}$ such that $G(\mathcal{X}) = S$, and let $\mathcal{Y}$ be a finite generating set of the S-module $\mathfrak{u}$ such that $\sharp\mathcal{Y} = \kappa(S, \mathfrak{u})$. Then, the group $G(\mathcal{X} \cup \mathcal{Y})$ contains $G(\mathcal{X}) = S$, and consequently,

$$G(\mathcal{X} \cup \mathcal{Y}) = G(S; \mathcal{Y}).$$

As seen in Example F.2, the Lie algebra $\mathfrak{g}(\mathcal{X} \cup \mathcal{Y})$ contains $\mathcal{Y}$. It follows that

$$\mathfrak{g}(\mathcal{X} \cup \mathcal{Y}) = \mathfrak{s} + \mathfrak{u} = \mathfrak{g}.$$

Since G is Zariski-connected, we conclude that $G(\mathcal{X} \cup \mathcal{Y}) = G$. This yields the desired inequality. □

As an immediate consequence of Lemma F.10(2), we obtain the following result in the reductive setting.

Proposition F.11 *Let G be a Zariski-connected, reductive algebraic group, and suppose that*

$$G = G_0 \cdot G_1 \cdots G_r$$

is the almost direct product, where G_0 is commutative, and the Lie algebras $\mathfrak{g}_i$ are of the form $\mathfrak{s}_i^{\oplus n_i}$, with each $\mathfrak{s}_i$ a simple Lie algebra not isomorphic to $\mathfrak{s}_j$ for $1 \leq i \neq j \leq r$. In what follows, ξ denotes any one of $\underline{\eta}$, η, or $\overline{\eta}$. Then we have

$$\xi(G) = \max_{0 \leq i \leq r} \xi(G_i).$$

Remark F.12 Let ξ denote any one of $\underline{\eta}$, η, or $\overline{\eta}$. When $\mathfrak{g}$ is the direct sum of n of a simple Lie algebra $\mathfrak{s}$ of a fixed type, the invariant $\xi(G)$ does not depend on n. We will give an explicit formula—for example, for $\overline{\eta}(G)$— in Propositions F.21 and F.26.

We now begin the proof of Theorem F.5. We first consider the elementary case in which G is commutative, addressed in Proposition F.13. We then prove Proposition F.16 which shows the result for $\underline{\eta}(G)$ and $\eta(G)$ in the general case where G is reductive. Finally, the results for $\overline{\eta}(G)$, when $\mathfrak{g}$ is the direct sum of a simple Lie algebras of a fixed type are given in Propositions F.21 and F.26. These complete the proof of Theorem F.5.

In the commutative case, the real algebraic group G decomposes as a direct sum of several copies of

$$\mathbf{SO}_2 := \left\{ \begin{pmatrix} x & -y \\ y & x \end{pmatrix} \middle| \, x^2 + y^2 = 1 \right\}, \quad \mathbf{G}_m := \left\{ \begin{pmatrix} x & 0 \\ 0 & y \end{pmatrix} \middle| \, xy = 1 \right\}.$$

The groups of real points of these algebraic groups are isomorphic, as Lie groups, to $\mathbb{T} := \{e^{it} \mid t \in \mathbb{R}\}$ and $\mathbb{R}^\times$, respectively. In what follows, we regard these Lie groups as equipped with the above real algebraic group structures, and we use the corresponding notation accordingly.

For $X = (X_1, \ldots, X_n) \in \mathbb{R}^n$ such that $X_1, \ldots, X_n$ are linearly independent over $\mathbb{Q}$, we define a countable subset of $\mathbb{R}$ by

$$\Lambda(X) := \bigcup_{\ell \in \mathbb{N}} \bigcup_{\lambda \in \mathbb{Z}^n \setminus \{0\}} \frac{\ell}{\sum_{j=1}^n \lambda_j X_j}. \tag{F.1}$$

Proposition F.13 *Let $G = \mathbb{T}^a \times (\mathbb{R}^\times)^b$. Suppose that real numbers $X_1, \ldots, X_{a+b}$ are linearly independent over $\mathbb{Q}$. Then*

$$(e^{\sqrt{-1}tX_1}, \ldots, e^{\sqrt{-1}tX_a}, e^{tX_{a+1}}, \ldots, e^{tX_{a+b}}) \in \mathbb{T}^a \times (\mathbb{R}^\times)^b$$

generates a Zariski-dense subgroup in G for any $t \in \mathbb{R} \setminus \pi\Lambda(X_1, \ldots, X_a)$. Furthermore, we have

$$\underline{\eta}(G) = 1, \quad \eta(G) = \overline{\eta}(G) = \begin{cases} 2 & (a > 0), \\ 1 & (a = 0). \end{cases}$$

Before proving Proposition F.13, we present two preliminary lemmas.

Lemma F.14 *Let $G = \mathbb{T}^a \times (\mathbb{R}^\times)^b$, and let* pr_i *denote the projection to the i-th factor for $i = 1, 2$. If G' is a real algebraic subgroup of G such that* $\mathrm{pr}_1(G') = \mathbb{T}^a$ *and* $\mathrm{pr}_2(G') = (\mathbb{R}^\times)^b$. *Then $G' = G$.*

Proof In our setting, every element g of G is semisimple, and its Jordan decomposition is given by $g = g_e g_h$, where $g_e = (\mathrm{pr}_1(g), 1)$ and $g_h = (1, \mathrm{pr}_2(g))$ are the elliptic and hyperbolic parts, respectively. Since the Jordan decomposition in the real algebraic subgroup G' is compatible with that of G, we have $G' = \mathrm{pr}_1(G') \times \mathrm{pr}_2(G')$. Hence, we conclude $G' = G$. □

Lemma F.15 *Suppose that real numbers $X_1, \ldots, X_n$ are linearly independent over $\mathbb{Q}$.*

(1) When $G = (\mathbb{R}^\times)^n$, $\exp(X) = (e^{X_1}, \ldots, e^{X_n}) \in G$ generates a Zariski-dense subgroup in G. In particular, $\underline{\eta}(G) = \eta(G) = \overline{\eta}(G) = 1$.
(2) When $G = \mathbb{T}^n$, $\exp(\pi\sqrt{-1}tX) \in G$ generates a Zariski-dense subgroup of G if $t \in \mathbb{R} \setminus \Lambda(X)$, where $\Lambda(X)$ is defined as in (F.1) *for $X = (X_1, \ldots, X_n)$. Furthermore, we have $\underline{\eta}(G) = 1$ and $\eta(G) = \overline{\eta}(G) = 2$.*

Proof (1). This is essentially Chevalley's lemma (e.g., [69, Lem. 2.A.1.2]).

In fact, suppose that a polynomial $f(x_1, \ldots, x_n) = \sum_{\alpha \in \mathbb{N}^n} a_\alpha x^\alpha$ vanishes on $\exp(\mathbb{Z}X)$, where $a_\alpha \in \mathbb{R}$ and $x^\alpha = x_1^{\alpha_1} \cdots x_n^{\alpha_n}$. This means that

$$\sum_{\alpha \in \mathbb{N}^n} a_\alpha \exp(\langle \alpha, X \rangle \ell) = 0 \quad \text{for all } \ell \in \mathbb{Z},$$

where $\langle \cdot, \cdot \rangle$ denotes the standard inner product on $\mathbb{R}^n$. Since the values $\langle \alpha, X \rangle$ are pairwise distinct for all $\alpha \in \mathbb{N}^n$ by the assumption on X, we deduce that $a_\alpha = 0$ by downward induction, starting from the largest $\langle \alpha, X \rangle$, and thus $f \equiv 0$. This shows $\underline{\eta}(G) = 1$. Since the assumption on X remains unchanged if we replace X with a non-zero scalar multiple tX, we conclude $\eta(G) = \overline{\eta}(G) = 1$.

(2). If $t \in \mathbb{R} \setminus \Lambda(X)$, then the set $\exp(\pi\sqrt{-1}\mathbb{Z}tX)$ is dense in $\mathbb{T}^n$ with respect to the usual topology, and hence also Zariski-dense. Therefore, we conclude that $\underline{\eta}(\mathbb{T}^n) = 1$.

Next, we show that $\eta(\mathbb{T}^n) \geq 2$. Let $X = (X_1, \ldots, X_n) \in \mathbb{R}^n$ be arbitrary. Then, there exists a positive number $t > 0$ as small as desired such that $tX_1 \in \mathbb{Q}\pi$. For such t, $G(t\{X\})$ is contained in the $1 \times \mathbb{T}^{n-1}$ because the projection of $\exp(\sqrt{-1}\mathbb{Z}tX) \subset \mathbb{T}^n$ to the first component has finite image. Hence, we have shown $\eta(\mathbb{T}^n) \geq 2$.

Finally, we show that $\overline{\eta}(\mathbb{T}^n) \leq 2$. We choose $X = (X_1, \ldots, X_n)$ and $Y = (Y_1, \ldots, Y_n)$ such that $X_1, \ldots, X_n, Y_1, \ldots, Y_n \in \mathbb{R}$ are $\mathbb{Q}$-linearly independent, and that $\Lambda(X) \cap \Lambda(Y) = \{0\}$. Then, for any positive real number t, at least one of the two elements $\exp(tX)$ and $\exp(tY)$ generates a Zariski-dense subgroup in G. Therefore, we conclude that $\overline{\eta}(\mathbb{T}^n) \leq 2$, and the proof of (2) is complete. □

Proof of Proposition F.13 Both assertions in Proposition F.13 follow immediately from Lemmas F.14 and F.15. □

We now consider the case where G is non-commutative.

Proposition F.16 *Let G be a Zariski-connected, non-commutative, real reductive algebraic group. Then we have $\underline{\eta}(G) = \eta(G) = 2$.*

To prove Proposition F.16, we introduce useful concepts and provide two lemmas.

Suppose that $\mathfrak{j}_{\mathbb{C}}$ is a Cartan subalgebra of a complex reductive Lie algebra $\mathfrak{g}_{\mathbb{C}}$. Let $\Delta(\mathfrak{g}_{\mathbb{C}}, \mathfrak{j}_{\mathbb{C}})$ denote the root system. We write

$$\mathfrak{g}_{\mathbb{C}} = \mathfrak{j}_{\mathbb{C}} \oplus \bigoplus_{\alpha \in \Delta(\mathfrak{g}_{\mathbb{C}}, \mathfrak{j}_{\mathbb{C}})} \mathfrak{g}_{\mathbb{C},\alpha}$$

for the root decomposition. Accordingly, for $X \in \mathfrak{g}_{\mathbb{C}}$, we decompose as $X = X_0 + \sum X_\alpha$.

Under this notation, we introduce the notion of $\mathfrak{j}$-*filled*, which will help to clarify the structure of the proof of Proposition F.16.

Definition F.17 We say that $X \in \mathfrak{g}_{\mathbb{C}}$ is $\mathfrak{j}_{\mathbb{C}}$-*filled* if it decomposes as $X = X_0 + \sum X_\alpha$ with $X_\alpha \neq 0$ for every $\alpha \in \Delta(\mathfrak{g}_{\mathbb{C}}, \mathfrak{j}_{\mathbb{C}})$. Let $\mathfrak{g}_{\mathbb{C}}(\mathfrak{j}_{\mathbb{C}})_{\mathrm{fill}}$ denote the set of $\mathfrak{j}_{\mathbb{C}}$-filled elements.

When $\mathfrak{g}$ is a real reductive Lie algebra and $\mathfrak{j}$ is a Cartan subalgebra, we write $\mathfrak{g}_{\mathbb{C}}$ and $\mathfrak{j}_{\mathbb{C}}$ for their complexifications, respectively. We say that an element X of $\mathfrak{g}$ is $\mathfrak{j}$-*filled* if $X \in \mathfrak{g}(\mathfrak{j})_{\mathrm{fill}} := \mathfrak{g} \cap \mathfrak{g}_{\mathbb{C}}(\mathfrak{j}_{\mathbb{C}})_{\mathrm{fill}}$.

Here are basic properties of $\mathfrak{j}$-filled elements:

Lemma F.18 *(1) The set $\mathfrak{g}(\mathfrak{j})_{\mathrm{fill}}$ is open dense in $\mathfrak{g}$ with respect to the usual topology.*
(2) If $X \in \mathfrak{g}(\mathfrak{j})_{\mathrm{fill}}$, then the Lie algebra generated by $\mathfrak{j}$ and X is equal to $\mathfrak{g}$.
(3) If $X \in \mathfrak{g}(\mathfrak{j})_{\mathrm{fill}}$, then $X \in \mathfrak{g}(\mathrm{Ad}(g)\mathfrak{j})_{\mathrm{fill}}$ if $g \in G$ is sufficiently close to the identity element e.

Proof Since $\mathfrak{g}_{\mathbb{C}}(\mathfrak{j}_{\mathbb{C}})_{\mathrm{fill}}$ is Zariski-open in $\mathfrak{g}_{\mathbb{C}}$, $\mathfrak{g}(\mathfrak{j})_{\mathrm{fill}}$ is open dense in $\mathfrak{g}$. Since every root space $\mathfrak{g}_{\mathbb{C},\alpha}$ is one-dimensional, the complex Lie algebra generated by $\mathfrak{j}$ and a $\mathfrak{j}$-filled element X is $\mathfrak{g}_{\mathbb{C}}$. Hence, the second statement holds. The last statement follows from the continuity of the root decomposition with respect to the choice of Cartan subalgebras. □

Lemma F.19 *Let $\mathfrak{g}$ be a real reductive Lie algebra. Then there exists a pair of elements X and Y satisfying the following properties: let $\mathfrak{j}_1$ be the centralizer of X in $\mathfrak{g}$, and $\mathfrak{j}_2$ be that of Y in $\mathfrak{g}$. Then*

(1) both $\mathfrak{j}_1$ and $\mathfrak{j}_2$ are Cartan subalgebras of $\mathfrak{g}$;
(2) X is $\mathfrak{j}_2$-filled, and Y is $\mathfrak{j}_1$-filled.

Let $\mathfrak{g}^{\mathrm{ss}}_{\mathrm{reg}}$ denote the set of semisimple, regular elements in $\mathfrak{g}$. Then $\mathfrak{g}^{\mathrm{ss}}_{\mathrm{reg}}$ contains an open dense subset of $\mathfrak{g}$.

Proof of Lemma F.19 We fix a Cartan subalgebra $\mathfrak{j}$, and take Y from $\mathfrak{g}^{ss}_{reg} \cap \mathfrak{g}(\mathfrak{j})_{\text{fill}}$. By Lemma F.18 (1), such a Y exists. The centralizer of Y in $\mathfrak{g}$, denoted by $\mathfrak{j}_2$, is a Cartan subalgebra of $\mathfrak{g}$ because $Y \in \mathfrak{g}^{ss}_{reg}$.

Since $Y \in \mathfrak{g}(\mathfrak{j})_{\text{fill}}$, by Lemma F.18 (3), there exists an open neighborhood V of e in G such that $Y \in \mathfrak{g}(\mathrm{Ad}(g)\mathfrak{j})_{\text{fill}}$ for any $g \in V$.

Since $\mathrm{Ad}(V)\mathfrak{j}$ contains a non-empty open subset of $\mathfrak{g}$, there exists $g \in V$ such that $\mathrm{Ad}(g)\mathfrak{j} \cap \mathfrak{g}^{ss}_{reg} \cap \mathfrak{g}(\mathfrak{j}_2)_{\text{fill}} \neq \emptyset$. We then choose X from this set.

Let $\mathfrak{j}_1$ denote the centralizer of X in $\mathfrak{g}$, which equals the Cartan subalgebra $\mathrm{Ad}(g)\mathfrak{j}$. Then X is $\mathfrak{j}_2$-filled, and Y is $\mathfrak{j}_1$-filled. Thus, the lemma is proved. □

We are ready to prove Proposition F.16.

Proof of Proposition F.16 Since G is non-commutative, it is clear that $2 \leq \underline{\eta}(G)$. Therefore, it suffices to show $\eta(G) \leq 2$.

We take two Cartan subalgebras $\mathfrak{j}_i$ $(i = 1, 2)$, $X \in \mathfrak{j}_1$, and $Y \in \mathfrak{j}_2$, as given in Lemma F.19. Let J_i be the Cartan subgroup of G with Lie algebras $\mathfrak{j}_i$ $(i = 1, 2)$.

By Lemma F.18 (1), we may replace X with a regular semisimple element $X' \in \mathfrak{j}_1$ such that X' is close enough to X that X' remains $\mathfrak{j}_2$-filled, while $G(t\{X'\}) = J_1$ for any $t \in \mathbb{R} \setminus \Lambda'$ by applying Proposition F.13 to the $\mathbb{R}$-torus J_1, where Λ' is a countable set of $\mathbb{R}$ depending on X', as given in (F.1).

Similarly, we may replace Y with a sufficiently close $Y' \in \mathfrak{j}_2$ such that Y' remains $\mathfrak{j}_1$-filled, while $G(tY') = J_2$ for any $t \in \mathbb{R} \setminus \Lambda''$, where Λ'' is a countable set of $\mathbb{R}$ depending on Y'. Since both Λ' and Λ'' are countable, we can take $s \in \mathbb{R}$, which is sufficiently close to 1, such that $s\Lambda' \cap \Lambda'' = \{0\}$. We set

$$\mathcal{X} := \{X', sY'\}.$$

Then, for any $t \in \mathbb{R} \setminus \{0\}$, at least one of the following holds:

$$G(t\{X'\}) = J_1 \quad \text{or} \quad G(t\{sY'\}) = J_2.$$

For instance, suppose that $G(t\{sY'\}) = J_2$. Then $G(t\{X', sY'\}) = G(J_2; t\{X'\})$. We write $X' = X' + \sum X'_\alpha$ for root decomposition. We take $H \in \mathfrak{j}_2$ such that $\alpha(H) \neq 0$ for any $\alpha \in \Delta(\mathfrak{g}_{\mathbb{C}}, \mathfrak{j}_{2,\mathbb{C}})$. Then there is an analytic curve $a \colon \mathbb{R} \to \mathfrak{g}$ such that $a(t) \in \mathfrak{g}(t\{X'\})$ for all $t \in \mathbb{R} \setminus \{0\}$ and $a(0) = [H, X'] = \sum_\alpha \alpha(H)X'_{\alpha'}$, as stated in Lemma 4.23. In particular, $a(0) \in \mathfrak{g}_{\text{fill}}$. Therefore, if t is sufficiently small, then $a(t) \in \mathfrak{g}_{\text{fill}}$. It follows from Lemma F.18 (2) that the Lie algebra $\mathfrak{g}(t\{X', sY'\})$ of $G(t\{X', sY'\})$ equals $\mathfrak{g}$.

Similarly in the case where $G(t\{X', sY'\}) = G(J_1; t\{sY'\})$, the proof works. Therefore, we have shown that $\mathfrak{g}(t\{X', sY'\}) = \mathfrak{g}$ in either case. Since G is Zariski-connected, we have $G(t\{X', sY'\}) = G$. Thus, the proposition is proved. □

We now determine $\overline{\eta}(G)$ for reductive groups G, starting with an exceptional case:

Lemma F.20 *If* $\mathfrak{g} = \mathfrak{su}(2)$, *then* $\overline{\eta}(G) = 3$.

Proof First, we show the inequality $2 < \overline{\eta}(G)$. Let X and $Y \in \mathfrak{g}$. Then there exists $t \in \mathbb{R}$ such that $\exp(tY) = 1$. Consequently, $G(t\{X, Y\}) = G(t\{X\})$ is abelian, which cannot coincide with G. Hence, we have shown that $3 \le \overline{\eta}(G)$.

Conversely, we can choose a basis $X_1, X_2, X_3 \in \mathfrak{g}$ such that, for any $t \neq 0$, at least two of $G(t\{X_i\})$ $(i = 1, 2, 3)$ are maximal tori in G. Therefore, $G(t\{X_1, X_2, X_3\}) = G$, showing $\overline{\eta}(G) \le 3$. □

Proposition F.21 *If $\mathfrak{g}$ is the direct sum of n copies of $\mathfrak{su}(2)$, then $\overline{\eta}(G) = 3$ for any $n \in \mathbb{N}_+$.*

Proof By Lemmas F.10 (1) and F.20, we have $\overline{\eta}(G) \le 3$. Let us prove the reverse inequality.

We write G as the almost direct product of groups $G = S^{(1)} \cdots S^{(n)}$, which corresponds to the direct sum decomposition of the Lie algebra $\mathfrak{g} \simeq \mathfrak{s}^{(1)} \oplus \cdots \oplus \mathfrak{s}^{(n)}$, where each $\mathfrak{s}^{(i)}$ is isomorphic to $\mathfrak{su}(2)$. We choose $X_1^{(i)}, X_2^{(i)}, X_3^{(i)}$ in $\mathfrak{s}^{(i)}$ for every $1 \le i \le n$, as in Lemma F.20, and define $X_j := \sum_{i=1}^n X_j^{(i)}$ for $j \in \{1, 2, 3\}$. Then, $G(t\{X_1, X_2, X_3\}) = G$ for all $t \neq 0$. □

The following lemma indicates that the case $G = SU(2)$ is the only exception.

Lemma F.22 *Suppose that G is a Zariski-connected, real simple algebraic group such that $\mathfrak{g}$ is not isomorphic to $\mathfrak{su}(2)$. There exist regular semisimple elements X and $Y \in \mathfrak{g}$ such that $G(t\{X, Y\}) = G$ for all $t \neq 0$. In particular, $\overline{\eta}(G) = 2$.*

For the proof, we need some notation and lemmas.

When $G_{\mathbb{C}}$ acts on $X_{\mathbb{C}}$, we denote by $(X_{\mathbb{C}})^g := \{x \in X_{\mathbb{C}} | gx = x\}$.

Lemma F.23 *Let $G_{\mathbb{C}}$ be a Zariski-connected, complex simple algebraic group, $Z_{G_{\mathbb{C}}}$ the center, $J_{\mathbb{C}}$ a Cartan subgroup, $H_{\mathbb{C}}$ a Zariski-connected algebraic subgroup containing $J_{\mathbb{C}}$, and $X_{\mathbb{C}} := G_{\mathbb{C}}/H_{\mathbb{C}}$.*

(1) $(X_{\mathbb{C}})^g = X_{\mathbb{C}}$ if and only if $g \in Z_{G_{\mathbb{C}}}$.

(2) We set

$$(X_{\mathbb{C}})_{\mathit{fix}} := \bigcup_{g \in J_{\mathbb{C}} \setminus Z_{G_{\mathbb{C}}}} (X_{\mathbb{C}})^g \tag{F.2}$$

Then $(X_{\mathbb{C}})_{\mathit{fix}}$ is a Zariski-closed subset with positive codimension in $X_{\mathbb{C}}$.

Proof (1). The condition $(X_{\mathbb{C}})^g = X_{\mathbb{C}}$ implies that g belongs to the proper normal subgroup $\bigcap_{\ell \in G_{\mathbb{C}}} \ell H_{\mathbb{C}} \ell^{-1}$. Since $G_{\mathbb{C}}$ is simple, we conclude $g \in Z_{G_{\mathbb{C}}}$. The converse statement is clear.

(2). Since $H_{\mathbb{C}}$ contains the Cartan subgroup $J_{\mathbb{C}}$, the defining equation $\exp(T)x = x$ in $X_{\mathbb{C}}$ for $T \in \mathfrak{j}_{\mathbb{C}}$ depends solely on the finite set

$$\Delta^T := \{\alpha \in \Delta(\mathfrak{g}_{\mathbb{C}}, \mathfrak{j}_{\mathbb{C}}) | e^{\alpha(T)} = 1\}.$$

Hence, the right-hand-side of (F.2) is a Zariski-closed subset with positive codimension in $X_{\mathbb{C}}$. □

For a fixed Cartan subgroup $J_{\mathbb{C}}$ of $G_{\mathbb{C}}$, we define

$$(G_{\mathbb{C}})_{\mathrm{fix}} := \bigcup_{J_{\mathbb{C}} \subset H_{\mathbb{C}}} \pi_{H_{\mathbb{C}}}^{-1}(G_{\mathbb{C}}/H_{\mathbb{C}})_{\mathrm{fix}}, \tag{F.3}$$

where $H_{\mathbb{C}}$ runs over the finite set of all Zariski-connected, maximal algebraic subgroups containing the Cartan subgroup $J_{\mathbb{C}}$, and $\pi_{H_{\mathbb{C}}} : G_{\mathbb{C}} \to G_{\mathbb{C}}/H_{\mathbb{C}}$ is the natural quotient map. Then, $(G_{\mathbb{C}})_{\mathrm{fix}}$ is a Zariski-closed subset of $G_{\mathbb{C}}$.

Lemma F.24 *Suppose* $Y \in \mathrm{Ad}(\ell^{-1})\mathfrak{j}_{\mathbb{C}}$ *where* $\ell \in G_{\mathbb{C}} \setminus (G_{\mathbb{C}})_{\mathit{fix}}$. *If* $\exp(Y) \notin Z_{G_{\mathbb{C}}}$, *then* $G_{\mathbb{C}}(J_{\mathbb{C}}; \{Y\}) = G_{\mathbb{C}}$.

Proof If this were not the case, there would exist a Zariski-connected maximal algebraic subgroup $H_{\mathbb{C}}$ such that $G(J_{\mathbb{C}}; \{Y\}) \subset H_{\mathbb{C}}$, where $Y = \mathrm{Ad}(\ell^{-1})X$ with $X \in \mathfrak{j}_{\mathbb{C}}$

Let $o \in G_{\mathbb{C}}/H_{\mathbb{C}}$ denote the origin. We observe that

$$\ell \cdot o \in \ell(G_{\mathbb{C}}/H_{\mathbb{C}})^{\exp(Y)} = (G_{\mathbb{C}}/H_{\mathbb{C}})^{\exp(X)}.$$

On the other hand, since $Y \in \mathfrak{j}_{\mathbb{C}}$ satisfies $\exp(Y) \notin Z_{G_{\mathbb{C}}}$, we have $\exp(X) \notin Z_{G_{\mathbb{C}}}$. It follows from the definition (F.3) that $(G_{\mathbb{C}}/H_{\mathbb{C}})^{\exp(X)} \subset (G_{\mathbb{C}}/H_{\mathbb{C}})_{\mathrm{fix}}$. This contradicts the assumption that $\ell \notin (G_{\mathbb{C}})_{\mathrm{fix}}$. □

Lemma F.25 *Let* G *be a Zariski-connected, real simple algebraic group such that* $\mathfrak{g} \neq \mathfrak{su}(2)$, *and let* $\mathfrak{j}$ *be a Cartan subalgebra of* $\mathfrak{g}$. *Then there exists a countable set* Ξ *in* $\mathfrak{j}$ *such that* $\exp(tY) \notin Z_G$ *for any* $t \in \mathbb{R} \setminus \{0\}$ *and for any* $Y \notin \mathbb{R}\Xi$.

Proof Let J be the Cartan subgroup of G with Lie algebra $\mathfrak{j}$. We note that the dimension of $\mathfrak{j}$ is greater that 1. Since the kernel of the exponential map $\exp : \mathfrak{j} \to J$ is a countable set, the lemma is clear. □

We are ready to prove Lemma F.22.

Proof of Lemma F.22 We shall find $X \in \mathfrak{j}$ and $Y \in \mathfrak{j}'$ in the order $\mathfrak{j} \Rightarrow Y \Rightarrow \mathfrak{j}' \Rightarrow X$ as follows.

First, we take a Cartan subalgebra $\mathfrak{j}$ of $\mathfrak{g}$. The set $(G_{\mathbb{C}})_{\mathrm{fix}}$ is defined, as in (F.3).

Next, we choose $Y \in \mathfrak{g}_{\mathrm{reg}}^{\mathrm{ss}}$ that satisfies the following three conditions, which are generic by Proposition F.26, Lemma F.25, and Proposition F.13, respectively:

(a) $\exp(tY) \notin Z_G$ for any $t \in \mathbb{R} \setminus \{0\}$;
(b) $Y \in \mathrm{Ad}(\ell^{-1})\mathfrak{j}$ for some $\ell \in G$ such that ℓ and ℓ^{-1} do not belong to $(G_{\mathbb{C}})_{\mathrm{fix}}$.
(c) $G(t\{Y\})$ is a Cartan subgroup of G for any $t \in \mathbb{R} \setminus \Lambda$, where Λ is a countable subset.

Let J' be the Cartan subgroup corresponding to $\mathfrak{j}' := \mathrm{Ad}(\ell^{-1})\mathfrak{j}$. Then, for any $t \in \mathbb{R} \setminus \Lambda$, we have $G(t\{Y\}) = J'$.

Third, we take $X \in \mathfrak{j}$ satisfying the following two conditions:

(a)' $\exp(tX) \notin Z_G$ for any $t \in \mathbb{R} \setminus \{0\}$;
(c)' $G(t\{X\}) = J$ for any $t \in \mathbb{R} \setminus \Lambda'$, where Λ' is a countable subset.

We already know that

(b)' $X \in \mathrm{Ad}(\ell)\mathfrak{j}'$, where $\ell \in G$ such that ℓ and ℓ^{-1} do not belong to $(G_{\mathbb{C}})_{\mathrm{fix}}$.

Rescaling X if necessary, we may and do assume that $\Lambda \cap \Lambda' = \{0\}$. Then, for any $t \neq 0$, at least one of the following holds: $G(t\{X\}) = J$ or $G(t\{Y\}) = J'$.

In the case where $G(t\{X\}) = J$, we have $G(t\{X, Y\}) = G(J; tY) = G$, where the last identity follows from Lemma F.24. On the other hand, in the case where $G(t; \{Y\}) = J'$, we have $G(t\{X, Y\}) = G(J'; tX) = G$, again by Lemma F.24. Therefore, we have shown $\overline{\eta}(G) \leq 2$.

Since the reverse inequality $2 \leq \overline{\eta}(G)$ is clear, the last assertion is also proved. □

Proposition F.26 *Suppose that G is a Zariski-connected real algebraic group such that $\mathfrak{g}$ is isomorphic to $\mathfrak{s}^n$, the direct sum of n copies of a simple Lie algebra $\mathfrak{s}$ which is not isomorphic to $\mathfrak{su}(2)$. Then there exist regular semisimple elements X and $Y \in \mathfrak{g}$ such that $G(t\{X, Y\}) = G$ for all $t \neq 0$. In particular, $\overline{\eta}(G) = 2$ for every $n \in \mathbb{N}_+$.*

Proof We write G as the almost direct product of groups

$$G = S_1 \cdots S_n,$$

which corresponds to the direct sum decomposition of the Lie algebra

$$\mathfrak{g} \simeq \mathfrak{s}_1 \oplus \cdots \oplus \mathfrak{s}_n,$$

where each $\mathfrak{s}_i$ is isomorphic to the same simple Lie algebra $\mathfrak{s}$. For $1 \leq i \leq n$, we choose Cartan subgroups J_i and J_i' of S_i, and regular semisimple elements $X_i, Y_i \in \mathfrak{s}_i$ as in Lemma F.22, satisfying the following conditions:

$$\begin{aligned} S(\{tX_i\}) &= J_i && \text{for any } t \in \mathbb{R} \setminus \Lambda_i, \\ S_i(\{tY_i\}) &= J_i' && \text{for any } t \in \mathbb{R} \setminus \Lambda_i', \\ S_i(J_i; \{tY_i\}) &= S_i && \text{for all } t \in \mathbb{R}, \\ S_i(J_i'; \{tX_i\}) &= S_i && \text{for all } t \in \mathbb{R}, \end{aligned}$$

where Λ_i and Λ_i' are countable sets such that $\Lambda_i \cap \Lambda_i' = \{0\}$. We define $X := X_1 + \cdots + X_n$ and $Y := Y_1 + \cdots + Y_n$. Then $G(t\{X, Y\})$ contains the subgroup S_i for every $1 \leq i \leq n$ and for each $t \in \mathbb{R} \setminus \{0\}$. Hence, it follows that $G(t\{X, Y\}) = G$ for all $t \in \mathbb{R} \setminus \{0\}$. □

Proof of Theorem F.5 By Proposition F.11, it suffices to show that the Lie algebra $\mathfrak{g}$ is either commutative or the direct sum of simple Lie algebras of the same type.

The commutative case is proved in Proposition F.13, with two exceptional cases given in Lemma F.15.

The formulas for $\underline{\eta}(G)$ and $\eta(G)$ in the non-commutative case are proved in Proposition F.16. The formula for $\overline{\eta}(G)$ in the non-commutative case is proved in Proposition F.26, with the exception case given in Proposition F.21.

Hence, the proof of Theorem F.5 is complete. □

References

1. Artin, M.: On the solutions of analytic equations. Invent. Math. **5**(4), 277–291 (1968)
2. Benoist, Y.: Actions propres sur les espaces homogènes réductifs. Ann. Math. (2) **144**(2), 315–347 (1996)
3. Berger, M.: Les espaces symétriques noncompacts. Ann. Sci. École Norm. Sup. (3) **74**, 85–177 (1957)
4. Beyrer, J., Kassel, F.: $\mathbb{H}^{p,q}$-convex cocompactness and higher higher teichmücller spaces. arXiv:2305.15031 (2023)
5. Bierstone, E., Milman, P.D.: Semianalytic and subanalytic sets. Inst. Hautes Études Sci. Publ. Math. (67), 5–42 (1988)
6. Bocheński, M., Tralle, A.: A solution of the problem of standard compact Clifford-Klein forms. arXiv:2403.10539v1 (2024)
7. Borel, A.: Linear Algebraic Groups, 2nd edn., Graduate Texts in Mathematics, vol. 126. Springer, New York (1991)
8. Burger, M., Iozzi, A., Wienhard, A.: Surface group representations with maximal Toledo invariant. Ann. Math. (2) **172**(1), 517–566 (2010)
9. Chevalley, C.: Théorie des Groupes de Lie. Tome II. Groupes algébriques, Actualités Scientifiques et Industrielles [Current Scientific and Industrial Topics], No. 1152, Hermann & Cie, Paris (1951)
10. Collingwood, D.H., McGovern, W.M.: Nilpotent Orbits in Semisimple Lie Algebras, Van Nostrand Reinhold Mathematics Series. Van Nostrand Reinhold Co., New York (1993)
11. Culler, M., Shalen, P.B.: Varieties of group representations and splittings of 3-manifolds. Ann. Math. (2) **117**(1), 109–146 (1983)
12. Eskin, A., McMullen, C.: Mixing, counting, and equidistribution in Lie groups. Duke Math. J. **71**(1), 181–209 (1993)
13. Flensted-Jensen, M.: Discrete series for semisimple symmetric spaces. Ann. Math. (2) **111**(2), 253–311 (1980)
14. Ghys, É.: Déformations des structures complexes sur les espaces homogènes de SL(2, $\mathbb{C}$). J. Reine Angew. Math. **468**, 113–138 (1995)
15. Goldman, W.M.: Nonstandard Lorentz space forms. J. Differential Geom. **21**(2), 301–308 (1985)
16. Goldman, W.M., Millson, J.J.: Local rigidity of discrete groups acting on complex hyperbolic space. Invent. Math. **88**(3), 495–520 (1987)
17. Goldman, W.M., Millson, J.J.: The deformation theory of representations of fundamental groups of compact Kähler manifolds. Inst. Hautes Études Sci. Publ. Math. (67), 43–96 (1988)
18. Guéritaud, F., Guichard, O., Kassel, F., Wienhard, A.: Anosov representations and proper actions. Geom. Topol. **21**(1), 485–584 (2017)
19. Guéritaud, F., Kassel, F.: Maximally stretched laminations on geometrically finite hyperbolic manifolds. Geom. Topol. **21**(2), 693–840 (2017)
20. Guichard, O.: Groupes plongés quasi isométriquement dans un groupe de Lie. Math. Ann. **330**(2), 331–351 (2004)

21. Johnson, D., Millson, J.J., Deformation spaces associated to compact hyperbolic manifolds, Discrete groups in geometry and analysis (New Haven, Conn., 1984), Progr. Math., vol. 67, Birkhäuser Boston, Boston, MA, pp. 48–106 (1987)
22. Kannaka, K.: Linear independence of generalized Poincaré series for anti–de Sitter 3-manifolds. SIGMA Symmetry Integrability Geom. Methods Appl. **17** (2021), Paper No. 042, 15
23. Kannaka, K.: Counting orbits of certain infinitely generated non-sharp discontinuous groups for the anti-de Sitter space. Selecta Math. (N.S.) **30**(1), 11 (2024)
24. Kannaka, K., Okuda, T., Tojo, K.: Zariski-dense discontinuous surface groups for reductive symmetric spaces. Symmetry in Geometry and Analysis (Festschrift in Honor of Toshiyuki Kobayashi), vol. 1, Progr. Math., pp. 321–354. Birkhäuser/Springer, Singapore (2025)
25. Kannaka, K., Tojo, K.: Hurwitz–Radon numbers and proper actions of semisimple Lie groups, in preparation
26. Kashiwara, M., Schapira, P.: Sheaves on Manifolds, Grundlehren der mathematischen Wissenschaften [Fundamental Principles of Mathematical Sciences], vol. 292. Springer, Berlin (1994), With a chapter in French by Christian Houzel, Corrected reprint of the 1990 original
27. Kassel, F.: Deformation of proper actions on reductive homogeneous spaces. Math. Ann. **353**(2), 599–632 (2012)
28. Kassel, F., Kobayashi, T.: Infinite multiplicity for poincaré series on certain standard locally homogeneous spaces, in preparation
29. Kassel, F., Kobayashi, T.: Poincaré series for non-Riemannian locally symmetric spaces. Adv. Math. **287**, 123–236 (2016)
30. Kassel, F., Kobayashi, T.: Invariant differential operators on spherical homogeneous spaces with overgroups. J. Lie Theory **29**(3), 663–754 (2019)
31. Kassel, F., Kobayashi, T.: Spectral analysis on pseudo-Riemannian locally symmetric spaces. Proc. Japan Acad. Ser. A Math. Sci. **96**(8), 69–74 (2020)
32. Kassel, F., Kobayashi, T., Spectral Analysis on Standard Locally Homogeneous Spaces, to appear in Lecture Notes in Mathematics **2367**, xi+116 pages (2025). https://doi.org/10.1007/978-981-96-1957-3
33. Kassel, F., Tholozan, N.: Sharpness of proper and cocompact actions on reductive homogeneous spaces. arXiv:2410.08179v1 (2024)
34. Kim, I., Pansu, P.: Flexibility of surface groups in classical simple Lie groups. J. Eur. Math. Soc. (JEMS) **17**(9), 2209–2242 (2015)
35. Klingler, B.: Local rigidity for complex hyperbolic lattices and Hodge theory. Invent. Math. **184**(3), 455–498 (2011)
36. Kneser, M.: Strong approximation, Algebraic Groups and Discontinuous Subgroups (Proc. Sympos. Pure Math., Boulder, Colo., 1965), Amer. Math. Soc., Providence, R.I. (1966), pp. 187–196
37. Kobayashi, T.: Proper action on a homogeneous space of reductive type. Math. Ann. **285**(2), 249–263 (1989)
38. Kobayashi, T.: Discontinuous groups acting on homogeneous spaces of reductive type, Representation theory of Lie groups and Lie algebras (Fuji-Kawaguchiko, 1990), World Sci. Publ., River Edge, NJ, 1992, pp. 59–75
39. Kobayashi, T.: On discontinuous groups acting on homogeneous spaces with noncompact isotropy subgroups. J. Geom. Phys. **12**(2), 133–144 (1993)
40. Kobayashi, T.: Discrete decomposability of the restriction of $A_{\mathfrak{q}}(\lambda)$ with respect to reductive subgroups and its applications. Invent. Math. **117**(2), 181–205 (1994)
41. Kobayashi, T.: Criterion for proper actions on homogeneous spaces of reductive groups. J. Lie Theory **6**(2), 147–163 (1996)
42. Kobayashi, T.: Discontinuous groups and Clifford-Klein forms of pseudo-Riemannian homogeneous manifolds, Algebraic and analytic methods in representation theory (Sønderborg, 1994), Perspect. Math., vol. 17. Academic Press, San Diego, CA, pp. 99–165 (1997)
43. Kobayashi, T.: Deformation of compact Clifford-Klein forms of indefinite-Riemannian homogeneous manifolds. Math. Ann. **310**(3), 395–409 (1998)

44. Kobayashi, T.: Discontinuous groups for non-Riemannian homogeneous spaces, Mathematics unlimited—2001 and beyond, pp. 723–747. Springer, Berlin (2001)
45. Kobayashi, T.: Intrinsic sound of anti–de Sitter manifolds, Lie theory and its applications in physics, Springer Proc. Math. Stat., vol. 191, pp. 83–99. Springer, Singapore (2016)
46. Kobayashi, T.: Bounded multiplicity theorems for induction and restriction. J. Lie Theory **32**(1), 197–238 (2022)
47. Kobayashi, T.: Conjectures on reductive homogeneous spaces, Mathematics going forward—collected mathematical brushstrokes, Lecture Notes in Math., vol. 2313. Springer, Cham, [2023] ©2023, pp. 217–231
48. Kobayashi, T., Nasrin, S.: Deformation of properly discontinuous actions of $\mathbb{Z}^k$ on $\mathbb{R}^{k+1}$. Internat. J. Math. **17**(10), 1175–1193 (2006)
49. Kobayashi, T., Nasrin, S.: Deformation space of discontinuous groups $\mathbb{Z}^k$ for a nilmanifold $\mathbb{R}^{k+1}$, Representation Theory of Groups and Extension of Harmonic Analysis (T. Kawazoe, ed.), Sūrikaisekikenkyūsho Kōkyūroku, no. 1467, pp. 101–111 (2006)
50. Kobayashi, T., Yoshino, T.: Compact Clifford-Klein forms of symmetric spaces—revisited, Pure Appl. Math. Q. **1**(3) (2005), Special Issue: In memory of Armand Borel. Part 2, 591–663
51. Kulkarni, R.S.: Proper actions and pseudo-Riemannian space forms. Adv. Math. **40**(1), 10–51 (1981)
52. Labourie, F.: Lectures on Representations of Surface Groups, Zurich Lectures in Advanced Mathematics, European Mathematical Society (EMS), Zürich (2013)
53. Martelli, B., Riolo, S., Slavich, L.: Compact hyperbolic manifolds without spin structures. Geom. Topol. **24**(5), 2647–2674 (2020)
54. Matsuki, T., Ōshima, T.: A description of discrete series for semisimple symmetric spaces, Group representations and systems of differential equations (Tokyo, 1982). Adv. Stud. Pure Math., vol. 4. North-Holland, Amsterdam, pp. 331–390 (1984)
55. Matsushima, Y., Murakami, S.: On vector bundle valued harmonic forms and automorphic forms on symmetric Riemannian manifolds. Ann. Math. (2) **78**, 365–416 (1963)
56. Millson, J.J.: On the first Betti number of a constant negatively curved manifold. Ann. Math. (2) **104**(2), 235–247 (1976)
57. Okuda, T.: Classification of semisimple symmetric spaces with proper $SL(2, \mathbb{R})$-actions. J. Differential Geom. **94**(2), 301–342 (2013)
58. Okuda, T.: Abundance of nilpotent orbits in real semisimple Lie algebras. J. Math. Sci. Univ. Tokyo **24**(3), 399–430 (2017)
59. Platonov, V., Rapinchuk, A.: Algebraic Groups and Number Theory, Pure and Applied Mathematics, vol. 139. Academic Press, Inc., Boston, MA (1994), Translated from the 1991 Russian original by Rachel Rowen
60. Raghunathan, M.S.: On the first cohomology of discrete subgroups of semisimple Lie groups. Amer. J. Math. **87**, 103–139 (1965)
61. Raghunathan, M.S.: Discrete Subgroups of Lie Groups, Ergebnisse der Mathematik und ihrer Grenzgebiete [Results in Mathematics and Related Areas], vol. Band 68. Springer, New York-Heidelberg (1972)
62. Selberg, A.: On discontinuous groups in higher-dimensional symmetric spaces, Contributions to function theory (Internat. Colloq. Function Theory, Bombay, 1960), Tata Inst. Fund. Res., Bombay, pp. 147–164 (1960)
63. Shafarevich, I.R.: Basic Algebraic Geometry, vols. 1, 2. Springer, Heidelberg (2013)
64. Shimura, G.: Arithmetic of Quadratic Forms. Springer Monographs in Mathematics. Springer, New York (2010)
65. Stallings, J.R.: On torsion-free groups with infinitely many ends. Ann. Math. (2) **88**, 312–334 (1968)
66. Tholozan, N.: Compact quotients of reductive homogeneous spaces and surface group representations, École Normale Supérieure – Université PSL (2021), Habilitation
67. Thurston, W.P.: Three-Dimensional Geometry and Topology. Vol. 1, Princeton Mathematical Series, vol. 35. Princeton University Press, Princeton, NJ (1997)

68. Tojo, K.: Classification of irreducible symmetric spaces which admit standard compact Clifford-Klein forms. Proc. Japan Acad. Ser. A Math. Sci. **95**(2), 11–15 (2019)
69. Wallach, N.R.: Real Reductive Groups. I, Pure and Applied Mathematics, vol. 132. Academic Press, Inc., Boston, MA (1988)
70. Weil, A.: On discrete subgroups of Lie groups. Ann. Math. (2) **72**, 369–384 (1960). II, ibid. **75**, 578–602 (1962)
71. Weil, A.: Remarks on the cohomology of groups. Ann. Math. (2) **80**, 149–157 (1964)
72. Wolpert, S.A.: Disappearance of cusp forms in special families. Ann. Math. (2) **139**(2), 239–291 (1994)

Non-Left-Orderability of Lattices in Higher-Rank Semisimple Lie Groups (After Deroin and Hurtado)

Dave Witte Morris

Abstract Let G be a connected, semisimple, real Lie group with finite centre, such that $\operatorname{rank}_{\mathbb{R}} G \geq 2$. B. Deroin and S. Hurtado recently proved the 30-year-old conjecture that no irreducible lattice in G has a left-invariant total order. (Equivalently, they proved that no such lattice has a nontrivial, orientation-preserving action on the real line.) We will explain many of the main ideas of the proof, by using them to prove the analogous result for lattices in p-adic semisimple groups. The p-adic case is easier, because some of the technical issues do not arise.

Keywords Orderable groups · Actions on the real line and the circle · Lattices in real Lie groups · Lattices in p-adic Lie

1 Introduction

Definition 1.1 ([16, p. 7]) A group Γ is *left-orderable* if it has a left-invariant total order. This means there is a transitive binary relation $\prec$, such that, for all $a, b, c \in \Gamma$,

- exactly one of the following is true: $a \prec b$, or $a = b$, or $a \succ b$, and
- if $a \prec b$, then $ca \prec cb$.

Understanding which groups are left-orderable is a longstanding problem in algebra [26, 31] and is also important in topology [11]. The following result proves a conjecture that was suggested informally (at least for the case where G is simple) by the current author more than 30 years ago, and became official when it was proposed independently by É. Ghys [22, p. 200] in 1999.

D. W. Morris (✉)
Department of Mathematics and Computer Science, University of Lethbridge, Lethbridge, AB, Canada
e-mail: dmorris@deductivepress.ca
URL: https://deductivepress.ca/dmorris

G. Prasad et al. (eds.), *Zariski Dense Subgroups, Number Theory and Geometric Applications*, Infosys Science Foundation Series,
https://doi.org/10.1007/978-981-95-3437-1_4

Theorem 1.2 (Deroin-Hurtado [14, Thm. 1.3]) *If Γ is an irreducible lattice in a connected, semisimple, real Lie group G with finite centre, and* $\operatorname{rank}_{\mathbb{R}} G \geq 2$, *then Γ is not left-orderable.*

Arithmetic subgroups of semisimple groups are lattices [30, Thm. 5.1.11, p. 86] (and the converse is true in higher real rank by the Margulis Arithmeticity Theorem [30, Thm. 5.2.1, p. 90]) so Theorem 1.2 can be restated in the following purely algebraic terms:

Corollary 1.3 (Deroin-Hurtado) *If* $\mathbf{G}$ *is an almost-simple algebraic group over* $\mathbb{Q}$, *such that* $\operatorname{rank}_{\mathbb{R}} \mathbf{G} \geq 2$, *then no finite-index subgroup of the arithmetic group* $\mathbf{G}(\mathbb{Z})$ *is left-orderable.*

The theorem also has the following reformulation:

Corollary 1.4 (Deroin-Hurtado [14, Thm. 1.3]) *Assume Γ is an irreducible lattice in a connected, semisimple, real Lie group G with finite centre, and* $\operatorname{rank}_{\mathbb{R}} G \geq 2$. *Then Γ has no nontrivial action on the real line by orientation-preserving homeomorphisms.*

Proof Suppose Γ has a nontrivial, orientation-preserving action on $\mathbb{R}$, and let N be the kernel of the action. Then Γ/N has a faithful, orientation-preserving action on $\mathbb{R}$, so it is not difficult to see that Γ/N is left-orderable [23, Thm. 6.8].

On the other hand, since no finite groups are left-orderable [16, Sect. 1.4.1, p. 27], we know that N is a normal subgroup of infinite index, so the Margulis Normal Subgroup Theorem [30, Thm. 17.1.1, p. 347] tells us that N is contained in the centre of G. So N is normal in G. Then Γ/N is a left-orderable, irreducible lattice in the semisimple Lie group G/N. This contradicts Theorem 1.2. □

Remark 1.5 If we do not require the homeomorphisms to preserve orientation, then there could be a nontrivial action: the map $x \mapsto -x$ is a homeomorphism of order 2, and it is possible that Γ has a nontrivial homomorphism to $\mathbb{Z}/2\mathbb{Z}$. However, it follows from Theorem 1.2 that every action of Γ on $\mathbb{R}$ (whether orientation-preserving or not) must be *virtually trivial*. This means that some finite-index subgroup of Γ acts trivially. (In fact, the subgroup will have index 1 or 2.)

By combining Theorem 1.2 with the following result, Deroin and Hurtado also obtained a strong result for actions on the circle S^1, not only actions on $\mathbb{R}$.

Theorem 1.6 (Ghys [22], cf. Burger-Monod [9, Cor. 1.5], [8, Cor. 1.7]) *Assume Γ is an irreducible lattice in a connected, semisimple, real Lie group G, and* $\operatorname{rank}_{\mathbb{R}} G \geq 2$. *If no almost-simple factor of G is locally isomorphic to* $\mathrm{SL}(2, \mathbb{R})$, *then every action of Γ on the circle S^1 has a finite orbit.*

Corollary 1.7 (Deroin-Hurtado [14]) *Assume Γ is an irreducible lattice in a connected, semisimple, real Lie group G, and* $\operatorname{rank}_{\mathbb{R}} G \geq 2$. *If no almost-simple factor of G is locally isomorphic to* $\mathrm{SL}(2, \mathbb{R})$, *then every action of Γ on the circle S^1 is virtually trivial.*

Proof By Theorem 1.6, we may let x be a point whose orbit is finite. Then its stabilizer Γ_x is a finite-index subgroup that acts on $S^1 \setminus \{x\}$, which is homeomorphic to $\mathbb{R}$. Then we see from Remark 1.5 that a finite-index subgroup of Γ_x must act trivially. □

Remark 1.8 The group $\mathrm{PSL}(2, \mathbb{R})$ acts faithfully on $\mathbb{R} \cup \{\infty\} \simeq S^1$ (by linear-fractional transformations). Therefore, if some almost-simple factor of G is locally isomorphic to $\mathrm{SL}(2, \mathbb{R})$, then every lattice in G has a highly nontrivial action on S^1 (because there is a surjective homomorphism $G \to \mathrm{PSL}(2, \mathbb{R})$). This shows that Corollary 1.7 would not be valid without the hypothesis that no factor is locally isomorphic to $\mathrm{SL}(2, \mathbb{R})$.

For simplicity, the statements of the above results assume that the centre of G is finite. There are counterexamples without this assumption, but Deroin and Hurtado proved that all of them arise from the fact that the action of $\mathrm{PSL}(2, \mathbb{R})$ on S^1 lifts to an orientation-preserving action of $\widetilde{\mathrm{SL}}(2, \mathbb{R})$ (the universal cover of $\mathrm{SL}(2, \mathbb{R})$) on $\mathbb{R}$. In particular:

Corollary 1.9 (Deroin-Hurtado [14, Thm. 1.1 and Sect. 2]) *Assume G is a connected, semisimple, real Lie group, and* $\mathrm{rank}_{\mathbb{R}} G \geq 2$. *An irreducible lattice Γ in G is left-orderable if and only if it is torsion-free and there is a surjective homomorphism $G \to \widetilde{\mathrm{SL}}(2, \mathbb{R})$.*

Remark 1.10 Theorem 1.2 is a huge advance over what was previously known. In particular:

(1) There was not a single example of a cocompact, torsion-free lattice Γ (in a connected, semisimple Lie group), for which it was known that no finite-index subgroup of Γ is left-orderable.
(2) For the noncocompact case, the theorem had been proved under the additional assumption that either $\mathrm{rank}_{\mathbb{Q}} \Gamma \geq 2$ [33] or G has more than one noncompact simple factor [27]. However, the theorem was not known to be true even for the special case where $G = \mathrm{SL}(3, \mathbb{R})$.

See [23, 29] for expository discussions of actions of lattices (and related groups) on $\mathbb{R}$ or S^1, and see [16, Chap. 4] for an introduction to the viewpoint on (non-) left-orderability that is used in the proof of Theorem 1.2.

Remark 1.11 The circle is a 1-dimensional manifold, so the Deroin-Hurtado Theorem is a part of the "Zimmer program," which includes the study of actions of lattices on low-dimensional manifolds. (However, unlike in Theorem 1.2, the actions are usually assumed to be differentiable, and are often C^∞.) The recent work of A. Brown, D. Fisher, and S. Hurtado [5–7] is a huge advance in the area (and seems to have been part of the inspiration for Deroin and Hurtado). See [18] for a survey of the Zimmer program, and see [10, 19] for expositions of the first Brown-Fisher-Hurtado paper.

In this paper, we will explain some of the main ideas in the proof of Theorem 1.2 (or, equivalently, Corollary 1.4). However, p-adic groups are easier than real ones, so we will actually prove the following:

Theorem 1.12 (Deroin-Hurtado) *If Γ is an irreducible lattice in a p-adic semisimple Lie group G, and* $\mathrm{rank}_{\mathbb{Q}_p} G \geq 2$, *then Γ has no nontrivial action on the real line by orientation-preserving homeomorphisms.*

Here is an outline of the paper.

- Section 2 recalls some standard definitions and notation, and a few basic results. (The reader is encouraged to skip this section, and refer back when necessary.)
- Section 3 sketches the proof of Theorem 1.12.
- Further explanation of several arguments can be found in Sect. 4. Some details are omitted, but specific references to the original work of Deroin and Hurtado [14] are provided.
- Our exposition of the proof of Theorem 1.12 relies on the simplifying assumption that $G = K\Gamma$ (see Assumption 3.7), but Sect. 5 provides a brief introduction to the main tool that allows this assumption to be eliminated (namely, the theory of harmonic functions).

Note 1.13 This is an expository paper. None of the nontrivial ideas or results are new, so the author does not claim credit for any of them, even if no reference is given. In particular, although Theorem 1.12 is not stated in [14], it follows directly from the methods there, so it can be considered to be implicit in the paper and should be attributed to Deroin and Hurtado.

On the other hand, the author takes responsibility for any errors and other deficiencies in the manuscript (of course).

2 Preliminaries

Measure Theory

Reminder 2.1 Let μ be a measure on a separable, metrizable topological space Y.

(1) We say that μ is a *probability measure* if $\mu(Y) = 1$.
(2) A measure μ' is *absolutely continuous* with respect to μ if, for every Borel set B, such that $\mu(B) = 0$, we also have $\mu'(B) = 0$.
(3) If μ' is absolutely continuous with respect to μ, then there is a non-negative measurable function $d\mu'/d\mu$ (called the *Radon-Nikodym derivative* of μ' with respect to μ), such that $\mu' = (d\mu'/d\mu)\,\mu$.
(4) If two measures are absolutely continuous with respect to each other, then they are in the same *measure class*. (This is an equivalence relation.)
(5) If $f: Y \to Z$ is a continuous function to a topological space Z, then we define a measure $f_*\mu$ on Z by
$$f_*\mu(B) = \mu\big(f^{-1}(B)\big).$$

Topological Groups

All groups in this paper (including G, K, Γ, and $\mathbb{R}$) are locally compact, which means there is a nonempty open set whose closure is compact.

Reminder 2.2 Assume H is a locally compact topological group.

(1) It is a basic fact in the theory of topological groups that H has a *Haar measure*, denoted m_H. This measure is left-invariant ($m_H(hB) = m_H(B)$ for all $h \in H$ and every Borel set B), finite on compact sets, and nonzero on nonempty open sets.
(2) H is *unimodular* if its Haar measure is bi-invariant. This means it is both left-invariant and right-invariant: $\mu_H(hB) = \mu_H(B) = \mu_H(Bh)$.
(3) If H is either a semisimple Lie group or a discrete group or a compact group, then H is unimodular.
(4) A discrete subgroup Γ of H is a *lattice* in H if there is an H-invariant probability measure $m_{H/\Gamma}$ on H/Γ.
(5) A lattice Γ is *irreducible* if ΓN is dense in H, for every closed, noncompact, normal subgroup N of H.
(6) A closed subgroup Γ of H is *cocompact* if H/Γ is compact.

Every discrete, cocompact subgroup is a lattice (because the Haar measure of every compact set is finite). The converse is not true in general, but the first part of the following result is a special case of the fact that it is true in p-adic groups:

Theorem 2.3 *Assume Γ is an irreducible lattice in a p-adic semisimple Lie group G, such that* $\operatorname{rank}_{\mathbb{Q}_p} G \geq 2$. *Then:*

(1) Γ is cocompact [28, Prop. IX.3.7, p. 313],
(2) the commutator subgroup $[\Gamma, \Gamma]$ *has finite index in Γ [28, Thm. IX.5.4, p. 325], and*
(3) if Γ acts on the circle S^1 by homeomorphisms, then the action has a finite orbit [34, Cor. 6.10].

The theory of roots provides a quite explicit description of the parabolic subgroups of every semisimple Lie group (see [3, 21.12, p. 234], for example). We will need the following consequence:

Lemma 2.4 *Assume P is a parabolic subgroup of a (real or p-adic) semisimple Lie group G. Then there is a split semisimple element a of P, such that:*

(1) $P = \big\{ h \in G \mid \{ a^n h a^{-n} \mid n \leq 0 \}$ *is bounded* $\big\}$, *and*
(2) if we let $U_a^- = \big\{ u \in G \mid a^n u a^{-n} \overset{n\to\infty}{\longrightarrow} \mathbf{1} \}$, *then*
  (a) the Lie algebra $\mathfrak{g}$ is the direct sum of the Lie algebra of P and the Lie algebra of U_a^-, so
  (b) $U_a^- P$ contains a neighbourhood of $\mathbf{1}$ in G.

Proof For concreteness, assume $G = \mathrm{SL}(3, \mathbb{Q}_p)$. Then it is well known (cf. [24, Sect. 21.3, pp. 134–136]) that every parabolic subgroup of G is conjugate to either

$$P_0 = G, \text{ or } P_1 = \begin{bmatrix} * & * & * \\ * & * & * \\ 0 & 0 & * \end{bmatrix}, \text{ or } P_2 = \begin{bmatrix} * & * & * \\ 0 & * & * \\ 0 & * & * \end{bmatrix}, \text{ or } P_3 = \begin{bmatrix} * & * & * \\ 0 & * & * \\ 0 & 0 & * \end{bmatrix}.$$

If $P = P_i$, we let $a = a_i$, where

$$a_0 = \mathbf{1}, \ a_1 = \mathrm{diag}(1/p, 1/p, p^2), \ a_2 = \mathrm{diag}(1/p^2, p, p), \ a_3 = \mathrm{diag}(1/p, 1, p).$$

It is easy to verify (1) with this choice of a. For example, if $i = 1$, then

$$a^n \begin{bmatrix} \mathsf{a} & \mathsf{b} & \mathsf{c} \\ \mathsf{d} & \mathsf{e} & \mathsf{f} \\ \mathsf{g} & \mathsf{h} & \mathsf{i} \end{bmatrix} a^{-n} = \begin{bmatrix} p^{-n} & 0 & 0 \\ 0 & p^{-n} & 0 \\ 0 & 0 & p^{2n} \end{bmatrix} \begin{bmatrix} \mathsf{a} & \mathsf{b} & \mathsf{c} \\ \mathsf{d} & \mathsf{e} & \mathsf{f} \\ \mathsf{g} & \mathsf{h} & \mathsf{i} \end{bmatrix} \begin{bmatrix} p^{n} & 0 & 0 \\ 0 & p^{n} & 0 \\ 0 & 0 & p^{-2n} \end{bmatrix}$$
$$= \begin{bmatrix} \mathsf{a} & \mathsf{b} & p^{-3n}\mathsf{c} \\ \mathsf{d} & \mathsf{e} & p^{-3n}\mathsf{f} \\ p^{3n}\mathsf{g} & p^{3n}\mathsf{h} & \mathsf{i} \end{bmatrix},$$

so we simply need to note that $\{\, p^{\mathsf{r}n}\mathsf{s} \mid n \leq 0 \,\}$ is bounded in $\mathbb{Q}_p$ if and only if either $\mathsf{r} \leq 0$ or $\mathsf{s} = 0$.

By combining the same calculation with the observation that $p^{\mathsf{r}n}\mathsf{s} \xrightarrow{n\to\infty} 0$ if and only if either $\mathsf{r} > 0$ or $\mathsf{s} = 0$, we see that $U_a^- = U_i$, where

$$U_0 = \{\mathbf{1}\}, \quad U_1 = \begin{bmatrix} 1 & 0 & 0 \\ 0 & 1 & 0 \\ * & * & 1 \end{bmatrix}, \quad U_2 = \begin{bmatrix} 1 & 0 & 0 \\ * & 1 & 0 \\ * & 0 & 1 \end{bmatrix}, \quad \text{and} \quad U_3 = \begin{bmatrix} 1 & 0 & 0 \\ * & 1 & 0 \\ * & * & 1 \end{bmatrix},$$

so (2a) clearly holds.

Now, define $f : U_a^- \times P \to G$ by $f(u, h) = uh$. We know from (2a) that the derivative of f at $(\mathbf{1}, \mathbf{1})$ is surjective, so the Inverse Function Theorem tells us that the image of f contains a neighbourhood of $\mathbf{1}$. This establishes (2b). □

Group Actions

Definition 2.5 Assume

- a topological group H acts by homeomorphisms on a metrizable topological space Y,
- μ_H is a probability measure on H, and
- μ_Y is a probability measure on Y.

Then:

(1) The action of H is *free* if $hy \neq y$ for all nonidentity $h \in H$ and all $y \in Y$.
(2) The measure μ_Y is H-*invariant* if $h_*\mu_Y = \mu_Y$ for all $h \in H$ (where $h_*\mu_Y$ was defined in Recall 2.1(5)). In other words, $\mu_Y(hB) = \mu_Y(B)$, for every $h \in H$ and every Borel set B.
(3) An H-invariant measure μ_Y is *ergodic* if, for every H-invariant Borel set B, either $\mu_Y(B) = 0$, or $\mu_Y(Y \setminus B) = 0$.
(4) The convolution $\mu_H * \mu_Y$ is the measure on Y defined by

$$\mu_H * \mu_Y = \int_H h_*\mu_Y \, d\mu_H(H), \quad \text{so} \quad (\mu_H * \mu_Y)(B) = \int_H \mu_Y(h^{-1}B) \, d\mu_H(h).$$

(5) We say that μ_Y is μ_H-*stationary* if $\mu_H * \mu_Y = \mu_Y$. More concretely, for every Borel set B, this means that we have

$$\mu_Y(B) = \int_H \mu_Y(h^{-1}B) \, d\mu_H(h).$$

In other words, the measure of every set is equal to the average of the measures of its translates.
(6) If μ_Y is H-invariant, then there is a unique probability measure η on the space $\mathcal{E}$ of ergodic H-invariant measures on Y, such that

$$\mu_Y = \int_{\mathcal{E}} \xi \, d\eta(\xi).$$

(See [32, Thm., p. 77], for example.) Measures in the support of η are called *ergodic components* of μ_Y.
(7) For $C > 0$, a function $f : \mathbb{R} \to \mathbb{R}$ is C-*Lipschitz* if

$$|f(x) - f(y)| \leq C\,|x - y| \quad \text{for all } x, y \in \mathbb{R}.$$

Dynamical Systems

Proposition 2.6 (cf. [17, Cor. 4.2 and Thm. 4.4, pp. 98–99]) *Assume $\mathbb{R}$ acts continuously on a compact, metrizable space Y. Then there is at least one $\mathbb{R}$-invariant probability measure on Y that is ergodic.*

Theorem 2.7 (Pointwise Ergodic Theorem [17, Thm. 2.30, p. 44]) *Assume a cyclic group $\langle g \rangle$ acts by homeomorphisms on a metrizable space Y, and μ is an ergodic $\langle g \rangle$-invariant probability measure on Y. Then, for every bounded, measurable function f on Y, and μ-a.e. $y \in Y$, we have*

$$\lim_{n\to\infty} \frac{1}{n} \sum_{i=1}^{n} f(g^i y) = \int_Y f \, d\mu.$$

Applying the Pointwise Ergodic Theorem to the ergodic components establishes a generalization to actions that are not assumed to be ergodic:

Corollary 2.8 (cf. [17, Thm. 2.30, p. 44]) *Assume a cyclic group $\langle g \rangle$ acts by homeomorphisms on a metrizable space Y, and μ is a $\langle g \rangle$-invariant measure on Y with $\mu(Y) < \infty$. Then, for every measurable function $f \colon Y \to [0, \infty)$, there is a g-invariant, measurable function $\overline{f} \colon Y \to [0, \infty]$, such that $\int_X \overline{f} \, d\mu = \int_X f \, d\mu$ and, for μ-a.e. $y \in Y$, we have*

$$\lim_{n\to\infty} \frac{1}{n} \sum_{i=1}^{n} f(g^i y) = \overline{f}(y).$$

Proposition 2.9 *Let Γ be an irreducible lattice in a (real or p-adic) semisimple Lie group G, and let P be a parabolic subgroup of G. Then:*

(1) The P-invariant probability measure on G/Γ is unique (namely, the G-invariant measure $m_{G/\Gamma}$ is the only P-invariant probability measure).
(2) G/Γ has no closed, nonempty, proper, P-invariant subsets. In other words, for all $y \in G/\Gamma$, the orbit Py is dense in G/Γ.

Idea of Proof Since P is parabolic, we know from Lemma 2.4(1) that there is a split semisimple element $a \in P$, such that

$$P = \big\{\, h \in G \mid \{\, a^n h a^{-n} \mid n \leq 0 \,\} \text{ is bounded} \,\big\}.$$

By the Moore Ergodicity Theorem [30, Exer. 11.2#11, p. 218], we also know that a is ergodic on G/Γ (with respect to the G-invariant probability measure $m_{G/\Gamma}$).

(2) Since a is ergodic, almost every a-orbit is dense. Hence, we can choose some $y' \approx y$, such that $\{\, a^n y' \mid n \geq 0 \,\}$ is dense in G/Γ. Let

$$U_a^- = \big\{\, u \in G \mid a^n u a^{-n} \xrightarrow{n\to\infty} \mathbf{1} \,\big\}.$$

Then $U_a^- P$ contains a neighbourhood of $\mathbf{1}$ in G, so we may write $y' = upy$ with $u \in U_a^-$ and $p \in P$.

For any $x \in G/\Gamma$, there is some large $n > 0$, such that $a^n y' \approx x$. Then

$$x \approx a^n y' = a^n upy = a^n u^- a^{-n} \cdot a^n py \approx \mathbf{1} \cdot a^n py \in Py.$$

(1) Let μ be an ergodic P-invariant probability measure on G/Γ, let f be a continuous function on G/Γ with compact support, and let

$$Y = \left\{ y \in G/\Gamma \;\middle|\; \frac{1}{n}\sum_{i=1}^{n} f(a^i y) \overset{n\to\infty}{\longrightarrow} \int_{G/\Gamma} f \, d\mu \right\}.$$

We have $\mu(Y) = 1$ by the Pointwise Ergodic Theorem (2.7). Since μ is P-invariant, this implies that for μ-a.e. $u \in Y$ and m_P-a.e. $p \in P$, we have $py \in Y$. Also, it is not difficult to see that $uy \in Y$ for every $u \in U_a^-$. This easily implies that $m_{G/\Gamma}(Y) > 0$. Hence, by the Pointwise Ergodic Theorem (2.7), there exists $y \in Y$, such that

$$\frac{1}{n}\sum_{i=1}^{n} f(a^i y) \overset{n\to\infty}{\longrightarrow} \int_{G/\Gamma} f \, dm_{G/\Gamma}.$$

So we must have $\int_{G/\Gamma} f \, d\mu = \int_{G/\Gamma} f \, dm_{G/\Gamma}$. Since f is an arbitrary continuous function with compact support, this implies $\mu = m_{G/\Gamma}$. □

3 Main Ideas of the Deroin-Hurtado Proof

In this section, we will present many of the main ideas in the Deroin-Hurtado paper [14]. (For readers interested in more details, references to specific statements in [14] will usually be provided. In many cases, additional details are also available in Sect. 4.) Our goal is to establish Theorem 1.12. For the reader's convenience, we reproduce the statement of this result:

Theorem 1.12 (Deroin-Hurtado) *If Γ is an irreducible lattice in a p-adic semisimple Lie group G, and* $\operatorname{rank}_{\mathbb{Q}_p} G \geq 2$, *then Γ has no nontrivial action on the real line by orientation-preserving homeomorphisms.*

Remark 3.2 The same proof applies when G is a finite product of p-adic semisimple Lie groups, for various primes p. Thus (in the algebraic spirit of Corollary 1.3), we can say: *Let*

- $\mathbf{G}$ *be an almost-simple algebraic group over a number field F,*
- *S be a finite set of places of F that includes all of the infinite places, and*
- *$\mathcal{O}_S$ be the ring of S-integers of F.*

If $\mathbf{G}(F_\infty)$ is compact for every infinite place ∞, and $\sum_{v \in S} \operatorname{rank}_{F_v} \mathbf{G} \geq 2$, *then no finite-index subgroup of the S-arithmetic group $\mathbf{G}(\mathcal{O}_S)$ is left-orderable.*

3.1 Assumptions and Notation

Notation 3.3 Throughout the remainder of this paper:

- p is a prime number,
- G is a p-adic semisimple Lie group, such that $\operatorname{rank}_{\mathbb{Q}_p} G \geq 2$,
- Γ is an irreducible lattice in G,
- K is a compact, open subgroup of G,
- P is a minimal parabolic subgroup of G, and
- A is a maximal $\mathbb{Q}_p$-split torus in P.

The proof of Theorem 1.12 is by contradiction, so we assume:

Assumption 3.4 Γ has a faithful action by orientation-preserving homeomorphisms of $\mathbb{R}$.

Remark 3.5

(1) The proof of Corollary 1.4 shows that any nontrivial, orientation-preserving action of Γ on $\mathbb{R}$ will become faithful after modding out a finite, normal subgroup of G. Therefore, even though Theorem 1.12 refers to a *nontrivial* action, no loss of generality results from Assumption 3.4's requirement that the action is *faithful*.

(2) The reader will not miss out on any of the main ideas if they assume that $G = \mathrm{SL}(3, \mathbb{Q}_p)$ or that $G = \mathrm{SL}(2, \mathbb{Q}_p) \times \mathrm{SL}(2, \mathbb{Q}_p)$. Even these two special cases were not known before the work of Deroin and Hurtado.

(3) If $G = \mathrm{SL}(3, \mathbb{Q}_p)$, we may let

$$K = \mathrm{SL}(3, \mathbb{Z}_p), \qquad P = \begin{bmatrix} * & * & * \\ 0 & * & * \\ 0 & 0 & * \end{bmatrix}, \qquad A = \begin{bmatrix} * & 0 & 0 \\ 0 & * & 0 \\ 0 & 0 & * \end{bmatrix}.$$

Note 3.6 By Theorem 2.3(1), we know that G/Γ is compact. On the other hand, since K is open, we also know that $K\backslash G$ is discrete. Therefore, the double-coset space $K\backslash G/\Gamma$ is both compact and discrete. So $K\backslash G/\Gamma$ must be finite.

Assumption 3.7 To simplify the proof, we will assume that the finite set $K\backslash G/\Gamma$ has only one element. This means:

$$G = K\Gamma. \tag{3.8}$$

(See Sect. 5 for some comments on the more complicated situation where $K\backslash G/\Gamma$ is a larger finite set.)

Remark 3.9 The conclusion of Note 3.6 is a key reason why the p-adic case is much easier than the real case. If G is a (noncompact) real semisimple Lie group and K is a compact subgroup of G, then $K \backslash G / \Gamma$ is a highly nontrivial manifold that plays a role in the proof of the Deroin-Hurtado Theorem (1.2).

Note 3.10 It is not difficult to see that the real line has no nontrivial, orientation-preserving homeomorphisms of finite order. Hence, Assumption 3.4 implies that Γ has no nontrivial elements of finite order [16, Sect. 1.4.1, p. 27]. Since every subgroup of Γ is discrete, and every discrete subgroup of a compact group is finite, this implies $\Gamma \cap K$ is trivial. By combining this with (3.8), we see that

$$\text{the natural map } (k, \gamma) \mapsto k\gamma \text{ is a } K\text{-equivariant homeomorphism } K \times \Gamma \simeq G. \tag{3.11}$$

Also, we can identify G/Γ with K.

Recall that a Borel measure μ on a topological space Y is a *probability measure* if $\mu(Y) = 1$ (see Recall 2.1(1)).

Notation 3.12 Let

(1) m_K be the Haar measure on K (normalized to be a probability measure), and
(2) μ_G be a nice probability measure on G whose support is compact and generates G. We assume [14, Defn. 3.6]:
   (a) μ_G is absolutely continuous with respect to the Haar measure m_G,
   (b) μ_G is *bi-K-invariant*, which means $\mu_G(B) = \mu_G(kB) = \mu_G(Bk)$ for every Borel set B and every $k \in K$, and
   (c) μ_G is *symmetric*, which means $\mu_G(B^{-1}) = \mu_G(B)$ for every Borel set B, where $B^{-1} = \{ b^{-1} \mid b \in B \}$.

Warning 3.13 The measure μ_G is definitely *not* the Haar measure on G (because the Haar measure on a noncompact group can never be a probability measure and cannot have compact support).

Definition 3.14 Define an atomic probability measure μ_Γ on Γ by $\mu_\Gamma(\{\gamma\}) = \mu_G(K\gamma)$. (Note that the support of μ_Γ is finite, since the support of μ_G is compact.) Since μ_G is K-invariant on the left (and the Haar measure on K is unique), this implies

$$\mu_G = m_K \times \mu_\Gamma,$$

where we are using the identification $G \simeq K \times \Gamma$ (see Note 3.10). We assume μ_G is chosen so that the support of μ_Γ generates Γ.

Remark 3.15 It is more difficult to construct the appropriate measure μ_Γ when Assumption 3.7 does not hold (see Definition 5.4(5)), and it is usually not true that μ_Γ has finite support.

3.2 The Almost-Periodic Space Z and the Induced G-Space X

The action of Γ on $\mathbb{R}$ obviously extends to an action of Γ on the one-point compactification $\mathbb{R} \cup \{\infty\}$. However, Γ fixes the point ∞ in this action. The following theorem shows that Γ acts with no fixed points on a much more sophisticated compactification of $\mathbb{R}$. (This is a general result about finitely generated groups that act on $\mathbb{R}$; it does not use our assumption that Γ is a lattice.) Deroin and Hurtado [14, Sect. 1.3.5] call this compactification the *almost-periodic space*.

Theorem 3.16 (see Theorem 4.1, [13, Thm. 1.1], [14, Thm. 5.4], [16, Cor. 4.2.11]) *There is a (nonempty) compact metrizable space Z, such that:*

(1) Γ acts on Z (by homeomorphisms) with no global fixed point,
(2) $\mathbb{R}$ has a continuous action on Z that is free,
(3) each $\mathbb{R}$-orbit is Γ-invariant, and
(4) the action of Γ on each $\mathbb{R}$-orbit is by orientation-preserving Lipschitz maps.

Remark 3.17

(1) Condition (1) states that the Γ-action has no fixed points. In fact, this implies that the action has no finite orbits (because, as was mentioned in Note 3.10, the real line has no nontrivial, orientation-preserving homeomorphisms of finite order).
(2) Condition (3) does not imply that the Γ-action commutes with the $\mathbb{R}$-action. In fact, the actions certainly do not commute (because Theorem 2.3(2) implies that Γ has no nontrivial homomorphism to $\mathbb{R}$).
(3) Condition (4) states that each element γ of Γ has a Lipschitz constant C_γ on each $\mathbb{R}$-orbit. In fact, the construction of Z yields a single constant C_γ that applies on every $\mathbb{R}$-orbit (but C_γ presumably depends on γ).

Notation 3.18 We will use additive notation for the action of $\mathbb{R}$ on Z:

$$\text{the image of a point } z \in Z \text{ by } t \in \mathbb{R} \text{ is denoted } z + t.$$

This additive notation allows us to also define subtraction for points that are in the same $\mathbb{R}$-orbit. Namely, for $z_1, z_2 \in Z$ and $t \in \mathbb{R}$:

$$\text{if } z_1 = z_2 + t, \text{ then we let } z_1 - z_2 := t.$$

Since the $\mathbb{R}$-action is assumed to be free, the difference $z_1 - z_2$ is well-defined (when it exists, i.e., whenever z_1 and z_2 are in the same $\mathbb{R}$-orbit).

Warning 3.19 For each $z \in Z$, the map $t \mapsto z + t$ is a homeomorphism from $\mathbb{R}$ onto the $\mathbb{R}$-orbit. However, we do not have a canonical identification of each $\mathbb{R}$-orbit with $\mathbb{R}$, because it is usually not possible to choose a well-defined base point in each $\mathbb{R}$-orbit to correspond to 0 (unless we invoke the Axiom of Choice, which will not yield a construction that is Borel measurable). Fortunately, the definition of subtraction does not require the choice of a base point.

A general technique (called "inducing" or "suspending" the given action) constructs an action of a group from an action of any closed subgroup. We now apply this to the action of Γ on Z.

Definition 3.20 ([14, Sect. 5.3], [35, p. 75]) Let

$$X = \mathrm{Ind}_\Gamma^G Z := (G \times Z)/\Gamma, \quad \text{where} \quad (h, z) * \gamma = (h\gamma, \gamma^{-1} z).$$

Then G acts on this quotient space by

$$g[(h, z)] = [(gh, z)],$$

where we use $[(h, z)]$ to denote the image of (h, z) under the natural quotient map from $G \times Z$ to $(G \times Z)/\Gamma$.

Note 3.21

(1) For $[(h, z)] \in X$ and $\gamma \in \Gamma$, we have

$$[(h\gamma, z)] = [(h\gamma \cdot \gamma^{-1}, \gamma z)] = [(h, \gamma z)].$$

(2) The natural map $(k, z) \mapsto [(k, z)]$ is a K-equivariant homeomorphism $K \times Z \simeq X$ (by (3.11)). This implies that X is compact (because K and Z are compact).
(3) The function $[(h, z)] \mapsto h\Gamma$ is a well-defined G-equivariant map $X \to G/\Gamma$. It gives X the structure of a fibre bundle over G/Γ with fibres homeomorphic to Z.
(4) For $g \in G$ and $k \in K$, there exist $k' \in K$ and $\gamma = \gamma(g, k) \in \Gamma$, such that $gk = k'\gamma$ (by (3.8)). Then, for all $z \in Z$, we have

$$g[(k, z)] = [(gk, z)] = [(k'\gamma, z)] = [(k', \gamma z)].$$

Hence, g acts on the entire fibre $[\{k\} \times Z]$ via the element $\gamma(g, k)$.

We have defined $z_1 - z_2$ when z_1 and z_2 are points in Z that are in the same $\mathbb{R}$-orbit. We extend this to points in X:

Notation 3.22 The action of $\mathbb{R}$ on Z can be extended to a free action of $\mathbb{R}$ on X via the identification $X \simeq K \times Z$: $[(k, z)] + t = [(k, z + t)]$. This yields a definition of $x - y$ for all $x, y \in X$, such that x and y are in the same $\mathbb{R}$-orbit.

Warning 3.23 For $g \in G$, it is usually *not* true that $[(g, z)] + t = [(g, z + t)]$. To see this, write $g = k\gamma$, with $k \in K$ and $\gamma \in \Gamma$ (by using (3.8)). Then:

$$\begin{aligned}[(g, z)] + t &= [(k\gamma, z)] + t = [(k, \gamma z)] + t = [(k, \gamma z + t)] \\ &= \big[\big(k\gamma, \gamma^{-1}(\gamma z + t)\big)\big] = \big[\big(g, \gamma^{-1}(\gamma z + t)\big)\big].\end{aligned}$$

This will rarely be equal to $[(g, z + t)]$, because Γ is not assumed to commute with the $\mathbb{R}$-action.

However, we do have the following weaker property, which is a consequence of Note 3.21(4) (and Theorem 3.16(4)):

Lemma 3.24 (cf. [14, Sect. 5.3.1]) *The action of G on X respects the $\mathbb{R}$-orbits, and the restriction of each element of G to an $\mathbb{R}$-orbit is Lipschitz (with a Lipschitz constant that does not depend on the choice of the orbit). More precisely, for all $g \in G$, there exists $C = C(g) \in \mathbb{R}^+$, such that, for all $x \in X$ and $t \in \mathbb{R}$, we have $g(x + t) \in gx + \mathbb{R}$ and*

$$|g(x + t) - gx| \le C\,|t|.$$

Furthermore, the action of G preserves the orientation of the $\mathbb{R}$-orbits: if $t > 0$, then $g(z + t) - g(z) > 0$.

3.3 The Probability Measures μ_Z, μ_X, and μ_X^P

Notation 3.25 Let

- μ_Z be an ergodic $\mathbb{R}$-invariant probability measure on Z (see Proposition 2.6 for the existence of such a measure), and
- $\mu_X = m_K \times \mu_Z$ (under the identification $X \simeq K \times Z$), so μ_X is a K-invariant probability measure on X.

Remark 3.26 See Example 5.7 for a definition of μ_X when $X \not\simeq K \times Z$.

Lipschitz functions on $\mathbb{R}$ are differentiable almost everywhere (and the derivative is bounded). Thus, we see from Lemma 3.24 that each element of G has a derivative along the $\mathbb{R}$-orbits (a.e.):

Definition 3.27 ([14, Sect. 5.2.2]) For $g \in G$ and μ_X-a.e. $x \in X$, we let

$$D_g^{\text{leaf}}(x) = \lim_{t \to 0} \frac{g(x + t) - g(x)}{t}.$$

We can similarly define $D_\gamma^{\text{leaf}}(z)$ for $\gamma \in \Gamma$ and μ_Z-a.e. $z \in Z$.

By definition, the measure μ_Z is $\mathbb{R}$-invariant, so moving along the $\mathbb{R}$-orbits preserves the measure. Since the Γ-action preserves these $\mathbb{R}$-orbits, any non-invariance of μ_Z under this action comes from distortion within the $\mathbb{R}$-orbits. Also, for any $\gamma \in \Gamma$, the action on each $\mathbb{R}$-orbit is Lipschitz, so the measure class of Lebesgue measure is preserved. This implies that $\gamma_* \mu_Z$ is in the same measure class as μ_Z. Furthermore:

Lemma 3.28 ([14, Lems. 5.5 and 5.10(4)]) *For every $\gamma \in \Gamma$, the Radon-Nikodym derivative $d\gamma_*\mu_Z/d\mu_Z$ is equal to $D^{\text{leaf}}_{\gamma^{-1}}$ (a.e.). Similarly, for all $g \in G$, we have*

$$\frac{dg_*\mu_X}{d\mu_X} = D^{\text{leaf}}_{g^{-1}} \quad (a.e.).$$

The almost-periodic space Z can be constructed so that the following holds. (See Definition 2.5(5) for the definition of "stationary.")

Proposition 3.29 (see Proposition 4.11, [14, Lem. 5.6]) *The measure μ_Z is μ_Γ-stationary.*

This has the following crucial consequence.

Corollary 3.30 (see Corollary 4.12, [14, Prop. 5.11]) *μ_X is μ_G-stationary.*

Recall that P is a minimal parabolic subgroup (see Notation 3.3 and perhaps also Remark 3.5(3)). It is known (for every G-action, not only the action on our space X) that all μ_G-stationary measures come from P-invariant measures, by averaging the K-translates:

Theorem 3.31 (Furstenberg [21, Thm. 2.1]) *There is a unique P-invariant probability measure μ_X^P on X, such that*

$$\mu_X = \int_K k_* \, \mu_X^P \, dm_K(k).$$

The remainder of the proof of the Deroin-Hurtado Theorem will show that the stationary measure μ_X is G-invariant (or, equivalently, that $\mu_X = \mu_X^P$). This will contradict the following result, and thereby complete the proof of Theorem 1.12, by establishing that Assumption 3.4 must be false, because it leads to a contradiction.

Proposition 3.32 ([14, Prop. 5.12]) *We have:*

(1) the stationary measure μ_Z is <u>not</u> Γ-invariant,
(2) the stationary measure μ_X is <u>not</u> G-invariant,
(3) the P-invariant measure μ_X^P is <u>not</u> K-invariant (so it is not G-invariant), and
(4) $\mu_X \neq \mu_X^P$.

Proof Translations are the only orientation-preserving homeomorphisms of $\mathbb{R}$ that preserve Lebesgue measure. Since the group of translations is abelian, but Γ is not abelian, it follows that the Lebesgue measure is not Γ-invariant (for any faithful action of Γ on $\mathbb{R}$). We will see that the proposition is an elaboration of this simple observation.

(1) Suppose μ_Z is Γ-invariant. Then for all $\gamma \in \Gamma$ we have

$$\frac{d\gamma_*\mu_Z}{d\mu_Z} = \frac{d\mu_Z}{d\mu_Z} = 1 \quad \text{a.e.},$$

so Lemma 3.28 tells us $D_\gamma^{\text{leaf}} = 1$ a.e. This implies that Γ acts by translations on each $\mathbb{R}$-orbit. However, the group of translations is abelian, and the abelianization of Γ is finite (see Theorem 2.3(2)). This contradicts Assumption 3.4, which states that the action of Γ is faithful.

(2) It can be shown that if μ_X is G-invariant, then μ_Z is Γ-invariant, which contradicts (1). Indeed, it is a general property of induced actions that μ_Z is Γ-invariant if and only if μ_X is G-invariant, but see Lemma 4.14 for a proof of the direction we need, that is specialized to our situation.

(4) Suppose $\mu_X = \mu_X^P$. Since μ_X^P is P-invariant, this assumption implies that μ_X is P-invariant. However, μ_X is also known to be K-invariant (see Notation 3.12(2)). These two subgroups generate all of G, so we conclude that μ_X is G-invariant. This contradicts (2).

(3) Suppose μ_X^P is K-invariant. This means $k_*\mu_X^P = \mu_X^P$ for all $k \in K$, so

$$\mu_X = \int_K k_*\mu_X^P \, dm_K(k) = \int_K \mu_X^P \, dm_K(k) = \mu_X^P.$$

This contradicts (4). □

3.4 Contraction on $\mathbb{R}$-Orbits and the Key Proposition

Definition 3.33 ([14, Sect. 7.1, p. 43]) We define $\chi_P \colon A \to \mathbb{R}$ by

$$\chi_P(a) = \int_X \log D_a^{\text{leaf}}(x) \, d\mu_X^P(x).$$

(There is a technical issue here: we know that $D_a^{\text{leaf}}(x)$ exists for μ_X-a.e. $x \in X$, but the definition assumes that it exists for μ_X^P-a.e. $x \in X$. See Lemma 4.15 for a proof that the derivative exists μ_X^P-a.e.)

Lemma 3.34 (see Lemma 4.16, [14, Sect. 7.1, p. 43]) *It follows from the Chain Rule (and the A-invariance of μ_X^P) that χ_P is a homomorphism.*

Notice that if $\chi_P(a) < 0$, then there is constant $C > 1$, such that if x and y are two nearby points in the same $\mathbb{R}$-orbit, then, on average,

$$|ax - ay| < \frac{|x - y|}{C}.$$

Applying a repeatedly should decrease the distance by an exponential factor, so one would expect the distance to tend to 0 in the limit. This is indeed true (if we start with two points that are not too far apart):

Lemma 3.35 (local contraction on $\mathbb{R}$-orbits, see Lemma 4.18, [14, Lem. 7.1]) *If $a \in A$, such that $\chi_P(a) < 0$, then for μ_X^P-a.e. $x \in X$, there exists $\epsilon = \epsilon(x) > 0$, such that*

$$a^n(x + \epsilon) \; - \; a^n(x - \epsilon) \; \to \; 0 \quad \text{as } n \to \infty.$$

With additional work, it can be shown that a contracts entire $\mathbb{R}$-orbits, not just ϵ-intervals. The resulting contraction property will be the key to the proof of Theorem 1.12.

Key Proposition 3.36 (Global contraction on $\mathbb{R}$-orbits, cf. [14, Prop. 7.7]) *For all $a \in A$, if $\chi_P(a) < 0$, then*

$$\text{for } \mu_X^P\text{-a.e. } x \in X, \text{ for all } y \in x + \mathbb{R}, \quad a^n x - a^n y \overset{n\to\infty}{\longrightarrow} 0.$$

Proof See Sect. 3.7 for an outline. More details can be found in Sect. 4.5. □

3.5 Invariance Under the Centralizer

Notation 3.37 For $a \in A$, let

$$U_a^+ = \left\{ u \in G \;\middle|\; \begin{array}{l} a^n u a^{-n} \to 1 \\ \text{as } n \to -\infty \end{array} \right\} \qquad \left(\begin{array}{c}\text{This is the } \textit{expanding horospherical} \\ \textit{subgroup} \text{ corresponding to } a.\end{array}\right)$$

and

$$C_G(a) = \{\, c \in G \mid ac = ca \,\} \qquad \text{(This is the } \textit{centralizer} \text{ of } a.)$$

Key Proposition 3.36 has the following consequence, which will be used to show (by a bootstrapping argument) that μ_X^P is G-invariant. This contradicts Proposition 3.32(3).

Corollary 3.38 (cf. [14, Prop. 8.3]) *If $a \in A$, such that $\chi_P(a) < 0$ and $U_a^+ \subseteq P$, then μ_X^P is $C_G(a)$-invariant.*

Note 3.39 Corollary 3.38 is not vacuous: there is an element a of A, such that $\chi_P(a) < 0$ and $U_a^+ \subseteq P$ (see Proposition 4.27).

Idea of Proof of Corollary 3.38 Let $c \in C_G(A)$. We wish to show $c_*\mu_X^P = \mu_X^P$.

The measure μ_X^P might not be ergodic for the action of a, so we should work with ergodic components of the measure, but, for simplicity, we ignore this technical issue. (See Remark 4.20 for a comment on this.) Then, by the Pointwise Ergodic Theorem (2.7), there exists $x \in X$, such that, for all $f \in C(X)$, we have

$$\lim_{n\to\infty} \frac{1}{n} \sum_{i=1}^{n} f(a^i x) = \int_X f \, d\mu_X^P. \tag{3.40}$$

We say that x is a "Birkhoff-generic point" for μ_X^P (w.r.t. a), and we call the average on the left-hand side a "Birkhoff average." Furthermore, since (3.40) holds for μ_X^P-a.e. $x \in X$, we see from Key Proposition 3.36 that we may assume

$$a^n x \;-\; a^n y \;\overset{n\to\infty}{\longrightarrow}\; 0 \quad \text{for all } y \in x + \mathbb{R}. \tag{3.41}$$

Since μ_X^P is P-invariant, the measure $c_*\mu_X^P$ is ${}^cP$-invariant, where ${}^cP = cPc^{-1}$ (a conjugate of P). Define $\pi \colon X \to G/\Gamma$ by $\pi\big([(h,z)]\big) = h\Gamma$, so π is a well-defined G-equivariant map (see Note 3.21(3)). Then $\pi_*(c_*\mu_X^P)$ is a ${}^cP$-invariant probability measure on G/Γ, so we see from Proposition 2.9(2) that its support is all of G/Γ. Since $\pi(cx)$ is Birkhoff generic for this measure (w.r.t. a), we conclude that the $\langle a \rangle$-orbit of $\pi(cx)$ is dense in G/Γ. Hence, we may choose $i \in \mathbb{Z}$, such that $\pi(a^i cx)$ is very close to $\pi(x)$.

Now, if we let

$$P_a^- = \left\{ g \in G \;\middle|\; \begin{array}{c} a^n g a^{-n} \text{ is bounded} \\ \text{for all } n \geq 0 \end{array} \right\},$$

then the Lie algebra $\mathfrak{g}$ is the direct sum of the Lie algebra of P_a^- and the Lie algebra of U_a^+. Hence, the conclusion of the preceding paragraph implies there exist small $g_- \in P_a^-$ and $u^+ \in U_a^+$, such that

$$\pi(a^i cx) = g^- u^+ \pi(x) = \pi(g^- u^+ x).$$

Writing $x = [(h,z)]$ with $h \in G$ and $z \in Z$, this means there exists $\gamma \in \Gamma$, such that

$$a^i ch\gamma = g^- u^+ h,$$

so (recalling that the $\mathbb{R}$-orbits in Z are Γ-invariant) we have

$$\begin{aligned} x &= c^{-1}a^{-i}a^i cx = c^{-1}a^{-i}[(a^i ch, z)] = c^{-1}a^{-i}[(a^i ch\gamma, \gamma^{-1}z)] \\ &= c^{-1}a^{-i}[(g^- u^+ h, \gamma^{-1}z)] \in c^{-1}a^{-i}[(g^- u^+ h, z)] + \mathbb{R} = c^{-1}a^{-i}g^- u^+ x + \mathbb{R}. \end{aligned} \tag{3.42}$$

Let $x_c = a^i c x$ and $x_0 = u^+ x$. We see from (3.42) and (3.41) that

$$a^n x - a^n(c^{-1} a^{-i} g^- x_0) \overset{n\to\infty}{\longrightarrow} 0.$$

Then, since $a^i c$ commutes with a^n, and is uniformly Lipschitz on the $\mathbb{R}$-orbits (see Lemma 3.24), we can multiply by $a^i c$ to conclude that

$$a^n x_c - a^n(g^- x_0) \overset{n\to\infty}{\longrightarrow} 0. \tag{3.43}$$

However:

- $d(a^n x_0, a^n x_c) \leq d(a^n x_0, a^n g^- x_0) + d(a^n x_c, a^n g^- x_0)$.
- Since $g_- \in P_a^-$, we know that $d(a^n x_0, a^n g^- x_0)$ stays bounded at about the size of g^-, so it is uniformly small.
- The action of $\mathbb{R}$ on Z is continuous (and Z is compact), so (3.43) implies $d(a^n x_c, a^n g^- x_0) \overset{n\to\infty}{\longrightarrow} 0$.

Therefore $a^n x_0$ is close to $a^n x_c$ for all n, so

x_0 and x_c have almost the same Birkhoff averages.

Now, we have another technical issue. Since $x_0 = u^+ x$ and x is Birkhoff generic for μ_X^P, we would like to be able to say that x_0 is Birkhoff generic for $u_*^+ \mu_X^P = \mu_X^P$. However, u^+ does not centralize a (far from it!), so this is not obvious. It is true for most values of u^+ though: since μ_X^P is U_a^+-invariant, one can show (for a.e. x) that the statement is true for $m_{U_a^+}$-a.e. choice of u^+ in U_a^+. By perturbing the generic point x, we can change the value of u^+ to get it into the good set (see [14, p. 50]). Therefore, we can assume that x_0 is indeed Birkhoff generic for μ_X^P.

On the other hand, x_c is Birkhoff generic for $c_* \mu_X^P$. Combining this with the conclusions of the preceding two paragraphs, we conclude that μ_X^P is ϵ-close to $c_* \mu_X^P$ (for every $\epsilon > 0$). By letting ϵ tend to 0, we conclude that $\mu_X^P = c_* \mu_X^P$. □

Remark 3.44 The assumption in Corollary 3.38 that $U_a^+ \subseteq P$ simply means that a is in a particular closed Weyl chamber in A.

3.6 Completion of the Proof by Propagating Invariance

We can now complete the proof of Theorem 1.12, by obtaining the following contradiction to Proposition 3.32(3). (This is where we use the assumption that $\operatorname{rank}_{\mathbb{Q}_p} G \geq 2$. If G has rank one, then $C_G(a') = C_G(A)$ for all nontrivial $a' \in A$, so Corollary 3.38 cannot show that μ_X^P is invariant under anything more than

$C_G(A)$.) The contradiction implies that our assumption (3.4) that there is a nontrivial, orientation-preserving action of Γ on $\mathbb{R}$ must be false. So it does indeed complete the proof of Theorem 1.12.

Corollary 3.45 ("Propagating Invariance" [14, Thm. 8.1 (and Sect. 9)]) *The measure μ_X^P is G-invariant.*

Idea of Proof For concreteness, let us assume $G = \mathrm{SL}(3, \mathbb{Q}_p)$. (See Corollary 4.21 for an explanation of the general case.) Let $\mathcal{W}_P$ be the closed Weyl chamber corresponding to P, so $U_a^+ \subseteq P$ for all $a \in \mathcal{W}_P$. We may assume P is upper triangular, as suggested in Remark 3.5(3). Let

$$\mathcal{H} = \{\, a \in A \mid \chi_P(a) < 0 \,\}.$$

This is an open halfspace in A.

Case 1 *Assume $\chi_P(a) < 0$ for all nontrivial $a \in \mathcal{W}_P$.*

Let a_1 and a_2 be on the two walls of $\mathcal{W}_P$, as pictured in Fig. 1a. Specifically, we may let

$$a_1 = \begin{bmatrix} 1/p & 1/p & p^2 \end{bmatrix} \qquad \text{and} \qquad a_2 = \begin{bmatrix} 1/p^2 & & \\ & p & \\ & & p \end{bmatrix}.$$

By the assumption of this case, we have $\chi_P(a_1) < 0$ and $\chi_P(a_2) < 0$. Also, we know $U_{a_1}^+ \subseteq P$ and $U_{a_2}^+ \subseteq P$, because $a_1, a_2 \in \mathcal{W}_P$. Indeed,

$$U_{a_1}^+ = \begin{bmatrix} 1 & & * \\ & 1 & * \\ & & 1 \end{bmatrix} \subset P \qquad \text{and} \qquad U_{a_1}^+ = \begin{bmatrix} 1 & * & * \\ & 1 & \\ & & 1 \end{bmatrix} \subset P.$$

So we see from Corollary 3.38 that μ_X^P is invariant under

$$C_G(a_1) = \begin{bmatrix} * & * & \\ * & * & \\ & & 1 \end{bmatrix} \qquad \text{and} \qquad C_G(a_2) = \begin{bmatrix} 1 & & \\ & * & * \\ & * & * \end{bmatrix}.$$

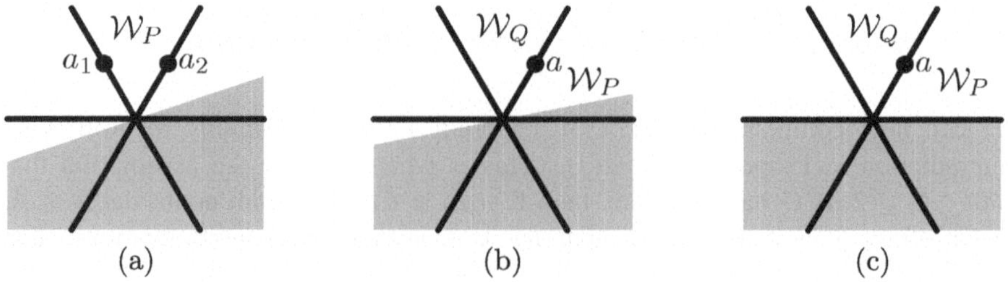

Fig. 1 The halfspace $\mathcal{H}$ is unshaded

These two centralizers generate G, so we conclude that μ_X^P is invariant under all of G.

Case 2 *The remaining case.*

We know that:

- $\mathcal{H}$ contains some element of $\mathcal{W}_P$ (by Note 3.39), and
- $\mathcal{H}$ does not contain every nontrivial element of $\mathcal{W}_P$ (by the assumption of this case).

So either the boundary of $\mathcal{H}$ passes through the interior of $\mathcal{W}_P$ (as in Fig. 1b), or the boundary of $\mathcal{H}$ contains one the boundary rays of $\mathcal{W}_P$, and the rest of $\mathcal{W}_P$ is in $\mathcal{H}$ (as in Fig. 1c). In either case, one of the two boundary rays of $\mathcal{W}_P$ is in $\mathcal{H}$.

Let $\mathcal{W}_Q$ be the Weyl chamber on the other side of this boundary ray, with corresponding minimal parabolic subgroup Q, and let a be a nontrivial point on the ray. Then $a \in \mathcal{H}$, which means $\chi_P(a) < 0$. (Also, $a \in \mathcal{W}_P$, so $U_a^+ \subseteq P$.) So we see from Corollary 3.38 that μ_X^P is $C_G(a)$-invariant. However, we also know by the definition of μ_X^P that it is P-invariant. So μ_X^P is invariant under the subgroup $\langle C_G(a), P\rangle$. This is a parabolic subgroup of G (because every subgroup that contains a parabolic subgroup is also a parabolic subgroup). In fact, since a is on the boundary between $\mathcal{W}_P$ and $\mathcal{W}_Q$, this subgroup is equal to $\langle P, Q\rangle$, and therefore contains Q. So μ_X^P is Q-invariant.

By the definition of μ_X^P, we also have $\mu_X = \int_K k_* \mu_X^P \, dm_K(k)$ (see Theorem 3.31). Hence, the uniqueness in Theorem 3.31 implies that $\mu_X^Q = \mu_X^P$. Then we see from Definition 3.33 that $\chi_Q = \chi_P$. Since it is clear from Fig. 1b, c that every nontrivial element of $\mathcal{W}_Q$ is in $\mathcal{H}$, we have $\chi_P(a) < 0$ for all nontrivial $a \in \mathcal{W}_Q$. Then, since $\chi_Q = \chi_P$, we conclude that the same inequality holds with χ_Q in the place of χ_P. Hence, Case 1 applies with the parabolic subgroup Q in the place of P. □

3.7 *Proof of Key Proposition 3.36*

Let $a \in A$, such that $\chi_P(a) < 0$. We know from Lemma 3.35 that the action of a contracts small intervals in $\mathbb{R}$-orbits (a.e.). In order to establish Key Proposition 3.36, we need to show that the contraction happens on the entire $\mathbb{R}$-orbit. To this end, we will employ contraction by generic elements of G, rather than only by elements of A.

For $g \in G$, we have $\mu_X(gX) = \mu_X(X) = 1$, so (by Lemma 3.28) we see that $\int_X D_g^{\text{leaf}} \, d\mu_X = 1$. As μ_X is not G-invariant (see Proposition 3.32(2)), we also know that D_g^{leaf} is not identically equal to 1 for all g, so, since log is a concave function, we conclude from Jensen's Inequality that

$$\int_G \int_X \log D_g^{\mathrm{leaf}}(x)\, d\mu_X(x)\, d\mu_G(g) < \log \int_G \int_X D_g^{\mathrm{leaf}}(x)\, d\mu_X(x)\, d\mu_G(g)$$
$$= \log \int_G 1\, d\mu_G \tag{3.46}$$
$$= 0.$$

This implies that (on average) applying a random element of G contracts distances that are small. Much like what we saw in Lemma 3.35 for a^n, this implies that repeated application of random elements of G will contract small intervals in the $\mathbb{R}$-orbits. However, we also have the following global contraction property of the Γ-action on each $\mathbb{R}$-orbit:

Proposition 3.47 (see Proposition 4.23, [14, Prop. 6.3]) *For every action of Γ on $\mathbb{R}$ without fixed points, there is a compact subset I of $\mathbb{R}$, such that for every $\epsilon > 0$ and all $s, t \in \mathbb{R}$, there exists $\gamma \in \Gamma$, such that*

$$|\gamma s - \gamma t| < \epsilon \text{ and } \{\gamma s, \gamma t\} \subset I.$$

By using this, it is possible to prove that repeated application of random elements of G yields contraction on entire $\mathbb{R}$-orbits, rather than only locally, and that the contraction is exponentially fast:

Proposition 3.48 (see Proposition 4.24, [14, Props. 6.4 and 6.8]) *There is a constant $\delta < 1$, such that, for μ_X-a.e. $x \in X$, for all $y \in \mathbb{R}x$, and for μ_G^∞-a.e. sequence $(g_i)_{i=1}^\infty$ of elements of G, we have*

$$|g_n g_{n-1} \cdots g_1 x \;-\; g_n g_{n-1} \cdots g_1 y| < \delta^n \quad \text{for all sufficiently large } n.$$

We have been proving the p-adic case in this paper, but the next part of the argument is easier to explain for real groups, so we will switch to that setting for the remainder of this section:

Assume, contrary to Notation 3.3, that
G is a connected real semisimple Lie group.

The group G acts by isometries on its associated symmetric space $K\backslash G$, which is a Riemannian manifold of nonpositive sectional curvature. (The p-adic case uses a Bruhat-Tits building, instead of a symmetric space.) Applying a random walk in G to a point in $K\backslash G$ yields a sequence of points, and it can be shown (without using our assumption that Γ acts on $\mathbb{R}$) that the negative curvature of $K\backslash G$ implies that this sequence almost surely moves along a geodesic, up to a sublinear error:

Theorem 3.49 (Karlsson-Margulis [25, Thm. 2.1]) *For μ_G^∞-a.e. sequence $(g_i)_{i=1}^\infty$ of elements of G, there is a geodesic a_t in $K\backslash G$, with $\mathsf{a}_0 = K$, such that*

$$\frac{d\big(K g_n g_{n-1} \cdots g_1,\ \mathsf{a}_n\big)}{n} \xrightarrow{n\to\infty} 0.$$

Definition 3.50 Fix a nonempty, open subset $\mathcal{O}$ of G. Then $\mathcal{O}$ generates G, so we can define the corresponding *word length*: for $g \in G$, we let

$$\ell_{\mathcal{O}}(g) = \min\{\, r \in \mathbb{N} \mid g \in \mathcal{O}^r \,\},$$

where $\mathcal{O}^r = \{\, h_1 h_2 \cdots h_r \mid h_i \in \mathcal{O} \text{ or } h_i^{-1} \in \mathcal{O} \,\}$.

Assumption 3.51 We assume the closure of $\mathcal{O}$ is compact.

Geodesics in $K \backslash G$ come from one-parameter subgroups of (K-conjugates of) A, and every element of A is K-conjugate to an element of $\mathcal{W}_P$, so Theorem 3.49 can be restated as follows:

Corollary 3.52 ([14, Thm. 3.18]) *For μ_G^∞-a.e. sequence $(g_i)_{i=1}^\infty$ of elements of G, there exist $k \in K$ and $\widehat{a_P} \in \mathcal{W}_P \smallsetminus \{\mathbf{1}\}$, such that*

$$\frac{\ell_{\mathcal{O}}(g_n g_{n-1} \cdots g_1\, k\, \widehat{a_P}^{\,-n})}{n} \overset{n\to\infty}{\longrightarrow} 0.$$

Furthermore, as $(g_i)_{i=1}^\infty$ varies, the element k is uniformly distributed in K (with respect to the Haar measure m_K).

We also have the following important technical result (which does not require our assumption that Γ acts on $\mathbb{R}$):

Proposition 3.53 (Benoist-Quint [2, Thm. 10.9(a,b), p. 158]) *The element $\widehat{a_P}$ is a well-defined element of $\mathcal{W}_P$ that is independent of $(g_i)_{i=1}^\infty$. (In fact, $\widehat{a_P}$ depends only on G, $\mathcal{W}_P$, and μ_G.)*

Combining these results has the following consequence:

Proposition 3.54 ([14, Prop. 7.6]) *For μ_X^P-a.e. $x \in X$ and every $y \in x + \mathbb{R}$, we have*

$$\widehat{a_P}^{\,n}\, x \;-\; \widehat{a_P}^{\,n}\, y \overset{n\to\infty}{\longrightarrow} 0.$$

Idea of Proof We will ignore some technical issues, including the difference between μ_X and μ_X^P. (Remark 4.25 explains how to get from μ_X to μ_X^P, by using the fact that U_a^+ is contained in P, and therefore preserves μ_X^P.)

Since μ_X is K-invariant, we see that, for m_K-a.e. $k \in K$, the conclusion of Proposition 3.48 holds with kx and ky in the place of x and y. Then the uniform distribution in the conclusion of Corollary 3.52 implies that we can choose k and $(g_i)_{i=1}^\infty$ so that the conclusions of Proposition 3.48 and Corollary 3.52 hold simultaneously. That is,

$$|g_n g_{n-1} \cdots g_1 kx \;-\; g_n g_{n-1} \cdots g_1 ky| \overset{n\to\infty}{\longrightarrow} 0 \text{ exponentially fast},$$

and the difference between $g_n g_{n-1} \cdots g_1\, k$ and $\widehat{a_P}^{\,n}$ is the product of a sublinear number of elements of $\mathcal{O}$. Since elements of $\mathcal{O}$ are uniformly Lipschitz, we conclude that $|\widehat{a_P}^{\,n} x \;-\; \widehat{a_P}^{\,n} y|$ also tends to 0 (exponentially fast). □

We have established that the single element $\widehat{a_P}$ has global contraction. We can use this to prove that the elements in a halfspace of A also contract entire $\mathbb{R}$-orbits. In other words, we now prove Key Proposition 3.36.

Proof of Key Proposition 3.36 Fix $a \in A$. For each $x \in X$, let

$$f(x) = \sup\{\, t \in \mathbb{R}^{\geq 0} \mid a^n(x+t) - a^n x \xrightarrow{n\to\infty} 0 \,\}.$$

Also let

$$X_0 = \{\, x \in X \mid f(x) < +\infty \,\}.$$

We wish to prove that $\mu_X^P(X_0) = 0$ if $\chi_P(a) < 0$.

For any $n \in \mathbb{Z}^+$, the element $\widehat{a_P}^n$ commutes with a (and is uniformly Lipschitz on the $\mathbb{R}$-orbits by Lemma 3.24). Therefore, if $x \in X_0$, then

$$\widehat{a_P}^n\big(x + f(x)\big) = \widehat{a_P}^n x + f(\widehat{a_P}^n x),$$

so

$$f(\widehat{a_P}^n x) \quad = \quad \widehat{a_P}^n\big(x + f(x)\big) \; - \; \widehat{a_P}^n x. \tag{3.55}$$

This is finite, so $\widehat{a_P}^n x \in X_0$; hence, X_0 is $\widehat{a_P}$-invariant.

By applying Corollary 2.8 to the action of $\widehat{a_P}$ on X_0, we obtain a certain function $\overline{f}\colon X_0 \to [0,\infty]$. From (3.55) and Proposition 3.54, we see that $f(\widehat{a_P}^n x) \xrightarrow{n\to\infty} 0$ for μ_X^P-a.e. $x \in X_0$. So the definition of $\overline{f}$ implies that $\overline{f}(x) = 0$ for μ_X^P-a.e. $x \in X_0$. Therefore

$$\int_{X_0} f \, d\mu_X^P = \int_{X_0} \overline{f} \, d\mu_X^P = \int_{X_0} 0 \, d\mu_X^P = 0,$$

so $f(x) = 0$ for μ_X^P-a.e. $x \in X_0$. However, if $\chi_P(a) < 0$, then we see from Lemma 3.35 that $f(x) > 0$ for μ_X^P-a.e. x. Thus, we must have $\mu_X^P(X_0) = 0$, as desired. □

4 Some Additional Explanations

For the interested reader, we provide some arguments that were omitted from Sect. 3.

4.1 Construction of the Almost-Periodic Space Z

Theorem 4.1 (Proof of Theorem 3.16, [13, Thm. 1.1], [14, Thm. 5.4]) *There is a (nonempty) compact metrizable space Z, such that:*

(1) Γ acts on Z (by homeomorphisms) with no global fixed point,

(2) $\mathbb{R}$ *has a continuous action on* Z *that is free,*
(3) each $\mathbb{R}$*-orbit is* Γ*-invariant, and*
(4) the action of Γ *on each* $\mathbb{R}$*-orbit is orientation preserving.*

Furthermore, two additional nice properties (4.9) *and* (4.10) *hold.*

Proof Stationary measures for an action can be messy, but B. Deroin, V. Kleptsyn, A. Navas, and K. Parwani (cf. [15, Prop. 8.1] or [16, Thm. 4.3.20]) showed that there is a nontrivial homomorphism $\phi\colon \Gamma \to \mathrm{Homeo}^+(\mathbb{R})$, such that the Lebesgue measure is stationary. For $s > t$, this implies

$$\begin{aligned} s - t = m_{\mathbb{R}}\big([s,t]\big) &= \int_\Gamma \gamma_* m_{\mathbb{R}}\big([t,s]\big)\, d\mu_\Gamma(\gamma) \\ &= \int_\Gamma m_{\mathbb{R}}\big(\big[\phi_\gamma(t), \phi_\gamma(s)\big]\big)\, d\mu_\Gamma(\gamma) = \int_\Gamma \big(\phi_\gamma(s) - \phi_\gamma(t)\big)\, d\mu_\Gamma(\gamma), \end{aligned}$$

so

$$\int_\Gamma \big(\phi_\gamma(t) - t\big)\, d\mu_\Gamma(\gamma) = \int_\Gamma \big(\phi_\gamma(s) - s\big)\, d\mu_\Gamma(\gamma).$$

This means that all points in $\mathbb{R}$ have the same mean displacement. However, the measure μ_Γ is symmetric (i.e., $\mu_\Gamma(\gamma) = \mu_\Gamma(\gamma^{-1})$), so steps to the right have exactly the same probability as steps to the left. Therefore, it is intuitively fairly clear (and was established in [15, proof of Prop. 8.1]) that is it impossible for all points to have a drift to the right (or for all points to have a drift to the left). Hence, the mean displacement must be 0, which means the following holds for $\varphi = \phi$:

$$\forall t \in \mathbb{R}, \quad \int_\Gamma \big(\varphi_\gamma(t) - t\big)\, d\mu_\Gamma(\gamma) \;=\; 0. \tag{4.2}$$

We now show that (4.2) implies the action on $\mathbb{R}$ is Lipschitz. To see this, let $s > t$, and fix some γ in the support of μ_Γ. Then

$$\begin{aligned} s - t &= \int_\Gamma \varphi_\lambda(s)\, d\mu_\Gamma(\lambda) - \int_\Gamma \varphi_\lambda(t)\, d\mu_\Gamma(\lambda) \quad (4.2) \\ &= \int_\Gamma \big(\varphi_\lambda(s) - \varphi_\lambda(t)\big)\, d\mu_\Gamma(\lambda) \\ &\geq \big(\varphi_\gamma(s) - \varphi_\gamma(t)\big)\, \mu_\Gamma(\gamma) \qquad \begin{pmatrix} \varphi_\lambda(s) - \varphi_\lambda(t) > 0 \text{ because the} \\ \text{action is orientation-preserving} \end{pmatrix}. \end{aligned}$$

Therefore, if we let $C = \max_{\gamma \in \operatorname{supp} \mu_\Gamma} 1/\mu_\Gamma(\gamma)$, then:

$$\text{(Lipschitz)} \quad \forall \gamma \in \operatorname{supp} \mu_\Gamma,\ \forall s, t \in \mathbb{R},\ |\varphi_\gamma(s) - \varphi_\gamma(t)| \leq C|s - t|. \tag{4.3}$$

Note that the Lipschitz constant C depends only on μ_Γ, so it is independent of φ.

Slightly lengthier arguments [14, proof of Thm. 5.4] show that (4.2) also implies the following properties (but the constant C may depend on φ):

$$\text{(bounded displacement) } \forall \gamma \in \operatorname{supp} \mu_\Gamma, \ \forall t \in \mathbb{R}, \ |\varphi_\gamma(t) - t| \leq C, \text{ and} \tag{4.4}$$

$$\text{(no approximate fixed points) } \max_{\gamma \in \operatorname{supp} \mu_\Gamma} |\varphi_\gamma(t) - t| \geq \frac{1}{C}. \tag{4.5}$$

Fix a constant C, such that (4.3), (4.4), and (4.5) all hold for $\varphi = \phi$, and let

$$Z := \left\{ \begin{array}{c|c} \text{nontrivial homomorphism} & \text{(4.2), (4.3), (4.4),} \\ \varphi : \Gamma \to \text{Homeo}^+(\mathbb{R}) & \text{and (4.5) hold} \end{array} \right\}.$$

Note that Z is nonempty (because it contains ϕ).

For each $\gamma \in \Gamma$, the bounded displacement condition implies that $\{\, \varphi_\gamma \mid \varphi \in Z \,\}$ is uniformly bounded on any compact interval, and the Lipschitz condition implies that this set of functions is equicontinuous. (Furthermore, the no approximate fixed points condition implies that the limit of any sequence in Z is nontrivial.) Therefore, we see from the Arzelà-Ascoli Theorem that Z is compact.

This completes the construction of the space Z. We now verify that it has the desired properties (1–4).

(2) The group $\mathbb{R}$ acts on Z by via conjugation by translations: for $s \in \mathbb{R}$, let

$$(s * \varphi)_\gamma(t) = \varphi_\gamma(s + t) - s.$$

(It is straightforward to check that if (4.2)–(4.5) hold for φ, then they also hold for $s * \varphi$.)

Suppose this $\mathbb{R}$-action is not free, so there is some nonzero $s \in \mathbb{R}$ and some $\varphi \in Z$, such that $s * \varphi = \varphi$. This means that for all $t \in \mathbb{R}$ and all $\gamma \in \Gamma$, we have

$$\varphi_\gamma(s + t) - s = \varphi_\gamma(t),$$

so the action of Γ centralizes the translation by s. Hence, the action of Γ on $\mathbb{R}$ factors through to a well-defined action on the quotient $\mathbb{R}/s\mathbb{Z}$. Then Theorem 2.3(3) tells us that this action must have a fixed point $t + s\mathbb{Z}$ (after replacing Γ with a finite-index subgroup). So the subset $t + s\mathbb{Z}$ of $\mathbb{R}$ is Γ-invariant. The group Γ must act on this set by order-preserving permutations, which means that it acts by translations. But the group of translations is abelian, and the abelianization of Γ is finite (see Theorem 2.3(2)). So we conclude that Γ fixes t (and also fixes all of the other points in $t + s\mathbb{Z}$). More precisely, $\varphi_\gamma(t) = t$ for every $\gamma \in \Gamma$. This contradicts the fact that there are no (approximate) fixed points.

(1) We define an action of Γ on Z by letting an element γ of Γ act on the point φ of Z via the action of the element $\varphi_\gamma(0)$ of $\mathbb{R}$, so

$$^{\gamma}\varphi = \varphi_\gamma(0) * \varphi. \tag{4.6}$$

More concretely:

$$^{\gamma}\varphi_\lambda(t) = \big(\varphi_\gamma(0) * \varphi\big)_\lambda(t) = \varphi_\lambda\big(t + \varphi_\gamma(0)\big) - \varphi_\gamma(0).$$

It may not be obvious that this defines an action, so we provide a verification. Noting that, for $\gamma \in \Gamma$ and $t \in \mathbb{R}$, we have

$$^{\gamma}(t * \varphi) = (t * \varphi)_\gamma(0) * (t * \varphi) = [\varphi_\gamma(0 + t) - t] * (t * \varphi) = \varphi_\gamma(t) * \varphi, \tag{4.7}$$

we see that

$$^{\gamma}\big(^{\lambda}\varphi\big) = {}^{\gamma}(\varphi_\lambda(0) * \varphi) = \varphi_\gamma\big(\varphi_\lambda(0)\big) * \varphi = \varphi_{\gamma\lambda}(0) * \varphi = {}^{\gamma\lambda}\varphi.$$

Suppose this Γ-action has a fixed point: ${}^{\gamma}\varphi = \varphi$ for every $\gamma \in \Gamma$. This means $\varphi_\gamma(0) * \varphi = \varphi$, so $\varphi_\gamma(0) = 0$ (because the $\mathbb{R}$-action is free). Since this holds for every $\gamma \in \Gamma$, we conclude that 0 is a fixed point of the action given by φ. Once again, we have a contradiction to the fact that there are no (approximate) fixed points.

(3) Since elements of Γ act on each point via an element of $\mathbb{R}$, it is clear that the each $\mathbb{R}$-orbit is Γ-invariant.

(4) If $s < t$, then $\varphi_\gamma(s) < \varphi_\gamma(t)$ (because φ_γ is orientation preserving). So we see from (4.7) that the action of Γ on each $\mathbb{R}$-orbit is orientation preserving. □

We now establish two nice properties of the almost-periodic space:

Lemma 4.8 *The almost-periodic space Z satisfies* (4.2) *on each $\mathbb{R}$-orbit:*

$$\forall z \in Z, \quad \int_\Gamma (\gamma z - z)\, d\mu_\Gamma(\gamma) = 0. \tag{4.9}$$

Therefore:

The action of Γ on each $\mathbb{R}$-orbit of Z is Lipschitz. (4.10)

In fact, for each $\gamma \in \Gamma$, there is a constant C, such that γ is C-Lipschitz on every $\mathbb{R}$-orbit.

Proof For $\varphi \in Z$, we have

$$\begin{aligned}
\int_\Gamma \big({}^\gamma\varphi - \varphi\big)\, d\mu_\Gamma(\gamma) &= \int_\Gamma \big(\varphi_\gamma(0) * \varphi - \varphi\big)\, d\mu_\Gamma(\gamma) && (4.6)\\
&= \int_\Gamma \varphi_\gamma(0)\, d\mu_\Gamma(\gamma) && \left(\begin{matrix}\text{(3.18): definition}\\ \text{of subtraction}\end{matrix}\right)\\
&= \int_\Gamma \big(\varphi_\gamma(0) - 0\big)\, d\mu_\Gamma(\gamma)\\
&= 0 && (4.2).
\end{aligned}$$

This establishes (4.9).

It was explained in the proof of Theorem 4.1 that this implies each element of the support of μ_Γ is Lipschitz on each $\mathbb{R}$-orbit (see (4.3)), and that the same Lipschitz constant can be used on all $\mathbb{R}$-orbits. Since the support generates Γ (see Definition 3.14), the conclusion can be extended to every $\gamma \in \Gamma$ (with a Lipschitz constant that depends on γ, but is the same on all $\mathbb{R}$-orbits). □

4.2 *The Measures μ_Z and μ_X are Stationary*

From (4.9), we obtain the following fact.

Proposition 4.11 (Proof of Proposition 3.29, [14, Lem. 5.6]) *The measure μ_Z is μ_Γ-stationary.*

Proof For a.e. $z \in Z$, we have

$$\begin{aligned}
\int_\Gamma \frac{d\gamma_*\mu_Z}{d\mu_Z}(z)\, d\mu_\Gamma(\gamma) &= \int_\Gamma \frac{d\gamma_*^{-1}\mu_Z}{d\mu_Z}(z)\, d\mu_\Gamma(\gamma)\\
&= \int_\Gamma D_\gamma^{\text{leaf}}(z)\, d\mu_\Gamma(\gamma)\\
&= \int_\Gamma \lim_{t\to 0} \frac{\gamma(z+t) - \gamma(z)}{t}\, d\mu_\Gamma(\gamma)\\
&= \lim_{t\to 0} \int_\Gamma \frac{\gamma(z+t) - \gamma(z)}{t}\, d\mu_\Gamma(\gamma) && \left(\begin{matrix}\text{the integral is}\\ \text{a finite sum}\\ \text{(see Defn. 3.14)}\end{matrix}\right)\\
&= \lim_{t\to 0} \int_\Gamma \frac{\big(\gamma(z+t) - (z+t)\big) - \big(\gamma(z) - z\big) + t}{t}\, d\mu_\Gamma(\gamma)
\end{aligned}$$

$$
\begin{aligned}
&= \lim_{t\to 0}\left[\frac{\int_\Gamma\big(\gamma(z+t)-(z+t)\big)\,d\mu_\Gamma(\gamma)}{t}\right.\\
&\qquad\qquad\left. - \frac{\int_\Gamma\big(\gamma(z)-z\big)\,d\mu_\Gamma(\gamma)}{t} + \int_\Gamma \frac{t}{t}\,d\mu_\Gamma(\gamma)\right]\\
&= \lim_{t\to 0}\left[\frac{0}{t}-\frac{0}{t}+1\right] \qquad (4.9)\\
&= 1.
\end{aligned}
$$

So

$$
\int_\Gamma \gamma_*\mu_Z\,d\mu_\Gamma(\gamma) = \int_\Gamma \frac{d\gamma_*\mu_Z}{d\mu_Z}\,\mu_Z\,d\mu_\Gamma(\gamma) = \left(\int_\Gamma \frac{d\gamma_*\mu_Z}{d\mu_Z}\,d\mu_\Gamma(\gamma)\right)\cdot\mu_Z = 1\cdot\mu_Z = \mu_Z. \qquad \square
$$

Corollary 4.12 (Proof of Corollary 3.30, [14, Prop. 5.11]) *μ_X is μ_G-stationary.*

Proof Let $[(k_0, z_0)] \in X$, with $k_0 \in K$ and $z_0 \in Z$. Then

$$
\begin{aligned}
&\int_G \frac{dg_*\mu_X}{d\mu_X}\big([(k_0,z_0)]\big)\,d\mu_G(g)\\
&= \int_G \frac{dg_*^{-1}\mu_X}{d\mu_X}\big([(k_0,z_0)]\big)\,d\mu_G(g)\\
&= \int_G D_g^{\text{leaf}}\big([(k_0,z_0)]\big)\,d\mu_G(g)\\
&= \int_G D_{gk_0}^{\text{leaf}}\big([(\mathbf{1},z_0)]\big)\cdot D_{k_0^{-1}}^{\text{leaf}}\big([(k_0,z_0)]\big)\,d\mu_G(g) && \text{(Chain Rule)}\\
&= \int_G D_{gk_0}^{\text{leaf}}\big([(\mathbf{1},z_0)]\big)\cdot 1\,d\mu_G(g)\\
&= \int_G D_g^{\text{leaf}}\big([(\mathbf{1},z_0)]\big)\,d\mu_G(g) && \left(\begin{array}{c}\mu_G \text{ is right}\\ K\text{-invariant}\\ \text{by (3.12(2b)}\end{array}\right)\\
&= \int_\Gamma\int_K D_{k\gamma}^{\text{leaf}}\big([(\mathbf{1},z_0)]\big)\,dm_K(k)\,d\mu_\Gamma(\gamma) && \text{(definition of } \mu_\Gamma)\\
&= \int_\Gamma\int_K D_\gamma^{\text{leaf}}\big([(\mathbf{1},z_0)]\big)\,dm_K(k)\,d\mu_\Gamma(\gamma) && \left(\begin{array}{c}\text{Chain Rule}\\ \text{and } D_k^{\text{leaf}}=1\end{array}\right)\\
&= \int_\Gamma D_\gamma^{\text{leaf}}\big([(\mathbf{1},z_0)]\big)\,d\mu_\Gamma(\gamma) && \left(\begin{array}{c}\text{integral of}\\ \text{a constant}\end{array}\right)\\
&= \int_\Gamma D_\gamma^{\text{leaf}}(z_0)\,d\mu_\Gamma(\gamma) && \Big(\gamma\big([(\mathbf{1},z)]\big) = [(\gamma,z)] = [(\mathbf{1},\gamma z)]\Big)\\
&= 1 && \text{(Proposition 4.11).} \qquad \square
\end{aligned}
$$

Warning 4.13 Corollary 4.12 is not a trivial consequence of Proposition 4.11: a stationary measure for an action of a lattice subgroup does not usually lead in an obvious way to a stationary measure for the induced action of the ambient group. However, D. Creutz [12, Sect. 3] has found a general construction.

Lemma 4.14 (Proof of Proposition 3.32(2)) *If μ_X is G-invariant, then μ_Z is Γ-invariant.*

Proof Given $\gamma \in \Gamma$ and $Z_0 \subseteq Z$, we wish to show $\mu_Z(Z_0) = \mu_Z(\gamma Z_0)$. Since K is open, there is an open neighbourhood K_0 of $\mathbf{1}$ in K, such that $\gamma K_0 \gamma^{-1} \subseteq K$. Then

$$\gamma[K_0 \times Z_0] = [\gamma K_0 \times Z_0] = [\gamma K_0 \gamma^{-1} \times \gamma Z_0],$$

so

$$\begin{aligned} m_K(K_0) \cdot \mu_Z(Z_0) &= \mu_X\big([K_0 \times Z_0]\big) && \text{(definition of } \mu_X) \\ &= \mu_X\big(\gamma[K_0 \times Z_0]\big) && \begin{pmatrix} \text{we are assuming that} \\ \mu_X \text{ is } G\text{-invariant} \end{pmatrix} \\ &= \mu_X\big([\gamma K_0 \gamma^{-1} \times \gamma Z_0]\big) && \begin{pmatrix} \text{definition of equivalence} \\ \text{relation on } G \times Z \end{pmatrix} \\ &= m_K(\gamma K_0 \gamma^{-1}) \cdot \mu_Z(\gamma Z_0) && \text{(definition of } \mu_X). \end{aligned}$$

However,

$$m_K(\gamma K_0 \gamma^{-1}) = m_G(\gamma K_0 \gamma^{-1}) = m_G(K_0) = m_K(K_0).$$

(The middle equality is because the semisimple group G is unimodular.) Therefore, by cancelling a factor of $m_K(K_0)$, we conclude that $\mu_Z(Z_0) = \mu_Z(\gamma Z_0)$. □

4.3 Contraction on the $\mathbb{R}$-Orbits

Lemma 4.15 (deal with the technical issue in Definition 3.33, cf. [14, Lem. 5.13]) *$D_a^{\text{leaf}}(x)$ exists for μ_X^P-a.e. $x \in X$.*

Proof Let

$$B = \{\, x \in X \mid D_a^{\text{leaf}}(x) \text{ does not exist} \,\} \quad \text{(the "bad" set).}$$

Suppose $[(k, z)] \in B$. By Assumption 3.7, we may write $ak = k'\gamma$ with $k' \in K$ and $\gamma \in \Gamma$. Then a acts on the entire $\mathbb{R}$-orbit $[(k, z + \mathbb{R})]$ via the element γ of Γ (see Note 3.21(4)). Since $[(k, z)] \in B$, this implies that $D_\gamma^{\text{leaf}}(z)$ does not exist. However, $K\gamma$ is open, so there is a neighbourhood $\mathcal{O}$ of $\mathbf{1}$ in K, such that $a\mathcal{O}k \subseteq K\gamma$. Then a acts via γ on every $\mathbb{R}$-orbit in $[\mathcal{O}k \times Z]$. This implies $\mathcal{O}[(k, z)] \subseteq B$.

For each $x \in B$, we have shown there is a neighbourhood $\mathcal{O}$ of $\mathbf{1}$ in K, such that $\mathcal{O}x \subseteq B$. If $\mu_X^P(B) > 0$, then, since the topology on K has a countable neighbourhood basis at $\mathbf{1}$, this implies there are a subset B' of positive measure, and a (small) neighbourhood $\mathcal{O}'$ of $\mathbf{1}$ in K, such that $\mathcal{O}'B' \subset B$. Then, since $\mu_X = m_K \times \mu_Z$, we have

$$\mu_X(B) \geq \mu_X(\mathcal{O}'B') \geq m_K(\mathcal{O}') \cdot \mu_Z\{\, z \in Z \mid \big(K \times \{z\}\big) \cap B' \neq \emptyset \,\} \neq 0.$$

This contradicts the fact that $D_a^{\text{leaf}}(x)$ exists for μ_X-a.e. $x \in X$. □

Lemma 4.16 (Proof of Lemma 3.34, [14, Sect. 7.1, p. 43]) *χ_P is a homomorphism.*

Proof We have

$$\begin{aligned}
\chi_P(a_1 a_2) &= \int_X \log D_{a_1 a_2}^{\text{leaf}}(x)\, d\mu_X^P(x) \\
&= \int_X \log D_{a_1}^{\text{leaf}}(a_2 x)\, d\mu_X^P(x) + \int_X \log D_{a_2}^{\text{leaf}}(x)\, d\mu_X^P(x) \quad \text{(Chain Rule)} \\
&= \int_X \log D_{a_1}^{\text{leaf}}(a_2 x)\, d\mu_X^P(x) + \chi_P(a_2).
\end{aligned}$$

Since $A \subseteq P$, we know that μ_X^P is a_2-invariant, so

$$\int_X \log D_{a_1}^{\text{leaf}}(a_2 x)\, d\mu_X^P(x) = \int_X \log D_{a_1}^{\text{leaf}}(x)\, d\mu_X^P(x) = \chi_P(a_1). \qquad \Box$$

Since the measure μ_Z is ergodic for the $\mathbb{R}$-action, it can be shown that the measure μ_X^P is ergodic for the action of P:

Lemma 4.17 ([14, Lem. 6.5]) *The P-invariant measure μ_X^P is ergodic.*

Idea of Proof Suppose μ_X^P is not ergodic, so $\mu_X^P = s\mu_1 + t\mu_2$ is a convex combination of two different P-invariant probability measures (with $st \neq 0$). Then

$$\mu_X = \int_K k_* \mu_X^P \, dm_K = s \int_K k_* \mu_1 \, dm_K + t \int_K k_* \mu_2 \, dm_K$$

is a convex combination of two G-stationary measures. Also, the uniqueness in Theorem 3.31 implies that neither of these stationary measures is equal to μ_X. So μ_X is not ergodic as a stationary measure.

Hence, there is a G-invariant function f that is not constant (a.e.). The Pointwise Ergodic Theorem (2.7) has an analogue for random walks with a stationary measure [14, Cor. 4.8] (cf. the proof of Proposition 4.27). Combining this with Proposition 4.24 (contraction on $\mathbb{R}$-orbits) implies that f is essentially constant on a.e. $\mathbb{R}$-orbit in X. Since the $\mathbb{R}$-action on Z is ergodic, this implies that $f\big([(h,z)]\big)$ is essentially independent of z. But the G-invariance implies that it is also essentially independent of h. So f is constant (a.e.). This is a contradiction. □

Lemma 4.18 (Proof of Lemma 3.35, [14, Lem. 7.1]) *If $a \in A$, such that $\chi_P(a) < 0$, then for μ_X^P-a.e. $x \in X$, there exists $\epsilon = \epsilon(x) > 0$, such that*

$$a^n(x-\epsilon) \; - \; a^n(x+\epsilon) \; \to \; 0 \quad as\; n \to \infty.$$

Proof For (small) $\eta > 0$, let

$$D_\eta(x) = \sup_{0<\epsilon<\eta} \log\left(\frac{a(x+\epsilon)-a(x)}{\epsilon}\right).$$

These functions are uniformly bounded, because we know that a and a^{-1} are uniformly Lipschitz on the $\mathbb{R}$-orbits (see Lemma 3.24). Also, the functions converge pointwise to D_a^{leaf} as $\eta \to 0$. By the Dominated Convergence Theorem, then

$$\textstyle\int_X D_\eta \, d\mu_X^P \overset{\eta\to 0}{\longrightarrow} \chi_P(a). \tag{4.19}$$

Now, a technical point. We will apply the Pointwise Ergodic Theorem (2.7) to the action of a. This requires an ergodic measure. We know from Lemma 4.17 that the measure μ_X^P is ergodic for the P-action, but we cannot expect it to be ergodic for the action of a. Therefore, we should consider the ergodic components of the measure. Fortunately, it can be shown that the integral in Definition 3.33 that defines $\chi_P(a)$ has the same value for each ergodic component [14, Prop. 7.2], so this is not a serious problem. Therefore, we will simply ignore it, and act as if the measure μ_X^P is ergodic for the action of a.

Note that, for $\epsilon, \delta > 0$, η small, and n large, we have

$$\begin{aligned}
\frac{1}{n}\log\left(\frac{a^n(x+\epsilon)-a^n x}{\epsilon}\right) &= \frac{1}{n}\big[\log\big(a^n(x+\epsilon)-a^n x\big] - \log\epsilon\big] \\
&= \frac{1}{n}\sum_{i=1}^{n}\log\left(\frac{a^i(x+\epsilon)-a^i x}{a^{i-1}(x+\epsilon)-a^{i-1}x}\right) && \begin{pmatrix}\text{telescoping}\\ \text{sum}\end{pmatrix} \\
&\le \frac{1}{n}\sum_{i=1}^{n} D_\eta(a^{i-1}x) && \begin{pmatrix}\text{if } a^{i-1}(x+\epsilon)-a^{i-1}x < \eta \\ \text{for } 1 \le i < n\end{pmatrix} \\
&< \delta + \int_X D_\eta \, d\mu_X^P && \begin{pmatrix}\text{Pointwise Ergodic}\\ \text{Theorem (2.7)}\end{pmatrix} \\
&< 2\delta + \chi_P(a) && (4.19).
\end{aligned}$$

So, for all large n, we have

$$a^n(x+\epsilon) - a^n x \; < \; \epsilon\, e^{nc} \qquad \text{where } c = 2\delta + \chi_P(a).$$

We can choose δ small enough that $2\delta + \chi_P(a) < 0$, so $e^{nc} \overset{n\to\infty}{\longrightarrow} 0$. Then we may choose ϵ small enough that $e^{nc} < \eta$ for all n. Then we conclude, by induction, that

$a^n(x+\epsilon) - a^n x < \eta$ for all n. So the above estimate applies for all n. Hence, we have shown that $a^n(x+\epsilon) - a^n x \overset{n\to\infty}{\longrightarrow} 0$ (exponentially fast).

A similar argument with negative ϵ completes the proof. □

4.4 μ_X^P is G-Invariant

Remark 4.20 (*Further details for the proof of Corollary* 3.38) The proof of Corollary 3.38 strongly uses the fact that μ_X^P is U_a^+-invariant, and also uses the fact that $\pi_*(c_*\mu_X^P)$ is a ${}^cP$-invariant probability measure, but the latter is needed only to conclude that the support of this measure is all of G/Γ.

Fortunately, both of these properties are true of the ergodic components. Firstly, it follows from the Mautner Lemma [1, Lem. II.3.6, p. 61] that if a probability measure is invariant under both a and its expanding horospherical subgroup U_a^+, then each a-ergodic component μ of the measure is invariant under U_a^+. Furthermore, it can be shown (by an argument very similar to the proof of Proposition 2.9(2)) that G/Γ has no nonempty, proper, closed subsets that are invariant under $\langle a, U_a^+\rangle$. This implies that the support of $\pi_*\mu$ is all of G/Γ.

Corollary 4.21 (Proof of all of Corollary 3.45, [14, Thm. 8.1 (and Sect. 9)]) *The measure μ_X^P is G-invariant.*

Proof Let $\mathcal{Q}_{\text{good}}$ be the set of all minimal parabolic subgroups Q containing A, such that μ_X^P is Q-invariant (or, equivalently, such that $\mu_X^Q = \mu_X^P$). Also let

$$\mathcal{H} = \{\, a \in A \mid \chi_P(a) < 0 \,\}.$$

The arguments given in the "Idea of proof" of Corollary 3.45 show that if $Q_1 \in \mathcal{Q}_{\text{good}}$, and $\mathcal{W}_{Q_2}$ is a Weyl chamber, such that $\mathcal{H}$ contains some nonidentity element of $\mathcal{W}_{Q_1} \cap \mathcal{W}_{Q_2}$, then $Q_2 \in \mathcal{Q}_{\text{good}}$. Therefore, if we let

$$A_{\text{good}} := \bigcup_{Q \in \mathcal{Q}_{\text{good}}} \mathcal{W}_Q,$$

it is not difficult to show that $\mathcal{H} \subseteq A_{\text{good}}$.

Furthermore, if w is any element of the Weyl group, such that $w(\mathcal{H}) \cap A_{\text{good}}$ is nonempty, then one can repeat the argument with $w(\mathcal{H})$ in the place of $\mathcal{H}$, to conclude that A_{good} also contains $w(\mathcal{H})$. For every element w of the Weyl group, this implies that either $w(A_{\text{good}}) = A_{\text{good}}$ or $w(A_{\text{good}})$ is disjoint from the interior of A_{good}.

Case 1 *Assume A_{good} is invariant under the Weyl group.*

Since the Weyl group acts transitively on the set of Weyl chambers, this implies that every Weyl chamber is in A_{good}, so $A_{\text{good}} = A$. This means that every minimal parabolic subgroup is in $\mathcal{Q}_{\text{good}}$, so μ_X^P is invariant under every minimal parabolic subgroup. Since the minimal parabolic subgroups generate G, we conclude that μ_X^P is G-invariant, as desired.

Case 2 *Assume* A_{good} *is* _not_ *invariant under the Weyl group.*

Recall that A_{good} contains the halfspace $\mathcal{H}$, and also that, for every element w of the Weyl group, either $w(A_{\text{good}}) = A$, or $w(A_{\text{good}})$ is disjoint from the interior of A_{good}. Then the assumption of this case implies that $A_{\text{good}} = \overline{\mathcal{H}}$ (the closure of $\mathcal{H}$) and, for every element of the Weyl group, either $w(\overline{\mathcal{H}}) = \overline{\mathcal{H}}$ or $w(\overline{\mathcal{H}}) = \overline{\mathcal{H}^{\text{op}}}$, the closure of the opposite halfspace $\{ a \in A \mid \chi_P(a) > 0 \}$.

The action of the Weyl group of a simple group is irreducible, so the only way this can happen is if:

- G has a simple factor G_1 with $\operatorname{rank}_{\mathbb{Q}_p} G_1 = 1$, and
- $\chi_P = c\,\alpha$, where α is the (unique) simple root of G_1 (for some ordering of the roots), and $c \in \mathbb{R}^+$.

Write $G = G_1 \times H$, and choose some $a \in A \cap G_1$, such that $\alpha(a) < 0$. Then a centralizes H, so we see from Corollary 3.38 that μ_X^P is H-invariant. The subgroup H is normal, so this implies that $k_*\mu_X^P$ is also H-invariant, for every $k \in K$:

$$h_*(k_*\mu_X^P) = k_*(k^{-1}hk)_*\mu_X^P = k_*\mu_X^P.$$

Since μ_X is the integral of these H-invariant measures (see Theorem 3.31), we conclude that μ_X is H-invariant. And we know that μ_X is K-invariant (by definition). So μ_X is invariant under KH. Thus, letting $\Gamma_0 := \Gamma \cap KH$, we know that μ_X is invariant under Γ_0. For every $\gamma \in \Gamma$, this implies that $D_\gamma^{\text{leaf}} = 1$ μ_X-a.e. (see Lemma 3.28), so γ commutes with the $\mathbb{R}$-flow. Therefore, Γ_0 acts by translations on each $\mathbb{R}$-orbit in Z. Since the group of translations is abelian, this implies that the commutator subgroup $[\Gamma_0, \Gamma_0]$ acts trivially on Z. On the other hand, Γ acts faithfully on Z. So we conclude that Γ_0 is abelian.

Let $\pi \colon G \to G_1$ be the projection with kernel H. Since Γ is irreducible, we know that $\pi(\Gamma)$ is dense in G_1. Since K is open, this implies that $\pi(\Gamma_0)$ is dense in $\pi(K)$, which is an open subgroup of G_1. Since nontrivial open subgroups are Zariski dense, we conclude that $\pi(\Gamma_0)$ is Zariski dense in G_1. Since G_1 is not abelian, this implies that $\pi(\Gamma_0)$ is not abelian, which contradicts the conclusion of the preceding paragraph. □

Remark 4.22 For real groups, Case 2 of the proof of Corollary 4.21 is much longer [14, Sect. 9, pp. 55–64]. This is because K is not open, so the intersection $\Gamma \cap KH$ may be abelian, or could even be trivial, so a different idea is needed.

4.5 More Proof of Key Proposition 3.36

Proposition 4.23 (Proof of Proposition 3.47, [14, Prop. 6.3]) *For every action of Γ on $\mathbb{R}$ without fixed points, there is a compact subset I of $\mathbb{R}$, such that for every $\epsilon > 0$ and all $x, y \in \mathbb{R}$, there exists $\gamma \in \Gamma$, such that*

$$|\gamma x - \gamma y| < \epsilon \text{ and } \{\gamma x, \gamma y\} \subset I.$$

Proof It is a general fact about orientation-preserving actions of finitely generated groups on the real line that one of the following possibilities must hold [15, Thm. 7.1]:

(1) the action has a fixed point, or
(2) there is a nontrivial homomorphism $\Gamma \to \mathbb{R}$, or
(3) there is a fixed-point-free homeomorphism φ of $\mathbb{R}$ that commutes with every element of Γ, or
(4) the action has the desired contraction property.

Since we are assuming that (1) does not hold, and Theorem 2.3(2) tells us that (2) does not hold, we may assume that (3) holds.

We may assume that φ is the translation $\varphi(t) = t + 1$. Then Γ acts on $\mathbb{R}/\mathbb{Z} = S^1$. By Theorem 2.3(3), this action has a finite orbit, so, by passing to a finite-index subgroup of G, we may assume that it has a fixed point. Assuming, without loss of generality, that the fixed point is $0 + \mathbb{Z}$, this implies that the set $\mathbb{Z}$ is Γ-invariant. Now, since the action of Γ is orientation-preserving, the action on $\mathbb{Z}$ must be by translations. However, the group of translations is isomorphic to the abelian group $\mathbb{Z}$, whereas the abelianization of Γ is finite (see Theorem 2.3(2) again). Hence, every point in $\mathbb{Z}$ is fixed by Γ, which contradicts an assumption of this proposition. □

Proposition 4.24 (Proof of Proposition 3.48, [14, Props. 6.4 and 6.8]) *There is a constant $\delta < 1$, such that, for μ_X-a.e. $x \in X$, for all $y \in \mathbb{R}x$, and for μ_G^∞-a.e. sequence $(g_i)_{i=1}^\infty$ of elements of G, we have*

$$|g_n g_{n-1} \cdots g_1 x \; - \; g_n g_{n-1} \cdots g_1 y| < \delta^n \quad \textit{for all sufficiently large } n.$$

Proof For convenience, let $\pi_n = g_n g_{n-1} \cdots g_1$. For any $\ell > 0$ and $x \in X$, it follows from Proposition 4.23 that if n is sufficiently large, then there is a positive probability that $\pi_n(x + \ell) - \pi_n x$ is small. Then an argument similar to the proof of Lemma 4.18 shows that this difference almost surely tends to 0. Hence, we see from the compactness of X that, for every $\ell, \epsilon > 0$, there exist $p > 0$ and $N \in \mathbb{N}$, such that, for every $x \in X$, we have

$$\mu_G^\infty\big(\{\, (g_i)_{i=1}^\infty \mid \pi_N(x + \ell) - \pi_N x \le \epsilon\}\big) \ge p.$$

For $x \in X$, let

$$C_x = \{\, g \in G \mid g(x + \ell) - g(x - \ell) < \epsilon \,\},$$

so $\mu_G^{*N}(C_x) \ge p$ (where μ_G^{*N} is the N-fold convolution $\mu_G * \mu_G * \cdots * \mu_G$).

We claim for each $x \in X$ and μ_G^∞-a.e. sequence $(g_i)_{i=1}^\infty$ of elements of G, there are infinitely many n, such that $\pi_{n+N}\pi_n^{-1} \in C_{\pi_n x}$. To see this, let $c(x)$ be the probability that this happens at least once. Note that:

- if $\pi_N \in C_x$, then we may let $n = 0$, and
- otherwise, we may ignore $n = 0, 1, 2, \ldots, N-1$, and start our search at $n = N$.

Therefore

$$c(x) \geq \mu_G^{*N}(C_x) + \int_{G \smallsetminus C_x} c(gx)\, d\mu_G^{*N}(g).$$

Therefore, the infinum c_- of $c(x)$ satisfies

$$c_- \geq p + (1-p)c_- = c_- + (1-c_-)p,$$

so $c_- = 1$. Hence, we have $c(x) = 1$ for all x. Then, since the random walk is a Markov process, we conclude that the event happens an infinite number of times.

Now, the Martingale Convergence Theorem tells us that for every $x \in X$, every $t \in \mathbb{R}$, and μ_G^∞-a.e. $(g_i)_{i=1}^\infty$, the sequence $\pi_n(x+t) - \pi_n(x-t)$ converges to a finite limit. For $\ell \in \mathbb{R}^+$, let $\mathcal{E}_\ell = \mathcal{E}_\ell(x,t)$ be the set of sequences for which this limit is strictly less than ℓ. Then, for n sufficiently large, the length of the interval is less than ℓ. Then, by the claim of the preceding paragraph, we know (almost surely) that there are infinitely many n, such that $\pi_{g_N g_{N-1} \cdots g_{n+1}} \in C_{\pi_n x}$. This implies

$$\pi_N(x+t) - \pi_N(x-t) < \epsilon.$$

Hence, since we know that the sequence $\pi_n(x+t) - \pi_n(x-t)$ converges, we conclude that it converges to 0.

An argument similar to the proof of Lemma 4.18 (but using (3.46) in place of the assumption that $\chi_P(a) < 0$) shows that the contraction is exponentially fast [14, Lem. 6.7 and Prop. 6.8]. □

As mentioned in Sect. 3, the next part of the argument is easier for real groups, so we will switch to that setting for the remainder of the section:

Assume, contrary to Notation 3.3, that
G is a connected real semisimple Lie group.

Remark 4.25 (*A technical point in the proof of Proposition* 3.54) For convenience, let $a = \widehat{a_P}$. Also let

$$X_0 = \{\, x \in X \mid \text{for all } u \in x + \mathbb{R}, \text{ we have } \widehat{a_P}^n\, x \;-\; \widehat{a_P}^n\, y \overset{n\to\infty}{\longrightarrow} 0 \,\}.$$

The "Idea of proof" of Proposition 3.54 established that $\mu_X(X_0) = 1$, and we wish to show that $\mu_X^P(X_0) = 1$.

The measure μ_X is K-invariant, so $k_*\mu_X(X_0) = 1$ for all $k \in K$. By Fubini's Theorem, we conclude, for μ_X-a.e. $x \in X$, that we have $kx \in X_0$ for m_K-a.e. $k \in K$. Since $\mu_X = \int_K k_*\mu_X^P \, dk$, we can change μ_X to μ_X^P:

$$\text{for } \mu_X^P\text{-a.e. } x \in X, \text{ for } m_K\text{-a.e. } k \in K, \text{ we have } kx \in X_0. \tag{4.26}$$

Fix μ_X^P-a.e. $x \in X$. Also fix a neighbourhood U of $\mathbf{1}$ in U_a^+, and let $\mathcal{O}^-$ be a small neighbourhood of $\mathbf{1}$ in the set

$$\{\, g \in G \mid \{a^n g a^{-n} \text{ is bounded}\,\}.$$

We see from (4.26) that for $m_{U_a^+}$-a.e. $u \in U$, we can choose some $g \in \mathcal{O}^-$, such that $gux \in X_0$. Letting $g_n := a^n g a^{-n}$, we have

$$|g_n a^n ux - g_n a^n uy| = |a^n gux - a^n guy| \stackrel{n\to\infty}{\longrightarrow} 0.$$

Since $\{g_n\}$ is bounded, this implies

$$|a^n ux - a^n uy| \stackrel{n\to\infty}{\longrightarrow} 0.$$

So $ux \in X_0$. This (and Fubini's Theorem) establishes that $u_*\mu_X^P(X_0) = 1$ for a.e. $u \in U$. However, μ_X^P is U-invariant (since $U \subseteq U_a^+ \subseteq P$). So we conclude that $\mu_X^P(X_0) = 1$.

We can also show that χ_P is nontrivial. More precisely:

Proposition 4.27 (Proof of Note 3.39, [14, Lem. 7.4]) *For $\widehat{a_P}$ as in Corollary 3.52, we have $\chi_P(\widehat{a_P}) < 0$ and $U_{\widehat{a_P}}^+ \subseteq P$.*

Idea of Proof Since $\widehat{a_P} \in \mathcal{W}_P$, we know that $U_{\widehat{a_P}}^+ \subseteq P$. Therefore, we only need to prove the other statement.

The Pointwise Ergodic Theorem (2.7) has an analogue for random walks with a stationary measure [14, Cor. 4.8]. (This was also mentioned in the proof of Lemma 4.17.) Applying this (and the Chain Rule) to the function $\log D_g^{\text{leaf}}(x)$ on $G \times X$ yields: for μ_G^∞-a.e. sequence $(g_i)_{i=1}^\infty$ of elements of G, and μ_X-a.e. $x \in X$, if we let $\pi_n = g_n g_{n-1} \cdots g_1$, then

$$\frac{1}{n} \log D_{\pi_n}^{\text{leaf}}(x) = \frac{1}{n} \sum_{i=0}^{n-1} \log D_{g_{i+1}}^{\text{leaf}}(\pi_i x) \stackrel{n\to\infty}{\longrightarrow} \int_G \int_X \log D_g^{\text{leaf}}(x) \, d\mu_X(x) \, d\mu_G(g) < 0, \tag{4.28}$$

where the final inequality is from (3.46).

For convenience, let $a = \widehat{a_P}$. As in the proof of Corollary 3.38, there is a technical issue that the measure μ_X^P might not be ergodic for the action of a, even though Lemma 4.17 tells us that it is ergodic for the action of P. An additional technical issue is that the above inequality is valid for μ_X-a.e. x, but the definition of $\chi_P(\widehat{a_P})$ uses the measure μ_x^P. We will ignore both of these issues. (Roughly speaking, they are handled by proving that statements we prove for x remain valid for kx, for a.e. $k \in K$.)

Assuming that μ_X^P is ergodic with respect to a, we have, for μ_X^P-a.e. $x \in X$:

$$\begin{aligned} \frac{1}{n} \log D_{a^n}^{\mathrm{leaf}}(x) &= \frac{1}{n} \sum_{k=0}^{n-1} \log D_a^{\mathrm{leaf}}(a^k x) && \text{(Chain Rule)} \\ &\overset{n\to\infty}{\longrightarrow} \int_X \log D_a^{\mathrm{leaf}}\, d\mu_X^P && \text{(Pointwise Ergodic Theorem (2.7))} \\ &= \chi_P(a) && \text{(definition of } \chi_P\text{).} \end{aligned} \tag{4.29}$$

Also, it follows from Corollary 3.52 that

$$\frac{\log D_{a^n}^{\mathrm{leaf}}(x) - \log D_{\pi_n}^{\mathrm{leaf}}(x)}{n} \overset{n\to\infty}{\longrightarrow} 0.$$

Therefore, if we ignore the technical issue that (4.29) is for μ_X^P-a.e. x, but (4.28) is for μ_X-a.e. x, we can equate the two (in the limit), and conclude that $\chi_P(a) < 0$. □

5 $K\Gamma$ Might Not Be All of G

We have been assuming that $G = K\Gamma$ (see Assumption 3.7). This makes it easy to extend any function f on Γ to a function f_G on G, simply by making it left K-invariant: $f_G(k\gamma) = f(\gamma)$. Without this assumption, there is usually no canonical way to extend a function on Γ to a function on G. However, we can do this for "harmonic" functions.

Definition 5.1 (*Furstenberg* [20, Defn. 4.1]) Assume μ_H is a probability measure on a topological group H, and $f\colon H \to \mathbb{R}$. We say that f is μ_H-*harmonic* if, for all $x \in H$, we have

$$f(x) = \int_H f(hx)\, d\mu_H(h).$$

That is, the value of f at any point is equal to the average (with respect to μ_H) of the values of f at certain other points.

Lemma 5.2 ([20, p. 363]) *Every μ_G-harmonic function f on G is left K-invariant. Therefore, f corresponds to a well-defined function $\overline{f}$ on $K\backslash G$.*

Proof For $k \in K$, we have

$$\begin{aligned} f(kx) &= \int_G f(gkx)\, d\mu_G(g) && (f \text{ is } \mu_G\text{-harmonic}) \\ &= \int_G f(gx)\, d\mu_G(g) && (\mu_G \text{ is right } K\text{-invariant (see Notation 3.12(2)))} \\ &= f(x) && (f \text{ is } \mu_G\text{-harmonic}). \end{aligned}$$ □

Remark 5.3 Choose some $x_0 \in G$, and construct the random walk $x_0, x_1, x_2, \ldots$ in G, starting at x_0, with law μ_G. (This means that, for every $n \geq 1$, the distribution of the nth step $x_n x_{n-1}^{-1}$ is given by the measure μ_G.) We define a stopping time

$$N = N(x_0, x_1, \ldots) = \min\{\, n \geq 0 \mid x_n \in K\Gamma \,\}.$$

Now, suppose $f \colon G \to \mathbb{R}$ is μ_G-harmonic. Then the sequence $\{f(x_n)\}_{n \geq 0}$ is a *martingale*. More precisely, $f(x_n)$ is the expected value of $f(x_{n+1})$, given $x_0, x_1, \ldots, x_n$. So it is not difficult to see that $f(x_0)$ is the expected value of $f(x_N)$ (where N is the above-mentioned stopping time, or any other stopping time that satisfies mild conditions). Since $x_N \in K\Gamma$ and f is left K-invariant (see Lemma 5.2), we can conclude that $f(x_0)$ is determined by the values of f on Γ. Since x_0 is arbitrary, this means that the value of f at every point of the group G is determined by its values on the subgroup Γ. Of course, this argument relies on the assumption that f is μ_G-harmonic.

Conversely, given an appropriate function on Γ, we will explain how to extend it to a μ_G-harmonic function that is defined on all of G.

Definition 5.4

(1) For each $x \in K\backslash G$, we define an atomic probability measure ω_x on $K\backslash G$ by:

$$\omega_x(y) = \mu_G(Khg^{-1}) \qquad \text{if } x = Kg \text{ and } y = Kh.$$

This is clearly independent of the choice of the representative h of y, and the following calculation shows that it is also independent of the choice of the representative g of x:

$$\mu_G\big(Kh(kg)^{-1}\big) = \mu_G\big(Khg^{-1}k^{-1}\big) = \mu_G\big(Khg^{-1}\big),$$

where the final equality follows from our assumption that μ_G is right K-invariant (see Notation 3.12(2b)).

(2) We consider the random walk $\omega = \omega(x_0)$ on $K\backslash G$ that starts at a point x_0, and has transition probability $\omega_x(y)$ from x to y.

(3) Let $\widehat{x} = K$ be the basepoint of $K\backslash G$. For $x \in K\backslash G$ and $\gamma \in \Gamma$, let $\mu_x(\gamma)$ be the probability that $\widehat{x}\gamma$ is the first point of $\widehat{x}\Gamma$ that is encountered by the random walk $\omega(x)$ that starts at x.
(4) Let $\mu_{K\backslash G}$ be the (atomic) measure on $K\backslash G$ that is induced by μ_G. More precisely, if $\pi \colon G \to K\backslash G$ is the natural quotient map, then $\mu_{K\backslash G} = \pi_* \mu_G$.
(5) Define a measure μ_Γ on Γ by:

$$\mu_\Gamma(\gamma) = \int_{K\backslash G} \mu_x(\gamma)\, d\mu_{K\backslash G}(x).$$

We will need an estimate of the following type:

Lemma 5.5 *We have $\mu_x(\Gamma) = 1$ for all $x \in K\backslash G$.*

Indeed, if we let $\ell(\gamma)$ be the word length of γ, with respect to some finite generating set of Γ, then there exists $\delta < 1$, such that

$$\sum_{\ell(\gamma)<n} \mu_x(\gamma) > 1 - \delta^n \qquad \textit{for all large } n.$$

Proof The random walk ω on $K\backslash G$ factors through to a random walk $\overline{\omega}$ on $K\backslash G/\Gamma$: for $x, y \in K\backslash G$, the transition probability from $x\Gamma$ to $y\Gamma$ is $\sum_{\gamma\in\Gamma} \omega_x(y\gamma)$. This is a random walk (or "Markov chain") on a finite set (see Note 3.6), so there is some $r \in \mathbb{N}$ and $\epsilon > 0$, such that, for every starting vertex, the walk has a probability $> \epsilon$ of reaching the point $\widehat{x}\Gamma$ within r steps. Therefore, the probability that the stopping time N is greater than n is less than $(1-\epsilon)^{n/r}$.

To complete the proof, combine this observation with the fact that, since Γ is cocompact, the word length metric ℓ_Γ on Γ is quasi-isometric to the metric on $\widehat{x}\Gamma$ that is obtained by restricting the metric on $K\backslash G$ [4, Prop. I.8.19, p. 140]. □

Proposition 5.6 ([14, Thm. 3.12]) *Every positive (or bounded) μ_Γ-harmonic function on Γ extends to a unique μ_G-harmonic function on G. Specifically, the μ_G-harmonic extension f_G of f is defined by:*

$$\overline{f_G}(x) = \sum_{\gamma\in\Gamma} \mu_x(\gamma)\, f(\gamma) \qquad \textit{for } x \in K\backslash G.$$

Proof To simplify the estimates, we assume that f is bounded. Then it is clear from Lemma 5.5 that the sum defining $\overline{f_G}(x)$ converges, so f_G is a function from G to $\mathbb{R}$.

The uniqueness of the extension is a consequence of the argument in Remark 5.3 (if the extension exists).

For $x = K\gamma \in \widehat{x}\Gamma$, the measure μ_x is the point mass at γ. So it is immediate from the definition of $\overline{f_G}$ that $\overline{f_G}(K\gamma) = f(\gamma)$. Therefore the formula for f_G does provide an extension of f.

To complete the proof, we show that f_G is μ_G-harmonic.

For $x_0 \notin \widehat{x}\,\Gamma$, we see from the definition that

$$\mu_{x_0}(\gamma) = \int_{K\backslash G} \mu_{x_1}(\gamma)\, d\mu_{K\backslash G}(x_1),$$

so

$$\overline{f_G}(x_0) = \sum_{\gamma\in\Gamma} \mu_{x_0}(\gamma)\, f(\gamma) = \int_{K\backslash G} \sum_{\gamma\in\Gamma} \mu_{x_1}(\gamma)\, f(\gamma)\, d\mu_G(x_1) = \int_{K\backslash G} \overline{f_G}(x_1)\, d\mu_G(x_1),$$

so f_G is μ_G-harmonic at the points of G that are in x_0.

To see that f_G is also μ_G-harmonic at points of $\widehat{x}\,\Gamma$, note that

$$\begin{aligned}
\overline{f_G}(K\lambda) &= f(\lambda) && (f_G \text{ is an extension of } f)\\
&= \sum_{\gamma\in\Gamma} \mu_\Gamma(\gamma)\, f(\gamma\lambda) && (f \text{ is } \mu_\Gamma\text{-harmonic})\\
&= \int_{K\backslash G} \sum_{\gamma\in\Gamma} \mu_x(\gamma)\, f(\gamma\lambda)\, d\mu_{K\backslash G}(x) && (\text{definition of } \mu_\Gamma)\\
&= \int_{K\backslash G} \sum_{\gamma\in\Gamma} \mu_{x\lambda}(\gamma\lambda)\, f(\gamma\lambda)\, d\mu_{K\backslash G}(x) && \left(\begin{array}{c}\text{definition of } \mu_x, \text{ and the}\\ \text{random walk } \omega \text{ is } G\text{-invariant}\end{array}\right)\\
&= \int_{K\backslash G} \sum_{\gamma\in\Gamma} \mu_{x\lambda}(\gamma)\, f(\gamma)\, d\mu_{K\backslash G}(x) && (\text{change of variables in the sum})\\
&= \int_{K\backslash G} \overline{f_G}(x\lambda)\, d\mu_{K\backslash G}(x) && (\text{definition of } \overline{f_G}). \qquad \square
\end{aligned}$$

Properties of harmonic functions (including Proposition 5.6) play key roles in the Deroin-Hurtado proof.

Example 5.7 Harmonic functions are the foundation of the construction of the measure μ_X when $G \neq K\Gamma$. Let Prob(Z) be the space of probability measures on Z, and define $f\colon \Gamma \to \text{Prob}(Z)$ by $f(\gamma) = \gamma_*^{-1}\mu_Z$. Then f is μ_Γ-harmonic (because μ_Z is μ_Γ-stationary, and μ_Γ is *symmetric*, i.e., $\mu_\Gamma(\gamma^{-1}) = \mu_\Gamma(\gamma)$). So it extends to a unique μ_G-harmonic function $f_G\colon G \to \text{Prob}(Z)$. Also note that f is right Γ-equivariant, in the sense that $f(\gamma\lambda) = \lambda_*^{-1} f(\gamma)$, so the uniqueness implies that f_G is also right Γ-equivariant. Hence, f_G represents a section of the bundle $\big(G \times \text{Prob}(Z)\big)/\Gamma \to G/\Gamma$. Integrating this section with respect to the G-invariant probability measure on G/Γ yields a probability measure μ_X on $(G \times Z)/\Gamma = X$. Since f_G is μ_G-harmonic (and the measure on G/Γ is G-invariant), the measure μ_X is μ_G-stationary.

Remark 5.8 Harmonic functions are left K-invariant (see Lemma 5.2), so the section f_G of Example 5.7 is constant on K. If $X \simeq K \times Z$ (as stated in Note 3.21(2)), this implies that $\mu_X = m_K \times \mu_Z$, which agrees with Notation 3.25.

Remark 5.9 (*cf.* [14, 5.3.2, 5.3.4, and 5.10(3)]) Here is a slightly different way to look at the construction of μ_X. Define $D\colon \Gamma \times Z \to \mathbb{R}^+$ by $D(\gamma, z) = D_\gamma^{\mathrm{leaf}}(z)$. For $\lambda, \gamma \in \Gamma$, we see from the Chain Rule that

$$D(\lambda\gamma, z) = D(\gamma, z)\, D(\lambda, \gamma z). \tag{5.10}$$

Since μ_Z is μ_Γ-stationary (see Proposition 4.11), this implies

$$\sum_{\lambda\in\Gamma} \mu_\Gamma(\lambda)\, D(\lambda\gamma, z) = D(\gamma, z) \sum_{\lambda\in\Gamma} \mu_\Gamma(\lambda)\, D(\lambda, \gamma z) = D(\gamma, z) \cdot 1 = D(\gamma, z).$$

This means that the function $\gamma \mapsto D(\gamma, z)$ is μ_Γ-harmonic, and therefore extends to a unique μ_G-harmonic function on G. Thus, we obtain a function $D_G\colon G \times Z \to \mathbb{R}^{\geq 0}$. (This function is μ_G-harmonic in the G-coordinate, and defined for μ_Z-a.e. z.) Let

$$\widetilde{\mu_X} = D_G \cdot (m_G \times \mu_Z),$$

so $\widetilde{\mu_X}$ is a measure on $G \times Z$ that is in the measure class of $m_G \times \mu_Z$.

For any $g \in G$, the uniqueness of D_G implies that (5.10) still holds if we replace D with D_G, and replace λ with g. Then for $\gamma \in \Gamma$ (acting on the right via $(g, z) \mapsto (g\gamma, \gamma^{-1}z)$), we can use the fact that $d\gamma_*^{-1}\mu_Z/d\mu_Z = D_G(\gamma, z)$ (see Lemma 3.28) to conclude that

$$\begin{aligned}
\frac{d\gamma_*\widetilde{\mu_X}}{d\widetilde{\mu_X}}(g, z) &= \frac{D_G(g\gamma^{-1}, \gamma z)}{D_G(g, z)} \cdot \frac{d\gamma_* m_G}{dm_G}(g) \cdot \frac{d\gamma_*^{-1}\mu_Z}{d\mu_Z}(z) \\
&= \frac{D_G(g\gamma^{-1}, \gamma z)}{D_G(g, z)} \cdot 1 \cdot D_G(\gamma, z) \\
&= 1.
\end{aligned}$$

This means that $\widetilde{\mu_X}$ is invariant for the action of Γ, so we can construct a well-defined measure μ_X on X from $\widetilde{\mu_X}$, by restricting the quotient map $G \times Z \to X$ to any fundamental domain of the action. The harmonicity of D in the G-coordinate implies that μ_X is μ_G-stationary.

The advantage of this approach is that it shows μ_X is in the same measure class as $m_G \times \mu_Z$ (restricted to a fundamental domain).

Acknowledgements I am grateful to two anonymous referees for helpful comments and corrections, and I thank B. Deroin, S. Hurtado, and J. Lécureux for informative discussions related to the proof of Theorem 1.2 and possible generalizations. I also acknowledge support of the Institut Henri Poincaré (UAR 839 CNRS-Sorbonne Université) and LabEx CARMIN (ANR-10-LABX-59-01).

References

1. Bekka, M.B., Mayer, M.: Ergodic Theory and Topological Dynamics of Group Actions on Homogeneous Spaces. Cambridge University Press, Cambridge (2000). ISBN 0-521-66030-0, MR 1781937
2. Benoist, Y., Quint, J.-F.: Random Walks on Reductive Groups. Springer, Cham (2016). ISBN 978-3-319-47719-0, MR 3560700
3. Borel, A.: Linear Algebraic Groups, 2nd edn. Springer, New York (1991). ISBN 0-387-97370-2, MR 1102012
4. Bridson, M., Haefliger, A.: Metric Spaces of Non-positive Curvature. Springer, Berlin (1999). ISBN 3-540-64324-9, MR 1744486
5. Brown, A., Fisher, D., Hurtado, S.: Zimmer's conjecture for actions of $\mathrm{SL}(m, \mathbb{Z})$. Invent. Math. 221(3), 1001–1060 (2020). MR 4132960
6. Brown, A., Fisher, D., Hurtado, S.: Zimmer's conjecture: subexponential growth, measure rigidity, and strong property (T). Ann. Math. (2)**196**(3), 891–940 (2022). MR 4502593
7. Brown, A., Fisher, D., Hurtado, S.: Zimmer's conjecture for non-uniform lattices: escape of mass and growth of cocycles (preprint) (2021). https://arxiv.org/abs/2105.14541
8. Burger, M.: An extension criterion for lattice actions on the circle. In: Geometry, Rigidity, and Group Actions. University of Chicago Press, Chicago, IL (2011), pp. 3–31. ISBN 978-0-226-23788-6, MR 2807827
9. Burger, M., Monod, N.: Bounded cohomology of lattices in higher rank Lie groups. J. Eur. Math. Soc. **1**, 199–235 (1999). MR 1694584
10. Cantat, S.: Progrès récents concernant le programme de Zimmer [d'après A. Brown, D. Fisher et S. Hurtado]. Astérisque, Séminaire Bourbaki. Vol. 2017/2018. Exposés 1136–1150 (2019), no. 414, Exp. No. 1136, 1–47. MR 4093195
11. Clay, A., Rolfsen, D.: Ordered Groups and Topology. American Mathematical Society, Providence, RI (2016). ISBN 978-1-4704-3106-8, MR 3560661
12. Creutz, D.: Stabilizers of stationary actions of lattices in semisimple groups. Groups Geom. Dyn. **18**(2), 551–570 (2024). MR 4729817
13. Deroin, B.: Almost-periodic actions on the real line. Enseign. Math. **59**, 183–194 (2013). MR 3113604
14. Deroin, B., Hurtado, S.: Non left-orderability of lattices in higher rank semi-simple Lie groups (preprint, 2020). arXiv:2008.10687
15. Deroin, B., Kleptsyn, V., Navas, A., Parwani, K.: Symmetric random walks on $\mathrm{Homeo}_+(\mathbb{R})$. Ann. Probab. **41**, 2066–2089 (2013). MR 3098067
16. Deroin, B., Navas, A., Rivas, C.: Groups, Orders, and Dynamics (preprint) (2014). arXiv:1408.5805
17. Einsiedler, M., Ward, T.: Ergodic Theory with a View Towards Number Theory. Springer, London (2011). ISBN 978-0-85729-020-5, MR 2723325
18. Fisher, D.: Groups acting on manifolds: around the Zimmer program. In: Geometry, Rigidity, and Group Actions. University of Chicago Press, Chicago, IL (2011), pp. 72–157. ISBN 978-0-226-23788-6, MR 2807830
19. Fisher, D.: Afterword: recent progress in the Zimmer program. In: Group Actions in Ergodic Theory, Geometry, and Topology. University of Chicago Press, Chicago, IL (2020), pp. 685–707. ISBN 978-0-226-56813-3, MR 4521508
20. Furstenberg, H.: A Poisson formula for semi-simple Lie groups. Ann. Math. (2) **77**, 335–386 (1963). MR 0146298
21. Furstenberg, H.: Noncommuting random products. Trans. Am. Math. Soc. **108**, 377–428 (1963). MR 0163345
22. Ghys, É.: Actions de réseaux sur le cercle. Invent. Math. **137**, 199–231 (1999). MR 1703323
23. Ghys, É.: Groups acting on the circle. Enseign. Math. (2) **47**(3-4), 329–407 (2001). MR 1876932
24. Humphreys, J.E.: Linear Algebraic Groups. Springer, New York (1975). ISBN 978-1-4684-9445-7, MR 0396773

25. Karlsson, A., Margulis, G.A.: A multiplicative ergodic theorem and nonpositively curved spaces. Comm. Math. Phys. **208**(1), 107–123 (1999). MR 1729880
26. Kopytov, V.M., Medvedev, N.Ya.: Right-Ordered Groups. Consultants Bureau, New York (1996). ISBN 0-306-11060-1, MR 1393199
27. Lifschitz, L., Morris, D.W.: Bounded generation and lattices that cannot act on the line. Pure Appl. Math. Q. **4**, Part 2, no. 1, 99–126 (2008). MR 2405997
28. Margulis, G.A.: Discrete Subgroups of Semisimple Lie Groups. Springer, Berlin (1991). ISBN 3-540-12179-X, MR 1090825
29. Morris, D.W.: Can lattices in $\mathrm{SL}(n, \mathbb{R})$ act on the circle?. In: Farb, B., Fisher, D. (eds.) Geometry, Rigidity, and Group Actions. University of Chicago Press, Chicago (2011), pp. 158–207. ISBN 9780226237886, MR 2807831
30. Morris, D.W.: Introduction to Arithmetic Groups. Deductive Press (2015). ISBN 978-0-9865716-0-2, MR 3307755
31. Mura, R., Rhemtulla, A.: Orderable Groups. Marcel Dekker, New York (1977). MR 0491396
32. Phelps, R.R.: Lectures on Choquet's Theorem. Springer, Berlin (2001). ISBN 3-540-41834-2, MR 1835574
33. Witte, D.: Arithmetic groups of higher $\mathbb{Q}$-rank cannot act on 1-manifolds. Proc. Am. Math. Soc. **122**, 333–340 (1994). MR 1198459
34. Witte, D., Zimmer, R.J.: Actions of semisimple Lie groups on circle bundles. Geom. Dedicata **87**, 91–121 (2001). MR 1866844
35. Zimmer, R.J.: Ergodic Theory and Semisimple Groups. Birkhäuser, Basel (1984). ISBN 978-1-4684-9488-4, MR 0776417. https://doi.org/10.1007/978-1-4684-9488-4

Mini Course on Uniform Stability of Higher-Rank Arithmetic Groups

Bharatram Rangarajan

Abstract These notes are based on lectures given by Alexander Lubotzky and the author in a mini-course at ICTS Bengaluru as part of the program *Zariski Dense Subgroups, Number Theory and Geometric Applications* in January 2024, and the results are from a sequence of joint works with Lev Glebsky and Nicolas Monod [20], and Francesco Fournier-Facio [16]. Lattices in higher-rank semisimple groups enjoy a number of rigidity properties like super-rigidity, quasi-isometric rigidity, first-order rigidity, and more. In these lectures, we will add another one: uniform stability. Namely, it will be shown that (most) such lattices Γ satisfy the following: Every finite-dimensional unitary "almost-representation" of Γ (with respect to a sub-multiplicative norm) is a small deformation of a true unitary representation. This extends a result of Kazhdan (1982) for amenable groups and or Burger–Ozawa–Thom (2013) for $SL(n, Z), n > 2$. Towards this goal, an elaborate cohomological theory is constructed capturing the obstruction to such stability, and it is shown that the vanishing of second cohomology implies stability in this setting. This cohomology can be roughly thought of as an asymptotic version of bounded cohomology, and sheds light on a question raised in [27] about a possible connection between vanishing of second bounded cohomology and Ulam stability. This machinery is then used to establish uniform stability of lamplighters and Thompson groups, and the main result about lattices in higher rank Lie groups.

Keywords Uniform stability · Bounded cohomology · Lattices in lie groups

1 Lecture 1: Introduction and Overview

Consider a semisimple group $G = \prod_{i=1}^{k} \mathbf{G}_i(K_i)$, where for $1 \leq i \leq k$, K_i is a local field, and $\mathbf{G}_i$ is an almost K_i-simple group. If the rank $\sum_{i=1}^{k} rk_{K_i}(\mathbf{G}_i) \geq 2$, such a G is referred to as a *higher rank* semisimple group. The class of irreducible lattices Γ in

B. Rangarajan (✉)
Zariski Dense Subgroups, Number Theory and Geometric Applications, ICTS, Bengaluru, India
e-mail: Bharatram.rangarajan@gmail.com

G. Prasad et al. (eds.), *Zariski Dense Subgroups, Number Theory and Geometric Applications*, Infosys Science Foundation Series,
https://doi.org/10.1007/978-981-95-3437-1_5

such groups G (referred to as *higher rank lattices*) form an interesting class of groups, which over the years, have been shown to satisfy many rigidity properties, such as local rigidity, Mostow strong-rigidity, Margulis super-rigidity (implying that they are arithmetic groups), Zimmer cocycle rigidity, quasi-isometric rigidity, first-order rigidity, etc. (see [2, 5, 14, 25] and the references therein). A common feature of the classical rigidity results is that such a higher rank lattice Γ has some clear family of representations, and all other representations are just easy variants of them.

The goal of this mini-course is to demonstrate another type of rigidity phenomena of these lattices. Before stating the exact formulation, let us recall that Margulis super-rigidity, while usually not formulated this way, also gives a full classification of all the finite dimensional unitary representations of a higher rank lattice Γ as above. Margulis super-rigidity implies that all such irreducible representations come from a combination of those that factor through finite quotients (and these are the only ones if Γ is a non-uniform lattice) and from the representations of Γ appearing naturally in its definition as an arithmetic group by Galois twisting (see Sect. 1.3 in [25]). The rigidity phenomenon studied here, which is called *uniform stability*, is the property that every unitary *almost-representation* of Γ is a small deformation of a unitary representation.

1.1 Uniform Stability of Groups

Let Γ be a discrete group and (G, d_G) be a metric group (where d_G is a bi-invariant metric on G). For $\epsilon > 0$, a map $\phi : \Gamma \to G$ is said to be an ϵ-almost homomorphism (or *ϵ-homomorphism*) if $d_G(\phi(xy), \phi(x)\phi(y)) \leq \epsilon$ for every $x, y \in \Gamma$. The value $\sup_{x,y\in\Gamma} d_G(\phi(xy), \phi(x)\phi(y))$ is called the *defect* of ϕ.

Let $\mathcal{G}$ be a family of metric groups. We say that Γ is *uniformly stable* with respect to $\mathcal{G}$ if for any $\epsilon > 0$, there exists $\delta = \delta(\epsilon)$ with $\lim_{\epsilon\to 0} \delta(\epsilon) = 0$ such that for any ϵ-homomorphism $\phi : \Gamma \to G$ (for $G \in \mathcal{G}$), there exists a homomorphism $\psi \in Hom(\Gamma, G)$ with $\sup_{x\in\Gamma} d_G(\phi(x), \psi(x)) \leq \delta$. In other words, Γ is uniformly stable with respect to $\mathcal{G}$ if any almost homomorphism of Γ to any group in the family $\mathcal{G}$ is close to a (true) homomorphism.

Questions of this nature were first raised and studied in [33–35], and of particular interest is the case when $\mathcal{G}$ is the family of unitary operators on Hilbert spaces and the metric is given by a norm (on the space of bounded operators), as studied in [10, 23]. Note that in this work, we will be interested solely in uniform stability, as opposed to *pointwise* stability, as studied in [3, 13] and the references therein.

The notion of uniform stability with respect to unitary operators on Hilbert spaces equipped with the operator norm is referred to in [10] as *strong Ulam stability*, while if we restrict the family to unitary operators on *finite-dimensional* Hilbert spaces, it is referred to as *Ulam stability*. In the pioneering work of Kazhdan [23] (and clarified further in [22, 32]), it is shown that

Theorem 1.1 ([23]) *Every (discrete) amenable group Γ is Ulam stable (in fact, even strongly Ulam stable).*

It is worth noting that the only known examples of strongly Ulam stable groups are amenable, and it is natural to ask if strong Ulam stability characterizes amenability. Let us mention at this point that one of the (innumerable) equivalent characterizations of amenability is given in terms of the vanishing of bounded cohomology with dual coefficients: Γ is amenable iff $\mathrm{H}^n_b(\Gamma, V) = 0$ for every dual Banach Γ-module V and $n > 0$. Here $\mathrm{H}^n_b(\Gamma, V)$ denotes the n-th *bounded* cohomology group of Γ with coefficients in the Banach Γ-module V. Kazhdan's proof does not use this result explicitly but does use a notion of ϵ-cocycles and approximate cohomology in degree 2.

Ulam stability was further studied in [10] where they show more examples (and non-examples) of Ulam stable groups. It is shown there that if a group contains a non-abelian free subgroup, then it is *not* strongly Ulam stable. In particular, this means that higher rank lattices are not strongly Ulam stable. On the positive side, they show:

Theorem 1.2 ([10]) *Let $\mathcal{O}$ be the ring of integers of a number field, S a finite set of primes, and $\mathcal{O}_S$ the corresponding localization. Then for every $n \geq 3$, $SL(n, \mathcal{O}_S)$ is Ulam stable.*

The proof of this result in [10] uses the fact that $SL(n, \mathcal{O}_S)$ (for $n \geq 3$) is boundedly generated by elementary matrices, and makes no reference to bounded cohomology (this result is further extended in [19] in the case of $n = 2$ when $\mathcal{O}_S$ has infinitely many units). However, note that for $\Gamma = SL(n, \mathcal{O}_S)$, $\mathrm{H}^2_b(\Gamma, V) = 0$ for every dual separable Γ-module V. In fact, it is shown in [8] that for every higher rank lattice Γ and any dual, separable Banach Γ-module V with $V^\Gamma = \{0\}$, $\mathrm{H}^2_b(\Gamma, V) = 0$. All this hints at a possible connection between bounded cohomology and Ulam stability, as raised by Monod in his ICM talk [27, Problem F], and serves as one of the starting points for our current work.

1.2 Main Results and Methods

The goal of this work is to generalize Theorems 1.1 and 1.2 to a wider class of groups (mainly, higher-rank lattices) and metrics. We shall consider the question of uniform stability with respect to the family $\mathfrak{U}$ of groups of unitary operators on *finite-dimensional* Hilbert spaces, with the metrics induced from submultiplicative norms on matrices (which we shall denote uniform $\mathfrak{U}$-stability or just "uniform stability" henceforth). These include the p-Schatten norms for $1 \leq p \leq \infty$ (and in particular, uniform $\mathfrak{U}$-stability subsumes Ulam stability). Furthermore, we shall actually obtain uniform stability *with a linear estimate*, which means that the distance of an almost homomorphism from a homomorphism is linearly bounded by its defect, and all our results are proved in this stronger notion of stability.

To this end, a new type of bounded cohomological theory is built to capture obstructions to uniform stability, so that uniform stability follows as a consequence of the vanishing of the second cohomology group in this theory. Our technique involves the following two main steps:

- **Defect Diminishing**: Expressing the problem of uniform stability as a homomorphism lifting problem, we can treat it as a culmination of intermediate lifts so that at each step, the lifting kernel is abelian. This is a uniform variant of defect diminishing that was introduced in [13] in the non-uniform setting, and is applicable when the relevant norms in the target groups are submultiplicative.
- **Asymptotic Cohomology**: Such a homomorphism lifting problem with abelian kernel naturally leads to a cohomological reformulation (as in [13]). However, unlike in the non-uniform setting where ordinary group cohomology comes up, in our uniform setting one would need to carefully construct a new cohomology theory such that the vanishing of the second cohomology group in this model implies (uniform) defect diminishing, and hence uniform stability.

The cohomological theory constructed here is an asymptotic variant of the bounded cohomology of the ultrapower ${}^*\Gamma$ with coefficients in a suitable ultraproduct Banach space $\mathcal{W}$, but *restricted to the internal objects in this universe*, which we shall call the *asymptotic cohomology* of Γ denoted $\mathrm{H}_a^\bullet(\Gamma, \mathcal{W})$.

Theorem 1.3 *Suppose* $\mathrm{H}_a^2(\Gamma, \mathcal{W}) = 0$, *then* Γ *is uniformly stable with a linear estimate.*

This new cohomology theory bears some similarity to the theory of bounded cohomology, and sometimes arguments there can be easily adapted to this model (for instance, we can show that $\mathrm{H}_a^2(\Gamma, \mathcal{W}) = 0$ for an amenable group Γ, immediately implying that amenable groups are Ulam-stable as in [22, 23, 32]), though at other times serious difficulties arise.

The groups Γ of particular interest are lattices in higher rank semisimple groups. Unlike some of the other rigidity results which sometimes hold for some lattices in rank one simple groups:

Proposition 1.4 *If* Γ *is a lattice in a semisimple group of rank* 1, *then* Γ *is not uniformly* $\mathfrak{U}$*-stable.*

It is shown in [18] that for such a Γ, $\mathrm{H}_b^2(\Gamma, \mathbb{R})$ is infinite dimensional. More precisely, Fujiwara constructs (many) non-trivial quasi-homomorphisms witnessing that the comparison map $c : \mathrm{H}_b^2(\Gamma, \mathbb{R}) \to H^2(\Gamma, \mathbb{R})$ is not injective. By exponentiation, such quasimorphisms yield almost homomorphisms to $U(1)$ that are not close to any homomorphism . In particular, a lattice of rank one is not uniformly U-stable, and to hope for uniform $\mathfrak{U}$-stability, the condition that rank of Γ is at least 2 is necessary.

For the main results, some definitions capturing properties of the class of semisimple groups of interest are needed. For a locally compact group G, we denote by $\mathrm{H}_b^\bullet(G, \mathbf{R})$ the *continuous* bounded cohomology of G with trivial coefficients.

- A locally compact group G is said to have the **$2\frac{1}{2}$property** (of vanishing bounded cohomology) if $\mathrm{H}_b^2(G, \mathbb{R}) = 0$ and $\mathrm{H}_b^3(G, \mathbb{R})$ is Hausdorff.
- Let G be a non-compact simple Lie group, and fix a minimal parabolic subgroup $P \leq G$. The group G is said to have **Property-$G(\mathcal{Q}_1, \mathcal{Q}_2)$** if there exist two proper parabolic subgroups Q_1 and Q_2 containing P, both having the $2\frac{1}{2}$-property, such that G is boundedly generated by the union $Q_1 \cup Q_2$. A semisimple group G is said to have Property-$G(\mathcal{Q}_1, \mathcal{Q}_2)$ if all its simple factors have Property-$G(\mathcal{Q}_1, \mathcal{Q}_2)$.

Note that if a semisimple group has Property-$G(\mathcal{Q}_1, \mathcal{Q}_2)$, then by definition, it must be of rank at least 2. But note that not all simple groups have the property (for instance, $SL_3(\mathbb{R})$). However, we will see that many classes of groups do have this property, for example, all simple groups (of rank at least 2) over $\mathbb{C}$ or over a non-archimedean field, and $SL_n(\mathbb{R})$ for $n \geq 4$.
We can now state our main result:

Theorem 1.5 *Let Γ be a lattice in a semisimple group G that has Property-$G(\mathcal{Q}_1, \mathcal{Q}_2)$. Then $\mathrm{H}_a^2(\Gamma, \mathcal{W}) = 0$, so in particular, Γ is uniformly stable.*

So the goal is to show that $\mathrm{H}_a^2(\Gamma, \mathcal{W}) = 0$ for Γ a lattice in a higher rank Lie group (satisfying certain conditions). For this, the results of [8] about the vanishing of bounded cohomology for such lattices can serve as an inspiration and a starting point. More specifically, the approach is inspired by a proof of [30] specifically in degree two, and a version of that result and proof technique are outlined below.

Theorem 1.6 ([30]) *Let G be a higher rank simple group, and $P \leq G$ be a minimal parabolic subgroup. Suppose G contains two proper parabolic subgroups Q_1 and Q_2 such that $P \subseteq Q_1 \cap Q_2$, G is generated by $Q_1 \cup Q_2$, and $\mathrm{H}_b^2(Q_1, \mathbb{R}) = \mathrm{H}_b^2(Q_2, \mathbb{R}) = 0$. Then for any lattice Γ in G and a dual separable Banach Γ-module W, $\mathrm{H}_b^2(\Gamma, W) = 0$.*

The proof of Theorem 1.6 proceeds in several steps briefly sketched below, and the corresponding steps and difficulties in the proof of Theorem 1.5 are also hinted at:

- The first step is to use an Eckman-Shapiro induction to construct a dual, separable, continuous Banach G-module V so that $\mathrm{H}_b^2(\Gamma, W) = \mathrm{H}_b^2(G, V)$, thus reducing the problem to showing that $\mathrm{H}_b^2(G, V) = 0$. A similar inductive procedure allows us to construct an ultraproduct Banach space $\mathcal{V}$ with an asymptotic action of G so that $\mathrm{H}_a^2(\Gamma, \mathcal{W}) = \mathrm{H}_a^2(G, \mathcal{V})$. Note that in the setting of asymptotic cohomology, we actually work with ultrapowers ${}^*\Gamma$ and ${}^*G$, and so ${}^*G/{}^*\Gamma$ is not locally compact. However, the restriction to internal objects allows us to carefully work out an induction procedure as needed. The induced module $\mathcal{V}$ also has an internal continuity property that is particularly helpful.
- Since P is amenable, the bounded cohomology $\mathrm{H}_b^\bullet(G, V)$ can be computed as the cohomology of the complex

$$0 \longrightarrow V^G \xrightarrow{\epsilon} L^\infty(G/P, V)^G \xrightarrow{d} L^\infty((G/P)^2, V)^G \xrightarrow{d} \dots$$

Furthermore, for a parabolic subgroup $P \leq Q \leq G$, the bounded cohomology $\mathrm{H}_b^\bullet(Q, V)$ can be computed as the cohomology of the complex

$$0 \longrightarrow V^Q \xrightarrow{\epsilon} L^\infty(G/P, V)^Q \xrightarrow{d} L^\infty((G/P)^2, V)^Q \xrightarrow{d} \dots$$

These steps too can be reworked in the asymptotic setting, analogous to the procedure in bounded cohomology theory, again thanks to the restriction on internality.

- The motivation behind the preceding step is that we have at our disposal the following double ergodicity theorem: Let V be a continuous G-module and $\alpha \in L^\infty\left((G/P)^2, V\right)^G$, then α is essentially constant. This theorem follows from the Mautner's lemma: let V be a continuous G-module and $N \leq G$ be a non-compact subgroup, then $V^N = V^G$. Both these results are particularly useful in the context of $\mathrm{H}_b^2(G, V)$.
  In our setting, there are particular difficulties in obtaining an analogous Mautner lemma due to the asymptotic nature of our model. One can overcome them for the specific Banach module $\mathcal{V}$ by applying a suitable correction to exact cocycles using structure results for the semisimple group G.
- For the parabolic subgroups Q_1 and Q_2 as in the hypothesis, an inflation-restriction sequence argument implies that $\mathrm{H}_b^2(Q_i, V) = \mathrm{H}_b^2(Q_i/N_i, V^{N_i})$ for N_i being the (amenable) radical of Q_i for $i \in \{1, 2\}$. By Mautner's lemma and the hypothesis that $\mathrm{H}_b^2(Q_i, \mathbb{R}) = 0$, one concludes that $\mathrm{H}_b^2(Q_i, V) = 0$.
  The analogous hypothesis in our asymptotic setting is Property-$G(\mathcal{Q}_1, \mathcal{Q}_2)$, where the conditions that $\mathrm{H}_b^2(Q_i, \mathbb{R}) = 0$ and $\mathrm{H}_b^3(Q_i, \mathbb{R})$ is Hausdorff together are used to conclude that $\mathrm{H}_a^2(\mathcal{Q}_i, \mathcal{V}) = 0$.
- Let $\omega \in L^\infty\left((G/P)^3, V\right)^G$ be a bounded 2-cocycle for G. Since $\mathrm{H}_b^2(Q_1, V) = \mathrm{H}_b^2(Q_2, V) = 0$, there exist $\alpha_1 \in L^\infty\left((G/P)^2, V\right)^{Q_1}$ and $\alpha_2 \in L^\infty\left((G/P)^2, V)\right)^{Q_2}$ such that $\omega = d\alpha_1 = d\alpha_2$. In particular, $\alpha_1 - \alpha_2$ is a 1-cocycle for $Q_1 \cap Q_2$. Since $P \leq Q_1 \cap Q_2$, $\alpha_1 - \alpha_2$ is a 1-cocycle for P as well. But since $\mathrm{H}_b^1(P, V) = 0$, one can show using the double ergodicity theorem, that $\alpha_1 = \alpha_2$ ($= \alpha$, say), implying that α is equivariant with respect to both Q_1 and Q_2, and hence is G-equivariant. Thus $\omega = d\alpha$ for $\alpha \in L^\infty\left((G/P)^2, V\right)^G$, proving that $\mathrm{H}_b^2(G, V) = 0$. This step too goes through in the asymptotic setting once one has the asymptotic variant of the ergodicity theorem used in the classical case.

While in this mini-course, we mainly focus on using the theory of asymptotic cohomology to prove uniform stability for lattices in semisimple groups, the framework and tools developed here have also been used in subsequent work [16] to prove uniform $\mathfrak{U}$-stability for other classes of groups such as lamplighter groups $\Gamma \wr \Lambda$ where Λ is infinite and amenable, as well as several groups of dynamical origin such as Thompson's group F. The techniques there too are analogous to corresponding vanishing results of bounded cohomology in [28], yet again highlighting the connections between the theories of bounded cohomology and asymptotic cohomology.

2 Lecture 2: Defect Diminishing and Cohomology

In this lecture, we being by studying the simpler *abelian* setting of $U(1)$-stability where the connections to cohomology are already apparent, and then describe the *defect diminishing* property which allows us to retain an abelian setting even in the case of stability over $U(n)$. This allows us to motivate the construction of asymptotic cohomology and reprove Kazhdan's theorem [23] in this framework.

2.1 $U(1)$-*Stability and Quasimorphisms*

Let us begin with a simpler setting, namely uniform stability of a discrete group Γ with respect to the abelian group $U(1)$. This section can be read independently of the rest.

Definition 2.7 For $\epsilon > 0$, a map $\phi : \Gamma \to U(1)$ is said to be an ϵ-homomorphism if

$$\sup_{x,y\in\Gamma} |\phi(xy) - \phi(x)\phi(y)| \leq \epsilon$$

The value $\sup_{x,y\in\Gamma} |\phi(xy) - \phi(x)\phi(y)|$ is called the *defect* of ϕ.

Definition 2.8 A group Γ is said to be uniformly $U(1)$-stable if for every $\delta > 0$, there exists $\epsilon > 0$ such that if $\phi : \Gamma \to U(1)$ is an ϵ-homomorphism, there exists a homomorphism $\psi : \Gamma \to U(1)$ such that $\sup_{x\in\Gamma} |\phi(x) - \psi(x)| < \delta$.

A closely related notion is that of a quasimorphism. A *quasimorphism* is a map $f : \Gamma \to \mathbb{R}$ such that there exists $D \geq 0$ such that for every $x, y \in \Gamma$,

$$|f(x) + f(y) - f(xy)| \leq D$$

Let $QM(\Gamma)$ denote the space of all quasimorphisms of Γ. A trivial example of a quasimorphism is obtained by perturbing a homomorphism by a bounded function, and such quasimorphisms form the subspace $Hom(\Gamma, \mathbb{R}) \oplus C_b(\Gamma, \mathbb{R})$ of $QM(\Gamma)$ (where $C_b(\Gamma, \mathbb{R})$ denotes the space of all bounded functions from Γ to $\mathbb{R}$) A quasimorphism that is not at a bounded distance from any homomorphism is called a *non-trivial quasimorphism*. In this setting, a question analogous to uniform stability is whether every quasimorphism is at a bounded distance from a homomorphism. That is, is $QM(\Gamma) = Hom(\Gamma, \mathbb{R}) \oplus C_b(\Gamma, \mathbb{R})$?

It is known that every quasimorphism class contains a unique *homogenous* quasimorphism (a quasimorphism f is said to be homogenous if for every $g \in \Gamma$ and $n \in \mathbb{N}$, $f(g^n) = nf(g)$). Suppose f is a homogenous quasimorphism of Γ that is not a homomorphism. Then its exponent $\mu := e^{2\pi i\epsilon f} : \Gamma \to U(1)$ is a function whose defect can be made arbitrarily small (by choosing $\epsilon \to 0$), but whose distance from homomorphisms is bounded below by a positive constant.

Proposition 2.9 ([10, 31]) *If* Γ *admits a non-trivial quasimorphism, then* Γ *is not uniformly* $U(1)$*-stable.*

Quasimorphisms are also closely related to group cohomology (this is well-known and classical, see [17, 27] for references). Observe that given a quasimorphism $f : \Gamma \to \mathbb{R}$, the map

$$df : \Gamma \times \Gamma \to \mathbb{R}$$

$$df(x, y) = f(x) + f(y) - f(xy)$$

is a 2-coboundary for Γ in $\mathbb{R}$, while also being a bounded function that satisfies the 2-cocycle condition (and hence a *bounded 2-cocycle*). This leads to the following characterization of quasimorphisms classes ([17, 27]):

Proposition 2.10 *The kernel, denoted* $E\,\mathrm{H}^2_b(\Gamma, \mathbb{R})$*, of the comparison map* $c : \mathrm{H}^2_b(\Gamma, \mathbb{R}) \to H^2(\Gamma, \mathbb{R})$ *is isomorphic to the space of quasimorphisms modulo the suabspace of trivial quasimorphisms.*

$$E\,\mathrm{H}^2_b(\Gamma, \mathbb{R}) \cong \frac{QM(\Gamma)}{C_b(\Gamma, \mathbb{R}) \oplus Hom(\Gamma, \mathbb{R})}$$

Corollary 2.11 *Suppose* $E\,\mathrm{H}^2_b(\Gamma, \mathbb{R}) \neq 0$*. Then* Γ *is not* $U(1)$*-stable (and hence, not Ulam stable either).*

Hence to show that Γ is not $U(1)$-stable, it is sufficient to show that the comparison map $c : \mathrm{H}^2_b(\Gamma, \mathbb{R}) \to H^2(\Gamma, \mathbb{R})$ is not injective. In [18], it is shown that for a lattice in a rank one semisimple Lie group, its comparison map has non-zero kernel, and hence.

Theorem 2.12 *Let* H *be a semisimple group (as in the introduction) of rank one, and let* Γ *be a lattice in* H*. Then* Γ *is not uniformly* $U(1)$*-stable.*

A partial converse to Corollary 2.11 is known as well, and is proved in [20].

Theorem 2.13 *Let* Γ *be a group such that* $\mathrm{H}^2_b(\Gamma, \mathbb{R})$ *is finite-dimensional and injects into* $H^2(\Gamma, \mathbb{R})$*. Then* Γ *is uniformly* $U(1)$*-stable.*

Combining this with the results of [8], the conclusion is:

Theorem 2.14 *If* Γ *is a higher rank lattice then it is uniformly* $U(1)$*-stable.*

2.2 *Defect Diminishing*

To precisely define the property of defect diminishing, it is helpful to work in the language of ultraproducts and non-standard analysis [21]. Let $\mathcal{U}$ be a non-principal ultrafilter on $\mathbb{N}$, which is fixed throughout. For an indexed collection $\{X_n\}_{n\in\mathbb{N}}$ of sets, let $\{X_n\}_{\mathcal{U}}$ denote its (algebraic) ultraproduct (which is $\prod_{\mathcal{U}} X_n := \prod_{n\in\mathbb{N}} X_n / \sim$

where for $\{x_n\}_{n\in\mathbb{N}}, \{y_n\}_{n\in\mathbb{N}} \in \prod_{n\in\mathbb{N}} X_n$, $\{x_n\}_{n\in\mathbb{N}} \sim \{y_n\}_{n\in\mathbb{N}}$ if $\{n : x_n = y_n\} \in \mathcal{U}$). If $X_n = X$ for every $n \in \mathbb{N}$, then $\prod_{\mathcal{U}} X$ is called the (algebraic) *ultrapower* of X, denoted $^*X$. Objects that arise as ultraproducts are said to be *internal* (while those that are not ultraproducts are called *external*).

For example, the ultrapower $^*\mathbb{R}$ of $\mathbb{R}$ is a non-archimedean ordered field called the *hyperreals*. It has two important subsets: the (external) set of **bounded** hyperreals, denoted $^*\mathbb{R}_b$, is the subset comprising elements $\{x_n\}_{\mathcal{U}} \in {^*\mathbb{R}}$ for which there exists $C \in \mathbb{R}_{\geq 0}$ such that $|x_n| \leq C$ for every $n \in \mathbb{N}$, and the (external) set of **infinitesimal** hyperreals, denoted $^*\mathbb{R}_{inf}$, is the subset comprising elements $\{x_n\}_{\mathcal{U}} \in {^*\mathbb{R}}$ such that for *every* real $\epsilon > 0$, there exists a large set $S \in \mathcal{U}$ such that $|x_n| < \epsilon$ for every $n \in S$. For $x, y \in {^*\mathbb{R}}$, denote by $x = O_{\mathcal{U}}(y)$ if $x/y \in {^*\mathbb{R}_b}$, and by $x = o_{\mathcal{U}}(y)$ if $x/y \in {^*\mathbb{R}_{inf}}$ (in particular,any bounded element $x \in {^*\mathbb{R}_b}$ is $x = O_{\mathcal{U}}(1)$ while any infinitesimal element $\epsilon \in {^*\mathbb{R}_{inf}}$ is $\epsilon = o_{\mathcal{U}}(1)$).

Returning to our setting, consider a sequence $\{\phi_n : \Gamma \to U(k_n)\}_{n\in\mathbb{N}}$. This can be used to construct an internal map $\{\phi_n\}_{\mathcal{U}} : {^*\Gamma} \to \prod_{\mathcal{U}} U(k_n)$, and allows us to redefine asymptotic homomorphisms and closeness to homomorphisms in the ultraproduct. Given two internal maps $\phi^{(1)} : {^*\Gamma} \to \prod_{\mathcal{U}} U(k_n)$ and $\phi^{(2)} : {^*\Gamma} \to \prod_{\mathcal{U}} U(k_n)$, let us denote by $dist(\phi^{(1)}, \phi^{(2)}) := \{dist(\phi_n^{(1)}, \phi_n^{(2)})\}_{\mathcal{U}}$. The maps $\phi^{(1)}$ and $\phi^{(2)}$ are said to be (internally) *asymptotically close* to each other if $dist(\phi^{(1)}, \phi^{(2)}) \in {^*\mathbb{R}_{inf}}$. An internal map $\psi : {^*\Gamma} \to \prod_{\mathcal{U}} G_n$ is said to be an *internal homomorphism* if there exists a sequence $\{\psi_n\}_{n\in\mathbb{N}}$ of homomorphisms such that $\psi = \{\psi_n\}_{\mathcal{U}}$.

- An internal map $\phi := \{\phi_n\}_{\mathcal{U}} : {^*\Gamma} \to \prod_{\mathcal{U}} U(k_n)$ is said to be (internally) **asymptotically close to an internal homomorphism** if there exists an internal homomorphism $\psi = \{\psi_n : \Gamma \to U(k_n)\}_{n\in\mathbb{N}}$ such that $dist(\phi, \psi) \in {^*\mathbb{R}_{inf}}$ (in other words, $\{D_{\Gamma,G_n}\}_{\mathcal{U}} \in {^*\mathbb{R}_{inf}}$).
- An internal map $\phi := \{\phi_n\}_{\mathcal{U}} : {^*\Gamma} \to \prod_{\mathcal{U}} U(k_n)$ is called an **internal asymptotic homomorphism** if $def(\phi) := \{def(\phi_n)\}_{\mathcal{U}} \in {^*\mathbb{R}_{inf}}$. For $\epsilon \in {^*\mathbb{R}_{inf}}$, we shall call an internal asymptotic homomorphism ϕ an internal ϵ-homomorphism if $def(\phi) = \epsilon$.
  (we similarly define internal $O_{\mathcal{U}}(\epsilon)$-homomorphisms).

The question of stability can be rephrased in this language as follows:

Lemma 2.15 *The group Γ is uniformly stable iff every internal asymptotic homomorphism $\phi : {^*\Gamma} \to \prod_{\mathcal{U}} U(k_n)$ for $G_n \in \mathcal{G}$ is asymptotically close to an internal homomorphism. The group Γ is $\mathcal{G}$-uniformly stable with a linear estimate iff for every internal asymptotic homomorphism $\phi : {^*\Gamma} \to \prod_{\mathcal{U}} U(k_n)$, there exists an internal homomorphism $\psi : {^*\Gamma} \to \prod_{\mathcal{U}} U(k_n)$ with $dist(\phi, \psi) = O_{\mathcal{U}}(def(\phi))$.*

One can also describe the question of stability as a homomorphism lifting problem. Let $\phi : {^*\Gamma} \to \prod_{\mathcal{U}} U(k_n)$ be an internal asymptotic homomorphism with defect $\epsilon \in {^*\mathbb{R}_{inf}}$. Consider the normal subgroup $B(\epsilon) := \{\{g_n\}_{\mathcal{U}} \in \prod_{\mathcal{U}} U(k_n) : \{\|g_n - I\|\}_{\mathcal{U}} = O_{\mathcal{U}}(\epsilon)\}$ (elements *bounded* by ϵ) of $\prod_{\mathcal{U}} U(k_n)$ and the quotient homomorphism $q_{B(\epsilon)} : \prod_{\mathcal{U}} U(k_n) \to \prod_{\mathcal{U}} U(k_n)/B(\epsilon)$ so that the composition $q_{B(\epsilon)} \cdot \phi : {^*\Gamma} \to \left(\prod_{\mathcal{U}} U(k_n)\right)/B(\epsilon)$ (which we shall henceforth denote $\tilde{\phi}$) is a homomorphism.

Lemma 2.16 *The group* Γ *is uniformly* $\mathcal{G}$*-stable with a linear estimate iff for every* $\epsilon \in {}^*\mathbb{R}_{inf}$*, every homomorphism* $\tilde{\phi} : {}^*\Gamma \to \prod_{\mathcal{U}} G_n / B(\epsilon)$ *that has an internal lift also has an internal lift homomorphism.*

Definition 2.17 The group Γ is said to have the **defect diminishing** property if for any internal asymptotic homomorphism $\phi : {}^*\Gamma \to \prod_{\mathcal{U}} U(k_n)$, there exists an asymptotic homomorphism $\psi : {}^*\Gamma \to \prod_{\mathcal{U}} U(k_n)$ such that

- The defect $def(\psi) = o_{\mathcal{U}}(def(\phi))$.
- The distance $dist(\phi, \psi) = O_{\mathcal{U}}(def(\phi))$.

While defect diminishing seems like a weaker requirement than stability, it is easy to show, using iteration and limits, that it implies stability.

Theorem 2.18 *The group* Γ *is uniformly stable with a linear estimate iff* Γ *has the defect diminishing property with respect to* $\mathcal{G}$.

However, the idea behind defining this notion is that reformulating stability in terms of defect diminishing allows us to reformulate stability as a homomorphism lifting problem with abelian kernel, further leading to a cohomological theory.
For $\epsilon \in {}^*\mathbb{R}_{inf}$, consider the subgroup $I(\epsilon) := \{\{g_n\}_{\mathcal{U}} \in \prod_{\mathcal{U}} U(k_n) : \{\|g_n - I\|\}_{\mathcal{U}} = o_{\mathcal{U}}(\epsilon)\}$ (note that $I(\epsilon)$ is normal in both $\prod_{\mathcal{U}} U(k_n)$ and $B(\epsilon)$), and let $q_{I(\epsilon)} : \prod_{\mathcal{U}} G_n \to \prod_{\mathcal{U}} G_n / I(\epsilon)$ be the canonical quotient homomorphism. Then, just like in Lemma 2.16, defect diminishing too can be reformulated as a homomorphism lifting problem.

Lemma 2.19 *The group* Γ *has the defect diminishing property with respect to* $\mathcal{G}$ *iff for every* $\epsilon \in {}^*\mathbb{R}_{inf}$ *and every homomorphism* $F : {}^*\Gamma \to \prod_{\mathcal{U}} U(k_n)/B(\epsilon)$ *that has an internal lift,* F *has an internal lift* $\hat{F} : {}^*\Gamma \to \prod_{\mathcal{U}} U(k_n)$ *such that* $q_{I(\epsilon)} \cdot \hat{F} : {}^*\Gamma \to \prod_{\mathcal{U}} U(k_n)/I(\epsilon)$ *is a homomorphism.*

The advantage here is that the submultiplicativity of the norm implies that:

Lemma 2.20 *The group* $B(\epsilon)/I(\epsilon)$ *is abelian.*

One can associate with ϕ an internal map $\rho_\phi : {}^*\Gamma \times \prod_{\mathcal{U}} U(k_n) \to \prod_{\mathcal{U}} U(k_n)$ given by

$$\rho_\phi(g)(u) := \phi(g) \cdot u \cdot \phi(g)^{-1} \tag{1}$$

For every $g \in {}^*\Gamma$ and $u \in B(\epsilon)$, $\rho_\phi(g)(u) \in B(\epsilon)$, while for $u \in I(\epsilon)$, $\rho_\phi(g)(u) \in I(\epsilon)$. Then

Lemma 2.21 *The internal map* $\rho_\phi : {}^*\Gamma \times \prod_{\mathcal{U}} U(k_n) \to \prod_{\mathcal{U}} U(k_n)$ *induces an action, denoted* $\tilde{\rho}_\phi$*, of* ${}^*\Gamma$ *on the abelian group* $B(\epsilon)/I(\epsilon)$.

The fact that $B(\epsilon)/I(\epsilon)$ is an abelian group with an action of ${}^*\Gamma$ will allow us to rephrase the lifting property of Lemma 2.19 in terms of the vanishing of a cohomology which is described next.

Remark 2.22 It is to be noted that the use of ultraproducts in this formulation is merely to capture the notion of defect diminishing in a succinct fashion and allow for a cohomological theory to be built in a traditional way. No non-trivial model-theoretic properties of ultrafilters are required, and the entire theory can be rebuilt redeveloped from scratch without the language of non-standard analysis too.

2.3 *Asymptotic Cohomology*

We shall now construct an asymptotic cocycle α_ϕ that takes values in $B(\epsilon)$ (and becomes a (true) cocycle in the quotient $B(\epsilon)/I(\epsilon)$).
Define the internal map $\alpha_\phi : {}^*\Gamma \times {}^*\Gamma \to \prod_{\mathcal{U}} U(k_n)$

$$\alpha_\phi(g_1, g_2) := \psi(g_1)\phi(g_2)\psi(g_1g_2)^{-1}$$

Note that by construction α_ϕ takes values in $B(\epsilon)$. Let $q_{B/I} : B(\epsilon) \to B(\epsilon)/I(\epsilon)$ be the canonical quotient homomorphism. Since α_ϕ takes values in $B(\epsilon)$, let $\tilde{\alpha}_\phi := q_{B/I} \cdot \alpha_\phi : {}^*\Gamma \times {}^*\Gamma \to B(\epsilon)/I(\epsilon)$. Then it is easy to check that $\tilde{\alpha}_\phi$ satisfies the following 2-cocycle condition:

Corollary 2.23 *For* $g_1, g_2, g_3 \in {}^*\Gamma$,

$$\tilde{\rho}_\phi(g_1) \cdot \tilde{\alpha}_\psi(g_2, g_3) - \tilde{\alpha}_\psi(g_1g_2, g_3) + \tilde{\alpha}_\psi(g_1, g_2g_3) - \tilde{\alpha}_\psi(g_1, g_2) = 0$$

Now suppose there exists an internal map $\beta : {}^*\Gamma \to \prod_{\mathcal{U}} G_n$ that takes values in $B(\epsilon)$ such that the following equation holds for all $g_1, g_2 \in {}^*\Gamma$:

$$\tilde{\alpha}_\phi(g_1, g_2) = \tilde{\beta}(g_1) + \rho_\phi(g_1) \cdot \tilde{\beta}(g_2) - \tilde{\beta}(g_1g_2)$$

Then the internal map $\psi : {}^*\Gamma \to \prod_{\mathcal{U}} U(k_n)$

$$\psi(g) := \phi(g)\beta(g)^{-1}$$

is also an internal lift of $\tilde{\phi}$ such that for every $g_1, g_2 \in {}^*\Gamma$, $\tilde{\alpha}_\psi(g_1, g_2) = 0$. In particular, this means that $\psi^\sim$ is the internal lift that we want.
The above discussion hints at a cohomological theory that captures the obstruction to such lifts. The idea is as follows: The map ϕ, with defect $O_{\mathcal{U}}(\epsilon)$, gives us a type of 2-cocycle of ${}^*\Gamma$ with coefficients in $B(\epsilon)/I(\epsilon)$, and if that cocycle happens to be a 1-couboundary, then ϕ can be corrected to obtain another lift that has defect $o_{\mathcal{U}}(\epsilon)$ and is still $O_{\mathcal{U}}(\epsilon)$-close to ϕ, thus implying the defect diminishing property.

To formally construct our cohomology, one first passes to the lie algebras so as to obtain asymptotic cocycles that take values in *Banach spaces* (rather than in a group). Let $\mathfrak{u}(k_n)$ be the Lie algebra of $U(k_n)$, and denote by $\mathcal{W}$ the ultraproduct $\prod_{\mathcal{U}} \mathfrak{u}(k_n)$. Note that $\mathcal{W}$ comes with a ${}^*\mathbb{R}$-valued norm, which we shall denote simply as $\|\cdot\|$, obtained as the ultraproduct of the respective norms of each $\mathrm{u}(k_n)$. The bounded elements of $\mathcal{W}$ shall be denoted $\mathcal{W}_b$ while the infinitesimal elements of $\mathcal{W}$ are denoted by $\mathcal{W}_{inf}$. The ultralimit $\mathcal{W}_b/\mathcal{W}_{inf}$ shall be denoted, as usual, by $\tilde{\mathcal{W}}$.

Define the logarithm ${}_\epsilon\log : \prod_{\mathcal{U}} U(k_n) \to \mathcal{W}$ and exponential map ${}_\epsilon\exp : \mathcal{W} \to \prod_{\mathcal{U}} U(k_n)$ (both scaled with respect to ϵ) as

$$ {}_\epsilon\log u := \frac{1}{\epsilon}\{\log u_{k_n}\}_{\mathcal{U}} $$

$$ {}_\epsilon\exp u := \{\exp \epsilon_{k_n} u_{k_n}\}_{\mathcal{U}} $$

Observe that the internal map ${}_\epsilon\log : \prod_{\mathcal{U}} U(k_n) \to \mathcal{W}$ when restricted to $B(\epsilon)$, takes values in $\mathcal{W}_b$, and elements in $I(\epsilon)$ are taken to $\mathcal{W}_{inf}$. This induces an isomorphism of the abelian groups $\tilde{\mathcal{W}}$ (which is also a real Banach space) and $B(\epsilon)/I(\epsilon)$.

Just like in Eq. 1, ${}^*\Gamma$ has an asymptotic action $\pi_\phi : {}^*\Gamma \times \mathcal{W} \to \mathcal{W}$ on $\mathcal{W}_b$ that induces an action on $\tilde{\mathcal{W}}$ given by

$$ \pi_\phi v := \phi(g) v \phi(g)^{-1} \tag{2} $$

Corresponding to the map $\alpha_\phi : {}^*\Gamma \times {}^*\Gamma \to \prod_{\mathcal{U}} U(k_n)$, define the internal map

$$ \alpha : {}^*\Gamma \times {}^*\Gamma \to \mathcal{W} $$

$$ \alpha(g_1, g_2) := {}_\epsilon\log \alpha_\phi(g_1, g_2) \tag{3} $$

It is clear that $\alpha : {}^*\Gamma \times {}^*\Gamma \to \mathcal{W}$ takes values only in $\mathcal{W}_b$. We shall denote by $\tilde{\alpha} : {}^*\Gamma \times {}^*\Gamma \to \tilde{\mathcal{W}}$ the map induced obtained by composing α with the canonical quotient map $\mathcal{W}_b \to \tilde{\mathcal{W}}$.

Lemma 2.24 *The map $\alpha : {}^*\Gamma \times {}^*\Gamma \to \mathcal{W}$ satisfies the following condition: for any $g_1, g_2, g_3 \in {}^*\Gamma$,*

$$ \pi_\phi(g_1)\alpha(g_2 h g_3) - \alpha(g_1 g_2, g_3) + \alpha(g_1, g_2 g_3) - \alpha(g_1, g_2) \in \mathcal{W}_{inf} $$

So transfering to the internal Lie algebra through the ${}_\epsilon\log$ map, we thus have a map $\alpha : {}^*\Gamma \times {}^*\Gamma \to \mathcal{W}$ which takes values in $\mathcal{W}_b$ and satisfies the 2-cocycle condition modulo $\mathcal{W}_{inf}$. Putting together what we discussed so far,

Proposition 2.25 *Suppose for every internal $\alpha : {}^*\Gamma \times {}^*\Gamma \to \mathcal{W}$ with $Im(\alpha) \subseteq \mathcal{W}_b$ that satisfies the 2-cocycle condition, there exists an internal $\beta : {}^*\Gamma \to \mathcal{W}$ taking values in $\mathcal{W}_b$ such that*

$$ \tilde{\alpha}(g_1, g_2) = \tilde{\beta}(g_1) + g_1 \cdot \tilde{\beta}(g_2) - \tilde{\beta}(g_1 g_2) $$

then Γ exhibits the defect diminishing property, and is therefore uniformly stable with a linear estimate.

Since we shall work with internal maps from $({}^*\Gamma)^2$ (or ${}^*\Gamma$) to $\mathcal{W}$ that take values in $\mathcal{W}_b$, it is helpful to describe such a map as an ultraproduct of bounded maps. An internal map $\alpha : {}^*\Gamma \times {}^*\Gamma \to \mathcal{W}$ with $Im(\alpha) \subseteq \mathcal{W}_b$ can be described as $\{\alpha_n\}_{\mathcal{U}}$ where for every $k \in \mathbb{N}$, $\alpha_n : \Gamma \times \Gamma \to \mathfrak{u}(k_n)$ is a bounded map, and such that $\{\|\alpha_n\|_\infty\}_{\mathcal{U}} \in {}^*\mathbb{R}_b$. In other words, the internal map α is the ultraproduct of bounded maps, and is also bounded as an ultraproduct.
From now on, we shall regard α as an element of $\prod_{\mathcal{U}} \ell^\infty(({}^*\Gamma)^2, \mathfrak{u}(k_n))$, which we shall henceforth denote $\mathcal{L}^\infty(({}^*\Gamma)^2, \mathcal{W})$ (understood as the space of internal maps from $({}^*\Gamma)^2$ to $\mathcal{W}$. In fact, α is actually an element of $\mathcal{L}^\infty(({}^*\Gamma)^2, \mathcal{W})_b$ since not only is it internally bounded, but also $\{\|\alpha_n\|_\infty\}_{\mathcal{U}} \in {}^*\mathbb{R}_b$. In general, we shall use the following notation:

- For $m \in \mathbb{N}$, the internal space $\mathcal{L}^\infty(({}^*\Gamma)^m, \mathcal{W})$ is defined as

$$\mathcal{L}^\infty(({}^*\Gamma)^m, \mathcal{W}) := \left\{\ell^\infty(\Gamma^m, \mathfrak{u}(k_n))\right\}_{\mathcal{U}}$$

- The (external) subspace of $\mathcal{L}^\infty(({}^*\Gamma)^m, \mathcal{W})$ comprising internal functions with bounded (supremum) norm will be denoted $\mathcal{L}_b^\infty(({}^*\Gamma)^m, \mathcal{W})$ while the (external) subspace of $\mathcal{L}_b^\infty(({}^*\Gamma)^m, \mathcal{W})$ comprising internal functions with infinitesimal (supremum) norm will be denoted $\mathcal{L}_{inf}^\infty(({}^*\Gamma)^m, \mathcal{W})$.
- The quotient $\mathcal{L}_b^\infty(({}^*\Gamma)^m, \mathcal{W})/\mathcal{L}_{inf}^\infty(({}^*\Gamma)^m, \mathcal{W})$ shall be denoted $\tilde{\mathcal{L}}^\infty(({}^*\Gamma)^m, \mathcal{W})$. This space comprises bounded maps from $({}^*\Gamma)^m$ to $\tilde{\mathcal{W}}$ that are induced from internal maps in $\mathcal{L}^\infty(({}^*\Gamma)^m, \mathcal{W})$. Note that $\tilde{\mathcal{L}}^\infty(({}^*\Gamma)^m, \mathcal{W})$ is a real Banach space.
- For a map $f \in \mathcal{L}_b^\infty(({}^*\Gamma)^m, \mathcal{W})$, we shall denote by $\tilde{f} \in \tilde{\mathcal{L}}^\infty(({}^*\Gamma)^m, \mathcal{W})$ the composition of f with the canonical quotient map $\mathcal{L}_b^\infty(({}^*\Gamma)^m, \mathcal{W}) \to \tilde{\mathcal{L}}^\infty(({}^*\Gamma)^m, \mathcal{W})$.

For convenience, we restate Theorem 2.25 in this notation:

Proposition 2.26 *Suppose for every $\alpha \in \mathcal{L}_b^\infty(({}^*\Gamma)^2, \mathcal{W})$ that satisfies the 2 cocycle condition Lemma 2.24, there exists a $\beta \in \mathcal{L}_b^\infty(({}^*\Gamma)^1, \mathcal{W})$ such that*

$$\tilde{\alpha}(g_1, g_2) = \tilde{\beta}(g_1) + g_1 \cdot \tilde{\beta}(g_2) - \tilde{\beta}(g_1 g_2)$$

Then Γ exhibits the defect diminishing property (and therefore uniformly stable with a linear estimate).

We conclude this lecture with a formal definition of asymptotic cohomology for locally compact groups.

Let $\mathcal{V} = \prod_{\mathcal{U}} V_n$ be an internal Banach space, with $\tilde{\mathcal{V}}$. It is said to be an *internal dual* Banach space if $\mathcal{V} = \prod_{\mathcal{U}} (V_n^\flat)^\sharp$ (that is, each V_n is the dual of a Banach space $V_n^\flat$).
Let $\pi : {}^*G \times \mathcal{V} \to \mathcal{V}$ be an internal isometry such that it induces an action $\tilde{\pi} : {}^*G \times \tilde{\mathcal{V}} \to \tilde{\mathcal{V}}$ of ${}^*G$ on the real Banach space $\tilde{\mathcal{V}}$. The internal Banach space $\mathcal{V}$, equipped with such an internal map π, is called an *asymptotic Banach ${}^*G$-module* with asymptotic ${}^*G$-representation π.

Consider the internal Banach space

$$\mathcal{L}^{\infty}\left(({}^*G)^m, \mathcal{V}\right) := \prod_{\mathcal{U}} L^{\infty}_{w*}(G^m, V_n)$$

For $f = \{f_n\}_{\mathcal{U}} \in \mathcal{L}^{\infty}\left(({}^*G)^m, \mathcal{V}\right)$, we denote by $\|f\|$ the hyperreal $\{\|f_n\|\}_{\mathcal{U}} \in {}^*\mathbb{R}$ and by $Im(f)$ the subset $\{Im(f_n)\}_{\mathcal{U}} \subseteq \mathcal{V}$ (here $Im(f_n)$ is the *essential image* of the weak-* measurable function f_n).
The internal map $\pi : {}^*G \times \mathcal{V} \to \mathcal{V}$ can be used to define the internal map $\tau^m : {}^*G \times \mathcal{L}^{\infty}(({}^*G)^m, \mathcal{V}) \to \mathcal{L}^{\infty}(({}^*G)^m, \mathcal{V})$ for each $m \geq 0$ defined as

$$(\tau^m(g)(f))(g_1, g_2, \ldots, g_m) := \pi(g) f(g^{-1}g_1, \ldots, g^{-1}g_m) \tag{4}$$

for every $g \in {}^*G$ and every $g_1, \ldots, g_m \in {}^*G$. This internal map $\tau^m : {}^*G \times \mathcal{L}^{\infty}(({}^*G)^m, \mathcal{V}) \to \mathcal{L}^{\infty}(({}^*G)^m, \mathcal{V})$ induces an action $\tilde{\tau}^m$ of ${}^*G$ on the ultralimit space $\tilde{\mathcal{L}}^{\infty}(({}^*G)^m, \mathcal{V})$. In particular, this means that $(\tau^m, \mathcal{L}^{\infty}(({}^*G)^m, \mathcal{V}))$ is an asymptotic Banach ${}^*G$-module. For simplicity, we shall denote the induced action of $g \in {}^*G$ on $\tilde{f} \in \tilde{\mathcal{L}}^{\infty}(({}^*G)^m, \mathcal{V})$ simply by $g \cdot \tilde{f}$. We shall denote by $\mathcal{L}^{\infty}(({}^*G)^m, \mathcal{V})^{\sim {}^*G}$ the set of elements $f \in \mathcal{L}^{\infty}(({}^*G)^m, \mathcal{V})$ such that $\tilde{f} \in \tilde{\mathcal{L}}^{\infty}(({}^*G)^m, \mathcal{V})^{{}^*G}$. Such an element f shall be referred to as *asymptotically* ${}^*G$*-equivariant*.

For each $m \geq 0$, define the internal map $d^m : \mathcal{L}^{\infty}(({}^*G)^m, \mathcal{V}) \to \mathcal{L}^{\infty}(({}^*G)^{m+1}, \mathcal{V})$

$$d^m f(g_0, \ldots, g_m) := \sum_{j=0}^{m} (-1)^j f(g_0, \ldots, \hat{g}_j, \ldots, g_m)$$

It is clear that $d^m \circ d^{m-1} = 0$ for every $m \geq 1$. Note that d^m induces a map

$$\tilde{d}^m : \tilde{\mathcal{L}}^{\infty}(({}^*G)^m, \mathcal{V}) \to \tilde{\mathcal{L}}^{\infty}(({}^*G)^{m+1}, \mathcal{V})$$

Furthermore, for $f \in \mathcal{L}^{\infty}(({}^*G)^m, \mathcal{V})^{\sim {}^*G}$, $d^m f \in \mathcal{L}^{\infty}(({}^*G)^{m+1}, \mathcal{V})^{\sim {}^*G}$. That is, $d^{\bullet}$ is an asymptotic ${}^*G$-morphism of asymptotic Banach ${}^*G$-modules.

Since the maps $\tilde{d}^m : \tilde{\mathcal{L}}^{\infty}(({}^*G)^m, \mathcal{V}) \to \tilde{\mathcal{L}}^{\infty}(({}^*G)^{m+1}, \mathcal{V})$ are ${}^*G$-equivariant for every $m \geq 1$, with $\tilde{d}^m \circ \tilde{d}^{m-1} = 0$, we have an asymptotic ${}^*G$-cochain complex

$$0 \xrightarrow{d^0} \mathcal{L}^{\infty}({}^*G, \mathcal{V}) \xrightarrow{d^1} \mathcal{L}^{\infty}(({}^*G)^2, \mathcal{V}) \xrightarrow{d^2} \mathcal{L}^{\infty}(({}^*G)^3, \mathcal{V}) \xrightarrow{d^3} \mathcal{L}^{\infty}(({}^*G)^4, \mathcal{V}) \xrightarrow{d^4} \ldots$$

and the induced ${}^*G$-cochain complex

$$0 \xrightarrow{\tilde{d}^0} \tilde{\mathcal{L}}^{\infty}({}^*G, \mathcal{V})^{{}^*G} \xrightarrow{\tilde{d}^1} \tilde{\mathcal{L}}^{\infty}(({}^*G)^2, \mathcal{V})^{{}^*G} \xrightarrow{\tilde{d}^2} \tilde{\mathcal{L}}^{\infty}(({}^*G)^3, \mathcal{V})^{{}^*G} \xrightarrow{\tilde{d}^3} \ldots$$

Definition 2.27 The **asymptotic cohomology group of G with coefficients in a dual asymptotic Banach $^*G$-module** $\mathcal{V}$, denoted $\mathrm{H}_a^\bullet(G, \mathcal{V})$, is defined to be the cohomology of the $^*G$-cochain complex

$$0 \xrightarrow{\tilde{d}^0} \tilde{\mathcal{L}}^\infty(^*G, \mathcal{V})^{*G} \xrightarrow{\tilde{d}^1} \tilde{\mathcal{L}}^\infty((^*G)^2, \mathcal{V})^{*G} \xrightarrow{\tilde{d}^2} \tilde{\mathcal{L}}^\infty((^*G)^3, \mathcal{V})^{*G} \xrightarrow{\tilde{d}^3} \dots$$

For instance, in the case of a discrete group Γ and the coefficients being the asymptotic Banach $^*\Gamma$-module $(\pi_\Gamma, \mathcal{W})$ where $\mathcal{W} = \prod_{\mathcal{U}} \mathfrak{u}_{k_n}$ and $\pi_\Gamma = \pi_\phi$, we will be interested in $\mathrm{H}_a^2(\Gamma, \mathcal{W})$ and the following can be used in lieu of Theorem 2.26:

Theorem 2.28 *Suppose* $\mathrm{H}_a^2(\Gamma, \mathcal{W}) = 0$. *Then* Γ *has the defect diminishing property (and is uniformly stable).*

Remark 2.29 Asymptotic cohomology shares some features with the more classical theory of *bounded cohomology* [26]. In a way, H_a can be thought of as a formulation of bounded cohomology, *with errors* designed to capture defect *diminishing* (as opposed to a single step of defect elimination as in the case of $U(1)$-stability).

2.4 Stability of Lamplighters and Thompson Groups

We begin with some general results about the acyclicity of amenable groups, transfer properties of asymptotic cohomology, and conclude with another application of our machinery in proving that lamplighters and Thompson groups are uniformly stable. Throughout our study henceforth, $\mathcal{V}$ is a dual asymptotic Banach $^*G$-module.

Definition 2.30 Let S be a regular G-space. A **conditional expectation** $\mathfrak{m} : L^\infty(G \times S) \to L^\infty(S)$ is a measurable norm one linear map such that

- $\mathfrak{m}(1_{G\times S}) = 1_S$
- For every $f \in L^\infty(G \times S)$ and every measurable set $A \subset S$, $\mathfrak{m}(f \cdot 1_{G\times A}) = \mathfrak{m}(f) \cdot 1_A$.

The G-action on S is said to be **Zimmer amenable** if there exists a G-equivariant conditional expectation $\mathfrak{m} : L^\infty(G \times S) \to L^\infty(S)$.

The significance of Zimmer amenability is that it allows us to compute $H_a^\bullet(G, \mathcal{V})$ using other cochain complexes, just as in bounded cohomology:

Theorem 2.31 *Let S be a regular G-space with a Zimmer-amenable action of G. Then* $\mathrm{H}_a^\bullet(G, \mathcal{V})$ *can be computed as the asymptotic cohomology of the asymptotic $^*G$-cochain complex*

$$0 \xrightarrow{d^0} \mathcal{L}^\infty(^*S, \mathcal{V}) \xrightarrow{d^1} \mathcal{L}^\infty((^*S)^2, \mathcal{V}) \xrightarrow{d^2} \mathcal{L}^\infty((^*S)^3, \mathcal{V}) \xrightarrow{d^3} \dots$$

We can now use the above theorem to obtain an asymptotic analogue of the classical result that amenable groups have vanishing bounded cohomology, and follows from the fact that if G is amenable, then the trivial space is Zimmer-amenable for G.

Corollary 2.32 *Let G be an amenable group and $\mathcal{V}$ be a dual asymptotic Banach $^*G$-module. Then for every $n \geq 1$, $\mathrm{H}^n_a(G, \mathcal{V}) = 0$.*

In particular, for a discrete amenable group Γ, $H^2_a(\Gamma, \mathcal{W}) = 0$, implying Kazhdan's result that amenable groups are uniformly stable.

Another useful transfer property between uniform stability of groups and their normal subgroups is obtained in [16]:

Theorem 2.33 ([16]) *Let $\Lambda \leq \Gamma$ be co-amenable, then if Λ is Ulam stable with a linear estimate, then so is Γ. Let $N \leq \Gamma$ be an amenable normal subgroup, then if Γ is Ulam stable with a linear estimate, then so is Γ/N.* ■

All these results can be used to show stability results for lamplighters $\Gamma \wr \Lambda$ where Λ is infinite and amenable, as well as several groups of dynamical origin such as the classical Thompson groups F, F', T and V.

Theorem 2.34 ([16]) *Let Γ, Λ be two countable groups, where Λ is infinite and amenable. Then $\Gamma \wr \Lambda$ is Ulam stable with a linear estimate. In particular, the Thompson's groups F, F', T and V are Ulam stable with a linear estimate.* ■

These results can be seen in light of the corresponding results on vanishing of bounded cohomology for such groups [29], and further illustrates connections between bounded cohomology and Ulam stability that deserves more scrutiny. In [16], the related notion of *approximability* of such groups, and their connections to stability is also studied.

3 Lecture 3: Vanishing of Second Cohomology for Higher-Rank Lattices

The main result of interest is in showing that higher-rank lattices are uniformly stable. This involves a sequence of steps that is described in the subsequent sections.

To begin with, consider the fact that for a lattice Γ in a locally compact group G, G is Zimmer-amenable as a regular Γ-space. Using Theorem 2.31, this serves as the starting point for an induction procedure to go from Γ to G which we shall soon describe.

Corollary 3.35 *Let $\Gamma \leq G$ be a lattice in a locally compact group G, and let $\mathcal{W}$ be a dual asymptotic Banach $^*\Gamma$-module. Then $\mathrm{H}^\bullet_a(\Gamma, \mathcal{W})$ can be computed as the asymptotic cohomology of the asymptotic $^*\Gamma$-cochain complex*

$$0 \xrightarrow{d^0} \mathcal{L}^\infty({}^*G, \mathcal{W}) \xrightarrow{d^1} \mathcal{L}^\infty(({}^*G)^2, \mathcal{W}) \xrightarrow{d^2} \mathcal{L}^\infty(({}^*G)^3, \mathcal{W}) \xrightarrow{d^3} \dots$$

The next corollary uses the fact that for a closed subgroup $Q \leq G$ (and in particular, when $Q = G$), and a closed amenable subgroup $P \leq G$, the space G/P is a regular Q-space that is Zimmer-amenable for the Q-action.

Corollary 3.36 *Let $P \leq G$ be a closed amenable subgroup of G, Q be a closed subgroup of G, and $\mathcal{V}$ be a dual asymptotic Banach ${}^*G$-module. Then $\mathrm{H}_a^\bullet(Q, \mathcal{V})$ can be computed as the asymptotic cohomology of the asymptotic ${}^*Q$-cochain complex*

$$0 \xrightarrow{d^0} \mathcal{L}^\infty(({}^*(G/P)), \mathcal{V}) \xrightarrow{d^1} \mathcal{L}^\infty(({}^*(G/P))^2, \mathcal{V}) \xrightarrow{d^2} \dots$$

The above corollary uses the fact that for a closed subgroup $Q \leq G$, the space G/P is a regular Q-space that is Zimmer-amenable for the Q-action.

From now on, we shall work with Γ being a lattice in a semisimple group G of the form $G = \prod_{i=1}^k \mathbf{G}_i(K_i)$ where for $1 \leq i \leq k$, K_i is a local field, and $\mathbf{G}_i$ is a connected, simply connected, almost K_i-simple group. Our goal is to use the results developed in the previous sections to prove that $\mathrm{H}_a^2(\Gamma, \mathcal{W}) = 0$ where $\mathcal{W} = \prod_{\mathcal{U}} \mathfrak{u}(k_n)$ with an asymptotic action π_Γ of ${}^*\Gamma$. In the next part of the lecture, we shall briefly sketch the main ingredients of the proof.

3.1 Step 1: Eckmann-Shapiro Induction

We shall now apply an induction procedure to construct a dual Banach ${}^*G$-module $\mathcal{V}$ such that $\mathrm{H}_a^\bullet(\Gamma, \mathcal{W}) = \mathrm{H}_a^\bullet(G, \mathcal{V})$. The space $\mathcal{V}$ shall be $\mathcal{L}^\infty({}^*D, \mathcal{W})$ where D is a Borelian left fundamental domain of Γ in G.
We shall next define an internal map

$$\pi_G : {}^*G \times \mathcal{L}^\infty({}^*D, \mathcal{W}) \to \mathcal{L}^\infty({}^*D, \mathcal{W})$$

$$\pi_G(g)(f)(z) := \pi_\Gamma(\gamma) f(x) \tag{5}$$

where $zg = \gamma x$ for $\gamma \in {}^*\Gamma$ and $x \in {}^*D$. Note that ${}^*G \times {}^*D \to {}^*D$ given by $(g, z) \mapsto x$, where $zg = \gamma x$, defines an internal right action of ${}^*G$ on ${}^*D$. So the above map π_G in Eq. 5 can be denoted as

$$\pi_G(g)(f)(z) := \pi_\Gamma(\gamma) f(zg)$$

This map π_G induces an action of ${}^*G$ on $\tilde{\mathcal{L}}^\infty({}^*D, \mathcal{W})$, which we shall denote $\tilde{\pi_G}$. In particular, note that this gives the internal Banach space $\mathcal{L}^\infty({}^*D, \mathcal{W})$ the structure of an asymptotic Banach ${}^*G$-module, with the asymptotic ${}^*G$-representation being π_G as defined above. It is also a dual asymptotic Banach ${}^*G$-module, and we have

Theorem 3.37 *For every $m \geq 0$, $\mathrm{H}_a^m(G, \mathcal{V}) \cong \mathrm{H}_a^m(\Gamma, \mathcal{W})$.*

It is often also advantageous to work with the (external) set $\mathcal{L}_b^\infty({}^*G, \mathcal{W})^{\sim {}^*\Gamma}$ which, upto infinitesimals, can be identified with the internal Banach space $\mathcal{V} := \mathcal{L}^\infty({}^*D, \mathcal{W})$.

3.2 Step 2: Asymptotic Mautner Property and Double Ergodicity

In this step, we use the property of bounded generation combined with amenability to obtain asymptotic versions of classical results on fixed points and double ergodicity.

Definition 3.38 Let $T \leq G$ be a closed subgroup and $M > 0$. The group G is said to be *M-boundedly generated by T-contracted elements* if any element $g \in G$, there exist $s_1, \ldots, s_m \in G$ with $m \leq M$, such that $g = s_1 s_2 \ldots s_m$, and each $s_i \in G$ is contracted by some element of T. More generally, for a family $\mathfrak{T}$ of closed subgroups of G, the group G is said to be **boundedly generated by $\mathfrak{T}$-contracted elements** if there exists $M > 0$ such that G is M-boundedly generated by T-contracted elements for every $T \in \mathfrak{T}$.

Remark 3.39 Suppose G is boundedly generated by $\mathfrak{T}$-contracted elements for a family $\mathfrak{T}$ of subgroups. Let $\mathcal{T} = \prod_{\mathcal{U}} T_n$ be an internal subgroup of ${}^*G$, with $T_n \in \mathfrak{T}$ for every $n \in \mathbb{N}$. Observe that there exists $M > 0$ such that any $g \in {}^*G$ can be expressed as $g = s_1 s_2 \ldots s_m$ such that each $s_i \in {}^*G$ is internally contracted by some element of $\mathcal{T}$.

Lemma 3.40 *Suppose G is boundedly generated by $\mathfrak{T}$-contracted elements, and let $\mathcal{T} = \prod_{\mathcal{U}} T_n$ be an internal subgroup of ${}^*G$, with $T_n \in \mathfrak{T}$ for every $n \in \mathbb{N}$. Then if $f \in \mathcal{L}_b^\infty({}^*D, \mathcal{W})^{\sim \mathcal{T}}$, then $f \in \mathcal{L}_b^\infty({}^*D, \mathcal{W})^{\sim {}^*G}$ (that is, if f is asymptotically fixed by all elements of $\mathcal{T}$, then it is asymptotically ${}^*G$-fixed).*

Lemma 3.41 *Let G be a semisimple group of the form $G = \prod_{i=1}^k \mathbf{G}_i(K_i)$ where for $1 \leq i \leq k$, K_i is a local field, and $\mathbf{G}_i$ is a connected, simply connected, almost K_i-simple K_i-isotropic group, and let $\mathfrak{T}$ be the family of maximal tori of the radicals of parabolic subgroups of G of the form $Q = \prod_{i=1}^k Q_i$, where each Q_i is a proper parabolic subgroup of $\mathbf{G}_i(K_i)$. Then G is boundedly generated by $\mathfrak{T}$-contracted elements.*

Combining these, we get the asymptotic version of Mautner's lemma that we will use:

Theorem 3.42 (Asymptotic Mautner Property) *Let G and a parabolic subgroup $Q \leq G$ be as in Lemma 3.41, and let $N \leq G$ be the radical of Q. Then $\mathcal{V}_b^{\sim {}^*N} = \mathcal{V}_b^{\sim {}^*G}$.*

We will also need an internal analogue of ergodicity (and double ergodicity) with coefficients (as in [26]). We recall the following classical definitions for comparison:

Definition 3.43 For a locally compact second countable group G and a regular G-space S, the G-action on S is said to be *ergodic* if any measurable G-invariant function $f : S \to \mathbb{R}$ is essentially constant. The G-action on S is said to be *doubly ergodic* if the diagonal G-action on $S \times S$ is ergodic.

There is no difference if $\mathbb{R}$ above is replaced by any dual separable Banach space: the G-action on S is ergodic (resp., double ergodic) iff for any dual separable Banach space E, any weak-* measurable G-*invariant* map $f : S \to E$ (resp, $f : S \times S \to E$) is essentially constant.
Such a definition can be strenghtened in the case of E that comes with an action of G. A G-action on a regular G-space S is said to be *ergodic with coefficients in a Banach G-module E* if any weak-$*$ measurable G-*equivariant* map $f : S \to E$ is essentially constant. The action is called *doubly ergodic with coefficients in E* if the corresponding condition holds for the diagonal G-action on $S \times S$.

In our setting, the goal is to show that for $f \in \mathcal{L}_b^\infty\left({}^*(G/P)^2, \mathcal{V}\right)^{\sim^* G}$, there exists $F \in \mathcal{V}_b$ so that $f - F \in \mathcal{L}_{inf}^\infty\left({}^*(G/P)^2, \mathcal{V}\right)$. In other words, we would like to prove an asymptotic version of double ergodicity of the action of G on G/P with coefficients being the asymptotic Banach ${}^*G$-module $\mathcal{V}$.

Theorem 3.44 (Double Ergodicity with Coefficients) *For $f \in \mathcal{L}_b^\infty\left({}^*(G/P)^2, \mathcal{V}\right)^{\sim^* G}$, there exists $v \in \mathcal{V}_b^{\sim^* G}$ so that $f - v \in \mathcal{L}_{inf}^\infty\left({}^*(G/P)^2, \mathcal{V}\right)$.*

3.3 Property $G(Q_1, Q_2)$ and the Main Theorem

We now sketch and discuss the proof of $H_a^2(G, \mathcal{V}) = 0$. We begin by noting that by that $\mathrm{H}_a^2(Q, \mathcal{V}) = 0$ for a proper parabolic subgroup Q of G as in Theorem 3.41 such that $\mathrm{H}_b^2(Q, \mathbb{R}) = 0$ and $\mathrm{H}_b^3(Q, \mathbb{R})$ is Hausdorff.
The following proposition is an application of Theorem 3.42:

Proposition 3.45 *Let G and a parabolic subgroup $Q \leq G$ be as in Theorem 3.41. Let $\omega \in \mathcal{L}_b^\infty(({}^*Q)^3, \mathcal{V})^{\sim^* Q}$ be an asymptotic 2-cocycle for Q with coefficients in $\mathcal{V}$. Then there exists an asymptotic 2-cocycle $\omega' \in \mathcal{L}_b^\infty(({}^*Q)^3, \mathcal{V})^{\sim^* Q}$ for Q that takes values in $\mathcal{V}_b^{\sim^* G}$ such that*

$$\tilde{\omega} - \tilde{\omega}' = \tilde{d}^1 \tilde{\alpha}_1$$

where $\alpha_1 \in \mathcal{L}_b^\infty(({}^*Q)^2, \mathcal{V})^{\sim^* Q}$.

Thus, the obstacle to the asymptotic 2-cocycle $\omega \in \mathcal{L}_b^\infty(({}^*Q)^3, \mathcal{V})^{\sim^* Q}$ being an asymptotic 2-coboundary is the asymptotic 2-cocycle $\omega' \in \mathcal{L}_b^\infty(({}^*Q)^3, \mathcal{V})^{\sim^* Q}$ that takes values in $\mathcal{V}_b^{\sim^* G}$. Under the right assumptions on Q, this obstacle too can be resolved:

Lemma 3.46 *Suppose $\mathrm{H}_b^2(Q, \mathbb{R}) = 0$ and $\mathrm{H}_b^3(Q, \mathbb{R})$ is Hausdorff. Then there exists $\alpha' \in \mathcal{L}_b^\infty\left(({}^*Q)^2, \mathcal{W}\right)^{{}^*Q}$ with $Im(\alpha') \subseteq \mathcal{W}_b^{\sim^*\Gamma}$ and $d^2\alpha' - \omega'' \in \mathcal{L}_{inf}^\infty\left(({}^*Q)^3, \mathcal{W}\right)$.*

In other words, assuming $\mathrm{H}_b^2(Q, \mathbb{R}) = 0$ and $\mathrm{H}_b^3(Q, \mathbb{R})$ is Hausdorff, we can conclude that $H_a^2(Q, \mathcal{V}) = 0$.

Theorem 3.47 *Let G and a parabolic subgroup $Q \leq G$ be as in Theorem 3.41, and suppose $\mathrm{H}_b^3(Q, \mathbb{R})$ is Hausdorff and $\mathrm{H}_b^2(Q, \mathbb{R}) = 0$. Then $\mathrm{H}_a^2(Q, \mathcal{V}) = 0$.*

This motivates the following definition: A locally compact group G has the **$2\frac{1}{2}$-property** if $\mathrm{H}_b^2(G, \mathbb{R})$ vanishes and $\mathrm{H}_b^3(G, \mathbb{R})$ is Hausdorff.

Definition 3.48 Let $G = \prod_{i=1}^{k} \mathbf{G}_i(K_i)$ be a semisimple group (where for $1 \leq i \leq k$, K_i is a local field, and $\mathbf{G}_i$ is a connected almost K_i-simple group), and let P be a minimal parabolic subgroup. Then G is said to have **Property-**$G(\mathcal{Q}_1, \mathcal{Q}_2)$ if there exist two parabolic subgroups Q_1 and Q_2 of G satisfying the following properties:

- Both Q_1 and Q_2 are of the form $Q_1 = \prod_{i=1}^{k} Q_{1,i}$ and $Q_2 = \prod_{i=1}^{k} Q_{2,i}$ with $Q_{1,i}$ and $Q_{2,i}$ being proper parabolic subgroups of $\mathbf{G}_i(K_i)$ for $1 \leq i \leq k$.
- Both Q_1 and Q_2 have the $2\frac{1}{2}$-property.
- The intersection $Q_1 \cap Q_2$ is a parabolic subgroup that contains the minimal parabolic subgroup P of G.
- The group G is boundedly generated by the union of Q_1 and Q_2.

Observe that the existence of two distinct proper parabolic subgroups immediately implies that if G has Property-$G(\mathcal{Q}_1, \mathcal{Q}_2)$, then each of its simple factors has rank at least 2.

With all the technical conditions and ingredients in place, we are ready to sketch the proof of the main theorem.

Theorem 3.49 *Let Γ be a lattice in a semisimple group $G = \prod_{i=1}^{k} \mathbf{G}_i(K_i)$ (where for $1 \leq i \leq k$, K_i is a local field, and $\mathbf{G}_i$ is a connected, simply connected, almost K_i-simple group) that has Property-$G(\mathcal{Q}_1, \mathcal{Q}_2)$. Then $\mathrm{H}_a^2(\Gamma, \mathcal{W}) = 0$.*

Proof We know that $\mathrm{H}_a^2(\Gamma, \mathcal{W}) = \mathrm{H}_a^2(G, \mathcal{V})$. Consider an asymptotic 2-cocycle

$$\omega \in \mathcal{L}_b^\infty(({}^*(G/P))^2, \mathcal{V})^{\sim^* G}$$

By Theorem 3.47, there exist $\alpha_1 \in \mathcal{L}_b^\infty(({}^*(G/P))^2, \mathcal{V})^{\sim^* Q_1}$ and $\alpha_2 \in \mathcal{L}_b^\infty(({}^*(G/P))^2, \mathcal{V})^{\sim^* Q_2}$ such that

$$\tilde{\omega} = \tilde{d}^2 \tilde{\alpha}_1 = \tilde{d}^2 \tilde{\alpha}_2$$

Since $P \subseteq Q_1 \cap Q_2$, note that $\alpha_1 - \alpha_2 \in \mathcal{L}_b^\infty(({}^*(G/P))^2, \mathcal{V})^{\sim^* P}$ is an asymptotic 1-cocyle for P. As P is amenable, $\mathrm{H}_a^1(P, \mathcal{V}) = 0$, and so there exists $\beta \in \mathcal{L}_b^\infty({}^*(G/P), \mathcal{V})^{\sim^* P}$ such that

$$\tilde{\alpha}_1 - \tilde{\alpha}_2 = \tilde{d}^1 \tilde{\beta}$$

We now use β to define the internal map

$$\beta_1 : ({}^*(G/P))^2 \to \mathcal{V}$$

$$\beta_1(g^*P, h^*P) := \pi_G(g)\beta(g^{-1}h^*P)$$

Note that $\beta_1 \in \mathcal{L}_b^\infty(({}^*(G/P))^2, \mathcal{V})^{\sim{}^*G}$, and by Lemma 3.44, there exists $F \in \mathcal{V}_b^{\sim{}^*G}$ so that $\beta_1 - F \in \mathcal{L}_{inf}^\infty(({}^*(G/P))^2, \mathcal{V})$. In particular, the same holds for β as well: $\beta - F \in \mathcal{L}_{inf}^\infty({}^*(G/P), \mathcal{V})$, implying that

$$\tilde{d}^1\tilde{\beta} = 0$$

This means that $\tilde{\alpha}_1 = \tilde{\alpha}_2$. Setting $\alpha := \alpha_1$, note that α is both asymptotically ${}^*Q_1$-equivariant and asymptotically ${}^*Q_2$-equivariant. Since G is boundedly generated by elements of Q_1 and Q_2, this implies that $\alpha \in \mathcal{L}_b^\infty(({}^*(G/P))^2, \mathcal{V})^{\sim{}^*G}$.

While the above theorem assumes that Γ is a lattice in a semisimple group G that is simply connected, we can extend this as follows:

Theorem 3.50 *Let Γ be a lattice in a semisimple group $G = \prod_{i=1}^k \mathbf{G}_i(K_i)$ (where for $1 \le i \le k$, K_i is a local field, and $\mathbf{G}_i$ is a connected almost K_i-simple group) that has Property-$G(\mathcal{Q}_1, \mathcal{Q}_2)$. Then Γ is uniformly $\mathfrak{U}$-stable with a linear estimate.*

Proof Let $\hat{G}$ be the universal cover of G, which is a semisimple simply connected group, and let $\hat{\Gamma}$ be the preimage of Γ in $\hat{G}$. Note that Γ and $\hat{\Gamma}$ are commensurable, hence it is sufficient to prove that $\hat{\Gamma}$ is is uniformly $\mathfrak{U}$-stable with a linear estimate. Note that since the fundamental group of G is abelian, it follows that $\hat{G}$ has Property-$G(\mathcal{Q}_1, \mathcal{Q}_2)$. Hence we can now apply Theorem 3.49 to $\hat{\Gamma} \le \hat{G}$ to conclude that $\mathrm{H}_a^2(\hat{\Gamma}, \mathcal{W}) = 0$, implying that $\hat{\Gamma}$ is uniformly $\mathfrak{U}$-stable with a linear estimate.

3.4 Groups with Property $G(Q_1, Q_2)$

We first list out groups that have the $2\frac{1}{2}$-property, and use that to construct groups having Property $G(Q_1, Q_2)$. We illustrate this idea in the case of groups over $\mathbb{C}$.

Theorem 3.51 *The following groups G have the $2\frac{1}{2}$-property:*

1. *$G = \mathbf{G}(k)$, where k is a non-Archimedean local field and $\mathbf{G}$ is a connected semisimple k-group.*
2. *$G = SL_n(\mathbb{R})$ for any $n \neq 2$.*
3. *$G = SL_n(\mathbb{C})$ for any n.*

In the complex case, since we know that $SL_n(\mathbb{C})$ has the $2\frac{1}{2}$-property for every n, we can use Dynkin diagrams to explicitly construct proper parabolic subgroups (which, modulo their amenable radical, would correspond to $SL_n(\mathbb{C})$ that we know has the $2\frac{1}{2}$-property) that together generate the group, to conclude that the simple group has Property-$G(\mathcal{Q}_1, \mathcal{Q}_2)$.

Proposition 3.52 *Let G be a simple, simply connected, complex Lie group of rank $n \geq 2$. Then G has Property-$G(\mathcal{Q}_1, \mathcal{Q}_2)$.*

Proof Recall that simple, simply connected Lie groups over $\mathbb{C}$ split, and hence are classified by Dynkin diagrams, so let X be the Dynkin diagram of G. Any subdiagram Y of X corresponds to a parabolic subgroup $Q \leq G$ containing a fixed minimal parabolic subgroup $P \leq G$, such that Q modulo its amenable radical is a semisimple group corresponding to the subdiagram Y. Furthermore, if the subdiagram Y is of type A_n (for $n \geq 1$), then the parabolic subgroup Q corresponding to Y, modulo its amenable radical, is $SL_{n+1}(\mathbb{C})$, and so, by Theorems 3.47 and 3.51, Q has the $2\frac{1}{2}$-property.

So constructing two proper parabolic subgroups Q_1 and Q_2, that both contain P and generate G, is equivalent to choosing two proper subdiagrams Y_1 and Y_2 of X that the union of the vertex sets of Y_1 and Y_2 is the vertex set of X. Furthermore, if we can ensure both these diagrams are of type A_n, then this gives us two proper parabolic subgroups with the $2\frac{1}{2}$-property, ensuring that G has Property-$G(\mathcal{Q}_1, \mathcal{Q}_2)$. We claim that for any connected Dynkin diagram X corresponding to a simple complex Lie algebra of rank at least 2, we can find two such subdiagrams Y_1 and Y_2 both of type A_n (for $n \geq 1$). This can be seen by considering each case separately, and is illustrated in the figure below. Observe that for the Dynkin diagram of type F_4, we construct subdiagrams Y_1 and Y_2 both of type A_2, while for a Dynkin diagram of any other type, we can always choose Y_1 and Y_2 to be of type A_1 and A_{m-1} (where m is the rank of the group G).

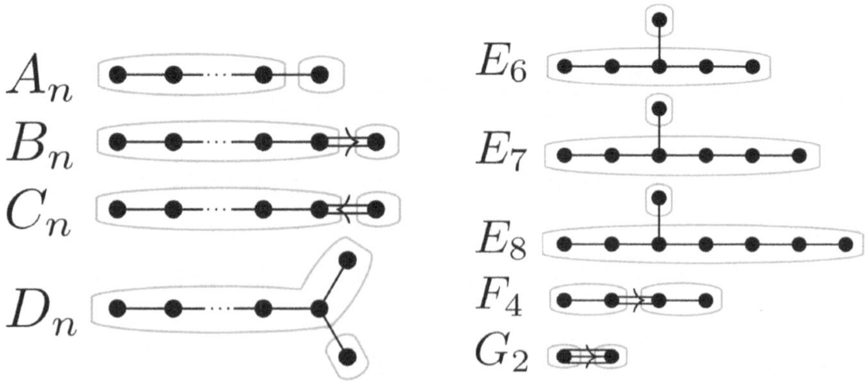

In the case of groups over $\mathbb{R}$, the situation is more involved. Firstly, note that even $SL_3(\mathbb{R})$ (which, by Theorem3.51, has the $2\frac{1}{2}$-property) does not have Property-$G(\mathcal{Q}_1, \mathcal{Q}_2)$ even though it has rank 2, because any proper parabolic subgroup of $SL_3(\mathbb{R})$, modulo its amenable radical, is $SL_2(\mathbb{R})$ for which $\mathrm{H}_b^2(SL_2(\mathbb{R}), \mathbb{R}) \neq 0$. However, since $SL_n(\mathbb{R})$ has the $2\frac{1}{2}$-property for $n \geq 3$, we can apply the proof technique of Proposition 3.52 in the case of certain *split* simple real Lie groups to show that:

Proposition 3.53 *Let G be a split simple real Lie group of type A_n (for $n \geq 3$), D_n (for $n \geq 3$), F_4, E_6, E_7 or E_8.Then G has Property-$G(\mathcal{Q}_1, \mathcal{Q}_2)$.*

Finally, in the case of a simple group G over a non-archimedean field, any parabolic subgroup Q has the $2\frac{1}{2}$-property. Thus, since G has rank at least 2, we can always find two proper parabolic subgroups, both having the $2\frac{1}{2}$-property, generating G. Hence,

Proposition 3.54 *The group $G = \mathbf{G}(k)$, where k is a non-Archimedean local field and $\mathbf{G}$ is a connected, simply connected, semisimple k-group of rank at least 2, has Property-$G(\mathcal{Q}_1, \mathcal{Q}_2)$.*

4 Conclusions and Open Questions

- **Prove a complete Ulam-stability result for higher rank lattices in semisimple Lie groups**.
  One could hope to extend our main results (specifically Theorem 3.49) by considering a larger family of groups for which the Property-$G(\mathcal{Q}_1, \mathcal{Q}_2)$ is known. However, this has its limitations since there are groups for which Property-$G(\mathcal{Q}_1, \mathcal{Q}_2)$ is simply not true. An important example of such a group is $SL_3(\mathbb{R})$, where for a maximal parabolic subgroup $Q \leq SL_3(\mathbb{R})$, $\mathrm{H}_b^2(Q, \mathbb{R}) = \mathrm{H}_b^2(SL_2(\mathbb{R}), \mathbb{R}) \neq 0$. We can ask if Property-$G(\mathcal{Q}_1, \mathcal{Q}_2)$ for the ambient group is even necessary for Ulam stability, or if it just happens to be an artifact of our proof technique.
- **Surjectivity/Injectivity of a comparison map** $\mathrm{H}_a^2(\Gamma, \mathcal{W}) \mapsto \mathrm{H}_b^2(\Gamma, \tilde{\mathcal{W}})$.
  There exists a natural forgetful map $\mathrm{H}_a^2(\Gamma, \mathcal{W}) \mapsto \mathrm{H}_b^2(\Gamma, \tilde{\mathcal{W}})$ (analogous to the comparison map $c : \mathrm{H}_b^2(\Gamma, W) \to H^2(\Gamma, W)$ for a Γ-module W). It is not immediate if this map is either surjective or injective. Suppose it were injective, then we could truly reduce the question of uniform stability with a linear estimate to the study of the second bounded cohomology group $\mathrm{H}_b^2(\Gamma, \tilde{W})$ which is a well-studied notion. For instance, it is known ([8]) that for a lattice Γ in a higher rank simple Lie group G, $\mathrm{H}_b^2(\Gamma, W) = 0$ for every dual separable Banach Γ-module W (note, however, that the Banach space ultraproduct $\tilde{W}$ is not separable unless it is finite dimensional).
- **Uniform versus non-uniform stability**
  All along in this paper, we have dealt with the question of uniform stability as opposed to non-uniform, or pointwise, stability. The connection between pointwise

stability and vanishing second cohomology is studied in [13, 24]. We stress that such stability results in the non-uniform setting is far from being known for most lattices. So far such results are known for many lattices in p-adic Lie groups (with respect to the Frobenius or p-Schatten norms for $p < \infty$), but almost nothing is known for lattices in real (or complex) Lie groups (see [7]). Moreover, when the family $M_n(\mathbb{C})$ is endowed with the operator norm (i.e. the p-Schatten norm for $p = \infty$) then it is known that the stability result is *not* true for most hgh rank lattices. This follows from the results in [11, 12] that show that if $H^{2j}(\Gamma, \mathbb{R}) \neq 0$ for some positive integer j, then Γ is not pointwise stable for the operator norm. Note that in our setting of uniform stability, the case of $p = \infty$ and $p < \infty$ are treated together without any problem.

- **Stability with respect to the (normalized) Hilbert-Schmidt norm**.
  In [6] it is shown that a lattice Γ that has Property (T) is not pointwise stable for the (normalized) Hilbert-Schmidt norm. In [1] it is shown that if a residually finite group is uniformly stable with respect to the (normalized) Hilbert-Schmidt norm, then it is virtually abelian. In particular, this means that no lattice in a non-compact semisimple Lie group is uniformly stable with respect to the (normalized) Hilbert-Schmidt norm.
  In both cases, the results still leave the possibility that higher rank lattices are *flexibly* stable (pointwise or uniform) with respect to the (normalized) Hilbert-Schmidt norm. For more on flexible stability, refer [4, 6].
- **Uniform stability with non-linear estimate**.
  Our machinery, whenever it can be applied, works to prove uniform stability with a linear estimate. We do not know of examples of groups that are uniformly stable, but without a linear estimate. It is interesting to compare this with [9] where it is shown that $\mathbb{Z}^2$ exhibits pointwise stability (with respect to Hamming metric on $Sym(n)$) but with a non-linear estimate. However, in the case of uniform stability in our setting, $\mathbb{Z}^2$ (being amenable) is uniformly stable for any submultiplicative norm on $U(n)$. The quantitative aspects of stability are an active line of current research.
- **Stability with respect to non-archimedean metrics**.
  A recent monograph of [15] studies stability with respect to p-adic groups. An interesting feature in this non-archimedean setting is that the ultrametric (strong triangle inequality) forces an equivalence between uniform and pointwise stability (for finitely presented groups). One can ask if further stability results in this setting too can be proved using the framework of asymptotic cohomology.

Acknowledgements I would like to thank my Ph.D. adviser, Alexander Lubotzky, for organizing the content of this mini-course and for his suggestions and comments on these notes.

References

1. Akhtiamov, D., Dogon, A.: On uniform Hilbert-Schmidt stability of groups. Proc. Am. Math. Soc. (2022)
2. Avni, N., Lubotzky, A., Meiri, C.: First order rigidity of non-uniform higher rank arithmetic groups. Invent. Math. **217**(1), 219–240 (2019)
3. Arzhantseva, G., Păunescu, L.: Almost commuting permutations are near commuting permutations. J. Funct. Anal. **269**(3), 745–757 (2015)
4. Becker, O., Chapman, M.: Stability of approximate group actions: uniform and probabilistic (2020). arXiv preprint arXiv:2005.06652
5. Brown, A., Fisher, D., Hurtado, S.: Zimmer's conjecture for actions of $sl(m, \mathbf{Z})$. Invent. Math. **221**(3), 1001–1060 (2020)
6. Becker, O., Lubotzky, A.: Group stability and property (T). J. Funct. Anal. **278**(1), 108298 (2020)
7. Bader, U., Lubotzky, A., Sauer, R., Weinberger, S.: Stability and instability of lattices in semisimple lie groups (in preparation) (2022)
8. Burger, M., Monod, N.: Bounded cohomology of lattices in higher rank Lie groups. J. Eur. Math. Soc. **1**(2), 199–235 (1999)
9. Becker, O., Mosheiff, J.: Abelian groups are polynomially stable. Int. Math. Res. Not. **2021**(20), 15574–15632 (2021)
10. Burger, M., Ozawa, N., Thom, A.: On Ulam stability. Israel J. Math. **193**, 109–129 (2013)
11. Connes, A., Gromov, M., Moscovici, H.: Conjecture de Novikov et fibrés presque plats. CR Acad. Sci. Paris Sér. I Math. **310**(5), 273–277 (1990)
12. Dadarlat, M.: Obstructions to matricial stability of discrete groups and almost flat k-theory. Adv. Math. **384**, 107722 (2021)
13. De Chiffre, M., Glebsky, L., Lubotzky, A., Thom, A.: Stability, cohomology vanishing, and non-approximable groups. In: Forum of Mathematics, Sigma, vol. 8. Cambridge University Press (2020)
14. Elek, G., Szabó, E.: Hyperlinearity, essentially free actions and l^2-invariants: the sofic property. Math. Ann. **332**(2), 421–441 (2005)
15. Fournier-Facio, F.: Ultrametric analogues of Ulam stability of groups (2021). arXiv preprint arXiv:2105.00516
16. Fournier-Facio, F., Rangarajan, B.: Ulam stability of lamplighters and Thompson groups (2023). arXiv preprint arXiv:2301.03970
17. Frigerio, R.: Bounded Cohomology of Discrete Groups, vol. 227. American Mathematical Society (2017)
18. Fujiwara, K.: The second bounded cohomology of a group acting on a Gromov-hyperbolic space. Proc. Lond. Math. Soc. **76**(1), 70–94 (1998)
19. Gamm, C.: ε-representations of groups and Ulam stability. Master's thesis, Georg-August-Universität Göttingen 21 (2011)
20. Glebsky, L., Lubotzky, A., Monod, N., Rangarajan, B.: Asymptotic cohomology and uniform stability for lattices in semisimple groups (2023). arXiv preprint arXiv:2301.00476
21. Goldblatt, R.: Lectures on the Hyperreals: An Introduction to Nonstandard Analysis, vol. 188. Springer Science & Business Media (2012)
22. Johnson, B.E.: Approximately multiplicative maps between Banach algebras. J. Lond. Math. Soc. **2**(2), 294–316 (1988)
23. Kazhdan, D.: On ϵ-representations. Israel J. Math. **43**(4), 315–323 (1982)
24. Lubotzky, A., Oppenheim, I.: Non p-norm approximated groups. J. d'Analyse Mathématique **141**(1), 305–321 (2020)
25. Margulis, G.A.: Discrete Subgroups of Semisimple Lie Groups, vol. 17. Springer Science & Business Media (1991)
26. Monod, N.: Continuous Bounded Cohomology of Locally Compact Groups, Number 1758. Springer Science & Business Media (2001)

27. Monod, N.: An invitation to bounded cohomology. In: Proceedings of the International Congress of Mathematicians, pp. 1183–1211. European Mathematical Society (2006)
28. Monod, N.: Lamplighters and the bounded cohomology of Thompson's group. Geom. Funct. Anal., 1–14 (2022)
29. Monod, N.: Lamplighters and the bounded cohomology of thompson's group. Geom. Funct. Anal. **32**(3), 662–675 (2022)
30. Monod, N., Shalom, Y.: Cocycle superrigidity and bounded cohomology for negatively curved spaces. J. Differ. Geom. **67**(3), 395–455 (2004)
31. Monod, N., Shalom, Y.: Orbit equivalence rigidity and bounded cohomology. Ann. Math., 825–878 (2006)
32. Shtern, A.I.: Quasi representations of amenable groups: results, errors, and hopes. Russ. J. Math. Phys. **20**(2), 239–253 (2013)
33. Turing, A.M.: Finite approximations to lie groups. Ann. Math. (2) **39**(1), 105–111 (1938)
34. Ulam, S.M.: A Collection of Mathematical Problems. Interscience Publishers (1960)
35. von Neumann, J.: Beweis des Ergodensatzes und des H-Theorems in der neuen Mechanik. Z. Phys. **57**(1), 30–70 (1929)

Prasad's Volume Formula and Its Applications

François Thilmany

Dedicated to Gopal Prasad on the occasion of his eightieth birthday.

Abstract After giving brief historical context, we present Prasad's famous formula for the volume of quotients of semisimple groups over local fields by arithmetic subgroups, faithfully to the original article [Volumes of S-arithmetic quotients of semisimple groups. *Publ. Math., Inst. Hautes Étud. Sci.*, 69:91–117, 1989]. We then survey the applications it has found in geometry, group theory, and arithmetic over the last four decades.

Keywords Prasad's volume formula · Tamagawa numbers · Lattices in Lie groups · Arithmetic groups

1 Introduction

The present article is a survey on the volume formula developed by Prasad for arithmetic quotients of semisimple groups over local fields.

In his influential paper [91], Prasad gave a very explicit formula for the covolume of principal arithmetic subgroups of a simple group G over a global field k. A principal S-arithmetic subgroup Γ of $\mathrm{G}(k)$ is determined by a collection $\{\mathrm{P}_v\}_{v\notin S}$ of compact open subgroups of $\mathrm{G}(k_v)$, one for each non-Archimedean place v outside the fixed set S. The volume of the quotient $(\prod_{v\in S}\mathrm{G}(k_v))/\,\Gamma$ is then expressed as a product of a global factor (composed of invariants of the global field k and of the form of the ambient group G), and of a local factor for each place $v \notin S$ (mainly depending of

F. Thilmany is supported by the FNRS under grant FC 4057, and by the FWO under grant 1221722N.

F. Thilmany (✉)
Departement Wiskunde, KU Leuven, Leuven, Belgium
e-mail: francois.thilmany@kuleuven.be

G. Prasad et al. (eds.), *Zariski Dense Subgroups, Number Theory and Geometric Applications*, Infosys Science Foundation Series,
https://doi.org/10.1007/978-981-95-3437-1_6

the size of the reduction of P_v to the residue field of k_v). We give the precise formula and explain its ingredients in Sect. 4. The proof of the volume formula relies on the knowledge of the Tamagawa number $\tau(G)$ of G, which we introduce in Sect. 3. In Sect. 2, for historical perspective, we recollect some earlier instances of formulas for the covolume of discrete subgroups.

Prasad's volume formula, in great part due to its explicit character, has had many applications to geometry and arithmetic group theory. Combined with the cohomological methods due to Rohlfs and with Margulis' arithmeticity theorems, the volume formula becomes a powerful tool to study the covolume of general lattices in semisimple groups over local fields. We briefly describe some of the major applications in Sects. 5–10 and 11.

2 A Brief Historical Perspective

2.1 Areas in the Hyperbolic Plane

The most basic situation, both from the point of view of geometry and Lie theory, is that of $\mathrm{PGL}_2(\mathbb{R})$ and its symmetric space, the real-hyperbolic plane $\mathcal{H}^2_{\mathbb{R}}$. In this situation, any discrete subgroup Γ of $\mathrm{PGL}_2(\mathbb{R})$ admits a Dirichlet fundamental domain F, which is an ideal polygon (i.e., a convex polygon with possibly infinitely many vertices, some of which may lie on the boundary of $\mathcal{H}^2_{\mathbb{R}}$). The covolume of Γ in $\mathrm{PGL}_2(\mathbb{R})$ is then proportional to the hyperbolic area of F. In turn, this area can be computed using any triangulation of F and the formula for the area of a hyperbolic triangle. If T is a triangle in $\mathcal{H}^2_{\mathbb{R}}$ with angles α, β, and γ, its hyperbolic area is given by

$$\mathrm{Area}(T) = \pi - \alpha - \beta - \gamma.$$

One can thus compute the volume of $\mathrm{PGL}_2(\mathbb{R})/\Gamma$ provided some understanding of the geometry of $\mathcal{H}^2_{\mathbb{R}}/\Gamma$.

Examples (i) Let $p, q, r \in \{2, 3, \ldots, \infty\}$ and let $T_{p,q,r}$ denote a triangle in $\mathcal{H}^2_{\mathbb{R}}$ with angles $\frac{\pi}{p}, \frac{\pi}{q}, \frac{\pi}{r}$ (such a triangle exists if and only if $\frac{1}{p} + \frac{1}{q} + \frac{1}{r} < 1$, and if so, is unique up to an isometry of $\mathcal{H}^2_{\mathbb{R}}$). The reflections across the sides of $T_{p,q,r}$ generate a discrete group $W_{p,q,r}$ in $\mathrm{PGL}_2(\mathbb{R})$, called the *$(p, q, r)$-triangle group*, which admits $T_{p,q,r}$ as fundamental domain. In consequence, the volume of $\mathrm{PGL}_2(\mathbb{R})/W_{p,q,r}$ is

$$\mathrm{Vol}(\mathrm{PO}_2(\mathbb{R})) \cdot \mathrm{Vol}(\mathcal{H}^2_{\mathbb{R}}/W_{p,q,r}) = 2\pi \cdot \mathrm{Area}(T_{p,q,r}) = 2\pi^2 \cdot \left(1 - \frac{1}{p} - \frac{1}{q} - \frac{1}{r}\right)$$

for this normalization of the Haar measure on $\mathrm{PGL}_2(\mathbb{R})$.

(ii) It is a familiar fact that $\mathrm{SL}_2(\mathbb{Z})$ acts on $\mathcal{H}^2_{\mathbb{R}}$ with fundamental domain $T_{3,3,\infty}$. Since $\mathrm{SL}_2(\mathbb{Z})$ contains the kernel $\{\pm 1\}$ of this action, the volume of $\mathrm{SL}_2(\mathbb{R})/\mathrm{SL}_2(\mathbb{Z})$ is

$$\mathrm{Vol}(\mathrm{SO}_2(\mathbb{R})) \cdot \mathrm{Vol}(\mathcal{H}^2_{\mathbb{R}} / \mathrm{SL}_2(\mathbb{Z})) = \pi \cdot \mathrm{Area}(T_{3,3,\infty}) = \frac{\pi^2}{3}.$$

Equivalently, one can observe that $\mathrm{PSL}_2(\mathbb{Z})$ has index 2 in (a conjugate of) $W_{2,3,\infty}$.

Such geometric considerations led Siegel [104] to prove that the orientation-preserving subgroup of $W_{2,3,7}$ achieves the smallest possible covolume among discrete groups in $\mathrm{PSL}_2(\mathbb{R})$: knowing that a discrete subgroup of $\mathrm{PSL}_2(\mathbb{R})$ has a polygonal fundamental domain, Siegel argued that in order to have smallest volume, this domain needs to have finitely many sides, in fact exactly 3; but the hyperbolic triangle with smallest area is precisely the (2, 3, 7)-triangle.

These ideas also appeared in Hurwitz's anterior work [55] on his automorphism theorem, which states that the automorphism group of a Riemann surface of genus g has size at most $84(g-1)$. The bound $84(g-1)$ can only be achieved when the Riemann surface is a cover of $\mathcal{H}^2_{\mathbb{R}} / W_{2,3,7}$.

2.2 *Minkowski's and Siegel's Formulae*

The first formula giving the covolume of a family of lattices in groups of increasing dimension appeared in the work of Minkowski [81]. Minkowski constructed a fundamental domain for the action of $\mathrm{SL}_n(\mathbb{Z})$ on the space of positive definite quadratic n-forms with discriminant ≤ 1, by taking the intersection of finitely many half-spaces with the set on which the discriminant is ≤ 1. He computed the volume of this fundamental domain, and in this way obtained for every $n \geq 2$ a formula for the covolume of $\mathrm{SL}_n(\mathbb{Z})$ in $\mathrm{SL}_n(\mathbb{R})$:

$$\mathrm{Vol}(\mathrm{SL}_n(\mathbb{R}) / \mathrm{SL}_n(\mathbb{Z})) = \frac{2}{n+1} \cdot \pi^{-\frac{n(n+1)}{4}} \cdot \Gamma\left(\frac{1}{2}\right) \cdots \Gamma\left(\frac{n}{2}\right) \cdot \zeta(2) \cdots \zeta(n),$$

which is not too far from the modern normalization $\zeta(2) \cdots \zeta(n)$.

It should be apparent by now that a meaningful normalization of the Haar measure on the ambient groups is necessary, otherwise values of volumes are senseless. This is especially important when comparing several lattices in multiple groups. We will see again in Sect. 5 how crucial the choice of an appropriate measure is.

Three decades later, Siegel [102] took interest in Minkowski's work and generalized his volume formula to an algebraic number field k. Let $\mathfrak{o}_k$ be the ring of integers of k, (r_k, s_k) its signature, $d_k = r_k + 2s_k$ its degree, D_k the absolute value of its discriminant, R_k its regulator, and ζ_k its zeta function. Siegel gave an analogous fundamental domain for the diagonal action of $\mathrm{GL}_n(\mathfrak{o}_k)$ on the space of $(\deg_{\mathbb{Q}} k)$-tuples consisting of r_k positive definite quadratic n-forms and $2s_k$ positive definite Hermitian n-forms, the product of whose discriminants is ≤ 1, and computed its invariant volume:

$$\mathrm{Vol}(\mathrm{G}()/\mathrm{GL}_n(\mathfrak{o}_k)) = \frac{2^{r_k+s_k}\pi^{s_k}}{n+1}\cdot D_k^{\frac{n^2-1}{2}} R_k\cdot\pi^{-d\frac{n(n+1)}{4}}\cdot 2^{-s_k\frac{n(n+1)}{2}}\cdot\prod_{i=1}^{n}\left(\Gamma\left(\frac{i}{2}\right)^{r_k}\Gamma(i)^{s_k}\right)\cdot\zeta_k(2)\cdots\zeta_k(n),$$

where $\mathrm{G}(\mathbb{R})$ is the kernel of the morphism $\mathrm{GL}_n(k\otimes_{\mathbb{Q}}\mathbb{R})\to\mathbb{R}^\times$ induced by the norm of the determinant $N\circ\det:\mathrm{GL}_n(k)\to k^\times$. Several arithmetic invariants of the field of definition k play a role in this formula, most notably the discriminant and the zeta function; this is not a quirk due to normalization, these invariants are key in the expression of covolumes of arithmetic groups as we will see in Sect. 4. Note that the formula above for GL_1 is reminiscent of the definition of the regulator.

In the same period, Siegel [103] also computed the covolume of $\mathrm{SL}_n(\mathbb{Z})$ in $\mathrm{SL}_n(\mathbb{R})$ using a completely different method. To a parameter $\lambda\in\mathbb{R}^\times$ and a continuous integrable function $f:\mathbb{R}^n\to\mathbb{R}$, Siegel associates a function $\mathcal{S}_\lambda f:\mathrm{SL}_n(\mathbb{R})/\mathrm{SL}_n(\mathbb{Z})\to\mathbb{R}$, *the Siegel transform of* f, through the formula:

$$\mathcal{S}_\lambda f(g\,\mathrm{SL}_n(\mathbb{Z})) = \lambda^n\sum_{v\in\mathbb{Z}^n} f(\lambda g v).$$

When $\lambda\to 0$, we have that $\mathcal{S}_\lambda f(g\,\mathrm{SL}_n(\mathbb{Z}))\to\int_{\mathbb{R}^n} f(gx)dx=\int_{\mathbb{R}^n} f(x)dx$. Thus, picking f positive with $\int_{\mathbb{R}^n} f(x)dx=1$ and μ an invariant measure on $\mathrm{SL}_n(\mathbb{R})/\mathrm{SL}_n(\mathbb{Z})$, setting aside convergence issues, we have

$$\lim_{\lambda\to 0}\int_{\mathrm{SL}_n(\mathbb{R})/\mathrm{SL}_n(\mathbb{Z})}\mathcal{S}_\lambda f(g)d\mu(g) = \int_{\mathrm{SL}_n(\mathbb{R})/\mathrm{SL}_n(\mathbb{Z})}\lim_{\lambda\to 0}\mathcal{S}_\lambda f(g)d\mu(g) = \mu(\mathrm{SL}_n(\mathbb{R})/\mathrm{SL}_n(\mathbb{Z})).$$

Siegel then studies the series

$$\chi(\lambda) = \sum_{\substack{v\in\mathbb{Z}^n\\ \text{primitive}}}\int_{\mathrm{SL}_n(\mathbb{R})/\mathrm{SL}_n(\mathbb{Z})} f(\lambda g v)d\mu(g),$$

establishing two crucial facts. The constant $\chi(1)$ can be computed inductively,

$$\chi(1) = \frac{n-1}{n}\cdot\mu_{n-1}(\mathrm{SL}_{n-1}(\mathbb{R})/\mathrm{SL}_{n-1}(\mathbb{Z}))$$

for a compatible choice of μ_m's on $\mathrm{SL}_m(\mathbb{R})/\mathrm{SL}_m(\mathbb{Z})$ (with $\mu=\mu_n$), and the relation $\chi(\lambda)=\lambda^{-n}\chi(1)$ holds. These allow to evaluate

$$\int_{\mathrm{SL}_n(\mathbb{R})/\mathrm{SL}_n(\mathbb{Z})}\mathcal{S}_\lambda f(g)d\mu(g) = \lambda^n\sum_{l=1}^{\infty}\chi(l\cdot\lambda) = \sum_{l=1}^{\infty} l^{-n}\chi(1) = \frac{n-1}{n}\cdot\zeta(n)\cdot\mu_{n-1}(\mathrm{SL}_{n-1}(\mathbb{R})/\mathrm{SL}_{n-1}(\mathbb{Z})),$$

which turns out to be independent of λ, and inductively,

$$\mu(\mathrm{SL}_n(\mathbb{R})/\mathrm{SL}_n(\mathbb{Z})) = \frac{1}{n}\cdot\zeta(n)\cdots\zeta(2).$$

2.3 Borel's Formula

We mention one last volume formula, this time for a family of non-split groups, due to Borel [12] (see also [73, 112] for a textbook treatment). Let B be a quaternion algebra over a number field k, and let (r_k, s_k), D_k, ζ_k be as above. Let $\mathcal{O}$ be a maximal order in B, and set $\mathcal{O}^1$ to be the multiplicative group of elements of norm 1 in $\mathcal{O}$. The group $\mathrm{SL}_{1,B}(k \otimes_{\mathbb{Q}} \mathbb{R})$ splits as a product $\mathrm{SL}_2(\mathbb{R})^a \times \mathrm{SL}_2(\mathbb{C})^{s_k} \times \mathrm{SU}_2(\mathbb{R})^{r_k - a}$ in which $\mathcal{O}^1$ embeds diagonally, where a depends on the ramification of B over the different real completions of k. Then for the product of the standard volume forms on $\mathcal{H}_{a,s_k} = (\mathcal{H}^2_{\mathbb{R}})^a \times (\mathcal{H}^2_{\mathbb{C}})^{s_k}$, Borel proved that

$$\mathrm{Vol}(\mathcal{H}_{a,s_k}/\mathcal{O}^1) = \frac{2(4\pi)^a}{(2\pi)^{2r_k}(8\pi^2)^{s_k}} \cdot D_k^{\frac{3}{2}} \cdot \zeta_k(2) \cdot \prod_{v \in R_f} (q_v - 1),$$

where R_f is the set of finite places of k over which B is ramified, and q_v denotes the size of the residue field of k_v.

Unlike Minkowski and Siegel, to prove his volume formula Borel did not construct fundamental domains. Instead, he used the fact that the Tamagawa number of $\mathrm{SL}_{1,B}$ equals 1, combined with local calculations. Note that Harder [50] had earlier given a formula for the covolume (with respect to the Euler–Poincaré measure) of the integral points of a Chevalley group scheme using the same approach. This approach also underlies the proof of Prasad's volume formula. In the next section, we will introduce Tamagawa numbers, and illustrate how they can be used to compute volumes using local data.

3 Tamagawa Numbers as Local-Global Principle for Volumes

3.1 Notation

The following notation will be used throughout this article. Let k be a global field. We denote by D_k the absolute value of the discriminant of k if k is a number field, and if k is a function field with constant field $\mathbb{F}_{q_k}$ and genus g_k, we set $D_k = q_k^{2g_k - 2}$. The class number of k will be denoted h_k. When k is a number field, the signature (that is, the number of real and complex places) of k will be denoted (r_k, s_k), so that $d_k = r_k + 2s_k$ is the degree of k over $\mathbb{Q}$.

Let V be the set of places of k, and V_∞, V_f denote the set of Archimedean places (possibly empty) and non-Archimedean places, respectively. For any subset S of V, we write S_∞, S_f for, respectively, $S \cap V_\infty$, $S \cap V_f$. For $v \in V$, k_v denotes the completion of k at v, and $|\ |_v$ the normalized absolute value on k_v. For $v \in V_f$, $\mathfrak{o}_v$ denotes the valuation ring of k_v, and $\mathfrak{f}_v$ the residue field of $\mathfrak{o}_v$, whose order is denoted

q_v. We will identify each place $v \in V_f$ with the valuation $v : k_v \to \mathbb{Z}$, for which we have $|x|_v = q_v^{-v(x)}$. For $v \in V_\infty$, we normalize $|x|_v = |x|$ if $k_v = \mathbb{R}$, and $|x|_v = |x|^2$ if $k_v = \mathbb{C}$. With these conventions, the product formula takes the form $\prod_{v \in V} |x|_v = 1$ for all $x \in k^\times$.

Recall, for example, that $\mathbb{Q}$ has exactly one Archimedean place, namely, (the class of) the standard absolute value, with respect to which the completion of $\mathbb{Q}$ is $\mathbb{R}$, so $r_\mathbb{Q} = d_\mathbb{Q} = 1$. The remaining places of $\mathbb{Q}$ are in bijection with the set of positive prime numbers: to each prime p corresponds to the p-adic valuation v_p given by $v_p(\frac{a}{b} \cdot p^n) = n$ when p does not divide the integer a nor b, so that $|\frac{a}{b} \cdot p^n|_{v_p} = p^{-n}$, and the product formula is now an immediate consequence of prime factorization. For background in algebraic number theory, we refer the reader to the textbooks [23] or [85].

3.2 Adèles

Recall that to the global field k, one associates the *ring* A_k *of adèles of* k, which is the restricted topological product $\prod'_{v \in V} k_v$ with respect to the collection of compact subrings $\{\mathfrak{o}_v \mid v \in V_f\}$. By construction, the topological ring A_k is locally compact, and k embeds inside A_k diagonally, while each completion k_v embeds in the v-th component. The image of k in A_k is discrete as a consequence of the product formula.

Let G be an affine algebraic k-group. Since A_k is a locally compact k-algebra, $\mathrm{G}(\mathrm{A}_k)$ is a locally compact group. Fixing an embedding $\mathrm{G} \to \mathrm{GL}_N$ identifies $\mathrm{G}(\mathrm{A}_k)$ with the restricted direct product $\prod'_{v \in V} \mathrm{G}(k_v)$ with respect to the compact subgroups $\mathrm{G}(k_v) \cap \mathrm{GL}_N(\mathfrak{o}_v)$ $v \in V_f$, in which $\mathrm{G}(k)$ embeds diagonally as a discrete subgroup.

3.3 Tamagawa's Observation

We call *gauge form* a non-zero left-invariant form on G defined over k and of degree $\dim(\mathrm{G})$. Any gauge form ω induces by integration a non-zero left-invariant measure $|\omega|_v$ on $\mathrm{G}(k_v)$ and $|\omega|_{\mathrm{A}_k}$ on $\mathrm{G}(\mathrm{A}_k)$. The latter coincides with the product measure $\prod_{v \in V} |\omega|_v$ under the identification $\mathrm{G}(\mathrm{A}_k) = \prod'_{v \in V} \mathrm{G}(k_v)$ described above. Since the space $\bigwedge^{\dim G} \mathrm{Lie}(G)$ of left-invariant forms of top degree is one-dimensional, any other gauge form on G is a scalar multiple $c\omega$ of ω. The measure $|c\omega|_{\mathrm{A}_k}$ induced on the adelic group then satisfies $|c\omega|_{\mathrm{A}_k} = (\prod_{v \in V} |c|_v) \cdot |\omega|_{\mathrm{A}_k} = |\omega|_{\mathrm{A}_k}$ by the product formula. In other words, the measure $|\omega|_{\mathrm{A}_k}$ is independent of the choice of ω. Its multiple $\tau = D_k^{-\dim \mathrm{G}/2} |\omega|_{\mathrm{A}_k}$ is called the *Tamagawa measure* of G; it is a special choice of Haar measure on $\mathrm{G}(\mathrm{A}_k)$, which would usually only be determined up to a scalar.

When the group G admits no non-trivial k-characters, $\mathrm{G}(\mathrm{A}_k)/\mathrm{G}(k)$ has finite invariant volume for the measure induced by τ (for k a number field, this is essentially a theorem of Borel and Harish-Chandra [13]). The quantity $\tau(\mathrm{G}(\mathrm{A}_k)/\mathrm{G}(k))$ is then called the *Tamagawa number of* G, and we will denote it by $\tau_k(\mathrm{G})$.

Weil provided a way to extend the definition of the Tamagawa number to include unimodular groups G which may admit characters, namely, by restricting to the subgroup of $\mathrm{G}(\mathrm{A}_k)$ of elements of adelic norm equal to 1. One can show that in this broader sense, the equality $\tau_k(\mathrm{GL}_1) = 1$ is equivalent to the class number formula.

In addition to introducing what is now called the Tamagawa measure, in 1959 Tamagawa made the illuminating observation that the Smith–Minkowski–Siegel mass formula [80, 100, 105] is equivalent to the fact that for an integral non-degenerate quadratic form q, $\tau_{\mathbb{Q}}(\mathrm{SO}_q) = 2$. Tamagawa posteriorly published his observation in [107]; the author could however not locate the original source.

Example Let $\mathrm{G} = \mathrm{G}_\mathrm{a}$ be the additive group over a number field k with ring of integers $\mathfrak{o}_k$. The canonical choice dx of gauge form on G_a in the usual coordinates $k[x] = k[\mathrm{G}_\mathrm{a}]$ induces on $\mathrm{G}_\mathrm{a}(k_v) = k_v$, respectively, the Lebesgue measure, twice the Lebesgue measure, the measure giving volume 1 to $\mathfrak{o}_v$, when k_v is $\mathbb{R}$, $\mathbb{C}$, or a non-Archimedean field.

Minkowski's geometry of numbers tells us that the action of $\mathfrak{o}_k$ by translations on $k_\infty = \prod_{v \in V_\infty} k_v$ admits a fundamental domain F, whose volume for the product measure $\prod_{v \in V_\infty} (dx)_v$ is precisely $D_k^{1/2}$. Then $F \times \prod_{v \in V_f} \mathfrak{o}_v$ is a fundamental domain for the action of the additive group k on A_k, hence

$$\tau(\mathrm{A}_k/k) = D_k^{-\frac{1}{2}} \cdot D_k^{\frac{1}{2}} \cdot \prod_{v \in V_f} dx_v(\mathfrak{o}_v) = 1.$$

The equality $\tau_k(\mathrm{G}_\mathrm{a}) = 1$ thus captures to the fact that $D_k^{\frac{1}{2}}$ is the covolume of $\mathfrak{o}_k$ in k_∞.

3.4 Weil's Conjecture on Tamagawa Numbers

Well aware of Tamagawa's observation, Weil formulated the following conjecture.

Conjecture (Weil) *If* G *is a simply connected semisimple k-group,*[1] *then* $\tau_k(\mathrm{G}) = 1$.

---

[1] In the present article, we will abbreviate "affine algebraic group over k" by "k-*group*" and "morphism of algebraic groups defined over k" by "k-*morphism*". We adopt the convention that (semi)simple and reductive k-groups are connected. Moreover, we call *(absolutely) simple* any k-group G whose normal k-subgroups (resp. k^{alg}-subgroups) have dimension 0 or dim G; this is sometimes called "quasi-simple" or "almost-simple" in the literature. A k-group G is *simply connected* when it has no proper finite étale covers, that is, any k-epimorphism $\iota : \tilde{\mathrm{G}} \to \mathrm{G}$ with finite étale kernel is an isomorphism.

Note that SO_q is not simply connected. Its universal cover is the double cover $Spin_q$, and $\tau_k(Spin_q) = \frac{1}{2}\tau_k(SO_q) = 1$,[2] so Tamagawa's observation actually supports Weil's conjecture.

Weil's conjecture is now known to hold, thanks to the successive work of many mathematicians; we summarize some of the main contributions. In the book [114] (a print of earlier notes of a course by Weil), Weil computed the Tamagawa number of most of the classical groups. Langlands [67] proved Weil's conjecture for all Chevalley groups over number fields, and Lai [66] generalized his methods to treat quasi-split groups. Following a proposal of Jacquet and Langlands, Kottwitz then proved that if G satisfies the Hasse principle, the Tamagawa number of any inner form of G equals $\tau_k(G)$. Since the Hasse principle is now known to hold for all simply connected semisimple k-groups,[3] the work of Kottwitz proves Weil's conjecture over number fields. Much more recently, Gaitsgory and Lurie [36] proved Weil's conjecture for groups over function fields, completing the other side of the picture. Note that Harder [51] had worked out the case of split groups over function fields earlier.

The appendix of [114], written by Ono, contains a relatively detailed history of Tamagawa's discovery, Weil's conjecture on Tamagawa numbers, and their computation for different groups up to that time. The reader can also consult the proceedings [14] for more background and details, and [90] for a textbook treatment. The contents of this section can be found in these works, in one form or another.

Knowing the value of the Tamagawa number of G allows us to express the volume of principal S-arithmetic subgroups in terms of local factors.

For the remainder of this section, let G be a simply connected semisimple k-group, and let S be a finite subset of V containing V_∞.

3.5 *Principal arithmetic subgroup*

Definition 3.1 A collection of compact open subgroups $(P_v)_{v\notin S}$ of $G(k_v)$ is called *coherent* if the subgroup $\prod_{v\in S} G(k_v) \times \prod_{v\notin S} P_v$ is open in $G(A_k)$. Let $(P_v)_{v\notin S}$ be such a collection. The subgroup

$$\Lambda(P_v)_{v\notin S} = G(k) \cap \Big(\prod_{v\in S} G(k_v) \times \prod_{v\notin S} P_v\Big)$$

of $G(k)$ is called the *principal S-arithmetic subgroup determined by* $(P_v)_{v\notin S}$.

For our purposes, the compact open subgroups $P_v \leq G(k_v)$ will often be taken to be *parahoric subgroups*. A subgroup of $G(k_v)$ is parahoric if it is the stabilizer of a

---

[2] See [88] for the values of τ_k along an isogeny class.

[3] Kneser [63, 64] proved the Hasse principle for classical groups over number fields. Harder [48, 49] proved it for exceptional groups other than E_8, as well as for groups of all types over function fields [52]. The last remaining case, type E_8 over a number field, was verified by Chernousov [24].

face in the Bruhat–Tits building of G over k_v. Parahoric subgroups have the advantage of admitting a reduction map $\mathrm{P}_v \to \overline{\mathrm{P}}_v(\mathfrak{f}_v)$ to an $\mathfrak{f}_v$-group $\overline{\mathrm{P}}_v$ associated with P_v via Bruhat–Tits theory (more precisely, $\overline{\mathrm{P}}_v$ is the special fiber of the $\mathfrak{o}_v$-group scheme G_v associated with $\mathrm{P}_v = \mathrm{G}_v(\mathfrak{o}_v)$). The reader interested in reviewing Bruhat–Tits theory can consult the original articles of Bruhat and Tits [17–19], the survey [110], or the novel treatment in Kaletha and Prasad's book [56].

As all compact open subgroups of $\mathrm{G}(k_v)$ are commensurable, any other choice of coherent collection will yield a principal S-arithmetic subgroup commensurable to $\Lambda(\mathrm{P}_v)_{v\notin S}$, and their covolumes will be related by the commensurability index.

3.6 From Tamagawa Number Back to Volume

Recall that when G is simply connected, G satisfies *strong approximation* with respect to a set of places S if and only if every simple factor of G is isotropic over k_v for some $v \in S$ (see [90, Sect. 7.4]). This means that, under these assumptions, $\mathrm{G}(k)$ is dense in $\mathrm{G}(\mathrm{A}_{k,S})$ (here $\mathrm{A}_{k,S}$ denotes the set of S-truncated adèles), hence

$$\Big(\prod_{v\in S}\mathrm{G}(k_v)\times\prod_{v\notin S}U_v\Big)\cdot\mathrm{G}(k)=\mathrm{G}(\mathrm{A}_k)$$

for any choice of sets $U_v \subset \mathrm{G}(k_v)$ for which $\prod_{v\in S}\mathrm{G}(k_v)\times\prod_{v\notin S}U_v$ has non-empty interior in $\mathrm{G}(\mathrm{A}_k)$.

If now $(\mathrm{P}_v)_{v\notin S}$ is a coherent collection of compact open subgroups, the canonical surjective $\mathrm{G}(k_S)$-equivariant continuous map

$$\begin{array}{c}\mathrm{G}(\mathrm{A}_k)/\mathrm{G}(k)=\Big(\big(\mathrm{G}(k_S)\times\prod_{v\notin S}\mathrm{P}_v\big)\cdot\mathrm{G}(k)\Big)\Big/\mathrm{G}(k)\xrightarrow{\sim}\big(\mathrm{G}(k_S)\times\prod_{v\notin S}\mathrm{P}_v\big)\big/\Lambda(\mathrm{P}_v)_{v\notin S}\\ \downarrow\\ \mathrm{G}(k_S)\big/\Lambda(\mathrm{P}_v)_{v\notin S}\end{array}$$

has fibers which are translates of $\prod_{v\notin S}\mathrm{P}_v$. Thus if ω is a gauge form on G defined over k (as in Sect. 3.3), we can compute

$$D_k^{\frac{\dim\mathrm{G}}{2}}\cdot\tau_k(\mathrm{G})=|\omega|_{\mathrm{A}_k}\big(\mathrm{G}(\mathrm{A}_k)/\mathrm{G}(k)\big)=|\omega|_S\big(\mathrm{G}(k_S)/\Lambda(\mathrm{P}_v)_{v\notin S}\big)\cdot\prod_{v\notin S}|\omega|_v(\mathrm{P}_v),$$

where $|\omega|_S$ is (induced by) the product measure $\prod_{v\in S}|\omega|_v$. Since P_v is compact open, $|\omega|_v(\mathrm{P}_v)$ is finite and non-zero.

In consequence, we get a formula for the covolume of the principal arithmetic subgroup $\Lambda(\mathrm{P}_v)_{v\notin S}$ in $\mathrm{G}(k_S)\cong\prod_{v\in S}\mathrm{G}(k_v)$, in terms of the Tamagawa number of G and of the volumes of the P_v with respect to $|\omega|_v$:

$$|\omega|_S\big(\mathrm{G}(k_S)/\Lambda(\mathrm{P}_v)_{v\notin S}\big) = D_k^{\frac{\dim \mathrm{G}}{2}} \cdot \tau_k(\mathrm{G}) \cdot \prod_{v\notin S} |\omega|_v(\mathrm{P}_v)^{-1}. \tag{T}$$

Of course, the difficulty now lies in understanding the values of the local factors $|\omega|_v(\mathrm{P}_v)$. By a clever renormalization of the measures $|\omega|_v$, Prasad was able to make explicit the product of the local factors, and establish the volume formula that we will present in the next section.

Example 3.2 For certain reasonable groups, the local factors can be made explicit by using a well-chosen gauge form ω. We illustrate this for SL_n over $\mathbb{Q}$.

Given a $n \times n$ matrix $x = (x_{ij})$, let $m(x)$ denote the (n, n)-minor of x. A good choice of gauge form for SL_n is

$$\omega = m(x)^{-1} \cdot dx_{11} \wedge \cdots \wedge dx_{n,n-1},$$

expressed in the local coordinates $x_{11}, x_{12}, \ldots, x_{n,n-1}$ on the Zariski-open set of SL_n defined by $m(x) \neq 0$. Note that on this open set, the (n, n)-entry of a matrix x is a polynomial in terms of $x_{11}, x_{12}, \ldots, x_{n,n-1}$, and $m(x)^{-1}$.

For p prime, let P_p be the compact open subgroup $\mathrm{SL}_n(\mathbb{Z}_p)$ of $\mathrm{SL}_n(\mathbb{Q}_p)$. Using the standard matrix representation of SL_n, it is clear that the collection $(\mathrm{P}_p)_{p\text{prime}}$ is coherent.

The reduction mod p map $\mathrm{P}_p \to \mathrm{SL}_n(\mathbb{F}_p)$ is surjective and has kernel $\mathrm{P}_p^1 = \mathrm{SL}_n(\mathbb{Z}_p) \cap (1 + p\mathrm{M}_n(\mathbb{Z}_p))$. Thus, $|\omega|_p(\mathrm{P}_p) = \#\,\mathrm{SL}_n(\mathbb{F}_p) \cdot |\omega|_p(\mathrm{P}_p^1)$. But for $x \in 1 + p\mathrm{M}_n(\mathbb{Z}_p)$, we have $|m(x)|_p = 1$, so $m(x)$ is non-zero. Reparametrizing $1 + p\mathrm{M}_n(\mathbb{Z}_p)$ by $x = 1 + py$ with $y \in \mathrm{M}_n(\mathbb{Z}_p)$, we see that there, ω takes the form:

$$\omega = \det(m(1 + py))^{-1} p^{n^2-1} dy_{11} \wedge \cdots \wedge dy_{n,n-1}.$$

This allows to compute

$$\begin{aligned} |\omega|_p(\mathrm{P}_p^1) &= \int_{\mathbb{Z}_p^{n^2-1}} |p^{n^2-1} \det(m(1 + py))^{-1}|_p dy_{11} \cdots dy_{n,n-1} \\ &= p^{1-n^2} \int_{\mathbb{Z}_p^{n^2-1}} dy_{11} \cdots dy_{n,n-1} = p^{1-n^2}, \end{aligned}$$

and in turn

$$|\omega|_p(\mathrm{P}_p) = p^{n-1} \cdot \prod_{i=0}^{n-2} (p^n - p^i) \cdot p^{1-n^2} = \prod_{i=2}^{n} (1 - p^{-i}).$$

The principal arithmetic subgroup defined by $(\mathrm{P}_p)_{p\notin S}$ is

$$\Lambda = \{g \in \mathrm{SL}_n(\mathbb{Q}) \mid g \in \mathrm{SL}_n(\mathbb{Z}_p) \text{ for every } p \notin S\} = \mathrm{SL}_n(\mathbb{Z}[S_f^{-1}]).$$

In view of the discussion above, we conclude that

$$|\omega|_S\Big(\prod_{v\in S}\mathrm{SL}_n(\mathbb{Q}_v)/\Lambda\Big)=\tau_{\mathbb{Q}}(\mathrm{SL}_n)\cdot\prod_{p\notin S}\prod_{i=2}^{n}(1-p^{-i})=\prod_{i=2}^{n}\prod_{p\notin S}(1-p^{-i}).$$

In particular, for $S=\{\infty\}$, we have the formula (by now leitmotiv)

$$|\omega|_\infty\big(\mathrm{SL}_n(\mathbb{R})/\,\mathrm{SL}_n(\mathbb{Z})\big)=\prod_{i=2}^{n}\prod_{p\text{ prime}}(1-p^{-i})=\prod_{i=2}^{n}\zeta(i).$$

4 Prasad's Volume Formula

Throughout this section, we assume that G is a simply connected absolutely simple group defined over the global field k. Fix a gauge form ω on G defined over k. As before, S is a fixed finite set of places of k containing V_∞.

Let $(\mathrm{P}_v)_{v\notin S}$ be a coherent collection of parahoric subgroups (see Definition 3.1), and for $v\in S_f$, complete it with $\mathrm{P}_v=\mathrm{I}_v$, any fixed Iwahori subgroup of $\mathrm{G}(k_v)$. (Recall that since G is simply connected, an Iwahori subgroup is the stabilizer of a chamber in the Bruhat–Tits building of G.) Let G_v be the $\mathfrak{o}_v$-group scheme associated with P_v by Bruhat–Tits theory. Recall that G_v is smooth, has generic fiber isomorphic to G over k_v, and $\mathrm{G}_v(\mathfrak{o}_v)=\mathrm{P}_v$.

4.1 *The Scaling Factors*

Assume for a moment that G is quasi-split. Then at each $v\in V_f$, $\mathrm{G}(k_v)$ admits a particularly nice parahoric subgroup $\mathcal{P}_v$, namely, we take $\mathcal{P}_v$ hyperspecial (if G splits over an unramified extension of k_v) or special of maximal volume (if G does not split over any unramified extension of k_v). Moreover, the choice of such $\mathcal{P}_v$ can be made so that the collection $(\mathcal{P}_v)_{v\in V_f}$ is coherent.

In order to give a manageable expression for the local factors in formula (T), at each non-Archimedean place v, Prasad scales the form ω by $c_v\in k_v^\times$ to arrange that $c_v\omega$ induces a gauge form on the $\mathfrak{o}_v$-group scheme $\mathcal{G}_v$ associated with $\mathcal{P}_v$ which is both defined over $\mathfrak{o}_v$ and whose reduction to the special fiber $\mathcal{G}_v\times_{\mathfrak{o}_v}\mathfrak{f}_v$ is not zero. This scalar c_v is unique up to multiplication by $\mathfrak{o}_v^\times$, hence $\gamma_v=|c_v|_v$ is well defined, and $|c_v\omega|_v=\gamma_v|\omega|_v$. If instead v is Archimedean, let $c_v\in\mathbb{R}^\times$ be such that the Haar measure induced by the form $c_v\omega$ on any maximal compact subgroup of $\mathrm{R}_{k_v/\mathbb{R}}\mathrm{G}(\mathbb{C})$ gives it measure 1. In particular, if $\mathrm{G}(k_v)$ is compact then $|c_v\omega|_v(\mathrm{G}(k_v))=1$.

The global effect of this rescaling is as follows:

Theorem ([91, Theorem 1.6])

$$\prod_{v \in V} \gamma_v = (D_\ell / D_k^{[\ell:k]})^{\frac{\mathfrak{s}(\mathrm{G})}{2}} \cdot \left(\prod_{v \in V_\infty} \left| \prod_{i=1}^{r} \frac{m_i!}{(2\pi)^{m_i+1}} \right|_v \right),$$

where ℓ is the splitting field of G*, $\mathfrak{s}(\mathrm{G})$ is an integer that depends on the root system of* G*, r is the absolute rank of* G*, and m_i are the exponents of the absolute Weyl group of* G.

The integer $\mathfrak{s}(\mathrm{G})$ is 0 (by convention) if G is split; otherwise, it is the number of short roots plus the number of short simple roots in the relative root system of G if the latter is reduced, and the number of all roots plus the number of simple roots in the relative root system of G if it is not reduced. For the convenience of the reader, we record this Lie-theoretic information in the table below.

Type	Diagram	Exponents $m_1, \ldots, m_r$	$\mathfrak{s}(\mathrm{G})$ $(= \mathfrak{s}(\mathcal{G}))$
A_r		1, 2, ..., r	$\begin{cases} \frac{1}{2}(r-1)(r+2) & r \text{ odd} \\ \frac{1}{2}r(r+3) & r \text{ even} \end{cases}$
B_r		1, 3, ..., $2r-1$	
C_r		1, 3, ..., $2r-1$	
D_r		1, 3, ..., $2r-3$, $r-1$	$2r-1$
E_6		1, 4, 5, 7, 8, 11	26
E_7	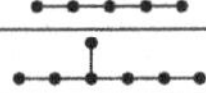	1, 5, 7, 9, 11, 13, 17	
E_8	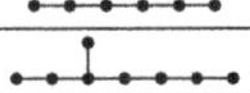	1, 7, 11, 13, 17, 19, 23, 29	
F_4		1, 5, 7, 11	
G_2		1, 5	

Now if G is not quasi-split, let $\mathcal{G}$ denote its quasi-split inner form. We make the choices of $\mathcal{P}_v$ and γ_v described above for $\mathcal{G}$ instead of G. Using an isomorphism $\varphi : \mathcal{G} \to \mathrm{G}$ defined over some Galois extension k' of k and such that ${}^\gamma\varphi \circ \varphi^{-1}$ is an inner automorphism for every $\gamma \in \mathrm{Gal}(k'/k)$, we then define the constants c_v as above, but for $\varphi^*(\omega)$ instead of ω; this does not depend on the choice φ. Similarly, the field ℓ and the constant $\mathfrak{s}(\mathrm{G})$ are defined using $\mathcal{G}$ instead of G.

4.2 Residue Field Datum and Volumes of Parahorics

For $v \in V_f$, let $\overline{\mathrm{M}}_v$ and $\overline{\mathcal{M}}_v$ denote the maximal reductive $\mathfrak{f}_v$-quotients of, respectively, $\overline{\mathrm{G}}_v$ and $\overline{\mathcal{G}}_v$, the fibers of G_v and $\mathcal{G}_v$ over the residue field $\mathfrak{f}_v$. Since $\mathfrak{f}_v$ is perfect, the unipotent radicals $\mathrm{R_u}(\overline{\mathrm{G}}_v)$ and $\mathrm{R_u}(\overline{\mathcal{G}}_v)$ are defined over $\mathfrak{f}_v$, so $\overline{\mathrm{M}}_v \cong \overline{\mathrm{G}}_v/\mathrm{R_u}(\overline{\mathrm{G}}_v)$ and $\overline{\mathcal{M}}_v \cong \overline{\mathcal{G}}_v/\mathrm{R_u}(\overline{\mathcal{G}}_v)$. Recall that for $v \in S_f$, we had set $\mathrm{P}_v = \mathrm{I}_v$; thus in this case the $\mathfrak{f}_v$-group $\overline{\mathrm{G}}_v$ happens to be solvable, and $\overline{\mathrm{M}}_v$ is a torus $\overline{\mathrm{T}}_v$ that can be identified with any maximal torus of $\overline{\mathrm{G}}_v$ via the quotient map.

The volumes of the parahorics P_v are given in terms of this datum as follows.

Proposition ([91, Proposition 2.10 (i)]) *For $v \in V_f$, we have*

$$\gamma_v |\omega|_v(\mathrm{P}_v) = \frac{\#\overline{\mathrm{M}}_v(\mathfrak{f}_v)}{q_v^{(\dim \overline{\mathrm{M}}_v + \dim \overline{\mathcal{M}}_v)/2}}.$$

4.3 Normalization of the Measure

For $v \in V_\infty$, let μ_v be the measure $|c_v\omega|_v = \gamma_v|\omega|_v$; in particular, $\mu_v(\mathrm{G}(k_v)) = 1$ whenever $\mathrm{G}(k_v)$ is compact. For $v \in S_f$, let μ_v be the *Tits measure*, that is, the Haar measure on $\mathrm{G}(k_v)$ for which the volume of the Iwahori subgroup I_v equals 1. The use of the Tits measure was suggested by Tits following his work with Kantor and Liebler on chamber-transitive actions on Bruhat–Tits buildings (see [57] and Sect. 9). Set $\mu_S = \prod_{v\in S} \mu_v$ to be the product measure on $\mathrm{G}(k_S) = \prod_{v\in S} \mathrm{G}(k_v)$.

With this setup, we can at last state the volume formula.

Theorem 4.1 (Prasad's volume formula [91, Theorem 3.7]) *With the notation introduced so far*

$$\mu_S(\mathrm{G}(k_S)/\Lambda(\mathrm{P}_v)_{v\notin S}) = D_k^{\frac{\dim \mathrm{G}}{2}} \left(D_\ell / D_k^{[\ell:k]}\right)^{\frac{\mathfrak{s}(\mathrm{G})}{2}} \cdot \left(\prod_{v\in V_\infty} \left|\prod_{i=1}^{r} \frac{m_i!}{(2\pi)^{m_i+1}}\right|_v\right) \cdot \prod_{v\in V_f} e(\mathrm{P}_v),$$

where the local factors are given by

$$e(\mathrm{P}_v) = \frac{q_v^{(\dim \overline{\mathrm{M}}_v + \dim \overline{\mathcal{M}}_v)/2}}{\#\overline{\mathrm{M}}_v(\mathfrak{f}_v)}, \quad v \in V_f.$$

(Recall that for $v \in S_f$, we have by convention $e(\mathrm{P}_v) = e(\mathrm{I}_v) = \frac{q_v^{(\dim \overline{\mathrm{T}}_v + \dim \overline{\mathcal{M}}_v)/2}}{\#\overline{\mathrm{T}}_v(\mathfrak{f}_v)}$.)

Proof From Equation (T) in Sect. 3.6, we know that

$$|\omega|_S\big(\mathrm{G}(k_S)/\Lambda(\mathrm{P}_v)_{v\notin S}\big) = D_k^{\frac{\dim \mathrm{G}}{2}} \cdot \tau_k(\mathrm{G}) \cdot \prod_{v\notin S} |\omega|_v(\mathrm{P}_v)^{-1}.$$

By construction of μ_S,

$$\mu_S\big(\mathrm{G}(k_S)/\Lambda(\mathrm{P}_v)_{v\notin S}\big) = \prod_{v\in V_\infty} \gamma_v \cdot \prod_{v\in S_f} |\omega|_v(\mathrm{I}_v)^{-1} \cdot |\omega|_S\big(\mathrm{G}(k_S)/\Lambda(\mathrm{P}_v)_{v\notin S}\big),$$

and using that $\tau_k(\mathrm{G}) = 1$ because G is simply connected, we deduce that

$$\mu_S\big(\mathrm{G}(k_S)/\Lambda(\mathrm{P}_v)_{v\notin S}\big) = D_k^{\frac{\dim \mathrm{G}}{2}} \cdot \prod_{v\in V_\infty} \gamma_v \cdot \prod_{v\in S_f} |\omega|_v(\mathrm{I}_v)^{-1} \cdot \prod_{v\notin S} |\omega|_v(\mathrm{P}_v)^{-1}$$
$$= D_k^{\frac{\dim \mathrm{G}}{2}} \cdot \prod_{v\in V} \gamma_v \cdot \prod_{v\in S_f} \big(\gamma_v|\omega|_v(\mathrm{I}_v)\big)^{-1} \cdot \prod_{v\notin S} \big(\gamma_v|\omega|_v(\mathrm{P}_v)\big)^{-1}.$$

Hence by the theorem in Sect. 4.1, it equals

$$D_k^{\frac{\dim \mathrm{G}}{2}} \cdot (D_\ell/D_k^{[\ell:k]})^{\frac{\mathfrak{s}(\mathrm{G})}{2}} \cdot \left(\prod_{v\in V_\infty} \left|\prod_{i=1}^{r} \frac{m_i!}{(2\pi)^{m_i+1}}\right|_v\right) \cdot \prod_{v\in V_f} \big(\gamma_v|\omega|_v(\mathrm{P}_v)\big)^{-1},$$

and finally by the proposition in Sect. 4.2, as promised

$$\mu_S\big(\mathrm{G}(k_S)/\Lambda(\mathrm{P}_v)_{v\notin S}\big) = D_k^{\frac{\dim \mathrm{G}}{2}} \cdot (D_\ell/D_k^{[\ell:k]})^{\frac{\mathfrak{s}(\mathrm{G})}{2}} \cdot \left(\prod_{v\in V_\infty} \left|\prod_{i=1}^{r} \frac{m_i!}{(2\pi)^{m_i+1}}\right|_v\right) \cdot \prod_{v\in V_f} e(\mathrm{P}_v). \qquad \square$$

Remark 4.2 The local factors $e(\mathrm{P}_v)$ of the Euler product are greater than 1 for every $v \in V_f$. In fact, in [91, Proposition 2.10], Prasad gives stronger bounds for $e(\mathrm{P}_v)$ in terms of q_v if G does not satisfy certain splitness conditions over k_v. These bounds and others of the same kind play an important role in many of the applications that we are about to discuss.

Remark 4.3 In the special case where k is a number field, $\mathrm{G} = \mathcal{G}$ is quasi-split, $S = V_\infty$, and $\mathrm{P}_v = \mathcal{P}_v$ is special, one has that $\overline{\mathrm{M}}_v = \overline{\mathcal{M}}_v$ is a simple $\mathfrak{f}_v$-group, and

$$e(\mathrm{P}_v) = e(\mathcal{P}_v) = \frac{q_v^{\dim \overline{\mathcal{M}}_v}}{\#\overline{\mathcal{M}}_v(\mathfrak{f}_v)}, \quad v \in V_f.$$

Using the formulas for the orders of the simple groups of Lie type, it follows that $\mathscr{X} = \prod_{v\in V_f} e(\mathcal{P}_v)$ is a product of values of certain L-functions (which depend on the form of $\mathcal{G}$ over k) at the integers $m_i + 1$ for $1 \le i \le r$. We record these values in the table below (see [56, Sect. 18.5.11]):

Type of the quasi-split group $\mathcal{G}$	Value of $\mathscr{Z}$
Split, any type	$\prod_{i=1}^{r} \zeta_k(m_i + 1)$
${}^2\mathrm{A}_r$ with r even	$\prod_{i=1}^{r/2} \left(\zeta_k(2i) L_{\ell/k}(2i+1)\right)$
${}^2\mathrm{A}_r$ with r odd	$\prod_{i=1}^{(r-1)/2} \left(\zeta_k(2i) L_{\ell/k}(2i+1)\right) \cdot \zeta_k(r+1)$
${}^2\mathrm{D}_r$	$\prod_{i=1}^{r-1} (\zeta_k(2i)) \cdot L_{\ell/k}(r)$
${}^2\mathrm{E}_6$	$\zeta_k(2) L_{\ell/k}(5) \zeta_k(6) \zeta_k(8) L_{\ell/k}(9) \zeta_k(12)$
${}^3\mathrm{D}_4$ or ${}^6\mathrm{D}_4$	$\zeta_k(2) L_{\ell/k}(4) \zeta_k(6)$

For a general group G and an arbitrary choice of parahorics P_v, the value of $\prod_{v \in V_f} e(\mathrm{P}_v)$ is an integer multiple of the corresponding one in the table.

Example 4.4 Consider again SL_n over $k = \mathbb{Q}$, any finite set $S \subset V$ containing $\{\infty\}$, and the hyperspecial parahorics $\mathrm{P}_p = \mathrm{SL}_n(\mathbb{Z}_p)$ for $p \notin S$. We have seen that the principal arithmetic subgroup they determine is $\mathrm{SL}_n(\mathbb{Z}[S_f^{-1}])$.

The group SL_n is split, hence $\ell = \mathbb{Q}$ and $\mathfrak{s}(\mathrm{SL}_n) = 0$. The exponents of the root system of type A_{n-1} are $1, \ldots, n-1$. The Bruhat–Tits group scheme associated with P_p for $p \notin S$ is simply $\mathrm{G}_p = \mathrm{SL}_n$ over $\mathbb{Z}_p$. In consequence, its special fiber is $\overline{\mathrm{G}}_p = \mathrm{SL}_n$ over $\mathbb{F}_p$, and we can compute

$$e(\mathrm{P}_p) = \frac{p^{n^2-1}}{\#\,\mathrm{SL}_n(\mathbb{F}_p)} = \frac{p^{n^2-1}}{p^{n-1}\prod_{i=0}^{n-2}(p^n - p^i)} = \prod_{i=2}^{n} \frac{1}{1-p^{-i}}.$$

For $p \in S$, $\mathrm{P}_p = \mathrm{I}_p$ reduces to the group $\overline{\mathrm{I}}_p$ of upper triangular unimodular $n \times n$ matrices over $\mathbb{F}_p$, whose maximal reductive quotient can be identified with the torus $\overline{\mathrm{T}}_p$ of diagonal unimodular $n \times n$ matrices over $\mathbb{F}_p$. Hence for $p \in S$,

$$e(\mathrm{P}_p) = e(\mathrm{I}_p) = \frac{p^{(n-1+n^2-1)/2}}{\#\overline{\mathrm{T}}_p(\mathbb{F}_p)} = \frac{p^{(n+2)(n-1)/2}}{(p-1)^{n-1}} = \frac{p^{n(n-1)/2}}{(1-p^{-1})^{n-1}} \geq p^{n(n-1)/2}.$$

Prasad's volume formula then immediately gives

$$\mu_S\big(\mathrm{SL}_n(\mathbb{R} \times \prod_{p \in S} \mathbb{Q}_p)/\,\mathrm{SL}_n(\mathbb{Z}[S_f^{-1}])\big) = \prod_{i=1}^{n-1} \frac{i!}{(2\pi)^{i+1}} \cdot \prod_{v \in S_f} \frac{p^{n(n-1)/2}}{(1-p^{-1})^{n-1}} \cdot \prod_{v \notin S} \prod_{i=2}^{n} \frac{1}{1-p^{-i}},$$

and in particular, when $S = \{\infty\}$

$$\mu_S(\mathrm{SL}_n(\mathbb{R})/\,\mathrm{SL}_n(\mathbb{Z})) = \prod_{i=1}^{n-1} \frac{i!}{(2\pi)^{i+1}} \cdot \prod_{i=2}^{n} \zeta(i).$$

The factor $\prod_{i=1}^{n-1} \frac{i!}{(2\pi)^{i+1}}$ is of course due to the choice of normalization for μ_∞, which is convenient in the presence of compact Archimedean factors.

Example 4.5 Consider now $\mathrm{SL}_{n,B}$, for B a quaternion algebra over a number field k. Let $S = V_\infty$ be the set of all Archimedean places of k, and choose P_v to be any coherent choice of special parahoric subgroups of $\mathrm{SL}_{n,B}(k_v)$ for $v \in V_f$. Then it turns out that the principal arithmetic subgroup determined by the P_v is the group $\mathcal{O}^1$ of elements of norm 1 of a maximal order $\mathcal{O}$ in B.

The group $\mathrm{SL}_{n,B}$ is an inner form of the split group SL_{2n}, so $\ell = k$. The reductions $\overline{\mathrm{G}}_v$ of P_v to $\mathfrak{f}_v$ have as maximal reductive quotient $\overline{\mathrm{M}}_v$ either SL_{2n} (if the quaternion algebra B splits over k_v) or the kernel of the determinant norm $\mathrm{GL}_n(\mathfrak{F}_v) \to \mathfrak{f}_v$, where $\mathfrak{F}_v$ is the quadratic extension of $\mathfrak{f}_v$ (if B is ramified at v). Thus,

$$e(\mathrm{P}_v) = \frac{q_v^{(2n)^2-1}}{\#\,\mathrm{SL}_{2n}(\mathfrak{f}_v)} = \prod_{i=2}^{2n} \frac{1}{1-q_v^{-i}}$$

if B splits over k_v, and otherwise

$$\begin{aligned} e(\mathrm{P}_p) &= \frac{q_v^{(2n^2-1+(2n)^2-1)/2}(q_v-1)}{\#\,\mathrm{GL}_n(\mathfrak{F}_v)} = \frac{q_v^{3n^2-1}(q_v-1)}{\prod_{i=0}^{n-1}(q_v^{2n}-q_v^{2i})} \\ &= \frac{q_v^{n^2-1}(q_v-1)}{\prod_{i=1}^{n}(1-q_v^{-2i})} = \prod_{i=1}^{n}(q_v^{2i-1}-1)\prod_{i=2}^{2n}\frac{1}{1-q_v^{-i}}. \end{aligned}$$

Prasad's volume formula now reads

$$\mu_\infty(\mathrm{SL}_{n,B}(k_\infty)/\mathcal{O}^1) = D_k^{\frac{4n^2-1}{2}} \cdot \left(\prod_{i=1}^{2n-1} \frac{i!}{(2\pi)^{i+1}}\right)^{[k:\mathbb{Q}]} \cdot \prod_{i=2}^{2n} \zeta_k(i) \cdot \prod_{v\in R_f}\left(\prod_{i=1}^{n}(q_v^{2i-1}-1)\right),$$

where R_f is the set of finite places of k over which B is ramified. In particular, for $n = 1$, we recover Borel's volume formula from Sect. 2:

$$\mu_\infty(\mathrm{SL}_{1,B}(k_\infty)/\mathcal{O}^1) = D_k^{\frac{3}{2}} \cdot \frac{1}{(2\pi)^{2[k:\mathbb{Q}]}} \cdot \zeta_k(2) \cdot \prod_{v\in R_f}(q_v-1).$$

Note in passing that if $k = \mathbb{Q}$ and B splits over $\mathbb{Q}_\infty = \mathbb{R}$, then $\mathcal{O}^1$ is a lattice in $\mathrm{SL}_{n,B}(\mathbb{R}) \cong \mathrm{SL}_{2n}(\mathbb{R})$ whose covolume is an integer multiple of that of $\mathrm{SL}_{2n}(\mathbb{Z})$, although $\mathcal{O}^1$ and $\mathrm{SL}_{2n}(\mathbb{Z})$ are not commensurable unless B splits over $\mathbb{Q}$.

5 A Finiteness Theorem for Arithmetic Groups of Bounded Covolume

It is known since the work of Wang [113] that in a fixed simple real Lie group not locally isomorphic to $SL_2(\mathbb{R})$ or $SL_2(\mathbb{C})$, there are only finitely many conjugacy classes of lattices whose covolume is less than a given bound. Motivated by his work [57] with Kantor and Liebler, Tits asked whether a similar result also holds for p-adic Lie groups, in fact not just for lattices in a single group, but also across a family of groups over different fields.

Of course, in order to compare covolumes in different groups, one needs a consistent way of normalizing the measure. Such a universal normalization exists for groups over non-Archimedean local fields: one can require the measure of the stabilizer of some (hence any) chamber in the Bruhat–Tits building to equal 1 (this is precisely the Tits measure that appeared in Sect. 4.3).

In response to Tits' question, Borel and Prasad [15] used the volume formula to prove the following striking finiteness theorem, which appeared right after Prasad's paper on the volume formula in Publ. Math. IHÉS.

Unless specified otherwise, in this section G will always denote an absolutely simple group (not necessarily simply connected) over a number field k. We fix for every $v \in V$ a Haar measure μ_v for $G(k_v)$ as in Sect. 4.3: for $v \in V_f$, μ_v is the Tits measure, giving volume 1 to the stabilizer of any chamber of the Bruhat–Tits building of G; for $v \in V_\infty$, the measure induced by μ_v on any maximal compact subgroup of $R_{k_v/\mathbb{R}}G(\mathbb{C})$ gives it measure 1. For any set S of places, μ_S is the product measure $\prod_{v\in S}\mu_v$.

5.1 *Finiteness theorem for arithmetic subgroups of bounded covolume [15, Theorem A]*

Theorem 5.1 *Let $b > 0$. There are only finitely many quadruples (k, G, S, Γ) with k, a number field; G, an absolutely simple k-group of absolute rank at least 2 up to k-isomorphism; S, a finite set of places of k containing V_∞; and Γ, an arithmetic subgroup of $G(k_S)$ up to conjugacy, such that $\mu_S(G(k_S)/\Gamma) \leq b$.*

The restriction on the absolute rank of G was later lifted, by filling in for type A_1 some of the steps of the proof of Borel and Prasad, see [8, Theorem 3.8].

Some conditional results of this kind are also available for the Euler–Poincaré measure, as well as over a function field; we refer the reader to [15, Sect. 7] and [16].

5.2 A Bound on the Index of a Principal Arithmetic Subgroup in Its Normalizer

Every arithmetic subgroup in a group without compact factors is by definition commensurable to a principal arithmetic subgroup. In fact, every maximal arithmetic subgroup Γ is the normalizer of an appropriate principal arithmetic subgroup Λ (this is *Rohlfs' maximality criterion*, see [15, Proposition 1.4 (iv)]). In order to bound the covolume of Γ, it thus suffices to compute the covolume of Λ and control the commensurability index $[\Gamma : \Lambda]$.

Borel and Prasad achieved the latter by using the action of the first Galois cohomology group of the center C of the universal cover of G (or flat cohomology if C is not reduced) on the local Dynkin diagrams Δ_v of G. This technique was invented by Rohlfs in [98], where he also proved the maximality criterion above. For simplicity, we will sketch the argument under the assumption that G is simply connected, but the bound given in the upcoming proposition is valid without this assumption.

From the exact sequence of k-groups

$$1 \to \mathrm{C} \to \mathrm{G} \xrightarrow{\varphi} \mathrm{Ad(G)} \to 1$$

follows for each $v \in V$ a longer exact sequence of Galois (or flat) cohomology sets

$$\begin{array}{ccccccccc} 1 \longrightarrow & \mathrm{C}(k) & \longrightarrow & \mathrm{G}(k) & \xrightarrow{\varphi} & \mathrm{Ad(G)}(k) & \xrightarrow{\delta} & \mathrm{H}^1(k,\mathrm{C}) & \longrightarrow & \mathrm{H}^1(k,\mathrm{G}) \\ & \downarrow & & \downarrow & & \downarrow & & \downarrow & & \downarrow \\ 1 \longrightarrow & \mathrm{C}(k_v) & \longrightarrow & \mathrm{G}(k_v) & \xrightarrow{\varphi} & \mathrm{Ad(G)}(k_v) & \xrightarrow{\delta_v} & \mathrm{H}^1(k_v,\mathrm{C}) & \longrightarrow & \mathrm{H}^1(k_v,\mathrm{G}). \end{array}$$

When k_v is non-Archimedean, $\mathrm{H}^1(k_v, \mathrm{G})$ vanishes by a theorem of Kneser. This means that δ_v induces an isomorphism $\mathrm{Ad(G)}(k_v)/\varphi(\mathrm{G}(k_v)) \cong \mathrm{H}^1(k_v, \mathrm{C})$ for each $v \in V_f$. Since G is simply connected, the action of $\mathrm{G}(k_v)$ on its Bruhat–Tits building is type-preserving. The action of $\mathrm{Ad(G)}(k_v)$ on the affine Dynkin diagram Δ_v of $\mathrm{G}(k_v)$ (via permutation of the types) thus factors through $\mathrm{Ad(G)}(k_v)/\varphi(\mathrm{G}(k_v)) \to \mathrm{Aut}(\Delta_v)$, which in turn gives a morphism $\xi_v : \mathrm{H}^1(k_v, \mathrm{C}) \to \mathrm{Aut}(\Delta_v)$. Denote Ξ_v the image of ξ_v, and $\mathrm{H}^1(k, \mathrm{C})_\xi$ the kernel of the morphism $\xi = (\xi_v)_{v \notin S} : \mathrm{H}^1(k, \mathrm{C}) \to \bigoplus_{v \notin S} \Xi_v$.

Let now Λ be a principal arithmetic subgroup of $\mathrm{G}(k_S)$ determined by parahorics $(\mathrm{P}_v)_{v \notin S}$. For $v \notin S$, set Θ_v to be the type of P_v, and Ξ_{Θ_v} to be the stabilizer of Θ_v in Ξ_v. Assume that the set S_i of places $v \in S$ for which the group G is isotropic over k_v is not empty. Let Γ be the normalizer of Λ in $\mathrm{G}(k_{S_i})$. Because Γ is also an arithmetic subgroup, its image $\varphi(\Gamma)$ lies in $\mathrm{Ad(G)}(k)$ (see [15, Proposition 1.2]). Thus the action of Γ on $\bigsqcup_{v \notin S} \Delta_v$ factors through

$$\Gamma/\Lambda \xrightarrow{\delta} \mathrm{H}^1(k, \mathrm{C})/\mathrm{H}^1(k, \mathrm{C})_\xi \xrightarrow{\xi} \bigoplus_{v \notin S} \Xi_v.$$

Since Γ normalizes Λ, its image in $\bigoplus_{v \notin S} \Xi_v$ lies in $\bigoplus_{v \notin S} \Xi_{\Theta_v}$. Moreover, the kernel of the map $\Gamma \to \mathrm{H}^1(k, \mathrm{C})$ induced by δ lies in $\Lambda \cdot \prod_{v \in S_i} \mathrm{C}(k_v)$ (see [15, Proposition 2.9]). Altogether, this proves the bound on $[\Gamma : \Lambda]$ we were looking for:

Proposition 5.2 ([15, Sect. 2.10]) *With* Λ *and* Γ *as above,*

$$[\Gamma : \Lambda] \leq \#(\mathrm{C}(k) \cap \Lambda)^{-1} \cdot \prod_{v \in S_i} \#\mathrm{C}(k_v) \cdot \#\mathrm{H}^1(k_v, \mathrm{C})_\xi \cdot \prod_{v \notin S} \#\Xi_{\Theta_v}.$$

One can in turn give a more explicit (but coarser) bound for the size of $\mathrm{H}^1(k_v, \mathrm{C})_\xi$; we refer the interested reader to [15, Propositions 5.1 and 5.6]. For inner forms this bound reads

$$\#\mathrm{H}^1(k_v, \mathrm{C})_\xi \leq h_k^\varepsilon \cdot n^{\varepsilon(r_k + s_k)} \cdot n^{\varepsilon \#(S_f \cup T)},$$

where h_k is the class number of k; the integers ε, n are such that $\mathrm{C} \cong \ker(\mathrm{GL}_1^\varepsilon \to \mathrm{GL}_1^\varepsilon : x \mapsto x^n)$; $r_k + s_k$ is the number of Archimedean places of k; and T is the set of places where G does not split. For outer forms, the class number of ℓ (the splitting field of the quasi-split inner form of G) comes into play, as well as the number of places where G is not quasi-split but does split over the maximal unramified extension.

5.3 *Bounds on the Arithmetical Invariants of k*

Both in the volume formula and in the previous proposition appear arithmetical invariants of the global field used to define the principal arithmetic subgroup Λ. When working with arbitrary arithmetic subgroups of a given group G over a local field, the field of definition may not be known. It is therefore very useful to have a control on the various arithmetic invariants appearing in the formulas in terms of simpler ones, for instance, in terms of the degree of k over $\mathbb{Q}$ when k is a number field, which will be assumed in this section. We remind the reader of some classical results in this direction.

Theorem (Minkowski's bound) *Let k be a number field with signature* (r_k, s_k) *and degree* $d_k = r_k + 2s_k$. *Then*

$$D_k^{\frac{1}{2}} \geq \left(\frac{\pi}{4}\right)^{s_k} \cdot \frac{d_k^{d_k}}{d_k!} \geq \left(\frac{\pi}{4}\right)^{\frac{d_k}{2}} \cdot \frac{e^{d_k}}{3\sqrt{d_k}} \quad \textit{(using Stirling's approximation).}$$

Odlyzko [86, 87] gave a series of lower bounds on the discriminant, which are better than Minkowski's in high degree; we record a sample of these bounds.

Theorem (Sample Odlyzko bound)

$$D_k \geq 60^{r_k} \cdot 22^{2s_k} \cdot e^{-81}.$$

Theorem (Upper bound in the Brauer–Siegel Theorem [101, Hilfssatz 1]) *Let h_k denote the class number of k, R_k its regulator, and w_k the number of roots of unity in k. Then*

$$h_k R_k \leq 2 w_k D_k \cdot \left(\frac{\pi}{12}\right)^{d_k}.$$

Lastly, we record a lower bound on the regulator established by Zimmert [117].

Theorem (Sample Zimmert bound)

$$R_k \geq 0.02 \cdot w_k \cdot e^{0.46 \cdot r_k + 0.1 \cdot s_k}.$$

These bounds or variants of them have been used, and at times are key, in the proof of the finiteness theorem (Theorem 5.1) and in most of the applications that we will describe in Sects. 6–9 and 10.

Example As an illustration, here is a very rough lower bound for the covolume of the normalizer Γ of a principal arithmetic group $\Lambda(\mathrm{P}_v)_{v \notin S}$ of a simple group G defined over a number field k that one deduces from Prasad's volume formula using the bounds in this section, and which only involves the degree d_k of k. For simplicity, we assume that G is an inner form of absolute rank at least 5. By the above,

$$[\Gamma : \Lambda] \leq \left(100\left(\frac{\pi}{12}\right)^{d_k} \cdot D_k \cdot n^{2\#S + \#T + d_k}\right)^{\varepsilon} \cdot \prod_{v \notin S} \#\Xi_{\Theta_v},$$

and as can be checked case by case (see, for instance, [15, Sect. 3.7]), when $r_v \geq 5$

$$\#\Xi_{\Theta_v}^{-1} \cdot e(\mathrm{P}_v) \geq 1 \text{ for } v \notin S \cup T; \quad n^{-\varepsilon} \cdot \#\Xi_{\Theta_v}^{-1} \cdot e(\mathrm{P}_v) \geq 1 \text{ for } v \in T.$$

Hence by Theorem 4.1,

$$\begin{aligned}\mu_S(\mathrm{G}(k_S)/\Gamma) &= [\Gamma : \Lambda]^{-1} \cdot D_k^{\frac{\dim \mathrm{G}}{2}} \left(D_\ell / D_k^{[\ell:k]}\right)^{\frac{\mathfrak{s}(\mathcal{G})}{2}} \cdot \left(\prod_{i=1}^{r} \frac{m_i!}{(2\pi)^{m_i+1}}\right)^{d_k} \cdot \prod_{v \in V_f} e(\mathrm{P}_v)\\ &\geq \left(100\left(\frac{\pi}{12}\right)^{d_k} n^{2\#S + d_k}\right)^{-\varepsilon} \left(\left(\frac{\pi}{4}\right)^{\frac{d_k}{2}} \frac{e^{d_k}}{3\sqrt{d_k}}\right)^{\frac{\dim \mathrm{G}}{2} - \varepsilon} \left(\prod_{i=1}^{r} \frac{m_i!}{(2\pi)^{m_i+1}}\right)^{d_k}.\end{aligned}$$

6 Hyperbolic Manifolds of Small Volume

Having at disposal the volume formula, one is drawn to go back to the context of Sect. 2 and start computing volumes of various quotients by arithmetic subgroups. We have already seen at the end of Sect. 4 that Prasad's volume formula recovers the different formulas presented in Sect. 2, but now arithmetic locally symmetric manifolds of any type are also on the menu.

In this section and in the next one, we give an overview of structural or classification results obtained using the volume formula concerning arithmetic manifolds whose volume is given or controlled by some bound. Typically, these results have the shape "if M is an arithmetic locally G-symmetric orbifold of volume less than b, then its fundamental group Γ has some specific form".

6.1 *Volumes of Arithmetic Real-Hyperbolic Orbifolds*

Using Prasad's volume formula (Theorem 4.1) and the techniques developed by Borel and Prasad to bound the index (Sect. 5.2), Belolipetsky [2] (for n even) and Belolipetsky and Emery [7] (for n odd) determined the non-compact arithmetic quotients of real-hyperbolic n-space $\mathcal{H}_{\mathbb{R}}^n$ of minimal volume, as well as the compact arithmetic quotients of $\mathcal{H}_{\mathbb{R}}^n$ of minimal volume among arithmetic quotients *of the first type*.

We summarize their results below, and refer the reader to [2, 7] for the exact description of the groups Γ_n^{nc}, Γ_n^{c} in each case. It turns out that for $n \geq 5$, the non-cocompact lattice Γ_n^{nc} has smaller covolume than the cocompact Γ_n^{c}.

Let Vol denote the usual hyperbolic volume on $\mathcal{H}_{\mathbb{R}}^n$. It is related to the normalization μ_∞ from Sect. 4.3 as follows: for any discrete subgroup Γ of $\mathrm{Spin}_{n,1}(\mathbb{R})$ containing the center, one has

$$\mathrm{Vol}(\mathcal{H}_{\mathbb{R}}^n/\Gamma) = \begin{cases} \dfrac{4(2\pi)^{n/2}}{(n-1)!!} \cdot \mu_\infty(\mathrm{Spin}_{n,1}(\mathbb{R})/\Gamma) = \dfrac{(2\pi)^{n/2}}{(n-1)!!} \cdot |\chi(\mathcal{H}_{\mathbb{R}}^n/\Gamma)| & \text{if } n \text{ is even,} \\[2ex] \dfrac{4\pi^{(n+1)/2}}{((n-1)/2)!} \cdot \mu_\infty(\mathrm{Spin}_{n,1}(\mathbb{R})/\Gamma) & \text{if } n \text{ is odd.} \end{cases}$$

For the purpose of the statement of the next theorem, a lattice Γ in $\mathrm{PO}_{n,1}(\mathbb{R})^\circ$ is *of the first type* (sometimes also called *of the simplest type*) if it is commensurable to an arithmetic subgroup of $\mathrm{PO}_{n,1}(\mathbb{R})$ constructed using the algebraic group PO_f, where f is a quadratic form defined over some number field k. In particular, f must have signature $(n, 1)$ at one Archimedean place of k, and signature $(n+1, 0)$ at every other Archimedean place.

A lattice in $\mathrm{PO}_{n,1}(\mathbb{R})^\circ$ is *of the second type* if it is constructed using the algebraic group PU_h instead, where h is a Hermitian form relative to a k-linear involution σ on a quaternion division algebra D over k whose fixed point is a three-dimensional subspace of D.

In terms of the classification of algebraic groups given by Tits [109], a k-form of $\mathrm{PO}_{n,1}$ yields a lattice of the first type if its Tits index is B_r, ${}^1\mathrm{D}_r^{(1)}$, or ${}^2\mathrm{D}_r^{(1)}$ and of the second type if its index is ${}^1\mathrm{D}_r^{(2)}$ or ${}^2\mathrm{D}_r^{(2)}$ (for r the floor of $(n+1)/2$).[4] In particular, when $n \geq 4$ is even, $\mathrm{PO}_{n,1}$ is of absolute type $\mathrm{B}_{n/2}$ and all arithmetic groups in $\mathrm{PO}_{n,1}(\mathbb{R})^\circ$ are of the first type.

Theorem ([2, 7]) *For every $n \geq 4$, there is a unique (up to conjugacy) non-cocompact arithmetic subgroup Γ_n^{nc} of $\mathrm{PO}_{n,1}(\mathbb{R})^\circ = \mathrm{Isom}^+(\mathcal{H}_\mathbb{R}^n)$ of smallest covolume among non-cocompact arithmetic subgroups. It is defined over $k_{\mathrm{nc}} = \mathbb{Q}$ and*

$$\mathrm{Vol}(\mathcal{H}_\mathbb{R}^n/\Gamma_n^{\mathrm{nc}}) = \begin{cases} \dfrac{4\cdot(2\pi)^r}{(2r-1)!!}\cdot\displaystyle\prod_{i=1}^{r}\left(\frac{(2i-1)!}{(2\pi)^{2i}}\zeta(2i)\right) & \text{if } n=2r \text{ with } r=0,1 \bmod 4,\\[2ex] \dfrac{2\cdot(2^r-1)(2\pi)^r}{(2r-1)!!}\cdot\displaystyle\prod_{i=1}^{r}\left(\frac{(2i-1)!}{(2\pi)^{2i}}\zeta(2i)\right) & \text{if } n=2r \text{ with } r=2,3 \bmod 4,\\[2ex] \dfrac{1}{2^{r-2}}\cdot\displaystyle\prod_{i=1}^{r-1}\left(\frac{(2i-1)!}{(2\pi)^{2i}}\zeta(2i)\right)\cdot\zeta(r) & \text{if } n=2r-1 \text{ with } r=1 \bmod 4,\\[2ex] \dfrac{(2^r-1)(2^{r-1}-1)}{3\cdot 2^{r-1}}\cdot\displaystyle\prod_{i=1}^{r-1}\left(\frac{(2i-1)!}{(2\pi)^{2i}}\zeta(2i)\right)\cdot\zeta(r) & \text{if } n=2r-1 \text{ with } r=3 \bmod 4,\\[2ex] \dfrac{3^{r-1/2}}{2^{r-1}}\cdot\displaystyle\prod_{i=1}^{r-1}\left(\frac{(2i-1)!}{(2\pi)^{2i}}\zeta(2i)\right)\cdot L_{\ell_{\mathrm{nc}}/\mathbb{Q}}(r) & \text{if } n=2r-1 \text{ with } r=2,4 \bmod 4, \end{cases}$$

where $\ell_{\mathrm{nc}} = \mathbb{Q}(\sqrt{-3})$, and $L_{\ell_{\mathrm{nc}}/\mathbb{Q}}$ is the L-function of the character associated with the quadratic extension $\ell_{\mathrm{nc}}/\mathbb{Q}$.

Theorem ([2, 7]) *For every $n \geq 4$, there is a unique (up to conjugacy) cocompact arithmetic subgroup Γ_n^{c} of $\mathrm{PO}_{n,1}(\mathbb{R})^\circ = \mathrm{Isom}^+(\mathcal{H}_\mathbb{R}^n)$ of smallest covolume among cocompact arithmetic subgroups of the first type.[5] It is defined over $k_{\mathrm{c}} = \mathbb{Q}[\sqrt{5}]$ and*

$$\mathrm{Vol}(\mathcal{H}_\mathbb{R}^n/\Gamma_n^{\mathrm{c}}) = \begin{cases} \dfrac{4\cdot 5^{r^2+r/2}(2\pi)^r}{(2r-1)!!}\cdot\displaystyle\prod_{i=1}^{r}\left(\frac{((2i-1)!)^2}{(2\pi)^{4i}}\zeta_{k_{\mathrm{c}}}(2i)\right) & \text{if } n=2r \text{ with } r \text{ even},\\[2ex] \dfrac{2\cdot 5^{r^2+r/2}(2\pi)^r(4r-1)}{(2r-1)!!}\cdot\displaystyle\prod_{i=1}^{r}\left(\frac{((2i-1)!)^2}{(2\pi)^{4i}}\zeta_{k_{\mathrm{c}}}(2i)\right) & \text{if } n=2r \text{ with } r \text{ odd},\\[2ex] \dfrac{5^{r^2-r/2}\cdot 11^{r-1/2}(r-1)!}{2^{2r-1}\pi^r}\cdot\displaystyle\prod_{i=1}^{r-1}\left(\frac{((2i-1)!)^2}{(2\pi)^{4i}}\zeta_{k_{\mathrm{c}}}(2i)\right)\cdot L_{\ell_{\mathrm{c}}/k_{\mathrm{c}}}(r) & \text{if } n=2r-1, \end{cases}$$

where ℓ_{c} is the quadratic field extension of k_{c} which is the splitting field of $X^4 - X^3 + 2X - 1$, and $L_{\ell_{\mathrm{c}}/k_{\mathrm{c}}}$ is the L-function of the character associated with the quadratic extension $\ell_{\mathrm{c}}/k_{\mathrm{c}}$.

---

[4] When $n = 3$ or $n = 7$, there is also an exceptional *third type*, stemming respectively from the isomorphism ${}^2\mathrm{D}_2 \cong {}^2(\mathrm{A}_1 \times \mathrm{A}_1)$ or from the trialitarian forms with Tits index ${}^6\mathrm{D}_4$; these exceptional types are formally unrelated. Note that groups with Tits index ${}^2\mathrm{D}_2$ are not absolutely simple.

[5] M. Belolipetsky informed the author that in [7], the treatment of cocompact lattices of the second type defined over $\mathbb{Q}$ was missing. Excluding those, Theorems 1 and 2 in [7] hold as stated here.

To put these results in context, we call back to mind the hyperbolic orbifolds of smallest volumes in dimensions $n = 2$ and 3. As was mentioned in Sect. 2.1, the case of $SL_2(\mathbb{R})$ was solved by Siegel using planar hyperbolic geometry: the smallest hyperbolic 2-orbifold is the quotient of $\mathcal{H}^2$ by the $(2, 3, 7)$-triangle group, with volume $\pi/42$. The case $n = 3$ took substantially more effort: more than 60 years after Siegel asked the question, Gehring, Marshall, and Martin [38, 78] determined the unique (up to conjugacy) smallest volume hyperbolic 3-orbifold. It is compact and its fundamental group is the hyperbolic reflection group with Coxeter diagram $\bullet\!-\!\bullet\overset{5}{-}\bullet\!-\!\bullet$, which is an arithmetic group defined using a form of $PO_{3,1}$ over the splitting field k_0 of $X^4 + 6X^3 + 12X^2 + 9X + 1$, a quartic field of discriminant -275. The volume of the corresponding orbifold is $\frac{275^{3/2}}{2^7\pi^6} \cdot \zeta_{k_0}(2)$. The non-compact orientable hyperbolic 3-orbifold of smallest volume had been determined previously by Meyerhoff [79]: under the identification $PO_{3,1}(\mathbb{R})^\circ \cong PGL_2(\mathbb{C})$, it is the quotient of $\mathcal{H}^3_{\mathbb{R}}$ by $PGL_2(\mathfrak{o})$, where $\mathfrak{o}$ is the ring of integers of $\mathbb{Q}(\sqrt{-3})$.

Note that the smallest hyperbolic 2-*manifold* has volume π, because the Euler–Poincaré characteristic of a hyperbolic surface is a strictly negative integer. The smallest orientable hyperbolic 3-*manifold* is the Weeks manifold (see [35]), which is arithmetic and cocompact but is not commensurable to the smallest orbifold of Gehring, Marshall, and Martin. This indicates that questions of minimal volume can be very subtle, as even the allowance of torsion or singular points can drastically change the answer.

Beyond dimension 3, Hild [54] determined the unique non-cocompact hyperbolic orbifolds of minimal volume in dimensions $5 \leq n \leq 9$ using crystallographic methods, and they are arithmetic, that is, they coincide with the quotients $\mathcal{H}^n_{\mathbb{R}}/\Gamma^{\mathrm{nc}}_n$ above. Subsequent work by Kellerhals [59], relying on the results of Belolipetsky and Emery, makes explicit the oriented non-cocompact arithmetic quotients of $\mathcal{H}^n_{\mathbb{R}}$ that achieve minimal volume (among oriented non-cocompact arithmetic quotients) for $n \in \{10, 11, 12, 13, 14, 16, 18\}$.

On a related note, Prasad's volume formula was also used by Emery and Kellerhals [32] to determine the three compact arithmetic 5-orbifolds with smallest volume, and by Emery [31] to show that there is no compact arithmetic hyperbolic n-manifold with $n \in \{6, 8, 10\}$ whose Euler–Poincaré characteristic has absolute value equal to 2.

For more on the topic of hyperbolic orbifolds of small volume, the interested reader can consult the more detailed surveys [5, 58].

6.2 *Volumes of Arithmetic Complex-Hyperbolic Orbifolds*

With the intent of finding the smallest complex-hyperbolic orbifolds, Emery and Stover [34] studied non-cocompact arithmetic subgroups of small covolume in $PU_{n,1}(\mathbb{R})$ for $n \geq 2$. It follows, for instance, from the classification [109] that such subgroups are constructed using the forms PU_h with h a Hermitian form of signature

$(n, 1)$ with respect to an imaginary quadratic extension ℓ of $\mathbb{Q}$.[6] Emery and Stover determined the smallest possible covolume in each commensurability class of non-cocompact arithmetic subgroups, and estimated the number of subgroups attaining this covolume. We record here the smallest covolume among all these.

Theorem ([34, Theorems 1.1–1.4]) *Let $n \geq 2$. If n is even, there are exactly two conjugacy classes of lattices Γ_n^{nc} with smallest covolume among non-cocompact arithmetic subgroups of $\mathrm{PU}_{n,1}(\mathbb{R})$. If $n = 1$, 3, or 5 modulo 8, there is exactly one conjugacy class of lattices Γ_n^{nc} with smallest covolume among non-cocompact arithmetic subgroups; for $n = 7$ modulo 8, there are either one or two conjugacy classes. In all cases, they are constructed using a Hermitian form with respect to the quadratic field extension $\ell_{\mathrm{nc}} = \mathbb{Q}(\sqrt{-3})$ of $k_{\mathrm{nc}} = \mathbb{Q}$. Their Euler–Poincaré characteristic is, respectively,*

$$\chi(\mathcal{H}_{\mathbb{C}}^n/\Gamma_n^{\mathrm{nc}}) = \begin{cases} \dfrac{n+1}{2^n} \cdot \displaystyle\prod_{i=1}^{n/2} \left(\zeta(1-2i) \cdot L_{\ell/\mathbb{Q}}(-2i)\right) & \textit{if } n \textit{ is even,} \\ \dfrac{(-1)^{(n+1)/2}(n+1)}{2^{n-1}} \cdot \displaystyle\prod_{i=1}^{(n-1)/2} \left(\zeta(1-2i) \cdot L_{\ell/\mathbb{Q}}(-2i)\right) \cdot \zeta(-n) & \textit{if } n \textit{ is odd.} \end{cases}$$

The Euler–Poincaré characteristic is related to the normalization of Sect. 4.3 via $\mu_\infty(\mathrm{SU}_{n,1}(\mathbb{R})/\Gamma) = (n+1)^{-2} \cdot |\chi(\mathcal{H}_{\mathbb{C}}^n/\Gamma)|$, for Γ any discrete subgroup of $\mathrm{SU}_{n,1}(\mathbb{R})$ containing the center. By the Chern–Gauss–Bonnet theorem, it relates to the complex-hyperbolic volume (normalized for the metric with holomorphic sectional curvature equal to -1) via $\mathrm{Vol}(\mathcal{H}_{\mathbb{C}}^n/\Gamma) = \frac{(-4\pi)^n}{(n+1)!} \cdot \chi(\mathcal{H}_{\mathbb{C}}^n/\Gamma)$. We refer the reader to [34] for the description of the arithmetic groups in question.

As far as the author is aware, beyond arithmetic groups the complex-hyperbolic manifolds of smallest volume are only known in dimension $n = 2$, where Hersonsky and Paulin [53] determined the smallest compact complex-hyperbolic surface, and Parker [89] determined the smallest non-compact complex-hyperbolic surface; they are of volume $8\pi^2$ and $8\pi^2/3$, respectively.

6.3 *Volumes of (Arithmetic) Quaternionic-Hyperbolic Orbifolds*

A few years ago, Emery and Kim investigated the last remaining case, that of quotients of the quaternionic-hyperbolic space.

[6] This fact is misreferenced in [34, Sect. 2.1]. However, it can be deduced from the classification by observing that (i) adjoint groups of outer type A are of the form PU_h for h a Hermitian form relative to an involution of the second kind on a division algebra D whose center ℓ is a quadratic extension of a base field k; (ii) if k is not $\mathbb{Q}$, then the form is always anisotropic at some Archimedean place of k hence any arithmetic subgroup constructed via this form is cocompact; and (iii) if $k = \mathbb{Q}$ but $D \neq \ell$, then because D splits over $\ell \otimes_{\mathbb{Q}} \mathbb{R} = \mathbb{C}$, the rank of PU_h over $\mathbb{R}$ will be at least $\deg_\ell D \cdot \mathrm{rank}_{\mathbb{Q}} \mathrm{PU}_h$ hence can equal 1 only when PU_h is anisotropic over $\mathbb{Q}$.

The group of orientation-preserving isometries of quaternionic-hyperbolic space $\mathcal{H}^n_{\mathbb{H}}$ is $\mathrm{PSp}_{n,1}(\mathbb{R})$, the group of $(n+1) \times (n+1)$ matrices with coefficients in Hamilton's quaternion algebra $\mathbb{H}$ preserving the standard Hermitian form (with respect to quaternion conjugation) of signature $(n, 1)$. Orientable quaternionic-hyperbolic orbifolds are thus in correspondence with the discrete subgroups of $\mathrm{PSp}_{n,1}(\mathbb{R})$.

Emery and Kim [33] determined the cocompact and non-cocompact quaternionic-hyperbolic orbifolds of smallest covolume, again using Prasad's volume formula. Note that arithmeticity holds for all groups of type C (by the work of Margulis [77] for higher rank, and of Corlette [25], and Gromov and Schoen [44] for groups of types C and F_4 of relative rank 1).

Theorem ([33, Theorems 1 and 2]) *For every $n \geq 2$, there is a unique (up to conjugacy) non-cocompact lattice Γ^{nc}_n of $\mathrm{Sp}_{n,1}(\mathbb{R})$ of smallest covolume among non-cocompact lattices. It is defined over $k_{\mathrm{nc}} = \mathbb{Q}$, the lattice is commensurable to the subgroup, not $\mathbb{Q}$, is commensurable to the subgroup of $\mathrm{Sp}_{n,1}(\mathbb{R})$ preserving $n+1$ copies of the Hurwitz order, and*

$$\chi(\mathcal{H}^n_{\mathbb{H}}/\Gamma^{\mathrm{nc}}_n) = \begin{cases} \dfrac{n+1}{2^{n+2}} \cdot \displaystyle\prod_{i=1}^{n+1}(2^i + (-1)^i) \cdot \prod_{i=1}^{n+1} \zeta(1-2i) & \text{if } n \text{ is even,} \\ \dfrac{n+1}{2^{n+2}} \cdot \displaystyle\prod_{i=1}^{(n+1)/2} (2^{2(2i-1)} - 1) \cdot \prod_{i=1}^{n+1} \zeta(1-2i) & \text{if } n \text{ is odd.} \end{cases}$$

Theorem ([33, Theorem 3]) *For every $n \geq 2$, there is a unique (up to conjugacy) cocompact lattice Γ^{c}_n of $\mathrm{Sp}_{n,1}(\mathbb{R})$ of smallest covolume among cocompact lattices. It is defined over $k_{\mathrm{c}} = \mathbb{Q}(\sqrt{5})$, the lattice is commensurable to the subgroup, not $\mathbb{Q}$, is commensurable to the subgroup of $\mathrm{Sp}_{n,1}(\mathbb{R})$ preserving $n+1$ copies of the icosian order, and*

$$\chi(\mathcal{H}^n_{\mathbb{H}}/\Gamma^{\mathrm{c}}_n) = \frac{n+1}{4^{n+1}} \cdot \prod_{i=1}^{n+1} \zeta_{k_{\mathrm{c}}}(1-2i).$$

The Euler–Poincaré characteristic is related to the normalization of Sect. 4.3 via $\mu_\infty(\mathrm{Sp}_{n,1}(\mathbb{R})/\Gamma) = 2^{-1}(n+1)^{-1} \cdot |\chi(\mathcal{H}^n_{\mathbb{H}}/\Gamma)|$, for Γ any discrete subgroup of $\mathrm{Sp}_{n,1}(\mathbb{R})$ containing the center. We refer to [33] for the precise description of Γ^{nc}_n and Γ^{c}_n.

Between Γ^{nc}_n and Γ^{c}_n, the latter has smaller covolume only for $n = 2$. Thus, except for $n \leq 2$, the smallest volume quaternionic-hyperbolic orbifold is not cocompact.

7 Classification of Fake Projective Planes

A *fake projective plane* is a smooth compact complex surface which has the same Betti numbers as the complex projective plane, but is not the complex projective plane. Any fake projective plane P is the quotient of the complex-hyperbolic plane

$\mathcal{H}^2_{\mathbb{C}}$ (the symmetric space of $\mathrm{PU}_{2,1}(\mathbb{R})$) by the fundamental group of P, which is by definition a torsion-free cocompact lattice of $\mathrm{PU}_{2,1}(\mathbb{R})$ with Euler–Poincaré characteristic equal to 3. By Mostow's strong rigidity theorem, the fundamental group of P uniquely determines its isometry class. In view of the last paragraphs of Sect. 6.2, fake projective planes have the smallest volume among compact complex-hyperbolic manifolds.

Mumford [84] constructed the first example of a fake projective plane, and proved that there can only be finitely many. Klingler [62] and Yeung [115, 116] independently proved that every fake projective plane is arithmetic, that is, its fundamental group is an arithmetic subgroup of $\mathrm{PU}_{2,1}(\mathbb{R})$. This allows us to put to work the methods of Sects. 4 and 5.

Let P be a fake projective plane and Π its fundamental group. By the Gauss–Bonnet theorem, the Euler–Poincaré characteristic of Π is proportional to the volume of P. More precisely, the preimage $\tilde{\Pi}$ of Π in $\mathrm{SU}_{2,1}(\mathbb{R})$ has Euler–Poincaré characteristic equal to 1, which amounts to $\mu_\infty(\mathrm{SU}_{2,1}(\mathbb{R})/\tilde{\Pi}) = 1/3$ with the normalization of Sect. 4.3 (also used in Sect. 5). One can thus recover the finiteness of the set of fake projective planes (in fact, of the set of complex arithmetic surfaces of Euler–Poincaré characteristic equal to 3) from Theorem 5.1. But the equality $\mu_\infty(\mathrm{SU}_{2,1}(\mathbb{R})/\tilde{\Pi}) = 1/3$ is much more informative, as it allowed Prasad and Yeung [92, 94] to classify the commensurability classes of fake projective planes using the volume formula. Note that there is a revised version of [92] available on theAll references made below are to this revised version. Let us also mention the Bourbaki Seminar [97] on this topic.

By the classification of semisimple algebraic groups [109], a form G of $\mathrm{SU}_{2,1}(\mathbb{R})$ over a number field k admitting a real place v_0 is given by the following datum: a quadratic extension ℓ of k over which v_0 ramifies, a simple algebra D of degree 3 over its center ℓ, an involution σ on D inducing the non-trivial automorphism of ℓ/k; the k-form G is then the special unitary group SU_h of norm 1 elements preserving the Hermitian form $h(x, y) = \sigma(x)y$ on $D \times D$. For a principal arithmetic subgroup of G to give a lattice in $\mathrm{SU}_{2,1}(\mathbb{R})$, the fields k and ℓ need to be, respectively, totally real and totally complex. (It turns out that G is uniquely determined by (k, ℓ, D), as different choices of σ yield k-isomorphic groups, see [92, Sect. 1.2].)

By estimating the different terms appearing in Prasad's volume formula, Prasad and Yeung show that every arithmetic subgroup Γ of $\mathrm{SU}_{2,1}(\mathbb{R})$ satisfying $\mu_\infty(\mathrm{SU}_{2,1}(\mathbb{R})/\Gamma) \leq 1/3$ normalizes a principal V_∞-arithmetic subgroup $\Lambda(\mathrm{P}_v)_{v \notin V_\infty}$ of a group G as above, where either $k = \mathbb{Q}$ and ℓ is one of 11 imaginary quadratic fields, or $k \neq \mathbb{Q}$ and (k, ℓ) is among 40 possible pairs (the precise lists can be found in [92, Proposition 3.5 and Sect. 8.2]).

With more in-depth analysis of these data, Prasad and Yeung then determine the precise list of those quadruples $(k, \ell, D, (\mathrm{P}_v)_{v \notin V_\infty})$ which do give rise to a fake projective plane. To rule out certain cases, they use that every simple algebra D giving rise to a fake projective plane must be a division algebra, a fact established by Cartwright and Steger [22, Theorem 2.1]. In fact, for each possible pair (k, ℓ), there is at most one central-simple ℓ-algebra D which gives rise to a fake projective plane (see [92, Sects. 4.5 and 9.2]).

Theorem (Classification of fake projective planes [92, Theorem 9.13]) *There are exactly 28 quadruples* $(k, \ell, D, (\mathrm{P}_v)_{v \notin V_\infty})$ *for which the normalizer of* $\Lambda(\mathrm{P}_v)_{v \notin V_\infty}$ *in* $\mathrm{G}(k_{v_0}) \cong \mathrm{SU}_{2,1}(\mathbb{R})$ *contains the preimage* $\tilde{\Pi}$ *of the fundamental group of a fake projective plane.*

They are distributed in nine commensurability classes, one for each pair of fields (k, ℓ) *in the following table.*

In particular, there are precisely nine commensurability classes of fake projective planes, each one uniquely determined by its trace field ℓ.

(k, ℓ)	$\mathscr{T}$	# of quadruples
$(\mathbb{Q}, \mathbb{Q}(\sqrt{-1}))$	$\{5\}$	2
	$\{2, 5\}$	1
$(\mathbb{Q}, \mathbb{Q}(\sqrt{-2}))$	$\{3\}$	2
	$\{2, 3\}$	1
$(\mathbb{Q}, \mathbb{Q}(\sqrt{-7}))$	$\{2\}$	2
	$\{2, 3\}$	2
	$\{2, 5\}$	2
$(\mathbb{Q}, \mathbb{Q}(\sqrt{-15}))$	$\{2\}$	4
$(\mathbb{Q}, \mathbb{Q}(\sqrt{-23}))$	$\{2\}$	2
$\mathcal{C}_2 : (\mathbb{Q}(\sqrt{5}), \mathbb{Q}(\sqrt{5}, \zeta_3))$	$\{\mathfrak{v}\}$	2
$\mathcal{C}_{10} : (\mathbb{Q}(\sqrt{2}), \mathbb{Q}(\sqrt{-7+4\sqrt{2}}))$	$\{\mathfrak{v}\}$	2
$\mathcal{C}_{18} : (\mathbb{Q}(\sqrt{6}), \mathbb{Q}(\sqrt{6}, \zeta_3))$	$\{\mathfrak{v}\}$	1
	$\{\mathfrak{v}, \mathfrak{v}_2\}$	2
$\mathcal{C}_{20} : (\mathbb{Q}(\sqrt{7}), \mathbb{Q}(\sqrt{7}, \zeta_4))$	$\{\mathfrak{v}\}$	1
	$\{\mathfrak{v}, \mathfrak{v}_3'\}$	1
	$\{\mathfrak{v}, \mathfrak{v}_3''\}$	1

In the table, the set $\mathscr{T}$ is the collection of the non-Archimedean places v of k where P_v is not maximal, along with the places v of k which do not ramify over ℓ and where P_v is not a hyperspecial parahoric. For $\mathcal{C}_2$, $\mathcal{C}_{10}$ and $\mathcal{C}_{20}$, $\mathfrak{v}$ is the unique place of k lying over 2; for $\mathcal{C}_{18}$, it is the unique place of k lying over 3. The place $\mathfrak{v}_2$ is the unique place of k lying over 2 for $\mathcal{C}_{18}$, and for $\mathcal{C}_{20}$, $\mathfrak{v}_3'$, $\mathfrak{v}_3''$ are the two places of k lying over 3. We refer to [92, Sects. 4–5, 7–9] for more information about the construction of G and $(\mathrm{P}_v)_{v \notin V_\infty}$ in terms of $\mathscr{T}$.

Using that for each class containing a fake projective plane there exists a (unique) non-Archimedean place $\mathfrak{v}$ of k where $\mathrm{G}_{k_\mathfrak{v}}$ is compact, Prasad and Yeung deduce that the abelianization of Π is non-trivial:

Corollary ([92, Theorem 10.1]) *If P is a fake projective plane with fundamental group Π, then* $\mathrm{H}_1(P, \mathbb{Z}) = \mathrm{H}_1(\Pi, \mathbb{Z}) = \Pi/[\Pi, \Pi]$ *is not trivial. In consequence, any smooth complex surface with the same integral homology groups as the complex projective plane is biholomorphic to it.*

Following up on the first version of this classification and assisted by computer, Cartwright and Steger [22, Theorem 2.2] determined the number of non-homeomorphic fake projective planes given by each quadruple $(k, \ell, D, (\mathrm{P}_v)_{v \notin V_\infty})$, and gave a set of generators for each of them, among other things. There are 50 fake projective planes up to homeomorphism (100 up to biholomorphism, as each one carries two complex structures). The reader can consult [22] for more details.

Very recently, using the work of Cartwright and Steger, Stover [106] studied the profinite completions of the 50 different fundamental groups of fake projective planes, and showed that among them, four pairs have isomorphic profinite completions, while the other 42 have distinct profinite completions.

Lastly, the existence of more general fake compact Hermitian symmetric spaces has been studied in [93, 95, 96], also using the volume formula. Arithmetic fake compact Hermitian symmetric spaces do not exist outside of types A_n, $1 \leq n \leq 4$. Inside of this range, A_3 is the only type for which it is not yet known whether they exist.

8 Lattices of Minimal Covolume in Semisimple Lie Groups

Taking a step back from the geometry of the symmetric space, we could also have asked from the start what lattices achieve the smallest covolume in the group itself. Both questions are obviously related: if one works with locally G-symmetric orbifolds, the volume of a quotient of the symmetric space of G is directly proportional to the covolume of its fundamental group in G, and the proportionality constant depends only on the normalization of the invariant measure on G versus on its symmetric space. If one restricts to locally symmetric manifolds, then one needs to filter this output in terms of torsion-free subgroups of G. Although theoretically straightforward, this last procedure can sometimes be a subtle matter in practice.

In this section, we present the groups in which the lattices of minimal covolume (at least among arithmetic ones) have been determined, along with a short description of their classification. By Margulis' arithmeticity theorem [77, Chap. IX], all irreducible lattices in groups of higher rank are arithmetic, so in those cases the classification becomes unconditional. In contrast, the question whether a lattice of minimal covolume in a simple group of rank 1 over $\mathbb{R}$ (or over another local field) is necessarily arithmetic, remains open:

Conjecture 8.1 (Folklore) *Let* G *be a semisimple* $\mathbb{R}$*-group. If* Γ *is a lattice of minimal covolume in* $\mathrm{G}(\mathbb{R})$*, then* Γ *is arithmetic.*

All the results discussed in Sect. 6 are in support of this conjecture, but so far no clear strategy to prove it has emerged. As highlighted by Belolipetsky in his ICM survey [5, Sect. 5], the known lower bounds for the covolume of non-necessarily-arithmetic lattices are decreasing super-exponentially in terms of the dimension of the ambient group, whereas the lower bound for arithmetic lattices deduced from the volume formula is increasing super-exponentially. Since the gap between these bounds is not substantiated by a non-arithmetic example, one wonders if the gap lies in our knowledge instead.

8.1 Real Lie Groups of Real Rank One

The works discussed in Sect. 6 give the lattices of minimal covolume in $\mathrm{PO}_{n,1}$ among arithmetic lattices of the first type, in $\mathrm{PU}_{n,1}$ among non-cocompact arithmetic lattices, and in $\mathrm{PSp}_{n,1}$ among all lattices. We refer the reader to that section and the references therein.

8.2 Chevalley Groups Over $\mathbb{F}_q((t))$

By establishing a quantitative version of the Kazhdan–Margulis theorem for Chevalley groups over $\mathbb{F}_q((t))$, Golsefidy classified the lattices of minimal covolume in the classical Chevalley groups over $\mathbb{F}_q((t))$. (It is conceivable that this theorem can also be proved using the volume formula.)

Theorem ([41, Main Theorem]) *Let* G *be a Chevalley group of rank at least* 2*, which is either of classical type or of type* E_6*. Suppose that* q *is a power of a prime* $p > 5$*, and that* $q > 9$ *when* G *is not of type* A*. Then* $\mathrm{G}(\mathbb{F}_q[t^{-1}])$ *is the unique lattice of minimal covolume in* $\mathrm{G}(\mathbb{F}_q((t)))$*, up to automorphism.*

In positive characteristic, let us also mention the preceding work of Lubotzky [69], who showed that the so-called *characteristic p modular group* $\mathrm{SL}_2(\mathbb{F}_q[t^{-1}])$ is a lattice of minimal covolume in $\mathrm{SL}_2(\mathbb{F}_q((t)))$. He observed that, contrary to the case of the (2, 3, 7)-triangle group in $\mathrm{SL}_2(\mathbb{R})$ (at the time, the only lattice of minimal covolume known among all lattices of a simple group), $\mathrm{SL}_2(\mathbb{F}_q[t^{-1}])$ is not cocompact. This led him to ask the following question.

Question (*Lubotzky*) In a semisimple Lie group over a local field, is a lattice of minimal covolume typically cocompact or non-cocompact?

Lubotzky also determined lattices of minimal covolume in $\mathrm{SL}_2(\mathbb{Q}_p)$ for every prime p (see [72] for extensions of $\mathbb{Q}_p$). As lattices in semisimple p-adic groups are always cocompact, the answer to Lubotzky's question in this case is unenlightening.

Using Prasad's volume formula (combined with techniques from his previous work), Golsefidy [43, Theorem 1] proved that any lattice of minimal covolume in a simply connected absolutely simple group of rank at least 2 over $\mathbb{F}_q((t))$ is non-cocompact. This answers Lubotzky's question in positive characteristic.

The available evidence, presented in Sects. 6 and 8, seems to indicate that lattices of minimal covolume in real Lie groups also tend to be non-cocompact, and that cocompact ones are low-dimensional exceptions.

8.3 Split Simple Groups Over $\mathbb{R}$

In his doctoral thesis, the author determined the lattices of minimal covolume in $\mathrm{SL}_n(\mathbb{R})$ for $n \geq 3$.

Theorem ([108]) *Let $n \geq 3$. Any lattice of minimal covolume in $\mathrm{SL}_n(\mathbb{R})$ is conjugate to $\mathrm{SL}_n(\mathbb{Z})$.*

As for many of the results presented in Sects. 6 and 7, the argument is based on Prasad's volume formula (Theorem 4.1); the methods of Rohlfs, Borel, and Prasad to control the index of principal arithmetic subgroups (Sect. 5.2); and a careful analysis of the arithmetic invariants appearing (using bounds in the nature of Sect. 5.3). One combines these techniques to successively determine the possible fields of definition, forms of algebraic groups, and maximal arithmetic subgroups that can achieve the prescribed covolume.

It is worth noting that lattices of minimal covolume in $\mathrm{SL}_n(\mathbb{R})$ for $n \geq 3$ are given by (the integer points of) the Chevalley group of type A_{n-1}. This was not the case for $\mathrm{SL}_2(\mathbb{R})$; in fact, $\mathrm{SL}_n(\mathbb{Z})$ is not cocompact, while the (2, 3, 7)-triangle group is.

Very recently, a second family of split simple real Lie groups, the symplectic groups, was studied by Džambić, Holm, and Köhl. Using the same methods, they determined that also in this case, lattices of minimal covolume are given by the Chevalley group of the same type.

Theorem ([28, Theorem 1.1]) *Let $n \geq 2$. Any lattice of minimal covolume in $\mathrm{Sp}_{2n}(\mathbb{R})$ is conjugate to $\mathrm{Sp}_{2n}(\mathbb{Z})$.*

Of course, as $\mathrm{Sp}_2 \cong \mathrm{SL}_2$, the same note can be made here as well.

9 Transitive Actions on Bruhat–Tits Buildings

Recall that when v is a non-Archimedean place of k, the Tits measure μ_v assigns volume 1 to any Iwahori subgroup of $\mathrm{G}(k_v)$ (see Sect. 4.3). Using the orbit map, μ_v is pushed forward on the set $\mathcal{C}$ of chambers of the Bruhat–Tits building of G over k_v to the counting measure, simply giving volume 1 to every chamber. Now if Γ is a discrete subgroup of $\mathrm{G}(k_v)$, one can associate to the quotient of $\mathcal{C}$ by the action of Γ a *combinatorial volume*

$$\mathrm{Vol}(\mathcal{C}/\Gamma) = \sum_{x \in \mathcal{C}/\Gamma} \frac{1}{\#\Gamma_x},$$

where x goes through a set of representatives of the orbits of Γ on $\mathcal{C}$, and Γ_x is the stabilizer of x in Γ. It can easily be shown (see, for instance, [1, Chap. 1]) that by the normalization of μ_v, we have in fact

$$\mathrm{Vol}(\mathcal{C}/\Gamma) = \mu_v(\mathrm{G}(k_v)/\Gamma).$$

In consequence, if Γ acts transitively on the set of chambers $\mathcal{C}$ of the building of G, then for any $x \in \mathcal{C}$,

$$\mu_v(\mathrm{G}(k_v)/\Gamma) = \frac{1}{\#\Gamma_x}$$

is the inverse of an integer; in particular, it is smaller than 1. Similarly, by comparing with the Haar measure on $\mathrm{G}(k_v)$ giving volume 1 to stabilizers of vertices in the building and accounting for the index of the stabilizer of a chamber inside the stabilizer of a vertex, any discrete vertex-transitive subgroup Γ will have a covolume in $\mathrm{G}(k_v)$ for μ_v which is a rational number (whose numerator is bounded by the denominator of that index, in lowest terms).

9.1 Vertex-Transitive Actions in Characteristic 0

Thus, discrete transitive subgroups of $\mathrm{G}(k_v)$ have small covolumes, and if they are arithmetic, one can use Prasad's volume formula to extract information about them. This is precisely the approach taken by Mohammadi and Golsefidy [82] to classify all vertex-transitive actions on Bruhat–Tits buildings of dimension at least 4, over local fields of characteristic 0.

If $\mathcal{B}$ is a locally finite irreducible affine building of dimension at least 3, Tits proved [111, Sect. 15 Corollaire] that $\mathcal{B}$ is the Bruhat–Tits building of an absolutely simple group G over a non-Archimedean local field F.

For such $\mathcal{B}$ to admit a vertex-transitive action, it is necessary that the Dynkin diagram of $\mathcal{B}$ is itself a vertex-transitive graph; this only happens if G is an inner form of a split group of type A_n. Given this assumption on the dimension of $\mathcal{B}$, G is of rank at least 4, hence every lattice in G is arithmetic by the work of Margulis.

Theorem ([82, Theorem B]) *Let $\mathcal{B}$ be a locally finite irreducible affine building of dimension at least* 4, *as above. Assume that there is a discrete subgroup Γ of* $\mathrm{Aut}\,\mathcal{B}$ *which acts transitively on the set of vertices of $\mathcal{B}$. Then $\mathcal{B}$ is the Bruhat–Tits building associated with an absolutely simple group* G *of type* A *over a non-Archimedean local field F.*

If F is of characteristic 0 *and* Γ *is a subgroup of* $\mathrm{Ad}(\mathrm{G})(F) \rtimes \mathrm{Aut}(F)$, *then* G *is the split* F*-group* SL_n *for* $n \in \{5, 6, 7, 8\}$, *and* F *is isomorphic to* $\mathbb{Q}_p$ *for an odd prime* p *congruent to either* 1 *modulo* 4, 1 *modulo* 3, *or* 1, 2, *or* 4 *modulo* 7. *Furthermore,* Γ *belongs to the finite list given in [82, Theorem A].*

In particular, there are no discrete vertex-transitive actions on Bruhat–Tits buildings over non-Archimedean local fields of characteristic 0 if the dimension of the building exceeds 7.

As a byproduct of their work, Mohammadi and Golsefidy [82, Appendix C] (following an observation by Serre [99, Sect. 3.7]) give a new proof of the Klingen–Siegel theorem [61]. The Klingen–Siegel theorem states that the values of the zeta function of a totally real number field k at strictly negative integers and of the L-function of totally complex quadratic extension l of k at even negative integers are rational numbers. They do so by interpreting products of these special values as the covolume of a well-chosen principal arithmetic subgroup Λ of a special linear or unitary group over k. As we have seen in the beginning of this section, this covolume is a rational number. Moreover, the size of its denominator can be controlled by the size of vertex stabilizers in Λ.

9.2 Vertex-Transitive Actions in Positive Characteristic

In an earlier work, Cartwright and Steger [21] exhibited for any prime power q and $n \geq 2$, a discrete subgroup of $\mathrm{PGL}_n(\mathbb{F}_q(\!(t)\!))$ acting simply transitively on the vertices of its Bruhat–Tits building. These are arithmetic subgroups constructed using forms of PGL_n coming from cyclic division algebras over $\mathbb{F}_q(t)$.

In his thesis [29], Ebrahimian layed the first steps of a classification of vertex-transitive actions over $\mathbb{F}_q(\!(t)\!)$. Using Prasad's volume formula, Ebrahimian showed that all vertex-transitive actions in positive characteristic arise from fields of genus 0, i.e., from the function fields $\mathbb{F}_q(t)$ for q a prime power. It is expected that by pushing these methods, one should also be able to obtain a classification of discrete vertex-transitive subgroups in positive characteristic. Note that Ebrahimian constructed vertex-transitive subgroups using special unitary groups over $\mathbb{F}_q(t)$, so Cartwright and Steger's examples do not account for all discrete vertex-transitive groups.

9.3 Chamber-Transitive Actions

Since their methods lie outside of the scope of this survey, we only remind the reader in passing that Kantor, Liebler, and Tits [57] classified the discrete chamber-transitive type-preserving actions on Bruhat–Tits buildings of dimension at least 2. There are finitely many such groups (up to conjugacy), and they only occur for groups over the local fields $\mathbb{Q}_2$, $\mathbb{Q}_3$, $\mathbb{F}_2((t))$, and $\mathbb{F}_{2^3}((t))$.

10 Counting Lattices by Covolume

As we have seen in Sect. 5, Wang [113] showed that in a given simple[7] real Lie group G not locally isomorphic to $\mathrm{SL}_2(\mathbb{R})$ or $\mathrm{SL}_2(\mathbb{C})$, there are only finitely many conjugacy classes of lattices whose covolume is less than a given bound b. It is thus legitimate to count them, or equivalently (in view of strong rigidity), to count locally G-symmetric orbifolds by volume. This counting problem is in a way a quantitative version of Wang's finiteness theorem (whose proof is not effective). Note that in $\mathrm{SL}_2(\mathbb{R})$ and $\mathrm{SL}_2(\mathbb{C})$, Borel [12] showed that nevertheless the number of conjugacy classes of arithmetic lattices with covolume less than b was finite.

Let $\mathrm{L}_G(x)$, resp. $\mathrm{L}_G^{\mathrm{tf}}(x)$ denote the number of conjugacy classes of lattices in G, torsion-free for the latter, with covolume less than x for some Haar measure on G fixed beforehand.

10.1 Hyperbolic Lattices

For hyperbolic manifolds of dimension $n \geq 4$, the asymptotic of $\mathrm{L}^{\mathrm{tf}}_{\mathrm{PO}_{n,1}(\mathbb{R})}$ was determined in [20]. There are strictly positive constants a, b such that

$$x^{ax} \leq \mathrm{L}^{\mathrm{tf}}_{\mathrm{PO}_{n,1}(\mathbb{R})}(x) \leq x^{bx}.$$

The lower bound is established by counting subgroups of a single non-arithmetic cocompact lattice in $\mathrm{PO}_{n,1}(\mathbb{R})$ (one constructed by Gromov and Piatetski-Shapiro); in particular, this argument does not extend to higher rank groups. The proof of the upper bound relies on the thick-thin decomposition.

Also using the thick-thin decomposition, Gelander [39] did extend the same upper bound to locally symmetric manifolds of dimension at least 4 excluding those of $\mathrm{SL}_3(\mathbb{R})$ and $\mathrm{SL}_2(\mathbb{R}) \times \mathrm{SL}_2(\mathbb{R})$, and to non-compact arithmetic locally symmetric orbifolds of dimension at least 4 (including those two excluded types), as well as to non-compact arithmetic hyperbolic 3-orbifolds.

[7] Although most results mentioned in this section are also available for irreducible lattices in semisimple groups, for simplicity we will stick with simple groups.

10.2 Arithmetic Lattices in Real Lie Groups

The counting problem can be decomposed into two parts: counting the number (A)$\mathrm{M}_G(x)$ of (arithmetic) maximal lattices in G of covolume less than x, and counting the number of subgroups of index at most n in a given maximal lattice (while controlling any overlap).

For the first part, Belolipetsky [4] gave an almost complete solution using Prasad's volume formula and the methods of Borel and Prasad. Write (A)$\mathrm{M}_G^{\mathrm{c},\mathrm{nc}}(x)$ for, respectively, the number of cocompact, non-cocompact maximal (arithmetic) lattices in G of covolume less than x. There are constants a, b, a', b', and for any $\varepsilon > 0$, a constant c such that for sufficiently large x:

$$x^a \le \mathrm{AM}_G^{\mathrm{c}}(x) \le x^{b \cdot c(\log x)^{\varepsilon}},$$
$$x^{a'} \le \mathrm{AM}_G^{\mathrm{nc}}(x) \le x^{b'}.$$

The second part is understood to a good extent via the theory of subgroup growth and its interplay with the congruence subgroup problem (see [70, Chaps. 6–8] on this subject). In summary, the number $c_n(\Gamma)$ of congruence subgroups of index at most n in an arithmetic subgroup Γ of G, if non-zero, satisfies for some constants a, b:

$$n^{\frac{a \log n}{\log \log n}} \le c_n(\Gamma) \le n^{\frac{b \log n}{\log \log n}}.$$

In case Γ has the congruence subgroup property, $c_n(\Gamma)$ of course accounts for all subgroups of index $\le n$.

It was initially conjectured in [20, 40] that the second part of this approach would dominate the counting problem for simple groups G with real rank ≥ 2, and hence $\mathrm{L}_G(x)$ in this case would conjecturally satisfy

$$x^{\frac{a \log x}{\log \log x}} \le \mathrm{L}_G(x) \le x^{\frac{b \log x}{\log \log x}},$$

for some (different) constants a, b.

But in a somewhat surprising twist, Belolipetsky and Lubotzky [10], using the infinite class field towers of Golod and Shafarevich, as well as Prasad's volume formula and accompanying techniques, showed that the growth rate is in fact

$$x^{a \log x} \le \mathrm{L}_G(x) \le x^{b \log x},$$

where the upper bound is conditional to the solution of the congruence subgroup problem (CSP) and the Margulis–Platonov (MP) conjecture (they are known to hold simultaneously for higher rank groups of type different from A_r, D_4, and E_6). Belolipetsky and Lubotzky [11] did confirm afterward that the conjectural bound above is correct for non-cocompact lattices in most groups, so cocompact lattices are solely responsible for the unexpected growth. In other words, when counted by covolume, most arithmetic lattices in a typical simple real Lie group are cocompact.

10.3 Arithmetic Lattices in Positive Characteristic

For Lie groups G of rank ≥ 2 over local fields of positive characteristic, Golsefidy [42, Theorems 1–3] determined the growth of the three analogous[8] quantities using a similar approach. After determining the congruence subgroup growth of Γ (in characteristic p, subgroup growth was only known for lattices constructed using Chevalley groups), and estimating the rate of M_G using Prasad's volume formula, Golsefidy deduces the rate of L_G conditional to (CSP) and (MP). There are constants a, a', b, b' depending on G and a'', b'' depending on Γ such that

$$\begin{aligned} x^{a \log x} &\leq \mathrm{L}_G(x) \leq x^{b \log x}, \\ x^{a'} &\leq \mathrm{M}_G(x) \leq x^{b' \log \log x}, \\ n^{a'' \log n} &\leq c_n(\Gamma) \leq n^{b'' \log n}. \end{aligned}$$

In particular, in this setting the congruence subgroup growth of an arbitrary lattice does dominate the count, contrary to the case of cocompact lattices in characteristic 0.

10.4 Minimal Number of Generators

Let us also briefly mention two related problems. First, the techniques of [10, 11] were recently used in [71] to give upper bounds on the minimal number $d(\Gamma)$ of generators of a non-cocompact lattice Γ in a simple Lie group of real rank at least 2 in terms of $\log \mathrm{Vol}(G/\Gamma)$, and lower bounds on $d(\Gamma)$ for certain cocompact lattices Γ constructed from the infinite class field towers.

10.5 Isospectral Lattices

The second concerns counting isospectral lattices. Since isospectral manifolds have the same volume, it makes sense to count the largest number $\mathrm{I}_G(x)$ of pairwise isospectral but non-conjugate lattices in G with covolume less than x. Belolipetsky and Linowitz [9] have shown that this number grows at a rate quite close to $\mathrm{L}_G(x)$: for G a simple Lie group of rank ≥ 2 over $\mathbb{R}$:

$$x^{\frac{a \log x}{(\log \log x)^2}} \leq \mathrm{I}_G(x).$$

[8] Note that in positive characteristic, $\mathrm{Aut}(G)/\mathrm{Inn}(G)$ is an infinite compact group, so lattices should really be counted up to automorphism, which does not change the answers for characteristic 0.

They used Prasad's volume formula to bound by d^{cd} the covolume of a^{d^2} non-conjugate isospectral subgroups, in terms of the degree d of a number field used to construct them via Sunada's method.

11 Miscellaneous Applications

In this last section, we gather various additional applications of Prasad's volume formula.

11.1 Motives of Reductive Groups

In [45], Gross attached to a reductive group G over a field k an Artin–Tate motive M, as follows. Assume for a moment that G is quasi-split. Let V be the space of primitive invariants of G (over some separable splitting field); V is the sum $\bigoplus_d V_d$ of its subspaces of degree d, where $d \in \{m_i + 1 \mid i = 1, \dots, r\}$ and the m_i's are the exponents of the absolute Weyl group of G (see Sect. 4.1). The Galois group $\mathrm{Gal}(k^{\mathrm{sep}}/k)$ acts on V, preserving the homogeneous pieces V_d, so one can view V_d as the Artin motive over k given by this representation of $\mathrm{Gal}(k^{\mathrm{sep}}/k)$. The *motive M associated with* G is then the direct sum of twists of the Artin motives

$$M = \bigoplus_{d \in \{m_i+1\}} V_d(1-d) = \bigoplus_{d \in \{m_i+1\}} V_d \otimes \mathbb{Q}(1)^{\otimes(1-d)},$$

where $\mathbb{Q}(1)$ denotes the Tate motive $\mathrm{H}_1(\mathrm{G_m})$ of rank 1 and weight -2 over k. If now G is an arbitrary reductive group, the motive associated with G is defined to be the motive of its quasi-split inner k-form $\mathcal{G}$.

When $k = k_v$ is a local field of characteristic 0 and the center of G is anisotropic, Gross [45, Theorem 8.1] showed that the value of the local L-function $L(M)$ of M at $s = 0$ satisfies the equation:

$$e(\mathrm{G}) \cdot c(\mathrm{G}) \cdot L(M)|_{s=0} \cdot \mu_v^{\mathrm{EP}} = \begin{cases} L(M^\vee(1))|_{s=0} \cdot \left| \prod_{i=1}^r \frac{m_i!}{(2\pi)^{m_i+1}} \right|_v \cdot |c_v \omega|_v & v \text{ Archimedean}, \\ L(M^\vee(1))|_{s=0} \cdot |c_v \omega|_v & v \text{ non-Archimedean}. \end{cases}$$

Here the motive $M^\vee(1) = \bigoplus_{d \in \{m_i+1\}} V_d(d)$ is the twisted dual of M, μ_v^{EP} is the Euler–Poincaré measure on $\mathrm{G}(k_v)$ introduced by Serre [99, Sect. 3], and $c_v \omega$ is a gauge form on G scaled as in Sect. 4.1. The sign $e(\mathrm{G})$ was introduced by Kottwitz [65] (it equals 1 when G is quasi-split), and $c(\mathrm{G})$ is the cohomological invariant given by

$$c(\mathrm{G}) = \frac{\#\mathrm{H}^1(k_v, \mathrm{T})}{\#\ker(\mathrm{H}^1(k_v, \mathrm{T}) \to \mathrm{H}^1(k_v, \mathrm{G}))},$$

where T is any maximal anisotropic torus of G.

Let now k be a number field, S be a finite set of places containing V_∞, and G be a reductive k-group. The S-integral L-function of M is the product $L_S(M)$ of the local L-functions of M over all places outside S. For $S = V_\infty$, $L_S(M)$ is the product of shifts by $1 - d$ of the Artin L-series of the representations V_d of $\mathrm{Gal}(k^{\mathrm{sep}}/k)$ mentioned above. Using the above equation, Gross gave a formula for the volume of $\mathrm{G}(\mathbb{A}_k)/\mathrm{G}(k)$ in terms of the value of $L_S(M)$ at $s = 0$.

Theorem ([45, Theorem 9.9]) *Let μ_S be the invariant measure on $\mathrm{G}(\mathbb{A}_k)$ given by the product of the local measures μ_v^{EP} for $v \in S$, and $L_v(M^\vee(1))|_{s=0} \cdot |c_v\omega|_v$ for $v \notin S$. Assume that the center of* G *is anisotropic over k. Then*

$$\prod_{v \in S} e(\mathrm{G} \times_k k_v) \cdot \prod_{v \in S} c(\mathrm{G} \times_k k_v) \cdot \mu_S(\mathrm{G}(\mathbb{A}_k)/\mathrm{G}(k)) = \tau_k(\mathrm{G}) \cdot L_S(M)|_{s=0}.$$

Note that if the center of G becomes isotropic over any place of S, both sides of the equality in the theorem degenerate to 0.

The proof of this theorem reduces to two main cases. First, when G is simply connected, absolutely simple, and quasi-split over k, the theorem is essentially equivalent to Prasad's volume formula. Second, when G is a one-dimensional torus, anisotropic over all Archimedean places of k and split by a totally complex quadratic extension of k, the theorem amounts to the class number formulae for G and $\mathrm{G_m}$.

11.2 Classifying One-Class Genera

Using the work described in Sect. 11.1 and its continuation [46], Gan, Hanke, and Yu [37, Proposition 2.13] gave an alternate proof of an exact formula due to Shimura for the mass of a maximal lattice in a quadratic or Hermitian space over a totally real number field, when the latter is totally definite.

This formula was then used by Hanke [47] and by Kirschmer [60] to classify the one-class genera of maximal integral lattices in a totally definite quadratic space of dimension $m \geq 3$, respectively, over $\mathbb{Q}$ and over a totally real number field of degree $d \geq 2$. There are 586 such genera (115 of them over $\mathbb{Q}$). The maximal possible degree d is 5, and the maximal possible value of m is, respectively, 10 over $\mathbb{Q}$ and 6 over other number fields.

11.3 Effective Equidistribution and Diameter of Arithmetic Orbifolds

Recently, Prasad's volume formula was used by Einsiedler, Margulis, Mohammadi, and Venkatesh [30] to prove an effective equidistribution theorem for adelic homogeneous spaces that does not require any splitting condition.

Let G be a semisimple algebraic group over a number field k, and set $X = \mathrm{G}(\mathrm{A}_k)/\mathrm{G}(k)$. Given datum $\mathscr{D} = (\mathrm{H}, \iota, g)$ consisting of a k-group H such that $\mathrm{H}(\mathrm{A}_k)/\mathrm{H}(k)$ has finite invariant volume, a k-morphism $\iota : \mathrm{H} \to \mathrm{G}$ with finite kernel, and $g \in \mathrm{G}(\mathrm{A}_k)$, one constructs an *algebraic homogeneous set*

$$Y_{\mathscr{D}} = g \cdot \iota(\mathrm{H}(\mathrm{A}_k)/\mathrm{H}(k)) \subset X.$$

It is canonically endowed with an action of $H_{\mathscr{D}} = g\iota(\mathrm{H}(\mathrm{A}_k))g^{-1} \leq \mathrm{G}(\mathrm{A}_k)$, and with the invariant probability measure $\mu_{\mathscr{D}}$ that is the push-forward of the normalized invariant measure on $\mathrm{H}(\mathrm{A}_k)/\mathrm{H}(k)$ via the map $x \mapsto g \cdot \iota(x)$.

In order to measure the size of an algebraic homogeneous set $Y_{\mathscr{D}}$ relative to X, Einsiedler, Margulis, Mohammadi, and Venkatesh define an *alternate notion of volume* as follows. To serve as a yardstick, fix an open pre-compact neighborhood Ω_0 of the identity in $\mathrm{G}(\mathrm{A}_k)$. Let $m_{\mathscr{D}}$ denote the measure on $H_{\mathscr{D}}$ which makes the orbit map $H_{\mathscr{D}} \to Y_{\mathscr{D}} : h \mapsto h \cdot g\mathrm{G}(k)$ locally measure-preserving (recall that $\mathrm{G}(k)$ is discrete in $\mathrm{G}(\mathrm{A}_k)$). Then

$$\mathrm{vol}(Y_{\mathscr{D}}) = m_{\mathscr{D}}(H_{\mathscr{D}} \cap \Omega_0)^{-1}$$

measures the inverse of the proportion of the $H_{\mathscr{D}}$-orbit $Y_{\mathscr{D}}$ that lifts to the fixed set Ω_0, hence gives an intrinsic measurement of the size or spread of $Y_{\mathscr{D}}$ in X. This quantity depends on the choice of Ω_0, but two choices give measurements which are bounded by positive multiples of each other. It is also polynomially related to the *discriminant of* $Y_{\mathscr{D}}$, see [30, Appendix B].

Theorem ([30, Theorem 1.5]) *Let $Y_{\mathscr{D}}$ be an algebraic homogeneous set given by datum $\mathscr{D} = (\mathrm{H}, \iota, g)$ such that $\iota(\mathrm{H})$ is a semisimple proper subgroup of* G *which is maximal among k^{alg}-subgroups of* G. *There are positive constants a, b depending only on k and* G *such that for every $y \in Y_{\mathscr{D}}$:*

$$\left| \int_{Y_{\mathscr{D}}} f \, \mathrm{d}\mu_{\mathscr{D}} - (Af)(y) \right| \leq a \, \mathrm{vol}(Y_{\mathscr{D}})^{-b} \mathcal{S}(f) \quad \textit{for all } f \in \mathrm{C}_{\mathrm{c}}^{\infty}(X).$$

Here A is the averaging operator on orbits of $\mathrm{G}(\mathrm{A}_k)^+$, the image of the points of the simply connected cover of G *(which is the usual average if* G *is simply connected, in general, Af is locally constant), and $\mathcal{S}(f)$ denotes a certain adelic Sobolev norm for which we refer the reader to [30, Appendix A].*

In the proof, the volume formula comes into the picture in two ways. First, it is used to relate this alternate notion of volume to the standard notion and to the discriminant. Second, it serves to *effectively find a good place*, that is, a place w of k with $q_w \leq c(\log \operatorname{vol}(Y_{\mathscr{D}}))^2$ (for some constant c depending on k and G), such that G, H are quasi-split over k_w, and Ω_0, resp. $\iota^{-1}(g^{-1}\Omega_0 g)$ contains a hyperspecial subgroup of $\mathrm{G}(k_w)$, resp. $\mathrm{H}(k_w)$. This place allows Einsiedler, Margulis, Mohammadi, and Venkatesh to obtain good dynamical properties without any splitting condition on G or H.

Applications of this theorem to the distribution of genera of quadratic forms and to property (τ) (its proof relies on (τ) only for SL_2) are given in [30].

The existence of a good place, combined with property (τ), was also used by Mohammadi, Golsefidy, and the author [83] to give small lifts of points in algebraic homogeneous sets, in a way which is polynomially effective in the inverse of the injectivity radius of the point and in the *volume* of the set in the sense above.

Let $\mathrm{G} = \mathrm{SL}_N$ over $k = \mathbb{Q}$. For any $\eta > 0$, let $B(\eta)$ denote the ball of radius η in $\mathfrak{sl}_N(\mathbb{R})$ with respect to, say, the maximum norm in the canonical basis, and set $\Omega_\eta = \exp(B(\eta)) \times \prod_{p \in V_f} \mathrm{SL}_N(\mathbb{Z}_p) \subset \mathrm{SL}_N(\mathrm{A}_{\mathbb{Q}})$. For $v \in V$ and $g \in \mathrm{SL}_N(\mathbb{Q}_v)$, let $|g|_{\mathrm{op}}$ be the maximum of the operator norm of g and g^{-1}, with respect to the maximum norm for the canonical basis of $\mathbb{Q}^N$, say. For the purpose of the next theorem, the *injectivity radius* of $x \in \mathrm{SL}_N(\mathrm{A}_{\mathbb{Q}})$ is the largest number $\eta = \operatorname{inj} x > 0$ such that the map $\Omega_\eta \to X : g \mapsto gx$ is injective.

Theorem ([83, Theorem 5.2]) *Let* $\mathrm{G} = \mathrm{SL}_N$ *over* $k = \mathbb{Q}$. *There are positive constants* a, b, c *depending only on* N, *and for any datum* $\mathscr{D} = (\mathrm{H}, \iota, 1)$ *with* H *a simply connected semisimple* $\mathbb{Q}$*-group, there exists a prime* $p \leq c(\log \operatorname{vol}(Y_{\mathscr{D}}))^2$ *such that the following holds. For any* $g \in \mathrm{H}(\mathrm{A}_{\mathbb{Q}})$, *there exists* $\gamma \in \mathrm{H}(\mathbb{Q})$ *such that* $g\gamma = h_1 h h_2$ *where* $h_1, h_2 \in \iota^{-1}(\Omega_{\operatorname{inj} \iota(g)})$ *and* $h \in \mathrm{H}(\mathbb{Q}_p)$ *satisfies*

$$|\iota(h)|_{\mathrm{op}} \preceq a \cdot (\operatorname{inj} \iota(g))^{-b} \cdot \operatorname{vol}(Y_{\mathscr{D}})^b.$$

On a related note, Belolipetsky proved the following inequality relating the diameter and the usual volume of compact arithmetic hyperbolic orbifolds.

Theorem ([6]) *There are constants* a, b *depending only on* n *such that for every compact arithmetic real- or complex-hyperbolic* n*-orbifold* M:

$$\operatorname{diam} M \leq \frac{a(\log \operatorname{Vol}(M)) + b}{h(M)},$$

where $h(M)$ *is the Cheeger constant of* M.

The proof relies on a bound for the injectivity radius in terms of $\log \operatorname{Vol}(M)$, which in turn is established using an estimate (deduced from Prasad's volume formula) of $\log \operatorname{Vol}(M)$ in terms of the degree d_k of the defining field k, and using Dobrowolski's bound for the Mahler measure of elements of k in terms of d_k.

11.4 Degrees of Strongly Special Subvarieties of Shimura Varieties

Recall that a Shimura datum (G, X) consists of a reductive $\mathbb{Q}$-group G and a $\mathrm{G}(\mathbb{R})$-conjugacy class X of $\mathbb{R}$-morphisms $u : \mathrm{Res}_{\mathbb{C}/\mathbb{R}}(\mathrm{G}_{\mathrm{m}}) \to \mathrm{G}$ subject to the Deligne conditions. Given a compact open subgroup K of $\mathrm{G}(\mathrm{A}_f)$ (here A_f is the ring of finite adèles of $\mathbb{Q}$), the set

$$\mathrm{Sh}_K(\mathrm{G}, X)(\mathbb{C}) = \mathrm{G}(\mathbb{Q})\backslash(X \times \mathrm{G}(\mathrm{A}_f)/K)$$

can be seen as the set of points of a quasi-projective variety $\mathrm{Sh}_K(\mathrm{G}, X)$ over $\mathbb{C}$, called the *Shimura variety of level K associated with the datum* (G, X). An irreducible subvariety V of $\mathrm{Sh}_K(\mathrm{G}, X)$ is called *special* when there is a Shimura datum (H, Y) and a monomorphism $f : \mathrm{H} \hookrightarrow \mathrm{G}$ mapping Y to X for which V is a component of a $\mathrm{G}(\mathrm{A}_f)$-translate of the image of the induced map $f : \mathrm{Sh}_L(\mathrm{H}, Y) \to \mathrm{Sh}_K(\mathrm{G}, X)$, for the compact subgroup $L = f^{-1}(K) \cap \mathrm{H}(\mathrm{A}_f)$ of $\mathrm{H}(\mathrm{A}_f)$. If in addition the image of H in Ad(G) is semisimple, then we say V is a *strongly special* subvariety of $\mathrm{Sh}_K(\mathrm{G}, X)$. A *special point* of $\mathrm{Sh}_K(\mathrm{G}, X)$ is then simply a special subvariety of dimension 0.

It is a standard fact that any special subvariety of $\mathrm{Sh}_K(\mathrm{G}, X)(\mathbb{C})$ contains a Zariski dense orbit of special points. The *André–Oort conjecture* asserts the converse: the Zariski closure of a set of special points in $\mathrm{Sh}_K(\mathrm{G}, X)(\mathbb{C})$ is a finite union of special subvarieties.

With the goal of deducing the André–Oort conjecture from the generalized Riemann hypothesis (GRH), Daw [26] related the degree of a special subvariety to its volume, then used Prasad's volume formula to give a lower bound for the degrees of strongly special subvarieties of a Shimura variety.

Theorem ([26, Theorem 1.4]) *Let* (G, X) *be a Shimura datum with* G *adjoint, and fix a connected component* X^+ *of* X. *Let* K *be a neat compact open subgroup of* $\mathrm{G}(\mathrm{A}_f)$ *which is the product of compact open subgroups* $K_p \leq \mathrm{G}(\mathbb{Q}_p)$.

There exist positive constants c and d such that if V is a strongly special subvariety of the Shimura variety $\mathrm{Sh}_K(\mathrm{G}, X)$ *defined by datum* (H, Y) *and compact open subgroup L, then*

$$\deg_{\mathscr{L}_K} V > c \cdot \Pi(\mathrm{H}, L)^d.$$

Here, the degree of V is computed with respect to the ample line bundle $\mathscr{L}_K$ on the Baily–Borel compactification of $\mathrm{Sh}_K(\mathrm{G}, X)$ induced by the line bundle of holomorphic forms of maximal degree on X, and the quantity $\Pi(\mathrm{H}, L)$ is the product of the finitely many primes p for which either H is not quasi-split over $\mathbb{Q}_p$, or H does not split over the maximal unramified extension of $\mathbb{Q}_p$, or L_p is not hyperspecial.

The lower bound in the theorem can replace a bound on Galois orbits due to Ullmo and Yafaev, and used in the breakthrough work of Klingler and Yafaev, to yield a new proof that GRH implies the André–Oort conjecture, which does not involve methods from ergodic theory.[9]

In subsequent work, Daw, Javanpeykar, and Kühne [27] generalized the previous theorem to positive-dimensional special subvarieties (still relying on Prasad's volume formula) and used it to give an effective upper bound on the degree of non-facteur maximal special subvarieties of $\mathrm{Sh}_K(\mathrm{G}, X)$.

11.5 Quantitative Estimates for Obstructions of Unitary Modular Varieties

Let k be an imaginary quadratic field and let h be a $k/\mathbb{Q}$-Hermitian form of signature $(1, n)$. The corresponding unitary group $\mathrm{U}_h(\mathbb{R})$ acts on the Hermitian symmetric domain $\mathscr{D}_h = \{v \in \mathbb{C}^{n+1} \mid h(v, v) > 0\}/\mathbb{C}^\times$, and for Γ an arithmetic subgroup of $\mathrm{U}_h(\mathbb{Q})$, the quotient $\mathscr{F}_h(\Gamma) = \mathscr{D}_h/\Gamma$ is called a *unitary modular variety*.

It turns out that given $a \in \mathbb{N}$, the canonical line bundle $K_{\mathscr{F}_h(\Gamma)}$ of $\mathscr{F}_h(\Gamma)$ can be decomposed into two parts:

$$K_{\mathscr{F}_h(\Gamma)} \sim_{\mathbb{Q}} \mathscr{M}(a) + ((n+1-a)\mathscr{L} - \Delta),$$

dictating, respectively, the reflective and the cusp obstructions for $\mathscr{F}_h(\Gamma)$ to be of general type (equivalently, for $K_{\mathscr{F}_h(\Gamma)}$ to be big).

Using Prasad's volume formula to bound the asymptotic growth of the dimension of the space of modular forms on $\mathscr{D}_h$ vanishing on the ramification divisors, Maeda [74, Theorem 1.1] proved, under adequate assumptions on h, that the line bundle $\mathscr{M}(a)$ is big if either n or a certain height of h is sufficiently large depending on $a > 0$, thereby lifting the reflective obstructions in this case.

11.6 Asymptotically Good Codes from Arithmetic Groups

In [75], Maire and Oggier introduced new families of codes, inspired by Lenstra's analogue of the Goppa codes, but using arithmetic groups instead of rings of integers in number fields.

Their construction roughly goes as follows. Let k be a number field, G a k-group, and Γ an arithmetic subgroup of $\mathrm{G}(k)$. Let S be a finite set of places of k, and for $v \in S$ let P_v be a parahoric subgroup of $\mathrm{G}(k_v)$ containing Γ. Assume that the characteristic of the residue field $\mathfrak{f}_v$ is the same for all $v \in S$, so that the reductions $\overline{\mathrm{P}}_v(\mathfrak{f}_v)$ of P_v for $v \in S$ embed in a common matrix algebra, say $\mathrm{M}_d(\mathbb{F}_q)$. The composition

[9] At the time of writing, it seems that an unconditional solution to the André–Oort conjecture has finally been reached, culminating with the recent preprint of Pila, Shankar, Tsimerman, Esnault, and Groechenig.

$\Theta : \Gamma \to \prod_{v \in S} \mathrm{P}_v \to \mathrm{M}_d(\mathbb{F}_q)^{\#S}$ can then be viewed as a code map, with message set any subset $\mathcal{M}$ of Γ on which Θ is injective, and code set $\mathcal{C} = \Theta(\mathcal{M})$ of block length $N = \#S$ on the alphabet $\mathcal{A} = \mathrm{M}_d(\mathbb{F}_q)$.

Suppose now that G admits no non-trivial k-character, so that Γ is a lattice in $G = \mathrm{G}(k \otimes_{\mathbb{Q}} \mathbb{R})$. If B is a compact subset of G, then it can easily be seen that there exists $g \in G$ such that

$$\#(gB \cap \Gamma) \geq \frac{\mu_\infty(B)}{\mu_\infty(G/\Gamma)}.$$

Having arranged that Θ be injective on $\mathcal{M} = gB \cap \Gamma$, this inequality can be used to estimate the rate $\frac{\log_q(\#\mathcal{C})}{d^2 N}$ of the code $\mathcal{C}$.

The case where $\mathrm{G} = \mathrm{G_a}$ is the additive group over k is essentially the situation considered by Lenstra [68], in which the rate of $\mathcal{C}$ depends on the discriminant of k. Maire and Oggier [75, Sect. 5.3] examined the multiplicative group $\mathrm{G_m}$ over k, where this time the rate of the code is dependent on the regulator of k. In the case of the group $\mathrm{SL}_{1,D}$ for D a central division k-algebra, Maire and Page [76, Proposition 20], used Prasad's volume formula to obtain a rather explicit bound for the rate of $\mathcal{C}$.

Recall that a sequence $(\mathcal{C}_i)_i$ of codes is called *asymptotically good* if both the rate of $\mathcal{C}_i$ and its minimum distance (for the normalized Hamming distance) are bounded away from 0. The estimate on the rate in the three instances just mentioned served to prove that there exist sequences of such codes which are asymptotically good. Notably, in all three cases, the construction of this asymptotically good sequence crucially used the existence of unramified towers of number fields with certain splitting conditions. We record here the result for $\mathrm{G} = \mathrm{SL}_{1,D}$.

Theorem ([76, Theorem 22]) *For any $d \geq 2$, there exists a sequence $(\mathcal{C}_i)_i$ of asymptotically good codes, each obtained from the construction above applied to the group Γ_i of elements of reduced norm* 1 *in a maximal order of a division algebra D_i of degree d over its center k_i, and having as fixed alphabet* $\mathrm{M}_d(\mathbb{F}_p)$ *with* $\log p = c \log d + O(\log \log d)$ *for some constant $c > 0$.*

Acknowledgements The author warmly thanks Alireza Salehi Golsefidy and Gopal Prasad for enlightening comments on this subject shared over the years. The author also thanks Misha Belolipetsky, Vincent Emery, Alireza Salehi Golsefidy, Matt Stover, and the anonymous referees for helpful comments made during the preparation of this article.

References

1. Bass, H., Lubotzky, A.: Tree lattices. With appendices by H. Bass, L. Carbone, A. Lubotzky, G. Rosenberg, and J. Tits., Prog. Math., vol. 176. Birkhäuser (2001)
2. Belolipetsky, M.: On volumes of arithmetic quotients of SO(1, n). Ann. Sc. Norm. Super. Pisa, Cl. Sci. (5) **3**(4), 749–770 (2004)
3. Belolipetsky, M.: Addendum: on volumes of arithmetic quotients of SO(1, n). Ann. Sc. Norm. Super. Pisa, Cl. Sci. (5) **6**(2), 263–268 (2007)

4. Belolipetsky, M.: Counting maximal arithmetic subgroups. With an appendix by J. Ellenberg and A. Venkatesh. Duke Math. J. **140**(1), 1–33 (2007)
5. Belolipetsky, M.: Hyperbolic orbifolds of small volume. In: Proceedings of the ICM, Seoul 2014. Vol. II: Invited Lectures, pp. 837–851. KM Kyung Moon Sa (2014)
6. Belolipetsky, M.: A bound for diameter of arithmetic hyperbolic orbifolds. Geom. Dedicata **214**, 295–302 (2021)
7. Belolipetsky, M., Emery, V.: On volumes of arithmetic quotients of $PO(n, 1)^{\circ}$, n odd. Proc. Lond. Math. Soc. (3) **105**(3), 541–570 (2012)
8. Belolipetsky, M., Gelander, T., Lubotzky, A., Shalev, A.: Counting arithmetic lattices and surfaces. Ann. Math. (2) **172**(3), 2197–2221 (2010)
9. Belolipetsky, M., Linowitz, B.: Counting isospectral manifolds. Adv. Math. **321**, 69–79 (2017)
10. Belolipetsky, M., Lubotzky, A.: Manifolds counting and class field towers. Adv. Math. **229**(6), 3123–3146 (2012)
11. Belolipetsky, M., Lubotzky, A.: Counting non-uniform lattices. Isr. J. Math. **232**(1), 201–229 (2019)
12. Borel, A.: Commensurability classes and volumes of hyperbolic 3-manifolds. Ann. Sc. Norm. Super. Pisa, Cl. Sci., IV. Ser. **8**, 1–33 (1981)
13. Borel, A., Harish-Chandra: Arithmetic subgroups of algebraic groups. Ann. Math. (2) **75**, 485–535 (1962)
14. Borel, A., Mostow, G.D. (eds.): Algebraic Groups and Discontinuous Subgroups. Proceedings of a Symposium in Pure Mathematics, vol. 9. American Mathematical Society (1966)
15. Borel, A., Prasad, G.: Finiteness theorems for discrete subgroups of bounded covolume in semi-simple groups. Publ. Math., Inst. Hautes Étud. Sci. **69**, 119–171 (1989)
16. Borel, A., Prasad, G.: Addendum to: Finiteness theorems for discrete subgroups of bounded covolume in semi-simple groups. Publ. Math., Inst. Hautes Étud. Sci. **71**, 173–177 (1990)
17. Bruhat, F., Tits, J.: Groupes réductifs sur un corps local. Publ. Math., Inst. Hautes Étud. Sci. **41**, 5–251 (1972)
18. Bruhat, F., Tits, J.: Groupes réductifs sur un corps local. II. Schémas en groupes. Existence d'une donnée radicielle valuée. Publ. Math., Inst. Hautes Étud. Sci. **60**, 1–194 (1984)
19. Bruhat, F., Tits, J.: Groupes algébriques sur un corps local. III. Compléments et applications à la cohomologie galoisienne. J. Fac. Sci., Univ. Tokyo, Sect. I A **34**, 671–698 (1987)
20. Burger, M., Gelander, T., Lubotzky, A., Mozes, S.: Counting hyperbolic manifolds. Geom. Funct. Anal. **12**(6), 1161–1173 (2002)
21. Cartwright, D.I., Steger, T.: A family of $\widetilde{A}_n$-groups. Isr. J. Math. **103**, 125–140 (1998)
22. Cartwright, D.I., Steger, T.: Enumeration of the 50 fake projective planes. C. R., Math., Acad. Sci. Paris **348**(1-2), 11–13 (2010). www.maths.usyd.edu.au/u/donaldc/fakeprojectiveplanes/
23. Cassels, J.W.S., Fröhlich, A., (eds.): Algebraic Number Theory, 2nd edn. London Mathematical Society (2010)
24. Chernousov, V.I.: On the Hasse principle for groups of type E_8. Sov. Math., Dokl. **39**(3), 592–596 (1989)
25. Corlette, K.: Archimedean superrigidity and hyperbolic geometry. Ann. Math. (2) **135**(1), 165–182 (1992)
26. Daw, C.: Degrees of strongly special subvarieties and the André-Oort conjecture. J. Reine Angew. Math. **721**, 81–108 (2016)
27. Daw, C., Javanpeykar, A., Kühne, L.: Effective estimates for the degrees of maximal special subvarieties. Sel. Math., New Ser. **26**(1), Paper No. 2, 31p. (2020)
28. Džambić, A., Holm, K., Köhl, R.: The Siegel Modular Group is the Lattice of Minimal Covolume in the Symplectic Group. J. Lie. Theory. Issn:0949-5932. **35**(3), 527–556 (2025)
29. Ebrahimian, R.: Steps Towards Characterizing Discrete Subgroups Acting Transitively on Vertices of a Bruhat-Tits Building. PhD thesis, Yale University (2010)
30. Einsiedler, M., Margulis, G.A., Mohammadi, A., Venkatesh, A.: Effective equidistribution and property (τ). J. Am. Math. Soc. **33**(1), 223–289 (2020)
31. Emery, V.: On compact hyperbolic manifolds of Euler characteristic two. Algebr. Geom. Topol. **14**(2), 853–861 (2014)

32. Emery, V., Kellerhals, R.: The three smallest compact arithmetic hyperbolic 5-orbifolds. Algebr. Geom. Topol. **13**(2), 817–829 (2013)
33. Emery, V., Kim, I.: Quaternionic hyperbolic lattices of minimal covolume. Forum Math. Sigma **10**, 19 (2022). Id/No e68
34. Emery, V., Stover, M.: Covolumes of nonuniform lattices in PU(n, 1). Am. J. Math. **136**(1), 143–164 (2014)
35. Gabai, D., Meyerhoff, R., Milley, P.: Minimum volume cusped hyperbolic three-manifolds. J. Am. Math. Soc. **22**(4), 1157–1215 (2009)
36. Gaitsgory, D., Lurie, J.: Weil's conjecture for function fields: volume I. Annals of Mathematics Studies, vol. 199. Princeton University Press (2019)
37. Gan, W.T., Hanke, J., Yu, J.-K.: On an exact mass formula of Shimura. Duke Math. J. **107**(1), 103–133 (2001)
38. Gehring, F.W., Martin, G.J.: Minimal co-volume hyperbolic lattices. I: the spherical points of a Kleinian group. Ann. Math. (2) **170**(1), 123–161 (2009)
39. Gelander, T.: Homotopy type and volume of locally symmetric manifolds. Duke Math. J. **124**(3), 459–515 (2004)
40. Goldfeld, D., Lubotzky, A., Nikolov, N., Pyber, L.: Counting primes, groups, and manifolds. Proc. Natl. Acad. Sci. USA **101**(37), 13428–13430 (2004)
41. Golsefidy, A.S.: Lattices of minimum covolume in Chevalley groups over local fields of positive characteristic. Duke Math. J. **146**(2), 227–251 (2009)
42. Golsefidy, A.S.: Counting lattices in simple Lie groups: the positive characteristic case. Duke Math. J. **161**(3), 431–481 (2012)
43. Golsefidy, A.S.: Lattices of minimum covolume are non-uniform. Isr. J. Math. **196**, 363–373 (2013)
44. Gromov, M., Schoen, R.: Harmonic maps into singular spaces and p-adic superrigidity for lattices in groups of rank one. Publ. Math., Inst. Hautes Étud. Sci. **76**, 165–246 (1992)
45. Gross, B.H.: On the motive of a reductive group. Invent. Math. **130**(2), 287–313 (1997)
46. Gross, B.H., Gan, W.T.: Haar measure and the Artin conductor. Trans. Am. Math. Soc. **351**(4), 1691–1704 (1999)
47. Hanke, J.: Enumerating maximal definite quadratic forms of bounded class number over Z in n >= 3 variables. Preprint, arXiv:1110.1876 [math.NT] (2011)
48. Harder, G.: Über die Galoiskohomologie halbeinfacher Matrizengruppen. I. Math. Z. **90**, 404–428 (1965)
49. Harder, G.: Über die Galoiskohomologie halbeinfacher Matrizengruppen II. Math. Z. **92**, 396–415 (1966)
50. Harder, G.: A Gauss-Bonnet formula for discrete arithmetically defined groups. Ann. Sci. Éc. Norm. Supér. **4**(4), 409–455 (1971)
51. Harder, G.: Chevalley groups over function fields and automorphic forms. Ann. Math. **2**(100), 249–306 (1974)
52. Harder, G.: Über die Galoiskohomologie halbeinfacher algebraischer Gruppen. III. J. Reine Angew. Math. **274–275**, 125–138 (1975)
53. Hersonsky, S., Paulin, F.: On the volumes of complex hyperbolic manifolds. Duke Math. J. **84**(3), 719–737 (1996)
54. Hild, T.: The cusped hyperbolic orbifolds of minimal volume in dimensions less than ten. J. Algebra **313**(1), 208–222 (2007)
55. Hurwitz, A.: Über algebraische Gebilde mit eindeutigen Transformationen in sich. Math. Ann. **41**, 403–442 (1893)
56. Kaletha, T., Prasad, G.: Bruhat-Tits theory. A new approach, New Mathematical Monographs, vol. 44. Cambridge University Press (2023)
57. Kantor, W.M., Liebler, R.A., Tits, J.: On discrete chamber-transitive automorphism groups of affine buildings. Bull. Am. Math. Soc., New Ser. **16**, 129–133 (1987)
58. Kellerhals, R.: Hyperbolic orbifolds of minimal volume. Comput. Methods Funct. Theory **14**(2–3), 465–481 (2014)

59. Kellerhals, R.: On minimal covolume hyperbolic lattices. Mathematics **5**(3), 16 (2017). Id/No 43
60. Kirschmer, M.: One-class genera of maximal integral quadratic forms. J. Number Theory **136**, 375–393 (2014)
61. Klingen, H.: Über die Werte der Dedekindschen Zetafunktion. Math. Ann. **145**, 265–272 (1962)
62. Klingler, B.: On the rigidity of certain fundamental groups, arithmeticity of complex hyperbolic lattices, and the "fake projective planes". Invent. Math. **153**(1), 105–143 (2003)
63. Kneser, M.: Hasse principle for H^1 of simply connected groups. In: Algebraic Groups and Discontinuous Subgroups, Proceedings of a Symposium in Pure Mathematics, vol. 9, pp. 143–148 (1966)
64. Kneser, M.: Lectures on Galois cohomology of classical groups. With an appendix by T. Springer, Lect. Math. Phys., vol. 47. Tata Institute of Fundamental Research (1969)
65. Kottwitz, R.E.: Sign changes in harmonic analysis on reductive groups. Trans. Am. Math. Soc. **278**, 289–297 (1983)
66. Lai, K.F.: Tamagawa number of reductive algebraic groups. Compos. Math. **41**, 153–188 (1980)
67. Langlands, R.P.: The volume of the fundamental domain for some arithmetical subgroups of Chevalley groups. In: Algebraic Groups and Discontinuous Subgroups, Proceedings of a Symposium in Pure Mathematics, vol. 9, pp. 143–148 (1966)
68. Lenstra, H.W.: Codes from algebraic number fields. In: Mathematics and Computer Science II. Fundamental Contributions in the Netherlands Since 1945. (Proceedings of a Symposium, Amsterdam), CWI Monogr., vol. 4, pp. 95–104 (1986)
69. Lubotzky, A.: Lattices of minimal covolume in SL_2: a nonarchimedean analogue of siegel's theorem $\mu \geq \pi/21$. J. Am. Math. Soc. **3**(4), 961–975 (1990)
70. Lubotzky, A., Segal, D.: Subgroup growth. Prog. Math., vol. 212. Birkhäuser (2003)
71. Lubotzky, A., Slutsky, R.: On the asymptotic number of generators of high rank arithmetic lattices. Mich. Math. J. **72**, 465–477 (2022)
72. Lubotzky, A., Weigel, T.: Lattices of minimal covolume in SL_2 over local fields. Proc. Lond. Math. Soc. (3) **78**(2), 283–333 (1999)
73. Maclachlan, C., Reid, A.W.: The arithmetic of hyperbolic 3-manifolds. Graduate Texts in Mathematics, vol. 219. Springer (2003)
74. Maeda, Y.: Reflective obstructions of unitary modular varieties. J. Algebra **647**, 341–399 (2024)
75. Maire, C., Oggier, F.: Maximal order codes over number fields. J. Pure Appl. Algebra **222**(7), 1827–1858 (2018)
76. Maire, C., Page, A.: Codes from unit groups of division algebras over number fields. Math. Z. **298**(1–2), 327–348 (2021)
77. Margulis, G.A.: Discrete subgroups of semisimple Lie groups, Ergeb. Math. Grenzgeb., 3. Folge, vol. 17. Springer (1991)
78. Marshall, T.H., Martin, G.J.: Minimal co-volume hyperbolic lattices. II: Simple torsion in a Kleinian group. Ann. Math. (2) **176**(1), 261–301 (2012)
79. Meyerhoff, R.: The cusped hyperbolic 3-orbifold of minimum volume. Bull. Am. Math. Soc., New Ser. **13**, 154–156 (1985)
80. Minkowski, H.: Untersuchungen über quadratische Formen: I. Bestimmung der Anzahl verschiedener Formen, welche ein gegebenes Genus enthält. Acta Math. **7**, 201–258 (1885)
81. Minkowski, H.: Diskontinuitätsbereich für arithmetische Äquivalenz. J. Reine Angew. Math. **129**, 220–274 (1905)
82. Mohammadi, A., Golsefidy, A.S.: Discrete subgroups acting transitively on vertices of a Bruhat-Tits building. Duke Math. J. **161**(3), 483–544 (2012)
83. Mohammadi, A., Golsefidy, A.S., Thilmany, F.: Diameter of homogeneous spaces: an effective account. Math. Ann. **385**(3–4), 1–40 (2023)
84. Mumford, D.: An algebraic surface with K ample, $(K^2) = 9$, $p_g = q = O$. Am. J. Math. **101**, 233–244 (1979)

85. Neukirch, J.: Algebraic Number Theory, Grundlehren Math. Wiss., vol. 322. Springer (1999)
86. Odlyzko, A.M.: Lower bounds for discriminants of number fields. Acta Arith. **29**, 275–297 (1976)
87. Odlyzko, A.M.: Lower bounds for discriminants of number fields. II. Tôhoku Math. J. (2) **29**, 209–216 (1977). https://www-users.cse.umn.edu/~odlyzko/unpublished/index.html
88. Ono, T.: On Tamagawa numbers. In: Algebraic Groups and Discontinuous Subgroups. Proceedings of a Symposium in Pure Mathematics, vol. 9, pp. 122–132 (1966)
89. Parker, J.R.: On the volumes of cusped, complex hyperbolic manifolds and orbifolds. Duke Math. J. **94**(3), 433–464 (1998)
90. Platonov, V., Rapinchuk, A.: Algebraic groups and number theory. Transl. from the Russian by Rachel Rowen, Pure and Applied Mathematics, vol. 139. Academic Press. Academic Press (1994)
91. Prasad, G.: Volumes of S-arithmetic quotients of semi-simple groups. With an appendix by M. Jarden and G. Prasad. Publ. Math., Inst. Hautes Étud. Sci. **69**, 91–117 (1989)
92. Prasad, G., Yeung, S.-K.: Fake projective planes. Invent. Math. **168**(2), 321–370 (2007). Revised version available at arXiv:math/0512115v7
93. Prasad, G., Yeung, S.-K.: Arithmetic fake projective spaces and arithmetic fake Grassmannians. Am. J. Math. **131**(2), 379–407 (2009)
94. Prasad, G., Yeung, S.-K.: Addendum to "Fake projective planes" Invent. Math. **168**, 321–370 (2007). Invent. Math. **182**(1), 213–227 (2010)
95. Prasad, G., Yeung, S.-K.: Nonexistence of arithmetic fake compact Hermitian symmetric spaces of type other than A_n ($n \leq 4$). J. Math. Soc. Japan **64**(3), 683–731 (2012)
96. Prasad, G., Yeung, S.-K.: Arithmetic fake compact Hermitian symmetric spaces of Type A_3. J. Math. Soc. Japan. **78**(1), 115–131 (2026). https://doi.org/10.2969/jmsj/92009200
97. Rémy, B.: Covolume of arithmetic S-groups and fake projective planes (after Mumford, Prasad, Klingler, Yeung, Prasad–Yeung). In: Séminaire Bourbaki. Volume 2007/2008. Exposés 982–996, pp. 83–130. Société Mathématique de France (2009)
98. Rohlfs, J.: Die maximalen arithmetisch definierten Untergruppen zerfallender einfacher Gruppen. Math. Ann. **244**, 219–231 (1979)
99. Serre, J.-P.: Cohomologie des groupes discrets. (Cohomology of discrete groups). Prospects Math., Ann. Math. Stud. **70**, 77–169 (1971)
100. Siegel, C.L.: Über die analytische Theorie der quadratischen Formen. Ann. Math. **2**(36), 527–606 (1935)
101. Siegel, C.L.: Über die Classenzahl quadratischer Zahlkörper. Acta Arith. **1**, 83–86 (1935)
102. Siegel, C.L.: The volume of the fundamental domain for some infinite groups. Trans. Am. Math. Soc. **39**, 209–218 (1936)
103. Siegel, C.L.: A mean value theorem in geometry of numbers. Ann. Math. **2**(46), 340–347 (1945)
104. Siegel, C.L.: Some remarks on discontinuous groups. Ann. Math. **2**(46), 708–718 (1945)
105. Smith, S.: On the orders and genera of quadratic forms containing more than three indeterminates. Second Notice. Lond. R. S. Proc. **16**, 197–208 (1868)
106. Stover, M.: Algebraic fundamental groups of fake projective planes. J. Eur. Math. Soc. (2024). To appear
107. Tamagawa, T.: Adèles. In: Algebraic Groups and Discontinuous Subgroups, Proceedings of a Symposium in Pure Mathematics, vol. 9, pp. 113–121 (1966)
108. Thilmany, F.: Lattices of minimal covolume in $\mathrm{SL}_n(\mathbb{R})$. Proc. Lond. Math. Soc. (3) **118**(1), 78–102 (2019)
109. Tits, J.: Classification of algebraic semisimple groups. In: Algebraic Groups and Discontinuous Subgroups, Proceedings of a Symposium in Pure Mathematics, vol. 9, pp. 33–62 (1966)
110. Tits, J.: Reductive groups over local fields. In: Automorphic Forms, Representations and L-Functions, Proceedings of a Symposium in Pure Mathematics, vol. 33(1), pp. 29–69 (1979)
111. Tits, J.: Immeubles de type affine. (Buildings of affine type). In: Buildings and the Geometry of Diagrams, Lect. 3rd 1984 Sess. C.I.M.E., Como/Italy, Lect. Notes Math., vol. 1181, pp. 159–190 (1986)

112. Voight, J.: Quaternion Algebras, Graduate Texts in Mathematics, vol. 288. Springer (2021)
113. Wang, H.-C.: Topics on totally discontinuous groups. Symmetric Spaces, short Courses presented at Washington Univ., Pure Appl. Math. **8**, 459–487 (1972)
114. Weil, A.: Adèles and algebraic groups. (Appendix 1: The case of the group G_2, by M. Demazure. Appendix 2: A short survey of subsequent research on Tamagawa numbers, by T. Ono), Prog. Math., vol. 23. Birkhäuser (1982)
115. Yeung, S.-K.: Integrality and arithmeticity of co-compact lattice corresponding to certain complex two-ball quotients of Picard number one. Asian J. Math. **8**(1), 107–130 (2004)
116. Yeung, S.-K.: Erratum to "Integrality and arithmeticity of co-compact lattice corresponding to certain complex two-ball quotients of Picard number one". Asian J. Math. **13**(2), 283–286 (2009)
117. Zimmert, R.: Ideale kleiner Norm in Idealklassen und eine Regulatorabschätzung. Invent. Math. **62**, 367–380 (1981)

Titles and Abstracts

Mini- courses

Brian Conrad (Stanford University (USA)) *Introduction to Bruhat-Tits theory*

Abstract. For connected reductive groups G over non-archimedean local fields k, Bruhat-Tits theory provides a metric space $B(G)$ (called the *building* of G) equipped with a $G(k)$-action that is a valuable tool for working with "large" compact open subgroups of $G(k)$. In particular, the conjugacy classes of maximal compact open subgroups (there is often more than one, unlike for connected Lie groups) can be understood through the study of $G(k)$-stabilizers of certain points in $B(G)$.

The understanding of the subgroup structure of $G(k)$ provided by Bruhat-Tits theory has made it a very powerful tool in number theory and representation theory. In these lectures, after some preliminaries about local fields and BN-pairs, we will give an axiomatic overview of the general theory (including some examples), discuss the utility of passage to "large" residue fields, and deduce some applications (such as useful group-theoretic decompositions). Further applications in representation theory will be discussed separately by Fintzen.

Pietro Corvaja (University of Udine (Italy)) *Bounded generation of linear groups and Diophantine approximation*

Abstract. We shall present different notions of bounded generation for Zariski-dense subgroups of linear algebraic groups; in particular, we shall treat the notion of exponential and purely exponential parametrization. After surveying on classical results, we shall show recent applications of Diophantine approximation techniques in the theory of linear groups which lead to a classification of groups admitting purely exponential parametrization or bounded generation.

G. Prasad et al. (eds.), *Zariski Dense Subgroups, Number Theory and Geometric Applications*, Infosys Science Foundation Series,
https://doi.org/10.1007/978-981-95-3437-1

Toshiyuki Kobayashi (The University of Tokyo (Japan)) *Global Analysis of Locally Symmetric Spaces with Indefinite Metric*

ABSTRACT. The local to global study of geometries was a major trend of twentieth century geometry, with remarkable developments achieved particularly in Riemannian geometry. In contrast, in areas such as pseudo-Riemannian geometry, familiar to us as the space-time of relativity theory, and more generally in pseudo-Riemannian geometry of general signature, surprising little was known about global properties of the geometry even if we impose a locally homogeneous structure. This theme has been developed rapidly in the last three decades.

In the series of lectures, I plan to discuss two topics by the general theory and some typical examples.

1. Global geometry: Properness criterion and its quantitative estimate for the action of discrete groups of isometries on reductive homogeneous spaces, existence problem of compact manifolds modeled on homogeneous spaces, and their deformation theory.

2. Spectral analysis: Construction of periodic L^2-eigenfunctions for the Laplacian with indefinite signature, stability question of eigenvalues under deformation of geometric structure, and spectral decomposition on the locally homogeneous space of indefinite metric.

Alex Lubotzky (Weizmann Institute (Israel)) and **Bharatram Rangarajan** (Hebrew University of Jerusalem (Israel)) *Uniform Stability of Higher-rank Arithmetic Groups*

ABSTRACT. Lattices in high-rank semisimple groups enjoy a number of rigidity properties like super-rigidity, quasi-isometric rigidity, first-order rigidity, and more. In these lectures, we will add another one: uniform (a.k.a. Ulam) stability. Namely, it will be shown that (most) such lattices D satisfy: every finite-dimensional unitary "almost-representation" of D (almost w.r.t. to a sub-multiplicative norm on the complex matrics) is a small deformation of a true unitary representation. This extends a result of Kazhdan (1982) for amenable groups and or Burger-Ozawa-Thom (2013) for $SL(n, \mathbb{Z})$, $n > 2$. The main technical tool is a new cohomology theory ("asymptotic cohomology") that is related to bounded cohomology in a similar way to the connection of the last one with ordinary cohomology. The vanishing of H^2 w.r.t. to a suitable module implies the above stability. The talks are based on a joint work of the speakers with L. Glebsky and N. Monod. See arXiv: 2301.00476.

TALKS

Mikhail Belolipetsky (IMPA (Brazil)) *Subspace stabilisers in hyperbolic lattices*

ABSTRACT. In a joint work with Nikolay Bogachev, Alexander Kolpakov and Leone Slavich we show that immersed totally geodesic m-dimensional suborbifolds of an n-dimensional arithmetic hyperbolic orbifold correspond to finite subgroups of the commensurator whenever m is sufficiently large. In particular, for $n = 3$ this condition includes all totally geodesic suborbifolds. We call such totally geodesic subspaces by finite centraliser subspaces (or fc-subspaces for short) and use them to formulate an arithmeticity criterion for hyperbolic lattices. We show that a hyperbolic orbifold is arithmetic if and only if it has infinitely many fc-subspaces, while in the non-arithmetic case the number of fc-subspaces is finite and bounded in terms of the volume. The case of special interest is that of exceptional trialitarian 7-dimensional orbifolds—we show that every such orbifold contains totally geodesic arithmetic hyperbolic 3-orbifolds of exceptional type.

Yves Benoist (University Paris-Saclay (France)) *Convolution and square on abelian groups*

ABSTRACT. The aim of this talk will be to construct functions on a cyclic group of odd order d whose "convolution square" is proportional to their square. For that, we will have to interpret the cyclic group as a subgroup of an abelian variety with complex multiplication and to use the modularity properties of their theta functions.

Indranil Biswas (Shiv Nadar University (India)) *Classification, reduction and stability of toric principal bundles*

ABSTRACT. Let X be a complex toric variety equipped with the action of an algebraic torus T, and let G be a complex linear algebraic group. We classify all T-equivariant principal G-bundles E over X and the morphisms between them. We then give a criterion for the equivariant reduction of the structure group of E to a Levi subgroup of G in terms of the automorphism group of the bundle. (With A. Dey, J. Dasgupta, B. Khan and M. Poddar.)

Maciej Bochenski (University of Warmia Mazury (Poland)) *Standard compact Clifford-Klein forms and Lie algebra decompositions*

ABSTRACT. Let G be a non-compact linear simple Lie group and $H \subset G$ a reductive subgroup. We say that the homogeneous space G/H admits a standard compact Clifford-Klein form if there exists a reductive subgroup $L \subset G$ such that L acts properly and co-compactly on G/H.

I will describe relations between real root decompositions of Lie triples $(\mathfrak{g}, \mathfrak{h}, \mathfrak{l})$ corresponding to standard compact Clifford-Klein forms, under the assumption that $\mathfrak{g}$ is not equal to $\mathfrak{h} + \mathfrak{l}$. This gives new classes of homogeneous spaces G/H which do not admit standard compact Clifford-Klein forms. For instance proper regular subalgebras $\mathfrak{h}$ of $\mathfrak{g}$ never generate homogeneous spaces G/H which admit standard compact Clifford-Klein forms (other than $\mathfrak{g} = \mathfrak{h} + \mathfrak{l}$).

Emmanuel Breuillard (University of Oxford (UK)) *Heights on character varieties, free subgroups and spectral gaps*

ABSTRACT. Given a finite set of elements in a semisimple algebraic group over the field of algebraic numbers, one can define its normalized height as a weighted sum over all places of its associated joint spectral radii. The height gap theorem asserts that this height is bounded away from zero provided the subgroup generated by the finite set is Zariski-dense. This result can be seen as a non-commutative analogue to the Lehmer or Bogomolov problem in diophantine number theory. In recent joint work with Oren Becker, we show how this can be used to deduce uniform spectral gaps for actions of Zariski-dense subgroups on homogeneous spaces of algebraic groups and establish the existence of a uniform lower bound on the first eigenvalue of Cayley graphs of certain finite simple groups of Lie type.

Emmanuel Breuillard (University of Oxford (UK)) *Random character varieties*

ABSTRACT. We study the G-character variety of a random group, where G is a semisimple complex Lie group. The typical behavior depends on the defect of the random presentation. We are able to describe what happens for all but an exponentially small probability of exceptions. In particular we compute the dimension and show the absolute irreducibility of the character variety in defect at least two. In defect one we also exhibit a phenomenon of Galois rigidity showing the finiteness of the character variety and proving lower bounds on its cardinality. The proofs are conditional on GRH via the use of effective Chebotarev type theorems and uniform mixing bounds for Cayley graphs of finite simple groups. This is joint work with Peter Varju and Oren Becker.

Ted Chinburg (University of Pennsylvania (USA)) *Linear representations of the Grothendieck-Teichmüller group*

ABSTRACT. I will speak on work with F. Bleher and A. Lubotzky arising from a construction by Lubotzky and Grunewald of linear representations of the automorphism group $\mathrm{Aut}(F_n)$ of a free group on n elements. It turns out that if one replaces F_n by its profinite completion $\hat{F}_n$, the construction of linear representations with large image of $\mathrm{Aut}(\hat{F}_n)$ becomes simpler and more general than the discrete case. This leads to a solution of an open problem from the 1990s having to do with constructing natural non-abelian representations of the so-called Grothendieck-Teichmüller group GT inside $\mathrm{Aut}(\hat{F}_2)$. I will discuss where GT comes from and current open problems in this area.

Jessica Fintzen (University of Bonn (Germany)) *Representations of p-adic groups*

ABSTRACT. The theory of Bruhat and Tits opened the door to studying and constructing representations of general p-adic reductive groups. I will give an overview of our understanding of the construction and category of representations of p-adic groups and indicate how it crucially relies on Bruhat–Tits theory and the Moy–Prasad filtration.

Tsachik Gelander (Hebrew University of Jerusalem (Israel)) *Things we can learn by looking at random manifolds*

ABSTRACT. The theory of invariant random subgroups (IRS), which has been developed quite rapidly during the last decade, has been very fruitful to the study of lattices and their asymptotic invariants. However, restricting to invariant measures limits the scope of problems that one can approach (in particular since the groups involved are highly non-amenable). It was recently realised that the more general notion of stationary random subgroups (SRS) is still very effective and opens paths to deal with questions which previously seemed out of reach.

In the talk I will describe various old and new results concerning arithmetic groups and general locally symmetric manifolds of finite as well as infinite volume that can be proved using "randomness," e.g.:

1. Kazhdan-Margulis minimal covolume theorem.
2. Most hyperbolic manifolds are non-arithmetic (joint work with A. Levit).
3. Higher rank manifolds of large volume have a large injectivity radius (joint with Abert, Bergeron, Biringer, Nikolov, Raimbault and Samet).
4. Margulis' infinite injectivity radius conjecture: For manifolds of rank at least 2, finite volume is equivalent to bounded injectivity radius (joint with M. Fraczyk).

Alireza Salehi Golsefidy (UCSD (USA)) *Random walks on group extensions*

ABSTRACT. Lindenstrauss and Varju asked the following questions: for every prime p, let S_p be a symmetric generating set of $G_p := \mathrm{SL}(2, F_p) \times \mathrm{SL}(2, F_p)$. Suppose the family of Cayley graphs $\{Cay(\mathrm{SL}(2, F_p), \mathrm{pr}_i(S_p))\}$ is a family of expanders, where pr_i is the projection to the i-th component. Is it true that the family of Cayley graphs $\{Cay(G_p, S_p)\}$ is a family of expanders? We answer this question and go beyond that by describing random-walks in various group extensions. (This is a joint work with Srivatsa Srinivas.)

Alireza Salehi Golsefidy (UCSD (USA)) *Closure of orbits of the pure mapping class group on the character variety*

ABSTRACT. For every surface S, the pure mapping class group G_S acts on the SL_2-character variety Ch_S of the fundamental group P of S. The character variety Ch_S is a scheme over the ring of integers. Classically this action on the real points $Ch_S(\mathbb{R})$ of the character variety has been studied in the context of the Teichmuller theory and $\mathrm{SL}(2, \mathbb{R})$-representations of P. In a seminal work, Goldman studied this action on a subset of $Ch_S(\mathbb{R})$ which comes from $\mathrm{SU}(2)$-representations of P. In this case, Goldman showed that if S is of genus $g > 1$ and zero punctures, then the action of G_S is ergodic. Previte and Xia studied this question from topological point of view, and when $g > 0$, proved that the orbit closure is as large as algebraically possible. Bourgain, Gamburd, and Sarnak studied this action on the F_p-points $Ch_S(F_p)$ of the character variety where S is a puncture torus. They conjectured that in this case, this action has only two orbits, where one of the orbits has only one point. Recently, this conjecture is proved for large enough primes by Chen. When S is an n-punture sphere, the finite orbits of this action on $Ch_S(\mathbb{C})$ are connected to the algebraic solutions of Painleve differential equations.

Marek Kaluba (Karlsruhe Institute for Technology (Germany)) and **Piotr Nowak** (IMPAN (Poland)) *Property (T) for automorphism groups of free groups*

ABSTRACT. Talk 1 (Kaluba) We will discuss the notion of positivity in rings, understood as being a sum of squares. This problem is classically relevant for polynomials, and in the context of property (T) it appears in the group ring and its augmentation ideal. We will discuss a computer-assisted approach, via semidefinite programming, of verifying whether an element of the group ring can be expressed as a sum of squares. An important feature of this method is that despite using numerical information, in certain situations it provides a rigorous proof of positivity. Together with a characterization of property (T) via positivity in the group ring this provides an approach to proving Kazhdan's property (T) for some groups.

Talk 2 (Nowak) We will use the methods described in the first talk to prove property (T) for automorphism groups of free groups. First we will consider a singular case of $\mathrm{Aut}(F_5)$ but the main focus will be the infinite case. We will describe an inductive procedure that allows to deduce property (T) for families of groups like $\mathrm{SL}_n(\mathbb{Z})$ and $\mathrm{Aut}(F_n)$ from positivity of a single group ring element. Using this inductive approach we will prove property (T) for $\mathrm{Aut}(F_n)$ for all $n \geq 6$, and reprove property (T) for $\mathrm{SL}_n(\mathbb{Z})$, $n \geq 3$. We will also discuss several consequences and applications, in particular we will show new estimates for Kazhdan constants.

Alex Kontorovich (Rutgers University (USA)) *Thin Orbits and Applications* (zoom)

ABSTRACT. We will discuss some recent progress on problems that can be reformulated as Diophantine questions on orbits of "thin" groups.

Rajesh Kulkarni (Michigan State University (USA)) *Moduli of binary cubic forms*

ABSTRACT. We discuss the moduli of binary cubic forms. We give a description of this moduli space in terms triples of an associated CM elliptic curve E, a degree-3 isogeny from E to E, and a point on E. We will also discuss an application of our construction.

Benjamin Linowitz (Oberlin College (USA)) *The systolic geometry of arithmetic locally symmetric spaces*

ABSTRACT. The systole of a compact Riemannian manifold M is the least length of a non-contractible loop on M. In this talk I will survey some recent work with S. Lapan and J. Meyer on the systolic geometry of arithmetic locally symmetric spaces, emphasizing lower bounds for systoles and systole growth along congruence covers.

Ben McReynolds (Purdue University (USA)) *What makes semisimple Lie groups special?* (zoom)

ABSTRACT. This talk will focus on a series of questions of the forum: What makes "blank" special? We will consider the cases when blank is a semisimple Lie group in the class of isometry groups of Hadamard spaces. The case when blank is a lattice in a semisimple Lie groups in the class of lattices in isometry groups of Hadamard spaces. The case when blank is a (finitely generated) subgroup of a lattice in a semisimple in the class of discrete subgroups of the semisimple Lie group. The focus will be through geometric properties and broad classes of geometric/topological invariants for the above objects.

Chen Meiri (Technion (Israel)) *Conjugacy width in higher rank arithmetic groups of orthogonal type* (zoom)

ABSTRACT. A basic question in group theory is what can we learn about a finitely generated and residually finite group from its profinite completion? In this talk, we will focus on the relation between the widths of conjugacy classes in a higher rank arithmetic group to the corresponding widths in the profinite completion of this group and explain the connection to the Congruence Subgroup Property. This is joint work with Nir Avni.

Yosuke Morita (Kyushu University (Japan)) *The existence problem of compact quotients of pseudo-Riemannian homogeneous spaces*

ABSTRACT. Let G/H be a homogeneous space. If a discrete subgroup of G acts properly and freely on G/H, the quotient space becomes a manifold locally modelled on G/H, and called a Clifford-Klein form. Since the late 1980s, the existence problem of compact Clifford-Klein forms of pseudo-Riemannian homogeneous spaces has been studied by various methods (e.g. Cartan projection of reductive Lie groups, rigidity theory in homogeneous dynamics, tempered unitary representations, de Rham and relative Lie algebra cohomology, Anosov representations). In this talk, I will explain my joint work-in-progress with Fanny Kassel (IHES) and Nicolas Tholozan (ENS) on a new necessary condition for the existence of compact Clifford-Klein forms. It is formulated in terms of the homotopy theory of sphere bundles and hence related to the Adams operations in KO-theory.

Dave Morris (University of Lethbridge (Canada)) *Arithmetic groups of higher real rank are not left-orderable (after Deroin and Hurtado)*

ABSTRACT. Bertrand Deroin and Sebastian Hurtado recently proved the 30-year-old conjecture that if $\mathbf{G}$ is an almost-simple algebraic $\mathbb{Q}$-group, and the real rank of $\mathbf{G}$ is at least two, then no arithmetic subgroup of $\mathbf{G}$ is left-orderable. We will discuss this theorem, and explain some of the main ideas of the proof, by illustrating them in the simpler case where the real field $\mathbb{R}$ is replaced with a p-adic field. Harmonic functions and continuous group actions are key tools.

Raman Parimala (Emory University (USA)) *Hasse principle for reductive groups over p-adic function fields*

ABSTRACT. In this talk we survey results on Hasse principle for homogeneous spaces under reductive groups over semiglobal fields, i.e., function fields of curves over complete discrete valued fields. We state some conjectures and explain recent progress on these conjectures for function fields of p-adic curves.

C.S. Rajan (Ashoka University (India)) *On non-commensurable isospectral locally symmetric spaces*

ABSTRACT. We give examples of non-commensurable but isospectral locally symmetric spaces, thereby completing the work of Lubotzky, Samuels and Vishne. The main step is to show that adelic conjugation of lattices in $SL(1, D)$ by the adjoint group preserves the spectrum, where D is a division algebra over a number field F (under some additional hypothesis), extending the work done by one of us, when D is quaternion. This is joint work with Sandeep Varma.

Igor Rapinchuk (Michigan State University (USA)) *On the Hasse principle for reductive algebraic groups over finitely generated fields*

ABSTRACT. One of the major results in the arithmetic theory of algebraic groups is the validity of the cohomological local-global (or Hasse) principle for simply-connected and adjoint semisimple groups over number fields. Over the last several years, there has been growing interest in studying Hasse principles for reductive groups over arbitrary finitely generated fields with respect to suitable sets of discrete valuations. In particular, we have conjectured that for divisorial sets, the corresponding Tate-Shafarevich set, which measures the deviation from the local-global principle, should be finite for all reductive groups. I will report on recent progress on this conjecture, focusing in particular on the case of algebraic tori as well as on connections to groups with good reduction.

Alan Reid (Rice University (USA)) *From representation rigidity to profinite rigidity*

ABSTRACT. A finitely generated residually finite group G is called *profinitely rigid* if for any other finitely generated residually finite group H, whenever the profinite completions of H and G are isomorphic, then H is isomorphic to G. In this talk we will discuss some recent work that constructs finitely presented groups that are profinitely rigid amongst finitely presented groups but not amongst finitely generated one. This builds on previous work that used "controlled SL(2)-representations" to construct examples of arithmetic lattices in $\mathrm{PSL}(2, \mathbb{R})$ and $\mathrm{PSL}(2, \mathbb{C})$ that are profinitely rigid.

Jinbo Ren (Xiamen University (China)) *On the construction of noncommensurable locally symmetric spaces*

ABSTRACT. In 2005, Lubotzky, Samuels and Vishne provided the first examples of *noncommensurable* compact isospectral locally symmetric spaces. These examples were constructed as quotients of the symmetric space $\mathrm{SL}_n(\mathbb{R})/\mathcal{K}$ associated to the group $\mathrm{SL}_n(\mathbb{R})$ by suitable arithmetic subgroups, and the proof of their isospectrality relied on some results of Harris and Taylor. In this talk, I will introduce a simpler proof of isospectrality of the relevant locally symmetric spaces which is based on a direct comparison of the Selberg trace formulas. The method is likely to be applicable to other types of locally symmetric spaces. This is joint work with Andrei Rapinchuk.

Sandip Singh (IITB (India)) *Hypergeometric groups and their arithmeticity*

ABSTRACT. A hypergeometric group is a subgroup of $\mathrm{GL}_n(\mathbb{C})$ generated by the companion matrices of two monic coprime polynomials of degree n. It arises as the monodromy group of a hypergeometric differential equation, and if the defining polynomials are also self-reciprocal and form a primitive pair, then its Zariski closure inside $\mathrm{GL}_n(\mathbb{C})$ is either a symplectic or an orthogonal group. In this talk, we will discuss the arithmeticity and thinness of the hypergeometric groups whose defining polynomials also have integer coefficients.

Matthew Stover (Temple University (USA)) *Central extensions of (arithmetic) lattices*

ABSTRACT. I will describe joint work with Domingo Toledo on residual finiteness for cyclic central extensions of fundamental groups of aspherical manifolds, its application to central extensions of certain arithmetic lattices, and discuss some open problems on residual properties of central extensions of lattices and their connections to the geometry of locally symmetric spaces.

Ramdorai Sujatha (UBC (Canada)) *Rational points on elliptic curves along anti-cyclotomic extensions*

ABSTRACT. Let K be an imaginary quadratic field and E/K be an elliptic curve with good ordinary reduction at an odd prime number p. We study the rational points of E along the anticyclotomic $\mathbb{Z}_p$-extension of K.

François Thilmany (Catholic University of Louvain (Belgium)) *Prasad's volume formula and its applications*

ABSTRACT. In 1989, G. Prasad established a hands-on formula to compute the covolume of S-arithmetic subgroups of semisimple groups over global fields. The formula has since then found many applications in the theory of arithmetic groups, starting with the contemporary Borel–Prasad finiteness theorem. In the talk, we aim to describe Prasad's volume formula and its contents. We will then survey some of its most striking applications, such as the finiteness theorem, the classification of fake projective planes, and the study of lattices of small covolume in various simple groups.

George Tomanov (University of Lyon (France)) *Orbit closures on homogeneous spaces and applications to number theory*

ABSTRACT. We study actions of algebraic tori on homogeneous spaces. A conjecture of Cassels and Swinnerton-Dyer about the purely real (homogeneous) forms formulated in 1955 and still open can be translated in terms of such actions. We prove that the natural generalization of the conjecture to the non-purely real forms is not valid. The proof is based on the complete description of the forms with discrete set of values at the integer vectors and non-representing non-trivially zero over the rational numbers.

Sandeep Varma (TIFR (India)) *Homogeneity results for invariant distributions on p-adic symmetric spaces*

ABSTRACT. Let G be the group of rational points of a connected reductive linear algebraic group defined over a nonarchimedean local field k. Under some conditions that are satisfied when the residue characteristic of k is large, S. DeBacker used Moy-Prasad theory to sharpen, refine and generalize some results of Waldspurger, establishing an explicit form of the Howe conjecture for G and proving the conjecture of T. Hales, A. Moy and G. Prasad on the range of validity for the Harish-Chandra–Howe local expansion for characters of irreducible admissible representations of G. We will report on joint work with J. Adler and E. Sayag, in which we pursue similar questions for p-adic symmetric spaces.

T.N. Venkataramana (ICTS (India)) *CSP and unipotent generators for higher rank nonuniform arithmetic groups*

ABSTRACT. Given a finite index subgroup Γ of $G(\mathbb{Z})$, where G is a $\mathbb{Q}$-simple algebraic group of $\mathbb{Q}$-rank at least one and $\mathbb{R}$-rank at least two, consider the group N generated by $\Gamma \cap U^+(\mathbb{Z})$ and $\Gamma \cap U^-(\mathbb{Z})$. Here U^+ is the unipotent radical of a minimal parabolic $\mathbb{Q}$-subgroup of G, and U^- is its opposite. An old result of Tits, Vaserstein, Raghunathan and myself says that N has finite index in $G(\mathbb{Z})$. The proof presented here is more streamlined and is much simpler.

The same proof shows the centrality of the congruence subgroup kernel, showing that the group $G(\mathbb{Z})$ has the congruence subgroup property.

SHORT COMMUNICATIONS

Pralay Chatterjee (IMSc, Chennai (India)) *Lower dimensional Betti numbers of homogeneous spaces of Lie groups*

ABSTRACT. We will present our joint work with I. Biswas and C. Maity on certain explicit descriptions of lower dimensional Betti numbers of homogeneous spaces of Lie groups. We will begin by recalling some of the earlier works in this subject and give our motivation. We will then describe our results and some of the applications in special cases.

Sourav Ghosh (Ashoka University (India)) *Finite step rigidity*

ABSTRACT. In this talk I will introduce Hitchin representations and show that only a finite part of the Jordan-Lyapunov spectra of a Hitchin representation is enough to completely determine it up to conjugacy classes.

Elyasheev Leibtag (Weizmann Institute (Israel)) *Images of Homomorphisms of Algebraic Groups*

ABSTRACT. Let G be an algebraic group over a local field k. We show that the image of $G(k)$ under every continuous homomorphism into a (Hausdorff) topological group is closed if and only if the center of $G(k)$ is compact. This is joint work with Uri Bader.

Arghya Mondal (Krea University (India)) *Property (T) for fiber and semidirect products*

ABSTRACT. Extension of a Property (T) group by another is a Property (T) group. It follows that Property (T) is closed under product. The same is true for fiber products of two homomorphisms from Property (T) groups, provided one of them splits. In general there are counterexamples. The special case of fiber product of two copies of a homomorphism reduces to the question of Property (T) for a semidirect product given the same for the non-normal factor. Here necessary and sufficient conditions can be given. This is a joint work with M. Mj. The general case of relative Property (T) for semidirect products, where the normal factor need not be abelian, is an ongoing work with S. Nayak.

Sumit Chandra Mishra (IISER Mohali, Punjab (India)) *Local-global principles for norm one tori and multinorm tori over semi-global fields*

ABSTRACT. A well-known result of Hasse states that the local-global principle holds for norms over number fields for cyclic extensions. In other words, if L/F is a cyclic extension of number fields then an element $\lambda \in F^\times$ is in the image of norm map $N_{L/F} : L^\times \to F^\times$ if and only if λ is in the image of the norm map locally everywhere, i.e., for completions associated to all archimedean and non-archimedean places of F. In this talk, we would consider local-global principles for norms and product of norms over fields which are function fields of curves over complete discretely valued fields, for example, $\mathbb{C}((t))(x)$.

Plinio Murillo (Fluminense Federal University (Brazil)) *On kissing number of hyperbolic manifolds* (zoom)

ABSTRACT. Motivated by a recent result of M. F. Bourque and B. Petri, we constructed a sequence of hyperbolic manifolds with a large number of closed geodesics of shortest length. The aim of this talk is to explain what "large" means, and how arithmetic groups enter in this context. This will involve results obtained in collaboration with Cayo Dória and Emanoel Freire.

Srinivasan Srimathy (TIFR (India)) *The genus of division algebras over discrete valued fields*

ABSTRACT. Given a field with a set of discrete valuations, we show how the genus of any division algebra over certain fields is related to the genus of some residue algebras at various valuations and the ramification data. Applications include showing triviality of the genus of quaternions over many fields such as higher local fields, function fields of curves over higher local fields and function fields of curves over real closed fields. When the base field is a function field of a curve over a global field with a rational point, the genus of any quaternion is related to the 2-torsion of the Tate-Shafarevich group of the Jacobian. As a consequence, when the curve is elliptic, the size of the genus can be computed directly using arithmetic data of the elliptic curve.

Itamar Vigdorovich (Weizmann Institute (Israel)) *Stationary dynamics on character spaces and applications to arithmetic groups*

ABSTRACT. To any group G is associated the space $\mathrm{Ch}(G)$ of all characters on G. After defining this space and discussing its interesting properties, I will turn to discuss dynamics on such spaces. Our main result is that the action of any arithmetic group on the character space of its amenable/solvable radical is stiff, i.e., any probability measure which is stationary under random walks must be invariant. This generalizes a classical theorem of Furstenberg for dynamics on tori. Relying on works of Bader, Boutonnet, Houdayer, and Peterson, this stiffness result is used to deduce dichotomy statements (and 'charmenability') for higher rank arithmetic groups pertaining to their normal subgroups, dynamical systems, representation theory and more. The talk is based on a joint work with Uri Bader.

Schedule of Talks

All talks are held in Ramanujan Lecture Hall

Monday January 1

10:30–11:20	**Toshiyuki Kobayashi** (The University of Tokyo (Japan)) Mini-course *Global Analysis of Locally Symmetric Spaces with Indefinite Metric* Lecture 1
11:20–12:00	Coffee break
12:00–12:50	**Indranil Biswas** (Shiv Nadar University (India)) *Classification, reduction and stability of toric principal bundles*
1:00–3:00	Lunch break
3:00–3:50	**Alex Lubotzky** (Weizmann Institute (Israel)) & **Bharatram Rangarajan** (Hebrew University of Jerusalem (Israel)) Mini-course *Uniform Stability of Higher-rank Arithmetic Groups* Lecture 1
4:00–4:30	Coffee break
4:30–5:20	**Yosuke Morita** (Kyushu University (Japan)) *The existence problem of compact quotients of pseudo-Riemannian homogeneous spaces*

Tuesday January 2

9:30–10:20	**Toshiyuki Kobayashi** (The University of Tokyo (Japan)) Mini-course *Global Analysis of Locally Symmetric Spaces with Indefinite Metric* Lecture 2
10:35–11:25	**Rajesh Kulkarni** (Michigan State University (USA)) *Moduli of binary cubic forms* (zoom)
11:30–12:00	Coffee break

G. Prasad et al. (eds.), *Zariski Dense Subgroups, Number Theory and Geometric Applications*, Infosys Science Foundation Series,
https://doi.org/10.1007/978-981-95-3437-1

12:00–12:50	**Maciej Bochenski** (University of Warmia Mazury (Poland)) *Standard compact Clifford-Klein forms and Lie algebra decompositions*
1:00–3:00	Lunch break
3:00–3:50	**Alex Lubotzky** (Weizmann Institute (Israel)) & **Bharatram Rangarajan** (Hebrew University of Jerusalem (Israel)) Mini-course *Uniform Stability of Higher-rank Arithmetic Groups* Lecture 2
4:00–4:30	Coffee break
4:30–5:20	**Igor Rapinchuk** (Michigan State University (USA)) *On the Hasse principle for reductive algebraic groups over finitely generated fields* (zoom)

Wednesday January 3

9:30–10:20	**Toshiyuki Kobayashi** (The University of Tokyo (Japan)) Mini-course *Global Analysis of Locally Symmetric Spaces with Indefinite Metric* Lecture 3
10:35–11:25	**Marek Kaluba** (Karlsruhe Institute for Technology (Germany)) & **Piotr Nowak** (IMPAN (Poland)) *Property (T) for automorphism groups of free groups* Lecture 1
11:30–12:00	Coffee break
12:00–12:50	**Alireza Salehi Golsefidy** (UCSD (USA)) *Random walks on group extensions*
1:00–3:00	Lunch break
3:00–3:50	**Alex Lubotzky** (Weizmann Institute (Israel)) & **Bharatram Rangarajan** (Hebrew University of Jerusalem (Israel)) Mini-course *Uniform Stability of Higher-rank Arithmetic Groups* Lecture 3
4:00–4:30	Coffee break
4:30–5:20	**Marek Kaluba** (Karlsruhe Institute for Technology (Germany)) and **Piotr Nowak** (IMPAN (Poland)) *Property (T) for automorphism groups of free groups* Lecture 2

Thursday January 4

9:30–10:20	**Raman Parimala** (Emory University (USA)) *Hasse principle for reductive groups over p-adic function fields*
10:35–11:25	**Ted Chinburg** (University of Pennsylvania (USA)) *Linear representations of the Grothendieck-Teichmüller group*
11:30–12:00	Coffee break

Short communications

12:00–12:20	**Srinivasan Srimathy** (TIFR (India)) *The genus of division algebras over discrete valued fields*
12:25–12:45	**Pralay Chatterjee** (IMSc, Chennai (India)) *Lower dimensional Betti numbers of homogeneous spaces of Lie groups*
12:50–1:10	**Arghya Mondal** (Krea University (India)) *Property (T) for fiber and semidirect products*

1:10–3:00	Lunch break
3:00–3:50	**Chen Meiri** (Technion (Israel)) *Conjugacy width in higher rank arithmetic groups of orthogonal type* (zoom)
4:00–4:30	Coffee break
4:30–5:20	**Matthew Stover** (Temple University (USA)) *Central extensions of (arithmetic) lattices*

Friday January 5

9:30–10:20	**Alex Kontorovich** (Rutgers University (USA)) *Thin Orbits and Applications* (zoom)
10:35–11:25	**Ramdorai Sujatha** (UBC (Canada)) *Rational points on elliptic curves along anticyclotomic extensions*
11:30–12:00	Coffee break
12:00–12:50	**Alireza Salehi Golsefidy** (UCSD (USA)) *Closure of orbits of the pure mapping class group on the character variety*
1:00–3:00	Lunch break
3:00–3:50	**Emmanuel Breuillard** (University of Oxford (UK)) *Heights on character varieties, free subgroups and spectral gaps*
4:00–4:30	Coffee break
4:30–5:20	**George Tomanov** (University of Lyon (France)) *Orbit closures on homogeneous spaces and applications to number theory*

Monday January 8

9:30–10:20	**Pietro Corvaja** (University of Udine (Italy)) Mini-course *Bounded generation of linear groups and Diophantine approximation* Lecture 1
10:35–11:25	**C.S. Rajan** (Ashoka University (India)) *On non-commensurable isospectral locally symmetric spaces*
11:30–12:00	Coffee break
12:00–12:50	**Emmanuel Breuillard** (University of Oxford (UK)) *Random character varieties*
1:00–3:00	Lunch break
3:00–3:50	**Brian Conrad** (Stanford University (USA)) Mini-course *Introduction to Bruhat-Tits theory* Lecture 1
4:00–4:30	Coffee break
4:30–5:20	**Alan Reid** (Rice University (USA)) *From representation rigidity to profinite rigidity*

Tuesday January 9

9:30–10:20	**Pietro Corvaja** (University of Udine (Italy)) Mini-course *Bounded generation of linear groups and Diophantine approximation* Lecture 2
10:35–11:25	**Jinbo Ren** (Xiamen University (China)) *On the construction of non-commensurable locally symmetric spaces* (zoom)

11:30–12:00	Coffee break
12:00–12:50	**Benjamin Linowitz** (Oberlin College (USA)) *The systolic geometry of arithmetic locally symmetric spaces*
1:00–3:00	Lunch break
3:00–3:50	**Brian Conrad** (Stanford University (USA)) Mini-course *Introduction to Bruhat-Tits theory* Lecture 2
4:00–4:30	Coffee break
4:30–5:20	**Ben McReynolds** (Purdue University (USA)) *What makes semisimple Lie groups special?* (zoom)

Wednesday January 10

9:30–10:20	**Pietro Corvaja** (University of Udine (Italy)) Mini-course *Bounded generation of linear groups and Diophantine approximation* Lecture 3
10:35–11:25	**T.N. Venkataramana** (ICTS (India)) *CSP and unipotent generators for higher rank nonuniform arithmetic groups*
11:30–12:00	Coffee break
12:00–12:50	**Sandip Singh** (IITB (India)) *Hypergeometric groups and their arithmeticity*
1:00–3:00	Lunch break
3:00–3:50	**Brian Conrad** (Stanford University (USA)) Mini-course *Introduction to Bruhat-Tits theory* Lecture 3
4:00–4:30	Coffee break
4:30–5:20	**Yves Benoist** (University Paris-Saclay (France)) *Convolution and square on abelian groups*

Thursday January 11

9:30–10:20	**Jessica Fintzen** (University of Bonn (Germany)) *Representations of p-adic groups* (zoom)
10:35–11:25	**François Thilmany** (Catholic University of Louvain (Belgium)) *Prasad's volume formula and its applications*
11:30–12:00	Coffee break

Short communications

12:00–12:20	**Sourav Ghosh** (Ashoka University (India))*Finite step rigidity*
12:30–12:50	**Elyasheev Leibtag** (Weizmann Institute (Israel)) *Images of Homomorphisms of Algebraic Groups*
1:00–3:00	Lunch break
3:00–3:50	**Sandeep Varma** (TIFR (India)) *Homogeneity results for invariant distributions on p-adic symmetric spaces*
4:00–4:30	Coffee break
4:30–5:20	**Mikhail Belolipetsky** (IMPA (Brazil)) *Subspace stabilisers in hyperbolic lattices* (zoom)

Friday January 12

9:30–10:20	**Tsachik Gelander** (Hebrew University of Jerusalem (Israel)) *Things we can learn by looking at random manifolds* (zoom)
10:35–11:25	**Dave Morris** (University of Lethbridge (Canada)) *Arithmetic groups of higher real rank are not left-orderable (after Deroin and Hurtado)* (zoom)
11:30–12:00	Coffee break

Short communications

12:00–12:20	**Sumit Chandra Mishra** (IISER Mohali, Punjab (India)) *Local-global principles for norm one tori and multinorm tori over semi-global fields*
12:25–12:45	**Plinio Murillo** (Fluminense Federal University (Brazil)) *On kissing number of hyperbolic manifolds* (zoom)
12:50–1:10	**Itamar Vigdorovich** (Weizmann Institute (Israel)) *Stationary dynamics on character spaces and applications to arithmetic groups*
1:10–3:00	Lunch break

The manufacturer's authorised representative in the EU is Springer Nature Customer Service Centre GmbH, Europaplatz 3, 69115 Heidelberg, Germany. If you have any concerns regarding our products, please contact ProductSafety@springernature.com

Printed and bound by CPI Group (UK) Ltd, Croydon, CR0 4YY
21/09/2026
02219986-0017